Grundlehren der mathematischen Wissenschaften 302

A Series of Comprehensive Studies in Mathematics

Herbert Lange   Christina Birkenhake

Complex Abelian Varieties

Springer-Verlag

Berlin Heidelberg New York
London Paris Tokyo
Hong Kong Barcelona
Budapest

Herbert Lange
Christina Birkenhake
Mathematisches Institut
Universität Erlangen-Nürnberg
Bismarckstr. 1 1/2
W-8520 Erlangen, FRG

Mathematics Subject Classification (1980): 14-02, 14KXX, 30F10, 32G20

ISBN 3-540-54747-9 Springer-Verlag Berlin Heidelberg New York
ISBN 0-387-54747-9 Springer-Verlag New York Berlin Heidelberg

Library of Congress Cataloging-in-Publication Data
Lange, H. (Herbert), 1943- Complex Abelian varieties / Herbert Lange, Christina Birkenhake. p. cm. -- (Grundlehren der mathematischen Wissenschaften ; 302)
Includes bibliographical references and index.
ISBN 0-387-54747-9 (Springer-Verlag New York : alk. paper).
ISBN 3-540-54747-9 (Springer-Verlag Berlin : alk. paper)
1. Abelian Varieties. 2. Riemann surfaces. I. Birkenhake, Christina. II. Title.
III. Series. QA564.L32 1992 516.3'53--dc20 92-23806

Printed in the United States of America

Typesetting: Camera ready by authors
41/3140-543210 - Printed on acid-free paper

Contents

Introduction

A *hyperelliptic integral* is by definition an integral of the form

$$\int_\gamma \frac{dz}{\sqrt{f(z)}} ,$$

where γ is a path in the complex plane $\mathbb{C}$ with coordinate z and $f(z) = (z-a_1)\cdots(z-a_d)$ with pairwise different constants a_i. If $d = \deg f$ is 1 or 2, an explicit integration by elementary functions is well known from calculus. If $d = 3$ or 4, integration is possible using elliptic functions. If however $d \geq 5$, no explicit integration is known in general.

The reason for this is the following: the differential $\omega = \frac{dz}{\sqrt{f(z)}}$ is not single valued, considered as a function on $\mathbb{C}$. Let C denote the compact Riemann surface associated to $\sqrt{f}$. By definition C is the double covering of the Riemann sphere $\mathbb{P}_1$, ramified at the points $a_1, \dots, a_d$ together with ∞ if d is odd. Now ω may be considered as a holomorphic differential on C. It is essentially the topological structure of C which causes the problem. The more complicated it is, the more difficult it is to integrate ω.

At the beginning of the 19th century the Norwegian mathematician Niels Henrik Abel (1802–1829) and the German mathematician Carl Gustav Jacob Jacobi (1804–1851) found a way to attack this problem. In geometric terms their method can be described as follows. The idea is to try to integrate not ω alone, but simultaneously the whole set of holomorphic differentials

$$\omega_i = z^{i-1} \frac{dz}{\sqrt{f(z)}} \quad \text{for} \quad i = 1, \dots, g = [\tfrac{d-1}{2}]$$

on C. For this, fix a point $p_0 \in C$ and consider the map

$$p \mapsto \left(\int_{p_0}^{p} \omega_1, \dots, \int_{p_0}^{p} \omega_g \right),$$

defined on a small neighbourhood U of p_0. Unfortunately this map cannot be extended to the whole of C, since the integrals depend on the path from p_0 to p. However, if we consider it modulo the values of integrals along all possible closed paths, we obtain a well-defined map. To be more precise, we consider $H_1(C, \mathbb{Z})$ to be the group of closed paths starting from p_0 modulo homology. The image of the map $H_1(C, \mathbb{Z}) \to \mathbb{C}^g$, $\gamma \mapsto (\int_\gamma \omega_1, \ldots, \int_\gamma \omega_g)$ is a lattice in $\mathbb{C}^g$, i.e. a discrete subgroup of rank $2g$. So the quotient

$$J(C) = \mathbb{C}^g / H_1(C, \mathbb{Z})$$

is a complex torus, called the *Jacobian variety* of C. It can be shown that $J(C)$ is isomorphic to a projective variety. A complex torus with this property is called an *abelian variety*. By construction the integration map $U \to \mathbb{C}^g$ discussed above extends to a holomorphic map

$$\alpha: C \to J(C)\,,\, p \mapsto \left(\int_{p_0}^{p} \omega_1, \ldots, \int_{p_0}^{p} \omega_g\right) \bmod H_1(C, \mathbb{Z})\ ,$$

called the *Abel-Jacobi map*. In these terms the integration of $(\omega_1, \ldots, \omega_g)$ is essentially equivalent to the following two steps:

(i) Determine the Jacobian variety $J(C)$.
(ii) Describe the Abel-Jacobi map $\alpha: C \to J(C)$.

Unfortunately only in very few cases can this be done explicitly (see Section 11.7). However, one can try to study the geometry of the pair $(J(C), \alpha)$, and this may be considered as a sort of substitute for step (i). This method is not restricted to hyperelliptic integrals, but works for holomorphic integrals on any compact Riemann surface.

Of course Abel and Jacobi did not express their results in the above geometric language. For them integration was entirely a matter of analysis. They worked with what we now call abelian functions, or rather abelian integrals. Some essential progress in the subject was made by Riemann. Even if he did not invent theta functions, he used them heavily in his investigations of abelian functions. Apart from Riemann, the analytic theory of abelian functions is mainly due to Weierstraß, Frobenius, Poincaré and Picard. For example, the fact that every abelian function is a quotient of theta functions was known to Weierstraß in 1878 and proved by Poincaré [1] in 1899.

Towards the end of the 19th century geometers started to study the theory of abelian and theta functions by geometric methods. Originally an "abelian variety" of dimension g meant a hypersurface of $\mathbb{P}_{g+1}$ given as the image of $\mathbb{C}^g$ under the map defined by $g+2$ suitable theta functions (see de Franchis [1] p. 53). Later this notion was extended to mean a projective variety given as the image of $\mathbb{C}^g$ under the map defined by any system of theta functions of the same type (see Lefschetz [2] p. 355). However, since these varieties often have unpleasant singularities and do not admit a group structure, the language of complex tori turned out to be more fruitful for this purpose. Picard seems to have been the first to express the theory of abelian functions in this language. But it was only after the fundamental paper of Lefschetz [1] that this point of view was generally accepted. The basics of the geometric theory of abelian varieties are largely due to Scorza, Rosati, and Lefschetz.

Today abelian varieties play an important role in several branches of mathematics. Number theory is certainly the subject in which they are most extensively applied. They are heavily used in class field theory, as well as for rationality and transcendence questions. In the theory of dynamical systems they represent a tool to solve certain Hamiltonian systems. Their importance in algebraic geometry lies in the fact that there are natural ways to associate to any smooth projective algebraic variety V an abelian variety X and investigate properties of V by studying X. Examples of this are the Picard variety, the Albanese variety, and certain intermediate Jacobians. Some problems in physics can be studied via abelian varieties. Let us mention only that theta functions are solutions of the heat equation in thermodynamics. A more recent branch of physics applying abelian varieties is string theory.

Apart from their importance in applications, geometric properties of abelian varieties are interesting for their own sake and this is the subject of the present book. The central topics are the projective embeddings of an abelian variety, their equations and geometric properties, discussed in Chapters 7 and 10. Moreover we construct several moduli spaces of abelian varieties with additional structure in Chapters 8 and 9 and give some applications for the theory of algebraic curves in Chapters 11 and 12.

The main tool for the proofs is the *theta group* of a line bundle L on X, introduced in Chapter 6. Implicitly it appears already in Chapter 3, where we construct a basis of canonical theta functions of $H^0(X, L)$ in such a way that the theta group of L acts essentially by permutation (see Theorem 3.2.7). Another tool is the notion of the *characteristic* of a nondegenerate line bundle on X. Characteristics represent a list of all line bundles in an algebraic equivalence class; their definition is a direct generalization of the classical notion of characteristics of theta functions. For more details concerning the contents we refer to the introductions at the start of every chapter.

In order to be able to present some more advanced results, we restrict ourselves to abelian varieties over the field $\mathbb{C}$ of complex numbers. The main advantage is that a line bundle on a complex torus $\mathbb{C}^g/\Lambda$ can be described by factors of automorphy on the universal covering $\mathbb{C}^g$. In Appendix B we give a short introduction and compile some results on factors of automorphy.

At the end of each chapter there is a section with exercises of varying degrees of difficulty. Many of the exercises are standard results which were not included in the text due to lack of space. Others contain theorems which we quote from research papers and which are intended as hints for further reading.

As for the prerequisites, the book is reasonably self-contained. We use however the basic language of algebraic geometry and complex analysis. In a few cases, in particular in the last chapters, we apply some deeper results, for which we have tried to give precise references. More details on the prerequisites are contained in the introductions to each chapter.

The history of abelian varieties is very intricate and we have not dared to give a proper account of it (for a good survey we refer to Krazer-Wirtinger [1], de Franchis [1] and Lefschetz [2]). However we try to give some hints about the origins of the material. One should be aware of the fact that many results have been discovered several times. The name attributed to a theorem is sometimes not the name of the person who first proved it. We apologize for incorrect or missing attributions.

There are several interesting topics concerning complex abelian varieties not covered in this book: for example rationality questions, families and degenerations of abelian varieties, compactifications of the moduli spaces, the Schottky problem, intermediate Jacobians and relations with Hodge theory. A reasonable presentation of each of these subjects would fill a whole volume for itself.

It is a pleasure to acknowledge the help and support we received from a number of people and institutions. In particular, we would like to thank W. Barth, H.-J. Brasch, W.-D. Geyer, J. Hilgert, G. Martens, M. Reid, W. Ruppert, C. Schoen, E. Sernesi, A. Silverberg, and R. Smith for many valuable comments and suggestions. We are very grateful to S. Schmiedl, who typed the TEX-file and produced the illustrations. We received financial support from the Deutsche Forschungs Gemeinschaft via the Schwerpunkt-programm "Komplexe Mannigfaltigkeiten" and the European Community via the Science project "Geometry of Algebraic Varieties" for which we would like to express our gratitude. Finally, the first author would like to thank the University of Warwick for hospitality.

Notation

$H^i(E)$	We use the same symbol to denote a vector bundle and its corresponding locally free sheaf. If E is a locally free sheaf and there is no ambiguity about the base space X of E, we write $H^i(E)$ instead of $H^i(X, E)$.
$h^i(E)$	dimension of the vector space $H^i(E)$.
$x \in D$	for a divisor D and a point x on a variety, $x \in D$ means that x is contained in the support of D.
$M(g \times g', \mathcal{R})$	module of $(g \times g')$-matrices over a ring $\mathcal{R}$.
$M_g(\mathcal{R})$	algebra of $(g \times g)$-matrices over a ring $\mathcal{R}$.
$\operatorname{diag}(x_1, \ldots, x_g)$	diagonal matrix with entries $x_1, \ldots, x_g$.
$\mathbb{1} = \mathbb{1}_g$	unit matrix of degree g.
$dx_1 \wedge \cdots \wedge \check{dx_\nu} \wedge \cdots \wedge dx_g$	$dx_1 \wedge \cdots \wedge dx_{\nu-1} \wedge dx_{\nu+1} \wedge \cdots \wedge dx_g$, the differential $(g-1)$-form with dx_ν omitted.
$\#S$	cardinality of a set S.
$e(\cdot)$	exponential function $z \mapsto e^z$.
$\mathfrak{S}_n$	symmetric group of degree n.
$\mathbb{C}_1$	the circle group $\{z \in \mathbb{C} \mid \lvert z \rvert = 1\}$.
$\bar{v}$	image of $v \in V \simeq \mathbb{C}^g$ under the projection map $\pi: V \to X = V/\Lambda$.
$\sim$	linear equivalence of divisors.
$\equiv$	algebraic equivalence of line bundles and divisors.
L^n	n-th tensor power of a line bundle L.
(L^g)	self-intersection number of a line bundle L on a g-dimensional complex torus.
$<S>$	vector space or group generated by a set S.
δ_{IJ}	Kronecker symbol for subsets $I, J \subset \{1, \ldots, n\}$.

Chapter 1

Complex Tori

A *lattice* in a complex vector space $\mathbb{C}^g$ is by definition a discrete subgroup of maximal rank in $\mathbb{C}^g$. It is a free abelian group of rank $2g$. A *complex torus* is a quotient $X = \mathbb{C}^g/\Lambda$ with Λ a lattice in $\mathbb{C}^g$. The complex torus X is a complex manifold of dimension g. It inherits the structure of a complex Lie group from the vector space $\mathbb{C}^g$. A meromorphic function on $\mathbb{C}^g$, periodic with respect to Λ, may be considered as a function on X. An *abelian variety* is a complex torus admitting sufficiently many meromorphic functions. Even though abelian varieties are the topic of this book, they will be introduced only in Chapter 4. In the first three chapters we study more generally arbitrary complex tori.

In his fundamental paper [1] mentioned already in the introduction Lefschetz derived, among other things, the most important topological properties of complex tori. This is the subject of the first three sections of this chapter. In Section 1 we show that a complex Lie group is a complex torus if and only if it is compact and connected, and we introduce period matrices. In Section 2 holomorphic maps between complex tori are studied: any such holomorphic map can be expressed in a unique way as the composite of a homomorphism and a translation. We introduce the rational and the analytic representation of homomorphisms. Moreover we study the most important class of homomorphisms, the isogenies. The singular cohomology groups of X are the subject of Section 3: it turns out that $H^n(X, \mathbb{Z})$ is a free abelian group of rank $\binom{2g}{n}$. Finally, in Section 4 a proof of the theorem of Hodge on the decomposition of the vector spaces $H^n(X, \mathbb{C})$ will be given. The proof is substantially easier than for a general compact Kähler manifold (see Griffiths-Harris [1]). Here the theory of Sobolev spaces is not necessary, and the Green operator can be defined explicitly using Fourier expansions.

As for prerequisites: apart from some basic topological results such as the Künneth formula and the universal coefficient theorem we use the theorems of de Rham and Dolbeault describing cohomology groups by differential forms. In Section 1 we apply the fact that the universal covering of a Lie group inherits a Lie group structure.

§ 1 Complex Tori

Let V denote a complex vector space of dimension g and Λ a lattice in V. By definition Λ is a discrete subgroup of rank $2g$ of V. The lattice Λ acts on V by addition. The quotient

$$X = V/\Lambda$$

is called a *complex torus*. According to Corollary A.7 it is a connected complex manifold. Moreover X is compact, since Λ is of maximal rank as a discrete subgroup of V and thus X is the image of a bounded subset of V. The addition in V induces the structure of an abelian complex Lie group on X. We write its group operation additively and call the map $\mu: X \times X \to X$, $\mu(x_1, x_2) = x_1 + x_2$ the *addition map*.

(1.1) Lemma. *Any connected compact complex Lie group X of dimension g is a complex torus.*

Proof. First we claim that X is abelian: consider the commutator map $\Phi(x,y) = xyx^{-1}y^{-1}$ and let U be a coordinate neighbourhood of the unit element 1 in X. For every $x \in X$ there exist open neighbourhoods V_x of x and W_x of 1 in X with $\Phi(V_x, W_x) \subseteq U$, since $\Phi(x,1) = 1 \in U$ and Φ is continuous. As X is compact, finitely many V_x cover X. Denoting by W the intersection of the corresponding finitely many open sets W_x, we get $\Phi(X,W) \subseteq U$. This implies $\Phi(X,W) = 1$, since holomorphic functions on compact manifolds are constant and $\Phi(1,x) = 1$ for every $x \in W$. As W is open and nonempty, this implies the assertion.

Let $\pi: V \to X$ be the universal covering map. The Lie group structure of X induces the structure of a simply connected complex Lie group on V, such that π is a homomorphism. Moreover V is abelian, since X is. Hence V is isomorphic to the vector space $\mathbb{C}^g$ (see Hochschild [1] Theorem 18.4.1). Finally the compactness of X implies that $\ker \pi$ is a lattice in V. □

For a complex torus $X = V/\Lambda$ the vector space V may be considered as the universal covering space of X. We will denote by

$$\pi: V \to X$$

the universal covering map. If v is an element of V, we often write $\bar{v}$ for its image $\pi(v)$ in X. Conversely, for $\bar{v} \in X$, the letter v will denote some representative of $\bar{v}$ in V. The kernel Λ of π can be identified with the fundamental group $\pi_1(X) = \pi_1(X, 0)$. Furthermore, since Λ is abelian, $\pi_1(X)$ is canonically isomorphic to the first homology group $H_1(X, \mathbb{Z})$. The torus X is locally isomorphic to V. This implies that V may be considered as the tangent space T_0X of X in 0. From the Lie theoretic point of view $\pi: V = T_0X \to X$ is just the exponential map.

As an example let us consider the case $g = 1$. Choosing a basis, we may identify V with the field of complex numbers $\mathbb{C}$. A lattice Λ in $\mathbb{C}$ is generated by 2 complex numbers λ_1 and λ_2 which are linearly independent over $\mathbb{R}$:

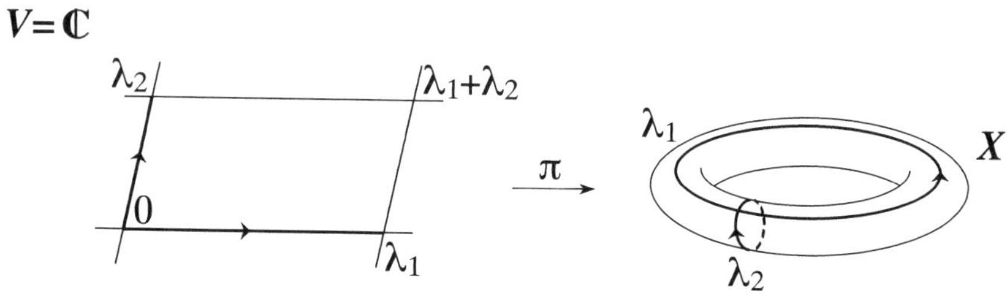

Identifying opposite sides of the parallelogram $0, \lambda_1, \lambda_1 + \lambda_2, \lambda_2$ we obtain the torus X. The images of the lines $\overline{0\lambda_1}$ and $\overline{0\lambda_2}$ are cycles on X also denoted by λ_1 and λ_2. Obviously λ_1 and λ_2 generate the group $H_1(X, \mathbb{Z})$. A 1-dimensional complex torus is called an *elliptic curve*. In Example 4.1.3 it will be shown that any elliptic curve admits the structure of an algebraic variety.

We return to the general case. In order to describe a complex torus $X = V/\Lambda$, choose bases $e_1, \cdots, e_g$ of V and $\lambda_1, \cdots, \lambda_{2g}$ of the lattice Λ. Write λ_i in terms of the basis $e_1, \ldots, e_g$: $\lambda_i = \sum_{j=1}^{g} \lambda_{ji} e_j$. The matrix

$$\Pi = \begin{pmatrix} \lambda_{11} & \cdots & \cdots & \lambda_{1,2g} \\ \vdots & & & \vdots \\ \lambda_{g1} & \cdots & \cdots & \lambda_{g,2g} \end{pmatrix}$$

in $M(g \times 2g, \mathbb{C})$ is called a *period matrix* for X. The period matrix Π determines the complex torus X completely, but certainly it depends on the choice of the bases for V and Λ. Conversely, given a matrix $\Pi \in M(g \times 2g, \mathbb{C})$, one may ask: Is Π a period matrix for some complex torus? The following proposition gives an answer to this question.

(1.2) Proposition. $\Pi \in M(g \times 2g, \mathbb{C})$ *is the period matrix of a complex torus if and only if the matrix* $P = \binom{\Pi}{\overline{\Pi}} \in M_{2g}(\mathbb{C})$ *is nonsingular, where* $\overline{\Pi}$ *denotes the complex conjugate matrix.*

Proof. Π is a period matrix if and only if the column vectors of Π span a lattice in $\mathbb{C}^g$, in other words, if and only if the columns are linearly independent over $\mathbb{R}$.

Suppose first that the columns of Π are linearly dependent over $\mathbb{R}$. Then there is an $x \in \mathbb{R}^{2g}$, $x \neq 0$, with $\Pi x = 0$, and we get $Px = 0$. This implies $\det P = 0$.

Conversely, if P is singular, there are vectors $x, y \in \mathbb{R}^{2g}$, not both zero, such that $P(x + iy) = 0$. But $\Pi(x + iy) = 0$ and $\Pi(x - iy) = \overline{\Pi(x + iy)} = 0$ imply $\Pi x = \Pi y = 0$. Hence the columns of Π are linearly dependent over $\mathbb{R}$. □

§ 2 Homomorphisms

There are two distinguished types of holomorphic maps between complex tori, namely homomorphisms and translations. We will see that every holomorphic map is a composition of one of each.

Let $X = V/\Lambda$ and $X' = V'/\Lambda'$ be complex tori of dimensions g and g'. A *homomorphism of* X *to* X' is a holomorphic map $f: X \to X'$, compatible with the group structures. The *translation by an element* $x_0 \in X$ is defined to be the holomorphic map $t_{x_0}: X \to X$, $x \mapsto x + x_0$.

(2.1) Proposition. *Let $h: X \to X'$ be a holomorphic map.*

a) There is a unique homomorphism $f: X \to X'$ such that $h = t_{h(0)}f$, i.e. $h(x) = f(x) + h(0)$ for all $x \in X$.

b) There is a unique $\mathbb{C}$-linear map $F: V \to V'$ with $F(\Lambda) \subset \Lambda'$ inducing the homomorphism f.

Proof. Define $f = t_{-h(0)}h$. We can lift the composed map $V \xrightarrow{\pi} X \xrightarrow{f} X'$ to a holomorphic map F into the universal covering V' of X'

$$\begin{array}{ccc} V & \xrightarrow{F} & V' \\ & {\scriptstyle f\pi}\searrow \quad \swarrow{\scriptstyle \pi'} & \\ & X' & \end{array}$$

in such a way that $F(0) = 0$. The diagram implies that for all $\lambda \in \Lambda$ and $v \in V$ we have $F(v + \lambda) - F(v) \in \Lambda'$. Thus the continuous map $v \mapsto F(v + \lambda) - F(v)$ is constant and we get $F(v + \lambda) = F(v) + F(\lambda)$ for all $\lambda \in \Lambda$ and $v \in V$. Hence the partial derivatives of F are $2g$-fold periodic and thus constant by Liouville's theorem. It follows that F is $\mathbb{C}$-linear and f is a homomorphism. The uniqueness of F and f is obvious. □

Under addition the set of homomorphisms of X into X' forms an abelian group denoted by $\mathrm{Hom}(X, X')$. Proposition 2.1 gives an injective homomorphism of abelian groups

$$\rho_a\colon \mathrm{Hom}(X,X') \to \mathrm{Hom}_{\mathbb{C}}(V,V'),\ f \mapsto F\ ,$$

the *analytic representation of* $\mathrm{Hom}(X,X')$. The restriction F_Λ of F to the lattice Λ is $\mathbb{Z}$-linear. F_Λ determines F and f completely. Thus we get an injective homomorphism

$$\rho_r\colon \mathrm{Hom}(X,X') \to \mathrm{Hom}_{\mathbb{Z}}(\Lambda,\Lambda'),\ f \mapsto F_\Lambda\ ,$$

the *rational representation of* $\mathrm{Hom}(X,X')$. We denote the extensions of ρ_a and ρ_r to $\mathrm{Hom}_{\mathbb{Q}}(X,X') := \mathrm{Hom}(X,X') \otimes_{\mathbb{Z}} \mathbb{Q}$ by the same letters. These will also be referred to as the analytic and rational representations. Since any subgroup of $\mathrm{Hom}_{\mathbb{Z}}(\Lambda,\Lambda') \simeq \mathbb{Z}^{4gg'}$ is isomorphic to $\mathbb{Z}^m$, the injectivity of ρ_r implies

(2.2) Proposition. $\mathrm{Hom}(X,X') \simeq \mathbb{Z}^m$ *for some* $m \le 4gg'$.

Let $X'' = V''/\Lambda''$ be a third complex torus. For $f \in \mathrm{Hom}(X,X')$ and $f' \in \mathrm{Hom}(X',X'')$ we have $\rho_a(f'f) = \rho_a(f')\rho_a(f)$. This follows immediately from the uniqueness statement in Proposition 2.1.b). In particular, if $X = X'$, ρ_a and ρ_r are representations of the ring $\mathrm{End}(X)$ respectively $\mathrm{End}_{\mathbb{Q}}(X) := \mathrm{End}(X) \otimes_{\mathbb{Z}} \mathbb{Q}$.

Suppose $\Pi \in M(g \times 2g, \mathbb{C})$ and $\Pi' \in M(g' \times 2g', \mathbb{C})$ are period matrices for X and X' with respect to some bases of V, Λ and V', Λ'. Let $f\colon X \to X'$ be a homomorphism. With respect to the chosen bases the representation $\rho_a(f)$ (respectively $\rho_r(f)$) is given by a matrix $A \in M(g' \times g, \mathbb{C})$ (respectively $R \in M(2g' \times 2g, \mathbb{Z})$. In terms of matrices the condition $\rho_a(f)(\Lambda) \subset \Lambda'$ means

$$A\Pi = \Pi' R\ . \tag{1}$$

Conversely, any two matrices $A \in M(g' \times g, \mathbb{C})$ and $R \in M(2g' \times 2g, \mathbb{Z})$ satisfying (1) define a homomorphism $X \to X'$. The following proposition shows, how ρ_a and ρ_r are related.

(2.3) Proposition. *The extended rational representation* $\rho_r \otimes 1\colon \mathrm{End}_{\mathbb{Q}}(X) \otimes \mathbb{C} \to \mathrm{End}_{\mathbb{C}}(\Lambda \otimes \mathbb{C}) = \mathrm{End}_{\mathbb{C}}(V \times V)$ *is equivalent to the direct sum of the analytic representation and its complex conjugate:* $\rho_r \otimes 1 \simeq \rho_a \oplus \overline{\rho_a}$.

Proof. Let Π denote the period matrix of X with respect to some bases of V and Λ. Suppose $f \in \mathrm{End}(X)$. If A and R are the matrices of $\rho_a(f)$ and $\rho_r(f)$ with respect to the chosen bases, we have by (1)

$$\begin{pmatrix} A & 0 \\ 0 & \overline{A} \end{pmatrix} \begin{pmatrix} \Pi \\ \overline{\Pi} \end{pmatrix} = \begin{pmatrix} \Pi \\ \overline{\Pi} \end{pmatrix} R\ .$$

This implies the assertion, since $\binom{\Pi}{\overline{\Pi}}$ is nonsingular by Proposition 1.2. □

Next we study the image $\operatorname{im} f$ and the kernel $\ker f$ of a homomorphism $f: X \to X'$ of complex tori.

(2.4) Proposition. *a)* $\operatorname{im} f$ *is a subtorus of* X'.

b) $\ker f$ *is a closed subgroup of* X. *The connected component* $(\ker f)_0$ *of* $\ker f$ *containing* 0 *is a subtorus of* X *of finite index in* $\ker f$.

Proof. This is a consequence of Lemma 1.1. However we will give a direct proof. Denote $F = \rho_a(f)$. As for a): since $\operatorname{im} f = F(V)/F(V) \cap \Lambda'$, we have to show that $F(V) \cap \Lambda'$ is a lattice in $F(V)$. But $F(V) \cap \Lambda'$ is discrete in $F(V)$ and generates $F(V)$ as an $\mathbb{R}$-vector space, since it contains $F(\Lambda)$. For b) we have only to show that $(\ker f)_0$ is a complex torus, since as a compact space $\ker f$ has only a finite number of connected components. F is a linear map, hence the connected component $F^{-1}(\Lambda')_0$ of $F^{-1}(\Lambda')$ containing 0 is a subvector space of V and $(\ker f)_0 = F^{-1}(\Lambda')_0 / \left(F^{-1}(\Lambda')_0 \cap \Lambda\right)$. Finally $(\ker f)_0$ being compact, the group $F^{-1}(\Lambda')_0 \cap \Lambda$ is a lattice in $F^{-1}(\Lambda')_0$. □

As an example consider the product $X \times X'$ of the complex tori $X = V/\Lambda$ and $X' = V'/\Lambda'$. It is again a complex torus: $X \times X' = V \times V'/\Lambda \times \Lambda'$. The projections of $X \times X'$ onto its factors and the natural embeddings of X (respectively X') into $X \times X'$ are homomorphisms of complex tori. Obviously the analytic (respectively rational) representation of these homomorphisms are just the projections and natural embeddings of the corresponding vector spaces (respectively lattices).

Next we define a special class of homomorphisms of complex tori, the isogenies. They will be of particular importance in the sequel. An *isogeny* of a complex torus X to a complex torus X' is by definition a surjective homomorphism $X \to X'$ with finite kernel. Obviously a homomorphism $X \to X'$ is an isogeny if and only if it is surjective and $\dim X = \dim X'$. If $\Gamma \subseteq X$ is a finite subgroup, the quotient X/Γ is a complex torus and the natural projection $p: X \to X/\Gamma$ is an isogeny. To see this, note that $\pi^{-1}(\Gamma) \subset V$ is a lattice containing Λ and $X/\Gamma = V/\pi^{-1}(\Gamma)$. Conversely it is clear that up to isomorphisms every isogeny is of this type.

By Proposition 2.4 every surjective homomorphism $f: X \to X'$ of complex tori factorizes canonically into a surjective homomorphism g with a complex torus as kernel and an isogeny h. This is the *Stein factorization* of f:

$$\begin{array}{ccccc} & & X/(\ker f)_0 & & \\ & {\scriptstyle g}\nearrow & & \searrow{\scriptstyle h} & \\ X & & \xrightarrow{\quad f \quad} & & X' \end{array} .$$

We define the *degree* $\deg f$ of a homomorphism $f\colon X \to X'$ to be the order of the group $\ker f$, if it is finite, and 0 otherwise. So for an isogeny we have

$$\deg f = (\Lambda' \colon \det \varrho_r(f)(\Lambda)) \ ,$$

the index of the subgroup $\varrho_r(f)(\Lambda)$ in Λ'. If moreover f is an endomorphism of X, then $\Lambda = \Lambda'$ and thus

$$\det f = \det \varrho_r(f) \ . \tag{2}$$

Note that $\det \varrho_r(f)$ is positive by Proposition 2.3. Equation (2) is valid vor an arbitrary endomorphism, since both sides are zero if f is not an isogeny.

Suppose f is surjective and $g\colon X' \to X''$ is a second homomorphism. Then we have for the degree of the composition $\deg gf = \deg f \cdot \deg g$. In particular, if f and g are isogenies, the composition gf is again an isogeny.

For any integer n define the homomorphism $n_X\colon X \to X$ by $x \mapsto nx$. If $n \neq 0$, its kernel X_n is called the *group of n-division points* of X.

(2.5) Proposition. $X_n \simeq (\mathbb{Z}/n\mathbb{Z})^{2g}$.

Proof. $\ker n_X = \frac{1}{n}\Lambda/\Lambda \simeq \Lambda/n\Lambda \simeq (\mathbb{Z}/n\mathbb{Z})^{2g}$. □

The proposition implies that for any $n \neq 0$ the homomorphism n_X is an isogeny of degree n^{2g}. In particular any complex torus is a divisible group. Another consequence is that $\operatorname{Hom}(X, X')$ is torsion free as an abelian group (see also Proposition 2.2). Hence $\operatorname{Hom}(X, X')$ can and will be considered as a subgroup of $\operatorname{Hom}_{\mathbb{Q}}(X, X')$. Moreover the definition of the degree of a homomorphism extends to $\operatorname{Hom}_{\mathbb{Q}}(X, X')$ by

$$\deg(rf) := r^{2g} \deg f$$

for any $r \in \mathbb{Q}$ and $f \in \operatorname{Hom}(X, X')$.

We will see that isogenies are “almost” isomorphisms. Define the *exponent* $e = e(f)$ of an isogeny f to be the exponent of the finite group $\ker f$. In other words $e(f)$ is the smallest positive integer n with $nx = 0$ for all x in $\ker f$.

(2.6) Proposition. *For any isogeny $f\colon X \to X'$ of exponent e there exists an isogeny $g\colon X' \to X$, unique up to isomorphisms, such that $gf = e_X$ and $fg = e_{X'}$.*

Proof. As $\ker f \subseteq \ker e_X$, there is a unique map $g\colon X' \to X$ such that $gf = e_X$. With e_X and f also g is an isogeny. The kernel of g is contained in the kernel of $e_{X'}$, since for every $x' \in \ker g$ there is an $x \in \ker e_X$ with $f(x) = x'$ and $ex' = ef(x) = f(ex) = 0$. Thus $e_{X'} = f'g$ for some isogeny $f'\colon X \to X'$ and we get $f'e_X = f'gf = e_{X'}f = fe_X$. This implies $f = f'$, since e_X is surjective. □

(2.7) Corollary. *a) Isogenies define an equivalence relation on the set of complex tori.*

b) An element in End(X) *is an isogeny if and only if it is invertible in* $\mathrm{End}_{\mathbb{Q}}(X)$.

Hence it makes sense to call two complex tori *isogenous*, if there is an isogeny between them.

§ 3 Cohomology of Complex Tori

The aim of this section is to compute the singular cohomology groups of complex tori with values in $\mathbb{Z}$. Let $X = V/\Lambda$ be a complex torus of dimension g. As a real manifold X is isomorphic to the product of the $2g$ circles $S_1 \simeq S_1^{(i)} = \lambda_i \mathbb{R}/\lambda_i \mathbb{Z}$, where $\lambda_1, \dots, \lambda_{2g}$ denotes a basis of the lattice Λ. By the Künneth formula this implies that $H_n(X, \mathbb{Z})$ and $H^n(X, \mathbb{Z})$ are free abelian groups of finite rank for all $n = 1, \dots, 2g$. According to Section 1 we have identifications

$$\pi_1(X) = H_1(X, \mathbb{Z}) = \Lambda \ .$$

So by the universal coefficient theorem there is a natural isomorphism $H^1(X, \mathbb{Z}) \simeq \mathrm{Hom}(\pi_1(X), \mathbb{Z})$ (see Greenberg-Harper [1], 23.28). The following lemma shows how the higher cohomology groups can be computed out of $H^1(X, \mathbb{Z})$.

(3.1) Lemma. *The canonical map* $\bigwedge^n H^1(X, \mathbb{Z}) \longrightarrow H^n(X, \mathbb{Z})$ *induced by the cup product is an isomorphism for every* $n \geq 1$.

This is just the Künneth formula generalized to n factors. For the convenience of the reader we give a proof using only the ordinary Künneth formula.

Proof. The assertion is a consequence of the compatibility of Künneth's formula with the cup product. To be more precise, we will show by induction on m, that the canonical homomorphisms

$$\gamma_m^n \colon \bigwedge^n H^1(S_1^m, \mathbb{Z}) \to H^n(S_1^m, \mathbb{Z})$$

are bijective for every $n \geq 1$. This is trivial for $m = 1$. Suppose $m > 1$ and γ_p^n is bijective for every $p < m$ (and of course for all n). According to the compatibility of Künneth's formula with the cup product the diagram

$$\begin{array}{ccc}
\bigoplus_{p+q=n} (\bigwedge^p H^1(S_1^{m-1}, \mathbb{Z}) \otimes \bigwedge^q H^1(S_1, \mathbb{Z})) & \xrightarrow{\gamma} & \bigoplus_{p+q=n} (H^p(S_1^{m-1}, \mathbb{Z}) \otimes H^q(S_1, \mathbb{Z})) \\
\| & & \\
\bigwedge^n (H^1(S_1^{m-1}, \mathbb{Z}) \oplus H^1(S_1, \mathbb{Z})) & & \downarrow \beta_m \\
\downarrow \alpha_m & & \\
\bigwedge^n H^1(S_1^m, \mathbb{Z}) & \xrightarrow{\gamma_m^n} & H^n(S_1^m, \mathbb{Z})
\end{array}$$

commutes, where $\gamma = \bigoplus_{p+q=n} \gamma_{m-1}^p \otimes \gamma_1^q$, and α_m and β_m denote the usual Künneth homomorphisms. The groups $H^p(S_1^{m-1}, \mathbb{Z})$ and $H^q(S_1, \mathbb{Z})$ are torsion free, so Künneth's formula implies that α_m and β_m are isomorphisms. Since γ is bijective by induction hypothesis, this completes the proof. □

By what we saw above we can identify $H^1(X, \mathbb{Z}) = \operatorname{Hom}(\Lambda, \mathbb{Z})$. Denoting by $\operatorname{Alt}^n(\Lambda, \mathbb{Z})$ the group of $\mathbb{Z}$-valued alternating n-forms $\bigwedge^n \operatorname{Hom}(\Lambda, \mathbb{Z})$ on Λ, we get as a consequence

(3.2) Corollary. *There is a canonical isomorphism* $H^n(X, \mathbb{Z}) \simeq \operatorname{Alt}^n(\Lambda, \mathbb{Z})$ *for every* $n \geq 1$.

(3.3) Corollary. $H_n(X, \mathbb{Z})$ *and* $H^n(X, \mathbb{Z})$ *are free* $\mathbb{Z}$*-modules of rank* $\binom{2g}{n}$ *for all* $n \geq 1$.

For an explicit basis of $H_n(X, \mathbb{Z})$ see Exercise 8.

The universal coefficient theorem yields $H^n(X, \mathbb{C}) = H^n(X, \mathbb{Z}) \otimes \mathbb{C}$. Denoting by $\operatorname{Alt}^n_{\mathbb{R}}(V, \mathbb{C})$ the group of $\mathbb{R}$-linear alternating n-forms on V with values in $\mathbb{C}$ and applying the canonical isomorphism $\operatorname{Alt}^n(\Lambda, \mathbb{Z}) \otimes \mathbb{C} = \operatorname{Alt}^n_{\mathbb{R}}(V, \mathbb{C})$ we get

(3.4) Corollary. *For every* $n \geq 1$ *there are canonical isomorphisms*

$$H^n(X, \mathbb{C}) \simeq \operatorname{Alt}^n_{\mathbb{R}}(V, \mathbb{C}) = \bigwedge^n \operatorname{Hom}_{\mathbb{R}}(V, \mathbb{C}) \simeq \bigwedge^n H^1(X, \mathbb{C}) \ .$$

De Rham's theorem states that integration of complex valued $\mathcal{C}^\infty$-forms induces an isomorphism

$$H^n(X, \mathbb{C}) \simeq H^n_{\mathrm{DR}}(X) = \frac{\{d\text{-closed } n\text{-forms}\}}{d\{(n-1)\text{-forms}\}} \ .$$

We will explain how, in the case of a complex torus, in every class of n-forms in $H^n_{\mathrm{DR}}(X)$ one can distinguish a uniquely determined representative: fix a basis $\lambda_1, \ldots, \lambda_{2g}$ of $\Lambda = H_1(X, \mathbb{Z})$ and denote by $x_1, \ldots, x_{2g}$ the corresponding real coordinate functions of V. A complex valued $\mathcal{C}^\infty$-form ω on X (respectively V) is called *(translation) invariant*, if $t_x^* \omega = \omega$ for all $x \in X$

(respectively $t_v^*\omega = \omega$ for all $v \in V$). Obviously the differentials $dx_1, \dots, dx_{2g}$ are invariant 1-forms on V. In particular, they are invariant with respect to translation by elements of Λ. Hence every dx_i is the pullback of a uniquely determined invariant 1-form on X via $\pi: V \longrightarrow X$. By abuse of notation we denote this 1-form on X also by dx_i.

Under the de Rham isomorphism the cohomology classes of $dx_1, \dots, dx_{2g}$ on X correspond to a basis of $H^1(X, \mathbb{C})$, since we have by construction $\int_{\lambda_i} dx_j = \delta_{ij}$. In particular the bases $dx_1, \dots, dx_{2g}$ and $\lambda_1, \dots, \lambda_{2g}$ are dual to each other. The cup product corresponds under the de Rham isomorphism to the exterior product of forms. Together with Corollary 3.4 this implies that the classes of the n-forms $dx_{i_1} \wedge \dots \wedge dx_{i_n}$, $i_1 < \dots < i_n$, form a basis of $H^n(X, \mathbb{C})$. Conversely it is obvious that every invariant differential form is a linear combination of these. In other words, they span the complex vector space of invariant n-forms on X. Combining everything we obtain

(3.5) Proposition. *Let $IF^n(X)$ denote the vector space of invariant n-forms on X. The de Rham isomorphism induces an isomorphism $H^n(X, \mathbb{C}) \simeq IF^n(X)$.*

§ 4 The Hodge Decomposition

In the last section we used the real structure of the complex torus $X = V/\Lambda$ to compute the cohomology groups $H^n(X, \mathbb{C})$. Here we want to show that the complex structure of X yields a direct sum decomposition of these vector spaces, the Hodge decomposition. We include a proof here, since in the case of a complex torus it is considerably easier than for a general compact Kähler manifold (see Griffiths-Harris [1]).

(4.1) Theorem. *a) For every $n \geq 0$ the de Rham and the Dolbeault isomorphisms induce an isomorphism*

$$H^n(X, \mathbb{C}) \simeq \bigoplus_{p+q=n} H^q(\Omega_X^p)$$

with Ω_X^p the sheaf of holomorphic p-forms on X.

b) For every pair (p, q) there is a natural isomorphism

$$H^q(\Omega_X^p) \simeq \bigwedge^p \Omega \otimes \bigwedge^q \overline{\Omega}$$

with $\Omega := \mathrm{Hom}_{\mathbb{C}}(V, \mathbb{C})$ and $\overline{\Omega} := \mathrm{Hom}_{\overline{\mathbb{C}}}(V, \mathbb{C})$, the group of $\mathbb{C}$-antilinear forms on V. In particular $h^p(\Omega_X^q) = h^q(\Omega_X^p)$.

The theorem is a consequence of the following propositions and corollaries. We start by describing the sheaf Ω^p_X. Identifying, as in Section 1, the vector space V with the complex tangent space $T_{X,0}$ of X at 0, the complex cotangent space $\Omega^1_{X,0}$ of X at 0 is $\Omega = \mathrm{Hom}_{\mathbb{C}}(V, \mathbb{C})$. For any $x \in X$ the translation t_{-x} induces a vector space isomorphism $dt_{-x}: T_{X,x} \to T_{X,0}$. Using the dual isomorphism $(dt_{-x})^*: \Omega = \Omega^1_{X,0} \to \Omega^1_{X,x}$, every p-covector $\varphi \in \bigwedge^p \Omega$ extends by

$$(\omega_\varphi)_x := \Big(\bigwedge^p (dt_{-x})^*\Big)\varphi$$

to a translation invariant holomorphic p-form ω_φ on X and the map $\varphi \mapsto \omega_\varphi$ defines a homomorphism of sheaves

$$\bigwedge^p \Omega \otimes_{\mathbb{C}} \mathcal{O}_X \longrightarrow \Omega^p_X \ . \tag{1}$$

Since the holomorphic forms, coming from a basis of $\bigwedge^p \Omega$, generate every fibre of Ω^p_X, this homomorphism is in fact an isomorphism. This proves

(4.2) Lemma. *Ω^p_X is a free $\mathcal{O}_X$-module of rank $\binom{g}{p}$.*

Choose a basis $e_1, \dots, e_g$ for V and denote by $v_1, \dots, v_g$ the corresponding complex coordinate functions on V. The differentials $dv_1, \dots, dv_g, d\bar{v}_1, \dots, d\bar{v}_g$ are linearly independent over $\mathbb{C}$. Hence by Proposition 3.5 they form a basis of the vector space $IF^1(X)$ of invariant 1-forms on X. Here again, as in Section 3, we denote the invariant forms on V and the corresponding forms on X by the same letter. For a multi-index $I = (i_1 < \dots < i_p)$ we write for abbreviation $dv_I = dv_{i_1} \wedge \ldots \wedge dv_{i_p}$ and $d\bar{v}_I = d\bar{v}_{i_1} \wedge \ldots \wedge d\bar{v}_{i_p}$. An element $\varphi \in IF^n(X)$ of the form

$$\varphi = \sum_{\substack{\#I=p \\ \#J=q}} \alpha_{IJ} dv_I \wedge d\bar{v}_J$$

with $\alpha_{IJ} \in \mathbb{C}$ and $p + q = n$, is called to be *of type* (p, q). This definition does not depend on the choice of the basis for V. Denoting by $IF^{p,q}(X)$ the vector space of all forms of type (p,q) in $IF^n(X)$ we get the direct sum decomposition

$$IF^n(X) = \bigoplus_{p+q=n} IF^{p,q}(X) \ .$$

Note that $IF^{p,0}(X)$ is just the space $H^0(\Omega^p_X)$ of global sections of the sheaf Ω^p_X. So $IF^{p,0}(X) \simeq \bigwedge^p \Omega$ by (1). Similarly there is an isomorphism $\bigwedge^q \overline{\Omega} \simeq IF^{0,q}(X)$ given by $\varphi \mapsto \omega_\varphi$ with $(\omega_\varphi)_x := \big(\bigwedge^q (dt_{-x})^*\big)\varphi$ for all $\varphi \in \bigwedge^q \overline{\Omega}$ and $x \in X$. So we get an isomorphism $\bigwedge^p \Omega \otimes \bigwedge^q \overline{\Omega} \simeq IF^{p,q}(X)$ by associating to every $\varphi_1 \otimes \varphi_2 \in \bigwedge^p \Omega \otimes \bigwedge^q \overline{\Omega}$ the invariant (p,q)-form $\omega_{\varphi_1 \otimes \varphi_2} := \omega_{\varphi_1} \wedge \omega_{\varphi_2}$. Combining this with the isomorphism of Proposition 3.5 we obtain

(4.3) Proposition. *There are natural isomorphisms*

$$H^n(X,\mathbb{C}) \simeq \bigoplus_{p+q=n} IF^{p,q}(X) \simeq \bigoplus_{p+q=n} \bigwedge^p \Omega \otimes \bigwedge^q \bar{\Omega}\,.$$

This is the first step in the proof of Theorem 4.1. In a second step we compute the cohomology groups $H^q(\Omega_X^p)$. For this denote by $\mathcal{A}_X^{p,q}$ the sheaf of complex valued $\mathcal{C}^\infty$-forms of type (p,q) on X. The definition of a $\mathcal{C}^\infty$-form of type (p,q) is analogous to the definition of an invariant form of type (p,q). According to Griffiths-Harris [1] p.42 the sheaf $\mathcal{A}_X^{p,q}$ is fine for every pair (p,q). Hence the Dolbeault resolution

$$0 \longrightarrow \Omega_X^p \longrightarrow \mathcal{A}_X^{p,0} \xrightarrow{\bar{\partial}} \mathcal{A}_X^{p,1} \xrightarrow{\bar{\partial}} \cdots \tag{2}$$

(see Griffiths-Harris [1] p.45) gives an isomorphism

$$H^q(\Omega_X^p) \simeq H^{p,q}(X) := \frac{\{\bar{\partial}\text{-closed } (p,q)\text{-forms}\}}{\bar{\partial}H^0(\mathcal{A}_X^{p,q-1})}\,. \tag{3}$$

It remains to show that for every pair (p,q) there is a canonical isomorphism $H^{p,q}(X) \xrightarrow{\sim} IF^{p,q}(X)$, since then the compositions

$$H^n(X,\mathbb{C}) \simeq \bigoplus_{p+q=n} IF^{p,q}(X) \simeq \bigoplus_{p+q=n} H^{p,q}(X) \simeq \bigoplus_{p+q=n} H^q(\Omega_X^p)$$

and
$$H^q(\Omega_X^p) \simeq H^{p,q}(X) \simeq \mathrm{IF}^{p,q}(X) \simeq \bigwedge^p \Omega \otimes \bigwedge^q \bar{\Omega}$$

give the assertions of Theorem 4.1.

We will do this by showing that $H^{p,q}(X)$ and $IF^{p,q}(X)$ both can be identified with the group of harmonic (p,q)-forms on X, which we are going to define next. For this we introduce a Kähler metric on X: any positive definite hermitian form on the vector space V determines a translation invariant hermitian metric ds^2 on V and consequently a translation invariant hermitian metric on X. We will work with the hermitian metric on X given by the identity matrix with respect to the chosen basis $e_1,\dots,e_g$ of V:

$$ds^2 = \sum_{\nu=1}^{g} dv_\nu \otimes d\bar{v}_\nu\,.$$

Its associated $(1,1)$-form $\omega = \frac{i}{2}\sum_{\nu=1}^{g} dv_\nu \wedge d\bar{v}_\nu$ is d-closed, which means, that ds^2 is a Kähler metric on X. In particular every complex torus is a Kähler manifold. For abbreviation denote the corresponding volume form by

$$dv = \frac{1}{g!}\bigwedge^g \omega = (-1)^{\frac{1}{2}g(g-1)}(\tfrac{i}{2})^g dv_1 \wedge \ldots \wedge dv_g \wedge d\bar{v}_1 \wedge \ldots \wedge d\bar{v}_g\,.$$

Moreover denote by $A^{p,q}$ the space of global sections of the sheaf $\mathcal{A}_X^{p,q}$, that is the vector space of complex valued $\mathcal{C}^\infty$-forms of type (p,q) on X. Every element of $A^{p,q}$ can be uniquely written as $\varphi = \sum \varphi_{IJ} dv_I \wedge d\bar{v}_J$ with complex valued $\mathcal{C}^\infty$-functions φ_{IJ} on X and where the sum runs over all multi-indices I of length p and J of length q. If $\psi = \sum \psi_{IJ} dv_I \wedge d\bar{v}_J$ is a second (p,q)-form, define

$$(\varphi, \psi) = \sum_{\substack{\#I=p \\ \#J=q}} \int_X \varphi_{IJ} \bar{\psi}_{IJ} dv \ .$$

This is the global inner product associated to the metric ds^2. It makes the vector space $A^{p,q}$ into a pre-Hilbert space. Define a linear operator $\bar{\delta}: A^{p,q+1} \longrightarrow A^{p,q}$ by

$$\bar{\delta}(\varphi dv_I \wedge d\bar{v}_K) = \sum_{\nu=1}^{q+1} (-1)^{p+\nu} \frac{\partial \varphi}{\partial v_{k_\nu}} dv_I \wedge d\bar{v}_{K-k_\nu} \ ,$$

where for the multi-index $K = (k_1, \ldots, k_{q+1})$ by $K - k_\nu$ we denote the multi-index $(k_1, \ldots, k_{\nu-1}, k_{\nu+1}, \ldots, k_{q+1})$. The following lemma shows in particular that

$$\triangle := \bar{\partial}\bar{\delta} + \bar{\delta}\bar{\partial}: A^{p,q} \to A^{p,q}$$

is the Laplace operator on $A^{p,q}$ associated to $\bar{\partial}$.

(4.4) Lemma. *Suppose* $\varphi dv_I \wedge d\bar{v}_J \in A_X^{p,q}$.

a) The operator $\bar{\delta}$ *is adjoint to the operator* $\bar{\partial}: A^{p,q} \to A^{p,q+1}$ *with respect to the inner product* $(\ , \ )$.

b) $\triangle(\varphi dv_I \wedge d\bar{v}_J) = - \sum_{\nu=1}^{g} \frac{\partial^2 \varphi}{\partial v_\nu \partial \bar{v}_\nu} dv_I \wedge d\bar{v}_J$.

The proof is an immediate computation using the product formula and Stokes' theorem. □

A form $\varphi \in A^{p,q}$ is called *harmonic*, if $\triangle \varphi = 0$. Let $\mathcal{H}^{p,q}$ denote the vector space of harmonic forms of type (p,q) on X. We want to show that $\mathcal{H}^{p,q} = IF^{p,q}(X)$. For this define a linear operator $H: A^{p,q} \longrightarrow A^{p,q}$ by

$$H(\varphi dv_I \wedge d\bar{v}_J) = \frac{1}{\mathrm{vol}(X)} (\int_X \varphi dv) dv_I \wedge d\bar{v}_J$$

where $\mathrm{vol}(X) := \int_X dv$. This is a projection of $A^{p,q}$ onto the vector space $IF^{p,q}(X)$. In particular $H^2 = H$. Moreover we have

$$\bar{\partial} H = H \bar{\partial} = 0 \ , \tag{4}$$

using $\frac{\partial \varphi}{\partial \bar{v}_\nu} dv = (-1)^{\frac{1}{2}g(g+1)+\nu-1} (\frac{i}{2})^g \, d(\varphi \, dv_1 \wedge \ldots \wedge \check{d\bar{v}_\nu} \wedge \ldots \wedge d\bar{v}_g)$ and Stokes' theorem.

The main point in the proof of the Hodge decomposition is the existence of an operator G, called *Green's operator*, which one can associate to H and $\triangle$. For a general Kähler manifold, G is constructed using the theory of Sobolev spaces. However, in the case of a complex torus G can be defined explicitly using Fourier expansions. This is the main simplification of the proof of the Hodge decomposition.

(4.5) Proposition. *There is a* $\mathbb{C}$*-linear operator* $G\colon A^{p,q} \longrightarrow A^{p,q}$ *satisfying* $\triangle G = G \triangle = 1 - H$ *and* $HG = GH = 0$.

Proof. For any $\varphi dv_I \wedge d\bar{v}_J \in A^{p,q}$ the complex valued $\mathcal{C}^\infty$-function $\pi^*\varphi$ on V is periodic with respect to the lattice Λ. Considering V as a real vector space $\pi^*\varphi$ is $2g$-fold periodic and thus admits a Fourier expansion. Let $x_1, \ldots, x_{2g}$ be the real coordinates of a vector x in V with respect to the real basis $e_1, \ldots, e_g, ie_1, \ldots, ie_g$ of V. In other words, if $x = (v_1, \ldots, v_g)$ with respect to $e_1, \ldots, e_g$, then $x_\nu = \operatorname{Re} v_\nu$ and $x_{g+\nu} = \operatorname{Im} v_\nu$ for $1 \le \nu \le g$. The group $\Lambda^\perp = \{y \in V \mid e(2\pi i \sum_{\nu=1}^{2g} \lambda_\nu y_\nu) = 1 \text{ for all } \lambda \in \Lambda\}$ is a lattice in V. Let

$$\pi^*\varphi(x) = \sum_{y \in \Lambda^\perp} a_y e(2\pi i \sum_{\nu=1}^{2g} x_\nu y_\nu)\ , \quad a_y \in \mathbb{C}\ ,$$

be the Fourier expansion of $\pi^*\varphi$. Define the operator $G\colon A^{p,q} \longrightarrow A^{p,q}$ by

$$G(\varphi dv_I \wedge d\bar{v}_J) = \varphi' dv_I \wedge d\bar{v}_J$$

where φ' is the unique complex valued $\mathcal{C}^\infty$-function on X with

$$\pi^*\varphi'(x) = \frac{1}{(2\pi)^2} \sum_{\substack{y \in \Lambda^\perp \\ y \neq 0}} \frac{a_y}{\sum y_\nu^2} e(2\pi i \sum_{\nu=1}^{2g} x_\nu y_\nu)\ .$$

In fact, $\pi^*\varphi'$ is a $\mathcal{C}^\infty$-function on V, since $\sum y_\nu^2 > 1$ for almost all $y \in \Lambda^\perp$, and thus the series converges uniformly on every compact subset of V. It is easy to verify that G satisfies the assertions. □

As a consequence we get the desired relation between harmonic forms and invariant forms on X.

(4.6) Corollary. $\mathcal{H}^{p,q} = IF^{p,q}(X)$, *that is* $\omega \in A^{p,q}$ *is harmonic if and only if* ω *is invariant.*

Proof. We have $IF^{p,q}(X) \subset \ker \triangle = \mathcal{H}^{p,q}$ by Lemma 4.4. The equation $G \triangle = 1 - H$ implies the converse inclusion. □

We need some further consequences of Proposition 4.5 concerning $\triangle, G, H$ and $\bar\partial$. First we claim that $\operatorname{im} \triangle = \ker H$. For the proof note that $\ker H \subset$

$\operatorname{im}\triangle$. This is a consequence of $\triangle G = 1$ on $\ker H$. Conversely, for any $\varphi = \triangle\psi$ we have $H\varphi = H\triangle\psi = H\triangle(\psi - H\psi) = (HG)\triangle^2\psi = 0$. This implies the assertion.

Since H is a projection, it follows that $A^{p,q}$ decomposes as

$$A^{p,q} = \ker H \oplus \operatorname{im} H = \operatorname{im}\triangle \oplus \operatorname{im} H \ .$$

Hence any $\varphi \in A^{p,q}$ is of form $\varphi = \triangle\psi_1 + H\psi_2$ for some $\psi_1, \psi_2 \in A^{p,q}$ and we get using (4) and Proposition 4.5

$$\begin{aligned} G\bar\partial\varphi &= G\bar\partial\triangle\psi_1 + G\bar\partial H\psi_2 \\ &= G\triangle\bar\partial\psi_1 = \bar\partial\psi_1 - H\bar\partial\psi_1 = \bar\partial\psi_1 - \bar\partial H\psi_1 \\ &= \bar\partial G\triangle\psi_1 = \bar\partial G\triangle\psi_1 + \bar\partial G H\psi_2 = \bar\partial G\varphi. \end{aligned}$$

Thus we have shown: $$\bar\partial G = G\bar\partial \ . \tag{5}$$

Finally we have

(4.7) Proposition. *H induces an isomorphism $H^{p,q}(X) \xrightarrow{\sim} IF^{p,q}(X)$.*

Proof. Consider the projection map $H\colon A^{p,q} \longrightarrow \operatorname{Im} H = IF^{p,q}(X)$. Its restriction to $\ker\bar\partial$ is surjective, since $\operatorname{Im} H \subset \ker\bar\partial$ by (4). On the other hand $H^{p,q}(X) = \ker\bar\partial/\bar\partial A^{p,q-1}$. So it remains to show that $\bar\partial A^{p,q-1} = \ker(H\colon \ker\bar\partial \longrightarrow IF^{p,q}(X))$. Applying Proposition 4.5 and (5) we have for $\varphi \in \ker H \cap \ker\bar\partial$

$$\varphi = H\varphi + G\triangle\varphi = G\triangle\varphi = G\bar\partial\,\bar\delta\varphi = \bar\partial G\bar\delta\varphi.$$

Hence $\varphi \in \bar\partial A^{p,q-1}$. The converse inclusion is a consequence of $H\bar\partial = 0$. □

This completes the proof of Theorem 4.1. □

Exercises

(1) Let $X = V/\Lambda$ be a complex torus.
- a) Show that X admits a complex subtorus of dimension g' if and only if there exists a subgroup $\Lambda' \subset \Lambda$ of rank $2g'$ such that the image of the canonical map $\Lambda' \otimes \mathbb{R} \to V$ is a $\mathbb{C}$-subvector space of V.
- b) Conclude from a) that any complex torus admits at most countably many complex subtori.
- c) Give an example of a complex torus of dimension ≥ 2 not admitting any nontrivial complex subtorus.

(2) Let $X = V/\Lambda$ be a complex torus of dimension g. Show that
a) there exist bases of V and Λ with respect to which the period matrix of X is of the form $(Z, \mathbb{1}_g)$ with $Z \in M_g(\mathbb{C})$ and $\det \operatorname{Im} Z \neq 0$,
b) $\det \begin{pmatrix} Z & \mathbb{1} \\ \bar{Z} & \mathbb{1} \end{pmatrix} = \det(2i \operatorname{Im} Z)$.

(3) For a complex torus X of dimension g: $\quad h^{1,1}(X) = \dim H^{1,1}(X) = g^2$.

(4) *(Real Tori)* Let V be an $\mathbb{R}$-vector space of dimension n and Λ a lattice in V. The quotient group $T = V/\Lambda$ has a unique structure of a real analytic manifold such that the canonical projection $p: V \to T$ is real analytic. T is a connected compact abelian real Lie group, called *real torus of dimension n*.
a) Any two real tori of dimension n are isomorphic as real Lie groups.
b) For any connected abelian real Lie group G of dimension n, there is an integer $m \leq n$ such that $G \simeq T \times \mathbb{R}^{n-m}$ with T a real torus of dimension m. In particular, any connected compact abelian real Lie group is a real torus and any simply connected abelian real Lie group is an $\mathbb{R}$-vector space.
c) Let S be a connected closed subgroup of a real torus T. Then S and T/S are real tori and $T \simeq S \times T/S$.

(5) Let X be a complex torus and $\varphi: X \to G$ a holomorphic map into a complex Lie group G. Show that the map $X \to G$, $x \mapsto \varphi(0)^{-1}\varphi(x)$ is a homomorphism of complex Lie groups.

(6) a) There is a bijection between the set of complex structures on the vector space $\mathbb{R}^{2g}$ and $\mathrm{GL}_{2g}(\mathbb{R})/\mathrm{GL}_g(\mathbb{C})$.
b) This induces a bijection between the set of isomorphism classes of complex tori of dimension g and the set of orbits in $\mathrm{GL}_{2g}(\mathbb{R})/\mathrm{GL}_g(\mathbb{C})$ under the natural action of $\mathrm{GL}_{2g}(\mathbb{Z})$.

(7) *(Pontryagin Product)* Let G be a real Lie group of dimension g. Let $\sigma: \Delta_p \to G$ and $\tau: \Delta_q \to G$ be singular p- respectively q-simplices. Here Δ_p denotes the standard p-simplex. If we divide the product $\Delta_p \times \Delta_q$ into $(p+q)$-simplices, then the map $\sigma \star \tau: \Delta_p \times \Delta_q \to G$, $\sigma \star \tau(s,t) = \sigma(s)\tau(t)$ is a singular $(p+q)$-chain. For singular p- and q-chains $\sigma = \sum m_i\sigma_i$ and $\tau = \sum n_j\tau_j$ define $\sigma \star \tau = \sum m_i n_j \sigma_i \star \tau_j$.
a) Show that the boundary operator ∂ satisfies

$$\partial(\sigma \star \tau) = (-1)^{\epsilon_1}\partial(\sigma) \star \tau + (-1)^{\epsilon_2}\sigma \star \partial(\tau),$$

where ϵ_1 and ϵ_2 are integers depending on p and q. Hence $\star$ induces a bilinear map $\star: H_p(G,\mathbb{Z}) \times H_q(G,\mathbb{Z}) \to H_{p+q}(G,\mathbb{Z})$, $[\sigma] \star [\tau] = [\sigma \star \tau]$, called the *Pontryagin product*. In terms of the multiplication map $\mu: G \times G \to G$ the Pontryagin product coincides with the composition $H_p(G,\mathbb{Z}) \times H_q(G,\mathbb{Z}) \xrightarrow{\times} H_{p+q}(G \times G, \mathbb{Z}) \xrightarrow{\mu_*} H_{p+q}(G,\mathbb{Z})$, where $\times$ denotes the exterior homology product.

b) For p-, q- and r-cycles σ, τ and λ, and the unit element $1 \in G$ show
   i) $[1] \star [\sigma] = [\sigma] \star [1]$,
   ii) $([\sigma] \star [\tau]) \star [\lambda] = [\sigma] \star ([\tau] \star [\lambda])$,
   iii) $[\sigma] \star [\tau] = (-1)^{pq} [\tau] \star [\sigma]$, if G is commutative.

c) Let $\iota: G' \rightarrow G$ be a Lie subgroup of dimension g'. Show that for any $[\sigma] \in H_p(G', \mathbb{Z})$ and $[\tau] \in H_{g'-p}(G', \mathbb{Z})$ and $[\lambda] \in H_{g-g'}(G, \mathbb{Z})$

$$(\iota_*[\sigma] \cdot (\iota_*[\tau] \star [\lambda]))_G = (-1)^{\epsilon}([\sigma] \cdot [\tau])_{G'}([G'] \cdot [\lambda])_{G},$$

where ϵ depends on g, g' and p and $(\,\cdot\,)_G$ and $(\,\cdot\,)_{G'}$ denote the intersection numbers in G and G' respectively.

d) Let G' and $[\lambda]$ be as above. Use c) to show that, if $[\sigma_1], \ldots, [\sigma_n]$ are linearly independent elements in $H_\bullet(G', \mathbb{Z})$, then the elements $\iota_*[\sigma_1], \ldots, \iota_*[\sigma_n]$, $\iota_*[\sigma_1] \star [\lambda], \ldots, \iota_*[\sigma_n] \star [\lambda]$ are linearly independent in $H_\bullet(G, \mathbb{Z})$ (see Pontryagin [1]).

(8) Let $X = V/\Lambda$ be a complex torus of dimension g and $\lambda_1, \ldots, \lambda_{2g}$ a basis of Λ. Via the identification $\Lambda = H_1(X, \mathbb{Z})$ of Section 3 the λ_i's can be considered as elements of $H_1(X, \mathbb{Z})$. Show that $\{\lambda_{i_1} \star \cdots \star \lambda_{i_p} \mid 1 \leq i_1 < \cdots < i_p \leq 2g\}$ is a basis of $H_p(X, \mathbb{Z})$ for any $1 \leq p \leq 2g$. (Hint: apply induction on p and use Exercise 7 d) and Corollary 3.3. A different proof will be given in Lemma 4.10.1.)

(9) Let X be a complex torus, $\mu: X \times X \rightarrow X$ the addition map and $p_i: X \times X \rightarrow X$, $i = 1, 2$, the natural projections. Show that a $\mathcal{C}^\infty$-1-form ω on X is translation-invariant if and only if $\mu^*\omega = p_1^*\omega + p_2^*\omega$.

(10) Let Γ be a free abelian group of finite rank. A *Hodge-structure of weight* 1 *on* Γ is a decomposition $\Gamma \otimes \mathbb{C} = H^{0,1} \oplus H^{1,0}$, where $H^{0,1}$ and $H^{1,0}$ are complex subvector spaces of $\Gamma \otimes \mathbb{C}$ with $H^{1,0} = \overline{H^{0,1}}$. Show that giving a Hodge structure of weight 1 on Γ is equivalent to giving a complex structure on the real torus $(\Gamma \otimes \mathbb{R})/\Gamma$, i.e. an isomorphism of real tori $(\Gamma \otimes \mathbb{R})/\Gamma \rightarrow X$, where X is a complex torus.

Chapter 2

Line Bundles on Complex Tori

In this chapter we describe the structure of the group $\mathrm{Pic}(X)$ of holomorphic line bundles on a complex torus $X = V/\Lambda$. The main result is the *Appell-Humbert Theorem*, which says that $\mathrm{Pic}(X)$ is an extension of the Néron-Severi group $\mathrm{NS}(X)$ by the group $\mathrm{Hom}(\Lambda, \mathbb{C}_1)$ of characters of Λ with values in the circle group $\mathbb{C}_1$. The group $\mathrm{NS}(X)$ turns out to be the group of hermitian forms H on V satisfying $\mathrm{Im}\, H(\Lambda, \Lambda) \subseteq \mathbb{Z}$. The theorem was proven for dimension 2 by Humbert [1] applying a result of Appell [1] and by Lefschetz [1] in general. The present formulation appears in Weil [3] and Mumford [2].

Holomorphic line bundles on a complex torus can be described in terms of factors of automorphy. This construction is basic for this chapter. We recall it in Appendix B for the convenience of the reader. An alternative statement of the Appell-Humbert Theorem is as follows: there is a canonical way to associate to any line bundle L on X a factor of automorphy. It is called the *canonical factor* of L. Canonical factors are a powerful and convenient tool for proofs. For example, they immediately yield the *Theorem of the Square* and the *Theorem of the Cube*. Other consequences are the construction of the *dual complex torus* $\widehat{X}$ parametrizing line bundles on X with first Chern class zero together with a universal line bundle $\mathcal{P}$ on $X \times \widehat{X}$, called the *Poincaré bundle*.

The chapter is organized as follows: In Section 1 we compute the first Chern class $c_1(L)$ of a line bundle L in terms of a factor of automorphy and determine the Néron-Severi group of X. Section 2 contains a proof of the Appell-Humbert Theorem. In Section 3 the main properties of canonical factors are derived and Sections 4 and 5 contain the constructions of the dual complex torus and the Poincaré bundle as well as some consequences. An important tool is the homomorphism $\phi_L: X \to \widehat{X}$ associated to a line bundle L. We show that two line bundles L_1 and L_2 on X are analytically equivalent if and only if $\phi_{L_1} = \phi_{L_2}$. Finally we give a criterion (Theorem 5.5) for an isogeny $f: X \to \widehat{X}$ to be of the form $f = \phi_L$.

As for prerequisites: the proof of Theorem 1.2 requires a good understanding of cocycles in group and Čech cohomology (the reader might find it hard). The basics for this are given in Appendix B. In the proof of Lemma 1.1 the $\bar\partial$-Poincaré lemma and the fact that $H^2(V, \mathbb{Z}) = 0$ are used. Finally, in Section 5 we apply Zariski's Main Theorem for normal complex analytic spaces.

§ 1 Line Bundles on Complex Tori

Let $X = V/\Lambda$ be a complex torus. In this section we want to introduce line bundles on X. We compute the first Chern class $c_1(L)$ for any holomorphic line bundle L on X in terms of a factor of automorphy of L. Moreover we determine the Néron-Severi group of X.

Denote the group of holomorphic line bundles on X by $\mathrm{Pic}(X)$. It can be identified with $H^1(X, \mathcal{O}_X^*)$ in a natural way. As we saw in Section 1.1, the canonical projection $\pi: V \to X$ is the universal covering and the lattice Λ may be identified with the fundamental group $\pi_1(X)$. In Appendix B we show that the line bundles on X, whose pullback to V is trivial, can be described by a factor of automorphy. The following lemma implies that this holds for every line bundle on X.

(1.1) Lemma. *Every holomorphic line bundle on a complex vector space V is trivial.*

Proof. From the exponential sequence $0 \longrightarrow \mathbb{Z} \longrightarrow \mathcal{O}_V \xrightarrow{e(2\pi i\cdot)} \mathcal{O}_V^* \longrightarrow 1$ we obtain the exact sequence $H^1(\mathcal{O}_V) \to H^1(\mathcal{O}_V^*) \to H^2(V, \mathbb{Z})$. But $H^1(\mathcal{O}_V) = 0$ by the $\bar\partial$-Poincaré lemma (see Griffiths-Harris [1] p.46), whereas one knows from Algebraic Topology that $H^2(V, \mathbb{Z}) = 0$. This implies the assertion. □

Combining Lemma 1.1 with Proposition B.1, we get a canonical isomorphism

$$\mathrm{Pic}(X) \simeq H^1(\Lambda, H^0(\mathcal{O}_V^*)) \,.$$

In other words, any holomorphic line bundle on X can be described by a factor of automorphy.

In order to introduce the first Chern class of a line bundle consider the exact sequence $0 \to \mathbb{Z} \to \mathcal{O}_X \to \mathcal{O}_X^* \to 1$ and its long cohomology sequence

$$\longrightarrow H^1(X, \mathbb{Z}) \longrightarrow H^1(\mathcal{O}_X) \longrightarrow H^1(\mathcal{O}_X^*) \xrightarrow{c_1} H^2(X, \mathbb{Z}) \longrightarrow \,.$$

The image $c_1(L)$ of a line bundle $L \in H^1(\mathcal{O}_X^*)$ in $H^2(X, \mathbb{Z})$ is called the *first Chern class of* L. According to Corollary 1.3.2 the groups $H^2(X, \mathbb{Z})$ and $\mathrm{Alt}^2(\Lambda, \mathbb{Z})$ are canonically isomorphic. Hence we may consider $c_1(L)$ as an alternating $\mathbb{Z}$-valued form on the lattice Λ. The following theorem shows how to compute $c_1(L)$ in terms of a factor of automorphy f of L. Note that any such f can be written in the form $f = e(2\pi i g)$ with a map $g\colon \Lambda \times V \to \mathbb{C}$, holomorphic in the second variable.

(1.2) Theorem. *There is a canonical isomorphism $H^2(X, \mathbb{Z}) \to \mathrm{Alt}^2(\Lambda, \mathbb{Z})$, which maps the first Chern class $c_1(L)$ of a line bundle L on X with factor of automorphy $f = e(2\pi i g)$ to the alternating form*

$$E_L(\lambda, \mu) = g(\mu, v + \lambda) + g(\lambda, v) - g(\lambda, v + \mu) - g(\mu, v)$$

for all $\lambda, \mu \in \Lambda$ and $v \in V$.

For the proof we need the following two lemmas. As usual let $Z^2(\Lambda, \mathbb{Z})$ denote the group of 2-cocycles on Λ with values in $\mathbb{Z}$.

(1.3) Lemma. *The map $\alpha\colon Z^2(\Lambda, \mathbb{Z}) \to \mathrm{Alt}^2(\Lambda, \mathbb{Z})$ defined by $\alpha F(\lambda, \mu) = F(\lambda, \mu) - F(\mu, \lambda)$ induces a canonical isomorphism*

$$\alpha\colon H^2(\Lambda, \mathbb{Z}) \to \mathrm{Alt}^2(\Lambda, \mathbb{Z}) .$$

Proof. A 2-cocycle $F \in Z^2(\Lambda, \mathbb{Z})$ is a map $F\colon \Lambda \times \Lambda \to \mathbb{Z}$ satisfying

$$\partial F(\lambda, \mu, \nu) := F(\mu, \nu) - F(\lambda + \mu, \nu) + F(\lambda, \mu + \nu) - F(\lambda, \mu) = 0$$

for all $\lambda, \mu, \nu \in \Lambda$. Hence

$$\begin{aligned}\alpha F(\lambda + \mu, \nu) - \alpha F(\lambda, \nu) - \alpha F(\mu, \nu) &= \\ = \partial F(\lambda, \nu, \mu) - \partial F(\nu, \lambda, \mu) - \partial F(\lambda, \mu, \nu) &= 0 ,\end{aligned}$$

so αF is an alternating bilinear form. Obviously $\alpha\colon Z^2(\Lambda, \mathbb{Z}) \to \mathrm{Alt}^2(\Lambda, \mathbb{Z})$ is a homomorphism of groups. Moreover for the group $B^2(\Lambda, \mathbb{Z})$ of 2-coboundaries of Λ with values in $\mathbb{Z}$ we have $\alpha(B^2(\Lambda, \mathbb{Z})) = 0$, since the elements $\partial f(\lambda, \mu) = f(\mu) - f(\lambda + \mu) + f(\lambda)$ of $B^2(\Lambda, \mathbb{Z})$ are symmetric in λ and μ. It follows that α descends to a homomorphism $H^2(\Lambda, \mathbb{Z}) \to \mathrm{Alt}^2(\Lambda, \mathbb{Z})$, which we also denote by α.

We first claim that α is surjective. To see this, note that for all $f, g \in \mathrm{Hom}(\Lambda, \mathbb{Z})$ the map $f \otimes g$ is in $Z^2(\Lambda, \mathbb{Z})$ and

$$\alpha(f \otimes g)(\lambda, \mu) = f \otimes g(\lambda, \mu) - f \otimes g(\mu, \lambda) = f \wedge g(\lambda, \mu) .$$

But the elements $f \wedge g$ generate $\bigwedge^2 \mathrm{Hom}(\Lambda, \mathbb{Z}) = \mathrm{Alt}^2(\Lambda, \mathbb{Z})$. Finally the groups $H^2(\Lambda, \mathbb{Z})$ and $\mathrm{Alt}^2(\Lambda, \mathbb{Z})$ both are free $\mathbb{Z}$-modules of the same rank by Corollaries 1.3.2 and 1.3.3 and Remark B.2. Hence α is also injective. □

Consider the exact sequence

$$0 \longrightarrow \mathbb{Z} = H^0(V, \mathbb{Z}) \longrightarrow H^0(\mathcal{O}_V) \overset{e(2\pi i \cdot)}{\longrightarrow} H^0(\mathcal{O}_V^*) \longrightarrow 1\,. \tag{1}$$

The lattice Λ acts on each of these cohomology groups in a compatible way, so that (1) induces a long exact cohomology sequence. In particular we get a connecting homomorphism $\delta: H^1(\Lambda, H^0(\mathcal{O}_V^*)) \to H^2(\Lambda, \mathbb{Z})$. By definition δ maps the cocycle $f = e(2\pi i g) \in Z^1(\Lambda, H^0(\mathcal{O}_V^*))$ to the 2-cocycle

$$\delta f(\lambda, \mu) = g(\mu, v + \lambda) - g(\lambda + \mu, v) + g(\lambda, v)$$

in $Z^2(\Lambda, \mathbb{Z})$, where $\lambda, \mu \in \Lambda$ and $v \in V$. Note that δf does not depend on the variable v, since f satisfies the *cocycle relation* $f(\lambda + \mu, v) = f(\mu + \lambda, v) = f(\mu, v + \lambda) f(\lambda, v)$ for all $\lambda, \mu \in \Lambda$ and $v \in V$ (see Appendix B).

The following lemma implies that the composed map $\alpha\delta: H^1(\Lambda, H^0(\mathcal{O}_V^*)) \to \mathrm{Alt}^2(\Lambda, \mathbb{Z})$ is just the homomorphism associating to a line bundle its first Chern class considered as an alternating 2-form on Λ. Denoting by ϕ_1 and ϕ_2 the canonical homomorphisms defined in Proposition B.1 and Remark B.2, we have

(1.4) Lemma. *ϕ_1 and ϕ_2 are isomorphisms and the following diagram is commutative*

$$\begin{array}{ccc} H^1(\Lambda, H^0(\mathcal{O}_V^*)) & \overset{\delta}{\longrightarrow} & H^2(\Lambda, \mathbb{Z}) \\ \downarrow{\scriptstyle \phi_1} & & \downarrow{\scriptstyle \phi_2} \\ H^1(\mathcal{O}_X^*) & \overset{c_1}{\longrightarrow} & H^2(X, \mathbb{Z}) \end{array} .$$

Proof. We will use the notation of Step I in the proof of Proposition B.1, where one has to replace $\pi_1(X)$ (resp. $\tilde{X}$) by Λ (resp. V). ϕ_1 is an isomorphism by Proposition B.1 and Lemma 1.1. A proof analogous to that of Proposition B.1 shows that ϕ_2 is also an isomorphism. It is easily checked that the diagram commutes, since the homomorphisms ϕ_1, ϕ_2 and c_1 are given as follows:

$(\phi_1 f)_{ij} = f(\lambda_{ij}, \pi_i^{-1})$ for $f \in Z^1(\Lambda, H^0(\mathcal{O}_V^*))$

$(c_1 h)_{ijk} = g_{jk} - g_{ik} + g_{ij}$ for $h = \{h_{ij} = e(2\pi i g_{ij})\}_I \in Z^1(X, \mathcal{O}_X^*)$ and

$(\phi_2 F)_{ijk} = F(\lambda_{ij}, \lambda_{jk}, \pi_i^{-1})$ for $F \in Z^2(\Lambda, H^0(V, \mathbb{Z})) = Z^2(\Lambda, \mathbb{Z})$

and for all $i, j, k \in I$. □

Proof of Theorem 1.2. The canonical isomorphism in question is the composed map $\alpha\phi_2^{-1}: H^2(X, \mathbb{Z}) \to \mathrm{Alt}^2(\Lambda, \mathbb{Z})$. In fact, according to Lemma 1.4 the cocycle δf represents the element $\phi_2^{-1} c_1(L)$ of $H^2(\Lambda, \mathbb{Z})$. An immediate computation gives $\alpha\phi_2^{-1} c_1(L) = \alpha\delta f = E_L$. □

One can show that the canonical isomorphism $\alpha\phi_2^{-1}\colon H^2(X,\mathbb{Z}) \to \mathrm{Alt}^2(\Lambda,\mathbb{Z})$ coincides with the canonical isomorphism of Corollary 1.3.2 (see Exercise 1 b)), but we do not need this fact. Its inverse isomorphism $\mathrm{Alt}^2(\Lambda,\mathbb{Z}) \to H^2(X,\mathbb{Z})$ extends to an isomorphism

$$\beta_2\colon \mathrm{Alt}^2(\Lambda,\mathbb{Z})\otimes\mathbb{C} = \mathrm{Alt}^2(\Lambda,\mathbb{C}) \to H^2(X,\mathbb{C})\ .$$

For both vector spaces we have a direct sum decomposition. On the one hand, $\mathrm{Alt}^2(\Lambda,\mathbb{C}) = \mathrm{Alt}^2_{\mathbb{R}}(V,\mathbb{C})$ can be identified with $\bigwedge^2\Omega\oplus(\Omega\otimes\overline{\Omega})\oplus\bigwedge^2\overline{\Omega}$, where $\Omega = \mathrm{Hom}_{\mathbb{C}}(V,\mathbb{C})$ and $\overline{\Omega} = \mathrm{Hom}_{\overline{\mathbb{C}}}(V,\mathbb{C})$ as in Section 1.4. On the other hand, there is the Hodge decomposition for $H^2(X,\mathbb{C})$. We want to show that the isomorphism β_2 respects these decompositions. To be more precise, we want to show that the following diagram is commutative

$$\begin{array}{ccccc}
\mathrm{Alt}^2(\Lambda,\mathbb{C}) & = & \mathrm{Alt}^2_{\mathbb{R}}(V,\mathbb{C}) & \simeq & \bigwedge^2\Omega\oplus(\Omega\otimes\overline{\Omega})\oplus\bigwedge^2\overline{\Omega} \\
{\scriptstyle\beta_2}\big\downarrow & & & & \big\downarrow{\scriptstyle\gamma_2} \\
H^2(X,\mathbb{C}) & \simeq & H^2_{\mathrm{DR}}(X) & = & H^{2,0}(X)\oplus H^{1,1}(X)\oplus H^{0,2}(X)\ .
\end{array} \tag{2}$$

Here γ_2 denotes the natural isomorphism of Proposition 1.4.3, which associates to every alternating form the corresponding invariant 2-form (see also Exercise 1). Recall that

$$\mathrm{Alt}^2_{\mathbb{R}}(V,\mathbb{C}) = \bigwedge^2\mathrm{Hom}_{\mathbb{R}}(V,\mathbb{C}) = \bigwedge^2(\Omega\oplus\overline{\Omega}) \quad \text{and}$$
$$H^2(X,\mathbb{C}) \simeq \bigwedge^2 H^1(X,\mathbb{C}) \simeq \bigwedge^2 H^1_{\mathrm{DR}}(X) = \bigwedge^2\big(H^{1,0}(X)\oplus H^{0,1}(X)\big)\ .$$

Hence for the commutativity of (2) it suffices to show that there is a corresponding commutative diagram on the H^1-level, all of whose isomorphisms are compatible with the wedge products, and this is a consequence of the subsequent lemma. Of course one could also compute the commutativity of (2) directly, but this is more complicated.

(1.5) Lemma. *The following diagram commutes*

$$\begin{array}{ccc}
H^1(\Lambda,\mathbb{C}) = \mathrm{Hom}_{\mathbb{R}}(V,\mathbb{C}) = & & \Omega\oplus\overline{\Omega} \\
{\scriptstyle\beta_1}\big\downarrow & & \big\downarrow{\scriptstyle\gamma_1} \\
H^1(X,\mathbb{C}) & \xleftarrow{\ \ \rho\ \ } & H^{1,0}(X)\oplus H^{0,1}(X)\ ,
\end{array}$$

where β_1 is the extension of the canonical isomorphism ϕ_1 of Remark B.2, γ_1 associates to every linear form the corresponding invariant 1-form and $\rho\colon H^1_{\mathrm{DR}}(X) \to H^1(X,\mathbb{C})$ is the de Rham isomorphism.

Proof. For the definition of the isomorphisms we will use the same notation as in the proof of Lemma 1.4. In particular we have for $f\in\mathrm{Hom}_{\mathbb{R}}(V,\mathbb{C})$

$$(\beta_1 f)_{ij} = f(\lambda_{ij}) \circ \pi_i^{-1} = f(\lambda_{ij})$$

for all $i, j \in I$ (see Remark B.2). Let $v_1, \ldots, v_g$ denote a system of complex coordinate functions for V. Then the linear form f can be uniquely written as $f = f_1 v_1 + \cdots + f_g v_g + f_1' \bar{v}_1 + \cdots + f_g' \bar{v}_g$ with $f_\nu, f_\nu' \in \mathbb{C}$, and

$$\gamma_1(f) = \sum_{\nu=1}^{g} f_\nu \, dv_\nu + \sum_{\nu=1}^{g} f_\nu' \, d\bar{v}_\nu \, .$$

To compute $\rho\gamma_1(f)$ note that the $\mathcal{C}^\infty$-function $g_i = f \circ \pi_i^{-1} \colon U_i \to \mathbb{C}$ satisfies $dg_i = \gamma_1(f) \big| U_i$ for all $i \in I$. Hence by definition of the de Rham isomorphism

$$\big(\rho\gamma_1(f)\big)_{ij} = g_j - g_i = f(\pi_j^{-1} - \pi_i^{-1}) = f(\lambda_{ij})$$

for all $i, j \in I$. This proves the assertion. □

In Theorem 1.2 we associated to every holomorphic line bundle on X an alternating $\mathbb{Z}$-valued form on Λ and thus via $\mathbb{R}$-linear extension an alternating form $V \times V \to \mathbb{R}$. Conversely, we will determine, which alternating forms come from line bundles in this way.

(1.6) Proposition. *For an alternating form $E \colon V \times V \to \mathbb{R}$ the following conditions are equivalent:*

i) There is a holomorphic line bundle L on X such that E represents the first Chern class $c_1(L)$.

ii) $E(\Lambda, \Lambda) \subseteq \mathbb{Z}$ and $E(iv, iw) = E(v, w)$ for all $v, w \in V$.

Proof. Consider the following diagram

$$\begin{array}{ccccc}
H^1(\mathcal{O}_X^*) & \xrightarrow{c_1} & H^2(X, \mathbb{Z}) & \longrightarrow & H^2(\mathcal{O}_X) \\
 & & \Big\downarrow{\scriptstyle \iota} & & \Big\| \\
 & & H^2(X, \mathbb{C}) & \xrightarrow{p} & H^{0,2}(X) \\
 & & \wr\Big\downarrow{\scriptstyle \beta_2^{-1}} & & \wr\Big\downarrow{\scriptstyle \gamma_2^{-1}} \\
 & & \bigwedge^2 \Omega \oplus (\Omega \otimes \bar{\Omega}) \oplus \bigwedge^2 \bar{\Omega} & \xrightarrow{p} & \bigwedge^2 \bar{\Omega}
\end{array}$$

where the upper line is part of the exact cohomology sequence of $0 \to \mathbb{Z} \to \mathcal{O}_X \to \mathcal{O}_X^* \to 0$, the map ι is the natural embedding, β_2 and γ_2 are the isomorphisms of diagram (2) and p denote the projection maps. We claim that the diagram commutes. To see this, it suffices to show that the homomorphism $H^2(X, \mathbb{Z}) \to H^2(\mathcal{O}_X)$ factorizes via the natural projection map $H^2(X, \mathbb{C}) = H^{2,0}(X) \oplus H^{1,1}(X) \oplus H^{0,2}(X) \to H^{0,2}(X)$. But this is an immediate consequence of the fact that this projection is induced by the natural projection map from the de Rham complex to the Dolbeault complex.

Now suppose $L \in H^1(\mathcal{O}_X^*)$ and $\beta_2^{-1}c_1(L) = E = E_1 + E_2 + E_3$ with $E_1 \in \bigwedge^2\Omega, E_2 \in \Omega \otimes \overline{\Omega}$ and $E_3 \in \bigwedge^2\overline{\Omega}$. But $E_1 = \overline{E_3}$, since E has values in $\mathbb{R}$, whereas according to the diagram above $E_3 = 0$. Hence $E = E_2$ such that E satisfies ii). The converse implication follows also from the above diagram. □

The following lemma shows that the alternating forms of Proposition 1.6 are just the imaginary parts of hermitian forms. Recall that a *hermitian form on* V is a map $H: V \times V \to \mathbb{C}$, which is $\mathbb{C}$-linear in the first component and satisfies $H(v, w) = \overline{H(w, v)}$ for all $v, w \in V$.

(1.7) Lemma. *There is a 1-1-correspondence between the set of hermitian forms* H *on* V *and the set of real valued alternating forms* E *on* V *satisfying* $E(iv, iw) = E(v, w)$, *given by*

$$E(v, w) = \operatorname{Im} H(v, w) \quad \text{and} \quad H(v, w) = E(iv, w) + iE(v, w)$$

for all $v, w \in V$.

Proof. Given E, the form H is hermitian, since

$$H(v, w) = E(iv, w) + iE(v, w) = -E(iw, -v) - iE(w, v) = \overline{H(w, v)}\ .$$

Conversely given H, the form $E = \operatorname{Im} H$ is alternating and $E(iv, iw) = \operatorname{Im} H(iv, iw) = \operatorname{Im} H(v, w) = E(v, w)$. □

Summing up, we can and will consider the first Chern class of a line bundle on X either as an alternating form or as a hermitian form on V. Define the *Néron-Severi group* $\operatorname{NS}(X)$ to be the image of the homomorphism $c_1: H^1(\mathcal{O}_X^*) \to H^2(X, \mathbb{Z})$. According to Proposition 1.6 and Lemma 1.7 we can identify $\operatorname{NS}(X)$ either with the group of $\mathbb{R}$-valued alternating forms E on V satisfying $E(\Lambda, \Lambda) \subseteq \mathbb{Z}$ and $E(iv, iw) = E(v, w)$, or with the group of hermitian forms H on V with $\operatorname{Im} H(\Lambda, \Lambda) \subseteq \mathbb{Z}$.

§ 2 The Appell-Humbert Theorem

Let $X = V/\Lambda$ be a complex torus. In the last section we saw that any holomorphic line bundle on X can be given by a factor of automorphy. Here we show that there is a canonical way to distinguish a factor in every class of $H^1(\Lambda, H^0(\mathcal{O}_V^*))$.

Consider the Néron-Severi group $\operatorname{NS}(X)$ as the group of hermitian forms $H: V \times V \to \mathbb{C}$ with $\operatorname{Im} H(\Lambda, \Lambda) \subseteq \mathbb{Z}$ and denote by $\mathbb{C}_1$ the circle group $\{z \in \mathbb{C} \mid |z| = 1\}$. A *semicharacter for* H is a map $\chi: \Lambda \to \mathbb{C}_1$ satisfying

$$\chi(\lambda + \mu) = \chi(\lambda)\chi(\mu)\, e(\pi i \operatorname{Im} H(\lambda, \mu)) \quad \text{for all } \lambda, \mu \in \Lambda.$$

By definition the characters on Λ with values in $\mathbb{C}_1$ are exactly the semicharacters for $0 \in \mathrm{NS}(X)$. Denote by $\mathcal{P}(\Lambda)$ the set of all pairs (H, χ) where $H \in \mathrm{NS}(X)$ and χ is a semicharacter for H. Clearly $\mathcal{P}(\Lambda)$ is a group with respect to the composition $(H_1, \chi_1) \circ (H_2, \chi_2) = (H_1 + H_2, \chi_1 \chi_2)$, and the following sequence is exact

$$1 \longrightarrow \mathrm{Hom}(\Lambda, \mathbb{C}_1) \xrightarrow{\iota} \mathcal{P}(\Lambda) \xrightarrow{p} \mathrm{NS}(X) \; .$$

Here $\iota(\chi) = (0, \chi)$ and $p(H, \chi) = H$. We will show that p is surjective. For this consider the map $\mathcal{P}(\Lambda) \to \mathrm{Pic}(X)$ defined in the following way: for $(H, \chi) \in \mathcal{P}(\Lambda)$ define $a = a_{(H,\chi)} \colon \Lambda \times V \to \mathbb{C}^*$ by

$$a(\lambda, v) = \chi(\lambda) e(\pi H(v, \lambda) + \tfrac{\pi}{2} H(\lambda, \lambda)) \; .$$

The map a is an element of $Z^1(\Lambda, H^0(\mathcal{O}_V^*))$, since it satisfies the cocycle relation:

$$\begin{aligned} a(\lambda + \mu, v) &= \chi(\lambda)\chi(\mu) e\big(\pi i \,\mathrm{Im}\, H(\lambda, \mu) + \pi H(v, \lambda{+}\mu) + \tfrac{\pi}{2} H(\lambda{+}\mu, \lambda{+}\mu)\big) \\ &= \chi(\lambda) e\big(\pi H(v + \mu, \lambda) + \tfrac{\pi}{2} H(\lambda, \lambda)\big) \chi(\mu) e\big(\pi H(v, \mu) + \tfrac{\pi}{2} H(\mu, \mu)\big) \\ &= a(\lambda, v + \mu) a(\mu, v) \end{aligned}$$

for all $v \in V$ and $\lambda, \mu \in \Lambda$. According to Proposition B.1 the cocycle $a_{(H,\chi)}$ determines a line bundle on X, which we denote by $L(H, \chi)$. By construction

$$L(H, \chi) \cong V \times \mathbb{C} / \Lambda$$

where Λ acts on $V \times \mathbb{C}$ by $\lambda \circ (v, t) = (v + \lambda, a_{(H,\chi)}(\lambda, v) t)$.

The aim is to show that for every line bundle L on X there is a unique pair $(H, \chi) \in \mathcal{P}(\Lambda)$ such that $L \simeq L(H, \chi)$.

(2.1) Lemma. *The map* $\mathcal{P}(\Lambda) \to \mathrm{Pic}(X)$, $(H, \chi) \mapsto L(H, \chi)$ *is a homomorphism of groups and the following diagram commutes*

$$\begin{array}{ccc} \mathcal{P}(\Lambda) & \searrow^{p} & \\ \downarrow & & \mathrm{NS}(X) \\ \mathrm{Pic}(X) & \nearrow_{c_1} & \end{array}$$

Proof. In order to see that the map is a homomorphism, it suffices to show $a_{(H_1,\chi_1)} a_{(H_2,\chi_2)} = a_{(H_1+H_2,\chi_1\chi_2)}$ for all $(H_1, \chi_1), (H_2, \chi_2) \in \mathcal{P}(\Lambda)$. But this is obvious from the definitions.

For the proof of the second statement we have to show that $c_1(L(H, \chi)) = H$ for all $(H, \chi) \in \mathcal{P}(\Lambda)$. Writing $\chi(\lambda) = e(2\pi i \varphi(\lambda))$ we have $a_{(H,\chi)}(\lambda, v) = e(2\pi i g(\lambda, v))$ with $g(\lambda, v) = \varphi(\lambda) - \frac{i}{2} H(v, \lambda) - \frac{i}{4} H(\lambda, \lambda)$. According to Theorem 1.2, the imaginary part of the hermitian form $c_1(L(H, \chi))$ is the alternating form

$$E_{L(H,\chi)}(\lambda,\mu) = g(\mu, v+\lambda) + g(\lambda, v) - g(\lambda, v+\mu) - g(\mu, v)$$
$$= \tfrac{1}{2i}(H(\lambda,\mu) - H(\mu,\lambda)) = \operatorname{Im} H(\lambda,\mu) .$$

Now Lemma 1.7 gives the assertion. □

The lemma implies in particular that $p: \mathcal{P}(\Lambda) \to \mathrm{NS}(X)$ is surjective. In order to show that $\mathcal{P}(\Lambda) \to \operatorname{Pic}(X)$ is an isomorphism, consider its restriction to $\operatorname{Hom}(\Lambda, \mathbb{C}_1)$. Let $\operatorname{Pic}^0(X)$ denote the kernel of $c_1: \operatorname{Pic}(X) \to \mathrm{NS}(X)$. So $\operatorname{Pic}^0(X)$ is the group of line bundles with first Chern class 0.

(2.2) Proposition. *The map $\mathcal{P}(\Lambda) \to \operatorname{Pic}(X)$ induces an isomorphism*

$$\operatorname{Hom}(\Lambda, \mathbb{C}_1) \xrightarrow{\sim} \operatorname{Pic}^0(X) .$$

Proof. Consider the commutative diagram

$$\begin{array}{ccccccc} H^1(X,\mathbb{Z}) & \to & H^1(\mathcal{O}_X) & \to & H^1(\mathcal{O}_X^*) & \xrightarrow{c_1} & H^2(X,\mathbb{Z}) \\ & & \uparrow p & & \| & & \\ & & H^1(X,\mathbb{C}) & \xrightarrow{\varepsilon} & H^1(\mathcal{O}_X^*) & & \end{array}$$

where the upper line is part of the cohomology sequence of the exponential sequence, p is the projection map associated to the Hodge decomposition $H^1(X,\mathbb{C}) \simeq H^1(\mathcal{O}_X) \oplus H^0(\Omega_X)$, and finally ε is the map associated to $e(2\pi i\,\cdot): \mathbb{C} \to \mathbb{C}^* \subseteq \mathcal{O}_X^*$. It follows that $\operatorname{Pic}^0(X) = \operatorname{im}\varepsilon$. In other words, any line bundle in $\operatorname{Pic}^0(X)$ can be given by a cocycle in $Z^1(X, \mathcal{O}_X^*)$ with constant coefficients. By the definition of the homomorphisms $H^1(\mathcal{O}_X^*) \to H^1(\Lambda, H^0(\mathcal{O}_V^*))$ and $H^1(X,\mathbb{C}) \to H^1(\Lambda,\mathbb{C})$ (see Remark B.2 in Appendix B) it is easy to see that the corresponding class in $H^1(\Lambda, H^0(\mathcal{O}_V^*))$ is represented by a factor of automorphy with constant coefficients.

We claim first that the homomorphism $\operatorname{Hom}(\Lambda, \mathbb{C}_1) \to \operatorname{Pic}^0(X)$ is surjective. To see this, let $L \in \operatorname{Pic}^0(X)$ and $f \in Z^1(\Lambda, H^0(\mathcal{O}_V^*))$ a factor for L with constant coefficients. Then $f: \Lambda \to \mathbb{C}^*$ is a homomorphism. Writing $f(\lambda) = e(2\pi i\, g(\lambda))$, we get

$$g(\lambda+\mu) \equiv g(\lambda) + g(\mu) \bmod \mathbb{Z}$$

for all $\lambda, \mu \in \Lambda$. Hence the imaginary part $\operatorname{Im} g: \Lambda \to \mathbb{R}$ is a homomorphism extending $\mathbb{R}$-linearly to a function $V \to \mathbb{R}$, denoted by the same symbol. Define a $\mathbb{C}$-linear form

$$l: V \to \mathbb{C} \quad \text{by} \quad v \mapsto \operatorname{Im} g(iv) + i \operatorname{Im} g(v) .$$

Since $e(2\pi i l) \in H^0(\mathcal{O}_V^*)$,

$$\chi_L(\lambda, v) := f(\lambda) e(2\pi i\, l(v) - 2\pi i\, l(v+\lambda))$$

is a cocycle in $Z^1(\Lambda, H^0(\mathcal{O}_V^*))$ equivalent to f. Moreover χ_L does not depend on v and takes values in $\mathbb{C}_1$, since $\chi_L(\lambda, v) = e(2\pi i\, g(\lambda) - 2\pi i\, l(\lambda)) = e(2\pi i(\operatorname{Re} g(\lambda) - \operatorname{Im} g(i\lambda))$ and $\operatorname{Re} g(\lambda) - \operatorname{Im} g(i\lambda)$ is real. The fact that f and $e(2\pi i l)$ are homomorphisms implies $\chi_L \in \operatorname{Hom}(\Lambda, \mathbb{C}_1)$, i.e. χ_L is a semicharacter for $0 \in \operatorname{NS}(X)$, and $L \simeq L(0, \chi_L)$. Thus the homomorphism $\operatorname{Hom}(\Lambda, \mathbb{C}_1) \to \operatorname{Pic}^0(X)$ is surjective.

To see that it is injective, suppose there is another homomorphism $\chi \in \operatorname{Hom}(\Lambda, \mathbb{C}_1)$ with $L \cong L(0, \chi)$. So χ_L and χ are equivalent cocycles in $Z^1(\Lambda, H^0(\mathcal{O}_V^*))$ and there is an $h \in H^0(\mathcal{O}_V^*)$ with $\chi_L(\lambda) = \chi(\lambda)h(v+\lambda)h(v)^{-1}$ for all $v \in V$ and $\lambda \in \Lambda$. Now $|\chi_L| = |\chi| = 1$ implies $|h(v+\lambda)| = |h(v)|$ for all $v \in V$ and $\lambda \in \Lambda$, such that h is bounded on V. By the theorem of Liouville h is constant and $\chi = \chi_L$. This implies the assertion. □

Summing up, we obtain the

(2.3) Appell-Humbert Theorem. *Let $X = V/\Lambda$ be a complex torus. There is a canonical isomorphism of exact sequences*

$$\begin{array}{ccccccccc}
1 & \longrightarrow & \operatorname{Hom}(\Lambda, \mathbb{C}_1) & \longrightarrow & \mathcal{P}(\Lambda) & \longrightarrow & \operatorname{NS}(X) & \longrightarrow & 0 \\
 & & \downarrow\simeq & & \downarrow\simeq & & \| & & \\
1 & \longrightarrow & \operatorname{Pic}^0(X) & \longrightarrow & \operatorname{Pic}(X) & \longrightarrow & \operatorname{NS}(X) & \longrightarrow & 0\ .
\end{array}$$

§ 3 Canonical Factors

Let $X = V/\Lambda$ be a complex torus. The Appell-Humbert Theorem tells us that there is a canonical way to associate a factor of automorphy to any line bundle on X. Namely, if $L \in \operatorname{Pic}(X)$ with hermitian form $c_1(L) = H$, one can distinguish a semicharacter χ for H such that $L \simeq L(H, \chi)$. The cocycle $a_L = a_{L(H,\chi)} \in Z^1(\Lambda, H^0(\mathcal{O}_V^*))$, defined by

$$a_L(\lambda, v) = \chi(\lambda)\, e(\pi H(v, \lambda) + \tfrac{\pi}{2} H(\lambda, \lambda))$$

for all $(\lambda, v) \in \Lambda \times V$ is called the *canonical factor (of automorphy) for L*. In this section we want to compile some properties of line bundles on X, which are proved using their canonical factors.

Let us start with some rules for working with semicharacters and canonical factors, which are frequently used in this book and follow immediately from the definitions.

(3.1) Lemma. *Let a_L be the canonical factor of $L = L(H,\chi) \in \mathrm{Pic}(X)$. For all $\lambda \in \Lambda, v, w \in V$ and $n \in \mathbb{Z}$ we have*

a) $\chi(n\lambda) = \chi(\lambda)^n$,

b) $a_L(\lambda, v+w) = a_L(\lambda, v)\, e(\pi H(w,\lambda))$,

c) $a_L(\lambda, v)^{-1} = a_L(-\lambda, v)\, e(-\pi H(\lambda,\lambda))$.

We want to study the behaviour of line bundles under pullbacks by holomorphic maps of complex tori. According to Proposition 1.2.1 every holomorphic map between complex tori is a composition of a homomorphism and a translation. We will consider both cases separately.

(3.2) Lemma. *For any $L = L(H,\chi) \in \mathrm{Pic}(X)$ and $\bar{v} \in X$ with representative $v \in V$*

$$t_{\bar{v}}^* L(H,\chi) = L\big(H, \chi e(2\pi i \operatorname{Im} H(v,\cdot))\big) \ .$$

Proof. The translation t_v on V induces the translation $t_{\bar{v}}$ on X. Moreover the induced map $t_{\bar{v}*}$ on the fundamental group Λ of X (as defined in Appendix B) is the identity. Hence, if a_L denotes the canonical factor of L, $(\mathrm{id}_\Lambda \times t_v)^* a_L$ is a factor for $t_{\bar{v}}^* L$ (see Appendix B), but not yet the canonical one. For $g(w) = e(-\pi H(w,v))$ the factor:

$$\begin{aligned}(\mathrm{id}_\Lambda \times t_v)^* a_L(\lambda, w) g(w+\lambda) g(w)^{-1} &= \\ = \chi(\lambda) e(\pi H(w+v,\lambda) + \tfrac{\pi}{2} H(\lambda,\lambda))\, e(-\pi H(w+\lambda, v) + \pi H(w,v)) \\ = \chi(\lambda) e(2\pi i \operatorname{Im} H(v,\lambda)) e(\pi H(w,\lambda) + \tfrac{\pi}{2} H(\lambda,\lambda))\end{aligned}$$

is equivalent to $(\mathrm{id}_\Lambda \times t_v)^* a_L$. Moreover it is the canonical factor for $t_{\bar{v}}^* L$, since $\chi e(2\pi i \operatorname{Im} H(v,\cdot))$ is a semicharacter for H. □

Comparing hermitian forms and semicharacters, we get as an immediate consequence the

(3.3) Theorem of the Square. *For all $\bar{v}, \bar{w} \in X$ and $L \in \mathrm{Pic}(X)$*

$$t_{\bar{v}+\bar{w}}^* L = t_{\bar{v}}^* L \otimes t_{\bar{w}}^* L \otimes L^{-1} \ .$$

Let $X' = V'/\Lambda'$ be a second complex torus and $f: X' \to X$ a homomorphism with analytic representation $F: V' \to V$ and rational representation $F_\Lambda: \Lambda' \to \Lambda$.

(3.4) Lemma. *For any $L(H,\chi) \in \mathrm{Pic}(X)$*

$$f^* L(H,\chi) = L(F^* H, F_\Lambda^* \chi) \ .$$

Proof. The induced maps between the universal coverings V' and V and the fundamental groups Λ' and Λ of X' and X are just the analytic and rational representation F and F_Λ. Hence $(F_\Lambda \times F)^* a_{L(H,\chi)} = a_{L(F^*H, F_\Lambda^*\chi)}$ is the canonical factor for f^*L (see Appendix B). □

As an example we apply Lemma 3.4 to the homomorphism $n_X: X \to X$ for any integer n.

(3.5) Proposition. *For every $L \in \operatorname{Pic}(X)$ and $n \in \mathbb{Z}$*

$$n_X^* L \simeq L^{\frac{n^2+n}{2}} \otimes (-1)_X^* L^{\frac{n^2-n}{2}} .$$

Proof. Recall the analytic representation $n_V: V \to V$, $v \mapsto nv$ of n_X. For $L = L(H, \chi)$ we get:

$$\begin{aligned} L^{\frac{n^2+n}{2}} \otimes (-1)_X^* L^{\frac{n^2-n}{2}} &= L\left(\tfrac{n^2+n}{2} H + \tfrac{n^2-n}{2}(-1)_V^* H, \chi^{\frac{n^2+n}{2}} \cdot (-1)_V^* \chi^{\frac{n^2-n}{2}}\right) \\ &= L(n^2 H, \chi(n\cdot)) \qquad \text{(by Lemma 3.1 a))} \\ &= L(n_V^* H, n_\Lambda^* \chi) \\ &= n_X^* L(H, \chi) . \qquad \text{(by Lemma 3.4)} \end{aligned}$$ □

A line bundle L on X is called *symmetric*, if $(-1)_X^* L \simeq L$. There are several reasons why it is convenient to work with symmetric line bundles. One reason is that they are easier to handle. For example, Proposition 3.5 simplifies as follows

(3.6) Corollary. *For every symmetric line bundle $L \in \operatorname{Pic}(X)$ and $n \in \mathbb{Z}$*

$$n_X^* L \simeq L^{n^2} .$$

Given a line bundle L in terms of H and χ, it is easy to decide whether L is symmetric:

(3.7) Proposition. *The line bundle $L = L(H, \chi)$ is symmetric if and only if $\chi(\Lambda) \subseteq \{\pm 1\}$.*

Proof. This is a direct consequence of $(-1)^* L(H, \chi) = L(H, \chi^{-1})$. □

§ 4 The Dual Complex Torus

In this section we define the dual $\widehat{X}$ of a complex torus X, prove some of its functorial properties and study the canonical map $\phi_L: X \to \widehat{X}$, associated to a line bundle L.

Let $X = V/\Lambda$ be a complex torus of dimension g. According to the Appell-Humbert Theorem the map $\mathrm{Hom}(\Lambda, \mathbb{C}_1) \to \mathrm{Pic}^0(X)$, $\chi \mapsto L(0,\chi)$ is an isomorphism of groups. The fact that $\mathrm{Hom}(\Lambda, \mathbb{C}_1) \simeq (\mathbb{R}/\mathbb{Z})^{2g}$ is a real torus suggests that $\mathrm{Pic}^0(X)$ may be given the structure of a complex torus.

Consider the $\mathbb{C}$-vector space $\overline{\Omega} = \mathrm{Hom}_{\overline{\mathbb{C}}}(V, \mathbb{C})$ of $\mathbb{C}$-antilinear forms $l: V \to \mathbb{C}$. The underlying real vector space of $\overline{\Omega}$ is canonically isomorphic to $\mathrm{Hom}_{\mathbb{R}}(V, \mathbb{R})$. The isomorphism is given by $l \mapsto k = \mathrm{Im}\, l$ with inverse map $k \mapsto l(v) = -k(iv) + ik(v)$. Hence the canonical $\mathbb{R}$-bilinear form

$$\langle\ ,\ \rangle: \overline{\Omega} \times V \to \mathbb{R}\ ,\ \langle l, v\rangle = \mathrm{Im}\, l(v)$$

is nondegenerate. This implies that

$$\widehat{\Lambda} := \{l \in \overline{\Omega} \mid \langle l, \Lambda\rangle \subseteq \mathbb{Z}\}$$

is a lattice in $\overline{\Omega}$, called the *dual lattice* of Λ. The quotient

$$\widehat{X} := \overline{\Omega}/\widehat{\Lambda}$$

is a complex torus of dimension g, the *dual complex torus*. Identifying V with the space of $\mathbb{C}$-antilinear forms $\overline{\Omega} \to \mathbb{C}$ by double antiduality, the nondegeneracy of $\langle\ ,\ \rangle$ implies that Λ is just the lattice in V dual to $\widehat{\Lambda}$. It follows that

$$\widehat{\widehat{X}} = X\ .$$

(4.1) Proposition. *The canonical homomorphism* $\overline{\Omega} \to \mathrm{Hom}(\Lambda, \mathbb{C}_1)$, $l \mapsto e(2\pi i\langle l, \cdot\rangle)$ *induces an isomorphism* $\widehat{X} \xrightarrow{\sim} \mathrm{Pic}^0(X)$.

According to Proposition 4.1, the group $\mathrm{Pic}^0(X)$ admits the structure of a complex torus in a natural way. We often identify the complex tori $\mathrm{Pic}^0(X)$ and $\widehat{X}$.

Proof. The nondegeneracy of the form $\langle\ ,\ \rangle$ implies that the map $\overline{\Omega} \to \mathrm{Hom}(\Lambda, \mathbb{C}_1)$ is surjective. By definition $\widehat{\Lambda}$ is precisely the kernel of this homomorphism. □

Let $X_\nu = V_\nu/\Lambda_\nu$ denote complex tori for $\nu = 1, 2$ and $f: X_1 \to X_2$ a homomorphism with analytic representation $F: V_1 \to V_2$. The (anti-)dual map $F^*: \overline{\Omega}_2 \to \overline{\Omega}_1$ associating to an antilinear form $l_2 \in \overline{\Omega}_2$ the antilinear form $l_2 \circ F \in \overline{\Omega}_1$ induces a homomorphism $\widehat{f}\colon \widehat{X}_2 \to \widehat{X}_1$, since $F^*\widehat{\Lambda}_2 \subseteq \widehat{\Lambda}_1$. According to Proposition 4.1 the following diagram commutes

$$\begin{array}{ccc} \widehat{X}_2 & \xrightarrow{\sim} & \mathrm{Pic}^0(X_2) \\ {\scriptstyle \widehat{f}}\downarrow & & \downarrow{\scriptstyle f^*} \\ \widehat{X}_1 & \xrightarrow{\sim} & \mathrm{Pic}^0(X_1) \end{array} \quad . \tag{1}$$

If $g\colon X_2 \to X_3$ is a second homomorphism of complex tori, then

$$\widehat{gf} = \widehat{f}\,\widehat{g}.$$

Moreover $\widehat{\mathrm{id}_X} = \mathrm{id}_{\widehat{X}}$ and $\widehat{\widehat{f}} = f$. Hence "$\widehat{\ }$" is a functor from the category of complex tori into itself. The following proposition says that this functor is exact.

(4.2) Proposition. *If $0 \to X_1 \to X_2 \to X_3 \to 0$ is an exact sequence of complex tori, the dual sequence $0 \to \widehat{X}_3 \to \widehat{X}_2 \to \widehat{X}_1 \to 0$ is also exact.*

Proof. Suppose $X_\nu = V_\nu/\Lambda_\nu$. Applying the serpent lemma, the induced sequence of lattices $0 \to \Lambda_1 \to \Lambda_2 \to \Lambda_3 \to 0$ is exact. As a sequence of free abelian groups it splits, so that

$$0 \to \mathrm{Hom}(\Lambda_3, \mathbb{C}_1) \to \mathrm{Hom}(\Lambda_2, \mathbb{C}_1) \to \mathrm{Hom}(\Lambda_1, \mathbb{C}_1) \to 0$$

is also exact. According to Proposition 4.1 and the Appell-Humbert Theorem this is just the assertion. □

(4.3) Proposition. *If $f\colon X_1 \to X_2$ is an isogeny of complex tori, the dual map $\widehat{f}\colon \widehat{X}_2 \to \widehat{X}_1$ is also an isogeny and its kernel is isomorphic to* $\mathrm{Hom}(\ker f, \mathbb{C}_1)$. *In particular* $\deg \widehat{f} = \deg f$.

Proof. Suppose $X_\nu = V/\Lambda_\nu$. We may assume that id_V is the analytic representation of f. By definition $\mathrm{id}_{\overline{\Omega}}$ is the analytic representation of $\widehat{f}$, so $\widehat{f}$ is an isogeny. According to diagram (1) and the Appell-Humbert Theorem

$$\ker \widehat{f} \simeq \ker(\mathrm{Hom}(\Lambda_2, \mathbb{C}_1) \to \mathrm{Hom}(\Lambda_1, \mathbb{C}_1)) \simeq \mathrm{Hom}(\Lambda_2/\Lambda_1, \mathbb{C}_1)$$

Since $\ker f \simeq \Lambda_2/\Lambda_1$, this implies the assertion. □

As a consequence of Propostion 4.3 we obtain a criterion for a line bundle to descend under an isogeny.

(4.4) Corollary. *Let $f\colon X_1 \to X_2$ be an isogeny of complex tori $X_\nu = V/\Lambda_\nu$ with analytic representation F. For a line bundle $L = L(H,\chi)$ on X_1 the following statements are equivalent:*

i) $L = f^*M$ *for some line bundle* $M \in \operatorname{Pic}(X_2)$.
ii) $\operatorname{Im} H(F^{-1}\Lambda_2, F^{-1}\Lambda_2) \subseteq \mathbb{Z}$.

Proof. Suppose $\operatorname{Im} H(F^{-1}\Lambda_2, F^{-1}\Lambda_2) \subseteq \mathbb{Z}$, that is $F^{-1*}H \in \operatorname{NS}(X_2)$. Choose an $\widetilde{M} \in \operatorname{Pic}(X_2)$ with $c_1(\widetilde{M}) = F^{-1^*}H$. Then $c_1(f^*\widetilde{M}) = H$, and thus $L \otimes f^*\widetilde{M}^{-1} \in \operatorname{Pic}^0(X_1)$. According to Proposition 4.3 the homomorphism $f^*: \operatorname{Pic}^0(X_2) \to \operatorname{Pic}^0(X_1)$ is surjective, so there is an $N \in \operatorname{Pic}^0(X_2)$ with $f^*N = L \otimes f^*\widetilde{M}^{-1}$. Now $M = \widetilde{M} \otimes N$ satisfies i). The converse implication is a consequence of Lemma 3.4. □

Let L be a line bundle on X. For any point $x \in X$ the line bundle $t_x^*L \otimes L^{-1}$ has first Chern class zero. So identifying $\widehat{X} = \operatorname{Pic}^0(X)$ we get a map

$$\phi_L: X \to \widehat{X}\,,\ x \mapsto t_x^*L \otimes L^{-1}$$

which according to the Theorem of the Square 3.3 is a homomorphism. In order to compute its analytic representation, suppose $L = L(H, \chi)$ and $X = V/\Lambda$.

(4.5) Lemma. *The map* $\phi_H: V \to \overline{\Omega}$, $\phi_H(v) = H(v, \cdot)$ *is the analytic representation of* ϕ_L.

Proof. Applying Lemma 3.2 we get

$$t_{\bar{v}}^*L \otimes L^{-1} = L\big(0, e(2\pi i \operatorname{Im} H(v, \cdot))\big) = L\big(0, e(2\pi i \langle \phi_H(v), \cdot\rangle)\big)\ .$$

Comparing this with the isomorphism $\widehat{X} \xrightarrow{\sim} \operatorname{Pic}^0(X) = \operatorname{Hom}(\Lambda, \mathbb{C}_1)$ of Proposition 4.1 gives the assertion. □

(4.6) Corollary. *a)* ϕ_L *depends only on the first Chern class* H *of* L.
b) $\phi_{L\otimes M} = \phi_L + \phi_M$ *for all* $L, M \in \operatorname{Pic}(X)$.
c) $\widehat{\phi_L} = \phi_L$ *under the natural identification* $\widehat{\widehat{X}} = X$
d) For any homomorphism $f: Y \to X$ *of complex tori the following diagram commutes*

$$\begin{array}{ccc} X & \xrightarrow{\phi_L} & \widehat{X} \\ {\scriptstyle f}\uparrow & & \downarrow{\scriptstyle \widehat{f}} \\ Y & \xrightarrow[\phi_{f^*L}]{} & \widehat{Y} \end{array}\ .$$

Proof. a), b) and d) follow immediately from the definition. As for c): recall that ϕ_H^* is the analytic representation of $\widehat{\phi_L}$. So the assertion follows from the fact that $\phi_H^* = \phi_H$ under the natural identification $\operatorname{Hom}_{\overline{\mathbb{C}}}(\overline{\Omega}, \mathbb{C}) = V$. □

Denote by $K(L)$ the kernel of ϕ_L. In order to describe $K(L)$, define

$$\Lambda(L) = \{v \in V \mid \operatorname{Im} H(v, \Lambda) \subseteq \mathbb{Z}\} .$$

Obviously $\Lambda(L) = \phi_H^{-1}(\widehat{\Lambda})$, which implies

$$K(L) = \Lambda(L)/\Lambda .$$

Since $K(L)$ and $\Lambda(L)$ depend only on the hermitian form H, we sometimes write $K(H)$ and $\Lambda(H)$ respectively for these groups. The following lemma compiles some elementary properties of $K(L)$.

(4.7) Lemma. *For any line bundle L on X*
a) $K(L \otimes P) = K(L)$ *for any* $P \in \operatorname{Pic}^0(X)$,
b) $K(L) = X$ *if and only if* $L \in \operatorname{Pic}^0(X)$,
c) $K(L^n) = n_X^{-1} K(L)$ *for any* $n \in \mathbb{Z}$,
d) $K(L) = n_X K(L^n)$ *for any* $n \in \mathbb{Z}$, $n \neq 0$.

Proof. a) and b) are immediate consequences of the definition and the Appell-Humbert Theorem. Suppose $L = L(H, \chi)$, then $L^n = L(nH, \chi^n)$ and thus $\Lambda(L^n) = \{v \in V \mid \operatorname{Im} H(nv, \Lambda) \subseteq \mathbb{Z}\} = \{\frac{1}{n}v \in V \mid v \in \Lambda(L)\}$. This implies c) and d). □

According to Lemma 4.5 the homomorphism ϕ_L associated to the line bundle L is an isogeny of complex tori if and only if the hermitian form $H = c_1(L)$ is nondegenerate. This suggests the following definition: a line bundle L on X is called *nondegenerate*, if the associated hermitian form $H = c_1(L)$ is nondegenerate, or equivalently if the alternating form $\operatorname{Im} H$ is nondegenerate (see Lemma 1.7). This implies

(4.8) Proposition. *A line bundle L on X is nondegenerate if and only if $K(L)$ is finite.*

(4.9) Proposition. $\deg \phi_L = \det(\operatorname{Im} H)$.

Proof. We may assume that H is nondegenerate, since otherwise both sides are zero. Then $\deg \phi_L = (\phi_H^{-1}(\widehat{\Lambda}) : \Lambda) = (\Lambda(L) : \Lambda) = \det(\operatorname{Im} H)$, using an elementary result of Linear Algebra. □

§ 5 The Poincaré Bundle

Let $X = V/\Lambda$ be a complex torus. According to Proposition 4.1 the points of the dual torus $\widehat{X}$ parametrize the isomorphism classes of line bundles in $\mathrm{Pic}^0(X)$. This suggests that there might be a line bundle on the product $X \times \widehat{X}$ inducing all line bundles of $\mathrm{Pic}^0(X)$. Such line bundles are called Poincaré bundles. To be more precise, we define: a holomorphic line bundle $\mathcal{P}$ on $X \times \widehat{X}$ satisfying

i) $\mathcal{P}|X \times \{L\} \simeq L$ for every $L \in \widehat{X}$, and
ii) $\mathcal{P}|\{0\} \times \widehat{X}$ is trivial

is called a *Poincaré bundle* for X. Note that condition i) is the relevant property of $\mathcal{P}$, whereas ii) serves for normalisation.

(5.1) Theorem. *There exists a Poincaré bundle $\mathcal{P}$ on $X \times \widehat{X}$, uniquely determined up to isomorphisms.*

Proof. Existence: Define a hermitian form $H: (V \times \Omega) \times (V \times \Omega) \to \mathbb{C}$ by

$$H((v_1, l_1), (v_2, l_2)) = \overline{l_2(v_1)} + l_1(v_2)\ .$$

According to Proposition 1.6 the form H is the first Chern class of some line bundle on $X \times \widehat{X}$, since $\mathrm{Im}\, H(\Lambda \times \widehat{\Lambda}, \Lambda \times \widehat{\Lambda}) \subseteq \mathbb{Z}$ by the definition of $\widehat{\Lambda}$. Furthermore define a map $\chi: \Lambda \times \widehat{\Lambda} \to \mathbb{C}_1$ by

$$\chi(\lambda, l_0) = e(\pi i\ \mathrm{Im}\, l_0(\lambda))\ .$$

Obviously χ is a semicharacter for H. So according to the Appell-Humbert Theorem the pair (H, χ) defines a line bundle $\mathcal{P}$ on $X \times \widehat{X}$.

We have to check the properties i) and ii). For this consider the canonical factor $a_{\mathcal{P}}: (\Lambda \times \widehat{\Lambda}) \times (V \times \Omega) \to \mathbb{C}^*$ of $\mathcal{P}$

$$a_{\mathcal{P}}((\lambda, l_0), (v, l)) = \chi(\lambda, l_0) e(\pi H((v, l), (\lambda, l_0)) + \tfrac{\pi}{2} H((\lambda, l_0), (\lambda, l_0)))\ .$$

As for i): suppose $L \in \widehat{X} = \mathrm{Pic}^0(X)$. There is an $l \in \Omega$, such that $L = L(0, e(2\pi i \,\mathrm{Im}\, l))$. The restriction $\mathcal{P}|X \times \{L\}$ is given by the factor $a_{\mathcal{P}}|\Lambda \times \{0\} \times V \times \{l\}$. But

$$a_{\mathcal{P}}((\lambda, 0), (v, l)) = \chi(\lambda, 0) e\big(\pi H\big((v, l), (\lambda, 0)\big) + \tfrac{\pi}{2} H\big((\lambda, 0), (\lambda, 0)\big)\big) = e(\pi \overline{l(\lambda)})$$

is equivalent to $a_{\mathcal{P}}((\lambda, 0), (v, l))\, e(\pi \overline{l(v + \lambda)})^{-1}\, e(\pi \overline{l(v)}) = e(2\pi i\ \mathrm{Im}\, l(\lambda))$, the canonical factor of L.

As for ii): the restriction of $\mathcal{P}|\{0\} \times \widehat{X}$ is given by the factor

$$a_{\mathcal{P}}((0,l_0),(0,l)) = 1 \quad \text{for all} \quad l_0 \in \widehat{\Lambda} \quad \text{and} \quad l \in \Omega\,,$$

the canonical factor of the trivial line bundle on $\{0\} \times \widehat{X}$.
Uniqueness: This is a direct consequence of the Seesaw Principle A.9. □

Note that by construction the Poincaré bundle is nondegenerate. It satisfies the following universal property.

(5.2) Proposition. *For any normal complex analytic space T and any line bundle $\mathcal{L}$ on $X \times T$ satisfying the conditions*
i) $\mathcal{L}|X \times \{t\} \in \mathrm{Pic}^0(X)$ for every $t \in T$, and
ii) $\mathcal{L}|\{0\} \times T$ is trivial,
there is a unique holomorphic map $\psi: T \to \widehat{X}$ such that $\mathcal{L} \simeq (\mathrm{id} \times \psi)^*\mathcal{P}$.

Note that set-theoretically the map ψ is given by $\psi(t) = \mathcal{L}|X \times \{t\}$. Note also that, if T is connected, in condition i) it suffices to assume $\mathcal{L}|X \times \{t_0\} \in \mathrm{Pic}^0(X)$ for some $t_0 \in T$.

Proof. Define a map $\psi: T \to \widehat{X}$ by $t \mapsto \mathcal{L}|X \times \{t\}$. We first claim that ψ is holomorphic. For this consider the line bundle $\mathcal{N} = p_{12}^*\mathcal{L} \otimes p_{13}^*\mathcal{P}^{-1}$ on $X \times T \times \widehat{X}$. Here p_{ij} is the projection of $X \times T \times \widehat{X}$ onto the i-th and j-th factor. The set $\Gamma = \big\{(t,L) \in T \times \widehat{X} \mid \mathcal{N}|X \times \{t,L\} \text{ is trivial} \big\}$ is Zariski closed in $T \times \widehat{X}$ by the Seesaw Theorem A.8 a). But Γ is the graph of ψ, since $\mathcal{N}|X \times \{t,L\} \simeq \mathcal{L}|X \times \{t\} \otimes L^{-1}$. It follows that the projection $p_1: \Gamma \to T$ is a bijective holomorphic map. Since T is normal, we can apply Zariski's main theorem to get that p_1 is biholomorphic. Therefore ψ is holomorphic. The fact that $\mathcal{L} \simeq (\mathrm{id} \times \psi)^*\mathcal{P}$ follows from the Seesaw Principle A.9.

As for the uniqueness of ψ, suppose $\psi': T \to \widehat{X}$ is another holomorphic map with $\mathcal{L} \simeq (\mathrm{id} \times \psi')^*\mathcal{P}$. Then

$$\psi(t) = (\mathrm{id} \times \psi)^*\mathcal{P}|X \times \{t\} = (\mathrm{id} \times \psi')^*\mathcal{P}|X \times \{t\} = \psi'(t)$$

for all $t \in T$, which completes the proof. □

We will use the Poincaré bundle to give a criterion for two line bundles on X to be analytically equivalent. Here we define the notion of analytic equivalence analogously to the notion of algebraic equivalence as follows: Line bundles L_1 and L_2 on X are called *analytically equivalent*, if there are a connected complex analytic space T, a line bundle $\mathcal{L}$ on $X \times T$ and points $t_1, t_2 \in T$ such that

$$\mathcal{L}|X \times \{t_\nu\} \simeq L_\nu$$

for $\nu = 1, 2$. It is an immediate consequence of the following criterion that this is in fact an equivalence relation.

(5.3) Proposition. *For line bundles L_1, L_2 on X the following statements are equivalent:*

i) L_1 *and* L_2 *are analytically equivalent,*
ii) $L_1 \otimes L_2^{-1} \in \operatorname{Pic}^0(X)$,
iii) $\phi_{L_1} = \phi_{L_2}$,
iv) $c_1(L_1) = c_1(L_2)$.

Proof. Suppose first $L_1 \otimes L_2^{-1} \in \operatorname{Pic}^0(X)$. Define $\mathcal{L} = \mathcal{P} \otimes p^* L_2$ on $X \times \widehat{X}$ where p denotes the projection onto the first factor. Since $\mathcal{L}|X \times \{L_1 \otimes L_2^{-1}\} \simeq L_1$ and $\mathcal{L}|X \times \{0\} \simeq L_2$, the line bundles L_1 and L_2 are analytically equivalent. This shows ii) $\Longrightarrow$ i).

As for i) $\Longrightarrow$ ii): suppose L_1 and L_2 are analytically equivalent and let the notation be as in the definition. It is easy to see that the map $T \to H^2(X, \mathbb{Z})$, $t \mapsto c_1(\mathcal{L}|X \times \{t\})$ is continuous. In fact, it is even constant, since T is connected and $H^2(X, \mathbb{Z})$ is a discrete group. Thus $c_1(L_1) = c_1(L_2)$, that is $L_1 \otimes L_2^{-1} \in \operatorname{Pic}^0(X)$.

Finally, the equivalence ii) $\Longleftrightarrow$ iii) follows immediately from Corollary 4.6 b) and Lemma 4.7 b), and ii) $\Longleftrightarrow$ iv) is a consequence of the Appell-Humbert Theorem. □

According to Proposition 5.3 the analytic equivalence classes of line bundles on X are just the classes of $\operatorname{Pic}(X)$ modulo its subgroup $\operatorname{Pic}^0(X)$. A class is determined uniquely by the corresponding hermitian form H in $\operatorname{NS}(X)$. It will be denoted by $\operatorname{Pic}^H(X)$. If moreover $L \in \operatorname{Pic}(X)$ is nondegenerate, we can say more:

(5.4) Corollary. *Suppose $L, L' \in \operatorname{Pic}(X)$ with L nondegenerate. L and L' are analytically equivalent if and only if $L' = t_x^* L$ for some $x \in X$.*

Proof. If L and L' are analytically equivalent, then $L' \otimes L^{-1} \in \operatorname{Pic}^0(X)$ by Proposition 5.3. Since L is nondegenerate, the map $\phi_L : X \to \operatorname{Pic}^0(X)$ is surjective and there is an $x \in X$ such that $L' \otimes L^{-1} = \phi_L(x) = t_x^* L \otimes L^{-1}$. The converse implication is obvious. □

Another application of the Poincaré bundle is the following criterion for a homomorphism $f : X \to \widehat{X}$ to be of the form ϕ_L for some line bundle L on X.

(5.5) Theorem. *Suppose $X = V/\Lambda$ is a complex torus and $f : X \to \widehat{X}$ a homomorphism with analytic representation $F : V \to \overline{\Omega}$. The following statements are equivalent:*

i) $f = \phi_L$ *for some line bundle* $L \in \operatorname{Pic}(X)$.
ii) The form $F : V \times V \to \mathbb{C}, (v, w) \mapsto F(v)(w)$ *is hermitian.*

For the proof we need the following lemma. Recall that X_n denotes the kernel of the multiplication map n_X.

(5.6) Lemma. *For a line bundle M on X and an integer n the following statements are equivalent:*

i) $M = L^n$ for some line bundle L on X.

ii) $X_n \subset K(M)$.

Proof. ii) $\Longrightarrow$ i): Suppose $X_n \subseteq K(M)$. If $M = L(H,\chi)$, this means that $\frac{1}{n}\Lambda \subseteq \Lambda(M) = \{v \in V \mid \operatorname{Im} H(v,\Lambda) \subseteq \mathbb{Z}\}$. Hence $\operatorname{Im}(nH)(\lambda_1,\lambda_2) \in \operatorname{Im} H(\lambda_1,\Lambda) \subseteq \mathbb{Z}$ for all $\lambda_1,\lambda_2 \in \frac{1}{n}\Lambda$, and thus by Corollary 4.4 there is an $\widetilde{M} \in \operatorname{Pic}(X)$ with $M^n = n_X^*\widetilde{M}$. According to Proposition 3.5 the line bundle $(M \otimes \widetilde{M}^{-n})^n = n_X^*\widetilde{M} \otimes \widetilde{M}^{-n^2}$ is an element of $\operatorname{Pic}^0(X)$. Hence also $M \otimes \widetilde{M}^{-n} \in \operatorname{Pic}^0(X)$. Since $\operatorname{Pic}^0(X) = \widehat{X}$ is a divisible group, there is an $N \in \operatorname{Pic}^0(X)$ with $M \otimes \widetilde{M}^{-n} \simeq N^n$. Now $L = \widetilde{M} \otimes N$ satisfies i).
i) $\Longrightarrow$ ii): Suppose $M \cong L^n$. By definition $\Lambda(M) = \Lambda(L^n) = \frac{1}{n}\Lambda(L)$, which implies ii). □

Proof of Theorem 5.5:. ii) $\Longrightarrow$ i): Suppose the form $F\colon V \times V \to \mathbb{C}$ is hermitian. Let M denote the pullback of the Poincaré bundle $\mathcal{P}$ under the homomorphism $(\mathrm{id}_X, f)\colon X \to X \times \widehat{X}$. We first claim that $2f = \phi_M$. If H denotes the hermitian form of $\mathcal{P}$ (see proof of Theorem 5.1), then by assumption

$$\begin{aligned}(v,w) \mapsto (\mathrm{id}_V, F)^* H(v,w) &= H((v,F(v)),(w,F(w))) \\ &= \overline{F(w)(v)} + F(v)(w) = 2F(v)(w)\end{aligned}$$

is the hermitian form of M. Since $(\mathrm{id}, F)^*\phi_H$ is the analytic representation of ϕ_M and $2F$ is the analytic representation of $2f$, this implies that $2f = \phi_M$.

According to Lemma 5.6 there is a line bundle $L \in \operatorname{Pic}(X)$ with $M = L^2$. By Corollary 4.6 b) we have $2\phi_L = \phi_{L^2} = 2f$. Since $\operatorname{Hom}(X,\widehat{X})$ is torsion free according to Proposition 1.2.2, this finally implies $f = \phi_L$.
i) $\Longrightarrow$ ii): If $f = \phi_L$, the analytic representation F is just the hermitian form $c_1(L)$ by definition of ϕ_L. □

Exercises

(1) Suppose $X = V/\Lambda$ is a complex torus of dimension g and $x_1,\ldots,x_{2g}$ are real coordinate functions with respect to an $\mathbb{R}$-basis $e_1,\ldots,e_{2g}$ of V.

a) Show that the following diagram commutes

$$\begin{array}{ccc} H^2(\Lambda,\mathbb{C}) & \xrightarrow{\alpha} & \operatorname{Alt}^2_{\mathbb{R}}(V,\mathbb{C}) \\ {\scriptstyle\phi_2}\downarrow & & \downarrow{\scriptstyle\gamma_2} \\ H^2(X,\mathbb{C}) & \xrightarrow{\varrho} & H^2_{\mathrm{DR}}(X) \end{array} .$$

Here α is the extension of the isomorphism of Lemma 1.3, ϕ_2 is the canonical isomorphism (see Remark B.2), ϱ is the de Rham isomorphism and γ_2 is the canonical isomorphism which sends the alternating form E to $\sum_{1\leq\nu<\mu\leq 2g} E(e_\mu, e_\nu)\, dx_\nu \wedge dx_\mu$.

b) Conclude that the isomorphism $\alpha\phi_2^{-1}\colon H^2(X,\mathbb{Z}) \to \mathrm{Alt}^2(\Lambda,\mathbb{Z})$ coincides with the isomorphism of Corollary 1.3.2.

(2) *(First Chern Class of a Line Bundle)* For a complex torus $X = V/\Lambda$ the composed map $c_1\colon H^1(X,\mathcal{O}_X^*) \to H^2(X,\mathbb{Z}) \hookrightarrow H^2(X,\mathbb{C}) \xrightarrow{\sim} H^2_{\mathrm{DR}}(X)$ is defined as follows: Suppose $L \in H^1(X,\mathcal{O}_X^*)$ is given by a cocycle $\{f_{\mu\nu}\}_I$ with respect to a covering $\{U_\mu\}_I$ of X. Choose $\mathcal{C}^\infty$-1-forms φ_μ on U_μ with $\frac{1}{2\pi i} d\log f_{\mu\nu} = \varphi_\nu - \varphi_\mu$. Then $c_1(L)$ is the global 2-form $d\,\varphi_\mu$.

a) Suppose $L = L(H,\chi)$ and $v_1,\ldots,v_g$ are complex coordinate functions with respect to a basis $e_1,\ldots,e_g$ of V. Show that:

$$c_1(L) = \tfrac{i}{2} \sum_{\nu,\mu=1}^{g} H(e_\nu, e_\mu)\, dv_\nu \wedge d\bar{v}_\mu \;.$$

b) Suppose L is of type $(d_1,\ldots,d_g)$ and $x_1,\ldots,x_g,y_1,\ldots,y_g$ are the real coordinate functions on V with respect to a symplectic basis $\lambda_1,\ldots,\lambda_g$, $\mu_1,\ldots,\mu_g$ of Λ for L (for the definition of the type and a symplectic basis see Section 3.1). Show that

$$c_1(L) = -\sum_{\nu=1}^{g} d_\nu dx_\nu \wedge dy_\nu \;.$$

(Hint: use the cocycle $\phi_1 a_L$, where a_L is the canonical factor of L and ϕ_1 is the isomorphism of Remark B.2.)

(3) Let $X = V/\Lambda$ be a complex torus. In terms of sheaves the canonical isomorphism $\phi_1\colon H^1(\Lambda, H^0(V,\mathcal{O}_V^*)) \xrightarrow{\sim} \mathrm{Pic}(X)$ can be described as follows: Let $f\colon \Lambda\times V \to \mathbb{C}^*$ denote a factor of automorphy and $[f]$ its class in the cohomology group $H^1(\Lambda, H^0(V,\mathcal{O}_V^*))$. The image $\phi_1[f]$ is the line bundle associated to the sheaf $\mathcal{L}$ defined by

$$\mathcal{L}(U) = \left\{ \vartheta\colon \pi^{-1}(U) \to \mathbb{C} \;\middle|\; \begin{array}{c} \vartheta \text{ holomorphic with } \vartheta(v+\lambda) = f(\lambda,v)\,\vartheta(v) \\ \text{for all } (\lambda,v) \in \Lambda\times\pi^{-1}(U) \end{array} \right\}$$

for any open $U \subset X$.

(4) *(The Néron-Severi Group of a Complex Torus)* Let $(Z,\mathbb{1}_g)$ be a period matrix of a complex torus X. According to Section 1 the Néron-Severi group $NS(X)$ may be considered as the group of alternating forms on $\mathbb{R}^{2g}$, integer valued on $\mathbb{Z}^{2g}$, which induce a hermitian form on $\mathbb{C}^g$ via the $\mathbb{R}$-linear isomorphism $\mathbb{R}^{2g} \to \mathbb{C}^g$, $x \mapsto (Z,\mathbb{1}_g)x$. Let E be an alternating form on $\mathbb{R}^{2g}$ with matrix $\begin{pmatrix} A & B \\ -{}^tB & C \end{pmatrix} \in M_{2g}(\mathbb{R})$. Show that the following conditions are equivalent:

i) $E \in \mathrm{NS}(X)$,
ii) $A, B, C \in M_g(\mathbb{Z})$ and $A - BZ + {}^tZ{}^tB + {}^tZCZ = 0$.

Hence the group of alternating matrices $\begin{pmatrix} A & B \\ -{}^tB & C \end{pmatrix} \in M_{2g}(\mathbb{Z})$ satisfying condition ii) is isomorphic to the Néron-Severi group $\mathrm{NS}(X)$.

In particular, if $g = 2$, $A = \begin{pmatrix} 0 & a \\ -a & 0 \end{pmatrix}$, $B = \begin{pmatrix} b & d \\ e & f \end{pmatrix}$ and $C = \begin{pmatrix} 0 & c \\ -c & 0 \end{pmatrix}$, condition ii) reads: $a, b, c, d, e, f \in \mathbb{Z}$ and $a + ez_{11} - bz_{12} + fz_{21} - dz_{22} + c \det Z = 0$.

(5) Let X be a complex torus of dimension g. The *Picard number* $\varrho(X)$ of X is by definition the rank of the Néron-Severi group $\mathrm{NS}(X)$.
Show $\rho(X) \leq h^{1,1}(X) = g^2$.

(6) *(Theorem of the Cube)* a) Let $X_\nu, \nu = 1, 2, 3$, be complex tori and L a line bundle on $X_1 \times X_2 \times X_3$ such that the restrictions of L to $X_1 \times X_2 \times \{0\}$, $X_1 \times \{0\} \times X_3$ and $\{0\} \times X_2 \times X_3$ are trivial. Show that L is trivial. (Hint: use canonical factors.)
b) Generalize this to products of $n > 3$ complex tori.

(7) Let X be a complex torus, Y be any complex manifold and $\varphi_\nu : Y \to X$, $\nu = 1, 2, 3$, holomorphic maps. Show that for any $L \in \mathrm{Pic}(X)$:

$$(\varphi_1 + \varphi_2 + \varphi_3)^* L \simeq$$
$$\simeq (\varphi_1 + \varphi_2)^* L \otimes (\varphi_1 + \varphi_3)^* L \otimes (\varphi_2 + \varphi_3)^* L \otimes \varphi_1^* L^{-1} \otimes \varphi_2^* L^{-1} \otimes \varphi_3^* L^{-1} .$$

(Hint: use the canonical factor for L.)

(8) Let $f : X = V/\Lambda \to Y = V/\Gamma$ be an isogeny of complex tori with analytic representation F and L a line bundle on X which descends to a line bundle M on Y.

a) Choose an isomorphism $\varphi : \overline{\Omega} \xrightarrow{\sim} V$. It induces an isomorphism $\operatorname{im} \phi_L \simeq V/\Lambda(L)$.
b) There is an isomorphism of groups: $\ker \widehat{f} \xrightarrow{\sim} \Lambda(L)/F^{-1}\Gamma(M)$. (Here $\Gamma(M)$ is defined analogously as $\Lambda(L)$ in Section 4.)

(9) Let $f : X \to Y$ be an isogeny of complex tori and L a line bundle on X which descends to a line bundle M_0 on Y.
a) Show that the set $\{M \in \mathrm{Pic}(Y) \mid f^* M \simeq L\}$ has a natural group structure with M_0 as identity element.
b) There is an isomorphism of groups:

$$\{M \in \mathrm{Pic}(Y) \mid f^* M \simeq L\} \xrightarrow{\sim} \Lambda(L)/\varrho_a(f)^{-1}\Lambda(M) .$$

(10) For any complex torus X and any integer $n \neq 0$ the homomorphism $L \mapsto n_X^* L$ on $\mathrm{Pic}(X)$ induces the n-th power map on $\widehat{X}$ and the n^2-th power map on $\mathrm{NS}(X)$.

(11) a) For complex tori X_1 and X_2 there is a canonical isomorphism $(X_1 \times X_2)\widehat{\ } \simeq \widehat{X}_1 \times \widehat{X}_2$.

b) Let $f_\nu: X_\nu \to Y_\nu$, $\nu = 1, 2$, be homomorphisms of complex tori. Show that $(f_1 \times f_2)\widehat{\ } = \widehat{f_1} \times \widehat{f_2}$ with respect to the canonical isomorphisms of a).

(12) a) For a complex torus X denote by $\Delta_X: X \to X \times X$, $x \mapsto (x, x)$ the diagonal map and by $\mu: X \times X \to X$, $(x, y) \mapsto x + y$ the addition map. Show that

$$\widehat{\mu} = \Delta_{\widehat{X}} .$$

b) Use a) to show that $(f+g)\widehat{\ } = \widehat{f} + \widehat{g}$ for homomorphisms $f, g: X \to Y$ of complex tori.

(13) Suppose $f: X \to Y$ is a homomorphism of complex tori of dimensions g and g' respectively. As usual, denote by $\varrho_a: \mathrm{Hom}(X, Y) \to M(g' \times g, \mathbb{C})$ and $\varrho_r: \mathrm{Hom}(X, Y) \to M(2g' \times 2g, \mathbb{Z})$ the analytic and rational representation.

a) $\varrho_a(\widehat{f}) = {}^t\overline{\varrho_a(f)}$,

b) $\varrho_r(\widehat{f}) = {}^t\varrho_r(f)$.

(14) Let L_ν be a line bundle on the complex torus X_ν for $\nu = 1, 2$ and denote by $p_\nu: X_1 \times X_2 \to X_\nu$ the natural projection. Show

$$\phi_{p_1^* L_1 \otimes p_2^* L_2} = \phi_{L_1} \times \phi_{L_2}: X_1 \times X_2 \to \widehat{X}_1 \times \widehat{X}_2 .$$

(15) Let X be a complex torus. Show that the pair $(\widehat{X}, \mathcal{P})$ is uniquely determined (up to isomorphisms) by the properties i) and ii) of the definition of the Poincaré bundle $\mathcal{P}$ and the Universal Property Proposition 5.2 for $\mathcal{P}$.

(16) Let X be a complex torus and $\kappa: X \to \widehat{\widehat{X}}$ the canonical isomorphism. Denote by $\mathcal{P}_X$ (respectively $\mathcal{P}_{\widehat{X}}$) the Poincaré bundle for X (respectively for $\widehat{X}$) and by s the canonical isomorphism $\widehat{X} \times X \to X \times \widehat{X}$ sending $(\hat{x}, x)$ to $(x, \hat{x})$. Show that $(1_{\widehat{X}} \times \kappa)^* \mathcal{P}_{\widehat{X}} \simeq s^* \mathcal{P}_X$ on $\widehat{X} \times X$.

(17) Let X be a complex torus and $\mathcal{P}$ the Poincaré bundle on $X \times \widehat{X}$. Denote by $p_1, p_2: X \times X \to X$ the natural projections and by $\mu: X \times X \to X$ the addition map.

a) Show that for any $L \in \mathrm{Pic}(X)$: $(1_X \times \phi_L)^* \mathcal{P} \simeq \mu^* L \otimes p_1^* L^{-1} \otimes p_2^* L^{-1}$.

b) Conclude that $L \in \mathrm{Pic}^0(X)$ if and only if $\mu^* L \simeq p_1^* L \otimes p_2^* L$.

Chapter 3

Cohomology of Line Bundles

In this chapter we compute the dimension of every cohomology group of every line bundle L on a complex torus $X = V/\Lambda$ (see Theorem 5.5). As a direct consequence we get a formula for the *Euler-Poincaré characteristic* $\chi(L)$ of L. The result is the Riemann-Roch Theorem. This approach to Riemann-Roch was first given in Deligne [1] and independently in Umemura [1].

The main steps of the proof are: the sections of L can be interpreted as theta functions on the vector space V. So if the hermitian form $H = c_1(L)$ is positive definite, one can compute $h^s(L)$ by considering Fourier expansions of the corresponding theta functions. If H is nondegenerate with $s > 0$ negative eigenvalues, one can apply a trick, in order to compute $h^s(L)$, which goes back to Wirtinger [2]: a suitable change of the complex structure of V defines in a canonical way a new line bundle M, which is positive definite and satisfies $h^s(L) = h^0(M)$. Furthermore, a vanishing theorem implies that all other cohomology groups of L are zero. Finally, the case that H is degenerate is reduced to the nondegenerate case by passing to an abelian variety of smaller dimension.

An important tool for the proof, used also in many other parts of the book, is the notion of characteristics to be introduced for every nondegenerate line bundle L on X: given a decomposition $\Lambda = \Lambda_1 \oplus \Lambda_2$ into subgroups Λ_1 and Λ_2 isotropic for the alternating form $\operatorname{Im} H$ associated to L, one can distinguish a line bundle L_0 among all line bundles in $\operatorname{Pic}^H(X)$. An element $c \in V$ is called the *characteristic* of L (with respect to the given decomposition of Λ), if $L \simeq t_{\bar{c}}^* L_0$. Using characteristics, all computations with theta functions in this chapter can be done in an intrinsic, coordinatefree way. In Chapter 8 we will see that in the principally polarized case this notion coincides with the classical notion of characteristics used for example in Krazer [1].

After the introduction of characteristics in Section 1 we derive some properties of theta functions in Section 2. The main result is Theorem 2.7, which gives an explicit basis of the vector space of canonical theta functions for any positive definite line bundle on X. It will play an important role in the theory of theta groups and Heisenberg groups in Chapter 6. Section 3 reduces the case of a positive semidefinite line bundle to a positive definite one. Section 4 contains the proof of the Vanishing Theorem of Mumford and Kempf. The remaining step in the computation of the cohomology groups, the above mentioned trick of Wirtinger, can be found in Section 5. Finally in Section 6 we deduce various forms of the Riemann-Roch Theorem.

In Section 4 we use Dolbeault's Theorem for the cohomology groups of a holomorphic line bundle as well as Serre duality.

§ 1 Characteristics

The notion of characteristics played an important role in the theory of theta functions at the end of the last century, as developed for example in Krazer [1]. Here we will generalize this notion and define characteristics for any nondegenerate line bundle on a complex torus. This will turn out to be decisive for the whole book.

Let $X = V/\Lambda$ be a complex torus of dimension g and L a line bundle on X with first Chern class H. Recall that H is a hermitian form on V, whose alternating form $E = \operatorname{Im} H$ is integer-valued on the lattice Λ. According to the elementary divisor theorem (see Frobenius [1] or Bourbaki [1] Alg.IX. 5.1 Th.1) there is a basis $\lambda_1, \dots, \lambda_g, \mu_1, \dots, \mu_g$ of Λ, with respect to which E is given by the matrix

$$\begin{pmatrix} 0 & D \\ -D & 0 \end{pmatrix},$$

where $D = \operatorname{diag}(d_1, \dots, d_g)$ with integers $d_\nu \geq 0$ satisfying $d_\nu | d_{\nu+1}$ for $\nu = 1, \dots, g-1$. The numbers $d_1, \dots, d_g$ are uniquely determined by E and Λ and thus by L. The vector $(d_1, \dots, d_g)$ as well as the matrix D are called the *type* of the line bundle L, and the basis $\lambda_1, \dots, \lambda_g, \mu_1, \dots, \mu_g$ is called a *symplectic* (or *canonical) basis* of Λ for L (or H or E respectively). Recall that the line bundle L is *nondegenerate* if the form H, and thus E, is nondegenerate. It is clear that this is equivalent to $d_\nu > 0$ for $\nu = 1, \dots, g$. A direct sum decomposition

$$\Lambda = \Lambda_1 \oplus \Lambda_2 \tag{1}$$

is called a *decomposition* for L (or H or E respectively) if Λ_1 and Λ_2 are isotropic with respect to the alternating form E. Such a decomposition always exists: If $\lambda_1, \dots, \lambda_g, \mu_1, \dots, \mu_g$ is a symplectic basis of Λ for L, then $\Lambda = \langle \lambda_1, \dots, \lambda_g \rangle \oplus \langle \mu_1, \dots, \mu_g \rangle$ is a decomposition for L. Conversely it follows immediately from the proof of the elementary divisor theorem that for every decomposition (1) for L there exists a symplectic basis $\lambda_1, \dots, \lambda_g, \mu_1, \dots, \mu_g$ of Λ for L with $\Lambda_1 = \langle \lambda_1, \dots, \lambda_g \rangle$ and $\Lambda_2 = \langle \mu_1, \dots, \mu_g \rangle$.

A decomposition

$$V = V_1 \oplus V_2 \tag{2}$$

with real subvector spaces V_1 and V_2, is called a *decomposition for* L (or H or E respectively), if $(V_1 \cap \Lambda) \oplus (V_2 \cap \Lambda)$ is a decomposition of Λ for L. Clearly a decomposition of V for L is a decomposition into maximal isotropic subvector spaces. Conversely, not every decomposition of V into maximal isotropic subvector spaces is a decomposition for L (see Exercise 1). According to the above definitions the decompositions of Λ for L are in one to one correspondence to the decompositions of V for L.

Let $H \in \mathrm{NS}(X)$ be nondegenerate. A decomposition $V = V_1 \oplus V_2$ for H leads to an explicit description of all line bundles L in $\mathrm{Pic}^H(X)$. For this define a map $\chi_0 : V \longrightarrow \mathbb{C}_1$ by

$$\chi_0(v) = e\big(\pi i E(v_1, v_2)\big) \ , \tag{3}$$

where $v = v_1 + v_2$ with $v_\nu \in V_\nu$. Using that V_1 and V_2 are isotropic with respect to E we find by an obvious computation

(1.1) Lemma. *For every* $v = v_1 + v_2$, $w = w_1 + w_2 \in V_1 \oplus V_2$

$$\chi_0(v + w) = \chi_0(v)\chi_0(w)e\big(\pi i E(v, w)\big)e\big(-2\pi i E(v_2, w_1)\big) \ .$$

Lemma 1.1 implies in particular that $\chi_0|\Lambda$ is a semicharacter for H. Let $L_0 = L(H, \chi_0)$ denote the corresponding line bundle. The data χ_0 and L_0 are uniquely determined by H and the chosen decomposition. In other words, the decomposition $V = V_1 \oplus V_2$ distinguishes a line bundle in $\mathrm{Pic}^H(X)$, namely L_0. With this notation

(1.2) Lemma. *Suppose H is a nondegenerate hermitian form on V and $V = V_1 \oplus V_2$ a decomposition for H.*

a) $L_0 = L(H, \chi_0)$ is the unique line bundle in $\mathrm{Pic}^H(X)$, *whose semicharacter is trivial on $\Lambda_\nu = V_\nu \cap \Lambda$ for $\nu = 1, 2$.*

b) For every $L = L(H, \chi)$ on X there is a point $c \in V$, uniquely determined up to translation by elements of $\Lambda(L)$, such that $L \simeq t_{\bar{c}}^* L_0$ or equivalently $\chi = \chi_0 e\big(2\pi i E(c, \cdot)\big)$.

Proof. a) follows immediately from the definition of χ_0 and Lemma 1.1. As for b): the existence of the point c is a direct consequence of Corollary 2.5.4 The uniqueness statement is a translation of the fact that $\ker \phi_L = \Lambda(L)/\Lambda$. The second assertion follows from Lemma 2.3.2. □

We call c a *characteristic of the line bundle L with respect to the chosen decomposition for L*. If we speak only of a characteristic c of L, we mean that a decomposition for L is fixed and that c is the characteristic of L with respect to it. We will see in Lemma 8.5.2 that this definition coincides with the classical notion of characteristics. Note that a characteristic is only defined for nondegenerate line bundles and determined only up to translations by elements of $\Lambda(L)$. Note also that 0 is a characteristic of L_0.

Recall the canonical factor of automorphy $a_L : \Lambda \times V \to \mathbb{C}^*$ of Section 2.3. As a first application of the notion of characteristics, we extend it to a function $V \times V \longrightarrow \mathbb{C}^*$, also denoted by a_L. This will turn out to be very useful for later computations in particular for the theory of Chapter 6. Suppose $L = L(H, \chi)$ is a nondegenerate line bundle on X and c is a characteristic for L with respect to the decomposition $V = V_1 \oplus V_2$. Define $a_L : V \times V \longrightarrow \mathbb{C}^*$ by

$$a_L(u, v) = \chi_0(u) e\big(2\pi i E(c, u) + \pi H(v, u) + \tfrac{\pi}{2} H(u, u)\big) . \qquad (4)$$

According to Lemma 1.1 its restriction to $\Lambda \times V$ is the canonical factor of automorphy $(\lambda, v) \mapsto \chi(\lambda) e(\pi H(v, \lambda) + \frac{\pi}{2} H(\lambda, \lambda))$ of L as defined in Section 2.3. So a_L is in fact an extension to $V \times V$ of the canonical factor of L. Certainly $a_L : V \times V \longrightarrow \mathbb{C}^*$ is not a factor of automorphy. The following technical lemma gives some properties of a_L, which turn out to be very useful subsequently.

(1.3) Lemma. *For all* $u = u_1 + u_2, v = v_1 + v_2,\ w \in V = V_1 \oplus V_2$

a) $a_L(u, v + w) = a_L(u, v) e\big(\pi H(w, u)\big)$,

b) $a_L(u + v, w) = a_L(u, v + w) a_L(v, w) e\big(2\pi i E(u_1, v_2)\big)$,

c) $a_L(u, v)^{-1} = a_L(-u, v) \chi_0(u)^{-2} e\big(-\pi H(u, u)\big)$,

d) $a_{L'}(u, v) = a_L(u, v) e\big(2\pi i E(w, u)\big)$ *with* $L' = t_w^* L$.

Proof. a), c) and d) are immediate consequences of the definition of a_L and Lemma 2.3.2 respectively. As for b), we have using Lemma 1.1:

$$\begin{aligned}
a_L(u + v, w) &= \\
&= \chi_0(u + v) e\big(2\pi i E(c, u + v) + \pi H(w, u + v) + \tfrac{\pi}{2} H(u + v, u + v)\big) \\
&= \chi_0(u) e\big(2\pi i E(c, u) + \pi H(v + w, u) + \tfrac{\pi}{2} H(u, u)\big) \\
&\qquad \cdot \chi_0(v) e\big(2\pi i E(c, v) + \pi H(w, v) + \tfrac{\pi}{2} H(v, v)\big) \\
&\qquad \cdot e\big(\pi i E(u, v) - 2\pi i E(u_2, v_1) + \tfrac{\pi}{2} H(u, v) - \tfrac{\pi}{2} H(v, u)\big) \\
&= a_L(u, v + w) a_L(v, w) e\big(2\pi i E(u_1, v_2)\big) .
\end{aligned}$$

□

(1.4) Lemma. *Let L be a nondegenerate line bundle on X and $\Lambda = \Lambda_1 \oplus \Lambda_2$ a decomposition for L with induced decomposition $V = V_1 \oplus V_2$.*

a) $\Lambda(L) = \Lambda(L)_1 \oplus \Lambda(L)_2$ with $\Lambda(L)_i = V_i \cap \Lambda(L)$ for $i = 1, 2$.

b) $K(L) = K_1 \oplus K_2$ with $K_i = \Lambda(L)_i/\Lambda_i$ for $i = 1, 2$.

c) $K_\nu \simeq \mathbb{Z}^g/D\mathbb{Z}^g = \bigoplus_{\mu=1}^g \mathbb{Z}/d_\mu\mathbb{Z}$ for $\nu = 1, 2$, if the line bundle L is of type $D = \operatorname{diag}(d_1, \ldots, d_g)$.

Proof. b) and c) are immediate consequences of a) and the elementary divisor theorem.

As for a), it suffices to show $\Lambda(L) \subseteq \Lambda(L)_1 \oplus \Lambda(L)_2$. Suppose $v \in \Lambda(L)$ and write $v = v_1 + v_2$ with $v_i \in V_i$ for $i = 1, 2$. By symmetry it suffices to show $v_1 \in \Lambda(L)$, i.e. $E(v_1, \lambda) \in \mathbb{Z}$ for all $\lambda \in \Lambda$. By definition the decomposition of V restricts to a decomposition of Λ, so $\lambda = \lambda_1 + \lambda_2$ with $\lambda_i \in \Lambda_i$ for every $\lambda \in \Lambda$. But $E(v_1, \lambda) = E(v_1, \lambda_1 + \lambda_2) = E(v_1, \lambda_2) = E(v, \lambda_2) \in \mathbb{Z}$. □

The decompositions of Lemma 1.4 a) and b) are also called *decompositions for L* (or H or E respectively). Clearly they are decompositions into maximal isotropic subspaces with respect to the induced alternating forms. Note that decompositions of Λ and $\Lambda(L)$ for L determine each other in such a way that $\Lambda_\nu \subset \Lambda(L)_\nu$ for $\nu = 1, 2$. Hence a decomposition of Λ induces a unique decomposition of $K(L)$ for L. But conversely, for any decomposition of $K(L)$ there might be many decompositions of Λ inducing it.

(1.5) Remark. One may ask, whether there are lattices in V containing Λ, for which a_L is still a factor of automorphy. In fact, it follows from Lemma 1.4 that $\Lambda(L)_1 \oplus \Lambda_2$ and $\Lambda_1 \oplus \Lambda(L)_2$ are lattices in V, and according to Lemma 1.3.b) the map a_L restricted to $(\Lambda(L)_1 \oplus \Lambda_2) \times V$ respectively $(\Lambda_1 \oplus \Lambda(L)_2) \times V$ satisfies the cocycle relation. In terms of line bundles and complex tori, this means: denote by

$$X_1 = V/(\Lambda(L)_1 \oplus \Lambda_2) = X/K_1 \text{ and } X_2 = V/(\Lambda_1 \oplus \Lambda(L)_2) = X/K_2$$

the corresponding complex tori and $p_\nu\colon X \longrightarrow X_\nu$, $\nu = 1, 2$, the isogenies induced by id_V. Then a_L determines line bundles M_ν on X_ν with $L = p_\nu^* M_\nu$ for $\nu = 1, 2$. By construction M_ν is the line bundle on X_ν with characteristic c with respect to $V = V_1 \oplus V_2$. Varying the characteristic c of L within $\Lambda(L)$, it is easy to see, how to obtain every descent M_ν of L to X_ν (see Exercise 2.9).

§ 2 Theta Functions

Let $X = V/\Lambda$ be a complex torus. As always denote by $\pi: V \longrightarrow X$ the canonical projection map. According to Lemma 2.1.1 the pullback π^*L of a line bundle L on X is trivial. On the other hand, the lattice Λ acts naturally on π^*L. Hence $H^0(L) \simeq \pi^*H^0(L)$ is isomorphic to $H^0(\mathcal{O}_V)^\Lambda$, the subspace of holomorphic functions on V invariant under the action of Λ. According to Appendix B this isomorphism depends on the choice of a factor of automorphy for L. To be more precise, suppose f is a factor of automorphy for L. Then $H^0(L)$ can be identified with the set of holomorphic functions $\vartheta: V \longrightarrow \mathbb{C}$ satisfying

$$\vartheta(v + \lambda) = f(\lambda, v)\vartheta(v)$$

for all $v \in V$ and $\lambda \in \Lambda$. These functions are called *theta functions for the factor* f.

In Section 2.2 we saw that for every line bundle L there is a canonical factor a_L. Correspondingly we call the theta functions for a_L *canonical theta functions for* L. In this section we determine the theta functions for L and compute the dimension $h^0(L)$ in the case that the hermitian form H of L is positive definite. For this it is convenient to introduce another factor for L, the *classical* factor of automorphy.

Suppose $L = L(H, \chi)$ is a nondegenerate line bundle on X. Denote $E = \operatorname{Im} H$ and fix a decomposition $V = V_1 \oplus V_2$ for L. Then

(2.1) Lemma. *V_2 generates V as a $\mathbb{C}$-vector space.*

Proof. $V_2 \cap iV_2$ is a complex subvector space of V, on which E vanishes identically. According to Lemma 2.1.7 also H vanishes on $V_2 \cap iV_2$. Since H is nondegenerate, $V_2 \cap iV_2 = \{0\}$ and thus $V = V_2 + iV_2$. This implies the assertion. □

The hermitian form H is symmetric on V_2, since its imaginary part E vanishes there. Denote by B the $\mathbb{C}$-bilinear extension of the symmetric form $H|V_2 \times V_2$. According to Lemma 2.1 the symmetric form B is defined on the whole of V. The following properties of H and B will be frequently applied.

(2.2) Lemma. *a)* $(H - B)(v, w) = \begin{cases} 0 & \textit{if} \quad (v, w) \in V \times V_2 \\ 2iE(v, w) & \quad (v, w) \in V_2 \times V \end{cases}.$

b) If H is positive definite, then $\operatorname{Re}(H - B)$ *is positive definite on* V_1.

Proof. As for a): $H - B = 0$ on $V \times V_2$, since H is $\mathbb{C}$-linear in its first component. Hence for $v \in V_2$ and $w \in V$

$$(H-B)(v,w) = \overline{H(w,v)}-B(w,v) = (H-B)(w,v)-2iE(w,v) = 2iE(v,w)\,.$$

As for b): since $V = V_2 + iV_2$, a vector $v \in V_1, v \neq 0$ can be uniquely written in the form $v = v_2 + iv_2'$ for some $v_2, v_2' \in V_2$ with $v_2' \neq 0$. Using a) we have

$$\begin{aligned}\mathrm{Re}(H-B)(v,v) &= \mathrm{Re}\big(2iE(v_2,v) - 2E(v_2',v)\big) = 2E(v,v_2') \\ &= 2E(iv_2',v_2') + 2iE(v_2',v_2') = 2H(v_2',v_2')\,.\end{aligned}$$

This implies the assertion. □

The bilinear form B enables us to introduce the classical factor of automorphy for L in a coordinate free way. Define $e_L\colon \Lambda \times V \longrightarrow \mathbb{C}^*$ by

$$e_L(\lambda,v) = \chi(\lambda)e\big(\pi(H-B)(v,\lambda) + \tfrac{\pi}{2}(H-B)(\lambda,\lambda)\big)\,.$$

With an immediate computation one checks that

$$e_L(\lambda,v) = a_L(\lambda,v)e\big(\tfrac{\pi}{2}B(v,v)\big)e\big(\tfrac{\pi}{2}B(v+\lambda,v+\lambda)\big)^{-1} \tag{1}$$

for all $(\lambda,v) \in \Lambda \times V$. This implies that e_L is a factor of automorphy for L, equivalent to the canonical factor a_L. It is called the *classical factor (of automorphy) for L*. Correspondingly the theta functions for the factor e_L are called *classical theta functions for L*. This terminology will be justified later (see Section 8.5). In fact, we will see that for a line bundle L of type $(1,\ldots,1)$ these functions are just the classical theta functions in the sense of Riemann (see Krazer [1]). The classical theta functions have the advantage to be periodic with respect to the subgroup Λ_2 of Λ.

Suppose L is of type $D = \mathrm{diag}(d_1,\ldots,d_g)$. Recall that $\left(\begin{smallmatrix}0 & D\\ -D & 0\end{smallmatrix}\right)$ is the matrix of E with respect to a symplectic basis of the lattice Λ. By definition $Pf(E) = \det D$ is called the *Pfaffian of E*. Note that $Pf(E)$ is independent of the choice of the symplectic basis. With this notation we have for any nondegenerate line bundle L on X

(2.3) Lemma. $h^0(L) \leq Pf(E)$.

Proof. Suppose L is of characteristic c and let $L_0 \in \mathrm{Pic}^H(X)$ be the line bundle of characteristic 0 with respect to a decomposition $V = V_1 \oplus V_2$ for L.
Step I: $h^0(L) = h^0(L_0)$.
Consider $H^0(L)$ and $H^0(L_0)$ as vector spaces of theta functions with respect to the classical factors e_L and e_{L_0}. We claim that the map $H^0(L) \longrightarrow H^0(L_0)$ defined by $\vartheta \mapsto \tilde{\vartheta} = e\big(\pi(H-B)(\cdot,c)\big)\vartheta(\cdot - c)$ is an isomorphism of vector spaces. For this it suffices to show that $\tilde{\vartheta}$ is a theta function for the factor

e_{L_0}. By Lemma 1.3 d) the factors e_L and e_{L_0} are related as $e_L(\lambda, v) = e_{L_0}(\lambda, v)e(2\pi i E(c,\lambda))$ for all $\lambda \in \Lambda$ and $v \in V$. Using this and Lemma 1.3 a) the assertion follows from an immediate computation.

Step II: $h^0(L_0) \leq Pf(E)$.

Suppose $\vartheta \in H^0(L_0)$. According to Lemma 2.2 a) and the definition of χ_0 we have $e_{L_0}(\lambda_2, v) = 1$ for all $v \in V$ and $\lambda_2 \in \Lambda_2$. This implies that ϑ is periodic with respect to Λ_2. Hence ϑ admits a Fourier expansion. By the properties of $(H - B)\colon V \times V \to \mathbb{C}$ given in Lemma 2.2 one easily sees that the Fourier series of ϑ can be written in the form

$$\vartheta(v) = \sum_{\lambda \in \Lambda(L)_1} \alpha_\lambda e(\pi(H - B)(v,\lambda))$$

for all $v \in V$ with uniquely determined complex coefficients α_λ. This follows also from Lemma 8.5.1 b), where we show that in terms of suitable coordinates $v = {}^t(v_1, \ldots, v_g)$ and $\lambda = {}^t(\lambda_1, \ldots, \lambda_g)$ we have $e(\pi(H - B)(v,\lambda)) = e(-2\pi i\, {}^t v \lambda)$. So this is the usual Fourier expansion.

The function ϑ satisfies $\vartheta(v + \lambda_1) = e_{L_0}(\lambda_1, v)\vartheta(v)$ for all $v \in V$ and $\lambda \in \Lambda_1$. But

$$\vartheta(v + \lambda_1) = \sum_{\lambda \in \Lambda(L)_1} \alpha_\lambda e(\pi(H - B)(\lambda_1,\lambda) + \pi(H - B)(v,\lambda))$$

and by definition of e_{L_0}

$$\begin{aligned} e_{L_0}(\lambda_1, v)\vartheta(v) &= \sum_{\lambda \in \Lambda(L)_1} \alpha_\lambda e_{L_0}(\lambda_1, 0) e(\pi(H - B)(v,\lambda_1) + \pi(H - B)(v,\lambda)) \\ &= \sum_{\lambda \in \Lambda(L)_1} \alpha_{\lambda - \lambda_1} e_{L_0}(\lambda_1, 0) e(\pi(H - B)(v,\lambda)) \,. \end{aligned}$$

Comparing coefficients gives $\alpha_{\lambda-\lambda_1} = \alpha_\lambda e_{L_0}(\lambda_1, 0)^{-1} e(\pi(H - B)(\lambda_1,\lambda)$ for all $\lambda \in \Lambda(L)_1$ and $\lambda \in \Lambda_1$. Hence ϑ is determined by the coefficients α_λ, where $\lambda \in \Lambda(L)_1$ runs over a set of representatives of $K(L)_1 = \Lambda(L)_1/\Lambda_1$. So with Lemma 1.4 we obtain $h^0(L_0) \leq (\Lambda(L)_1 : \Lambda_1) = Pf(E)$. □

For the rest of this section we assume that the first Chern class H of L is a positive definite hermitian form. Such a line bundle is called *positive definite.* We will see that for a positive definite line bundle the inequality of Lemma 2.3 is even an equality. In fact, we will construct an explicit basis of canonical theta functions of $H^0(L)$.

Suppose $c \in V$ is a characteristic of L with respect to the decomposition $V = V_1 \oplus V_2$. Define a function $\vartheta^c\colon V \longrightarrow \mathbb{C}$ by

$$\begin{aligned} \vartheta^c(v) = e\big(&-\pi H(v,c) - \tfrac{\pi}{2} H(c,c) + \tfrac{\pi}{2} B(v + c, v + c)\big) \\ &\cdot \sum_{\lambda \in \Lambda_1} e\big(\pi(H - B)(v + c,\lambda) - \tfrac{\pi}{2}(H - B)(\lambda,\lambda)\big) \,. \end{aligned} \tag{2}$$

(2.4) Lemma. *ϑ^c is a canonical theta function for $L = t_{\bar{c}}^* L_0$.*

Proof. Step I: ϑ^c is holomorphic on V.
It suffices to show that the function

$$f(v) = \sum_{\lambda \in \Lambda_1} \left| e\left(\pi(H-B)(v,\lambda) - \tfrac{\pi}{2}(H-B)(\lambda,\lambda)\right)\right|$$

converges uniformly on every compact subset of V. For this choose a norm map $\|\cdot\|: V \longrightarrow \mathbb{R}$ with $\|\Lambda\| \subset \mathbb{Z}$. Since $\operatorname{Re}(H-B)$ is positive definite on V_1 by Lemma 2.2 b), there is an $R > 0$ such that

$$\left| e\left(\tfrac{\pi}{2}(H-B)(\lambda,\lambda)\right)\right| \geq e(R\|\lambda\|^2)$$

for all $\lambda \in \Lambda_1$. Moreover for every $r > 0$ there is an $R' > 0$ such that

$$\left| e\left(\pi(H-B)(v,\lambda)\right)\right| \leq e(R'\|\lambda\|)$$

for all $v \in V$ with $\|v\| \leq r$. It follows that

$$f(v) \leq \sum_{\lambda \in \Lambda_1} e(R'\|\lambda\| - R\|\lambda\|^2) \leq k\Big(\sum_{n \in \mathbb{Z}} e(R'n - Rn^2)\Big)^{2g} < \infty$$

for all $v \in V$ with $\|v\| \leq r$ and some constant $k > 0$. This implies the assertion.
Step II: ϑ^c satisfies: $\vartheta^c(v+\lambda) = a_L(\lambda, v)\vartheta^c(v)$ for all $(\lambda, v) \in \Lambda \times V$.
For convenience suppose first that $L = L_0$. Consider the decomposition $\lambda = \lambda_1 + \lambda_2$ with $\lambda_\nu \in \Lambda_\nu$. With immediate computations applying Lemmas 1.2 a) and 2.2 a) one checks that $\vartheta^0(v + \lambda_i) = a_{L_0}(\lambda_i, v)\vartheta^0(v)$ for $i = 1, 2$ and all $v \in V$. Then using the cocycle relation the assertion follows for ϑ^0.

In order to prove the assertion in the case of an arbitrary characteristic $c \in V$ notice that by definition ϑ^c and ϑ^0 are related as follows

$$\vartheta^c(v) = e\left(-\pi H(v,c) - \tfrac{\pi}{2}H(c,c)\right)\vartheta^0(v+c)\ . \tag{3}$$

Using this we obtain

$$\begin{aligned}
\vartheta^c(v+\lambda) &= e\left(-\pi H(v+\lambda,c) - \tfrac{\pi}{2}H(c,c)\right)a_{L_0}(\lambda, v+c)\vartheta^0(v+c) \\
&= e\left(2\pi i E(c,\lambda)\right)a_{L_0}(\lambda,v)\vartheta^c(v) \qquad \text{(by Lemma 1.3 a) and (3))} \\
&= a_{t_{\bar{c}}^* L_0}(\lambda,v)\vartheta^c(v) \qquad \text{(by Lemma 1.3 d))}\ .
\end{aligned}$$

This completes the proof. □

The proof actually shows more: it works even for $\lambda = \lambda_1 + \lambda_2 \in \Lambda_1 \oplus \Lambda(L)_2$. Hence, if M_2 denotes the line bundle on $X_2 = V/\Lambda_1 \oplus \Lambda(L)_2$ of Remark 1.5, we have shown

(2.5) Corollary. *ϑ^c is a canonical theta function for M_2.*

We can use ϑ^c to construct more canonical theta functions for L: for any $\overline{w} \in K(L) = \Lambda(L)/\Lambda$ define

$$\vartheta^c_{\overline{w}} = a_L(w, \cdot)^{-1} \vartheta^c(\cdot + w) \tag{4}$$

where w as usual denotes some representative of $\overline{w}$ in $\Lambda(L)$. It is easy to see that this definition does not depend on the choice of the representative w. In particular we have $\vartheta^c_{\overline{0}} = \vartheta^c$. For a generalization of this formula see Exercise 2.

(2.6) Corollary. *$\vartheta^c_{\overline{w}}$ is a canonical theta function for L for every $\overline{w} \in K(L)$.*

Proof.
$$\begin{aligned} \vartheta^c_{\overline{w}}(v + \lambda) &= a_L(w, v + \lambda)^{-1} \vartheta^c(v + \lambda + w) \\ &= e\big(\pi H(w, \lambda) - \pi H(\lambda, w)\big) a_L(\lambda, v) a_L(w, v)^{-1} \vartheta^c(v + w) \\ &\qquad \text{(by Lemmas 2.4 and 1.3 a))} \\ &= e\big(2\pi i E(w, \lambda)\big) a_L(\lambda, v) \vartheta^c_{\overline{w}}(v) \\ &= a_L(\lambda, v) \vartheta^c_{\overline{w}}(v)\ . \qquad \text{(since } w \in \Lambda(L)) \end{aligned}$$
□

The set of theta functions $\vartheta^c_{\overline{w}}$ corresponding to a suitably chosen subset of $K(L)$ forms a basis of $H^0(L)$. This is the contents of the following theorem, the main result of this section.

(2.7) Theorem. *Suppose $L = L(H, \chi)$ is a positive definite line bundle on X and let c be a characteristic with respect to a decomposition $V = V_1 \oplus V_2$ for L. Then the set $\{\vartheta^c_{\overline{w}} \mid \overline{w} \in K(L)_1\}$ is a basis of the vector space $H^0(L)$ of canonical theta functions for L.*

Note that the basis $\{\vartheta^c_{\overline{w}}\}$ is uniquely determined by the choice of the decomposition for L and the characteristic c. This basis will be an essential tool in Chapter 6. Moreover applying $\# K(L)_1 = Pf(E)$ (by Lemma 1.4), we get as a consequence

(2.8) Corollary. *$h^0(L) = Pf(E)$ for any positive definite $L = L(H, \chi)$.*

Proof of Theorem 2.7. Using the definition of $\vartheta^c_{\overline{w}}$ and Lemma 1.3, equation (3) generalizes to

$$\vartheta^c_{\overline{w}}(v) = e\big(-\pi H(v, c) - \tfrac{\pi}{2} H(c, c)\big) \vartheta^0_{\overline{w}}(v + c) \tag{5}$$

for all $\overline{w} \in K(L)_1$. Hence it suffices to prove the Theorem in the case $c = 0$.

According to equation (1) the function $e(-\frac{\pi}{2}B)\vartheta_w^c$ is a classical theta function and thus periodic with respect to Λ_2. Let $w_1, \dots, w_N \in \Lambda(L)_1$ denote a set of representatives of $K(L)_1 = \Lambda(L)_1/\Lambda_1$. In view of Lemma 2.3 it suffices to show that the functions $e(-\frac{\pi}{2}B)\vartheta_w^0$, $\nu = 1, \dots, N$ are linearly independent. We will do this by comparing the coefficients of their Fourier series. For all $v \in V$ and $1 \leq \nu \leq N$ we have according to (2) and (4)

$$\begin{aligned}
e\big(-\tfrac{\pi}{2}B(v,v)\big)\vartheta_{w_\nu}^0(v) &= a_{L_0}(w_\nu, v)^{-1} e\big(-\tfrac{\pi}{2}B(v,v) + \tfrac{\pi}{2}B(v+w_\nu, v+w_\nu)\big) \\
&\quad \cdot \sum_{\lambda\in\Lambda_1} e\big(\pi(H-B)(v+w_\nu,\lambda) - \tfrac{\pi}{2}(H-B)(\lambda,\lambda)\big) \\
&= \sum_{\lambda\in\Lambda_1} e\big(-\pi(H-B)(v,w_\nu) - \tfrac{\pi}{2}(H-B)(w_\nu,w_\nu) - \tfrac{\pi}{2}(H-B)(\lambda,\lambda) - \\
&\qquad + \pi(H-B)(w_\nu,\lambda) + \pi(H-B)(v,\lambda)\big) \\
&= \sum_{\lambda\in\Lambda_1} e\big(-\tfrac{\pi}{2}(H-B)(\lambda-w_\nu,\lambda-w_\nu) + \tfrac{\pi}{2}(H-B)(w_\nu,\lambda) - \tfrac{\pi}{2}(H-B)(\lambda,w_\nu) + \\
&\qquad + \pi(H-B)(v,\lambda-w_\nu)\big) \\
&= \sum_{\lambda\in\Lambda_1} e\big(-\tfrac{\pi}{2}(H-B)(\lambda-w_\nu,\lambda-w_\nu) + \pi i E(w_\nu,\lambda)\big) e\big(\pi(H-B)(v,\lambda-w_\nu)\big) \\
&= \sum_{\lambda\in\Lambda_1-w_\nu} e\big(-\tfrac{\pi}{2}(H-B)(\lambda,\lambda)\big) e\big(\pi(H-B)(v,\lambda)\big)\ .
\end{aligned}$$

From this the assertion is obvious: a general Fourier series for the lattice Λ_2 is of the form $\sum_{\lambda\in\Lambda(L)_1} \alpha_\lambda e\big(\pi(H-B)(v,\lambda)\big)$ (see proof of Lemma 2.3). But the sum in the series of $e(-\frac{\pi}{2}B)\vartheta_{w_\nu}^0$ runs only over the coset $\Lambda_1 - w_\nu$. Since these cosets are pairwise disjoint in $\Lambda(L)_1$, the functions $e(-\frac{\pi}{2}B)\vartheta_{w_\nu}^0$, $\nu = 1, \dots, N$, are linearly independent. □

As a consequence of equation (5) we have the following corollary, which will be very useful subsequently.

(2.9) Corollary. *Suppose $L = L(H,\chi)$ and $L' = L(H,\chi')$ are positive definite line bundles with characteristics c and c'. Let $\tau: V \to \mathbb{C}^*$ be the holomorphic function*

$$\tau(v) = e\big(\pi i \operatorname{Im} H(c',c) - \pi H(v, c'-c) - \tfrac{\pi}{2}H(c'-c, c'-c)\big)\ .$$

Then $\vartheta \mapsto \tau \cdot t_{c'-c}^*\vartheta$ defines an isomorphism of the $\mathbb{C}$-vector spaces of canonical theta functions $H^0(L) \to H^0(L')$.

In particular this homomorphism is in diagonal form with respect to the bases of $H^0(L)$ and $H^0(L')$ of Theorem 2.7.

§ 3 The Positive Semidefinite Case

A line bundle on X is called *positive semidefinite*, if its first Chern class is a positive semidefinite hermitian form. In this section we want to generalize Corollary 2.8 and compute $h^0(L)$ for any positive semidefinite line bundle L. To do this we need some preliminaries.

Let $L = L(H, \chi)$ be a line bundle on $X = V/\Lambda$. According to Section 2.4 the line bundle L determines a homomorphism $\phi_L: X \longrightarrow \widehat{X}$ with kernel $K(L) = \Lambda(L)/\Lambda$. Denote by $K(L)_0$ and $\Lambda(L)_0$ the connected components containing 0 of $K(L)$ and $\Lambda(L)$ respectively. The group

$$K(L)_0 = (\ker \phi_L)_0 = \Lambda(L)_0/\Lambda(L)_0 \cap \Lambda$$

is a subtorus of X.

(3.1) Lemma. $\Lambda(L)_0 = \{v \in V | H(v, V) = 0\}$, *the radical of* H.

Proof. The assertion is a consequence of the fact that $\Lambda(L)_0$ is the kernel of the analytic representation $v \mapsto \phi_H(v) = H(v, \cdot)$ of ϕ_L (see Lemma 2.4.5).□

Denote by $p: X \longrightarrow \overline{X} = X/K(L)_0$ the natural map. We want to give a criterion for the line bundle L to descend to $\overline{X}$.

(3.2) Lemma. *There is a line bundle* $\overline{L}$ *on* $\overline{X}$ *with* $L \simeq p^*\overline{L}$ *if and only if* $L|K(L)_0$ *is trivial. If* $\overline{L}$ *exists, it is nondegenerate and* $h^0(L) = h^0(\overline{L})$.

Proof. By definition $\overline{X} = \overline{V}/\overline{\Lambda}$, where $\overline{V} = V/\Lambda(L)_0$ and $\overline{\Lambda} = \Lambda/\Lambda(L)_0 \cap \Lambda$. According to Appendix B the line bundle $L(H, \chi)$ descends to $\overline{X}$ if and only if H descends to $\overline{V}$ and χ descends to $\overline{\Lambda}$. But this is the case if and only if $H|\Lambda(L)_0 = 0$ and $\chi|\Lambda(L)_0 \cap \Lambda = 1$, i.e. $L|K(L)_0$ is trivial.

Suppose now $L|K(L)_0$ is trivial. By construction $\overline{L}$ is nondegenerate. Certainly $h^0(\overline{L}) \leq h^0(L)$. But even equality holds, since otherwise L would admit a section whose restriction to $K(L)_0$ is nontrivial. □

Using this one can compute $h^0(L)$ for any positive semidefinite line bundle on X.

(3.3) Theorem. *For a positive semidefinite line bundle* $L = L(H, \chi)$ *on* X *and* $E = \operatorname{Im} H$

$$h^0(L) = \begin{cases} Pfr(E) & \text{if} \quad L|K(L)_0 \text{ is trivial} \\ 0 & \phantom{\text{if}} \quad L|K(L)_0 \text{ is nontrivial.} \end{cases}$$

Here $Pfr(E)$ denotes the *reduced Pfaffian* of the alternating form E, that is $Pfr(E) = \prod_{\nu=1}^{s} d_\nu$, if L is of type $(d_1, \ldots, d_s, 0, \ldots, 0)$ with $s \geq 1$ and $d_s \neq 0$, and $Pfr(E) = 1$, if $s = 0$.

Proof. Suppose first that L is trivial on $K(L)_0$. By the previous lemma L descends to a nondegenerate line bundle $\overline{L}$ on $\overline{X}$ with $h^0(L) = h^0(\overline{L})$. Denote by $\overline{E}$ the alternating form of $\overline{L}$. By construction $Pf(\overline{E}) = Pfr(E)$, such that $h^0(L) = Pfr(E)$ according to Corollary 2.8.

Suppose now $L|K(L)_0$ is nontrivial, that is the semicharacter χ restricts to a nontrivial character on $\Lambda(L)_0 \cap \Lambda$. Let ϑ be a canonical theta function for L. For any $w \in V$ the function $t_w^* \vartheta$ is holomorphic on V. Since the canonical factor a_L of L restricted to $\big(\Lambda(L)_0 \cap \Lambda\big) \times V$ is $\chi|\Lambda(L)_0 \cap \Lambda$ the function $t_w^*\vartheta$ satisfies

$$t_w^*\vartheta(v+\lambda) = \vartheta(v+w+\lambda) = a_L(\lambda, v+w)\vartheta(v+w) = \chi(\lambda) t_w^*\vartheta(v)$$

for all $\lambda \in \Lambda(L)_0 \cap \Lambda$ and $v \in \Lambda(L)_0$. Hence $t_w^*\vartheta$ is bounded and thus constant on $\Lambda(L)_0$. Since $\chi(\lambda_0) \neq 1$ for some $\lambda_0 \in \Lambda(L)_0 \cap \Lambda$, this implies $t_w^*\vartheta \equiv 0$ on $\Lambda(L)_0$. In particular $\vartheta(w) = 0$, implying $\vartheta \equiv 0$, since w was arbitrary. □

§ 4 The Vanishing Theorem

Let again L denote an arbitrary line bundle on $X = V/\Lambda$. The aim of this chapter is to compute $h^q(L)$ for all q. In the last section we determined $h^0(L)$ for positive semidefinite line bundles L. As a further step we will prove here a vanishing theorem due to Mumford and Kempf (see Kempf [1]). The proof uses the theory of harmonic forms with values in L.

To this end we first recall some definitions and the relations between $H^q(L)$ and the vector space $\mathcal{H}^q(L)$ of harmonic forms with values in L: tensoring the Dolbeault resolution of $\mathcal{O}_X$ (see 1.4(2) with $p = 0$) with L over $\mathcal{O}_X$, we obtain the exact sequence

$$0 \to L \to \mathcal{A}_X^{0,0}(L) \xrightarrow{\bar\partial} \mathcal{A}_X^{0,1}(L) \xrightarrow{\bar\partial} \mathcal{A}_X^{0,2}(L) \to \cdots$$

where $\mathcal{A}_X^{0,q}(L) = \mathcal{A}_X^{0,q} \otimes_{\mathcal{O}_X} L$ denotes the sheaf of $\mathcal{C}^\infty$-forms of type $(0,q)$ on X with values in L. For simplicity we write $\bar\partial$ instead of $\bar\partial \otimes \mathrm{id}$. Denote by $A^{0,q}(L)$ the vector space of global sections of $\mathcal{A}_X^{0,q}(L)$. We get the complex

$$0 \to H^0(L) \to A^{0,0}(L) \xrightarrow{\bar\partial} A^{0,1}(L) \xrightarrow{\bar\partial} A^{0,2}(L) \to \cdots .$$

Its cohomology groups $H^{0,q}(L)$ are called *Dolbeault cohomology groups of* L (see Griffiths-Harris [1] p.150).

A *hermitian metric on the line bundle* L is a hermitian inner product on each fibre $L(x)$ of L depending differentiably on $x \in X$. We will see that such a metric together with a suitable Kähler metric on X induces a global inner product on the vector spaces $A^{0,q}(L)$ in a natural way.

Assume that we have already such a global inner product $(\ ,\ )$ on $A^{0,q}(L)$. Let $\bar{\delta}\colon A^{0,q+1}(L) \to A^{0,q}(L)$ denote the adjoint operator of $\bar{\partial}$ with respect to $(\ ,\ )$ and $\Delta = \bar{\partial}\bar{\delta} + \bar{\delta}\bar{\partial}\colon A^{0,q}(L) \to A^{0,q}(L)$ the corresponding Laplacian. We denote by $\mathcal{H}^q(L)$ the subvector space $\ker \Delta$ of $A^{0,q}(L)$ of *harmonic forms with values in* L with respect to Δ. As in Griffiths-Harris [1] p.152 one can generalize the Hodge Theorem on harmonic forms (Corollary 1.4.6 combined with Proposition 1.4.7) to forms with values in L: there is an isomorphism $H^{0,q}(L) \simeq \mathcal{H}^q(L)$. On the other hand by Dolbeault's Theorem (see Griffiths-Harris [1] p.100) we have an isomorphism $H^q(L) \simeq H^{0,q}(L)$. Summarizing, we obtain an isomorphism

$$H^q(L) \simeq \mathcal{H}^q(L)\ . \tag{1}$$

Our first aim is to give an explicit description of the harmonic forms with values in L. Suppose $L = L(H,\chi)$. We start by defining a hermitian metric on L. For this consider the elements of $A^{0,0}(L)$ as $\mathcal{C}^\infty$-functions $f\colon V \to \mathbb{C}$ satisfying

$$f(v+\lambda) = a_L(\lambda, v) f(v) \tag{2}$$

for $(\lambda, v) \in \Lambda \times V$. They are called *differentiable theta functions for* L. Define for $f, g \in A^{0,0}(L)$

$$\langle f, g\rangle(v) = f(v)\overline{g(v)}e\big(-\pi H(v,v)\big)\ . \tag{3}$$

Obviously $\langle f, g\rangle$ is a $\mathcal{C}^\infty$-function on V, periodic with respect to Λ. Hence $\langle\ ,\ \rangle\colon A^{0,0}(L) \times A^{0,0}(L) \to A^{0,0}$ defines a hermitian metric on L.

Next we define a Kähler metric ds^2 on the complex torus X: fix a basis $e_1, \dots, e_g$ of V with respect to which the matrix of the hermitian form $H = c_1(L)$ is diagonal and denote by $v_1, \dots, v_g$ the corresponding coordinate functions of V. Choose positive real numbers $k_1, \dots, k_g$. Then

$$ds^2 = \sum_{\nu=1}^{g} k_\nu dv_\nu \otimes d\bar{v}_\nu$$

defines a Kähler metric on X (see also Section 1.4). Later in this section we will choose the coefficients k_ν suitably, in order to assure that the metric ds^2 of X is compatible with the hermitian form H. The volume element corresponding to ds^2 is

$$dv = (\tfrac{i}{2})^g \prod_{\nu=1}^{g} k_\nu dv_1 \wedge d\bar{v}_1 \wedge \dots \wedge dv_g \wedge d\bar{v}_g$$

and the global inner product $(\ ,\ )\colon A^{0,0}(L) \times A^{0,0}(L) \to \mathbb{C}$ associated to the metrics is given by

$$(f,g) = \int_X \langle f,g\rangle dv\ .$$

In analogy to the elements of $A^{0,0}(L)$ any $\omega \in A^{0,q}(L)$ may be considered as a $\mathcal{C}^\infty$-form of type $(0,q)$ on the vector space V satisfying $t_\lambda^*\omega = a_L(\lambda,\cdot)\omega$ for all $\lambda \in \Lambda$. Clearly ω can be uniquely written in the form $\sum_I \varphi_I d\bar v_I$, where the sum is to be taken over all multi-indices $I = (i_1 < \cdots < i_q)$, and where each $\varphi_I \in A^{0,0}(L)$. Let $\sum_I \psi_I d\bar v_I$ be another $(0,q)$-form in $A^{0,q}(L)$. Define a global inner product on $A^{0,q}(L)$ by

$$\Big(\sum_I \varphi_I d\bar v_I, \sum_I \psi_I d\bar v_I\Big) = \sum_I k^I(\varphi_I,\psi_I)$$

where $k^I = \prod_{\nu\in I} k_\nu^{-1}$.

Denote for abbreviation $\partial_\nu = \frac{\partial}{\partial v_\nu}$ and $\bar\partial_\nu = \frac{\partial}{\partial \bar v_\nu}$. Since $a_L(\lambda,v)$ is holomorphic in v, the partial derivative $\bar\partial_\nu$ is a linear operator of $A^{0,0}(L)$ into itself. This shows that the differential operator $\bar\partial\colon A^{0,q}(L) \to A^{0,q+1}(L)$ is given by

$$\bar\partial(\varphi d\bar v_I) = \sum_{\nu=1}^{g} \bar\partial_\nu \varphi d\bar v_\nu \wedge d\bar v_I\ .$$

Let $\bar\delta_\nu\colon A^{0,0}(L) \to A^{0,0}(L)$ and $\bar\delta\colon A^{0,q+1}(L) \to A^{0,q}(L)$ denote the adjoint operators of $\bar\partial_\nu$ and $\bar\partial$ with respect to the chosen inner products.

(4.1) Lemma. *Let* $\varphi \in A^{0,0}(L)$.
a) $\bar\delta_\nu\varphi = -\partial_\nu\varphi + \pi h_\nu \bar v_\nu \varphi$
b) $\bar\delta(\varphi d\bar v_J) = \sum_{\nu=1}^{q+1}(-1)^{\nu-1}k_{j_\nu}^{-1}\bar\delta_{j_\nu}\varphi d\bar v_{J-j_\nu}$ *for* $J = (j_1 < \cdots < j_{q+1})$.

Proof. Denote $\delta'_\nu\varphi = -\partial_\nu\varphi + \pi h_\nu \bar v_\nu\varphi$. We have to show $(\bar\partial_\nu\varphi,\psi) = (\varphi,\delta'_\nu\psi)$ for all $\varphi,\psi \in A^{0,0}(L)$. But $\langle\bar\partial_\nu\varphi,\psi\rangle - \langle\varphi,\delta'_\nu\psi\rangle = \bar\partial_\nu\langle\varphi,\psi\rangle$ and $\int_X \bar\partial_\nu\langle\varphi,\psi\rangle dv = \left(\frac{i}{2}\right)^g \prod_{\nu=1}^g k_\nu \int_X d\big(\langle\varphi,\psi\rangle dv_1\wedge d\bar v_1\wedge\cdots\wedge \check{d\bar v_\nu}\wedge\cdots\wedge d\bar v_g\big) = 0$ by Stokes' theorem. This proves a).

For b) it suffices to check that

$$(\bar\partial(\psi d\bar v_{J-j_\mu}),\varphi d\bar v_J) = \Big(\psi d\bar v_{J-j_\mu}, \sum_{\nu=1}^{q+1}(-1)^{\nu-1}k_{j_\nu}^{-1}\bar\delta_{j_\nu}\varphi d\bar v_{J-j_\nu}\Big)\ ,$$

and this is a consequence of a). □

Using Lemma 4.1, we are in position to compute the Laplacian operator $\Delta = \bar\delta\bar\partial + \bar\partial\bar\delta\colon A^{0,q}(L) \to A^{0,q}(L)$.

(4.2) Proposition. *For all $\varphi d\bar{v}_I \in A^{0,q}(L)$ and $I = (i_1 < \cdots < i_q)$*

$$\Delta(\varphi d\bar{v}_I) = \sum_{\nu=1}^{g} k_\nu^{-1} \bar{\delta}_\nu \bar{\partial}_\nu \varphi d\bar{v}_I + \pi \sum_{\nu=1}^{q} k_{i_\nu}^{-1} h_{i_\nu} \varphi d\bar{v}_I \ .$$

Proof. First we compute $\bar{\delta}$ for the form $\varphi d\bar{v}_\mu \wedge d\bar{v}_I$. Suppose $i_{\lambda-1} < \mu < i_\lambda$ and denote $J = (j_1 < \cdots < j_{q+1}) = (i_1 < \cdots < i_{\lambda-1} < \mu < i_\lambda < \cdots < i_q)$. Then we get using Lemma 4.1

$$\begin{aligned}
\bar{\delta}(\varphi d\bar{v}_\mu \wedge d\bar{v}_I) = \bar{\delta}((-1)^{\lambda-1}\varphi d\bar{v}_J) &= \sum_{\nu=1}^{q+1} (-1)^{\lambda+\nu} k_{j_\nu}^{-1} \bar{\delta}_{j_\nu} \varphi d\bar{v}_{J-j_\nu} \\
&= \sum_{\nu=1}^{\lambda-1} (-1)^{\lambda+\nu} k_{i_\nu}^{-1} \bar{\delta}_{i_\nu} \varphi d\bar{v}_{J-i_\nu} + k_\mu^{-1} \bar{\delta}_\mu \varphi d\bar{v}_{J-\mu} \\
&\quad + \sum_{\nu=\lambda}^{q} (-1)^{\lambda+\nu-1} k_{i_\nu}^{-1} \bar{\delta}_{i_\nu} \varphi d\bar{v}_{J-i_\nu} \\
&= k_\mu^{-1} \bar{\delta}_\mu \varphi d\bar{v}_I + \sum_{\nu=1}^{q} (-1)^\nu k_{i_\nu}^{-1} \bar{\delta}_{i_\nu} \varphi d\bar{v}_\mu \wedge d\bar{v}_{I-i_\nu} \ .
\end{aligned}$$

Using that $\bar{\delta}_{i_\nu} \bar{\partial}_\mu - \bar{\partial}_\mu \bar{\delta}_{i_\nu} = 0$ for $i_\nu \neq \mu$ and $\bar{\partial}_{i_\nu} \bar{\delta}_{i_\nu} \varphi = \bar{\delta}_{i_\nu} \bar{\partial}_{i_\nu} \varphi + \pi h_{i_\nu} \varphi$ we obtain

$$\begin{aligned}
\Delta(\varphi d\bar{v}_I) &= \bar{\delta}\Big(\sum_{\substack{\mu=1 \\ \mu\notin I}}^{g} \bar{\partial}_\mu \varphi d\bar{v}_\mu \wedge d\bar{v}_I\Big) + \bar{\partial}\Big(\sum_{\nu=1}^{q} (-1)^{\nu-1} k_{i_\nu}^{-1} \bar{\delta}_{i_\nu} \varphi d\bar{v}_{I-i_\nu}\Big) \\
&= \sum_{\substack{\mu=1 \\ \mu\notin I}}^{g} k_\mu^{-1} \bar{\delta}_\mu \bar{\partial}_\mu \varphi d\bar{v}_I + \sum_{\nu=1}^{q} \sum_{\substack{\mu=1 \\ \mu\notin I}}^{g} (-1)^\nu k_{i_\nu}^{-1} \bar{\delta}_{i_\nu} \bar{\partial}_\mu \varphi d\bar{v}_\mu \wedge d\bar{v}_{I-i_\nu} \\
&\quad + \sum_{\nu=1}^{q} \sum_{\mu=1}^{g} (-1)^{\nu-1} k_{i_\nu}^{-1} \bar{\partial}_\mu \bar{\delta}_{i_\nu} \varphi d\bar{v}_\mu \wedge d\bar{v}_{I-i_\nu} \\
&= \sum_{\substack{\mu=1 \\ \mu\notin I}}^{g} k_\mu^{-1} \bar{\delta}_\mu \bar{\partial}_\mu \varphi d\bar{v}_I + \sum_{\nu=1}^{q} k_{i_\nu}^{-1} (\bar{\delta}_{i_\nu} \bar{\partial}_{i_\nu} \varphi + \pi h_{i_\nu} \varphi) d\bar{v}_I \ .
\end{aligned}$$

This implies the assertion. □

The basis of V was chosen in such a way that the hermitian form H of L is in diagonal form. Denoting by r and s the numbers of the positive and negative eigenvalues of H respectively, we may assume that $H(v, w) = \sum_{\nu=1}^{r+s} h_\nu v_\nu \bar{w}_\nu$ with $h_\nu = +1$ for $\nu \leq r$ and $h_\nu = -1$ for $r + 1 \leq \nu \leq r + s$. Now choose the real numbers k_ν defining ds^2 as follows:

$$k_\nu = \begin{cases} \frac{1}{s+1} & \text{if} \quad \nu \leq r \\ 1 & \phantom{\text{if}} \quad \nu > r \end{cases} \ .$$

Denote $R_I = \#I \cap \{1, \ldots, r\}$ and $S_I = \#I \cap \{r+1, \ldots, r+s\}$ for any multi-index I. Then we have

(4.3) Proposition. *$\big(\Delta(\varphi d\bar{v}_I), \varphi d\bar{v}_I\big) \geq \pi\big((s+1)R_I - S_I\big)\big(\varphi d\bar{v}_I, \varphi d\bar{v}_I\big)$ for every $\varphi d\bar{v}_I \in A^{0,q}(L)$.*

Proof: $\big(\Delta(\varphi d\bar{v}_I), \varphi d\bar{v}_I\big) =$

$$= \sum_{\nu=1}^{g} k_\nu^{-1}(\bar{\delta}_\nu \bar{\partial}_\nu \varphi d\bar{v}_I, \varphi d\bar{v}_I) + \pi\big((s+1)R_I - S_I\big)(\varphi d\bar{v}_I, \varphi d\bar{v}_I) ,$$

which gives the assertion, since $(\bar{\delta}_\nu \bar{\partial}_\nu \varphi d\bar{v}_I, \varphi d\bar{v}_I) = (\bar{\partial}_\nu \varphi d\bar{v}_I, \bar{\partial}_\nu \varphi d\bar{v}_I) \geq 0$ for all ν. □

Proposition 4.2 shows in particular that the Laplacian Δ acts on the subvector spaces of $A^{0,q}(L)$ of monomial forms $\varphi d\bar{v}_I$ for every fixed multi-index I. So its kernel, the subvector space $\mathcal{H}^q(L)$ of harmonic forms in $A^{0,q}(L)$, decomposes as

$$\mathcal{H}^q(L) = \bigoplus_{\#I=q} \mathcal{H}_I^q(L) , \tag{4}$$

where $\mathcal{H}_I^q(L)$ denotes the subvector space of $\mathcal{H}^q(L)$ of monomial forms $\varphi d\bar{v}_I$.

(4.4) Corollary. *$\mathcal{H}_I^q(L) = 0$ if $R_I > 0$.*

Proof. Suppose $\varphi d\bar{v}_I \in \mathcal{H}_I^q(L)$. Applying Proposition 4.3 we get

$$0 = (0, \varphi d\bar{v}_I) = \big(\Delta(\varphi d\bar{v}_I), \varphi d\bar{v}_I\big) \geq \pi\big((s+1)R_I - S_I\big)(\varphi d\bar{v}_I, \varphi d\bar{v}_I) .$$

Now $((s+1)R_I - S_I) \geq 1$ and $(\varphi d\bar{v}_I, \varphi d\bar{v}_I) \geq 0$ imply $\varphi d\bar{v}_I = 0$. □

Using the isomorphism $H^q(L) \simeq \mathcal{H}^q(L)$ mentioned at the beginning of this section (see equation (1)), we deduce from Corollary 4.4 the *Vanishing Theorem of Mumford-Kempf.*

(4.5) Theorem. *Suppose L is a line bundle on X, whose hermitian form has r positive and s negative eigenvalues. Then*

$$H^q(L) = 0 \quad \textit{for} \quad q > g - r \quad \textit{and} \quad q < s .$$

Proof. If $q > g - r$, every multi-index I of length q intersects $\{1, \ldots, r\}$, i.e. $R_I > 0$. Thus $H^q(L) = \mathcal{H}^q(L) = 0$ by Corollary 4.4 and (4).

According to Lemma 1.4.2 the canonical line bundle $K_X = \Omega_X^g$ of X is trivial. Hence Serre Duality (Hartshorne [1] III 7.7) says $H^q(L) \simeq H^{g-q}(L^{-1})$.

Since $c_1(L^{-1}) = -c_1(L)$, the hermitian form of L^{-1} has s positive eigenvalues. Thus $H^{g-q}(L^{-1}) = 0$ for $g-q > g-s$ by the first part of the proof. This completes the proof. □

For any nondegenerate line bundle L the *index* $i(L)$ *of* L is defined to be the number of negative eigenvalues of the (nondegenerate) hermitian form $H = c_1(L)$. We get as a special case Mumford's Vanishing Theorem (Mumford [1] p.150]) saying: $H^q(L) = 0$ *unless* $q = i(L)$. For a different proof see Exercise 6.

§ 5 Cohomology of Line Bundles

Let the notation and assumptions be as in the last section. In particular r (respectively s) is the number of positive (respectively negative) eigenvalues of the hermitian form H of the line bundle L. In order to complete the computation of the dimensions of the cohomology groups of L, it remains to determine $h^q(L)$ for $s \le q \le g-r$. The following Lemma shows that for this it suffices to consider $H^s(L)$.

(5.1) Lemma. $h^q(L) = \binom{g-r-s}{q-s} h^s(L)$ *for* $s \le q \le g-r$.

Proof. Step I: $\mathcal{H}^q(L) = \bigoplus \mathcal{H}^q_I(L)$ *where the direct sum runs over all multi-indices* $I = (i_1 < \cdots < i_q)$ *with* $R_I = 0$ *and* $S_I = s$.
According to equation 4.(4) and Corollary 4.4 it remains to verify $\mathcal{H}^q_I(L) = 0$ for $R_I = 0$ and $S_I < s$.

Suppose first that $R_I = 0$ and $S_I \le s$. By Proposition 4.2 we have

$$\Delta(\varphi\, d\bar{v}_I) = \psi\, d\bar{v}_I \quad \text{with} \quad \psi = \sum_{\nu=1}^{g} k_\nu^{-1}\bar{\delta}_\nu \bar{\partial}_\nu \varphi - \pi S_I \varphi \,.$$

If $J = I \cap \{r+1, \ldots, r+s\}$, $R_I = 0$ and $S_J = S_I$, such that $\Delta(\varphi\, d\bar{v}_J) = \psi\, d\bar{v}_J$. Hence the map

$$\mathcal{H}^q_I(L) \to \mathcal{H}^{S_I}_J(L), \quad \varphi\, d\bar{v}_I \mapsto \varphi\, d\bar{v}_J \tag{1}$$

is an isomorphism. Moreover, if we assume $S_I < s$, then Theorem 4.5 implies that $\mathcal{H}^{S_I}_J(L) = 0$. This completes the proof of Step I.
Step II: Using Step I twice and the isomorphism (1) we have with $J = (r+1 < \cdots < r+s)$

$$\begin{aligned} H^q(L) \simeq \mathcal{H}^q(L) &= \bigoplus_{\substack{\#I=q \\ R_I=0,\, S_I=s}} \mathcal{H}^q_I(L) = \bigoplus_{\substack{\#I=q \\ R_I=0,\, S_I=s}} \mathcal{H}^s_J(L) \\ &\simeq \mathcal{H}^s_J(L) \otimes \mathbb{C}^{\binom{g-r-s}{q-s}} \simeq H^s(L) \otimes \mathbb{C}^{\binom{g-r-s}{q-s}} \end{aligned}$$

□

The idea for computing $h^s(L)$ is to associate to the line bundle L on X a complex torus Y and a positive semidefinite line bundle M on Y, for which there exists an isomorphism $H^s(X,L) \simeq H^0(Y,M)$. Then the problem is reduced to Theorem 3.3.

Recall that $\Lambda(L)_0$ is the radical of H (see Lemma 3.1). In particular by the choice of the basis $e_1,\dots,e_g$ of V, the radical $\Lambda(L)_0$ is the subvector space of V spanned by $e_{r+s+1},\dots,e_g$. We denote by V_+ (respectively V_-) the subvector space of V spanned by $e_1,\dots,e_r$ (respectively $e_{r+1},\dots,e_{r+s}$). Then V decomposes as

$$V = V_+ \oplus V_- \oplus \Lambda(L)_0 \ .$$

Let W be the underlying real vector space of V, and j the complex structure on W defining V. So j is the $\mathbb{R}$-linear map $j\colon W \to W$ given by multiplication by $i = \sqrt{-1}$ in V. Certainly the direct sum decomposition of V induces a direct sum decomposition

$$W = W_+ \oplus W_- \oplus W_0 \tag{2}$$

over $\mathbb{R}$. Define a new complex structure $k\colon W \to W$ by

$$k(w) = \begin{cases} j(w) & \text{if} \quad w \in W_+ \oplus W_0 \\ -j(w) & \phantom{\text{if}} \quad w \in W_- \end{cases} .$$

Let U denote the complex vector space defined by k and W. Then $\{e_1,\dots,e_g\}$ is also a basis for U and the corresponding complex coordinates $u_1,\dots,u_g$ satisfy

$$u_\nu = \begin{cases} v_\nu & \text{if} \quad \nu \in \{1,\dots,r,r+s+1,\dots,g\} \\ \bar{v}_\nu & \phantom{\text{if}} \quad \nu \in \{r+1,\dots,r+s\} \end{cases} . \tag{3}$$

The lattice Λ in V does not depend on the complex structure, so Λ is also a lattice in U and $Y = U/\Lambda$ is a complex torus of dimension g. Recall that $L = L(H,\chi)$. According to Lemma 2.1.7

$$\widetilde{H}(v,w) = \operatorname{Im} H\big(k(v),w\big) + i \operatorname{Im} H(v,w)$$

is a hermitian form on U. By construction it is positive semidefinite and satisfies $\operatorname{Im}\widetilde{H}(\Lambda \times \Lambda) \subseteq \mathbb{Z}$. Moreover χ is a semicharacter for $\widetilde{H}$, since $\operatorname{Im}\widetilde{H} = \operatorname{Im} H$. Let M denote the line bundle on Y defined by $\widetilde{H}$ and χ: $M = L(\widetilde{H},\chi)$.

We claim that there is an isomorphism between $H^s(L)$ and $H^0(M)$. To see this, we will work again on the level of harmonic forms with values in L and M respectively. As we saw in the proof of Lemma 5.1, we have $\mathcal{H}^s(L) = \mathcal{H}^s_J(L)$, where $J = \{r+1,\dots,r+s\}$. Denote

$$A^{0,s}_J(L) = \{\varphi \, d\bar{v}_J | \varphi \in A^{0,0}(L)\} \ .$$

In order to define an isomorphism $A_J^{0,s}(L) \to A^{0,0}(M)$, consider the function $f: W \to \mathbb{C}^*$, $f(w) = e(\pi \widetilde{H}(w_-, w_-))$, where $w = w_+ + w_- + w_0$ is the decomposition according to (2).

(5.2) Lemma. *The map* $A_J^{0,s}(L) \xrightarrow{\sim} A^{0,0}(M)$, $\varphi \, d\bar{v}_J \mapsto \varphi f$ *is an isomorphism of* $\mathbb{C}$*-vector spaces.*

Note that $\varphi \, d\bar{v}_J$ is a form on V, whereas the function φf is defined on the vector space U. However this makes sense, since the underlying real vector space is W in both cases.

Proof. By definition we have to show that $\varphi(w + \lambda) = a_L(\lambda, w) \, \varphi(w)$ if and only if $(\varphi f)(w + \lambda) = a_M(\lambda, w)(\varphi f)(w)$ for all $w \in W$ and $\lambda \in \Lambda$. Using the definition of $\widetilde{H}$ and the fact that the decomoposition (2) is orthogonal for $\widetilde{H}$ and H, we have

$$\begin{aligned}
&a_M(\lambda, w) f(w + \lambda)^{-1} f(w) \\
&= \chi(\lambda) \, e\big(\pi \widetilde{H}(w, \lambda) + \tfrac{\pi}{2} \widetilde{H}(\lambda, \lambda) - \pi \widetilde{H}(w_- + \lambda_-, w_- + \lambda_-) + \pi \widetilde{H}(w_-, w_-)\big) \\
&= \chi(\lambda) \, e\big(\pi \widetilde{H}(w_+, \lambda_+) - \pi \widetilde{H}(\lambda_-, w_-) + \tfrac{\pi}{2} \widetilde{H}(\lambda_+, \lambda_+) - \tfrac{\pi}{2} \widetilde{H}(\lambda_-, \lambda_-)\big) \\
&= \chi(\lambda) \, e\big(\pi H(w_+, \lambda_+) + \pi H(w_-, \lambda_-) + \tfrac{\pi}{2} H(\lambda_+, \lambda_+) + \tfrac{\pi}{2} H(\lambda_-, \lambda_-)\big) \\
&= a_L(\lambda, w) \; .
\end{aligned}$$

This implies the assertion. □

(5.3) Theorem. *The isomorphism of Lemma 5.2 induces an isomorphism of* $\mathbb{C}$*-vector spaces*

$$H^s(L) \xrightarrow{\sim} H^0(M) \; .$$

Proof. It remains to show that the isomorphism of Lemma 5.2 restricts to an isomorphism on the subvector spaces of harmonic forms $\mathcal{H}_J^s(L) \xrightarrow{\sim} \mathcal{H}^0(M)$. In other words we have to show that $\varphi \, d\bar{v}_J$ is a harmonic form with values in L if and only if φf is a harmonic differentiable theta function for M.

The function φf is harmonic if and only if $\frac{\partial}{\partial \bar{u}_\nu}(\varphi f) = 0$ for $\nu = 1, \ldots, g$. Since $f(u) = e(\pi \sum_{\nu=r+1}^{r+s} u_\nu \bar{u}_\nu)$ we obtain applying (3)

$$\frac{\partial}{\partial \bar{u}_\nu}(\varphi f) = \frac{\partial \varphi}{\partial \bar{u}_\nu} f + \varphi \frac{\partial f}{\partial \bar{u}_\nu} = \begin{cases} (\frac{\partial \varphi}{\partial v_\nu} + \pi \bar{v}_\nu \varphi) f & \text{if} \quad \nu \in \{r+1, \ldots, r+s\} \\ \frac{\partial \varphi}{\partial \bar{v}_\nu} f & \phantom{\text{if}} \quad \nu \notin \{r+1, \ldots, r+s\} \; . \end{cases}$$

Hence it suffices to show that $\varphi \, d\bar{v}_J$ is harmonic if and only if $\frac{\partial \varphi}{\partial v_\nu} + \pi \bar{v}_\nu \varphi = 0$ for $r + 1 \leq \nu \leq r + s$ and $\frac{\partial \varphi}{\partial \bar{v}_\nu} = 0$ otherwise. But this is an immediate consequence of Lemma 4.1, since a form ω is harmonic if and only if $\bar{\partial} \omega = \bar{\delta} \omega = 0$. □

(5.4) Corollary. $h^s(L) = \begin{cases} Pfr(E) & \textit{if} \quad L|K(L)_0 \quad \textit{is trivial} \\ 0 & \quad L|K(L)_0 \quad \textit{is nontrivial}\,. \end{cases}$

This is a consequence of Theorems 3.3 and 5.3. The only thing we have to check is that $L|K(L)_0$ is trivial if and only if $M|K(M)_0$ is trivial. But this is obvious, since the semicharacters of L and M coincide. □

Combining everything we get as a final result

(5.5) Theorem. *Let L be a line bundle on a complex torus X of dimension g, whose first Chern class H has r positive and s negative eigenvalues. Then*

$$h^q(L) = \begin{cases} \binom{g-r-s}{q-s} Pfr(E) & \textit{if} \quad s \leq q \leq g-r \quad \textit{and } L|K(L)_0 \textit{ is trivial} \\ 0 & \quad \textit{otherwise}\,. \end{cases}$$

§ 6 The Riemann-Roch Theorem

For a line bundle L on the g-dimensional complex torus $X = V/\Lambda$ we denote as usual by $\chi(L) = \sum_{\nu=0}^{g}(-1)^\nu h^\nu(L)$ the *Euler-Poincaré characteristic* of L. The Riemann-Roch Theorem is a formula for $\chi(L)$. Since we computed the dimensions of all cohomology groups of L, we can derive it by just adding.

(6.1) Analytic Riemann-Roch Theorem. *Let L be a line bundle on X, whose first Chern class H has s negative eigenvalues. Then*

$$\chi(L) = (-1)^s \ Pf(E)\ .$$

In other words, if L is of type $D = \operatorname{diag}(d_1, \dots, d_g)$ and s is the number of negative eigenvalues of $c_1(L)$, then $\chi(L) = (-1)^s d_1 \cdot \ldots \cdot d_g$. In particular, for degenerate L we have $d_g = 0$, so $\chi(L) = 0$.

Proof. If $L|K(L)_0$ is nontrivial, all cohomology groups vanish according to Theorem 5.5 and $Pf(E) = 0$, since L necessarily is degenerate in this case. So we may assume that $L|K(L)_0$ is trivial. Suppose r is the number of positive eigenvalues of H. Then Theorem 5.5 yields

$$\chi(L) = \sum_{q=s}^{g-r} (-1)^q \binom{g-r-s}{q-s} Pfr(E)$$

$$= \begin{cases} (-1)^s \ Pf(E) & \text{if} \quad g = r+s \\ 0 & \quad g > r+s\,. \end{cases}$$

This implies the assertion. □

Since $\deg \phi_L = \det E = Pf(E)^2$, we obtain as an immediate consequence

(6.2) Corollary. *$\chi(L)^2 = \deg \phi_L$ for every $L \in \operatorname{Pic}(X)$.*

The following geometric version of Riemann-Roch expresses the Euler-Poincaré characteristic of a line bundle in terms of its self-intersection number. For this recall the *self-intersection number* (L^g) of a line bundle L on X

$$(L^g) := \int_X \bigwedge^g c_1(L) .$$

Here the first Chern class $c_1(L)$ is considered to be a 2-form on X via the de Rham isomorphism $H^2(X, \mathbb{C}) \simeq H^2_{\mathrm{DR}}(X)$ (see also Exercise 2.2).

(6.3) Geometric Riemann-Roch Theorem. *For any line bundle L on X*

$$\chi(L) = \tfrac{1}{g!}(L^g) .$$

For the proof we need the following two lemmas. The first lemma computes the first Chern class $c_1(L) \in H^2_{\mathrm{DR}}(X)$ of L. Suppose $L = L(H, \chi)$. Choose a symplectic basis $\lambda_1, \ldots, \lambda_g, \mu_1, \ldots, \mu_g$ of Λ for $E = \operatorname{Im} H$ and denote by $x_1, \ldots, x_g, y_1, \ldots, y_g$ the corresponding real coordinate functions of V.

(6.4) Lemma. *If L is of type $D = \operatorname{diag}(d_1, \ldots, d_g)$, then*

$$c_1(L) = -\sum_{\nu=1}^{g} d_\nu dx_\nu \wedge dy_\nu .$$

Proof. By definition the canonical isomorphism $\gamma_2 \colon \operatorname{Alt}^2_{\mathbb{R}}(V, \mathbb{C}) \to H^2_{\mathrm{DR}}(X)$ sends the alternating form E to the 2-form $\gamma_2(E) = -\sum_{\nu=1}^{g} d_\nu dx_\nu \wedge dy_\nu$ (see Exercise 2.1). This implies the assertion, since $\gamma_2(E) = c_1(L)$ according to Lemma 2.1.4 and the commutative diagram 2.1(2). (For a different proof of the lemma see Exercise 2.2.) □

The lemma implies in particular that for degenerate line bundles L we have $\bigwedge^g c_1(L) = 0$ and thus $(L^g) = 0$. So for the proof of Theorem 6.3 it remains to consider nondegenerate line bundles.

The minus sign in Lemma 6.4 arises from the particular choice of the coordinate functions $x_1, \ldots, y_g$. The corresponding orientation is not always positive, as the following lemma shows.

(6.5) Lemma. *If L is nondegenerate of index s and $x_1, \dots, x_g, y_1, \dots, y_g$ are the real coordinate functions of V corresponding to a symplectic basis of Λ for L, then*

$$\int_X dx_1 \wedge dy_1 \wedge \dots \wedge dx_g \wedge dy_g = (-1)^{g+s} .$$

Proof. We have to show that the volume form $(-1)^{g+s} dx_1 \wedge dy_1 \wedge \dots \wedge dx_g \wedge dy_g$ represents the natural positive orientation of the complex vector space V. For this recall that $\mu_1, \dots, \mu_g$ is a $\mathbb{C}$-basis of V and denote by $v_1, \dots, v_g$ the corresponding complex coordinate functions. By definition the volume form $(\frac{i}{2})^g dv_1 \wedge d\bar{v}_1 \wedge \dots \wedge dv_g \wedge d\bar{v}_g$ gives the natural positive orientation of V. In order to compare both volume forms, let $(Z, \mathbb{1}_g)$ denote the period matrix of X with respect to the chosen bases. Since

$$(\tfrac{i}{2})^g dv_1 \wedge d\bar{v}_1 \wedge \dots \wedge dv_g \wedge d\bar{v}_g = (-1)^g \det(\operatorname{Im} Z) dx_1 \wedge dy_1 \wedge \dots \wedge dx_g \wedge dy_g ,$$

it remains to show $(-1)^s \det(\operatorname{Im} Z) > 0$. To see this, suppose $Y \in M_g(\mathbb{C})$ is the matrix of the hermitian form H of L with respect to the basis $\mu_1, \dots, \mu_g$. Then ${}^t(Z, \mathbb{1}) Y (\overline{Z}, \mathbb{1})$ is the matrix of H with respect to the $\mathbb{R}$-basis $\lambda_1, \dots, \lambda_g, \mu_1, \dots, \mu_g$ and

$$\begin{pmatrix} 0 & D \\ -D & 0 \end{pmatrix} = \operatorname{Im} \begin{pmatrix} {}^tZY\overline{Z} & {}^tZY \\ Y\overline{Z} & Y \end{pmatrix},$$

$\lambda_1, \dots, \mu_g$ being a symplectic basis of $E = \operatorname{Im} H$. So Y is real and $D = (\operatorname{Im} {}^tZ)Y$. Since by assumption Y has s negative eigenvalues, this completes the proof. □

Theorem 6.3 is a direct consequence of the previous lemmas and Theorem 6.1, since $\bigwedge^g c_1(L) = (-1)^g g!\, d_1 \cdot \dots \cdot d_g dx_1 \wedge dy_1 \wedge \dots \wedge dx_g \wedge dy_g$ with the notation as above. □

As a consequence we get a formula for the Euler-Poincaré characteristic of the pullback of a line bundle. Let $f \colon X' = V'/\Lambda' \to X = V/\Lambda$ be a surjective homomorphism of complex tori. Then we have

(6.6) Corollary. $\chi(f^*L) = \deg f\, \chi(L)$.

Proof. If f is not an isogeny, f^*L is degenerate and thus both sides of the equation are 0. Suppose f is an isogeny. Then according to Section 1.2, we have $\deg f = (\Lambda : \rho_a(f)(\Lambda'))$. So we get using the transformation formula and the Geometric Riemann-Roch Theorem

$$g!\, \chi(f^*L) = (f^*L^g) = \int_{X'} \bigwedge^g c_1(f^*L)$$

$$= (\Lambda : \rho_a(f)(\Lambda')) \int_X \bigwedge^g c_1(L) = \deg f\, (L^g) = \deg f\, g!\, \chi(L) . \quad \square$$

Exercises

(1) Let $L(H,\chi)$ be a nondegenerate line bundle on a complex torus $X = V/\Lambda$. Give an example of a decomposition of V into maximal isotropic subvector spaces with respect to $\operatorname{Im} H$, which is not a decomposition for L.

(2) Let $L = L(H,\chi)$ be a positive definite line bundle on a complex torus $X = V/\Lambda$. Suppose L is of characteristic c with respect to a decomposition $\Lambda = \Lambda_1 \oplus \Lambda_2$ for L.

a) Show that for any $u, v = v_1 + v_2, w = w_1 + w_2 \in V$:

$$a_L(v,u)^{-1} a_L(w, u+v-w) = e(2\pi i \operatorname{Im} H(w_1, w_2 - v_2)) a_L(v-w,u)^{-1}$$

b) Use a) to generalize equation (4) of Section 2, to show for any $v, w \in \Lambda(L)_1$:

$$\vartheta^c_{\bar{v}} = a_L(v-w,\cdot)^{-1}\, \vartheta^c_{\bar{w}}(\cdot + v - w)\ .$$

(3) With the notation of Section 2 show the following generalization of Corollary 2.5: for any $\overline{w} \in K(L)_1$ the function $\vartheta^c_{\overline{w}}$ is a canonical theta function for $t^*_{\overline{w}} M_2$, where M_2 is the descent to $X_2 = X/K(L)_2$ of the line bundle L (see Remark 1.5).

(4) Let $L = L(H,\chi)$ be a line bundle on a complex torus X and $K(L)_0$ the connected component of $K(L)$ containing 0. Show that for all $q \geq s$ there is an isomorphism $H^q(L) \simeq H^s(L) \otimes H^{q-s}(\mathcal{O}_{K(L)_0})$, where s denotes the number of negative eigenvalues of H.

(5) Let $L = L(H,\chi)$ be a nondegenerate line bundle of index i on a complex torus $X = V/\Lambda$. Let V_+ (respectively V_-) denote the sum of the eigenspaces of H with positive (respectively negative) eigenvalues. Show that $H^i(L)$ can be identified with the space of $\mathcal{C}^\infty$-theta functions with respect to the canonical factor of L, holomorphic on V_+ and anti-holomorphic on V_-. (Hint: use the methods of Section 5.)

(6) *(Mumford's Index Theorem)* Let L_0 be a positive definite line bundle and L any nondegenerate line bundle on a complex torus X. Consider $\chi(L_0^n \otimes L)$ as a polynomial in n.

a) The polynomial $\chi(L_0^n \otimes L)$ has only real roots.

b) The index $i(L)$ is the number of positive roots of $\chi(L_0^n \otimes L)$. (Hint: use the Analytic Riemann-Roch Theorem.)

(7) *(Cohomology of the Poincaré bundle)* Let X be a complex torus of dimension g. Show that for the Poincaré bundle $\mathcal{P}$ on $X \times \widehat{X}$

$$h^q(\mathcal{P}) = \begin{cases} \mathbb{C} & \text{if} \quad q = g \\ 0 & \phantom{\text{if}} \quad q \neq g \end{cases} .$$

(Hint: Show that $\mathcal{P}$ is nondegenerate of index g.)

(8) *(Poincaré's Reducibility Theorem for Complex Tori)*

a) Let X be a complex torus admitting a nondegenerate line bundle. For any complex subtorus Y of X there exists a complex subtorus Z of X such that $Y \cap Z$ is finite and $Y + Z = X$. In other words, X is isogenous to $Y \times Z$.

b) Consider the complex tori $X = \mathbb{C}^2/\Pi\mathbb{Z}^4$ with $\Pi = \begin{pmatrix} i & \sqrt{2} & 1 & 0 \\ 0 & i & 0 & 1 \end{pmatrix}$ and $Y = \mathbb{C}/(i,1)\mathbb{Z}^2$ embedded as a complex subtorus of X via $z \mapsto (z, 0)$. Show that X admits no complex subtorus Z such that $Y \times Z$ is isogenous to X. Together with a) this implies that X does not admit any nondegenerate line bundle.

(9) *(Riemann-Roch for Vector Bundles on Complex Tori)* Let X be a complex torus of dimension g and E a holomorphic vector bundle of rank r on X with Chern polynomial $c_t(E) = \sum_{i=0}^r c_i(E)t^i \in H^{\bullet}_{\mathrm{DR}}(X)[t]$. Writing $c_t(E) = \prod_{i=1}^r (1 + a_i t)$, where $a_1, \dots, a_r$ are just formal symbols, the *Chern character* of E is defined as $\mathrm{ch}(E) = \sum_{i=1}^r \sum_{k=0}^\infty \frac{a_i^k}{k!}$. Denote by $\mathrm{ch}(E)_g$ the image in $\mathbb{C}$ of the component of degree g of $\mathrm{ch}(E)$ under the natural isomorphism $H^{2g}_{\mathrm{DR}}(X) \to \mathbb{C}$, $\omega \mapsto \int_X \omega$.

a) Deduce from the general Hirzebruch-Riemann-Roch formula (see Hirzebruch [1])

$$\chi(E) = \sum_{i=1}^{g} (-1)^i h^i(E) = \mathrm{ch}(E)_g$$

(Hint: use that the tangent bundle on X is trivial.)

b) Write $c_i = c_i(E)$ and $c_i = 0$, if $i > r$. Then

$$\begin{aligned}
\text{for } g = 1: &\quad \chi(E) = c_1, \\
\text{for } g = 2: &\quad \chi(E) = \tfrac{1}{2}(c_1^2 - 2c_2), \\
\text{for } g = 3: &\quad \chi(E) = \tfrac{1}{3!}(c_1^3 - 3c_1c_2 + 3c_3), \\
\text{for } g = 4: &\quad \chi(E) = \tfrac{1}{4!}(c_1^4 - 4c_1^2c_2 + 4c_1c_3 + 2c_2^2 - 4c_4)\ .
\end{aligned}$$

c) Deduce the Geometric Riemann-Roch Theorem: $\chi(L) = \frac{(L^g)}{g!}$ for all line bundles L on X.

Chapter 4

Abelian Varieties

An *abelian variety* is by definition a complex torus admitting a positive definite line bundle. The Riemann Relations are necessary and sufficient conditions for a complex torus to be an abelian variety. They were given by Riemann in the special case of the Jacobian variety of a curve (see Chapter 11). For the general statement we refer to Poincaré-Picard [1] and Frobenius [2], although it was apparently known to Riemann and Weierstraß. Another characterization of abelian varieties is due to Lefschetz [1] p. 367: a complex torus is an abelian variety if and only if it admits the structure of an algebraic variety. Lefschetz showed that if L is a positive definite line bundle on a complex torus X, then L^n is very ample for any $n \geq 3$, i.e. the map $\varphi_{L^n}: X \to \mathbb{P}_N$ associated to the line bundle L^n is an embedding.

Apart from these results this chapter contains a complete answer to the question of whether the square L^2 of a positive definite line bundle L is very ample or not. The Decomposition Theorem 3.1 reduces this problem to two cases, a polarization without fixed component and an irreducible principal polarization. In the first case L^2 is very ample, as was shown by Ohbuchi [1] and in the latter case L^2 embeds the corresponding Kummer variety. The question of whether a line bundle L is very ample if it is not of the form M^n for some $n \geq 2$, is up to now only known in dimensions $g \leq 2$. For this we refer to Chapter 10.

In Section 1 we introduce the notion of a polarized abelian variety. Section 2 contains a proof of the Riemann Relations. In Section 3 we prove the Decomposition Theorem. Section 4 contains the main results about the Gauss map of a divisor on an abelian variety: let L be a positive definite line bundle on X. For any reduced divisor D in the linear system of L the image of the Gauss map is not contained in a hyperplane (Proposition 4.1). If D is moreover irreducible, the Gauss map is even dominant (Proposition 4.2). The first statement will be applied in the proofs of the above mentioned theorems of Lefschetz (Theorem 5.1) and Ohbuchi (Theorem 5.5). The second statement will be applied for the above mentioned embedding of the Kummer variety (Theorem 8.1). In Section 6 we study symmetric line bundles: the main result is the Inverse Formula 6.4. It describes the action of the involution $(-1)_X$ on the vector space $H^0(L)$ in terms of the basis of canonical theta functions of Theorem 3.2.7. As a consequence we obtain a formula for the dimensions of the eigenspaces for the eigenvalues ± 1. In Section 7 we study symmetric divisors. In Section 9 we compile some facts concerning maps into abelian varieties. In particular we show that any rational map of a normal variety to an abelian variety is everywhere defined. Finally, in the last two sections we introduce the Pontryagin product and give a proof of the Theorem of Lieberman 11.1, saying that numerical and homological equivalence of algebraic cycles on abelian varieties coincide.

The rest of this book deals with abelian varieties rather than complex tori. Since abelian varieties are algebraic, we switch (in Section 5) from the language of complex manifolds to the language of algebraic varieties. We assume the reader to be familiar only with the basic notions and results of algebraic geometry. In this chapter, for example, we apply the language of linear systems, for which we refer to Hartshorne [1] II, 7. In Section 11 we use the Lefschetz decomposition of the cohomology group $H^p(X, \mathbb{C})$.

§ 1 Polarized Abelian Varieties

Let $X = V/\Lambda$ be a complex torus. A *polarization* on X is by definition the first Chern class $H = c_1(L)$ of a positive definite line bundle L on X. By abuse of notation we sometimes consider the line bundle L itself as a polarization. The type of L (see Section 3.1) is called the *type of the polarization.* A polarization is called *principal* if it is of type $(1, \ldots, 1)$. An *abelian variety* is by definition a complex torus X admitting a polarization $H = c_1(L)$. The pair (X, H) is called a *polarized abelian variety.* Again we often write (X, L) instead of (X, H).

A polarization L on X defines an isogeny $\phi_L: X \to \widehat{X}$ (see Section 2.4). Conversely, according to Theorem 2.5.5 an isogeny $\varphi: X \to \widehat{X}$ is of the form ϕ_L for some polarization L on X if its analytic representation $V \to \overline{\Omega} = \mathrm{Hom}_{\overline{\mathbb{C}}}(V, \mathbb{C})$ is a positive definite hermitian form. Hence one could define equivalently a polarization to be an isogeny $X \to \widehat{X}$ with this property.

A *homomorphism of polarized abelian varieties* $f: (Y, M) \to (X, L)$ is a homomorphism of complex tori $f: Y \to X$ such that $f^*c_1(L) = c_1(M)$. According to Proposition 2.5.3 this means that f^*L and M are analytically equivalent. Note that f necessarily has a finite kernel, since otherwise f^*L would be degenerate. Conversely, if $f: Y \to X$ is a homomorphism of complex tori with finite kernel and L a polarization on X, then f^*L defines a polarization on Y, which is called the *induced polarization.* This proves

(1.1) Proposition. *a) A complex subtorus of an abelian variety is an abelian variety.*

b) A complex torus isogenous to an abelian variety is an abelian variety.

As a special case of b) we note that the dual complex torus $\widehat{X}$ of an abelian variety X is also an abelian variety. Therefore it makes sense to speak of the *dual abelian variety.*

The question, whether a polarization on X descends via an isogeny, is answered by Corollary 2.4.4. Moreover we have

(1.2) Proposition. *Every polarization is induced by a principal polarization via an isogeny.*

Proof. Let (X, L) be a polarized abelian variety of type $(d_1, \dots, d_g)$ and let $p_1 \colon X \to X_1$ be the isogeny of Remark 3.1.5. As we saw, there is an $M_1 \in \operatorname{Pic}(X_1)$ with $L = p_1^* M_1$. According to Lemma 3.1.4 the isogeny p_1 is of degree $\prod d_\nu$. From the Riemann-Roch Theorem and Corollary 3.6.6 we obtain $\prod d_\nu = \chi(L) = \prod d_\nu \cdot \chi(M_1)$, that is $\chi(M_1) = 1$. Applying Riemann-Roch again this implies the assertion, since M_1 is positive definite. □

As a first example we will show that every elliptic curve is an abelian variety.

(1.3) Example. Suppose $X = \mathbb{C}/\Lambda$ is an elliptic curve (see Section 1.1). Without loss of generality we may assume that $\{1, z\}$, with a complex number z, $\operatorname{Im} z > 0$, is a basis for Λ. Define $H \colon \mathbb{C} \times \mathbb{C} \to \mathbb{C}$ by $H(v, w) = \frac{v \cdot \bar{w}}{\operatorname{Im} z}$. It is easy to check that H is a hermitian form with $\operatorname{Im} H(\Lambda, \Lambda) \subseteq \mathbb{Z}$, so $H \in \operatorname{NS}(X)$. Since H is positive definite, X is an abelian variety. So every elliptic curve is an abelian variety. However not every complex torus of dimension ≥ 2 is an abelian variety. For examples see Exercise 10.7.

Let $X = V/\Lambda$ be an abelian variety of dimension g and $L \in \operatorname{Pic}(X)$ a polarization. The line bundle L induces in the usual way a meromorphic map $\varphi_L \colon X \to \mathbb{P}_N$ defined as follows: if $\sigma_0, \dots, \sigma_N$ is a basis of $H^0(L)$, then

$$\varphi_L(x) = (\sigma_0(x) : \cdots : \sigma_N(x)) ,$$

whenever $\sigma_\nu(x) \neq 0$ for some ν. Choosing a factor of automorphy f for L we may consider $H^0(L)$ as the vector space of theta functions on V with respect to f. Let $\vartheta_0, \dots, \vartheta_N$ denote the basis of theta functions for the factor f. Then the map φ_L is given by

$$\varphi_L(\bar{v}) = (\vartheta_0(v) : \cdots : \vartheta_N(v)) ,$$

whenever defined. Note that φ_L does not depend on the choice of the factor f, however it depends on the choice of the basis of $H^0(L)$. Changing this basis means modifying φ_L by a projective transformation of $\mathbb{P}_N$.

We want to study the map φ_L. In particular we want to give sufficient conditions for φ_L to be an embedding. For this it turns out to be convenient to use the language of linear systems of divisors (see Hartshorne [1]). Let $|L|$ denote the complete linear system associated to the line bundle L. Recall that the symbol $\sim$ indicates linear equivalence of divisors. The following lemma is a generalization of the Theorem of the Square 2.3.3. It turns out to be an important tool for studying the map φ_L.

(1.5) Lemma. *For $\bar{v}_1, \dots, \bar{v}_n \in X$ with $\sum_{\nu=1}^n \bar{v}_\nu = 0$ and $D \in |L|$*

$$\sum_{\nu=1}^{n} t_{\bar{v}_\nu}^* D \sim nD \qquad \textit{or equivalently} \qquad \bigotimes_{\nu=1}^{n} t_{\bar{v}_\nu}^* L \simeq L^n .$$

Proof. Suppose $L = L(H, \chi)$. Using Lemma 2.3.2 we have

$$\bigotimes_{\nu=1}^{n} t_{\bar{v}_\nu}^* L \simeq L\Big(nH, \prod_{\nu=1}^{n} \chi\, e\big(2\pi i \operatorname{Im} H(v_\nu, \cdot)\big)\Big) \simeq L(nH, \chi^n) \simeq L^n . \qquad \square$$

Another easy remark, which will be applied many times, is the following: by definition $t_x^* D = D - x$, so

$$y \in t_x^* D \Longleftrightarrow x \in t_y^* D \tag{1}$$

for every $x, y \in X$. As a first application we get

(1.6) Proposition. *If L is a positive definite line bundle on X of type $(d_1, \dots, d_g)$ and $d_1 \geq 2$, then φ_L is a holomorphic map.*

According to Lemmas 2.5.6 and 3.1.4 the assumption $d_1 \geq 2$ means that L is the d_1-th power of some positive definite line bundle on X.

Proof. Suppose $x \in X$. We have to show that there exists a divisor in the linear system $|L|$ not containing x. Choose $M \in \operatorname{Pic}(X)$ with $L \simeq M^{d_1}$. Since M is positive definite, $|M|$ contains a divisor D. By continuity of the addition map there exist points $x_1, \dots, x_{d_1 - 1} \notin t_x^* D$ such that $x_{d_1} := -\sum_{\nu=1}^{d_1 - 1} x_\nu$ is not contained in $t_x^* D$. Lemma 1.5 implies $\sum_{\nu=1}^{d_1} t_{x_\nu}^* D \sim d_1 D \in |L|$. By choice of $x_1, \dots, x_{d_1}$ and (1) we have $x \notin \sum_{\nu=1}^{d_1} t_{x_\nu}^* D$. $\square$

Another consequence of Lemma 1.5 is

(1.7) Proposition. *For any positive definite line bundle L on X the general member of $|L|$ is reduced.*

Proof. Suppose $D = nE + F \in |L|$ with $n \geq 2$ and $E > 0, F \geq 0$. According to Lemma 1.5 we have $nE \sim \sum_{\nu=1}^{n} t_{x_\nu}^* E$ for all $x_1, \ldots, x_n \in X$ satisfying $\sum_{\nu=1}^{n} x_\nu = 0$. This implies the assertion, since we can choose $n - 1$ of these points arbitrarily in X. □

Finally we need

(1.8) Lemma. *Let $L \in \mathrm{Pic}(X)$ be positive definite. There is an open dense set $U \subset |L|$ such that for every $D \in U$ the identity $t_x^* D = D$ only holds for $x = 0$.*

Proof. Suppose $0 \neq x \in X$ and $t_x^* D = D$ for an open dense set of divisors D in $|L|$ and hence for all D in $|L|$. In particular $t_x^* L \simeq L$ and x is contained in the finite group $K(L)$. The point x generates a finite subgroup G of X, say of order $n \geq 2$. Let $Y = X/G$ and $p: X \to Y$ the canonical projection map. The assumption implies that every $D \in |L|$ descends to an effective divisor E on Y. In particular $L \simeq p^* \mathcal{O}_Y(E)$. On the other hand, according to Proposition 2.4.3 there are only finitely many line bundles on Y, say $M_1, \ldots, M_n$, such that $L \simeq p^* M_\nu$. Now $H^0(L) = \bigcup_{\nu=1}^{N} p^* H^0(M_\nu)$ implies $H^0(L) = p^* H^0(M_\nu)$ for some ν and Riemann-Roch gives the contradiction $h^0(M_\nu) = h^0(L) = n h^0(M_\nu) > h^0(M_\nu)$. □

§ 2 The Riemann Relations

Before we proceed, we work out in terms of period matrices, what it means that a complex torus is an abelian variety.

Let $X = V/\Lambda$ be a complex torus of dimension g. Choose a basis $e_1, \ldots, e_g$ of V and $\lambda_1, \ldots, \lambda_{2g}$ of Λ and let Π be the corresponding period matrix. With respect to these bases we have

$$X = \mathbb{C}^g / \Pi \mathbb{Z}^{2g} .$$

The aim of this section is to prove the following

(2.1) Theorem. *X is an abelian variety if and only if there is a nondegenerate alternating matrix $A \in M_{2g}(\mathbb{Z})$ such that*

i) $\Pi A^{-1} \, {}^t\Pi = 0$,
ii) $i \Pi A^{-1} \, {}^t\overline{\Pi} > 0$.

The conditions i) and ii) are called *Riemann Relations.* By definition the complex torus X is an abelian variety if and only if X admits a polarization. It turns out that A is the matrix of the alternating form defining the polarization.

For the proof we start with an arbitrary nondegenerate alternating form E on Λ. Denote by A its matrix with respect to the basis $\lambda_1, \dots, \lambda_{2g}$. Define $H: \mathbb{C}^g \times \mathbb{C}^g \to \mathbb{C}$ by

$$H(u,v) = E(iu,v) + iE(u,v) \ .$$

Here E denotes also the extension of the alternating form E to $\Lambda \otimes \mathbb{R} = \mathbb{C}^g$. Now the theorem is a direct consequence of the following two lemmas which work out conditions for H to be a positive definite hermitian form.

(2.2) Lemma. *H is an hermitian form on $\mathbb{C}^g$ if and only if* $\Pi A^{-1}\,{}^t\Pi = 0$.

Proof. According to Lemma 2.1.7 the form H is hermitian if and only if $E(iu, iv) = E(u,v)$ for all $u, v \in \mathbb{C}^g$. In order to analise this condition in terms of matrices define

$$I = \begin{pmatrix} \Pi \\ \overline{\Pi} \end{pmatrix}^{-1} \begin{pmatrix} i\mathbb{1} & \\ & -i\mathbb{1} \end{pmatrix} \begin{pmatrix} \Pi \\ \overline{\Pi} \end{pmatrix} \ .$$

The matrix I satisfies $i\Pi = \Pi I$. Since $E(\Pi x, \Pi y) = {}^t x A y$ for all $x, y \in \mathbb{R}^{2g}$, the form H is hermitian if and only if ${}^t I A I = A$ or equivalently

$$\begin{pmatrix} i\mathbb{1} & \\ & -i\mathbb{1} \end{pmatrix} \left(\begin{pmatrix} \Pi \\ \overline{\Pi} \end{pmatrix} A^{-1}\,{}^t\!\begin{pmatrix} \Pi \\ \overline{\Pi} \end{pmatrix} \right)^{-1} \begin{pmatrix} i\mathbb{1} & \\ & -i\mathbb{1} \end{pmatrix} = \left(\begin{pmatrix} \Pi \\ \overline{\Pi} \end{pmatrix} A^{-1}\,{}^t\!\begin{pmatrix} \Pi \\ \overline{\Pi} \end{pmatrix} \right)^{-1} .$$

Comparing the $g \times g$-blocks of both sides one sees that this is the case if and only if $\Pi A^{-1}\,{}^t\Pi = 0$. □

In order to complete the proof of Theorem 2.1 we compute the matrix of H with respect to the basis $e_1, \dots, e_g$ under the assumption that H is hermitian.

(2.3) Lemma. *Suppose the form H is hermitian. Then* $2i(\overline{\Pi} A^{-1}\,{}^t\Pi)^{-1}$ *is the matrix of H with respect to the given basis. In particular H is positive definite if and only if* $i\Pi A^{-1}\,{}^t\overline{\Pi} > 0$.

Proof. Write $u = \Pi x$ and $v = \Pi y$ with $x, y \in \mathbb{R}^{2g}$. With the notation as in the proof of Lemma 2.2 and using $\Pi A^{-1}\,{}^t\Pi = 0$ we get

$$\begin{aligned} E(iu,v) &= {}^t x\,{}^t I A y = {}^t\!\begin{pmatrix} u \\ \bar{u} \end{pmatrix} \begin{pmatrix} i\mathbb{1} & \\ & -i\mathbb{1} \end{pmatrix} \left(\begin{pmatrix} \Pi \\ \overline{\Pi} \end{pmatrix} A^{-1}\,{}^t\!\begin{pmatrix} \Pi \\ \overline{\Pi} \end{pmatrix} \right)^{-1} \begin{pmatrix} v \\ \bar{v} \end{pmatrix} \\ &= {}^t\!\begin{pmatrix} u \\ \bar{u} \end{pmatrix} \begin{pmatrix} 0 & i(\overline{\Pi} A^{-1}\,{}^t\Pi)^{-1} \\ -i(\Pi A^{-1}\,{}^t\overline{\Pi})^{-1} & 0 \end{pmatrix} \begin{pmatrix} v \\ \bar{v} \end{pmatrix} \\ &= i\,{}^t u (\overline{\Pi} A^{-1}\,{}^t\Pi)^{-1} \bar{v} - i\,{}^t\bar{u} (\Pi A^{-1}\,{}^t\overline{\Pi})^{-1} v \ . \end{aligned}$$

Similarly one computes $E(u,v) = {}^t u(\overline{\Pi} A^{-1}\,{}^t\Pi)^{-1}\bar{v} + {}^t\bar{u}(\Pi A^{-1}\,{}^t\overline{\Pi})^{-1} v$. So $H(u,v) = E(iu,v) + iE(u,v) = 2i\,{}^t u(\overline{\Pi} A^{-1}\,{}^t\Pi)^{-1}\bar{v}$. □

Finally, given a polarized abelian variety $(X = V/\Lambda, L)$ we want to outline the Riemann Relations in terms of a symplectic basis for L. Suppose $\Pi = (\Pi_1, \Pi_2)$, with $\Pi_i \in M_g(\mathbb{C})$, is the period matrix of X with respect to a basis $e_1, \dots, e_g$ of V and a symplectic basis $\lambda_1, \dots, \lambda_{2g}$ of Λ for L. If L is of type $D = \operatorname{diag}(d_1, \dots, d_g)$, then by definition $\begin{pmatrix} 0 & D \\ -D & 0 \end{pmatrix}$ is the matrix of the alternating form of L with respect to $\lambda_1, \dots, \lambda_{2g}$ and Riemann's Relations read

$$i) \quad \Pi_2 D^{-1}\,{}^t\Pi_1 - \Pi_1 D^{-1}\,{}^t\Pi_2 = 0\ ,$$

$$ii) \quad i\Pi_2 D^{-1}\,{}^t\overline{\Pi}_1 - i\Pi_1 D^{-1}\,{}^t\overline{\Pi}_2 > 0\ .$$

§ 3 The Decomposition Theorem

Let (X, L), $L \in \operatorname{Pic}(X)$, be a polarized abelian variety. In this section we decompose (X, L) as a polarized abelian variety. This will turn out to be convenient for studying the associated map $\varphi_L : X \to \mathbb{P}_N$.

The linear system $|L|$ has a unique decomposition

$$|L| = |M| + F_1 + \cdots + F_r\ , \tag{1}$$

where $|M|$ is the moving part of $|L|$ and $F_1 + \cdots + F_r$ is the decomposition of the fixed part of $|L|$ into irreducible components. Note that $F_\nu \neq F_\mu$ for $\nu \neq \mu$ by Proposition 1.7. Denote

$$N_\nu = \mathcal{O}_X(F_\nu)$$

for $\nu = 1, \dots, r$. The line bundles M and $N_1, \dots, N_r$ are positive semidefinite with $h^0(M) > 1$ and $h^0(N_\nu) = 1$ for $\nu = 1, \dots, r$. So according to Theorem 3.3.3 the restrictions of M and N_ν to the subtori $K(M)_0$ and $K(N_\nu)_0$ of X respectively are trivial. Denote by $p_M : X \to X_M := X/K(M)_0$ and $p_{N_\nu} : X \to X_{N_\nu} := X/K(N_\nu)_0$ the canonical projections. Then Lemma 3.3.2 provides positive definite line bundles $\overline{M}$ on X_M and $\overline{N}_\nu$ on X_{N_ν} with

$$M = p_M^* \overline{M} \quad \text{and} \quad N_\nu = p_{N_\nu}^* \overline{N}_\nu \quad \text{for } \nu = 1, \dots, r.$$

The pairs $(X_M, \overline{M})$ and $(X_{N_\nu}, \overline{N}_\nu)$ are polarized abelian varieties. In particular, the $\overline{N}_\nu$'s define principal polarizations on the abelian varieties X_{N_ν}, since $h^0(\overline{N}_\nu) = h^0(N_\nu) = 1$. Consider the product $X_M \times X_{N_1} \times \cdots \times X_{N_r}$ and denote by q_M and q_{N_ν} the projections of $X_M \times X_{N_1} \times \cdots \times X_{N_r}$ onto its factors. With this notation we can state

(3.1) Decomposition Theorem. *The map*

$$(p_M, p_{N_1}, \dots, p_{N_r}) \colon X \to X_M \times X_{N_1} \times \dots \times X_{N_r}$$

is an isomorphism of polarized abelian varieties (X, L) *and* $(X_M \times X_{N_1} \times \dots \times X_{N_r}, q_M^* M \otimes q_{N_1}^* N_1 \otimes \dots \otimes q_{N_r}^* N_r)$.

For the proof we need some preliminaries: generalizing the definition of the self-intersection number in Section 3.6 we define the *intersection number* $(L_1 \cdot \ldots \cdot L_g)$ of the line bundles $L_1, \dots, L_g$ on X by

$$(L_1 \cdot \ldots \cdot L_g) = \int_X c_1(L_1) \wedge \dots \wedge c_1(L_g) \,.$$

If $L_1 \simeq \dots \simeq L_\nu$ and $L_{\nu+1} \simeq \dots \simeq L_g$, we write $(L_1^\nu \cdot L_g^{g-\nu})$ instead of $(L_1 \cdot \ldots \cdot L_1 \cdot L_g \cdot \ldots \cdot L_g)$. Moreover, since the intersection number depends only on the first Chern classes, it makes sense to define

$$(H_1 \cdot \ldots \cdot H_g) = (L_1 \cdot \ldots \cdot L_g)$$

for hermitian forms $H_\nu \in \mathrm{NS}(X)$ and line bundles $L_\nu \in \mathrm{Pic}^{H_\nu}(X)$. In the sequel we freely apply some elementary properties of intersection numbers, for which we refer to Griffiths-Harris [1]. Furthermore we need

(3.2) Lemma. *Let* L_1 *and* L_2 *be line bundles on* X.

a) If L_1 *and* L_2 *are positive (semi-)definite, then* $(L_1^\nu \cdot L_2^{g-\nu}) \underset{(=)}{>} 0$ *for* $\nu = 0, \dots, g$.

b) If L_2 *is positive definite and* $(L_1^\nu \cdot L_2^{g-\nu}) > 0$ *for* $\nu = 0, \dots, g$, *then* L_1 *is positive definite.*

More generally one can show that $(L_1 \cdot \dots \cdot L_g) \geq 0$ for any positive semidefinite line bundles $L_1, \dots, L_g$ on X, but we do not need this fact.

Proof. Suppose $X = V/\Lambda$ and H_i is the hermitian form on V associated to L_i for $i = 1, 2$. Since two hermitian forms can be diagonalized simultaneously, there is a basis of V, with respect to which $H_1 = \mathrm{diag}(h_1, \dots, h_g)$ and $H_2 = \mathrm{diag}(k_1, \dots, k_g)$ with real numbers h_i and k_i, which are nonnegative, if L_1 and L_2 are positive semidefinite, and positive in the positive definite case. Denoting by $v_1, \dots, v_g$ the complex coordinate functions with respect to the chosen basis we have $c_1(L_1) = \frac{i}{2} \sum_{\mu=1}^g h_\mu dv_\mu \wedge d\bar{v}_\mu$ and $c_1(L_2) = \frac{i}{2} \sum_{\mu=1}^g k_\mu dv_\mu \wedge d\bar{v}_\mu$ (see Exercise 2.2). Then

$$\bigwedge^\nu c_1(L_1) = (\tfrac{i}{2})^\nu \sum_{i_1, \dots, i_\nu = 1}^g h_{i_1} \cdot \dots \cdot h_{i_\nu} dv_{i_1} \wedge d\bar{v}_{i_1} \wedge \dots \wedge dv_{i_\nu} \wedge d\bar{v}_{i_\nu} \,.$$

A similar formula holds for $\bigwedge^{g-\nu} c_1(L_2)$. Using this we get by the definition of the intersection numbers

$$(L_1^\nu \cdot L_2^{g-\nu}) =$$
$$= \int_X (\tfrac{i}{2})^g \sum_{\sigma\in\mathfrak{S}_g} h_{\sigma(1)} \cdot \ldots \cdot h_{\sigma(\nu)} k_{\sigma(\nu+1)} \cdot \ldots \cdot k_{\sigma(g)} dv_1 \wedge d\bar{v}_1 \wedge \cdots \wedge dv_g \wedge d\bar{v}_g$$
$$= c \sum_{\sigma\in\mathfrak{S}_g} h_{\sigma(1)} \cdot \ldots \cdot h_{\sigma(\nu)} k_{\sigma(\nu+1)} \cdot \ldots \cdot k_{\sigma(g)} \overset{\geq}{(=)} 0$$

with $c = \int_X (\frac{i}{2})^g dv_1 \wedge d\bar{v}_1 \wedge \cdots \wedge dv_g \wedge d\bar{v}_g$. The constant c is positive, since the volume element $(\frac{i}{2})^g dv_1 \wedge d\bar{v}_1 \wedge \cdots \wedge dv_g \wedge d\bar{v}_g$ represents the natural positive orientation of V. This shows a).

As for b): we may assume that $k_1 = \cdots = k_g = 1$. Let s_ν denote the elementary symmetric polynomial of degree ν. We compute as above

$$\frac{1}{\nu!(g-\nu)!}(L_1^\nu \cdot L_2^{g-\nu}) = c \sum_{1\leq i_1<\cdots<i_\nu\leq g} h_{i_1} \cdot \ldots \cdot h_{i_\nu} = c\, s_\nu(h_1, \ldots, h_g) \,.$$

So $s_\nu(h_1, \ldots, h_g) > 0$ by assumption.

Since $\prod_{\nu=1}^g (t - h_\nu) = \sum_{\nu=0}^g (-1)^\nu s_\nu(h_1, \ldots, h_g) t^{g-\nu}$ and the alternating coefficients of the polynomial on the right hand side are positive, the roots $h_1, \ldots, h_g$ are also positive. Hence L_1 is positive definite. □

As an immediate consequence we obtain an improvement of the *Nakai-Moishezon Criterion* (see Hartshorne [1], Appendix A, Theorem 5.1) in the special case of an abelian variety.

(3.3) Corollary. *Suppose L_0 is a line bundle defining a polarization on X. A line bundle L on X defines a polarization if and only if $(L^\nu \cdot L_0^{g-\nu}) > 0$ for $\nu = 1, \ldots, g$.*

For the proof of the Decomposition Theorem we need another lemma. Let M_ν denote a line bundle on X with $h^0(M_\nu) \geq 1$ for $\nu = 1, 2$. In particular M_ν is positive semidefinite. According to Theorem 3.3.3 and Lemma 3.3.2 the line bundle M_ν descends to a positive definite line bundle $\overline{M}_\nu$ on $X_{M_\nu} := X/K(M_\nu)_0$ via the natural projections $p_{M_\nu}: X \to X_{M_\nu}$ for $\nu = 1, 2$.

(3.4) Lemma. *Suppose $M_1 \otimes M_2$ is positive definite and the homomorphism $(p_{M_1}, p_{M_2}): X \to X_{M_1} \times X_{M_2}$ is not surjective and has finite kernel. Then*

$$h^0(M_1 \otimes M_2) \geq h^0(M_1) + h^0(M_2)$$

Proof. Writing $g_\nu = \dim X_{M_\nu}$ for $\nu = 1, 2$, we have by assumption $g < g_1 + g_2$. Then

$$h^0(M_1 \otimes M_2) = \tfrac{1}{g!}\big((M_1 \otimes M_2)^g\big) = \tfrac{1}{g!}\big((p_{M_1}^*\overline{M}_1 \otimes p_{M_2}^*\overline{M}_2)^g\big)$$
$$= \sum_{\nu=g-g_2}^{g_1} \Big(\tfrac{(p_{M_1}^*\overline{M}_1)^\nu}{\nu!} \cdot \tfrac{(p_{M_2}^*\overline{M}_2)^{g-\nu}}{(g-\nu)!}\Big) ,$$

since the intersection products $\big((p_{M_1}^*\overline{M}_1)^\nu\big) = p_{M_1}^*(\overline{M}_1^\nu)$ and $\big((p_{M_2}^*\overline{M}_2)^{g-\nu}\big)$ $= p_{M_2}^*(\overline{M}_2^{g-\nu})$ vanish for $\nu > g_1$ respectively for $g-\nu > g_2$. For the summand with index $\nu = g_1$ we have

$$\Big(\tfrac{(p_{M_1}^*\overline{M}_1)^{g_1}}{g_1!} \cdot \tfrac{(p_{M_2}^*\overline{M}_2)^{g-g_1}}{(g-g_1)!}\Big)$$
$$= \Big(\tfrac{(p_{M_1}^*\overline{M}_1)^{g_1}}{g_1!(g-g_1)!} \cdot \Big\{(p_{M_2}^*\overline{M}_2)^{g-g_1} + \sum_{\mu=1}^{g-g_1} \tbinom{g-g_1}{\mu}(p_{M_1}^*\overline{M}_1)^\mu \cdot (p_{M_2}^*\overline{M}_2)^{g-g_1-\mu}\Big\}\Big)$$

(again, since the intersection product $\big((p_{M_1}^*\overline{M}_1)^{g_1+\mu}\big)$ vanishes for $\mu > 0$)

$$= \Big(\tfrac{(p_{M_1}^*\overline{M}_1)^{g_1}}{g_1!} \cdot \tfrac{(M_1 \otimes M_2)^{g-g_1}}{(g-g_1)!}\Big)$$
$$= h^0(\overline{M}_1)\frac{\big(p_{M_1}^*(\text{point})\cdot(M_1\otimes M_2)^{g-g_1}\big)}{(g-g_1)!}$$

(by Riemann-Roch, since $\overline{M}_1$ is positive definite)

$$= h^0(\overline{M}_1)\frac{\big((M_1 \otimes M_2|K(M_1)_0)^{g-g_1}\big)}{(g-g_1)!}$$
$$= h^0(\overline{M}_1) \cdot h^0(M_1 \otimes M_2|K(M_1)_0)$$

(again by Riemann-Roch, since $M_1 \otimes M_2$ is positive definite)

$$\geq h^0(\overline{M}_1) = h^0(M_1) .$$

Similarly we have for the summand with index $\nu = g - g_2$

$$\Big(\tfrac{(p_{M_1}^*\overline{M}_1)^{g-g_2}}{(g-g_2)!} \cdot \tfrac{(p_{M_2}^*\overline{M}_2)^{g_2}}{g_2!}\Big) \geq h^0(M_2) .$$

Now by assumption $g_1 \neq g - g_2$. Since all other summands are non negative by Lemma 3.2, this implies the assertion. □

Proof of Theorem 3.1. Define $L_r = M \otimes N_1 \otimes \cdots \otimes N_{r-1}$. According to the decomposition (1) we have $L \simeq L_r \otimes N_r$ with $h^0(L) = h^0(L_r) \geq 1$. Hence L_r descends via the natural projection $p_{L_r}: X \to X_{L_r}$ to a positive definite line bundle $\overline{L}_r$ on $X_{L_r} := X/K(L_r)_0$ such that $h^0(L_r) = h^0(\overline{L}_r)$. Denote by q_{L_r} and q_{N_r} the projections of $X_{L_r} \times X_{N_r}$ onto its factors.

We claim that $(p_{L_r}, p_{N_r}): (X, L) \to (X_{L_r} \times X_{N_r}, q_{L_r}^*\overline{L}_r \otimes q_{N_r}^*\overline{N}_r)$ is an isomorphism of polarized abelian varieties.

Suppose we have proven this. Applying this argument to $(X_{L_r}, \overline{L}_r)$ instead of (X, L) we can split off the principal polarized abelian variety

$(X_{N_{r-1}}, N_{r-1})$ in the same way. Repeating this process we finally obtain the asserted decomposition.

For the proof of the claim it suffices to show that (p_{L_r}, p_{N_r}) is an isomorphism. The kernel $K(L_r)_0 \cap K(N_r)_0$ of (p_{L_r}, p_{N_r}) is finite, since it is contained in the finite group $K(L)$. If (p_{L_r}, p_{N_r}) would not be surjective, we could apply Lemma 3.4 to get

$$h^0(L) \geq h^0(L_r) + h^0(N_r) = h^0(L) + 1 ,$$

a contradiction. Hence (p_{L_r}, p_{N_r}) is an isogeny. By definition we have $L = (p_{L_r}, p_{N_r})^*(q_{L_r}^* \overline{L}_r \otimes q_{N_r}^* \overline{N}_r)$. Applying Riemann-Roch, Corollary 3.6.6 and the Künneth formula we get

$$\begin{aligned} \tfrac{1}{\deg(p_{L_r}, p_{N_r})}(L^g) &= \left((q_{L_r}^* \overline{L}_r \otimes q_{N_r}^* \overline{N}_r)^g\right) = g!\, h^0(q_{L_r}^* \overline{L}_r \otimes q_{N_r}^* \overline{N}_r) \\ &= g!\, h^0(\overline{L}_r)\, h^0(\overline{N}_r) = g! h^0(L) = (L^g), \end{aligned}$$

which implies the assertion. □

We conclude this section with a result, which is an improvement of *Bertini's theorem.* This is another consequence of Lemma 3.4. Let L denote a positive definite line bundle on X without fixed components. Bertini's theorem states that the general member of the linear system $|L|$ is irreducible if only the dimension of the image of the map $\varphi_L : X \to \mathbb{P}_N$ is at least two. The following theorem says that this is true even if it is one.

(3.5) Theorem. *For a positive definite line bundle L on X without fixed components the general member of $|L|$ is irreducible.*

Proof. Assume the contrary, that is $L = M_1 \otimes M_2$ with line bundles M_ν on X such that $h^0(M_\nu) \geq 2$ for $\nu = 1, 2$ and the map $|M_1| \times |M_2| \to |L|$, given by $(D_1, D_2) \mapsto D_1 + D_2$, is surjective. Let the notation be as in Lemma 3.4. Then the map $(p_{M_1}, p_{M_2}) : X \to X_{M_1} \times X_{M_2}$ has finite kernel, since $\ker(p_{M_1}, p_{M_2}) = K(M_1)_0 \cap K(M_2)_0 \subseteq K(L)$.

If (p_{M_1}, p_{M_2}) would not be surjective, then by Lemma 3.4 and the fact that $\dim |M_1| + \dim |M_2| \geq \dim |L|$

$$h^0(L) \geq h^0(M_1) + h^0(M_2) \geq h^0(L) + 1 ,$$

a contradiction. Hence (p_{M_1}, p_{M_2}) is an isogeny.

Denote by q_{M_1} and q_{M_2} the projections of $X_{M_1} \times X_{M_2}$ onto its factors, such that $L \simeq (p_{M_1}, p_{M_2})^*(q_{M_1}^* \overline{M}_1 \otimes q_{M_2}^* \overline{M}_2)$. Then Künneth's formula implies

$$
\begin{aligned}
h^0(M_1)+h^0(M_2)-1 \geq h^0(L) &= \tfrac{1}{g!}(L^g) \\
&= \tfrac{1}{g!}\deg(p_{M_1},p_{M_2})\big((q_{M_1}^*\overline{M}_1 \otimes q_{M_2}^*\overline{M}_2)^g\big) \\
&= \deg(p_{M_1},p_{M_2})h^0(X_{M_1}\times X_{M_2}, q_{M_1}^*\overline{M}_1 \otimes q_{M_2}^*\overline{M}_2) \\
&= \deg(p_{M_1},p_{M_2})h^0(X_{M_1},\overline{M}_1)h^0(X_{M_2},\overline{M}_2) \\
&= \deg(p_{M_1},p_{M_2})h^0(M_1)h^0(M_2) \geq h^0(M_1)h^0(M_2)\ .
\end{aligned}
$$

Hence

$$
1 \leq (h^0(M_1)-1)(h^0(M_2)-1) = h^0(M_1)h^0(M_2)-h^0(M_1)-h^0(M_2)+1 \leq 0\ ,
$$

a contradiction. □

§ 4 The Gauss Map

Let $X = V/\Lambda$ be an abelian variety of dimension g and L a positive definite line bundle on X. Suppose $D \in |L|$ is a reduced divisor (according to Proposition 1.7 the linear system $|L|$ always contains such a divisor). By D_s we denote the smooth part of the support of D. Then for every $\overline{w} \in D_s$ the tangent space $T_{D,\overline{w}}$ is a $(g-1)$-dimensional vector space and its translation to zero is a well defined $(g-1)$-dimensional subvector space of $T_{X,0} = V$. Consider $H^0(L)$ as vector space of theta functions on V. Then there is a theta function $\vartheta \in H^0(L)$, uniquely determined up to a constant, such that $\pi^*D = (\vartheta)$.

Let $v_1,\dots,v_g$ denote the coordinate functions with respect to some complex basis of V. The equation of the tangent space $T_{D,\overline{w}}$ at a point $\overline{w} \in D$ is $\sum_{\nu=1}^g \frac{\partial\vartheta}{\partial v_\nu}(w)(v_\nu - w_\nu) = 0$. So the 1-dimensional subspace of the dual vector space V^* determined by $T_{D,\overline{w}}$ is generated by the vector $(\frac{\partial\vartheta}{\partial v_1}(w),\dots,\frac{\partial\vartheta}{\partial v_g}(w))$ (in coordinates with respect to the dual basis). The *Gauss map* $G\colon D_s \to \mathbb{P}_{g-1} = \mathbb{P}(V^*)$ of D is defined by

$$
G(\overline{w}) = \big(\tfrac{\partial\vartheta}{\partial v_1}(w) : \dots : \tfrac{\partial\vartheta}{\partial v_g}(w)\big)\ .
$$

Obviously G is a holomorphic map, neither depending on the choice of ϑ nor on the choice of the factor for L. In this section we derive some properties of the Gauss map G, which will be applied in the sequel.

(4.1) Proposition. *For a reduced divisor $D \in |L|$, with L a positive definite line bundle on X, the image of the Gauss map is not contained in a hyperplane.*

Proof. Assume the contrary. This means there is a nonzero tangent vector $t \in V$ contained in $T_{D,\bar{v}}$ for every $\bar{v} \in D$. We may choose the basis of V in such a way that $t = (1, 0, \dots, 0)$. Moreover we may assume that the function ϑ corresponding to D is a theta function with respect to the canonical factor $a_L = a_{L(H,\chi)}$ for L. Then the assumption means that $\frac{\partial \vartheta}{\partial v_1}(v) = 0$ for all $v \in V$ with $\vartheta(v) = 0$. Since D is reduced, the function

$$f = \frac{1}{\vartheta} \frac{\partial \vartheta}{\partial v_1}$$

is holomorphic on V, and the functional equation of ϑ translated to f is

$$f(v + \lambda) = f(v) + \pi H(t, \lambda)$$

for all $v \in V, \lambda \in \Lambda$. This implies that df is the pullback of a holomorphic differential on X and according to Proposition 1.3.5 we may assume that $df = \sum_{\nu=1}^{g} \alpha_\nu dv_\nu$ for some $\alpha_\nu \in \mathbb{C}$. Integrating we obtain $f = \sum_{\nu=1}^{g} \alpha_\nu v_\nu + c$, where c is a constant. Inserting this into the functional equation of f, we get $\sum_{\nu=1}^{g} \alpha_\nu \lambda_\nu = \pi H(t, \lambda)$, where $\lambda_1, \dots, \lambda_g$ denote the coordinates of λ with respect to the given basis. Hence

$$f(v) = \pi H(t, v) + c.$$

Since f is holomorphic and H is nondegenerate and $\mathbb{C}$-antilinear in the second variable, this implies $t = 0$, a contradiction. □

If moreover the divisor D is irreducible, one can say more.

(4.2) Proposition. *For any irreducible reduced divisor $D \in |L|$, with L a positive definite line bundle on X, the Gauss map is dominant.*

For an example of a divisor, for which the Gauss map is not dominant, compare Exercise 5.

Proof. For any $x \in D$ denote by B_x the maximal complex subtorus of X with $x + B_x \subset D$. Since X admits only countably many complex subtori (see Exercise 1.1), and D is irreducible, there is a complex subtorus B of X such that $B_x = B$ for almost all $x \in D$ and thus $D + B = D$. In particular this implies $B \subset K(L) = K(\mathcal{O}_X(D))$. But by assumption $K(L)$ is finite, so $B = 0$.

Suppose now that G is not dominant. Then $\operatorname{im} G$ is contained in a subvariety Y of codimension 1 in $\mathbb{P}_{g-1}$ and all fibres of G are of dimension ≥ 1. Let $\overline{w}$ denote a general point of D_s, where general means that $\overline{w}$ is smooth in the fibre $G^{-1}(G(\overline{w}))$ and $G(\overline{w})$ is smooth in Y. By what we said above, it suffices to show that $B_{\overline{w}} \neq 0$.

We may choose the basis of V in such a way that $G(\overline{w}) = (1:0:\cdots:0)$. As above we denote by $v_1, \ldots, v_g$ the corresponding coordinate functions and by ϑ a theta function corresponding to D. Since $\frac{\partial\vartheta}{\partial v_1}(w) \neq 0$, we may apply the implicit function theorem to get that $\pi^* D$ is given locally around w by an equation

$$v_1 + f(v_2, \ldots, v_g) = 0 .$$

It follows that for every vector v near w in the inverse image under $\pi: V \to X$ of the fibre $G^{-1}(G(\overline{w}))$

$$G(\bar{v}) = \left(1 : \tfrac{\partial f}{\partial v_2}(v) : \cdots : \tfrac{\partial f}{\partial v_g}(v)\right) = (1:0:\cdots:0) . \tag{1}$$

This implies that

$$\frac{\partial f}{\partial v_\nu} = 0, \quad \nu = 2, \ldots, g \tag{2}$$

are equations for $\pi^{-1}G^{-1}(G(\overline{w}))$ locally around w. If $x_1, \ldots, x_g$ denote the given homogenous coordinates in $\mathbb{P}_{g-1}$, then $z_\nu = \frac{x_\nu}{x_1}$, $\nu = 2, \ldots, g$, is a set of affine coordinates of $\mathbb{P}_{g-1} - \{x_1 = 0\} \simeq \mathbb{C}^{g-1}$, such that $G(\overline{w}) = (0, \ldots, 0)$. By assumption the variety Y is smooth near $G(\overline{w})$. So we may assume that (after a suitable linear transformation of $z_2, \ldots, z_g$) it is given locally around $G(\overline{w})$ by an equation

$$z_g = h(z_2, \ldots, z_g) , \tag{3}$$

with a power series h vanishing of order ≥ 2 in $G(\overline{w}) = (0, \ldots, 0)$. Note that applying the same linear transformation to $v_2, \ldots, v_g$, the equations (1) and (2) remain valid. So by definition of the Gauss map and (3) we get

$$\frac{\partial f}{\partial v_g} = \frac{\partial f}{\partial v_1} \cdot h\left(\frac{\partial f}{\partial v_2} \Big/ \frac{\partial f}{\partial v_1}, \ldots, \frac{\partial f}{\partial v_g} \Big/ \frac{\partial f}{\partial v_1}\right)$$

near w. Hence for $\nu = 2, \ldots, g$

$$\begin{aligned}\frac{\partial}{\partial v_g}\frac{\partial f}{\partial v_\nu} &= \frac{\partial}{\partial v_\nu}\frac{\partial f}{\partial v_g} \\ &= \frac{\partial^2 f}{\partial v_\nu \partial v_1} \cdot h\left(\frac{\partial f}{\partial v_2} \Big/ \frac{\partial f}{\partial v_1}, \ldots, \frac{\partial f}{\partial v_g} \Big/ \frac{\partial f}{\partial v_1}\right) + \frac{\partial f}{\partial v_1} \cdot \frac{\partial}{\partial v_\nu} h\left(\frac{\partial f}{\partial v_2} \Big/ \frac{\partial f}{\partial v_1}, \ldots, \frac{\partial f}{\partial v_g} \Big/ \frac{\partial f}{\partial v_1}\right) = 0\end{aligned}$$

on $\pi^{-1}G^{-1}G(\overline{w})$ near w be equation (2), since h vanishes of order ≥ 2 in $(0, \ldots, 0)$. This means that $\pi^{-1}G^{-1}(G(\overline{w}))$ is invariant under translations in direction of v_g. Denoting by A the complex subtorus of X generated by $\{\bar{v} \mid v_1 = \cdots = v_{g-1} = 0\}$ we get

$$\overline{w} + A \subset G^{-1}(G(\overline{w})) + A \subset G^{-1}(G(\overline{w})) \subset D .$$

This completes the proof, since $0 \neq A \subset B_{\overline{w}}$. □

§ 5 Projective Embeddings

Let (X, L) be a polarized abelian variety of type $(d_1, \dots, d_g)$. In this section we study the map φ_L. In Section 1 we saw that φ_L is a holomorphic map, if $d_1 \geq 2$. Here we show that φ_L is an embedding, if $d_1 \geq 3$, and give a criterion for this to happen in case $d_1 = 2$.

(5.1) Theorem of Lefschetz. *If L is a positive definite line bundle on X of type $(d_1, \dots, d_g)$ with $d_1 \geq 3$, then $\varphi_L : X \to \mathbb{P}_N$ is an embedding.*

Proof. We have to show (i) that φ_L is injective and (ii) that the differential $d\varphi_{L,x}$ is injective for every $x \in X$.

As for (i): assume $y_1, y_2 \in X$ with $\varphi_L(y_1) = \varphi_L(y_2)$. So $y_1 \in D$ if and only if $y_2 \in D$ for any $D \in |L|$. According to Lemma 2.5.6 there is a positive definite $M \in \operatorname{Pic}(X)$ with $L \simeq M^{d_1}$. By Proposition 1.7 and Lemma 1.8 there is a reduced divisor $D_M \in |M|$ such that $t_x^* D_M = D_M$ only for $x = 0$.

Suppose $x_1 \in t_{y_1}^* D_M$. By continuity of the addition map and since $d_1 \geq 3$, there are $x_2, \dots, x_{d_1} \in X$ with $x_1 + \dots + x_{d_1} = 0$ such that $y_2 \notin t_{x_\nu}^* D_M$ for $\nu = 2, \dots, d_1$. Since $\sum_{\nu=1}^{d_1} t_{x_\nu}^* D_M$ is a divisor in $|L|$ (see Lemma 1.5) containing y_1, we have by assumption $y_2 \in \sum_{\nu=1}^{d_1} t_{x_\nu}^* D_M$ and hence $y_2 \in t_{x_1}^* D_M$ by construction. So $x_1 \in t_{y_2}^* D_M$ and this holds for an arbitrary $x_1 \in t_{y_1}^* D_M$. Since D_M is reduced, we obtain $t_{y_1}^* D_M \subset t_{y_2}^* D_M$ and thus $t_{y_1}^* D_M = t_{y_2}^* D_M$, the situation being symmetric. Applying Lemma 1.8 we conclude $y_1 = y_2$.

As for (ii): suppose $t \neq 0$ is a tangent vector at $x \in X$. It suffices to show that there is a divisor $D \in |L|$ passing through x such that t is not tangent at D in x.

Assume the contrary, that is t is tangent at D in x for all $D \in |L|$ containing x. Fix a reduced divisor $D_M \in |M|$. For $x_1 \in t_x^* D_M$ we can choose as above $x_2, \dots, x_{d_1} \in X$ with $x_1 + \dots + x_{d_1} = 0$ such that $x \notin t_{x_\nu}^* D_M$ for $\nu = 2, \dots, d_1$. Since $x \in \sum_{\nu=1}^{d_1} t_{x_\nu}^* D_M \in |L|$, we have by assumption that t is tangent at the divisor $\sum_{\nu=1}^{d_1} t_{x_\nu}^* D_M$ in x and hence t is tangent at $t_{x_1}^* D_M$ in x. This holds for all $x_1 \in t_x^* D_M$. Hence t is tangent to D_M at all points of D_M. But this implies that the image of the Gauss map for D_M is contained in a hyperplane, contradicting Proposition 4.1. □

Before we study the case $d_1 = 2$, we deduce some consequences. Recall that by definition a line bundle L is *ample*, if L^n is very ample for some $n \geq 1$, that is if φ_{L^n} is an embedding. We have the following characterizations for a line bundle to be ample.

(5.2) Proposition. *For a line bundle L on X the following statements are equivalent.*

i) L is ample.

ii) L is positive definite.

iii) $H^0(L) \neq 0$ and $K(L)$ is finite.

iv) $H^0(L) \neq 0$ and $(L^g) > 0$.

Proof. i) $\Longrightarrow$ ii) Suppose φ_{L^n} is an embedding for some $n \geq 1$. It follows that $h^0(L^n) \neq 0$ and L^n is positive semidefinite according to Theorem 3.4.5. But L^n and as such L itself is even positive definite, since otherwise φ_{L^n} would not be injective by Theorem 3.3.3.

ii) $\Longrightarrow$ iii) is an immediate consequence of Corollary 3.2.8 and the definition of $K(L)$.

iii) $\Longrightarrow$ iv) L is nondegenerate, since $K(L)$ is finite. According to Theorem 3.4.5 we have $h^q(L) \neq 0$ only for $q = 0$. Hence the assertion is a consequence of the Geometric Riemann-Roch Theorem 3.6.3.

iv) $\Longrightarrow$ i) From the Riemann-Roch Theorems together with Theorem 3.5.5 we get $h^0(L) = \chi(L) = d_1 \cdot \cdots \cdot d_g > 0$, where $(d_1, \ldots, d_g)$ is the type of L. Consequently L is positive definite and thus ample by Theorem 5.1. □

As a consequence we obtain the familiar version of Lefschetz's theorem:

(5.3) Corollary. *If $L \in \operatorname{Pic}(X)$ is ample, L^n is very ample for $n \geq 3$.*

Proposition 5.2 leads to the following criterion for a complex torus X to be an abelian variety. Recall that the transcendence degree of the field of meromorphic functions on X is called the *algebraic dimension* $a(X)$ of X. Necessarily $a(X) \leq \dim X$ as for any connected compact complex manifold.

(5.4) Theorem. *For a complex torus X the following conditions are equivalent*

i) X is an abelian variety,

ii) X admits the structure of a projective algebraic variety,

iii) $a(X) = \dim X$.

Proof. i) $\Longrightarrow$ ii): Recall that by definition an abelian variety is a complex torus admitting a positive definite line bundle. According to Proposition 5.2 this means that X is an abelian variety if and only if it can be analytically embedded into projective space. But by the Theorem of Chow A.3 any closed analytic subvariety of $\mathbb{P}_n$ is algebraic.

It remains to show iii) $\Longrightarrow$ i), the implication ii) $\Longrightarrow$ iii) being trivial. Suppose $f_1, \ldots, f_g$, with $g = \dim X$, are algebraically independent meromorphic functions on X. Denote by D_i the polar divisor of f_i and $L = \mathcal{O}_X(D_1 + \cdots + D_g)$. Let σ be a section in $H^0(L)$ corresponding to the divisor $D_1 + \cdots + D_g$. For every i there exists a uniquely determined section $\sigma_i \in H^0(L)$ such that $f_i = \frac{\sigma_i}{\sigma}$. The line bundle L is positive semidefinite, since $h^0(L) > 0$. Let $p: X \to \overline{X} = X/K(L)_0$ be the natural map of Section 3.3. According to Lemma 3.3.2 there is a positive definite line bundle $\overline{L}$ on $\overline{X}$ such that $L = p^*\overline{L}$, again since $h^0(L) > 0$. Moreover $p^*: H^0(\overline{L}) \to H^0(L)$ is an isomorphism . This implies that $f_i = p^*h_i$ for some meromorphic functions h_i on $\overline{X}$. Certainly $h_1, \ldots, h_g$ are algebraically independent and $g \le a(\overline{X}) \le \dim \overline{X} \le \dim X = g$. So p is an isomorphism, i.e. L is positive definite. □

In the sequel we will deal exclusively with abelian varieties. Hence, according to Appendix A, we can work either in the analytic or the algebraic category. It seems more natural to us to consider abelian varieties as algebraic varieties rather than analytic varieties and we will do this without further noticing. For the necessary identifications we refer to Appendix A.

In the last part of this section we study the map $\varphi: X \to \mathbb{P}_n$ corresponding to an ample line bundle of type $(2, d_2, \ldots, d_g)$ on X. According to Lemma 2.5.6 such a line bundle is of the form L^2 with an ample line bundle L on X. Let us start slightly more generally with an arbitrary ample line bundle L on X. In order to study the map φ_{L^2}, we decompose the polarized abelian variety (X, L) applying the Decomposition Theorem 3.1:

$$(X, L) \simeq (X_M \times X_{N_1} \times \cdots \times X_{N_r}, q_M^* M \otimes q_{N_1}^* N_1 \otimes \cdots \otimes q_{N_r}^* N_r). \quad (1)$$

Here the line bundle M on X_M is a polarization without fixed components and the line bundles N_ν define irreducible principal polarizations on X_{N_ν} for $\nu = 1, \ldots, r$. Let $\varphi_{M^2}: X_M \to \mathbb{P}_{n_0}$ and $\varphi_{N_\nu^2}: X_{N_\nu} \to \mathbb{P}_{n_\nu}$ for $\nu = 1, \ldots, r$ be the corresponding holomorphic maps (see Proposition 1.6) and denote by $\psi: \mathbb{P}_{n_0} \times \mathbb{P}_{n_1} \times \cdots \times \mathbb{P}_{n_r} \to \mathbb{P}_n$ the Segre embedding. Then, via the decomposition (1), the holomorphic map $\varphi_{L^2}: X \to \mathbb{P}_n$ decomposes as follows

$$\varphi_{L^2} = \psi(\varphi_{M^2} \times \varphi_{N_1^2} \times \cdots \times \varphi_{N_r^2}) ,$$

and for the investigation of the map φ_{L^2} we are reduced to the cases

I: $L = M$, a polarization without fixed components,
II: $L = N_1$, an irreducible principal polarization.

In this section we will study the case I, whereas case II will be treated in Section 8. The following theorem was proved by Ramanan [1] for a generic abelian variety and by Obuchi [1] in general. Here we follow the proof given in Lange-Narasimhan [1].

(5.5) Theorem. *If L is an ample line bundle without fixed components, then L^2 is very ample.*

Proof. As in the proof of the theorem of Lefschetz we have to show (i) that φ_{L^2} is injective and (ii) that the differential $d\varphi_{L^2,x}$ is injective for every $x \in X$.

As for (i): assume $y_1, y_2 \in X$ with $\varphi_{L^2}(y_1) = \varphi_{L^2}(y_2)$. So $y_1 \in D$ if and only if $y_2 \in D$ for all $D \in |L^2|$. According to Proposition 1.7, Lemma 1.8 and Theorem 3.5 there is an irreducible reduced $D_1 \in |L|$ such that $t_x^* D_1 = D_1$ only for $x = 0$. Again since $|L|$ has no fixed component, there is an irreducible $D_2 \in |L|$ with $D_2 \neq (-1)^* t_{y_1+y_2}^* D_1$.

For $x \in t_{y_1}^* D_1$ we have $y_1 \in t_x^* D_1 \subset t_x^* D_1 + t_{-x}^* D_2 \in |L^2|$. By assumption this implies $y_2 \in t_x^* D_1 + t_{-x}^* D_2$, which in turn is equivalent to $x \in t_{y_2}^* D_1 + (-1)^* t_{y_2}^* D_2$. This holds for every $x \in t_{y_1}^* D_1$. Since D_1 is reduced, it follows that $t_{y_1}^* D_1 \subset t_{y_2}^* D_1 + (-1)^* t_{y_2}^* D_2$ or equivalently

$$D_1 = t_{-y_1}^* t_{y_1}^* D_1 \subset t_{-y_1+y_2}^* D_1 + (-1)^* t_{y_1+y_2}^* D_2 \ .$$

Since $D_1 \neq (-1)^* t_{y_1+y_2}^* D_2$ by construction, and the divisors D_1 and D_2 are irreducible, it follows that $D_1 = t_{-y_1+y_2}^* D_1$. By assumption on D_1 this implies $y_1 = y_2$.

As for (ii): suppose $t \neq 0$ is a tangent vector at $x \in X$. It suffices to show that there is a divisor $D \in |L^2|$ containing x such that t is not tangent at D in x.

Assume the contrary: the vector t is tangent at D in x for all $D \in |L^2|$ containing x. According to Proposition 1.7, Theorem 3.5, and the assumption that $|L|$ has no fixed component there are irreducible reduced divisors D_1 and D_2 in $|L|$ with $t_x^* D_1 \neq (-1)^* t_x^* D_2$. Let $D_{1,s}$ denote the smooth part of D_1. For any $y \in t_x^* D_{1,s} - (-1)^* t_x^* D_2$ we have

$$x \in t_y^* D_{1,s} \subset t_y^* D_1 + t_{-y}^* D_2 \in |L^2| \ .$$

By assumption this implies that t is tangent to $t_y^* D_1 + t_{-y}^* D_2$ in x. But x is not contained in $t_{-y}^* D_2$, since otherwise $y \in (-1)^* t_x^* D_2$. Hence t is tangent to $t_y^* D_{1,s}$ in x. This is true for every y in an open dense subset of $t_x^* D_{1,s}$. So it implies that t is tangent to D_1 in every point of D_1. But this means that the image of the Gauss map for D_1 is contained in a hyperplane, contradicting Proposition 4.1. □

§ 6 Symmetric Line Bundles

As we saw in the last section, for the investigation of the map φ_{L^2} it suffices to consider the cases that L has no fixed components and that L is an irreducible principal polarization. We treated the first case in Theorem 5.5. For the latter case it turns out to be convenient to assume that L is a symmetric line bundle. In order to show that this is no restriction, we begin this section by studying how the map φ_L changes, when L moves in its algebraic equivalence class.

The definition of *algebraic equivalence of line bundles* is analogous to the definition of analytic equivalence of line bundles (see Section 2.5), only the parameter space T has to be algebraic. In particular algebraically equivalent line bundles are analytically equivalent. So Corollary 2.5.4 implies that ample line bundles on an abelian variety are algebraically equivalent if and only if they differ by a translation. Hence we have to compare φ_L and $\varphi_{t_x^*L}$. In fact, the following lemma follows immediately from the definitions.

(6.1) Lemma. *For an ample line bundle L on X and a point $x \in X$ the following diagram commutes up to an automorphism of $\mathbb{P}_n$*

$$\begin{array}{ccc} X & \searrow^{\varphi_{t_x^*L}} & \\ {\scriptstyle t_x}\downarrow & & \mathbb{P}_n \\ X & \nearrow_{\varphi_L} & \end{array}$$

Notice that the maps φ_L and $\varphi_{t_x^*L}$ depend on the choice of bases of $H^0(L)$ and $H^0(t_x^*L)$. If one chooses the bases in a compatible way, the diagram actually commutes: let $\vartheta_0, \dots, \vartheta_n$ be a basis of $H^0(L)$, then $t_x^*\vartheta_0, \dots, t_x^*\vartheta_n$ is a basis of $H^0(t_x^*L)$ according to Corollary 3.2.9 and we have $\varphi_{t_x^*L} = \varphi_L \circ t_x$.

Lemma 6.1 implies that the image $\overline{X}$ of φ_L in $\mathbb{P}_n$ does not depend on L itself, but only on the algebraic equivalence class of L. This reflects the fact that the map φ_L and the chosen point 0 of the group X are independent of each other. So, in order to investigate the projective variety $\overline{X}$ in $\mathbb{P}_n$, we may choose the line bundle L suitably within its algebraic equivalence class. Good candidates for this are the symmetric line bundles on X, introduced in Section 2.3. The next lemma implies that the space $\mathrm{Pic}^H(X)$ contains a symmetric line bundle for any $H \in \mathrm{NS}(X)$.

Recall from Proposition 2.3.7 that a line bundle $L = L(H, \chi)$ is symmetric if and only if its semicharacter χ has values in $\{\pm 1\}$. If L is nondegenerate, a decomposition $\Lambda = \Lambda_1 \oplus \Lambda_2$ for L distinguishes a line bundle in the algebraic equivalence class $\mathrm{Pic}^H(X)$, namely the bundle $L_0 = L(H, \chi_0)$

of characteristic 0 (see Section 3.1). The semicharacter χ_0 was defined by $\lambda = \lambda_1 + \lambda_2 \mapsto e(\pi i \operatorname{Im} H(\lambda_1, \lambda_2))$, where $\lambda_\nu \in \Lambda_\nu$, $\nu = 1, 2$. This shows that L_0 is symmetric.

For any $H \in \mathrm{NS}(X)$ denote by $\mathrm{Pic}_s^H(X)$ the set of symmetric line bundles in the algebraic equivalence class $\mathrm{Pic}^H(X)$. The sets $\mathrm{Pic}_s^H(X)$ have the following structure.

(6.2) Lemma. *a)* $\mathrm{Pic}_s^0(X)$ *is a vector space of dimension* $2g$ *over* $\mathbb{Z}/2\mathbb{Z}$.
b) For any nonzero $H \in \mathrm{NS}(X)$ *the set* $\mathrm{Pic}_s^H(X)$ *is a principal homogeneous space over* $\mathrm{Pic}_s^0(X)$.

Proof. a) follows from the fact that $\mathrm{Pic}_s^0(X)$ is just the set of 2-division points in the dual abelian variety.

For b) we first claim that $\mathrm{Pic}_s^H(X) \neq \emptyset$. By what we have said above, $\mathrm{Pic}_s^H(X)$ is nonempty for any nondegenerate H. So it remains to consider the case that H is degenerate. As it was shown in Section 3.3, there is a surjective homomorphism of abelian varieties $p\colon X \to \overline{X}$ and a nondegenerate $\overline{H} \in \mathrm{NS}(\overline{X})$ with $H = p^*\overline{H}$. With $\overline{L}_0 \in \mathrm{Pic}_s^{\overline{H}}(\overline{X})$ also the pullback $p^*\overline{L}_0$ is symmetric. So $\mathrm{Pic}_s^H(X) \neq \emptyset$.

Now suppose $L \in \mathrm{Pic}_s^H(X)$. Obviously the bijective map $\mathrm{Pic}^0(X) \to \mathrm{Pic}^H(X)$, $P \mapsto L \otimes P$ induces a bijection $\mathrm{Pic}_s^0(X) \to \mathrm{Pic}_s^H(X)$, defining the structure of a principal homogeneous space. □

If H is nondegenerate, we can interpret the action of $\mathrm{Pic}_s^0(X)$ on $\mathrm{Pic}_s^H(X)$ in terms of the characteristics with respect to the chosen decomposition: let $L_0 \in \mathrm{Pic}_s^H(X)$ be the line bundle with characteristic 0 as above. One easily sees that the line bundle with characteristic c

$$t_{\bar{c}}^* L_0 = L\big(H, \chi_0 e\big(2\pi i \operatorname{Im} H(c, \cdot)\big)\big) \tag{1}$$

is symmetric if and only if the character $e(2\pi i \operatorname{Im} H(c, \cdot))$ on Λ takes only values in $\{\pm 1\}$. This is the case if and only if $c \in \frac{1}{2}\Lambda(H)$. Hence the 2^{2g} line bundles $t_{\bar{c}}^* L_0$ with characteristic c in $\frac{1}{2}\Lambda(H)$ (modulo $\Lambda(H)$) build up the principal homogeneous space $\mathrm{Pic}_s^H(X)$ and the action of $\mathrm{Pic}_s^0(X)$ on $\mathrm{Pic}_s^H(X)$ is induced by the map $\frac{1}{2}\Lambda(H) \to \mathrm{Pic}_s^H(X)$, $c \mapsto t_{\bar{c}}^* L_0$.

Let L be any symmetric line bundle on X. A biholomorphic map $\varphi : L \to L$ is called *isomorphism of* L *over* $(-1)_X$, if the diagram

$$\begin{array}{ccc} L & \xrightarrow{\varphi} & L \\ \downarrow & & \downarrow \\ X & \xrightarrow{(-1)_X} & X \end{array}$$

commutes and for every $x \in X$ the induced map $\varphi(x)\colon L(x) \to L(-x)$ is $\mathbb{C}$-linear. Here $L(x)$ denotes the fibre of L over the point x and $\varphi(x)$ is the restriction of L to $L(x)$. The isomorphism φ is called *normalized*, if the induced map $\varphi(0) : L(0) \to L(0)$ is the identity.

(6.3) Lemma. *Any symmetric line bundle $L \in \operatorname{Pic}(X)$ admits a unique normalized isomorphism $(-1)_L : L \to L$ over $(-1)_X$.*

Proof. The biholomorphic map $(-1) \times 1 : V \times \mathbb{C} \to V \times \mathbb{C}$ is an isomorphism of the trivial line bundle on V over $(-1)_V$. Since L is symmetric, its canonical factor a_L satisfies $a_L(-\lambda, -v) = a_L(\lambda, v)$ for all $(\lambda, v) \in \Lambda \times V$ according to Lemma 2.3.4. This implies that the action of Λ on $V \times \mathbb{C}$ via a_L defining L (see Section 2.2) is compatible with $(-1) \times 1$. Hence $(-1) \times 1$ descends to an isomorphism $(-1)_L : L \to L$ over $(-1)_X$. Certainly $(-1)_L$ is normalized, because $(-1) \times 1$ induces the identity on the fibre $\{0\} \times \mathbb{C}$. The uniqueness of $(-1)_L$ follows from the fact that any two automorphisms of L differ by a nonzero constant. □

Suppose now $L = L(H, \chi)$ is an ample symmetric line bundle on X. The normalized isomorphism $(-1)_L$ induces an involution on the vector space of canonical theta functions for L

$$(-1)_L^* : H^0(L) \to H^0(L) \, , \quad \vartheta \mapsto (-1)_V^* \vartheta \, .$$

Denote by $H^0(L)_+$ and $H^0(L)_-$ the eigenspaces of the involution $(-1)_L^*$. For the computation of the dimensions $h^0(L)_+$ and $h^0(L)_-$ we need to work out, how $(-1)_L$ acts on $H^0(L)$.

(6.4) Inverse Formula. *Let $\{\vartheta_{\overline{w}}^c \mid \overline{w} \in K(L)_1\}$ denote the basis of $H^0(L)$ of Theorem 3.2.7 and $c = c_1 + c_2$ the decomposition of the characteristic $c \in \frac{1}{2}\Lambda(L)$ of L. Then*

$$(-1)_L^* \vartheta_{\overline{w}}^c = e(4\pi i \operatorname{Im} H(w + c_1, c_2)) \vartheta^c_{-\overline{w} - 2\overline{c}_1} \, .$$

In particular, if L is of characteristic 0, then $(-1)_L^* \vartheta_{\overline{w}}^0 = \vartheta_{-\overline{w}}^0$ for all $\overline{w} \in K(L)_1$.

Proof. One easily sees from its definition that the theta function ϑ_0^0 is even. So for all $v \in V$

$$(-1)_L^* \vartheta_{\overline{w}}^c(v) = \vartheta_{\overline{w}}^c(-v) = e(\pi H(v, c) - \tfrac{\pi}{2} H(c, c)) \vartheta_{\overline{w}}^0(-v + c)$$

(by equation 3.2(5))

$$= e(\pi H(v, c) - \tfrac{\pi}{2} H(c, c)) a_{L_0}(w, -v + c)^{-1} \vartheta_0^0(-v + c + w)$$

(by equation 3.2(4))

$$= e(\pi H(v,c) - \tfrac{\pi}{2} H(c,c)) a_{L_0}(w, -v+c)^{-1} \vartheta_0^0(v - c - w)$$

(since ϑ_0^0 is even)

$$= e(\pi H(v,c) - \tfrac{\pi}{2} H(c,c)) a_{L_0}(w, -v+c)^{-1}$$
$$a_{L_0}(-2c_2, v - c_1 + c_2 - w)\vartheta_0^0(v - c_1 + c_2 - w)$$

(by Corollary 3.2.5)

$$= e(\pi H(v,c) - \tfrac{\pi}{2} H(c,c)) a_{L_0}(w, -v+c)^{-1}$$
$$a_{L_0}(-2c_2, v - c_1 + c_2 - w) a_{L_0}(-w - 2c_1, v + c)\vartheta^0_{-\overline{w}-2\overline{c}_1}(v+c)$$

(by equation 3.2(4))

$$= e(2\pi H(v,c)) a_{L_0}(w, -v+c)^{-1} a_{L_0}(-2c_2, v - c_1 + c_2 - w)$$
$$a_{L_0}(-w - 2c_1, v+c)\vartheta^c_{-\overline{w}-2\overline{c}_1}(v)$$

(by equation 3.2(5)).

Now using the definition of a_{L_0} and the fact that $E(w, c_1) = 0$ (c_1 and w both are contained in the subspace V_1, which is isotropic for E), one easily deduces the assertion. □

(6.5) Proposition. *Let $L \in \operatorname{Pic}^H(X)$ be an ample symmetric line bundle on X of characteristic c with respect to a decomposition of $\Lambda(L)$ for L. Write $c = c_1 + c_2$ and define $S = \{\overline{w} \in K(L)_1 \mid 2\overline{w} = -2\overline{c}_1\}$ and $S^\pm = \{\overline{w} \in S \mid e(4\pi i \operatorname{Im} H(w + c_1, c_2)) = \pm 1\}$. Then*

$$h^0(L)_\pm = \tfrac{1}{2}\big(h^0(L) - \#S\big) + \#S^\pm \ .$$

Proof. Let $\{\vartheta^c_{\overline{w}}, \overline{w} \in K(L)_1\}$ denote the basis of $H^0(L)$ of Theorem 3.2.7. Define for any $\overline{w} \in K(L)_1$

$$\theta^\pm_{\overline{w}} = e\big(-4\pi i \operatorname{Im} H(w + c_1, c_2)\big)\vartheta^c_{\overline{w}} \pm \vartheta^c_{-\overline{w}-2\overline{c}_1} \ .$$

It follows immediately from the Inverse Formula that $\theta^+_{\overline{w}}$ is an even function and $\theta^-_{\overline{w}}$ is odd. Since $\{\theta^+_{\overline{w}}, \theta^-_{\overline{w}}, \overline{w} \in K(L)_1\}$ spans the vector space $H^0(L)$, the theta functions $\theta^+_{\overline{w}}, \overline{w} \in K(L)_1$ span $H^0(L)_+$. By definition

$$\theta^+_{-\overline{w}-2\overline{c}_1} = e\big(4\pi i \operatorname{Im} H(w + c_1, c_2)\big)\theta^+_{\overline{w}} \ .$$

So for $\overline{w} \in K(L)_1 - S$ the functions $\theta^+_{\overline{w}}$ and $\theta^+_{-\overline{w}-2\overline{c}_1}$ are linearly dependent over $\mathbb{C}$. Moreover for $\overline{w} \in S$ we have

$$\theta^+_{\overline{w}} = \Big(e\big(-4\pi i \operatorname{Im} H(w + c_1, c_2)\big) + 1\Big)\vartheta^c_{\overline{w}} = \begin{cases} 2\vartheta^c_{\overline{w}} & \text{if} \quad \overline{w} \in S^+ \\ 0 & \phantom{\text{if}} \quad \overline{w} \in S^- \ . \end{cases}$$

To see this note that S is the disjoint union of the sets S^+ and S^-, since $\overline{w} \in S$ implies $2w + 2c_1 \in \Lambda$ such that $e(-4\pi i \operatorname{Im} H(w + c_1, c_2)) = e(-\pi i \operatorname{Im} H(2w + 2c_1, 2c_2)) = \pm 1$.

Choosing for every $\overline{w} \in K(L)_1 - S$ one function out of $\{\theta^+_{\overline{w}}, \theta^+_{-\overline{w}-2\overline{c}_1}\}$, these functions together with the functions $\vartheta^c_{\overline{w}}$ for $\overline{w} \in S^+$ form obviously a basis for $H^0(L)_+$. Noting that $\# K_1 = h^0(L)$ this implies the assertion. □

One can determine the sets S and S^+ in terms of the characteristic and the type of the line bundle L. In this way we get explicit formulas for $h^0(L)_+$ and $h^0(L)_-$. For the general case compare Exercise 11 or Birkenhake-Lange [2]. Here we only treat the most important case, the line bundle of characteristic 0.

(6.6) Corollary. *Let L_0 denote the ample line bundle of type $(d_1, \ldots, d_g)$, with characteristic 0 with respect to some decomposition for L_0. Suppose $d_1, \ldots, d_s$ are odd and $d_{s+1}, \ldots, d_g$ are even, then*

$$h^0(L_0)_\pm = \tfrac{1}{2} h^0(L_0) \pm 2^{g-s-1} \ .$$

For the proof note that $S = S^+ = K(L_0)_1 \cap X_2$ and $\#(K(L_0)_1 \cap X_2) = 2^{g-s}$, since $K(L_0)_1 \simeq \oplus_{r=1}^g \mathbb{Z}/d_r\mathbb{Z}$. □

It is easy to derive analogous formulas for any symmetric line bundle L with $h^0(L) > 0$ using the reduction to the ample case of Section 3.3.

§ 7 Symmetric Divisors

Let $X = V/\Lambda$ be an abelian variety of dimension g. A divisor D on X is called *symmetric*, if $(-1)^*_X D = D$. The main result of this section is Proposition 7.5, where we compute the number of 2-division points, at which a symmetric divisor has even or odd multiplicity.

Let D be a symmetric divisor on X. Certainly the line bundle $L = \mathcal{O}_X(D)$ is also symmetric. Suppose L is ample. As we saw in Section 4, the divisors D in the linear system $|L|$ correspond one to one to canonical theta functions ϑ for L modulo $\mathbb{C}^*$ via $\pi^* D = (\vartheta)$. In order to determine the theta functions corresponding to symmetric divisors, we observe that, if ϑ is an even or odd theta function, the corresponding divisor D is symmetric. The following lemma shows that the converse is also true.

(7.1) Lemma. *For $D \in |L|$ and $\vartheta \in H^0(L)$ with $\pi^* D = (\vartheta)$ the following conditions are equivalent:*

i) D is symmetric.

ii) $\vartheta \in H^0(L)_+$ or $\vartheta \in H^0(L)_-$.

Proof. If one considers $H^0(L)$ as the space of sections of the line bundle L, the statement is obvious by the construction of the normalized isomorphism $(-1)_L$. For convenience of the reader we also include a proof in terms of canonical theta functions. It suffices to show i) $\Longrightarrow$ ii): Since D is symmetric, there is a nowhere vanishing holomorphic function ε_D on V such that $\vartheta(-v) = \varepsilon_D(v)\vartheta(v)$ for all $v \in V$. On the other hand $a_L(\lambda, v) = a_L(-\lambda, -v)$ for all $v \in V$ and $\lambda \in \Lambda$, since the line bundle L is symmetric. Hence we have

$$\begin{aligned} \varepsilon_D(v+\lambda)\vartheta(v) &= \varepsilon_D(v+\lambda)\vartheta(v+\lambda)a_L(\lambda, v)^{-1} \\ &= \vartheta(-v-\lambda)a_L(-\lambda, -v)^{-1} \\ &= \vartheta(-v) = \varepsilon_D(v)\vartheta(v) \end{aligned}$$

for all $v \in V$ and $\lambda \in \Lambda$. This means ε_D is $2g$-fold periodic on V. So ε_D is constant by Liouville's theorem. Since $(-1)_V$ is an involution, $\varepsilon_D = +1$ or $\varepsilon_D = -1$. □

The lemma shows in particular that for any symmetric $L \in \operatorname{Pic}(X)$ with $h^0(L) > 0$ there is an effective symmetric divisor D with $L = \mathcal{O}_X(D)$.

For an arbitrary, not necessarily effective divisor D on X denote by $\operatorname{mult}_x(D)$ the multiplicity of D at a point $x \in X$. A symmetric divisor D is called *even* (respectively *odd*), if $\operatorname{mult}_0(D)$ is even (respectively odd). If D is moreover effective and ϑ a corresponding theta function, $\operatorname{mult}_{\bar{v}}(D)$ is just the subdegree of the Taylor expansion of ϑ in $v \in V$, and D is even (respectively odd) if and only if ϑ is even (respectively odd).

(7.2) Proposition. *Let $L = L(H, \chi)$ be a symmetric line bundle. For any symmetric divisor D on X with $L = \mathcal{O}_X(D)$ we have*

$$(-1)^{\operatorname{mult}_x(D)} = \chi(\lambda)(-1)^{\operatorname{mult}_0(D)}$$

for every 2-division point $x = \pi(\frac{1}{2}\lambda) \in X_2$.

Proof. Without loss of generality we may assume that D is effective. Let $\vartheta \in H^0(L)$ be a corresponding canonical theta function. Then we have

$$\vartheta(-v) = (-1)^{\operatorname{mult}_0(D)}\vartheta(v)$$

for all $v \in V$. On the other hand, $\operatorname{mult}_x(D) = \operatorname{mult}_0(t_x^* D)$ and according to Corollary 3.2.9 $\widetilde{\vartheta} := e(-\frac{\pi}{2}H(\cdot, \lambda))\vartheta(\cdot + \frac{1}{2}\lambda))$ is a canonical theta function for $t_x^* L$ corresponding to $t_x^* D$. For every $v \in V$

$$
\begin{aligned}
(-1)^{\mathrm{mult}_x(D)}\tilde{\vartheta}(v) &= \tilde{\vartheta}(-v) \\
&= e\big(\tfrac{\pi}{2}H(v,\lambda)\big)\vartheta(-v+\tfrac{1}{2}\lambda) \\
&= (-1)^{\mathrm{mult}_0(D)}e\big(\tfrac{\pi}{2}H(v,\lambda)\big)\vartheta(v+\tfrac{1}{2}\lambda-\lambda) \\
&= (-1)^{\mathrm{mult}_0(D)}e\big(\tfrac{\pi}{2}H(v,\lambda)\big) \\
&\quad\cdot\chi(-\lambda)e\big(\pi H(v+\tfrac{1}{2}\lambda,-\lambda)+\tfrac{\pi}{2}H(\lambda,\lambda)\big)\vartheta(v+\tfrac{1}{2}\lambda) \\
&= (-1)^{\mathrm{mult}_0(D)}\chi(\lambda)\tilde{\vartheta}(v)\ ,
\end{aligned}
$$

since $\chi(\lambda)=\chi(-\lambda)$. This implies the assertion. □

Our next aim is to compute the number of 2-division points $x \in X_2$ with even (respectively odd) multiplicity $\mathrm{mult}_x(D)$. We will see that this number is independent of the choice of D within the set of even (respectively odd) symmetric divisors linearly equivalent to D. For this we apply the theory of quadratic forms over the $\mathbb{Z}/2\mathbb{Z}$-vector space X_2. For any $H \in \mathrm{NS}(X)$ define a map $e^H\colon X_2 \times X_2 \to \{\pm 1\}$ by

$$e^H(\bar{v},\bar{w}) = e\big(\pi i \operatorname{Im} H(2v,2w)\big)\ .$$

This definition does not depend on the choice of the representatives of $\bar{v}$ and $\bar{w}$ in V and e^H takes values in $\{\pm 1\}$, since $\operatorname{Im} H(\Lambda\times\Lambda) \subseteq \mathbb{Z}$. So e^H is a symmetric bilinear form on the $2g$-dimensional $\mathbb{Z}/2\mathbb{Z}$-vector space X_2. By definition a *quadratic form associated to* e^H is a map $q\colon X_2 \to \{\pm 1\}$ satisfying

$$q(x)q(y)q(x+y) = e^H(x,y) \tag{1}$$

for all $x, y \in X_2$. Every element of $\mathrm{Pic}_s^H(X)$ induces such a quadratic form: suppose $L = L(H,\chi)$ is a symmetric line bundle. We define

$$q_L(\bar{v}) = \chi(2v) \tag{2}$$

for every $\bar{v} \in X_2$. The map $q_L\colon X_2 \to \{\pm 1\}$ is well defined, since χ takes only values in $\{\pm 1\}$. Moreover the defining equation for a semicharacter, $\chi(\lambda)\chi(\mu) = \chi(\lambda+\mu)e(\pi i \operatorname{Im} H(\lambda,\mu))$ for every $\lambda,\mu \in \Lambda$, translates just to equation (1). Thus q_L is a quadratic form associated to e^H. According to Lemma 6.2 we get in this way 2^{2g} quadratic forms for e^H.

(7.3) Remark. It is easy to see that the quadratic form q_L coincides with the form e_*^L defined in Mumford [1] (see Exercise 13). For another definition of q_L see Exercise 12.

We need the following elementary lemma on quadratic forms in characteristic 2.

(7.4) Lemma. *Let U be a $\mathbb{Z}/2\mathbb{Z}$-vector space of dimension $2g$ and suppose that $e: U \times U \to \{\pm 1\}$ is a symmetric bilinear form of rank $2s$ with radical $K = \{u \in U \mid e(u, \cdot) \equiv 1\}$. Suppose $q: U \to \{\pm 1\}$ is a quadratic form associated to e.*

a) If $q|K$ is trivial, then either
i) $\#q^{-1}(1) = 2^{2g-s-1}(2^s+1)$ and $\#q^{-1}(-1) = 2^{2g-s-1}(2^s-1)$ or
ii) $\#q^{-1}(1) = 2^{2g-s-1}(2^s-1)$ and $\#q^{-1}(-1) = 2^{2g-s-1}(2^s+1)$.

b) If $q|K$ is nontrivial, then $\#q^{-1}(1) = \#q^{-1}(-1) = 2^{2g-1}$.

Proof. Step I: Suppose e is nondegenerate, i.e. $s = g$ and $K = \{0\}$: According to the elementary divisor theorem (see Bourbaki [1] Alg. IX.5.1 Th. 1) there is a basis $u_1, \dots, u_g, u_1', \dots, u_g'$ of U such that $e(u_i, u_j) = e(u_i', u_j') = 1$ and $e(u_i, u_j') = (-1)^{\delta_{ij}}$ for $1 \le i, j \le g$.

Suppose first that $g = 1$: Then

$$q(u_1)q(u_1')q(u_1 + u_1') = e(u_1, u_1') = -1 \ .$$

Hence $\#q^{-1}(1) = 3$ or 1 and $\#q^{-1}(-1) = 1$ or 3, since $q(0) = 1$ in any case.

Now suppose $g > 1$ and that the assertion holds for all $g' < g$. Define subvector spaces $U_{g-1} = \langle u_i, u_i' | i = 2, \dots, g\rangle$ and $U_1 = \langle u_1, u_1'\rangle$. The restrictions $q_{g-1} = q|U_{g-1}$ and $q_1 = q|U_1$ are quadratic forms with nondegenerate associated bilinear forms $e|U_{g-1} \times U_{g-1}$ and $e|U_1 \times U_1$. Any $v \in U$ decomposes uniquely as $v = v_{g-1} + v_1$ with $v_{g-1} \in U_{g-1}$ and $v_1 \in U_1$. Since $e(v_{g-1}, v_1) = 1$, we get $q(v) = q_{g-1}(v_{g-1})q_1(v_1)$. It follows that

$$\#q^{-1}(1) = \#q_{g-1}^{-1}(1)\#q_1^{-1}(1) + \#q_{g-1}^{-1}(-1)\#q_1^{-1}(-1) \ .$$

If we are in case i) for q_{g-1} and q_1, then

$$\#q^{-1}(1) = 2^{g-2}(2^{g-1}+1) \cdot 3 + 2^{g-2}(2^{g-1}-1) \cdot 1 = 2^{g-1}(2^g+1) \ .$$

Hence $\#q^{-1}(-1) = 2^{2g} - \#q^{-1}(1) = 2^{g-1}(2^g - 1)$ and we are again in case i). Similarly one checks the other possibilities to see that one ends up in case i) or in case ii).

Step II: Suppose e is trivial, i.e. $s = 0$: If q is trivial, then $\#q^{-1}(1) = 2^{2g}$ and we are in case i) of a). Otherwise q is a surjective homomorphism $U \to \{\pm 1\}$, which implies the assertion.

Step III: Suppose $0 < s < g$: Let W denote an orthogonal complement of K in U with respect to the bilinear form e. The restriction $q_W = q|W$ is a quadratic form as in Step I and $q_K = q|K$ is a quadratic form as in Step II. Every $u \in U$ admits a unique decomposition $u = w + k$ with $w \in W$ and $k \in K$ such that $q(u) = q_W(w) \cdot q_K(k)$. It follows that

$$\#q^{-1}(1) = \#q_W^{-1}(1)\#q_K^{-1}(1) + \#q_W^{-1}(-1)\#q_K^{-1}(-1) \ .$$

Inserting the results of Step I and Step II for q_W and q_K, we obtain the assertion. □

For any not necessarily effective symmetric divisor $D \neq 0$ on X, define

$$X_2^+(D) = \{x \in X_2 \mid \operatorname{mult}_x(D) \equiv 0 \pmod 2\} \quad \text{and}$$
$$X_2^-(D) = \{x \in X_2 \mid \operatorname{mult}_x(D) \equiv 1 \pmod 2\} .$$

Obviously X_2 is the disjoint union of $X_2^+(D)$ and $X_2^-(D)$. One can compute the cardinalities of these sets.

(7.5) Proposition. *Let D be a nontrivial symmetric divisor on X and $L = \mathcal{O}_X(D)$. Suppose $L = L(H, \chi)$ is of type $(d_1, \dots, d_g)$ with $d_1, \dots, d_s$ odd and $d_{s+1}, \dots, d_g$ even.*

a) If $\chi | 2\Lambda(L) \cap \Lambda$ is trivial, then either
  i) $\#X_2^+(D) = 2^{2g-s-1}(2^s + 1)$ and $\#X_2^-(D) = 2^{2g-s-1}(2^s - 1)$, or
  ii) $\#X_2^+(D) = 2^{2g-s-1}(2^s - 1)$ and $\#X_2^-(D) = 2^{2g-s-1}(2^s + 1)$.

b) If $\chi | 2\Lambda(L) \cap \Lambda$ is nontrivial, then $\#X_2^+(D) = \#X_2^-(D) = 2^{2g-1}$.

Proof. According to Propsition 7.2 we have for the quadratic form q_L defined in (2)

$$q_L(x) = (-1)^{\operatorname{mult}_x(D) - \operatorname{mult}_0(D)}$$

for all $x \in X_2$ and hence $q_L^{-1}(1) = X_2^+(D)$, if D is even, and $q_L^{-1}(1) = X_2^-(D)$, if D is odd. It is easy to see that the rank of the bilinear form e^H is s. So in order to apply Lemma 7.4, it remains to show that q_L is trivial on the radical K of e^H if and only if $\chi | 2\Lambda(L) \cap \Lambda$ is trivial. Let $\pi \colon V \to X$ denote the canonical projection, then by definition of e^H

$$\pi^{-1}(K) = \{v \in \tfrac{1}{2}\Lambda \mid \operatorname{Im} H(v, \Lambda) \subseteq \mathbb{Z}\} = \Lambda(L) \cap \tfrac{1}{2}\Lambda .$$

Since $q_L(\bar{v}) = \chi(2v)$, the form q_L is trivial on K if and only if χ is trivial on $2(\Lambda(L) \cap \frac{1}{2}\Lambda) = 2\Lambda(L) \cap \Lambda$. □

In case a) of the Proposition there is still an ambiguity. In order to decide, which of the possibilities i) or ii) holds, one has to consider the characteristic of the line bundle and the parity of the multiplicity of the divisor D in 0. A general formula would be messy. We give here a precise statement only in the case, which we need later. Let L be a nondegenerate symmetric line bundle of type $(d_1, \dots, d_g)$. As always let $L_0 = L(H, \chi_0) \in \operatorname{Pic}^H(X)$ denote the line bundle of characteristic 0 with respect to some decomposition for H.

(7.6) Corollary. *Suppose d_1 is even and D is a symmetric divisor on X with $\mathcal{O}_X(D) = L$.*

a) If $L = L_0$, then $\#X_2^+(D) = \begin{cases} 2^{2g} & \text{if } D \text{ is even} \\ 0 & \text{if } D \text{ is odd} \end{cases}$.

b) If $L \neq L_0$, then $\#X_2^+(D) = \#X_2^-(D) = 2^{2g-1}$.

In particular, this includes the case of a Kummer polarization $(2, \dots, 2)$, which will be studied in the next section.

Proof. With the notation of Proposition 7.5 we have $s = 0$ and $2\Lambda(L) \cap \Lambda = \Lambda$. By assumption the alternating form $\operatorname{Im} H$ on Λ is divisible by 2, such that $\chi_0(\lambda) = e(\pi i \operatorname{Im} H(\lambda_1, \lambda_2)) = 1$ for all $\lambda = \lambda_1 + \lambda_2 \in \Lambda$. Hence in case $L = L_0$ Proposition 7.5 a) gives $\#X_2^+(D) = 2^{2g}$ or 0. Since by definition $0 \in X_2^+(D)$ for an even divisor D, this proves a). As for b): suppose $L = t_{\bar{c}}^* L_0$ with $c \in \frac{1}{2}\Lambda(L) - \Lambda(L)$. One immediately sees that the semicharacter $\chi_0 e(\pi i \operatorname{Im} H(2c, \cdot))$ of L is nontrivial on Λ. So Proposition 7.5 b) gives the assertion. □

A formula for the cardinality of $X_2^+(D)$ in the case that d_g is odd is given in Exercise 14.

(7.7) Remark. Let L be a symmetric line bundle on X of type $(d_1, \dots, d_g)$ with $d_1, \dots, d_s$ odd and $d_{s+1}, \dots, d_g$ even. Denote by X_2^+ (respectively X_2^-) the set of 2-division points $x \in X_2$ such that the normalized isomorphism $(-1)_L$ acts on the fibre $L(x)$ by multiplication with $+1$ (respectively -1). Using Lemma 7.3 and Exercise 12 one can show

a) if $q_L | K(L) \cap X_2$ is trivial, then either

i) $\#X_2^+ = 2^{2g-s-1}(2^s + 1)$ *and* $\#X_2^- = 2^{2g-s-1}(2^s - 1)$, *or*

ii) $\#X_2^+ = 2^{2g-s-1}(2^s - 1)$ *and* $\#X_2^- = 2^{2g-s-1}(2^s + 1)$.

b) if $q_L | K(L) \cap X_2$ is nontrivial, then $\#X_2^+ = \#X_2^- = 2^{2g-1}$.

§ 8 Kummer Varieties

As we saw in Section 5, for the investigation of the map φ_{L^2} it suffices to consider the cases that L has no fixed components, which we treated in Theorem 5.5, and that L is an irreducible principal polarization. In this section we study the latter case.

Let $X = V/\Lambda$ be an abelian variety of dimension g. The *Kummer variety* associated to X is defined to be the quotient

$$K_X = X/\langle(-1)_X\rangle \,.$$

From Appendix A we deduce that K_X is an algebraic variety of dimension g over $\mathbb{C}$, smooth apart from 2^{2g} singular points of multiplicity 2^{g-1}, the images of the 2-division points of X under the natural map $p: X \to K_X$.

Let $L = L(H, \chi)$ be an ample symmetric line bundle on X defining an irreducible principal polarization. Since $\chi(\Lambda) \subseteq \{\pm 1\}$ by Proposition 2.3.7, the semicharacter χ^2 of L^2 is identically 1 on Λ. This implies that L^2 is of characteristic 0 with respect to any decomposition for L. According to Corollary 6.6 all theta functions in $H^0(L^2)$ are even. Hence there is a map $\psi = \psi_{L^2}: K_X \to \mathbb{P}_{2^g-1}$ such that the following diagram commutes

$$\begin{array}{ccc} X & \xrightarrow{\varphi=\varphi_{L^2}} & \mathbb{P}_{2^g-1} \\ & {\scriptstyle p}\searrow \quad \nearrow{\scriptstyle \psi} & \\ & K_X & \end{array}$$

(8.1) Theorem. *If $L \in \operatorname{Pic}(X)$ is symmetric and defines an irreducible principal polarization on X, then $\psi: K_X \to \mathbb{P}_{2^g-1}$ is an embedding.*

Proof. Denote by Θ the unique (necessarily symmetric) divisor in the linear system $|L|$.

Step I: ψ is injective: Suppose $x, y \in X$ with $x \neq \pm y$. We have to show that there is a divisor $D \in |L^2|$ with $x \in D$ and $y \notin D$.

Since L is a principal polarization, $\phi_L: X \to \widehat{X}$ is an isomorphism and consequently $\Theta \neq t^*_{y-x}\Theta$ and $\Theta \neq t^*_{-(x+y)}\Theta$. Hence there is an element $z \in \Theta$ with $z \notin t^*_{y-x}\Theta \cup t^*_{-(x+y)}\Theta$. Consider the divisor

$$D = t^*_{z-x}\Theta + t^*_{x-z}\Theta \in |L^2| \ .$$

It satisfies $x = z + (x - z) \in t^*_{z-x}\Theta \subset D$. Moreover $y \notin t^*_{z-x}\Theta$, since otherwise $z \in t^*_{y-x}\Theta$, and $y \notin t^*_{x-z}\Theta$, since otherwise $z \in t^*_{-(x+y)}(-1)^*\Theta = t^*_{-(x+y)}\Theta$. Hence $y \notin D$.

Step II: The differential $d\psi_q$ is injective for all smooth points $q \in K_X$. Suppose $x \in X$, $2x \neq 0$, and $t \neq 0$ is a tangent vector to X at x. We have to show that there is a divisor $D \in |L^2|$ containing x such that t is not tangent to D at x.

According to Proposition 4.1 the image of the Gauß map for Θ is not contained in a hyperplane. This implies that there is a point $y \in \Theta$ with $y \notin t^*_{-2x}\Theta$ such that t is not tangent to Θ at y. The divisor

$$D = t^*_{y-x}\Theta + t^*_{x-y}\Theta \in |L^2|$$

contains x, since $x = y + (x - y) \in t^*_{y-x}\Theta \subset D$. By choice of y the vector t is not tangent to $t^*_{y-x}\Theta$ at x. Moreover $x \notin t^*_{x-y}\Theta$, since otherwise $y \in t^*_{-2x}(-1)^*\Theta = t^*_{-2x}\Theta$. Thus t is not tangent to D at x.

Step III: The differential $d\psi_q$ is injective for a singular point $q \in K_X$. Without loss of generality we may assume that q is the image of $0 \in X$, i.e. $q = p(0)$. Since we identified $T_{X,0}$ with V and q is an ordinary double point of K_X, the tangent space of K_X at q can be identified with the symmetric product S^2V. If $v_1, \dots, v_g$ denote coordinate functions of V, $\{\frac{\partial}{\partial v_1}, \dots, \frac{\partial}{\partial v_g}\}$ is a basis of $T_{X,0} = V$ and $\{\frac{\partial^2}{\partial v_\nu \partial v_\mu} \mid 1 \le \nu \le \mu \le g\}$ is a basis of $T_{K_X,q} = S^2V$.

According to Grothendieck [1] n° 221 Corollaire 5.3 the tangent space of $\mathbb{P}_{2^g-1} = P(H^0(L^2)^*)$ at a point P (considered as a hyperplane in $H^0(L^2)$) is $T_{\mathbb{P}_{2^g-1},P} = \mathrm{Hom}(P, H^0(L^2)/P)$. We have to show that the natural map

$$d\psi_q : S^2V \to \mathrm{Hom}\big(\varphi(0), H^0(L^2)/\varphi(0)\big)$$

is injective. It is defined as follows: choose an isomorphism $H^0(L^2)/\varphi(0) \simeq \mathbb{C}$ such that $\mathrm{Hom}(\varphi(0), H^0(L^2)/\varphi(0)) = \mathrm{Hom}(\varphi(0), \mathbb{C})$. By definition $\varphi(0) = \{\vartheta \in H^0(L^2) \mid \vartheta(0) = 0\}$ and we have

$$d\psi_q\Big(\sum_{\nu \le \mu} \alpha_{\nu\mu} \frac{\partial^2}{\partial v_\nu \partial v_\mu}\Big) = \begin{cases} \varphi(0) & \longrightarrow \quad \mathbb{C} \\ \vartheta & \longmapsto \quad \sum\limits_{\nu\le\mu} \alpha_{\nu\mu} \frac{\partial^2 \vartheta}{\partial v_\nu \partial v_\mu}(0) \end{cases} .$$

Given $0 \ne \sum_{\nu\le\mu} \alpha_{\nu\mu} \frac{\partial^2}{\partial v_\nu \partial v_\mu} \in S^2V$, we have to show that there is a theta function $\vartheta \in \varphi(0)$ with $\sum \alpha_{\nu\mu} \frac{\partial^2 \vartheta}{\partial v_\nu \partial v_\mu}(0) \ne 0$. For this denote by Q the quadric in $P(V^*)$ defined by $\sum \alpha_{\nu\mu} \frac{\partial}{\partial v_\nu} \frac{\partial}{\partial v_\mu} = 0$. Since the Gauss map $G: \Theta_s \to P(V^*)$ is dominant by Proposition 4.2, there is an element $\overline{y} \in \Theta_s$ such that $G(\overline{y}) \notin Q$. In other words

$$\sum_{\nu\le\mu} \alpha_{\nu\mu} \frac{\partial\theta}{\partial v_\nu}(y) \frac{\partial\theta}{\partial v_\mu}(y) \ne 0 ,$$

where θ denotes a theta function associated to Θ. Now consider the theta function

$$\vartheta = \theta(\cdot + y)\,\theta(\cdot - y) \in H^0(L^2) .$$

It is an element of $\varphi(0)$, since $\vartheta(0) = \theta(y)\,\theta(-y) = 0$. Moreover we have

$$\begin{aligned} \frac{\partial^2\vartheta}{\partial v_\nu \partial v_\mu}(0) &= \frac{\partial\theta}{\partial v_\nu}(y)\frac{\partial\theta}{\partial v_\mu}(-y) + \frac{\partial\theta}{\partial v_\nu}(-y)\frac{\partial\theta}{\partial v_\mu}(y) \\ &= \begin{cases} -2\frac{\partial\theta}{\partial v_\nu}(y)\frac{\partial\theta}{\partial v_\mu}(y) & \text{if } \theta \text{ is even} \\ +2\frac{\partial\theta}{\partial v_\nu}(y)\frac{\partial\theta}{\partial v_\mu}(y) & \phantom{\text{if }} \theta \text{ is odd} \end{cases} . \end{aligned}$$

This implies that $\sum_{\nu\le\mu} \alpha_{\nu\mu} \frac{\partial^2\vartheta}{\partial v_\nu \partial v_\mu}(0) = \pm 2 \sum_{\nu\le\mu} \alpha_{\nu\mu} \frac{\partial\theta}{\partial v_\nu}(y) \frac{\partial\theta}{\partial v_\mu}(y) \ne 0$. This completes the proof of the theorem. □

Let L be again an arbitrary ample symmetric line bundle on X. Combining the Decomposition Theorem 3.1, Theorem 5.5 and the previous theorem we obtain the following result on the map φ_{L^2}: Let

$$(X, L) = (X_1, L_1) \times \cdots \times (X_r, L_r)$$

denote the decomposition of the polarized abelian variety (X, L) into a product of irreducible polarized abelian varieties. To be precise, we assume that L is isomorphic to the exterior tensor product of the line bundles L_ν, $\nu = 1, \ldots, r$. In particular all L_ν's are symmetric. Suppose that (X_ν, L_ν) is principally polarized for $\nu = 1, \ldots, s$ and is not principally polarized for $\nu = s+1, \ldots, r$. For $\nu = 1, \ldots, s$ let $p_\nu: X_\nu \to K_{X_\nu}$ be the natural projection onto the Kummer variety K_{X_ν} associated to X_ν. Define

$$K = K_{X_1} \times \cdots \times K_{X_s} \times X_{s+1} \times \cdots \times X_r$$

and $p = p_1 \times \cdots \times p_s \times \mathrm{id}_{X_{s+1}} \times \cdots \times \mathrm{id}_{X_r}: X \to K$. Then $\varphi = \varphi_{L^2}$ factors as

$$\begin{array}{ccc} X & \xrightarrow{\varphi} & \mathbb{P}_n \\ & {\scriptstyle p}\searrow \quad \nearrow{\scriptstyle \psi} & \\ & K & \end{array}$$

with a holomorphic mapping ψ. Thus we obtain as a consequence of Theorems 5.5 and 8.1:

(8.2) Theorem. *ψ is an embedding.*

We observe that φ is of degree 2^s onto its image. In particular, if none of the components (X_ν, L_ν) is principally polarized, φ is an embedding. Using Lemma 6.1, Theorem 8.2 generalizes easily to an arbitrary (not necessarily symmetric) ample line bundle L. One has only to modify the projection map $p: X \to K$ slightly.

§ 9 Morphisms into Abelian Varieties

In this section we compile some properties of morphisms of algebraic varieties into abelian varieties. First we need the

(9.1) Rigidity Lemma. *Let $f: Y \times Z \to X$ be a morphism of algebraic varieties. Suppose Y is complete. If*

$$f\big(\{y_0\} \times Z\big) = x_0 = f\big(Y \times \{z_0\}\big)$$

for some $y_0 \in Y$, $z_0 \in Z$ and $x_0 \in X$, then f is constant.

Proof. Let $U \subset X$ be an open affine neighbourhood of x_0 and $q: Y \times Z \to Z$ the natural projection. The variety Y being complete implies that the set $A := qf^{-1}(X - U)$ is closed in Z. Note that a point z is contained in $Z - A$ if and only if $f(Y \times \{z\}) \subset U$. In particular $z_0 \in Z - A$. Hence $Z - A$ is open and dense in Z. Using the fact that any morphism of a complete variety into an affine variety is constant, we conclude that the image of $Y \times \{z\}$ under f is a point whenever $z \in Z - A$. So $f\big(Y \times \{z\}\big) = f\big(\{y_0\} \times \{z\}\big) = x_0$ for all $z \in Z - A$. This implies the assertion since $Y \times (Z - A)$ is open and dense in $Y \times Z$. □

(9.2) Corollary. *Let Y and Z be algebraic varieties, one of them complete, and X an abelian variety. If $f: Y \times Z \to X$ is a morphism with $f(Y \times \{z_0\}) = 0$ for some $z_0 \in Z$, then there is a uniquely determined morphism $g: Z \to X$ such that $f(y, z) = g(z)$ for all $(y, z) \in Y \times Z$.*

Proof. Choose $y_0 \in Y$ and define $f' : Y \times Z \to X$, $f'(y, z) = f(y, z) - f(y_0, z)$. Then $f'(\{y_0\} \times Z) = 0 = f'(Y \times \{z_0\})$ and by the Rigidity Lemma $f' \equiv 0$. Thus $g(z) := f(y_0, z)$ satisfies the assertion. □

(9.3) Corollary. *Let Y and Z be algebraic varieties, one of them complete, and let X be an abelian variety. Suppose $f: Y \times Z \to X$ is a morphism with $f(y_0, z_0) = 0$ for some $y_0 \in Y$, $z_0 \in Z$. Then there are uniquely determined morphisms $g: Y \to X$ and $h: Z \to X$ with $g(y_0) = 0 = h(z_0)$ such that for all $(y, z) \in Y \times Z$*

$$f(y, z) = g(y) + h(z) \ .$$

Proof. Define $f': Y \times Z \to X$ by $f'(y, z) = f(y, z) - f(y, z_0) - f(y_0, z)$. Since $f'(\{y_0\} \times Z) = 0 = f'(Y \times \{z_0\})$, the Rigidity Lemma gives $f' \equiv 0$. So $g(y) := f(y, z_0)$ and $h(z) := f(y_0, z)$ satisfy the assertion. The uniqueness of g and h follows by fixing $y = y_0$ and $z = z_0$ respectively in the equation $f(y, z) = g(y) + h(z)$. □

Recall that a *rational map* $f: Y \dashrightarrow X$ of complex algebraic varieties is an equivalence class of pairs (U, f_U) with an open dense subset U of Y and a morphism $f_U: U \to X$. Such pairs (U, f_U) and (V, f_V) are equivalent if $f_U | U \cap V = f_V | U \cap V$. The map f is called to be *defined in a point* $y \in Y$, if there is a pair (U, f_U) as above with $y \in U$.

Let $\mathbb{C}(Y)$ denote as usual the field of rational functions on Y. A rational map $f: Y \dashrightarrow X$ induces a homomorphism of local rings $\mathcal{O}_{X,x} \to \mathbb{C}(Y)$ for every $x \in X$. It is easy to see that f is defined at a point $y \in Y$ if and only if there is a $x \in X$ such that the induced map $\mathcal{O}_{X,x} \to \mathbb{C}(Y)$ factorizes via $\mathcal{O}_{Y,y}$. Note that then necessarily $x = f(y)$.

(9.4) Theorem. *Any rational map $f: Y \dashrightarrow X$ from a normal variety Y to an abelian variety X is defined on the whole of Y.*

Proof. First recall that f is not defined at most in a subvariety of codimension ≥ 2, since Y is normal and X is complete (see Grothendieck-Dieudonné [1] Corollaire 8.2.12). Assume f is not defined at one point. It suffices to show that there is a subvariety D of codimension 1 in Y such that f is not defined on the whole of D.

Consider the rational map $F: Y \times Y \dashrightarrow X$,

$$F(y_1, y_2) = f(y_1) - f(y_2)$$

whenever f is defined in y_1 and y_2.

We claim that F is defined in a point (y, y) if and only if f is defined in y. For the proof suppose F is defined in $(y_0, y_0) \in Y \times Y$ and let (U, F_U) be a representative of F with $(y_0, y_0) \in U$. There is a point $y_1 \in Y$, at which f is defined, such that y_0 is contained in the open set $V = \{y \in Y \mid (y_1, y) \in U\}$. Then $f(y) = f(y_1) - F(y_1, y)$ for all $y \in V$; in particular f is defined in y_0. The converse implication is obvious.

Suppose now f is not defined at a point $y \in Y$. Then F is not defined at (y, y). By what we have said above, this means that the homomorphism $\phi : \mathcal{O}_{X,0} \to \mathbb{C}(Y \times Y)$ induced by F satisfies

$$\operatorname{im} \phi \not\subset \mathcal{O}_{Y \times Y,(y,y)} \tag{1}$$

Since Y is normal, $\mathcal{O}_{Y \times Y,(y,y)}$ is the subring of all functions φ in $\mathbb{C}(Y \times Y)$ which are defined at (y, y). So by (1) there is a $\varphi \in \operatorname{im} \phi \subset \mathbb{C}(Y \times Y)$ such that (y, y) is contained in the polar divisor $(\varphi)_\infty$. Consider the intersection D of $(\varphi)_\infty$ with the diagonal Δ in $Y \times Y$. It is of codimension ≤ 1 in $\Delta \simeq Y$, since Δ is a complete intersection. Clearly F is not defined at every $(y', y') \in D$. Hence by what we have said above, f is not defined at all y' contained in a subvariety of codimension ≤ 1 in Y. This completes the proof. □

(9.5) Proposition. *Every rational map $f: \mathbb{P}_N \dashrightarrow X$ from the projective space to an abelian variety X is constant.*

In particular an abelian variety does not contain any rational curves.

Proof. According to Theorem 9.4 the map f is defined everywhere. Since any two points in $\mathbb{P}_N$ are joined by a line $\mathbb{P}_1$, it suffices to show that any morphism $f: \mathbb{P}_1 \to X$ is constant.

Let $X = V/\Lambda$ and $v_1, \ldots, v_g$ a basis of V. Then $dv_1, \ldots, dv_g$ is a basis of holomorphic differentials of X. Since $\mathbb{P}_1$ does not admit any global holomorphic differential, we have $f^* dv_\nu = 0$ for $\nu = 1, \ldots, g$. Thus f is constant. □

§ 10 The Pontryagin Product

Let $X = V/\Lambda$ be an abelian variety of dimension g. The addition on X induces a multiplication on the homology ring $H_\bullet(X,\mathbb{Z})$, the Pontryagin product. In this section we show that the Pontryagin product is dual to the cup product in $H^\bullet(X,\mathbb{Z})$. As a first application we derive a formula, which expresses the homology class of an ample divisor by a symplectic basis of $\Lambda = H_1(X,\mathbb{Z})$ in terms of the Pontryagin product. The results of this section generalize immediately to any nondegenerate line bundle on a complex torus.

Let $\times\colon H_p(X,\mathbb{Z}) \times H_q(X,\mathbb{Z}) \to H_{p+q}(X\times X,\mathbb{Z})$ denote the exterior homology product. The addition map $\mu\colon X\times X \to X$ induces a homomorphism $\mu_*\colon H_{p+q}(X\times X,\mathbb{Z}) \to H_{p+q}(X,\mathbb{Z})$. The *Pontryagin product* on X is defined to be the composition

$$\star\colon H_p(X,\mathbb{Z}) \times H_q(X,\mathbb{Z}) \xrightarrow{\times} H_{p+q}(X\times X,\mathbb{Z}) \xrightarrow{\mu_*} H_{p+q}(X,\mathbb{Z}) \ .$$

For the definition of the Pontryagin product in terms of cycles see Exercise 1.7. Recall that the exterior homology product is anti-commutative, i.e. $\sigma\times\tau = (-1)^{pq}\tau\times\sigma$ for all $\sigma\in H_p(X,\mathbb{Z})$ and $\tau\in H_q(X,\mathbb{Z})$ (see Greenberg-Harper [1] Corollary 29.29). This implies that also the Pontryagin product is anti-commutative:

$$\sigma\star\tau = (-1)^{p+q}\tau\star\sigma \ .$$

Let L be an ample line bundle on X. Fix a symplectic basis $\lambda_1,\dots,\lambda_{2g}$ of $\Lambda = H_1(X,\mathbb{Z})$ for L and denote by $x_1,\dots,x_{2g}$ the corresponding real coordinate functions on V. As we saw in Section 1.3, the differentials $dx_1,\dots,dx_{2g}$ define a basis of $H^1(X,\mathbb{Z})$, dual to $\lambda_1,\dots,\lambda_{2g}$, i.e. $\int_{\lambda_i} dx_j = \delta_{ij}$. In this and the following section a multi-index always is an ordered multi-index in $\{1,\dots,2g\}$. For $I = (i_1 < \cdots < i_p)$ we write $\lambda_I = \lambda_{i_1}\star\cdots\star\lambda_{i_p}$ and $dx_I = dx_{i_1}\wedge\cdots\wedge dx_{i_p}$. According to Proposition 1.3.5 the set $\{dx_I \mid \#I = p\}$ is a basis of $H^p(X,\mathbb{Z})$. This will be used to prove the following lemma.

(10.1) Lemma. *The set $\{\lambda_I \mid \#I = p\}$ is a basis of $H_p(X,\mathbb{Z})$, dual to the basis $\{dx_I \mid \#I = p\}$ of $H^p(X,\mathbb{Z})$ with respect to the natural isomorphism $H_p(X,\mathbb{Z}) \to H^p(X,\mathbb{Z})^*$, $\sigma\mapsto\int_\sigma$.*

Proof. We have to show that $\int_{\lambda_I} dx_J = \delta_{IJ}$ for all multi-indices I and J of length p. We will do this by induction on p. For $p=1$ this is clear. Assume the assertion is proved for all multi-indices of length p and let $I = (i_1 < \cdots < i_p < i_{p+1})$ and $J = (j_1 < \cdots < j_{p+1})$. Denote by I_p the multi-index $(i_1 < \cdots < i_p)$. If $p_i\colon X\times X \to X$ is the natural projection onto the i-th factor, then $p_1 + p_2 = \mu$ is the addition map on X and

$$\int\limits_{\lambda_I} dx_J = \int\limits_{\mu(\lambda_{I_p}\times\lambda_{i_{p+1}})} dx_J = \int\limits_{\lambda_{I_p}\times\lambda_{i_{p+1}}} \mu^* dx_J$$

$$= \int\limits_{\lambda_{I_p}\times\lambda_{i_{p+1}}} \bigwedge_{\nu=1}^{p+1} (p_1^* dx_{j_\nu} + p_2^* dx_{j_\nu})$$

$$= \int\limits_{\lambda_{I_p}\times\lambda_{i_{p+1}}} \sum_{\nu=1}^{p+1} (-1)^{p+1-\nu} p_1^* dx_{J-j_\nu} \wedge p_2^* dx_{j_\nu} \quad \text{(the other summands}$$

vanish by Fubini's Theorem, since $\lambda_{i_{p+1}}$ is a 1-cycle on X)

$$= \sum_{\nu=1}^{p+1} (-1)^{p+1-\nu} \int\limits_{\lambda_{I_p}} dx_{J-j_\nu} \cdot \int\limits_{\lambda_{i_{p+1}}} dx_{j_\nu} = \delta_{IJ}\ .$$

For the last equation we used the fact that $\delta_{I_p, J-j_\nu} \cdot \delta_{i_{p+1}, j_\nu} = 0$ unless $\nu = p+1$, since the multi-indices are ordered. □

According to the lemma, the map $D\colon H_p(X,\mathbb{Z}) \to H^p(X,\mathbb{Z})$, defined by $\lambda_I \mapsto dx_I$, is an isomorphism for every p. Note that this isomorphism depends not only on the line bundle L, but also on the choice of the symplectic basis. However, it shows that the Pontryagin product in homology is dual to the cup product in cohomology. To be more precise, the following diagram is commutative, as one immediately checks from the definition of the maps

$$\begin{array}{ccc} H_p(X,\mathbb{Z}) \times H_q(X,\mathbb{Z}) & \xrightarrow{\star} & H_{p+q}(X,\mathbb{Z}) \\ {\scriptstyle D\times D}\downarrow & & \downarrow{\scriptstyle D} \\ H^p(X,\mathbb{Z}) \times H^q(X,\mathbb{Z}) & \xrightarrow{\wedge} & H^{p+q}(X,\mathbb{Z}) \end{array} .$$

Let $P\colon H_p(X,\mathbb{Z}) \to H^{2g-p}(X,\mathbb{Z})$ denote the isomorphism induced by Poincaré Duality. We want to compute P explicitly in terms of the bases $\{\lambda_I\}$ and $\{dx_I\}$. If $I = (i_1 < \cdots < i_p)$ is a multi-index in $\{1,\ldots,2g\}$, denote by $I^o = (i_1^o < \cdots < i_{2g-p}^o)$ the multi-index, for which as a set $I \cup I^o = \{1,\ldots,2g\}$, and define the sign $\varepsilon(I)$ of I by

$$\varepsilon(I)\, dx_I \wedge dx_{I^o} = dx_1 \wedge dx_{g+1} \wedge \cdots \wedge dx_g \wedge dx_{2g}\ .$$

(10.2) Lemma. $P(\lambda_I) = (-1)^{g+p}\varepsilon(I)\, dx_{I^o}$ *for every multi-index* I *of length* p.

Proof. By definition of Poincaré Duality we have $\int_X P(\lambda_I) \wedge \varphi = \int_{\lambda_I} \varphi$ for any p-form φ on X. Since $\int_{\lambda_I} dx_J = \delta_{IJ}$ and the differentials dx_J with $\#J = p$ form a basis of $H^p(X,\mathbb{Z})$, the Poincaré dual of λ_I is necessarily of the form $P(\lambda_I) = c\,dx_{I^o}$ for some $c \in \mathbb{C}^*$. It remains to compute the constant c. According to Lemma 3.6.5 we have $\int_X dx_1 \wedge dx_{g+1} \wedge \cdots \wedge dx_g \wedge dx_{2g} = (-1)^g$. So

$$1 = \int_X P(\lambda_I) \wedge dx_I = c \int_X dx_{I^o} \wedge dx_I$$
$$= c\,(-1)^p \varepsilon(I) \int_X dx_1 \wedge dx_{g+1} \wedge \cdots \wedge dx_g \wedge dx_{2g} = c\,(-1)^{g+p}\varepsilon(I) \ ,$$

which implies the assertion. □

Recall that $H_\bullet(X,\mathbb{Z})$ is a ring with respect to the intersection product (see Griffiths-Harris [1]). For $\sigma \in H_p(X,\mathbb{Z})$ and $\tau \in H_q(X,\mathbb{Z})$ we denote by $\sigma \cdot \tau$ their intersection product in $H_{p+q-2g}(X,\mathbb{Z})$. If in particular $\tau \in H_{2g-p}(X,\mathbb{Z})$, then $\sigma \cdot \tau$ is an element of $H_0(X,\mathbb{Z})$, which is canonically isomorphic to $\mathbb{Z}$. The *intersection number* $(\sigma \cdot \tau)$ is by definition the image of $\sigma \cdot \tau$ in $\mathbb{Z}$. If V is a p-cycle on X, we denote by $\{V\}$ its homology class in $H_p(X,\mathbb{Z})$. Note that we consider the cycles λ_i already as homology classes via the natural identification $\Lambda = H_1(X,\mathbb{Z})$.

Suppose D is a divisor in the linear system $|L|$. If L is of type $(d_1,\ldots,d_g)$, then we have

(10.3) Lemma. $(\lambda_i \star \lambda_j \cdot \{D\}) = -d_i\delta_{g+i,j}$ *for all* $i \leq j$.

Proof. We may assume $i < j$, since $\lambda_i \star \lambda_i = 0$. According to Griffiths-Harris [1] p. 141 the Poincaré dual of $\{D\}$ is the first Chern class $c_1(L)$ of L. Recall from Lemma 3.6.4 that $c_1(L) = -\sum_{\nu=1}^g d_\nu dx_\nu \wedge dx_{g+\nu}$. Since the intersection product in homology is Poincaré dual to the cup product in cohomology, this implies using Lemma 10.1

$$(\lambda_i \star \lambda_j \cdot \{D\}) = \int_X P(\lambda_i \star \lambda_j) \wedge c_1(L) = \int_{\lambda_i \star \lambda_j} c_1(L)$$
$$= -\sum_{\nu=1}^{g} d_\nu \int_{\lambda_i\star\lambda_j} dx_\nu \wedge dx_{g+\nu} = -d_i\delta_{i+g,j} \ . \qquad \square$$

Preserving the notation of above, we can state the main result of this section.

(10.4) Theorem. *For all* $0 \leq p \leq g$

$$\{D\}^{g-p} = (-1)^p (g-p)! \sum_S \Big(\prod_{\nu \notin S} d_\nu\Big) \lambda_{s_1} \star \lambda_{g+s_1} \star \cdots \star \lambda_{s_p} \star \lambda_{g+s_p} \ .$$

Here the sum is to be taken over all subsets $S = \{s_1,\ldots,s_p\}$ *of* $\{1,\ldots,g\}$.

Proof. Since the intersection product in homology is Poincaré dual to the cup product in cohomology, we have by Lemma 3.6.4

$$\{D\}^{g-p} = P^{-1}\Big(\bigwedge^{g-p} c_1(L)\Big) = (-1)^{g-p} P^{-1}\Big(\bigwedge^{g-p} \sum_{\nu=1}^{g} d_\nu dx_\nu \wedge dx_{g+\nu}\Big)$$

for all $0 \le p \le g$. Now the assertion follows by an easy computation using Lemma 10.2. □

The following corollary lists the most important cases.

(10.5) Corollary. *a)* $\{D\}^0 = \{X\} = (-1)^g \lambda_1 \star \lambda_{g+1} \star \cdots \star \lambda_g \star \lambda_{2g}$

b) $\{D\} = (-1)^{g-1} \sum_{\nu=1}^{g} d_\nu \, \lambda_1 \star \lambda_{g+1} \star \cdots \star \check{\lambda}_\nu \star \check{\lambda}_{g+\nu} \star \cdots \star \lambda_g \star \lambda_{2g}$

c) $\{D\}^{g-1} = -(g-1)! \sum\limits_{\nu=1}^{g} d_1 \cdot \ldots \cdot \check{d}_\nu \cdot \ldots \cdot d_g \, \lambda_\nu \star \lambda_{g+\nu}$

d) $(\{D\}^g) = d_1 \cdot \ldots \cdot d_g \, g! = (L^g)$.

The last equation uses the Riemann-Roch Theorem 3.6.3. □

§ 11 Homological Versus Numerical Equivalence

Let L be an ample line bundle on an abelian variety $X = V/\Lambda$. In this section we prove a theorem, due to Liebermann [1], saying that numerical and homological equivalence on an abelian variety coincide. It will be applied in Chapters 11 and 12. The crucial point of the proof is to express the operator $\mathbf{\Lambda}$, associated to the Kähler metric on X given by L, in terms of the Pontryagin product.

First recall some facts about algebraic cycles: an *algebraic cycle* V *on* X *with coefficients in* $\mathbb{Q}$ is by definition a finite formal sum

$$\mathsf{V} = \sum r_i \mathsf{V}_i$$

with rational numbers r_i and algebraic subvarieties V_i of X, which we assume to be all of the same dimension. If $\dim \mathsf{V}_i = p$, then V is also called an *algebraic p-cycle.* Any algebraic p-cycle V defines a homology class $\{\mathsf{V}\}$ in $H_{2p}(X, \mathbb{Q})$ in a natural way. This gives a $\mathbb{Q}$-vector space homomorphism from the vector space of algebraic p-cycles into the homology group $H_{2p}(X, \mathbb{Q})$. Two algebraic p-cycles V_1 and V_2 are called *homologically equivalent*, if $\{\mathsf{V}_1\} = \{\mathsf{V}_2\}$. They are called *numerically equivalent*, if the intersection numbers satisfy $(\{\mathsf{V}_1\} \cdot \{\mathsf{W}\}) = (\{\mathsf{V}_2\} \cdot \{\mathsf{W}\})$ for any algebraic $(g-p)$-cycle W on X. Obviously homological equivalence implies numerical equivalence. The aim of this section is to prove that the converse also holds.

(11.1) Theorem. *Two algebraic cycles on an abelian variety are homologically equivalent if and only if they are numerically equivalent.*

For the proof we need some definitions from Kähler theory as well as the Lefschetz decomposition. Since it is valid for arbitrary compact Kähler manifolds, we omit the proof here, but refer to Griffiths-Harris [1] and Wells [1].

Consider the first Chern class $c_1(L)$ as a translation invariant $(1,1)$-form on X. Since L is ample and translation invariant forms are closed, $c_1(L)$ defines a Kähler metric ds^2 on X (see Section 1.4). Choose a basis of V with corresponding complex coordinate functions $v_1, \ldots, v_g$ in such a way that $c_1(L) = \frac{i}{2}\sum_{\nu=1}^{g} dv_\nu \wedge d\bar{v}_\nu$. Let x_ν and $x_{g+\nu}$ denote the real and imaginary part of v_ν for $\nu = 1, \ldots, g$. With respect to these coordinates $c_1(L)$ is given as

$$c_1(L) = \sum_{\nu=1}^{g} dx_\nu \wedge dx_{g+\nu} ,$$

and the corresponding volume element is

$$dv = \frac{1}{g!}\bigwedge^{g} c_1(L) = dx_1 \wedge dx_{g+1} \wedge \cdots \wedge dx_g \wedge dx_{2g} .$$

By definition of the self-intersection number of a line bundle and the Riemann-Roch Theorem 3.6.3 we have

$$\int_X dv = \frac{1}{g!}\int_X \bigwedge^{g} c_1(L) = \frac{1}{g!}(L^g) = \chi(L) .$$

Recall the $*$-operator on $H^{\bullet}(X,\mathbb{C})$ induced by the Kähler structure. Suppose $I = (i_1 < \cdots < i_p)$ is a multi-index in $\{1,\ldots,2g\}$. Then $*\, dx_I$ is defined to be the uniquely determined translation invariant $(2g-p)$-form on X such that

$$dx_I \wedge *\, dx_I = dv .$$

Since the invariant differentials dx_I with $\#I = p$ form a basis of $H^p(X,\mathbb{C})$, $\mathbb{C}$-linear extension gives a $\mathbb{C}$-linear operator $*\colon H^p(X,\mathbb{C}) \to H^{2g-p}(X,\mathbb{C})$. As in the last section define the complementary index $I^o = (i_1^o < \cdots < i_{2g-p}^o)$ and the sign $\varepsilon(I)$ by $\varepsilon(I)\, dx_I \wedge dx_{I^o} = dv$. An immediate computation gives

$$*\, dx_I = \varepsilon(I)\, dx_{I^o} \quad \text{and} \quad **\, dx_I = (-1)^p dx_I \tag{1}$$

for any multi-index $I = (i_1 < \cdots < i_p)$ in $\{1,\ldots,2g\}$. Define as usual linear operators

$$\mathbf{L}\colon H^p(X,\mathbb{C}) \to H^{p+2}(X,\mathbb{C}),\ \varphi \mapsto c_1(L) \wedge \varphi, \ \text{ and}$$

$$\mathbf{\Lambda}\colon H^p(X,\mathbb{C}) \to H^{p-2}(X,\mathbb{C}),\ \mathbf{\Lambda} = *^{-1}\,\mathbf{L}\, * .$$

Moreover, consider the Hodge decomposition $H^p(X,\mathbb{C}) = \bigoplus_{\mu+\nu=p} H^{\mu,\nu}(X)$ (see Section 1.4), and let $p_{\mu,\nu}\colon H^p(X,\mathbb{C}) \to H^{\mu,\nu}(X)$ denote the projection map. Then a third linear operator on $H^p(X,\mathbb{C})$ is defined by

$$J\colon H^p(X,\mathbb{C}) \to H^p(X,\mathbb{C})\,,\; J = \sum_{\mu+\nu=p} i^{\mu-\nu} p_{\mu,\nu}\,.$$

An element $\varphi \in H^p(X,\mathbb{C})$ is called *primitive*, if $\mathbf{\Lambda}\varphi = 0$.

(11.2) Lefschetz Decomposition. *a) Any $\varphi \in H^p(X,\mathbb{C})$ can be written uniquely in the form*

$$\varphi = \sum_\nu \mathbf{L}^\nu \varphi_{p-2\nu}$$

with primitive elements $\varphi_{p-2\nu} \in H^{p-2\nu}(X,\mathbb{C})$. Here the sum runs over all nonnegative integers $\nu \geq p-g$.

b) Each $\varphi_{p-2\nu}$ in a) can be expressed in the form

$$\varphi_{p-2\nu} = \sum_{\mu,\nu} r_{\mu,\nu} \mathbf{L}^\mu \mathbf{\Lambda}^{\mu+\nu} \varphi$$

with rational numbers $r_{\mu,\nu}$.

c)
$$* \mathbf{L}^\nu \varphi_{p-2\nu} = (-1)^{\frac{1}{2}p(p+1)+\nu} \tfrac{\nu!}{(g-p+\nu)!} \mathbf{L}^{g-p+\nu} J\varphi_{p-2\nu}$$
for all $\varphi_{p-2\nu}$ in a).

For the proof see Griffiths-Harris [1] p. 122 and Wells [1] Theorems V, 3.12 and V, 3.16. The idea is to show that the Lie algebra over $\mathbb{C}$ generated by $\mathbf{L}$ and $\mathbf{\Lambda}$ is isomorphic to the Lie algebra $sl_2(\mathbb{C})$ and to use the representation theory of $sl_2(\mathbb{C})$.

The next proposition shows how the $\mathbf{\Lambda}$-operator can be expressed in terms of the $*$-operator and the Pontryagin product $\star$. As in the last section denote by $P\colon H_p(X,\mathbb{C}) \to H^{2g-p}(X,\mathbb{C})$ the Poincaré duality isomorphism.

(11.3) Proposition. $\mathbf{\Lambda}\varphi = \frac{1}{\chi(L)} P\big(P^{-1}(* c_1(L)) \star P^{-1}\varphi\big)$ *for all $\varphi \in H^\bullet(X,\mathbb{C})$.*

Proof. For any $\varphi \in H^\bullet(X,\mathbb{C})$ we have by definition of $\mathbf{\Lambda}$ and equation (1)

$$\mathbf{\Lambda}\varphi = *^{-1} \mathbf{L} * \varphi = *^{-1}(c_1(L) \wedge * \varphi) = *^{-1}(** c_1(L) \wedge * \varphi)\,.$$

Hence it suffices to show that

$$*^{-1}(*\varphi_1 \wedge *\varphi_2) = \tfrac{1}{\chi(L)} P(P^{-1}\varphi_1 \star P^{-1}\varphi_2) \tag{2}$$

for all $\varphi_1, \varphi_2 \in H^\bullet(X,\mathbb{C})$. Let $I = (i_1 < \cdots < i_p)$ and $J = (j_1 < \cdots < j_q)$ be multi-indices in $\{1,\ldots,2g\}$ and define $\epsilon_{I,J} \in \{0,\pm 1\}$ by $dx_I \wedge dx_J = \epsilon_{I,J} dx_{I\cup J}$. Here $I \cup J$ denotes the ordered multi-index with elements out of the set $I \cup J$. So $\epsilon_{I,J} = 0$, if $I \cap J \neq \emptyset$. Then

$$\begin{aligned}
^{-1}(\,dx_I \wedge *\,dx_J) &= \varepsilon(I)\varepsilon(J) *^{-1}(dx_{I^o} \wedge dx_{J^o}) \\
&= (-1)^{p+q}\varepsilon(I)\varepsilon(J)\epsilon_{I^o,J^o} *\, dx_{I^o\cup J^o} \\
&= (-1)^{p+q}\varepsilon(I)\varepsilon(J)\,\epsilon_{I^o,J^o}\varepsilon(I^o \cup J^o)dx_{I\cap J}\ .
\end{aligned}$$

In order to compute the right hand side of equation (2), denote by $\lambda_1, \ldots, \lambda_{2g}$ the basis of $H_1(X, \mathbb{C})$ which is dual to $dx_1, \ldots, dx_{2g}$ with respect to the natural pairing. Note that $\lambda_1, \ldots, \lambda_{2g}$ is not necessarily a symplectic basis of $\Lambda = H_1(X, \mathbb{Z})$ for L. But with the same proof as for Lemma 10.2 one shows

$$P(\lambda_I) = P(\lambda_{i_1} \star \cdots \star \lambda_{i_p}) = (-1)^p \varepsilon(I)\,\frac{1}{\chi(L)} dx_{I^o}\ .$$

So we have

$$\begin{aligned}
\tfrac{1}{\chi(L)} P(P^{-1}dx_I \star P^{-1}dx_J) &= \chi(L)(-1)^{p+q}\varepsilon(I^o)\varepsilon(J^o)P(\lambda_{I^o} \star \lambda_{J^o}) \\
&= \chi(L)\varepsilon(I)\varepsilon(J)\,\epsilon_{I^o,J^o}P(\lambda_{I^o\cup J^o}) \\
&= (-1)^{p+q}\varepsilon(I)\varepsilon(J)\,\epsilon_{I^o,J^o}\varepsilon(I^o \cup J^o)dx_{I\cap J}\ .
\end{aligned}$$

Since the dx_I's form a basis of $H^\bullet(X, \mathbb{C})$, this completes the proof. □

Proof of Theorem 11.1. Let A denote the ring of algebraic cohomology classes in $H^\bullet(X, \mathbb{Q}) \subset H^\bullet(X, \mathbb{C})$, i.e. the Poincaré duals of homology classes of algebraic cycles on X. Recall from Griffiths-Harris [1] p. 163 that any algebraic cohomology class in $H^{2p}(X, \mathbb{C})$ is of type (p, p) with respect to the Hodge decomposition.

Step I: $*\,A \subset A$.

Suppose $\varphi \in A \cap H^{2p}(X, \mathbb{C})$. We have to show $*\,\varphi \in A$. Note first that $*\,c_1(L) = \alpha \bigwedge^{g-1} c_1(L)$ for some $\alpha \in \mathbb{Q}$, since $c_1(L) \wedge *\,c_1(L) = \alpha dv = \frac{\alpha}{g!} \bigwedge^g c_1(L)$. In particular $*\,c_1(L) \in A$. We deduce from Proposition 11.3 that $\mathbf{\Lambda} A \subset A$. Moreover by definition $\mathbf{L} A \subset A$.

Now consider the Lefschetz Decomposition $\varphi = \sum_\nu L^\nu \varphi_{2p-2\nu}$. According to 11.2.b) and what we have said above, all $\varphi_{2p-2\nu} \in A$. Since moreover $J|A = \mathrm{id}_A$, part c) of the Lefschetz Decomposition shows that $*\,\varphi \in A$.

Step II: Suppose V is an algebraic p-cycle on X, numerically equivalent to zero. We have to show that its class $\{\mathsf{V}\}$ is zero in $H_{2p}(X, \mathbb{C})$. With the notation from above its Poincaré dual $P\{\mathsf{V}\}$ is an element of $A \cap H^{2g-2p}(X, \mathbb{C})$. According to Step I we have $*\,P\{\mathsf{V}\} \in A$. Hence there is an algebraic $(g-p)$-cycle W with rational coefficients on X such that $P\{\mathsf{W}\} = *\,P\{\mathsf{V}\}$. It follows that

$$\begin{aligned}
0 = (\{\mathsf{V}\} \cdot \{\mathsf{W}\}) &= \int_X P\{\mathsf{V}\} \wedge P\{\mathsf{W}\} \\
&= \int_X P\{\mathsf{V}\} \wedge *\,P\{\mathsf{V}\} = \int_X P\{\mathsf{V}\} \wedge *\,\overline{P\{\mathsf{V}\}}\ ,
\end{aligned}$$

since $P\{\mathsf{V}\} \in H^{2g-2p}(X,\mathbb{Q})$ is invariant under complex conjugation. This immediately implies that $P\{\mathsf{V}\} = 0$ and thus $\{\mathsf{V}\} = 0$. □

Exercises

(1) Suppose X is an abelian variety with period matrix $\Pi \in M(g \times 2g, \mathbb{C})$ and $A \in M_{2g}(\mathbb{Z})$ the alternating matrix defining a polarization as in Theorem 2.1. There is a matrix $\Lambda \in M(g \times 2g, \mathbb{C})$ such that $A = {}^t\Pi\Lambda - {}^t\Lambda\Pi$.

(2) Suppose X and X' are abelian varieties with period matrices $\Pi \in M(g \times 2g, \mathbb{C})$ and $\Pi' \in M(g' \times 2g', \mathbb{C})$. There is a nontrivial homomorphism $X \to X'$ if and only if there is a matrix $Q \neq 0$ in $M(2g' \times 2g, \mathbb{Q})$ with $\Pi' Q\, {}^t\Pi = 0$ (Hint: use Exercise 1).

(3) *(Real Riemann Matrices according to H. Weyl [1])* Let $X = V/\Lambda$ be an abelian variety of dimension g and $\Pi \in M(g \times 2g, \mathbb{C})$ a period matrix for X. Suppose $A \in M_{2g}(\mathbb{Z})$ is an alternating matrix satisfying the Riemann Relations
i) $\Pi A^{-1}\, {}^t\Pi = 0$ and ii) $i\Pi A^{-1}\, {}^t\overline{\Pi} > 0$. Define

$$R = R(\Pi) = \begin{pmatrix} \Pi \\ \overline{\Pi} \end{pmatrix}^{-1} \begin{pmatrix} -i\mathbb{1}_g & 0 \\ 0 & i\mathbb{1}_g \end{pmatrix} \begin{pmatrix} \Pi \\ \overline{\Pi} \end{pmatrix}$$

a) R is independent of the chosen basis of V.
b) The matrix R satisfies the following properties
   1) R is real,
   2) $R^2 = -\mathbb{1}_{2g}$,
   3) AR is positive definite and symmetric.
c) Conversely, for any $R \in M_{2g}(\mathbb{R})$ satisfying 1), 2) and 3) there is a period matrix Π satisfying i) and ii) with $R = R(\Pi)$. The matrix Π is uniquely determined by R up to multiplication by a nonsingular matrix from the left.

A matrix $R \in M_{2g}(\mathbb{R})$ with 1), 2) and 3) is called *real Riemann matrix for* X. The main advantage of a real Riemann matrix is that an endomorphism of X may be described in a simpler way:

d) A matrix $M \in M_{2g}(\mathbb{Z})$ is the rational representation of an endomorphism of X if and only if $RM = MR$ for some real Riemann matrix R for X.

(4) Let $X = V/\Lambda$ be an abelian variety of dimension g and D a reduced effective divisor on X. Show that the Gauss map $G\colon D_s \to \mathbb{P}_{g-1}$ is given as follows: If $z_1, \dots, z_g$ denote complex coordinate functions on V and $\omega_\nu = dz_1 \wedge \dots \wedge dz_{\nu-1} \wedge dz_{\nu+1} \wedge \dots \wedge dz_g$, $\nu = 1, \dots, g$, then

$$G(\bar{v}) = \big(\omega_1(v) : \dots : \omega_g(v)\big)$$

for every $\bar{v} \in D_s$ with representative $v \in V$.

(5) Let $X = E_1 \times \cdots \times E_g$ be a product of elliptic curves. Consider the divisor $D = \sum_{\nu=1}^{g} E_1 \times \cdots \times E_{\nu-1} \times \{0\} \times E_{\nu+1} \times \cdots \times E_g$ on X. Show that the image of the Gauss map $G\colon D_s \to \mathbb{P}_{g-1}$ consists of g points spanning $\mathbb{P}_{g-1}$.

(6) *(Generalized Gauss Map)* Let $X = V/\Lambda$ be an abelian variety of dimension g and Y a subvariety of dimension n. For any smooth point y of Y the translation to the origin of the tangent space at Y in y is an n-dimensional subvector space of $T_{X,0} = V$. This defines a holomorphic map G of the smooth part Y_s of Y into the Grassmannian $\mathrm{Gr}(n, V)$ of n-dimensional subvector spaces of V. If the canonical sheaf of Y is an ample line bundle, G is generically one to one. (See Ran [2].)

(7) *(The maximal quotient abelian variety X_a of the complex torus X)* Let X be a complex torus. Recall that for a line bundle L on X the connected component of $K(L) = \ker \phi_L$ containing 0 is denoted by $K(L)_0$.

a) Show that $K(L_1 \otimes L_2) = K(L_1) \cap K(L_2)$ for any positive semidefinite $L_1, L_2 \in \mathrm{Pic}(X)$.
b) Conclude that there is a positive semidefinite line bundle L_a on X such that $K(L_a)_0 \subseteq K(L)_0$ for all positive semidefinite $L \in \mathrm{Pic}(X)$.
c) Show that $X_a := X/K(L_a)_0$ is the maximal abelian quotient variety of X.
d) (*Universal Property of X_a*) Denote by $p\colon X \to X_a$ the natural projection. For any homomorphism $f\colon X \to Y$ into an abelian variety Y there exists a unique homomorphism $g\colon X_a \to Y$ such that $f = gp$.
e) The homomorphism $p\colon X \to X_a$ induces an isomorphism between the divisor groups of X and X_a.
f) The homomorphism $p\colon X \to X_a$ induces an isomorphism between the fields of meromorphic functions on X and X_a.
g) Give an example of a complex torus $X \neq 0$ with $X_a = 0$.

(8) Let X be a complex torus of dimension g. The following statements are equivalent:

(i) X is an abelian variety.
(ii) For every $L \in \mathrm{Pic}(X)$ there is a divisor D on X such that $L = \mathcal{O}_X(D)$.

(9) Let $X = V/\Lambda$ be a complex torus of algebraic dimension $a(X)$. Consider $\mathrm{NS}(X)$ as the group of hermitian forms on V, whose imaginary part is integer valued on Λ. Show

a) $a(X) = \max\{\mathrm{rank}\, H \mid H \in \mathrm{NS}(X)\,,\ H \geq 0\}$
(Hint: use Section 3.3. and Theorem 5.4)
b) $\rho(X) = \mathrm{rk}\, \mathrm{NS}(X) = 0$ implies $a(X) = 0$.

(10) Let X be a simple abelian variety, i.e. X admits no nontrivial abelian subvariety. Show that any algebraic subvarieties V and W of X with $\dim V + \dim W \geq \dim X$ have a nonempty intersection.

(11) Let H be a polarization of type $D = \operatorname{diag}(d_1, \dots, d_g)$ on $X = V/\Lambda$ with $d_1, \dots, d_s$ odd and $d_{s+1}, \dots, d_g$ even, and $L \in \operatorname{Pic}_s^H(X)$. Suppose L is of characteristic $c = c_1 + c_2 \in \frac{1}{2}\Lambda(L)$ with respect to the decomposition defined by a symplectic basis $\lambda_1, \dots, \mu_g$ of Λ. The symplectic basis induces a homomorphism

$$\psi : K(L) \to (\mathbb{Z}^g/D\mathbb{Z}^g)^2/2(\mathbb{Z}^g/D\mathbb{Z}^g)^2 = (\mathbb{Z}/2\mathbb{Z})^{2(g-s)} .$$

Use Proposition 6.5 to show that

$$h^0(L)_\pm = \begin{cases} \frac{1}{2}h^0(L) & \psi(2\bar{c}) \neq 0 \\ \frac{1}{2}h^0(L) \pm 2^{g-s-1} & \text{if} \quad \psi(2\bar{c}) = 0 \text{ and } e\big(4\pi i \operatorname{Im} H(c_1, c_2)\big) = 1 \\ \frac{1}{2}h^0(L) \mp 2^{g-s-1} & \psi(2\bar{c}) = 0 \text{ and } e\big(4\pi i \operatorname{Im} H(c_1, c_2)\big) = -1 . \end{cases}$$

(12) Let L be a symmetric line bundle on X. For any $x \in X_2$ the normalized isomorphism $(-1)_L$ induces an automorphism $(-1)_L(x)$ of the fibre $L(x)$, which is multiplication by a constant denoted by $e_*^L(x) \in \mathbb{C}$.

a) e_*^L is a map on X_2 with values in $\{\pm 1\}$.

b) e_*^L coincides with q_L, the quadratic form defined in Section 7.

(13) Let L be a symmetric line bundle on an abelian variety $X = V/\Lambda$ with $L^n = \mathcal{O}_X$ for some $n \in \mathbb{Z}$. Suppose D is a divisor of X with $\mathcal{O}_X(D) = L$. Then there is a rational function g on X such that $(g) = n^*D$.

a) Show that for any $x \in X_n$

$$q_L^{(n)}(x) := \tfrac{g(x+y)}{g(y)}$$

is an n-th root of unity independent of the choice of D and of the point $y \in X$.

b) Suppose $L = L(0, \chi)$. Show that

$$q_L^{(n)}(\bar{v}) = \chi(nv)$$

for all $\bar{v} \in X_n$.

c) The map

$$q^{(n)} \colon X_n \times \widehat{X}_n \to \mu_n, \quad (x, L) \mapsto q_L^{(n)}(x)$$

is a nondegenerate pairing. In particular $q_L^{(2)}$ coincides with the quadratic form q_L defined in Section 7.

(14) Let H be a polarization on an abelian variety $X = V/\Lambda$ of type $(d_1, \dots, d_g)$ with d_g odd.

a) $2\Lambda(H) \cap \Lambda = 2\Lambda$.

b) There are $2^{g-1}(2^g \pm 1)$ symmetric line bundles $L \in \operatorname{Pic}_s^H(X)$ such that

$$\# X_2^+(D) = 2^{g-1}(2^g \pm 1)$$

for all even symmetric divisors D on X with $\mathcal{O}(D) = L$.

c) If L is of characteristic zero, and D a symmetric divisor with $L = \mathcal{O}(D)$,

$$\#X_2^{\pm}(D) = \begin{cases} 2^{g-1}(2^g \pm 1) & \text{if} \quad D \text{ is even} \\ 2^{g-1}(2^g \mp 1) & \phantom{\text{if}} \quad D \text{ is odd.} \end{cases}$$

(15) Give a proof of Remark 7.8.

(16) Let (X, H) be a principally polarized abelian variety of dimension g. A subset of 2-division points $A \subset X_2$ is called *azygetic* if $e^H(x+y, x+z) = -1$ for all pairwise different points $x, y, z \in A$. For the definition of $e^H(\cdot, \cdot)$ see Section 7.

a) Show that $\#A \leq 2g+2$.

An azygetic subset $A \subset X_2$ with $2g+2$ elements is called *fundamental system* (see Krazer [1], p. 283).

b) There exist exactly $\frac{2^{2g+g^2}}{(2g+2)!}(2^{2g}-1)(2^{2g-2}-1)\cdots(2^2-1)$ fundamental systems.

c) Suppose X is an abelian surface, $L \in \operatorname{Pic}^H(X)$ is of characteristic zero and D the unique divisor in the linear system $|L|$. The set $X_2^-(D)$ (see Section 7) is a fundamental system. Moreover $\sum_{x \in X_2^-(D)} x = 0$.

Chapter 5

Endomorphisms of Abelian Varieties

In Chapter 1 we saw that the ring of endomorphisms $\operatorname{End}(X)$ of a complex torus X is a free abelian group of finite rank. This implies that $\operatorname{End}_{\mathbb{Q}}(X)$ is a finite dimensional $\mathbb{Q}$-algebra. If moreover X is an abelian variety, any polarization L induces an anti-involution $f \mapsto f'$ on $\operatorname{End}_{\mathbb{Q}}(X)$, called the *Rosati involution*. It is the adjoint operator with respect to the hermitian form $c_1(L)$.

The main result of this chapter is Theorem 1.8, which says that the Rosati involution is positive. This means that the symmetric bilinear form $(f,g) \to \operatorname{Tr}_r(f'g)$ on $\operatorname{End}_{\mathbb{Q}}(X)$ is positive definite. Apparently this metric was introduced by Severi [1]. Sometimes it is called the *Weil metric*, since Weil used it in [1] as an essential tool in his proof of the Riemann hypothesis for algebraic curves. Most results of this chapter are consequences of Theorem 1.8. For example, we deduce that $\operatorname{End}_{\mathbb{Q}}(X)$ is a semisimple $\mathbb{Q}$-algebra and that the automorphism group of any polarized abelian variety is finite.

An abelian variety is called *simple* if it does not admit any nontrivial abelian subvariety. The endomorphism algebra of a simple abelian variety is a skew field of finite dimension over $\mathbb{Q}$ admitting a positive anti-involution. Albert classified in [1] and [2] the pairs $(F, \,')$ with F a skew field of finite dimension over $\mathbb{Q}$ and $'$ a positive anti-involution of F. We present a proof of his results in the last part of this chapter. In Chapter 9 we will see that almost all pairs $(F, \,')$ can be realized as endomorphism algebras of a simple abelian variety.

In Section 1 we introduce the Rosati involution and prove the positivity of the Weil metric. In Section 2 we show that a principal polarization induces an isomorphism between the Néron-Severi group $\operatorname{NS}(X)$ and the group $\operatorname{End}^s(X)$ of endomorphisms of X which are symmetric under the Rosati involution. Moreover, Theorem 2.5 identifies the set of polarizations of X within $\operatorname{End}^s(X)$. Theorem 3.2 shows how the abelian subvarieties of a given abelian variety X are reflected in the endomorphism algebra $\operatorname{End}_{\mathbb{Q}}(X)$: a polarization on X induces a bijection between the set of abelian subvarieties of X and the set of symmetric idempotents of $\operatorname{End}_{\mathbb{Q}}(X)$. An immediate consequence is Poincaré's Complete Reducibility Theorem. In Section 4 we show that any two algebraic cycles V and W of complementary dimension in X define an endomorphism $\delta(V,W)$ of the abelian variety X. It depends only on the algebraic equivalence classes of V and W. In the special case of a curve C and a divisor D the endomorphism $\delta(C,D)$ will be of particular importance in the theory of Jacobians and Prym varieties in Chapters 11 and 12. Finally in Section 5 we prove Albert's results mentioned above on the classification of skew fields of finite dimension over $\mathbb{Q}$ with positive anti-involution.

In Section 5 we use some classical results of algebra and number theory, the Skolem-Noether Theorem on central simple algebras, the approximation theorem, Hilbert's Satz 90, and a result on the Brauer group of a number field.

§ 1 The Rosati Involution

Let $X = V/\Lambda$ be an abelian variety of dimension g. In Section 1.2 we introduced the endomorphism algebra $\mathrm{End}_{\mathbb{Q}}(X)$ and its analytic and rational representations ϱ_a and ϱ_r. In this section we show that every polarization on X induces an anti-involution, called the Rosati involution, and a positive definite bilinear form on $\mathrm{End}_{\mathbb{Q}}(X)$.

Fix a polarization L on X. It induces an isogeny $\phi_L\colon X \to \widehat{X}$ depending only on the class of L in $\mathrm{NS}(X)$. The exponent $e(L)$ of the finite group $K(L) = \ker \phi_L$ is called the *exponent of the polarization* L. According to Proposition 1.2.6 there exists a unique isogeny $\psi_L\colon \widehat{X} \to X$ such that $\psi_L \phi_L = e(L)_X$ and $\phi_L \psi_L = e(L)_{\widehat{X}}$, the multiplications by the integer $e(L)$ on X and $\widehat{X}$ respectively. Thus ϕ_L has an inverse in $\mathrm{Hom}_{\mathbb{Q}}(\widehat{X}, X)$, namely

$$\phi_L^{-1} = \tfrac{1}{e(L)} \psi_L \ .$$

We write the elements of $\mathrm{End}_{\mathbb{Q}}(X)$ in the form rf with $f \in \mathrm{End}(X)$ and $r \in \mathbb{Q}$. Define the dual of $rf \in \mathrm{End}_{\mathbb{Q}}(X)$ to be $\widehat{rf} = r\widehat{f} \in \mathrm{End}_{\mathbb{Q}}(\widehat{X})$. Consider the map $'\colon \mathrm{End}_{\mathbb{Q}}(X) \to \mathrm{End}_{\mathbb{Q}}(X)$ defined by

$$f' = \phi_L^{-1} \widehat{f} \phi_L \ .$$

Using Exercise 2.12 one checks that it satisfies

$$(rf + sg)' = rf' + sg'$$
$$(fg)' = g'f' \quad \text{and} \quad f'' = f$$

for all $f, g \in \mathrm{End}_{\mathbb{Q}}(X)$ and $r, s \in \mathbb{Q}$. So $'$ is an anti-involution on $\mathrm{End}_{\mathbb{Q}}(X)$, called the *Rosati (anti-)involution* with respect to the polarization L.

Suppose $L = L(H, \chi)$ and $E = \operatorname{Im} H$. The following proposition shows that the Rosati involution is the adjoint operator with respect to the hermitian form H as well as with respect to the alternating form E.

(1.1) Proposition. *Suppose* $f \in \operatorname{End}_{\mathbb{Q}}(X)$.

a) $E(\rho_r(f)(\lambda), \mu) = E(\lambda, \rho_r(f')(\mu))$ *for all* $\lambda, \mu \in \Lambda$.
b) $H(\rho_a(f)(v), w) = H(v, \rho_a(f')(w))$ *for all* $v, w \in V$.

Proof. Recall from Section 2.4 that the canonical bilinear form $\langle\, , \rangle : \overline{\Omega} \times V \to \mathbb{R}$, $\langle l, v \rangle = \operatorname{Im} l(v)$ is nondegenerate, that $\phi_H : V \to \overline{\Omega}$, $v \to H(v, \cdot)$ is the analytic representation of ϕ_L, and that $\varrho_a(\widehat{f}) = \varrho_a(f)^*$. This implies $\rho_a(f') = \phi_H^{-1} \rho_a(f)^* \phi_H$. Hence for all $v, w \in V$

$$\begin{aligned} E\big(\rho_a(f')(v), w\big) &= \langle \phi_H\big(\rho_a(f')(v)\big), w\rangle = \langle \rho_a(f)^* \phi_H(v), w\rangle \\ &= \langle \phi_H(v), \rho_a(f)(w)\rangle = E\big(v, \rho_a(f)(w)\big) \ . \end{aligned}$$

Since $\rho_r(f) = \rho_a(f)|\Lambda$ and $f'' = f$, this implies a). Moreover

$$\begin{aligned} H\big(\rho_a(f)(v), w\big) &= E\big(i\,\rho_a(f)(v), w\big) + i\,E\big(\rho_a(f)(v), w\big) \\ &= E\big(\rho_a(f)(i\,v), w\big) + i\,E\big(\rho_a(f)(v), w\big) \\ &= E\big(i\,v, \rho_a(f')(w)\big) + i\,E\big(v, \rho_a(f')(w)\big) \\ &= H\big(v, \rho_a(f')(w)\big) \ . \end{aligned}$$

This implies b). □

For any $f \in \operatorname{End}_{\mathbb{Q}}(X)$ the characteristic polynomial P_f^r of the rational representation $\varrho_r(f)$ is

$$P_f^r(t) = \det\big(t\, \mathrm{id}_\Lambda - \varrho_r(f)\big) \ .$$

It is a monic polynomial in t of degree $2g$ with rational coefficients. Similarly the characteristic polynomial P_f^a of the analytic representation $\varrho_a(f)$

$$P_f^a(t) = \det\big(t\, \mathrm{id}_V - \varrho_a(f)\big)$$

is a monic polynomial in t of degree g with complex coefficients. The polynomials P_f^a and P_f^r are related as follows:

(1.2) Proposition. *For any* $f \in \operatorname{End}_{\mathbb{Q}}(X)$

a) $P_f^r = P_f^a \cdot \overline{P_f^a}$
b) $P_f^r(n) = \deg(n_X - f)$ *for all* $n \in \mathbb{Z}$.

Proof. a) is a consequence of Proposition 1.2.3 which states that $\varrho_r \simeq \varrho_a \oplus \bar{\varrho}_a$. As for b): Recall from Section 1.2 that the degree of an endomorphism is equal to the determinant of its rational representation. Hence $\deg(n_X - f) = \det \varrho_r(n_X - f) = \det\big(n \operatorname{id}_\Lambda - \varrho_r(f)\big) = P_f^r(n)$. □

Suppose $f \in \operatorname{End}_{\mathbb{Q}}(X)$ and

$$P_f^r(t) = \sum_{\nu=0}^{2g} (-1)^\nu r_\nu t^{2g-\nu} \quad \text{and} \quad P_f^a(t) = \sum_{\nu=0}^{g} (-1)^\nu a_\nu t^{g-\nu}$$

with coefficients $r_\nu \in \mathbb{Q}$, $r_0 = 1$ and $a_\nu \in \mathbb{C}$, $a_0 = 1$. The *rational* and the *analytic trace* of f are defined by

$$\operatorname{Tr}_r(f) = r_1 \quad \text{and} \quad \operatorname{Tr}_a(f) = a_1 \ .$$

Similarly the *rational* and the *analytic norm* of f are defined by

$$N_r(f) = r_{2g} \quad \text{and} \quad N_a(f) = a_g \ .$$

As an immediate consequence of Proposition 1.2 we get

(1.3) Corollary. *For any* $f \in \operatorname{End}_{\mathbb{Q}}(X)$

a) $N_r(f) = |N_a(f)|^2 = \deg(f)$,

b) $\operatorname{Tr}_r(f) = 2 \operatorname{Re} \operatorname{Tr}_a(f)$.

The analytic trace and norm of f and f' are related as follows:

(1.4) Lemma. *For any* $f \in \operatorname{End}_{\mathbb{Q}}(X)$ *we have* $P_{f'}^a(t) = \overline{P_f^a(t)}$. *In particular*

$$\operatorname{Tr}_a(f') = \overline{\operatorname{Tr}_a(f)} \quad \textit{and} \quad N_a(f') = \overline{N_a(f)} \ .$$

Proof. It suffices to prove the first assertion. Using Exercise 2.13 we have

$$\begin{aligned} P_{f'}^a(t) &= \det(t \operatorname{id}_V - \rho_a(\phi_L^{-1})\rho_a(\hat{f})\rho_a(\phi_L)) \\ &= \det(\rho_a(\phi_L)^{-1}(t \operatorname{id}_{\overline{\Omega}} - \rho_a(\hat{f}))\rho_a(\phi_L)) \\ &= \det(t \operatorname{id}_{\overline{\Omega}} - \rho_a(\hat{f})) \\ &= \overline{\det(t \operatorname{id}_{\mathbb{C}^g} - {}^t\rho_a(f))} = \overline{P_f^a(t)} \end{aligned}$$ □

Before we proceed, we compute the rational trace of an endomorphism f of X in terms of intersection numbers. For this we need some further notation: For any line bundle M on X define $D_M(f)$ to be the line bundle

$$D_M(f) = (f + 1_X)^* M \otimes f^* M^{-1} \otimes M^{-1} \ .$$

(1.5) Proposition. $(M^g) \operatorname{Tr}_r(f) = g \cdot (D_M(f) \cdot M^{g-1})$.

Proof. Comparing first the Chern classes one easily checks that for all integers n we have $(n_X - f)^*M \equiv D_M(f)^{-n} \otimes f^*M \otimes M^{n^2}$. So we get for the self-intersection number

$$\begin{aligned}(((n_X - f)^*M)^g) &= ((D_M(f)^{-n} \otimes f^*M \otimes M^{n^2})^g) \\ &= (M^g)n^{2g} - g(D_M(f) \cdot M^{g-1})n^{2g-1} + \cdots .\end{aligned}$$

On the other hand, according to Corollary 3.6.6 we have $\chi((n_X - f)^*M) = \deg(n_X - f)\chi(M)$ and Riemann-Roch and Proposition 1.2 give

$$(((n_X - f)^*M)^g) = P_f^r(n)(M^g) .$$

Comparing coefficients gives the assertion. □

As above let L be a polarization on X with Rosati involution $'$. According to Lemma 1.4 and Corollary 1.3 b)

$$(f, g) \mapsto \operatorname{Tr}_r(f'g) = \operatorname{Tr}_a(f'g) + \operatorname{Tr}_a(g'f) .$$

defines a symmetric bilinear form on $\operatorname{End}_{\mathbb{Q}}(X)$ with values in $\mathbb{Q}$. We claim that the associated quadratic form $f \mapsto \operatorname{Tr}_a(f'f)$ is positive definite. To see this we give, more generally, a geometric interpretation of the coefficients of the polynomial $P_{f'f}^r$.

(1.6) Lemma. *For all* $f \in \operatorname{End}(X)$ *and* $n \in \mathbb{Z}$

$$\chi(L)\, P_{f'f}^a(n) = \chi(f^*L^{-1} \otimes L^n) .$$

Proof. According to Corollary 2.4.6 we have $\phi_{f^*L} = \widehat{f}\phi_L f$. Applying the Riemann-Roch Theorem and Proposition 1.2 we get

$$\begin{aligned}\chi(f^*L^{-1} \otimes L^n)^2 &= \deg \phi_{f^*L^{-1} \otimes L^n} = \deg(n\phi_L - \widehat{f}\,\phi_L f) \\ &= \deg(n\phi_L - \phi_L f'f) = \deg \phi_L \deg(n_X - f'f) \\ &= \chi(L)^2 P_{f'f}^r(n) = \left(\chi(L)\, P_{f'f}^a(n)\right)^2 .\end{aligned}$$

The last equation follows from Lemma 1.4, since $f'f$ is symmetric with respect to the Rosati involution. Hence $\chi(f^*L^{-1} \otimes L^n) = \pm\chi(L)\, P_{f'f}^a(n)$ as polynomials in n. But for large n both sides are positive, since L is ample. □

We obtain the following geometric interpretation of the coefficients of the polynomial $P_{f'f}^a$.

(1.7) Corollary. *Suppose* $f \in \operatorname{End}(X)$ *and* $P^a_{f'f}(t) = \sum_{\nu=0}^{g}(-1)^\nu a_\nu t^{g-\nu}$. *For* $\nu = 0, \ldots, g$

$$a_\nu = \binom{g}{\nu}\frac{(f^*L^\nu \cdot L^{g-\nu})}{(L^g)} \geq 0 \ .$$

Proof. Applying Riemann-Roch we conclude from the previous lemma

$$P^a_{f'f}(n) = \frac{((f^*L^{-1} \otimes L^n)^g)}{(L^g)} = \sum_{\nu=0}^{g}(-1)^\nu \binom{g}{\nu}\frac{(f^*L^\nu \cdot L^{g-\nu})}{(L^g)} n^{g-\nu} \ ,$$

and the equality of the coefficients holds. All intersection numbers are non-negative by Lemma 4.3.2. □

For any nonzero endomorphism f of X the line bundle $L|\operatorname{im} f$ is ample and $f^*: H^0(L|\operatorname{im} f) \to H^0(f^*L)$ is injective. Hence there is a nontrivial effective divisor D on X with $f^*L = \mathcal{O}_X(D)$ and we get

$$\operatorname{Tr}_r(f'f) = 2a_1 = 2g\frac{(f^*L \cdot L^{g-1})}{(L^g)} = 2g\frac{((L|D)^{g-1})}{(L^g)} > 0 \ ,$$

since $L|D$ is ample on D. Thus we get as a consequence

(1.8) Theorem. $(f, g) \mapsto \operatorname{Tr}_r(f'g)$ *is a positive definite symmetric bilinear form on the* $\mathbb{Q}$*-vector space* $\operatorname{End}_{\mathbb{Q}}(X)$.

Finally we give some applications of Theorem 1.8.

(1.9) Corollary. *The group of automorphisms of any polarized abelian variety* (X, L) *is finite.*

Note that the group of automorphisms of an (unpolarized) abelian variety may be of infinite order. For an example consider the product $X \times X$ of an abelian variety X with itself.

Proof. Suppose f is an automorphism of (X, L). Then $f^*L \otimes L^{-1} \in \operatorname{Pic}^0(X)$ so that $\phi_L = \phi_{f^*L} = \widehat{f}\phi_L f$. We deduce that $f'f = 1$. Consequently $f \in \operatorname{End}(X) \cap \{\varphi \in \operatorname{End}(X) \otimes_{\mathbb{Z}} \mathbb{R} | \operatorname{Tr}_a(\varphi'\varphi) = g\}$. Since the group $\operatorname{End}(X)$ is discrete in $\operatorname{End}(X) \otimes \mathbb{R}$ (see Proposition 1.2.2) and since moreover the set $\{\varphi \in \operatorname{End}(X) \otimes_{\mathbb{Z}} \mathbb{R} | \operatorname{Tr}_a(\varphi'\varphi) = g\}$ is compact according to Theorem 1.8, this intersection is finite. □

(1.10) Corollary. *Let* f *be an automorphism of a polarized abelian variety* (X, L) *and* $n \geq 3$ *an integer. If* $f|X_n = \operatorname{id}_{X_n}$, *then* $f = 1_X$.

Proof. Assume the contrary, i.e. $f \neq 1_X$. According to Corollary 1.9 the automorphism f has finite order. By eventually passing to a power of f we may assume that f is of order p for some prime p. Since the only unipotent automorphism of (X, L) is the identity, there is an eigenvalue ξ of f which is a primitive p-th root of unity. By assumption $X_n \subset \ker(1_X - f)$. Hence there is a $g \in \operatorname{End}(X)$ such that $ng = 1_X - f$. This implies that there is an algebraic integer η, namely an eigenvalue of g, such that

$$n\eta = 1 - \xi .$$

Applying the norm of the field extension $\mathbb{Q}(\xi)|\mathbb{Q}$ we get

$$n^{p-1} N_{\mathbb{Q}(\xi)|\mathbb{Q}}(\eta) = N_{\mathbb{Q}(\xi)|\mathbb{Q}}(1 - \xi) = (1 - \xi) \cdot \ldots \cdot (1 - \xi^{p-1}) = p .$$

This is impossible, since p is a prime and $n \geq 3$. □

According to Corollary 1.10 the restriction to X_n induces an embedding $\operatorname{Aut}(X, L) \hookrightarrow \operatorname{Aut}_{\mathbb{Z}/n\mathbb{Z}}(X_n) = \operatorname{GL}_{2g}(\mathbb{Z}/n\mathbb{Z})$ for any $n \geq 3$. This gives an easy bound for the order of the group of automorphisms of a polarized abelian variety.

§ 2 Polarizations

Recall that by definition a polarization on an abelian variety $X = V/\Lambda$ is a class of an ample line bundle L in $\operatorname{NS}(X)$. By abuse of notation we often write L instead of its class in $\operatorname{NS}(X)$. If L is of type $(d_1, \ldots, d_g)$, we define the *degree of the polarization* L to be the product $d_1 \cdot \ldots \cdot d_g$. In this section we study the subset of $\operatorname{NS}(X)$ of polarizations of a given degree. The aim is to give a formula for the number of isomorphism classes of such polarizations.

Fix a polarization L_0 on X. The inverse $\phi_{L_0}^{-1}$ of $\phi_{L_0}: X \to \widehat{X}$ exists in $\operatorname{Hom}_{\mathbb{Q}}(\widehat{X}, X)$. Hence for every line bundle L on X the product $\phi_{L_0}^{-1}\phi_L$ is an element of $\operatorname{End}_{\mathbb{Q}}(X)$ depending only on the class of L in $\operatorname{NS}(X)$. Denoting $\operatorname{NS}_{\mathbb{Q}}(X) = \operatorname{NS}(X) \otimes_{\mathbb{Z}} \mathbb{Q}$ the polarization L_0 induces in this way a homomorphism of abelian groups

$$\operatorname{NS}_{\mathbb{Q}}(X) \to \operatorname{End}_{\mathbb{Q}}(X) , \ L \mapsto \phi_{L_0}^{-1}\phi_L .$$

Consider the Rosati involution $f \mapsto f'$ on $\operatorname{End}_{\mathbb{Q}}(X)$ with respect to the polarization L_0. An element $f \in \operatorname{End}_{\mathbb{Q}}(X)$ is called *symmetric* (with respect to L_0), if $f' = f$. Let $\operatorname{End}_{\mathbb{Q}}^s(X)$ (respectively $\operatorname{End}^s(X)$) denote the subset of $\operatorname{End}_{\mathbb{Q}}(X)$ (respectively $\operatorname{End}(X)$) of symmetric elements. $\operatorname{End}^s(X)$ is an additive group and $\operatorname{End}_{\mathbb{Q}}^s(X)$ is a $\mathbb{Q}$-vector space and we have

$$\operatorname{End}_{\mathbb{Q}}^s(X) \simeq \operatorname{End}^s(X) \otimes_{\mathbb{Z}} \mathbb{Q} .$$

(2.1) Proposition. *a) Let L_0 be a polarization on X. The map*

$$\varphi: \mathrm{NS}_{\mathbb{Q}}(X) \to \mathrm{End}^s_{\mathbb{Q}}(X)\,,\ L \mapsto \phi_{L_0}^{-1}\phi_L$$

is an isomorphism of $\mathbb{Q}$-vector spaces.

b) If L_0 is a principal polarization, φ restricts to an isomorphism of groups

$$\varphi: \mathrm{NS}(X) \to \mathrm{End}^s(X)\ .$$

Proof. According to Lemma 2.4.7 b) the map φ is injective. Hence it suffices to show that $f \in \mathrm{End}_{\mathbb{Q}}(X)$ is in the image of φ if and only if f is symmetric with respect to L_0. But $f \in \mathrm{im}\varphi$ means that $\phi_{L_0} f = \phi_L$ for some $L \in \mathrm{Pic}(X)$. According to Theorem 2.5.5 this is the case if and only if the bilinear form $(v,w) \mapsto \varrho_a(\phi_{L_0} f)(v,w) = H_0(\rho_a(f)(v),w)$ is hermitian, where $H_0 = c_1(L_0)$. By Proposition 1.1 b) the form $H_0(\rho_a(f)(\cdot),\cdot)$ is hermitian if and only if

$$H_0\big(\rho_a(f)(v),w\big) = \overline{H_0\big(\rho_a(f)(w),v\big)} = \overline{H_0\big(w,\rho_a(f')(v)\big)} = H_0\big(\rho_a(f')(v),w\big).$$

Since H_0 is nondegenerate, this is fulfilled if and only if $f' = f$. This completes the proof of a). For b) we only note that for a principal polarization the map ϕ_{L_0} is an isomorphism, that is $\phi_{L_0}^{-1}\,\phi_L \in \mathrm{End}(X)$ for all L. □

(2.2) Remarks. a) From the proof of the proposition one easily deduces the inverse map $\varphi^{-1}: \mathrm{End}^s_{\mathbb{Q}}(X) \to \mathrm{NS}_{\mathbb{Q}}(X)$: If $f \in \mathrm{End}^s_{\mathbb{Q}}(X)$, the element $\varphi^{-1}(f) \in \mathrm{NS}_{\mathbb{Q}}(X)$ is uniquely determined by $\phi_{L_0} f \in \mathrm{Hom}_{\mathbb{Q}}(X,\widehat{X})$, according to Theorem 2.5.5, since the form $(v,w) \mapsto \varrho_a(\phi_{L_0} f)(v,w) = H_0(\varrho_a(f)(v),w)$ is hermitian on V.

b) Suppose S is a commutative subring of $\mathrm{End}^s_{\mathbb{Q}}(X)$. The multiplication on S induces a multiplication " $\circ$ " on its preimage $\tilde{S}$ in $\mathrm{NS}_{\mathbb{Q}}(X)$, considered as a subspace of $\mathrm{Hom}_{\mathbb{Q}}(X,\widehat{X})$. For $\phi_1,\phi_2 \in \tilde{S}$ we have $\phi_1 \circ \phi_2 = \phi_2\phi_{L_0}^{-1}\phi_1$. □

Suppose $f = \varphi(L)$ is a symmetric endomorphism. The following proposition gives a geometric interpretation for the coefficients of the analytic characteristic polynomial P^a_f in terms of L.

(2.3) Proposition. *Let $f = \phi_{L_0}^{-1}\,\phi_L \in \mathrm{End}^s_{\mathbb{Q}}(X)$ with characteristic polynomial $P^a_f(t) = \sum_{\nu=0}^{g}(-1)^\nu a_\nu t^{g-\nu}$. Then*

$$d_0\, a_\nu = \tfrac{(L_0^{g-\nu}\cdot L^\nu)}{(g-\nu)!\nu!} \quad \textit{for} \quad \nu = 1,\ldots,g\ ,$$

where d_0 denotes the degree of the polarization L_0.

Proof. Applying Riemann-Roch and Proposition 1.2 we get

$$\begin{aligned}\chi(L_0^n \otimes L^{-1})^2 &= \deg \phi_{L_0^n \otimes L^{-1}} = \deg(n\phi_{L_0} - \phi_L) \\ &= \deg \phi_{L_0} \deg(n_X - \phi_{L_0}^{-1} \phi_L) = d_0^2 \deg(n_X - f) \\ &= d_0^2 P_f^r(n) = d_0^2 \left(P_f^a(n)\right)^2 .\end{aligned}$$

The last equation follows from Lemma 1.4, since f is symmetric. Now the Euler-Poincaré characteristic $\chi(L_0^n \otimes L^{-1})$ is positive for large n as L_0 is ample. So

$$\chi(L_0^n \otimes L^{-1}) = d_0 P_f^a(n) .$$

On the other hand we get by Riemann-Roch

$$\chi_0(L_0^n \otimes L^{-1}) = \tfrac{1}{g!}\left((L_0^n \otimes L^{-1})^g\right) = \sum_{\nu=0}^{g} (-1)^\nu \tfrac{(L_0^{g-\nu} \cdot L^\nu)}{(g-\nu)!\nu!} n^{g-\nu} .$$

Comparing coefficients gives the assertion. □

One can use this proposition to determine the subset of $\mathrm{NS}(X)$ of polarizations of a given degree in terms of the endomorphism algebra. An endomorphism in $\mathrm{End}(X)$ is called *totally positive*, if the zeros of its characteristic polynomial P_f^a are all positive.

(2.4) Theorem. *For a principal polarization L_0 on X the isomorphism $\varphi: \mathrm{NS}(X) \to \mathrm{End}^s(X)$ induces a bijection between the sets of*

a) polarizations of degree d on X, and

b) totally positive symmetric endomorphisms with analytic norm d of X.

Proof. According to Corollary 4.3.3 the line bundle L defines a polarization of degree d on X if and only if $(L_0^{g-\nu} \cdot L^\nu) > 0$ for $\nu = 1, \ldots, g-1$ and $(L^g) = d\,g!$. By Proposition 2.3 the alternating coefficients of the analytic characteristic polynomial of $f = \phi_{L_0}^{-1}\phi_L$ are positive. On the other hand, P_f^a has real zeroes by Lemma 1.4, since f is symmetric. This implies the assertion. □

(2.5) Remark. If L_0 is an arbitrary not necessarily principal polarization we have a similar statement: Call an element $l \in \mathrm{NS}_{\mathbb{Q}}(X)$ a *polarization*, if ml is represented by an ample line bundle on X for a suitable integer $m > 0$. Then the map $\varphi: \mathrm{NS}_{\mathbb{Q}}(X) \to \mathrm{End}^s_{\mathbb{Q}}(X)$ induces a bijection between the sets of

a) polarizations in $\mathrm{NS}_{\mathbb{Q}}(X)$, and

b) totally positive symmetric elements in $\mathrm{End}^s_{\mathbb{Q}}(X)$.

Pulling back a line bundle by an endomorphism of X defines an action of $\mathrm{End}(X)$ on $\mathrm{NS}(X)$. Given a principal polarization L_0 this induces an action of $\mathrm{End}(X)$ on $\mathrm{End}^s(X)$ via the diagram

$$\begin{array}{ccc} \mathrm{End}(X) \times \mathrm{NS}(X) & \longrightarrow & \mathrm{NS}(X) \\ \big\downarrow {\scriptstyle (\mathrm{id},\varphi)} & & \big\downarrow {\scriptstyle \varphi} \\ \mathrm{End}(X) \times \mathrm{End}^s(X) & \xrightarrow{\ \tau\ } & \mathrm{End}^s(X) \end{array}$$

(2.6) Lemma. *$\tau(\alpha, f) = \alpha' f \alpha$ for $\alpha \in \mathrm{End}(X)$ and $f \in \mathrm{End}^s(X)$.*

Proof. By Proposition 2.1 there is an $L \in \mathrm{NS}(X)$ with $f = \phi_{L_0}^{-1}\phi_L$. Using Corollary 2.4.6 c) we get

$$\begin{aligned} \varphi(\alpha^* L) &= \phi_{L_0}^{-1}\phi_{\alpha^* L} = \phi_{L_0}^{-1}\widehat{\alpha}\phi_L\alpha \\ &= (\phi_{L_0}^{-1}\widehat{\alpha}\phi_{L_0})(\phi_{L_0}^{-1}\phi_L)\alpha = \alpha' f \alpha\ . \end{aligned}$$ □

Two polarizations L and L' on X are called *isomorphic*, if there is an automorphism α of X such that $L' = \alpha^* L$ in $\mathrm{NS}(X)$. This defines an equivalence relation on the set of polarizations of given degree on X. Using Theorem 2.4 and Lemma 2.6 one can translate this equivalence relation into terms of $\mathrm{End}^s(X)$.

(2.7) Corollary. *For a principal polarization L_0 on X the isomorphism $\varphi\colon \mathrm{NS}(X) \to \mathrm{End}^s(X)$ induces a bijection between the sets of*

a) isomorphism classes of polarizations of degree d on X, and

b) equivalence classes of totally positive symmetric endomorphisms with analytic norm d with respect to the equivalence relation:

$$f_1 \sim f_2 \Longleftrightarrow f_1 = \alpha' f_2 \alpha \quad \textit{for some} \quad \alpha \in \mathrm{End}(X)\ .$$

(2.8) Remark. One can use Corollary 2.7 in order to determine the set of isomorphism classes of polarizations of degree d explicitly in special cases (see Exercises 12 and 13). By a theorem of Narasimhan and Nori this set is always finite (see Exercise 11).

§ 3 Norm-Endomorphisms and Symmetric Idempotents

In this section we describe the set of abelian subvarieties of an abelian variety X in terms of the endomorphism algebra $\mathrm{End}_{\mathbb{Q}}(X)$. Given a polarization L on X we associate to every abelian subvariety Y of X an endomorphism N_Y, the norm-endomorphism, and a symmetric idempotent ε_Y. We will see that the symmetric idempotents are in one to one correspondence to the abelian subvarieties of X. This leads to a criterion for an endomorphism to be a norm-endomorphism. One of the various consequences is that $\mathrm{End}_{\mathbb{Q}}(X)$ is a semisimple $\mathbb{Q}$-algebra.

Let (X, L) be a polarized abelian variety and Y an abelian subvariety of X with canonical embedding $\iota : Y \hookrightarrow X$. Define the *exponent* of the abelian subvariety Y to be the exponent $e(\iota^* L)$ of the induced polarization on Y and write $e(Y) = e(\iota^* L)$. We have (as in Section 1) the isogeny

$$\psi_{\iota^* L} = e(Y)\phi_{\iota^* L}^{-1} : \widehat{Y} \to Y .$$

With this notation define the *norm-endomorphism of X associated to Y (with respect to L)* by

$$N_Y = \iota \psi_{\iota^* L} \hat{\iota} \phi_L ,$$

i.e. as the composition $X \xrightarrow{\phi_L} \widehat{X} \xrightarrow{\hat{\iota}} \widehat{Y} \xrightarrow{\psi_{\iota^* L}} Y \xrightarrow{\iota} X$. The name norm-endomorphism comes from the theory of Jacobian varieties. In fact, it is a generalization of the usual notion of a norm-endomorphism associated to a covering of algebraic curves (see Section 12.3).

(3.1) Lemma. *For any abelian subvariety Y of X*

$$N_Y' = N_Y \quad \textit{and} \quad N_Y^2 = e(Y) N_Y ,$$

where $'$ denotes the Rosati involution with respect to the polarization L.

Proof. $N_Y' = \phi_L^{-1}(\widehat{\phi_L \iota \psi_{\iota^* L} \hat{\iota}})\phi_L = N_Y$, since $\widehat{\phi_L} = \phi_L$ and $\widehat{\psi_{\iota^* L}} = \psi_{\iota^* L}$ by Corollary 2.4.6. The second assertion follows by a similar computation using $\hat{\iota}\phi_L \iota = \phi_{\iota^* L}$. □

We will show that these conditions characterize norm-endomorphisms. For this note that for the norm-endomorphism N_Y the element

$$\varepsilon_Y := \tfrac{1}{e(Y)} N_Y = \iota \phi_{\iota^* L}^{-1} \hat{\iota} \phi_L$$

of $\mathrm{End}_{\mathbb{Q}}(X)$ satisfies $\varepsilon_Y' = \varepsilon_Y$ and $\varepsilon_Y^2 = \varepsilon_Y$. In other words, given a polarization L on X, we associate to every abelian subvariety Y of X a symmetric idempotent ε_Y in $\mathrm{End}_{\mathbb{Q}}(X)$. Conversely, if ε is a symmetric idempotent in $\mathrm{End}_{\mathbb{Q}}(X)$, there is an integer $n > 0$ such that $n\varepsilon \in \mathrm{End}(X)$. Define

$$X^{\varepsilon} = \mathrm{im}(n\varepsilon) \ .$$

Certainly this definition does not depend on the choice of n. Thus to every symmetric idempotent ε we associate an abelian subvariety X^{ε} of X.

(3.2) Theorem. *The assignments $\varphi: Y \mapsto \varepsilon_Y$ and $\psi: \varepsilon \mapsto X^{\varepsilon}$ are inverse to each other and give a bijection between the sets of*
a) abelian subvarieties of X, and
b) symmetric idempotents in $\mathrm{End}_{\mathbb{Q}}(X)$.

Proof. By definition we have $\psi\varphi(Y) = Y$ for any abelian subvariety Y of X. It remains to show that ψ is injective. Suppose that ε_1 and ε_2 are symmetric idempotents in $\mathrm{End}_{\mathbb{Q}}(X)$ with $X^{\varepsilon_1} = X^{\varepsilon_2}$. We have to show that $\varepsilon_1 = \varepsilon_2$.

Choose a positive integer n such that $f_i = n\varepsilon_i \in \mathrm{End}(X)$. Then $f_1^2 = nf_1$ and $f_2^2 = nf_2$. This means that f_ν is multiplication by n on $X^{\varepsilon_\nu} = \mathrm{im}\, f_\nu$ implying $f_2 f_1 = nf_1$ and $f_1 f_2 = nf_2$. So $(f_1 - f_2)^2 = nf_1 - nf_2 - nf_1 + nf_2 = 0$ and hence

$$\mathrm{Tr}_r\big((f_1 - f_2)'(f_1 - f_2)\big) = \mathrm{Tr}_r\big((f_1 - f_2)^2\big) = \mathrm{Tr}_r(0) = 0 \ .$$

According to Theorem 1.8 this implies $f_1 = f_2$ and thus $\varepsilon_1 = \varepsilon_2$. □

As a direct consequence we obtain the following criterion for an endomorphism to be a norm-endomorphism.

(3.3) Corollary. *For $f \in \mathrm{End}(X)$ and $Y = \mathrm{im}\, f$ the following statements are equivalent*
i) $f = N_Y$,
ii) $f' = f$ and $f^2 = e(Y)f$.

In general it is not easy to compute the exponent $e(Y)$. So Corollary 3.3 is not very useful in practice. In case of a principal polarization L we have a better criterion. Recall that an endomorphism $f \neq 0$ is called *primitive*, if $f = ng$ for some $g \in \mathrm{End}(X)$ holds only for $n = \pm 1$. Equivalently f is primitive if and only if its kernel does not contain a subgroup X_n of n-division points of X for some $n \geq 2$.

(3.4) Norm-endomorphism Criterion. *Let L be a principal polarization on X. For $f \in \mathrm{End}(X)$ the following statements are equivalent*
i) $f = N_Y$ for some abelian subvariety Y of X.
ii) The following three conditions hold

a) f is either primitive or $f = 0$.
b) $f = f'$,
c) $f^2 = ef$ for some positive integer e.

Proof. It suffices to show that the norm-endomorphism of a nontrivial abelian subvariety Y of the principally polarized abelian variety (X, L) is primitive. Since ϕ_L is an isomorphism and $\iota: Y \hookrightarrow X$ is an embedding, it suffices to show that the kernel of $\psi_{\iota^* L}\widehat{\iota}$ does not contain $\widehat{X}_n$ for any $n \geq 2$. But $(\psi_{\iota^* L}\widehat{\iota})\widehat{\ } = \iota\psi_{\iota^* L}$ does not contain $\widehat{Y}_n$ for any $n \geq 2$ by definition of $\psi_{\iota^* L}$ and since ι is an embedding. This implies the assertion. □

Theorem 3.2 has some important applications. First note that the set of symmetric idempotents in $\mathrm{End}_{\mathbb{Q}}(X)$ admits a canonical involution, namely

$$\varepsilon \mapsto 1 - \varepsilon \ .$$

So by Theorem 3.2 the polarization L of X induces a canonical involution on the set of abelian subvarieties of X:

$$Y \mapsto Z := X^{1-\varepsilon_Y} \ .$$

We call Z the *complementary abelian subvariety of* Y *in* X (with respect to the polarization L). Of course Y is also the complementary abelian subvariety of Z in X. Hence it makes sense to call (Y, Z) a *pair of complementary abelian subvarieties of* X (with respect to the polarization L). In general the exponents $e(Y)$ of Y and $e(Z)$ of Z are different (for an example see Section 12.1). However, if L is a principal polarization, then $e(Y) = e(Z)$ (see Corollary 12.1.2)

It follows immediately from the definitions and Lemma 3.1 that the norm-endomorphisms N_Y and N_Z satisfy the following properties

$$N_Y|Y = e(Y)1_Y \tag{1}$$
$$N_Y|Z = 0 \tag{2}$$
$$N_Y N_Z = 0 \tag{3}$$
$$e(Y)N_Z + e(Z)N_Y = e(Z)e(Y)1_X \ . \tag{4}$$

This leads to

(3.5) Poincaré's Reducibility Theorem. *Let (X, L) be a polarizied abelian variety and (Y, Z) a pair of complementary abelian subvarieties of X. Then the map*

$$(N_Y, N_Z): X \to Y \times Z$$

is an isogeny.

Proof. The map (N_Y, N_Z) has finite kernel, since by (4) the kernel of (N_Y, N_Z) consists of $e(Y)e(Z)$-division points. In order to show that it is surjective, suppose $(y,z) \in Y \times Z$. There are $y_1, z_1 \in X$ such that

$$y = N_Y\big(e(Y)e(Z)y_1\big) \quad \text{and} \quad z = N_Z\big(e(Y)e(Z)z_1\big) \ .$$

Thus
$$(N_Y, N_Z)\big(e(Z)N_Y(y_1) + e(Y)N_Z(z_1)\big) = (y,z) \ .$$
□

Poincaré's Reducibility Theorem has several important consequences.

(3.6) Corollary. *For any pair (Y,Z) of complementary abelian subvarieties of X the addition map $\mu: Y \times Z \to X$, $\mu(y,z) = y+z$ is an isogeny.*

Proof. Equation (4) means $\mu(e(Z)1_Y \times e(Y)1_Z)(N_Y, N_Z) = e(Y)e(Z)1_X$. So with (N_Y, N_Z), $((e(Z)1_Y \times e(Y)1_Z)$, and $e(Y)e(Z)1_X$ also μ is an isogeny.
□

An abelian variety X is called *simple*, if it does not contain any abelian subvariety apart from X and 0. By induction one immediately obtains

(3.7) Poincaré's Complete Reducibility Theorem. *Given an abelian variety X there is an isogeny*

$$X \to X_1^{n_1} \times \cdots \times X_r^{n_r}$$

with simple abelian varieties X_ν not isogenous to each other. Moreover the abelian varieties X_ν and the integers n_ν are uniquely determined up to isogenies and permutations.

(3.8) Corollary. $\mathrm{End}_{\mathbb{Q}}(X)$ *is a semisimple $\mathbb{Q}$-algebra. To be more precise: if $X \to X_1^{n_1} \times \cdots \times X_r^{n_r}$ is an isogeny as in the previous corollary, then*

$$\mathrm{End}_{\mathbb{Q}}(X) \simeq M_{n_1}(F_1) \oplus \cdots \oplus M_{n_r}(F_r) \ ,$$

where $F_\nu = \mathrm{End}_{\mathbb{Q}}(X_\nu)$ are skew fields of finite dimension over $\mathbb{Q}$.

Proof. Without loss of generality we may assume $X = X_1^{n_1} \times \cdots \times X_r^{n_r}$. Since $\mathrm{Hom}(X_\nu^{n_\nu}, X_\mu^{n_\mu}) = 0$ for $\nu \neq \mu$, we obtain

$$\mathrm{End}_{\mathbb{Q}}(X) = \bigoplus_{\nu=1}^{r} \mathrm{End}_{\mathbb{Q}}(X_\nu^{n_\nu}) \ .$$

Certainly $\mathrm{End}_{\mathbb{Q}}(X_\nu^{n_\nu})$ equals the ring of $(n_\nu \times n_\nu)$-matrices with entries in $\mathrm{End}_{\mathbb{Q}}(X_\nu)$. For the simple abelian variety X_ν every nonzero endomorphism is an isogeny and hence invertible in $\mathrm{End}_{\mathbb{Q}}(X)$. This proves that $\mathrm{End}_{\mathbb{Q}}(X)$ is a skew field over $\mathbb{Q}$. It is of finite dimension by Proposition 1.2.2. □

(3.9) Corollary. *For any abelian variety X the Néron-Severi group* $\mathrm{NS}(X)$ *is a free abelian group of finite rank.*

This is a consequence of Corollary 3.8, Proposition 2.1 and the fact that $\mathrm{NS}(X)$ is torsion free. It also follows from the injectivity of the map $\mathrm{NS}(X) \to \mathrm{Hom}(X, \widehat{X})$, $L \mapsto \phi_L$ (see Proposition 2.5.3), and the property of $\mathrm{Hom}(X, \widehat{X})$ to be a free $\mathbb{Z}$-module of finite rank. □

For any symmetric idempotent ε in $\mathrm{End}_{\mathbb{Q}}(X)$ one can compute the dimension of the corresponding abelian subvariety X^{ε}:

(3.10) Corollary. $\dim X^{\varepsilon} = \mathrm{Tr}_a(\varepsilon)$.

Proof. Denote $Y = X^{\varepsilon}$ and let Z be the complementary abelian subvariety of X. Using (1) and (2) we see that the following diagram is commutative

$$\begin{array}{ccc} X & \xrightarrow{(N_Y, N_Z)} & Y \times Z \\ {\scriptstyle N_Y}\downarrow & & \downarrow \begin{pmatrix} e(Y)1_Y & 0 \\ 0 & 0 \end{pmatrix} \\ X & \xrightarrow{(N_Y, N_Z)} & Y \times Z \end{array} .$$

By Poincaré's Reducibility Theorem (N_Y, N_Z) is an isogeny, so we have in $\mathrm{End}_{\mathbb{Q}}(X)$:

$$N_Y = (N_Y, N_Z)^{-1} \begin{pmatrix} e(Y)1_Y & 0 \\ 0 & 0 \end{pmatrix} (N_Y, N_Z) .$$

This gives $\mathrm{Tr}_a(\varepsilon) = \frac{1}{e(Y)} \mathrm{Tr}_a(N_Y) = \frac{1}{e(Y)} \mathrm{Tr}_a\left(\begin{pmatrix} e(Y)1_Y & 0 \\ 0 & 0 \end{pmatrix}\right) = \dim Y$. □

§ 4 Endomorphisms Associated to Cycles

Let X be an abelian variety of dimension g and denote by V and W algebraic cycles on X of complementary dimension. There is a canonical way to associate to the pair (V, W) an endomorphism $\delta(\mathsf{V}, \mathsf{W})$ of X. Following Morikawa [1] and Matsusaka [1], we will show that $\delta(\mathsf{V}, \mathsf{W})$ depends only on the algebraic equivalence classes of V and W.

In this section we consider only *algebraic cycles* V with coefficients in $\mathbb{Z}$, i.e. finite formal sums

$$\mathsf{V} = \sum r_i \mathsf{V}_i$$

with integers r_i and algebraic subvarieties V_i of X, which we assume to be all of the same dimension. If $\dim \mathsf{V}_i = p$, then V is also called an *algebraic p-cycle*. Let $\mathsf{W} = \sum s_i \mathsf{W}_i$ be an algebraic q-cycle on X. The cycles V and W are said to *intersect properly*, if $\mathsf{V}_i \cap \mathsf{W}_j$ is either of dimension $p+q-g$ or empty, whenever $r_i \neq 0 \neq s_j$.

(4.1) Moving Lemma. *Let* V *be an algebraic* p*-cycle and* W *an algebraic* q*-cycle on* X*. There is an open dense subset* U *in* X *such that* V *and* $t_x^*\mathsf{W}$ *intersect properly for all* $x \in U$.

Proof. We may assume that V and W are subvarieties of X. Consider the difference map $d: \mathsf{V} \times \mathsf{W} \to X$, $(v,w) \mapsto w - v$. The fibre of d over any $x \in X$ is

$$d^{-1}(x) \simeq \mathsf{V} \cap t_x^*\mathsf{W} \ .$$

Since d is a closed morphism, there is an open dense subset U of X such that $d^{-1}(x)$ is either of dimension $p+q-g$ (if d is surjective) or empty for all $x \in U$ (see Hartshorne [1] Exercise II, 3.22). □

We need a version of the Moving Lemma with parameters.

(4.2) Lemma. *Let* T *be an algebraic variety and* Z *an algebraic cycle on* $X \times T$ *intersecting* $X \times \{t\}$ *properly for any* $t \in T$*. Let* $\mathsf{Z}(t)$ *be the cycle on* X *defined by* $\mathsf{Z}(t) \times \{t\} = \mathsf{Z} \cdot (X \times \{t\})$*. For any algebraic cycle* W *on* X *there is an open dense subset* $U \subset T \times X$ *such that* $\mathsf{Z}(t)$ *and* $t_x^*\mathsf{W}$ *intersect properly for all* $(t,x) \in U$.

The proof is analogous to the proof of the Moving Lemma. Instead of the difference map d one uses the morphism $\mathsf{W} \times \mathsf{Z} \to T \times X$, $(w,x,t) \mapsto (t, w-x)$. □

Let V and W be algebraic cycles on X of complementary dimension. Suppose V and W intersect properly, then the usual intersection product $\mathsf{V} \cdot \mathsf{W}$ is a 0-cycle on X, i.e. $\mathsf{V} \cdot \mathsf{W} = \sum_{i=1}^n r_i x_i$ with points x_i on X and integers r_i. Define

$$S(\mathsf{V} \cdot \mathsf{W}) = r_1 x_1 + \cdots + r_n x_n \in X \ ,$$

where the sum means addition in X. Note that S is symmetric and bilinear, i.e. $S(\mathsf{V} \cdot \mathsf{W}) = S(\mathsf{W} \cdot \mathsf{V})$ and $S(\mathsf{V} + \mathsf{V}', \mathsf{W}) = S(\mathsf{V}, \mathsf{W}) + S(\mathsf{V}', \mathsf{W})$ for cycles V and V' both intersecting W properly.

Let now (V, W) be an arbitrary pair of algebraic cycles of complementary dimension on X. The pair (V, W) induces an endomorphism $\delta(\mathsf{V}, \mathsf{W})$ of X in the following way. According to the Moving Lemma 4.1 the cycle V intersects $t_x^*\mathsf{W}$ properly for all x of an open dense subset of X. So $x \mapsto S(\mathsf{V} \cdot t_x^*\mathsf{W})$ defines a rational map $X \to X$ which according to Theorem 4.9.4 extends to a morphism $S: X \to X$. By Proposition 1.2.1 there is an endomorphism

$\delta(\mathsf{V},\mathsf{W})$ of X and a point $c \in X$, both uniquely determined by S, such that $\delta(\mathsf{V},\mathsf{W}) = S - c$. So for $\delta(\mathsf{V},\mathsf{W})\colon X \to X$ we have

$$\delta(\mathsf{V},\mathsf{W})(x) = S(\mathsf{V}\cdot t_x^*\mathsf{W}) - c$$

whenever V intersects $t_x^*\mathsf{W}$ properly. The bilinearity of S implies

$$\delta(\mathsf{V}+\mathsf{V}',\mathsf{W}) = \delta(\mathsf{V},\mathsf{W}) + \delta(\mathsf{V}',\mathsf{W}) \quad \text{and}$$
$$\delta(\mathsf{V},\mathsf{W}+\mathsf{W}') = \delta(\mathsf{V},\mathsf{W}) + \delta(\mathsf{V},\mathsf{W}')$$

for all algebraic cycles V,V' and W,W' of complementary dimension on X. Note that in the special case that V intersects W properly we have $c = S(\mathsf{V}\cdot\mathsf{W})$, i.e.

$$\delta(\mathsf{V},\mathsf{W})(x) = S\big(\mathsf{V}\cdot(t_x^*\mathsf{W} - \mathsf{W})\big)$$

whenever defined. The next proposition shows that we always may assume that V intersects W properly.

(4.3) Proposition. $\delta(\mathsf{V},\mathsf{W}) = \delta(\mathsf{V}',\mathsf{W})$ *for any algebraically equivalent algebraic p-cycles* V *and* V' *and any algebraic $(g-p)$-cycle* W *on* X.

Proof. Without loss of generality we may assume that V intersects W properly. By the definition of algebraic equivalence we may assume that there is a normal algebraic variety T and an algebraic cycle Z in $X \times T$ intersecting $X \times \{t\}$ properly for every $t \in T$ such that

$$Z\cdot(X\times\{t_0\}) = \mathsf{V}\times\{t_0\} \quad \text{and} \quad Z\cdot(X\times\{t_1\}) = \mathsf{V}'\times\{t_1\}$$

for some $t_0, t_1 \in T$. For any $t \in T$ define the p-cycle V_t by $Z\cdot(X\times\{t\}) = \mathsf{V}_t\times\{t\}$. According to Lemma 4.2 there exists an open dense subset U of $T\times X$, such that V_t intersects $t_x^*\mathsf{W}$ properly for every $(t,x)\in U$. Since $\mathsf{V} = \mathsf{V}_{t_0}$ intersects W properly by assumption, we may assume that $(t_0, 0) \in U$. Passing eventually to a smaller subset, we may assume that V_t intersects also W properly for every $(t,x)\in U$. In other words with $(t,x)\in U$ also $(t,0)\in U$. So

$$\phi(t,x) := S\big(\mathsf{V}_t\cdot(t_x^*\mathsf{W} - \mathsf{W})\big)\ ,$$

for all $(t,x)\in U$, defines a rational map $\phi\colon T\times X \to X$ which by Theorem 4.9.4 is everywhere defined. We have $\phi(t,0) = S(\mathsf{V}_t\cdot(\mathsf{W}-\mathsf{W})) = 0$ for any $(t,0)\in U$ and thus for all $t\in T$. Hence by Corollary 4.9.2 the morphism ϕ does not depend on T. In particular $\delta(\mathsf{V},\mathsf{W}) = \phi(t_0,\cdot) = \phi(t_1,\cdot) = \delta(\mathsf{V}',\mathsf{W})$. □

Recall that for arbitrary algebraic cycles V and W of complementary dimension $(\mathsf{V}\cdot\mathsf{W})$ denotes the intersection number of V and W. If V and $t_x^*\mathsf{W}$ intersect properly, then $(\mathsf{V}\cdot\mathsf{W})$ is the degree of the 0-cycle $\mathsf{V}\cdot t_x^*\mathsf{W}$.

(4.4) Lemma. $\delta(\mathsf{V},\mathsf{W}) + \delta(\mathsf{W},\mathsf{V}) = -(\mathsf{V}\cdot\mathsf{W})1_X.$

Proof. We may assume that V and W intersect properly. Then for all x of an open dense subset of X

$$\begin{aligned}\delta(\mathsf{V},\mathsf{W})(x) &= S(\mathsf{V}\cdot t_x^*\mathsf{W}) - S(\mathsf{V}\cdot\mathsf{W}) = S(t_{-x}^*\mathsf{V}\cdot\mathsf{W}) - (\mathsf{V}\cdot\mathsf{W})x - S(\mathsf{W}\cdot\mathsf{V})\\ &= S(\mathsf{W}\cdot(t_{-x}^*\mathsf{V}-\mathsf{V})) - (\mathsf{V}\cdot\mathsf{W})x = -\delta(\mathsf{W},\mathsf{V})(x) - (\mathsf{V}\cdot\mathsf{W})x\ .\end{aligned}$$

So Theorem 4.9.4 implies the assertion. □

Combining Proposition 4.3 and Lemma 4.4 we obtain

(4.5) Corollary. *The homomorphism* $\delta(\mathsf{V},\mathsf{W})$ *depends only on the algebraic equivalence classes of* V *and* W.

One can show that $\delta(\mathsf{V},\mathsf{W})$ depends only on the numerical equivalence classes of V and W (see Matsusaka [1]), but we do not need this fact.

(4.6) Lemma. $\delta(\mathsf{V}_0,\mathsf{V}_1\cdot\ldots\cdot\mathsf{V}_r) = \sum_{i=1}^r \delta(\mathsf{V}_0\cdot\mathsf{V}_1\cdot\ldots\cdot\check{\mathsf{V}}_i\cdot\ldots\cdot\mathsf{V}_r,\mathsf{V}_i)$ *for all algebraic cycles* $\mathsf{V}_0,\ldots,\mathsf{V}_r$ *on* X *with* $\sum_{i=0}^r \dim\mathsf{V}_i = rg$.

Here the notation $\check{\mathsf{V}}_i$ means that the cycle $\check{\mathsf{V}}_i$ has to be omitted in the intersection product. Moreover by $\mathsf{V}_0\cdot\ldots\cdot\check{\mathsf{V}}_i\cdot\ldots\cdot\mathsf{V}_r$ for $i=0,\ldots,r$ we mean any cycle in the algebraic equivalence class of the corresponding intersection product. The assumption on the dimension implies that the cycles V_i and $\mathsf{V}_0\cdot\ldots\cdot\check{\mathsf{V}}_i\cdot\ldots\cdot\mathsf{V}_r$ are of complementary dimension for all $0\le i\le r$. So all endomorphisms in the formula are well defined.

Proof. Passing eventually to suitable translations we may assume that V_i and $\mathsf{V}_0\cdot\ldots\check{\mathsf{V}}_i\cdot\ldots\cdot\mathsf{V}_r$ intersect properly. Suppose first $r=2$. Then for a general $x\in X$

$$\begin{aligned}\delta(\mathsf{V}_0,\mathsf{V}_1\cdot\mathsf{V}_2)(x) &= S(\mathsf{V}_0\cdot(t_x^*\mathsf{V}_1\cdot t_x^*\mathsf{V}_2 - \mathsf{V}_1\cdot\mathsf{V}_2))\\ &= S(\mathsf{V}_0\cdot(t_x^*\mathsf{V}_1\cdot t_x^*\mathsf{V}_2 - t_x^*\mathsf{V}_1\cdot\mathsf{V}_2)) + S(\mathsf{V}_0\cdot(t_x^*\mathsf{V}_1\cdot\mathsf{V}_2 - \mathsf{V}_1\cdot\mathsf{V}_2))\\ &= \delta(\mathsf{V}_0\cdot\mathsf{V}_1,\mathsf{V}_2)(x) + \delta(\mathsf{V}_0\cdot\mathsf{V}_2,\mathsf{V}_1)(x)\ .\end{aligned}$$

This proves the assertion for $r=2$. The general case follows by induction. □

(4.7) Proposition. $\delta(D^r,D^{g-r}) = -\frac{g-r}{g}(D^g)1_X$ *for any divisor* D *on* X *and* $0\le r\le g$.

Proof. Using Lemma 4.6 and Lemma 4.4 we have

$$\delta(D,D^{g-1}) = (g-1)\delta(D^{g-1},D) = (g-1)(-\delta(D,D^{g-1}) - (D^g)1_X)$$

implying the assertion for $r = 1$. Using this and again Lemmas 4.6 and 4.4 we get for every $0 \leq r \leq g$

$$\begin{aligned}\delta(D^r, D^{g-r}) &= (g-r)\delta(D^{g-1}, D) = (g-r)(-\delta(D, D^{g-1}) - (D^g)1_X) \\ &= (g-r)(\tfrac{g-1}{g} - 1)(D^g)1_X = -\tfrac{g-r}{g}(D^g)1_X \ . \qquad \square\end{aligned}$$

§ 5 The Endomorphism Algebra of a Simple Abelian Variety

Let X be a simple abelian variety of dimension g and L a polarization on X. According to Corollary 3.8 the algebra $F = \mathrm{End}_{\mathbb{Q}}(X)$ is a skew field of finite dimension over $\mathbb{Q}$. The Rosati involution $f \mapsto f'$ with respect to the polarization L is an anti-involution on F such that $f \mapsto \mathrm{Tr}_r(f'f) = 2\,\mathrm{Tr}_a(f'f)$ is a positive definite quadratic form on F (see Theorem 1.8). We want to investigate the possibilities for such pairs $(F, \ ')$.

First we express the quadratic form $\mathrm{Tr}_r(f'f)$ in terms of $(F, \ ')$. Let K denote the center of the skew field F. The degree $[F : K]$ of F over K is a square, say d^2. The characteristic polynomial of any $f \in F$ over K is a d'th power of a polynomial $t^d - a_1 t^{d-1} + \cdots + (-1)^d a_0 \in K[t]$, called the *reduced characteristic polynomial of f over K*. In these terms the *reduced trace of f over K* is defined as

$$\mathrm{tr}_{F|K}(f) = a_1 \ .$$

For any subfield $k \subseteq K$ we define the *reduced trace of f over k* by

$$\mathrm{tr}_{F|k}(f) = \mathrm{tr}_{K|k}(\mathrm{tr}_{F|K}(f)) \ ,$$

where $\mathrm{tr}_{K|k}$ denotes the usual trace for the field extension $K|k$.

We claim that with $f \mapsto \mathrm{Tr}_r(f'f)$ also the quadratic form $f \mapsto \mathrm{tr}_{F|\mathbb{Q}}(f'f)$ is positive definite on $F = \mathrm{End}_{\mathbb{Q}}(X)$. To see this note that by Corollary 1.7 and Theorem 1.8 the alternating coefficients a_ν of the analytic characteristic polynomial of $f'f$ ($\neq 0$) are all nonnegative rational numbers with $a_1 > 0$. It follows that the zeros of the minimal polynomial of $f'f$ over $\mathbb{Q}$ are all positive. This immediately implies the assertion.

We need some preliminaries. For abbreviation we call an anti-involution $x \mapsto x'$ on a semisimple algebra over $\mathbb{Q}$ or $\mathbb{R}$ *positive*, if the quadratic form $\mathrm{tr}(x'x)$ is positive definite. Here tr denotes the reduced trace over $\mathbb{Q}$ or $\mathbb{R}$. It is well known that any finite dimensional simple $\mathbb{R}$-algebra is isomorphic to either $M_r(\mathbb{R})$ or $M_r(\mathbb{C})$ or $M_r(\mathbb{H})$ for some r, where $\mathbb{H}$ denotes the skew field of Hamiltonian quaternions. For any of these algebras there is a natural anti-involution, namely

$$x^* = \begin{cases} {}^t x & \text{for} \quad M_r(\mathbb{R}) \\ {}^t\bar{x} & \quad M_r(\mathbb{C}) \quad \text{and} \quad M_r(\mathbb{H})\ . \end{cases} \tag{1}$$

Here $x \mapsto \bar{x}$ means complex (respectively quaternion) conjugation. The following lemma shows that up to isomorphisms * is the unique positive anti-involution.

(5.1) Lemma. *For any simple* $\mathbb{R}$*-algebra* A *of finite dimension with positive anti-involution* $x \mapsto x'$ *there is an isomorphism* φ *of* A *onto one of the matrix algebras as above such that for every* $x \in A$

$$\varphi(x') = \varphi(x)^*\ .$$

Proof. We may assume that A is either $M_r(\mathbb{R})$ or $M_r(\mathbb{C})$ or $M_r(\mathbb{H})$ for some r. The two anti-involutions $x \mapsto x'$ and $x \mapsto x^*$ on A differ by an automorphism of A. By the Skolem-Noether Theorem (see Jacobson [1] II, p. 222) this implies that there is an $a \in A$ such that

$$x' = a^{-1}x^*a\ .$$

It suffices to show that $a = \pm b^*b$ for some $b \in A$, since then the isomorphism $\varphi\colon x \mapsto bxb^{-1}$ satisfies $\varphi(x') = \varphi(x)^*$ for every $x \in A$.

Since $x = x'' = a^{-1}a^*x(a^{-1}a^*)^{-1}$ for any $x \in A$, the element $\lambda := a^{-1}a^*$ is in the center of A, that is in $\mathbb{R}$ or $\mathbb{C}$. Moreover we have $|\lambda| = 1$, since $a = a^{**} = \bar{\lambda}\lambda a$.

We claim that we may assume that $\lambda = 1$ and thus $a^* = a$. For the proof suppose first that $A \in M_r(\mathbb{C})$ or $M_r(\mathbb{H})$ and $\lambda \neq 1$. There is a $\mu \in \mathbb{C}$ with $\lambda = \mu^2$. Replacing a by μa gives the assertion in this case.

If $A = M_r(\mathbb{R})$ we proceed as follows. Suppose $\lambda = -1$. Then a is an alternating matrix and thus $a = c^* \begin{pmatrix} \begin{smallmatrix} 0 & 1 \\ -1 & 0 \end{smallmatrix} & 0 \\ 0 & * \end{pmatrix} c$ for some $c \in M_r(\mathbb{R})$. With $x_0 = c^{-1}\operatorname{diag}(1, -1, 0, \ldots, 0)c$ we get

$$0 < \operatorname{tr}(x_0'x_0) = \operatorname{tr}\left(c^{-1}\begin{pmatrix} \begin{smallmatrix} 0 & 1 \\ -1 & 0 \end{smallmatrix} & 0 \\ 0 & * \end{pmatrix}^{-1} c^{*-1}x_0^*c^*\begin{pmatrix} \begin{smallmatrix} 0 & 1 \\ -1 & 0 \end{smallmatrix} & 0 \\ 0 & * \end{pmatrix} cx_0\right)$$

$$= \operatorname{tr}\left(\begin{pmatrix} \begin{smallmatrix} 0 & 1 \\ -1 & 0 \end{smallmatrix} & 0 \\ 0 & * \end{pmatrix}^{-1} (cx_0c^{-1})^* \begin{pmatrix} \begin{smallmatrix} 0 & 1 \\ -1 & 0 \end{smallmatrix} & 0 \\ 0 & * \end{pmatrix} (cx_0c^{-1})\right) = -2\ ,$$

a contradiction. Hence $a^* = a$, which completes the proof of the claim.

From Linear Algebra we know that there is a matrix $b \in A$ such that

$$a = b^* \operatorname{diag}(\varepsilon_1, \ldots, \varepsilon_r)b$$

with $\varepsilon_\nu \in \{\pm 1\}$ for $1 \leq \nu \leq r$. It remains to show that $\varepsilon_\mu = \varepsilon_\nu$ for $1 \leq \mu, \nu \leq r$.

Suppose $\varepsilon_\mu \neq \varepsilon_\nu$. Denoting by $e(\mu,\nu)$ the matrix, whose (μ,ν)-th entry is 1 and the others are 0, we have

$$\begin{aligned} 0 &< \operatorname{tr}\big(((b^{-1}e(\mu,\nu)b)'((b^{-1}e(\mu,\nu)b)\big) \\ &= \operatorname{tr}\big(\operatorname{diag}(\varepsilon_1,\ldots,\varepsilon_r)\, e(\nu,\mu)\ \operatorname{diag}(\varepsilon_1,\ldots,\varepsilon_r)\, e(\mu,\nu)\big) = \varepsilon_\mu\varepsilon_\nu = -1\ , \end{aligned}$$

a contradiction. So $\varphi(x) = bxb^{-1}$ satisfies the assertion of the Lemma. □

Recall that $(F,\ ')$ denotes a skew field of finite dimension over $\mathbb{Q}$ with positive anti-involution $x \mapsto x'$. The anti-involution $x \mapsto x'$ restricts to an involution on the center K of F, whose fixed field we denote by K_0.

(5.2) Lemma. *K_0 is a totally real number field, i.e. every embedding $K_0 \hookrightarrow \mathbb{C}$ factorizes via $\mathbb{R}$.*

Proof. Let $\sigma_1,\ldots,\sigma_r$ be the real embeddings and $\sigma_{r+1},\bar\sigma_{r+1},\ldots,\sigma_{r+s},\bar\sigma_{r+s}$ the nonreal complex embeddings of K_0 and assume $s \geq 1$. The approximation theorem (see v.d. Waerden [1] II.p.234) implies that for any $\varepsilon > 0$ there is an $x \in K_0$ such that $|\sigma_\nu(x)| < \varepsilon$ for $1 \leq \nu < r+s$ and $|\sigma_{r+s}(x) - i| < \varepsilon$. For small ε the term $2\operatorname{Re}\sigma_{r+s}(x^2) \approx -2$ is dominant in

$$\operatorname{tr}_{K_0|\mathbb{Q}}(x'x) = \operatorname{tr}_{K_0|\mathbb{Q}}(x^2) = \sum_{\nu=1}^{r} \sigma_\nu(x^2) + 2\sum_{\nu=1}^{s} \operatorname{Re}\sigma_{r+\nu}(x^2)\ .$$

Hence $0 < \operatorname{tr}_{F|\mathbb{Q}}(x'x) = \operatorname{tr}_{K_0|\mathbb{Q}}(\operatorname{tr}_{F|K_0}(x'x)) = [F : K_0]\operatorname{tr}_{K_0|\mathbb{Q}}(x^2) < 0$, a contradiction. □

The pair $(F,\ ')$ (respectively the anti-involution $'$) is called to be *of the first kind*, if the anti-involution is trivial on K, that is $K = K_0$, and *of the second kind* otherwise. We first consider the case that $(F,\ ')$ is of the first kind.

Recall that the skew field F is called a quaternion algebra over K, if it is of dimension 4 over its center K. A quaternion algebra F over K admits a canonical anti-involution, namely $x \mapsto \bar{x} = \operatorname{tr}_{F|K}(x) - x$.

(5.3) Theorem. *Let F be a skew field of characteristic 0 and $x \mapsto x'$ an anti-involution of F with center K. $(F,\ ')$ is a skew field of finite dimension over $\mathbb{Q}$ with positive anti-involution of the first kind if and only if K is a totally real number field and one of the following cases holds:*

a) $F = K$ and $x' = x$ for all $x \in F$,

b) F is a quaternion algebra over K and for every embedding $\sigma\colon K \hookrightarrow \mathbb{R}$

$$F \otimes_\sigma \mathbb{R} \simeq M_2(\mathbb{R})\ .$$

Moreover there is an element $a \in F$ with $a^2 \in K$ totally negative such that the anti-involution $x \mapsto x'$ is given by $x' = a^{-1}\bar{x}a$.

c) F is a quaternion algebra over K and for every embedding $\sigma: K \hookrightarrow \mathbb{R}$

$$F \otimes_\sigma \mathbb{R} \simeq \mathbb{H} .$$

Moreover the anti-involution $x \mapsto x'$ is given by $x' = \bar{x}$.

An algebra F as in b) is called a *totally indefinite quaternion algebra*, and an algebra F as in c) is called a *totally definite quaternion algebra.*

Proof. In Steps I—IV we show that a skew field $(F, \,' \,)$ of finite dimension over $\mathbb{Q}$ with positive anti-involution of the first kind is of type (a), (b) or (c). In step V we will prove the converse.

Step I: The anti-involution $x \mapsto x'$ on F may be considered as an isomorphism between F and its opposite algebra F^{op}, defined by the product $x \circ y = yx$ on the K-vector space F. Since the elements in the Brauer group $\text{Br}(K)$ corresponding to F and F^{op} are inverse to each other, F has order 1 or 2 considered as an element in $\text{Br}(K)$. Using a result on the Brauer group (see Jacobson [1] II, Theorem 9.23) this implies that the rank of F over K is either $1^2 = 1$ or $2^2 = 4$. It follows that F is either K, and we are in case (a), or a quaternion algebra over K.

Step II: Suppose F is a quaternion algebra over K. As in the proof of Lemma 5.1 there is an $a \in F$ such that

$$x' = a^{-1}\bar{x}a \quad \text{with} \quad \bar{a} = \lambda a \text{ and } |\lambda| = 1$$

for all $x \in F$. Since K is totally real, $\lambda = \pm 1$. In particular $a^2 \in K$. Denote by $\sigma_1, \ldots, \sigma_e$ the different embeddings of K and write $\mathbb{R}_{\sigma_\nu} = \mathbb{R}$ when $\mathbb{R}$ is considered as a K-algebra via σ_ν. Then

$$F \otimes_\mathbb{Q} \mathbb{R} \simeq F \otimes_K (K \otimes_\mathbb{Q} \mathbb{R}) \simeq F \otimes_K (\mathbb{R}_{\sigma_1} \times \cdots \times \mathbb{R}_{\sigma_e}) = \mathop{\times}_{\nu=1}^{e} F \otimes_{\sigma_\nu} \mathbb{R} ,$$

with
$$F \otimes_{\sigma_\nu} \mathbb{R} = M_2(\mathbb{R}) \quad \text{or} \quad \mathbb{H}$$

for $1 \leq \nu \leq e$. Denote by x_ν the image of $x \in F$ in $F \otimes_{\sigma_\nu} \mathbb{R}$. The anti-involutions $x \mapsto x'$ and $x \mapsto \bar{x}$ extend in a natural way to anti-involutions $x_\nu \mapsto x'_\nu$ and $x_\nu \mapsto \bar{x}_\nu$ on $F \otimes_{\sigma_\nu} \mathbb{R}$ for $1 \leq \nu \leq e$. Note that in the case $F \otimes_{\sigma_\nu} \mathbb{R} = \mathbb{H}$ the anti-involution $x_\nu \mapsto \bar{x}_\nu$ coincides with the usual Hamiltonian conjugation, whereas in the case $F \otimes_{\sigma_\nu} \mathbb{R} = M_2(\mathbb{R})$ the matrix $\bar{x}_\nu = \text{tr}_{F\otimes_{\sigma_\nu}\mathbb{R}/\mathbb{R}}(x_\nu) - x_\nu$ is the adjoint to the matrix x_ν.

Since F is dense in $F \otimes_{\sigma_\nu} \mathbb{R}$, we have $\text{tr}(x'_\nu x_\nu) \geq 0$ for every $x_\nu \in F \otimes_{\sigma_\nu} \mathbb{R}$ by continuity. But the nullspace of this quadratic form must be a rational subspace, since it is the orthogonal complement of the whole space. Hence it is 0 and $\text{tr}(x'_\nu x_\nu)$ is positive definite on $F \otimes_{\sigma_\nu} \mathbb{R}$ for all $1 \leq \nu \leq e$.

Step III: Either $F \otimes_{\sigma_\nu} \mathbb{R} = M_2(\mathbb{R})$ for all $1 \leq \nu \leq e$ or $F \otimes_{\sigma_\nu} \mathbb{R} = \mathbb{H}$ for all $1 \leq \nu \leq e$. Moreover in the latter case $(F, {}')$ is of type (c).
According to Lemma 5.1 we can identify $F \otimes_{\sigma_\nu} \mathbb{R}$ with $M_2(\mathbb{R})$ respectively $\mathbb{H}$ in such a way that

$$x'_\nu = x^*_\nu$$

for all $x_\nu \in F \otimes_{\sigma_\nu} \mathbb{R}$ and every $1 \leq \nu \leq e$. For a and λ as in Step II we get

$$x^*_\nu = a_\nu^{-1} \bar{x}_\nu a_\nu \quad \text{and} \quad \bar{a}_\nu = \lambda a_\nu \ . \tag{2}$$

Suppose $F \otimes_{\sigma_\mu} \mathbb{R} = \mathbb{H}$ for some $1 \leq \mu \leq e$. By (1) the two canonical anti-involutions agree, that is $x^*_\mu = \bar{x}_\mu$ for all $x_\mu \in F \otimes_{\sigma_\mu} \mathbb{R}$. Hence $a_\mu \in \mathbb{R}$, implying $\lambda = +1$, and thus $a \in K$. It follows that

$$x' = a^{-1} \bar{x} a = \bar{x} \ .$$

Assuming $F \otimes_{\sigma_\nu} \mathbb{R} = M_2(\mathbb{R})$ for some $\nu \neq \mu$ we get

$$\operatorname{tr}(x'_\nu x_\nu) = \operatorname{tr}(\bar{x}_\nu x_\nu) = 2 \det(x_\nu) \tag{3}$$

for all $x_\nu \in M_2(\mathbb{R})$, a contradiction.
Step IV: If $F \otimes_{\sigma_\nu} \mathbb{R} = M_2(\mathbb{R})$ for $1 \leq \nu \leq e$, then a^2 is totally negative.
In this case $\lambda = -1$, since the canonical involution $x_\nu \mapsto \bar{x}_\nu$ is not positive as we saw in (3). So we have $\bar{a}_\nu = -a_\nu$ for every $1 \leq \nu \leq e$. Using (2) we get $a^*_\nu = -a_\nu$ and thus

$$a_\nu = \begin{pmatrix} 0 & \alpha_\nu \\ -\alpha_\nu & 0 \end{pmatrix}$$

for some $\alpha_\nu \in \mathbb{R}$. This implies that $a_\nu^2 = -\alpha_\nu^2 \mathbb{1}_2$ for $\nu = 1, \ldots, e$. So a^2 is totally negative.
Step V: The algebras $(F, {}')$ of type (a), (b) and (c) are of finite dimension over $\mathbb{Q}$ with positive anti-involution of the first kind.
The assertion is obvious in the cases (a) and (c). To verify this for the case (b) we have to show that for any $a \in F$ with $a^2 \in K$ totally negative, the anti-involution $x \mapsto x' = a^{-1} \bar{x} a$ is positive. Let $\sigma_1, \ldots, \sigma_e \colon K \to \mathbb{R}$ denote the real embeddings. For $x \in F$ we denote again by x_ν its image in $F \otimes_{\sigma_\nu} \mathbb{R} = M_2(\mathbb{R})$. By assumption we have $a_\nu^2 = k_\nu \mathbb{1}_2$ with $k_\nu < 0$. An elementary matrix calculation shows that this means

$$a_\nu = \begin{pmatrix} \alpha & \beta \\ \gamma & -\alpha \end{pmatrix} \quad \text{with} \quad \alpha^2 + \beta\gamma = k_\nu < 0 \ .$$

By definition $x^*_\nu = \begin{pmatrix} 0 & 1 \\ -1 & 0 \end{pmatrix} \bar{x}_\nu \begin{pmatrix} 0 & 1 \\ -1 & 0 \end{pmatrix}^{-1}$ and thus

$$x'_\nu = (\begin{pmatrix} 0 & 1 \\ -1 & 0 \end{pmatrix} a_\nu)^{-1} x^*_\nu \begin{pmatrix} 0 & 1 \\ -1 & 0 \end{pmatrix} a_\nu \ .$$

But $\begin{pmatrix} 0 & 1 \\ -1 & 0 \end{pmatrix} a_\nu = \begin{pmatrix} \gamma & -\alpha \\ -\alpha & -\beta \end{pmatrix}$ is symmetric and either positive definite or negative definite, as $\det(\begin{pmatrix} 0 & 1 \\ -1 & 0 \end{pmatrix} a_\nu) = -\alpha^2 - \beta\gamma > 0$. It follows that

$$\operatorname{tr}(x'_\nu x_\nu) = \operatorname{tr}((\begin{pmatrix} 0 & 1 \\ -1 & 0 \end{pmatrix} a_\nu)^{-1} x^*_\nu \begin{pmatrix} 0 & 1 \\ -1 & 0 \end{pmatrix} a_\nu x_\nu) > 0 ,$$

since the product of a positive definite symmetric matrix with a positive semidefinite symmetric matrix in $M_2(\mathbb{R})$ has positive trace. This completes the proof of the theorem. □

Suppose now that $(F, {}')$ is of the second kind, that is the anti-involution $x \mapsto x'$ does not act trivially on the center K of F. According to Lemma 5.2 its fixed field K_0 is totally real. Moreover we have

(5.4) Lemma. *If $(F, {}')$ is a skew field of finite dimension over $\mathbb{Q}$ with positive anti-involution of the second kind, then its center K is totally complex, that is no embedding $K \hookrightarrow \mathbb{C}$ factors via $\mathbb{R}$, and the restriction of the anti-involution to K is complex conjugation.*

Proof. Assume K is not totally complex. Then there is an embedding $\sigma_1: K \hookrightarrow \mathbb{R}$. Let $\sigma_2: K \hookrightarrow \mathbb{R}$ denote the embedding defined by

$$\sigma_2(x) = \sigma_1(x')$$

for all $x \in K$ and denote the remaining embeddings by $\sigma_3, \ldots, \sigma_e: K \hookrightarrow \mathbb{C}$. According to the approximation theorem (see v.d. Waerden [1] II.p.234) for any $\varepsilon > 0$ there is an $x \in K$ with $|\sigma_1(x) + 1| < \varepsilon$, $|\sigma_2(x) - 1| < \varepsilon$, and $|\sigma_\nu(x)| < \varepsilon$ for $3 \le \nu \le e$. For small ε the dominant term of $\operatorname{tr}_{K|\mathbb{Q}}(x'x) = \sum_{\nu=1}^e \sigma_\nu(x'x)$ is $\sigma_1(x'x) + \sigma_2(x'x) = 2\sigma_1(x)\sigma_2(x) \approx -2$. Hence

$$\operatorname{tr}_{F|\mathbb{Q}}(x'x) = [F : K] \operatorname{tr}_{K|\mathbb{Q}}(x'x) < 0 ,$$

contradicting the positivity of the anti-involution $x \mapsto x'$. Hence K is totally complex.

Moreover by Lemma 5.2 complex conjugation induces an involution on K with fixed field K_0, implying that it coincides with the involution $x \mapsto x'$ on K. □

Denoting by $\overline{F}$ the complex conjugate K-algebra of F and by $\overline{F}^{\mathrm{op}}$ the K-algebra opposite to $\overline{F}$, we may consider the anti-involution $x \mapsto x'$ as an isomorphism $F \to \overline{F}^{\mathrm{op}}$ of K-algebras. However, not every such isomorphism corresponds to an anti-involution. In other words, a necessary condition for the existence of an anti-involution of the second kind on F is that F is isomorphic to $\overline{F}^{\mathrm{op}}$.

Conversely, suppose $\tau: F \to F^{\mathrm{op}}$ is any isomorphism. Since τ^2 is an automorphism of F over K, by the Skolem-Noether Theorem there is a $c \in F$ such that

$$\tau^2(x) = c^{-1}xc$$

for all $x \in F$. The following proposition gives a criterion for the existence of an anti-involution on F in terms of c.

(5.5) Proposition. *For a skew field F of finite dimension over $\mathbb{Q}$ with center a totally complex quadratic extension K of a totally real number field K_0 the following conditions are equivalent:*

i) There exists an anti-involution of the second kind on F.

ii) There exists an isomorphism $\tau: F \to F^{\mathrm{op}}$ such that $\tau^2(x) = c^{-1}xc$ for all $x \in F$ and some $c \in F$ implies $c\tau(c) \in N_{K|K_0}(K^*)$.

Proof. By what we said above we have only to show the implication ii) $\Longrightarrow$i). We may assume that $c \notin K$, since otherwise τ is already an anti-involution of the second kind. By assumption there is a $\lambda \in K$ such that $c\tau(c) = \bar{\lambda}\lambda$. Define a map $F \to F$ by $x \mapsto \tilde{x} = (\lambda + c)\tau(x)(\lambda + c)^{-1}$. This is an anti-involution, since

$$\begin{aligned}
\tilde{\tilde{x}} &= (\lambda + c)\,\tau(\lambda + c)^{-1}\tau^2(x)\,\tau(\lambda + c)\,(\lambda + c)^{-1} \\
&= (\lambda + c)\,\tau(\lambda + c)^{-1}c^{-1}xc\,\tau(\lambda + c)\,(\lambda + c)^{-1} \\
&= (\lambda + c)\,(\lambda + c)^{-1}\bar{\lambda}^{-1}x\bar{\lambda}(\lambda + c)\,(\lambda + c)^{-1} = x\,.
\end{aligned}$$ □

The next theorem shows that there is a positive anti-involution on F whenever there is any anti-involution of the second kind. Furthermore it classifies all positive anti-involutions on F.

(5.6) Theorem. *Let F be a skew field of finite dimension over $\mathbb{Q}$ with center a totally complex quadratic extension K of a totally real number field K_0. Moreover suppose that F admits an anti-involution $x \mapsto \tilde{x}$ of the second kind. Then there exists a positive anti-involution $x \mapsto x'$ of the second kind and for every embedding $\sigma: K \hookrightarrow \mathbb{C}$ an isomorphism*

$$\varphi: F \otimes_\sigma \mathbb{C} \xrightarrow{\sim} M_d(\mathbb{C})$$

such that $x \mapsto x'$ extends via φ to the canonical anti-involution $X \mapsto {}^t\overline{X}$ on $M_d(\mathbb{C})$.

Any other positive anti-involution on F is of the form

$$x \mapsto ax'a^{-1}$$

with $a \in F$, $a' = a$ and such that $\varphi(a \otimes 1)$ is a positive definite hermitian matrix in $M_d(\mathbb{C})$ for every embedding $\sigma: K \to \mathbb{C}$.

Proof. Step I: Denote by σ_0 the restriction of σ to K_0. Then we can identify $F \otimes_\sigma \mathbb{C} = F \otimes_\sigma (K \otimes_{\sigma_0} \mathbb{R}) = F \otimes_{\sigma_0} \mathbb{R}$ in such a way that the anti-involution $x \otimes \alpha \mapsto \tilde{x} \otimes \bar{\alpha}$ on $F \otimes_\sigma \mathbb{C}$ translates to the anti-involution $x \otimes r \mapsto \tilde{x} \otimes r$ on $F \otimes_{\sigma_0} \mathbb{R}$. Hence there is an isomorphism

$$\psi\colon F \otimes_{\sigma_0} \mathbb{R} \xrightarrow{\sim} M_d(\mathbb{C})$$

such that the anti-involution $x \mapsto \tilde{x}$ on F extends via ψ to an anti-involution on $M_d(\mathbb{C})$, which by the Skolem-Noether Theorem is of the form

$$X \mapsto A^{-1}\,{}^t\overline{X}A$$

for some $A \in \mathrm{Gl}_d(\mathbb{C})$. From the proof of Lemma 5.1 we see that we may assume ${}^t\overline{A} = A$.

Hence A is contained in the set $\{B \in M_d(\mathbb{C}) | A^{-1}\,{}^t\overline{B}A = B\}$. On the other hand, if U denotes the K_0-vector space $\{b \in F | \tilde{b} = b\}$, we have

$$\psi(U \otimes_{\sigma_0} \mathbb{R}) = \{B \in M_d(\mathbb{C}) | A^{-1}\,{}^t\overline{B}A = B\} .$$

Since U is dense in $U \otimes_{\sigma_0} \mathbb{R}$, there is an $a \in U$ such that $\psi(a \otimes 1)$ is arbitrarily close to A. The map

$$x \mapsto x' = a\tilde{x}a^{-1}$$

is again an anti-involution, since $\tilde{a} = a$. Its extension to $M_d(\mathbb{C})$ is

$$X \mapsto \psi(a \otimes 1)A^{-1}\,{}^t\overline{X}A\psi(a \otimes 1)^{-1} .$$

This is a positive anti-involution, since $X \mapsto {}^t\overline{X}$ is a positive anti-involution on $M_d(\mathbb{C})$ and $\psi(a \otimes 1)A^{-1}$ is arbitrarily close to $\mathbb{1}_d$. Thus we have shown that $x \mapsto x'$ is a positive anti-involution on F. According to Lemma 5.1 there is an isomorphism $\varphi\colon F \otimes_\sigma \mathbb{C} \xrightarrow{\sim} M_d(\mathbb{C})$ as claimed.

Step II: By the Skolem-Noether Theorem any positive anti-involution on F is of the form

$$x \mapsto ax'a^{-1}$$

with $a \in F$. As in the proof of Lemma 5.1 we see that $a' = \lambda a$ for some $\lambda \in K$ with $\bar{\lambda}\lambda = 1$. Applying Hilbert's Satz 90 (see Jacobson [1] I, Theorem 4.31) there is a $\mu \in K$ such that $\lambda = \bar{\mu}\mu^{-1}$. Replacing a by $\mu^{-1}a$ we see that we may assume $a' = a$.

Hence for $A = \varphi(a \otimes 1)$ we have ${}^t\overline{A} = A$ and it remains to show that A is positive definite. For the hermitian matrix A there is a unitary matrix $T \in M_d(\mathbb{C})$ such that ${}^t\overline{T}AT = \mathrm{diag}(r_1, \dots, r_d)$ for some $r_\nu \in \mathbb{R}$, $1 \le \nu \le d$. But for all matrices $X = (x_{ij}) \in M_d(\mathbb{C})$, $X \ne 0$, we have

$$
\begin{aligned}
0 &< \operatorname{tr}\bigl(A^t(TX\,{}^t\overline{T})A^{-1}TX\,{}^t\overline{T}\bigr)\\
&= \operatorname{tr}\bigl(({}^t\overline{T}AT)\,{}^t\overline{X}({}^t\overline{T}AT)^{-1}X\bigr)\\
&= \operatorname{tr}\bigl(\operatorname{diag}(r_1,\dots,r_d)\,{}^t\overline{X}\,\operatorname{diag}(r_1,\dots,r_d)^{-1}X\bigr)\\
&= \sum_{i,j=1}^{d} |x_{ij}|^2 \frac{r_j}{r_i}\,.
\end{aligned}
$$

Hence A or $-A$ is positive definite. Since we may replace a by $-a$, this completes the proof. □

Recall the situation of the beginning of this section: let X be a simple abelian variety of dimension g with polarization L. Then $F = \operatorname{End}_{\mathbb{Q}}(X)$ is a skew field of finite dimension over $\mathbb{Q}$ with positive anti-involution $x \mapsto x'$, the Rosati involution with respect to L. In Theorems 5.3 and 5.6 we have seen the possibilities for such a pair $(F,\ ')$.

As above let K denote the center of F and K_0 the fixed field of the anti-involution restricted to K. Denote

$$[F:K]=d^2,\quad [K:\mathbb{Q}]=e,\quad [K_0:\mathbb{Q}]=e_0 \quad \text{and} \quad \operatorname{rk}\operatorname{NS}(X)=\varrho\,.$$

Then we have the following restrictions for these values:

(5.7) Proposition.

$F = \operatorname{End}_{\mathbb{Q}}(X)$	d	e_0	ϱ	*restriction*
totally real number field	1	e	e	$e\mid g$
totally indefinite quaternion algebra	2	e	$3e$	$2e\mid g$
totally definite quaternion algebra	2	e	e	$2e\mid g$
$(F,\ ')$ *of the second kind*	d	$\frac{1}{2}e$	e_0d^2	$e_0d^2\mid g$

Proof. The values of d and e_0 follow from Theorems 5.3 and 5.6. In order to compute the Picard number ϱ recall from Proposition 2.1 that $\varrho = \dim_{\mathbb{R}} \operatorname{End}^s_{\mathbb{Q}}(X) \otimes_{\mathbb{Q}} \mathbb{R}$. Obviously $\varrho = e$ in the totally real number field case. In the case that $\operatorname{End}_{\mathbb{Q}}(X)$ is a totally indefinite quaternion algebra we have $\operatorname{End}_{\mathbb{Q}}(X) \otimes_{\mathbb{Q}} \mathbb{R} \simeq \times_{\nu=1}^{e} M_2(\mathbb{R})$ by Theorem 5.3 and Lemma 5.1 such that the anti-involution translates to transposition on the factors. So $\varrho = 3e$. Similarly in the totally definite quaternion algebra case $\operatorname{End}_{\mathbb{Q}}(X) \otimes_{\mathbb{Q}} \mathbb{R} \simeq \times_{\nu=1}^{e} \mathbb{H}$ such that the anti-involution translates to quaternion conjugation on the factors and thus $\varrho = e$. Finally in the last case, by Theorem 5.6 we have an isomorphism $\operatorname{End}_{\mathbb{Q}}(X) \otimes_{\mathbb{Q}} \mathbb{R} \simeq \times_{\nu=1}^{e_0} M_d(\mathbb{C})$ carrying the anti-involution to $X \mapsto {}^t\overline{X}$ on every factor and thus $\varrho = e_0 d^2$.

As for the restrictions, note first that $\dim_{\mathbb{Q}} F = ed^2$ divides $2g$, since $\operatorname{End}_{\mathbb{Q}}(X)$ admits a faithful representation in the vector space $\Lambda \otimes \mathbb{Q}$ and thus $\Lambda \otimes \mathbb{Q}$ is a vector space over the skew field F. This gives the restrictions for the last 3 lines.

It remains to show that $e|g$, if $(\mathrm{End}_{\mathbb{Q}}(X), {}') = (K, \mathrm{id}_K)$. For this consider the isomorphism $\varphi: \mathrm{NS}_{\mathbb{Q}}(X) \to \mathrm{End}^s_{\mathbb{Q}}(X) = \mathrm{End}_{\mathbb{Q}}(X) = K$ from Proposition 2.1. Define a map $p: \mathrm{End}(X) \to \mathbb{Q}$ by

$$p(f) = \frac{\chi(\varphi^{-1}(f))}{\chi(L)} .$$

According to the Geometric Riemann-Roch Theorem p is a homogeneous polynomial function of degree g on the $\mathbb{Z}$-module $\mathrm{End}(X)$. Hence, setting $p(\frac{f}{n}) = n^{-g}p(f)$, we can extend it to a homogeneous polynomial function of degree g on the whole $\mathbb{Q}$-vector space K, which we also denote by p.

We claim that $p: K \to \mathbb{Q}$ is multiplicative. For the proof let $f_1, f_2 \in \mathrm{End}(X)$. Applying Remark 2.2 b) and Corollary 3.6.2 we have

$$\begin{aligned} p(f_1f_2)^2 &= \frac{\deg(\varphi^{-1}(f_1f_2))}{\deg \phi_L} \\ &= \frac{\deg(\varphi^{-1}(f_1)\phi_L^{-1}\varphi^{-1}(f_2))}{\deg \phi_L} \\ &= \frac{\deg \varphi^{-1}(f_1)}{\deg \phi_L} \cdot \frac{\deg \varphi^{-1}(f_2)}{\deg \phi_L} = p(f_1)^2 p(f_2)^2 . \end{aligned}$$

Since p is a polynomial function and $p(1f) = +p(1)p(f)$ for all $f \in \mathrm{End}^s_{\mathbb{Q}}(X)$, this proves the claim.

Finally it is well known that the degree of any multiplicative homogeneous polynomial function on the $\mathbb{Q}$-vector space K is divisible by the dimension of K over $\mathbb{Q}$ (see Mumford [2] Lemma p. 179). Hence $e = [K : \mathbb{Q}]$ divides g. □

Theorems 5.3, 5.6, and Proposition 5.7 provide necessary conditions for a skew field F with positive anti-involution $'$ to be the endomorphism algebra of a simple abelian variety. In Chapter 9 we will study the converse question and construct abelian varieties for a given endomorphism algebra. It turns out that, apart from some exceptions, all of the above pairs $(F, {}')$ can be realized.

Exercises

(1) For a square free integer $d \geq 1$ consider the imaginary quadratic field $\mathbb{Q}(\sqrt{-d})$ with maximal order $\mathfrak{o}$. If $\{1, \omega\}$ denotes the usual basis of $\mathfrak{o}$ and f a positive integer, then $\mathfrak{o}_f = \mathbb{Z} \oplus f\omega\mathbb{Z}$ is a lattice in $\mathbb{C}$ and $E_f = \mathbb{C}/\mathfrak{o}_f$ is an elliptic curve. Show that $\mathrm{End}(E_f) = \mathfrak{o}_f$. In particular, if $f \geq 2$, then $\mathrm{End}(E_f)$ is not a maximal order in $\mathbb{Q}(\sqrt{-d})$.

(2) Let X_ν be an abelian variety with polarization L_ν of degree d_ν for $\nu = 1,2$. Then $p_1^* L_1 \otimes p_2^* L_2$ is a polarization on $X_1 \times X_2$ of degree $d_1 d_2$.

(3) *(Zarhin's trick)* Let L be a polarization of exponent e on an abelian variety $X = V/\Lambda$.

a) Suppose there is an $f \in \mathrm{End}(X)$ such that

i) $f(K(L)) \subseteq K(L)$

ii) $\varrho_a(f'f)|\frac{1}{e}\Lambda \equiv -\mathrm{id}_{\frac{1}{e}\Lambda} \pmod{\Lambda}$, where $'$ denotes the Rosati involution with respect to L.

Then $X \times \widehat{X}$ is principally polarized.

b) Conclude that for any abelian variety X the abelian variety $(X \times \widehat{X})^4$ is principally polarized.

(4) Let f be an automorphism of order n of an abelian variety of dimension g. Show that $\varphi(n) \leq 2g$, where φ is the Euler function of elementary number theory. In particular, $n \leq 6$ for $g = 1$, $n \leq 12$ for $g = 2$ and $n \leq 18$ for $g = 3$.

(5) Let X be an abelian variety. There exists an integer $n = n(X)$ such that for any abelian subvariety Y of X there exists an abelian subvariety Z of X with $Y + Z = X$ and $\# Y \cap Z \leq n$ (see Bertrand [1]).

(6) Let X be an abelian variety and L a line bundle on X. For any $f_1, f_2 \in \mathrm{End}(X)$ define a line bundle on X by

$$D_L(f_1, f_2) = (f_1 + f_2)^* L \otimes f_1^* L^{-1} \otimes f_2^* L^{-1} .$$

Note that $D_L(f, 1_X) = D_L(f)$ as defined in Section 1.

a) $D_L \colon \mathrm{End}(X) \times \mathrm{End}(X) \to \mathrm{Pic}(X)$ is symmetric and bilinear.

b) The map D_L depends only on the class of L in $\mathrm{NS}(X)$: $D_L = D_{L \otimes P}$ for all $P \in \mathrm{Pic}^0(X)$.

c) $\phi_{D_L(f_1,f_2)} = \widehat{f_1} \phi_L f_2 + \widehat{f_2} \phi_L f_1$.

d) Suppose $M \in \mathrm{Pic}(X)$ defines a principal polarization on X, then $D_M(\phi_M^{-1} \widehat{f_1} \phi_L f_2) = D_L(f_1, f_2)$.

(7) Let X be an abelian variety of dimension g and $L \in \mathrm{Pic}(X)$ a polarization. Then $\mathrm{Tr}_r(f_1' f_2) = \frac{g}{(L^g)}(D_L(f_1, f_2) \cdot L^{g-1})$ for all $f_1, f_2 \in \mathrm{End}(X)$. (Hint: generalize the proof of Proposition 1.5)

(8) Show that for line bundles on an abelian variety numerical equivalence and algebraic equivalence coincide. (Hint: suppose $L \sim_{\mathrm{num}} 0$, then $f^* L \sim_{\mathrm{num}} 0$ for any endomorphism f of X. Conclude $0 = (D_L(f_1, f_2) \cdot M^{g-1})$ for some polarization $M \in \mathrm{Pic}(X)$ and all $f_1, f_2 \in \mathrm{End}(X)$. Use Exercises 13 a) and 14) to deduce $\mathrm{Tr}_r(\phi_M^{-1} \widehat{f_1} \phi_L f_2) = 0$. Inserting $f_1 = 1_X$ and $f_2 = \phi_M^{-1} \phi_L$ gives $\phi_L = 0$.)

(9) Let $X \subset \mathbb{P}_n$ be a projectively normal embedding of an abelian variety of dimension g and $f \in \text{End}(X)$. We say that f *can be described by forms of degree* d on an open dense set $U \subseteq X$, if there are polynomials $p_0, \ldots, p_n \in \mathbb{C}[X_0, \ldots, X_n]$, homogeneous of degree d and not all zero such that for every $x \in U$ we have $f(x) = (p_0(x) : \cdots : p_n(x))$. We say that f *can be described completely by forms of degree* d, if there is an open covering $\{U_i\}_{i \in I}$ of X such that f can be described by forms of degree d on every U_i.

a) Show that f can be described completely be forms of degree d if and only if the line bundle $f^*\mathcal{O}_X(-1) \otimes \mathcal{O}_X(d)$ is base-point free.

b) Let $a_1, \ldots, a_g \in \mathbb{R}$ be the eigenvalues of the analytic representation of $f'f$, where f' is the Rosati adjoint of f. If $d > \max(a_1, \ldots, a_g) + 1$, then f can be described completely by forms of degree d. In particular, the multiplication map d_X can be described completely by forms of degree $d^2 + 2$.

c) The addition map $\mu: X \times X \to X \hookrightarrow X \times X$ can be described completely by quadrics if and only if $\mathcal{O}_X(1)$ is symmetric (see Lange [3]).

(10) Let X be an abelian variety of dimension g. Recall from Exercise 2.5 that $\rho(X) \leq g^2$. The following conditions are equivalent

i) $\rho(X) = g^2$.

ii) X is isogeneous to E^g with an elliptic curve E with complex multiplication.

iii) X admits a period matrix $\Pi \in M(g \times 2g, K)$ with K an imaginary quadratic field.

iv) X is isomorphic to a product $E_1 \times \cdots \times E_g$ with pairwise isogeneous elliptic curves with complex multiplication.

(Hint for iii) $\iff$ iv): use Exercise 10.5.)

(11) *(Narasimhan-Nori Theorem)* Let X be an abelian variety and d a positive integer. The number of isomorphism classes of polarizations of degree d is finite. (Hint: show that $\text{Aut}(X)$ is an arithmetic group and apply a result of Borel (see Borel[1], Théorème 9.11, Narasimhan-Nori [1]).)

(12) *(The number $\pi(X)$ of isomorphism classes of principal polarizations of X)* Let X be an abelian variety of dimension g with $\text{End}(A) = \mathfrak{o}$ the maximal order of a totally real number field K of degree g over $\mathbb{Q}$. Let U denote the group of units in $\mathfrak{o}$ and U^+ the subgroup of totally positive units. If h denotes the class number and h^+ the narrow class number of K (e.g. the order of the factor group of the group of ideals modulo the subgroup of totally positive principal ideals), then we have for the number of isomorphism classes $\pi(X)$ of principal polarizations of X: $\pi(X) = 0$ or $\pi(X) = \#U^+/U^2 = \frac{h^+}{h}$. (See Lange [2].)

(13) Let the notation be as in Exercise 9 and assume that X admits a principal polarization.

a) If $\sigma_0, \ldots, \sigma_{g-1}$ denote the real embeddings $K \hookrightarrow \mathbb{R}$, and if $\eta_1, \ldots, \eta_{g-1}$ is a system of fundamental units of $\mathfrak{o}$ and $\eta_0 = -1$, then

$$\pi(X) = 2^{g-\mathrm{rk}M} ,$$

where M denotes the matrix $(\operatorname{sign}\sigma_i(\eta_j))$.

b) Conclude the following result of Humbert [1]: If $g = 2$, then

$$\pi(X) = \begin{cases} 1 \\ 2 \end{cases} \text{if } \operatorname{Aut}(X) \text{ contains } \begin{matrix} \text{an} \\ \text{no} \end{matrix} \text{ element of negative norm.}$$

c) If $g = 3$, we have $\pi(X) = 1, 2$ or 4. Suppose $K = \mathbb{Q}(\omega)$. Show that $\pi(X) = 2$, if ω is the root of $x^3 + 12x^2 + 32x - 1 = 0$ with $0 < \omega < \frac{1}{2}$. Show that $\pi(X) = 4$, if ω is the root of $x^3 - 12x^2 + 26x - 1 = 0$.

d) Let $g = 4$ and $K = \mathbb{Q}(\sqrt{6}, \sqrt{7})$. Show that $\pi(X) = 2$. (Hint: use the set of fundamental units $\eta_1 = 8 + 3\sqrt{7}, \eta_2 = \sqrt{6} + \sqrt{7}$ and $\eta_3 = \frac{1}{2}(6 + 6\sqrt{6} + 2\sqrt{7} + \sqrt{42})$.)

(14) Give an example of an abelian surface X admitting two principal polarizations L_0 and L_1 such that $(L_0 \cdot L_1) \geq 3$. (Hint: use Exercise 13 b) and Proposition 2.3.)

(15) Let (X, L) be a polarized abelian variety. Consider the canonical isomorphism $\varphi\colon \mathrm{NS}_{\mathbb{Q}}(X) \to \mathrm{End}_{\mathbb{Q}}(X)$ of Proposition 2.1. Show that for any abelian subvariety Y of X with norm-endomorphism N_Y with respect to L:

$$\varphi^{-1}(N_Y) = e(Y)^{-1} N_Y^* L .$$

(16) *(A Second Proof for Corollary 3.3.)* Let L be a polarization on the abelian variety X and $f \in \mathrm{End}(X)$, $Y = \mathrm{im} f$, with $f' = f$ and $f^2 = e(i_Y^* L) f$. Denote $Z = \phi_L(Y)$ and $\iota_Z\colon Z \to \widehat{X}$ the canonical embedding.

a) There is a $\varphi \in \mathrm{Hom}(X, Z)$ such that

$$\phi_L f = \iota_Z \varphi . \tag{1}$$

Show that $e(L)\phi_L^{-1}\widehat{\varphi}(\widehat{Z}) = Y$.

b) There is a $\psi \in \mathrm{Hom}(\widehat{Z}, Y)$ such that

$$e(L)\phi_L^{-1}\widehat{\varphi} = \iota_Y \psi . \tag{2}$$

Show that $$\phi_L^{-1}\iota_Z\widehat{\psi} = e(L)\, e(\iota_Y^* L)\, \iota_Y\, \phi_{\iota_Y^* L}^{-1} . \tag{3}$$

c) Use (1), (2) and (3) to conclude that $f = N_Y$ (see Birkenhake-Lange [4]).

(17) Let E be an elliptic curve with $\mathrm{End}(E) = \mathbb{Z}$ and $X = E \times E$. The line bundle $L = p_1^*\mathcal{O}_E(0) \otimes p_2^*\mathcal{O}_E(0)$ defines a principal polarization on X.

a) There is a canonical isomorphism $\varphi\colon \mathrm{End}(X) \xrightarrow{\sim} M_2(\mathbb{Z})$ such that for the Rosati involution $'$ (with respect to L)

$$\varphi(f') = {}^t\varphi(f) \ .$$

b) The preimages under φ of

$$\tfrac{1}{1+r^2}\begin{pmatrix} 1 & r \\ r & r^2 \end{pmatrix}, \quad \tfrac{1}{1+r^2}\begin{pmatrix} r^2 & -r \\ -r & 1 \end{pmatrix},$$

where $r \in \mathbb{Q}$ arbitrary, are all the symmetric idempotents $\neq 0, 1$ in $\mathrm{End}_{\mathbb{Q}}(X)$.

(18) Let (X, L) be a polarized abelian variety and $\varepsilon_1, \ldots, \varepsilon_{r+s}$ are symmetric idempotents of $\mathrm{End}_{\mathbb{Q}}(X)$ with $\sum_{\nu=1}^r \varepsilon_\nu = \sum_{\nu=1}^s \varepsilon_{r+\nu}$. Show that the abelian varieties $\times_{\nu=1}^r X^{\varepsilon_\nu}$ and $\times_{\nu=1}^s X^{\varepsilon_{r+\nu}}$ are isogenous.

(19) Let X be an abelian variety of dimension g and Y and Y' abelian subvarieties of dimension p.

a) For any subvariety Z of dimension $g-p$ the homomorphism $\delta(Y, Z)\colon X \to Y$ is surjective.

b) If Y is numerical equivalent to Y', then $Y = Y'$. (Hint: use a) and the Remark after Corollary 4.5.)

(20) Suppose X is an abelian variety of dimension g with $\mathrm{End}_{\mathbb{Q}}(X)$ a commutative field. Let K_0 be the maximal totally real subfield and $m = \frac{g}{[K_0:\mathbb{Q}]}$. Any nontrivial line bundle L on X is nondegenerate of index $i(L) = \nu m$ for some integer $0 \leq \nu \leq [K_0 : \mathbb{Q}]$. Moreover, for any of these values there is a line bundle of this index. In particular, if $\mathrm{End}(X) = \mathbb{Z}$, any nontrivial line bundle is of index 0 or g.

Chapter 6

Theta and Heisenberg Groups

For the introductory remarks of this chapter let us assume that L is a very ample line bundle on an abelian variety $X = V/\Lambda$ and $\varphi_L: X \hookrightarrow \mathbb{P}_N$ the associated embedding. Recall the group $K(L)$ consisting of all $x \in X$ with $t_x^* L \simeq L$. We will see that the translations of X by elements of $K(L)$ extend to linear automorphisms of $\mathbb{P}_N$. In fact, $K(L)$ is the largest group of translations with this property. This leads to a projective representation $\varrho: K(L) \to \mathrm{PGL}_N(\mathbb{C})$, with respect to which the embedding φ_L is equivariant. It will be an important tool in the investigation of the geometric properties of the embedded abelian variety $\varphi_L(X)$ in $\mathbb{P}_N$.

It is not possible to lift ϱ to an ordinary representation of $K(L)$. However ϱ lifts to an ordinary representation $\tilde{\varrho}$ of the extension $\mathcal{G}(L)$ of $K(L)$, defined by the following pullback diagram

$$\begin{array}{ccccccccc} 1 & \longrightarrow & \mathbb{C}^* & \longrightarrow & \mathcal{G}(L) & \longrightarrow & K(L) & \longrightarrow & 0 \\ & & \| & & \downarrow{\scriptstyle\tilde{\varrho}} & & \downarrow{\scriptstyle\varrho} & & \\ 1 & \longrightarrow & \mathbb{C}^* & \longrightarrow & \mathrm{GL}_{N+1}(\mathbb{C}) & \longrightarrow & \mathrm{PGL}_N(\mathbb{C}) & \longrightarrow & 0 \end{array} .$$

The group $\mathcal{G}(L)$ is called the *theta group* of L. Since the representation $\tilde{\varrho}$ can be described explicitly in terms of canonical theta functions, this leads to a description of ϱ.

Although the representation $\tilde{\varrho}$ has already been used occasionally in the early days of the investigation of theta functions (see e.g. Krazer [1]), the first systematic account was given by Mumford. In [1] Mumford introduced the theta group $\mathcal{G}(L)$ and its representation $\tilde{\varrho}$ on $H^0(L)$ to study abelian varieties over an arbitrary algebraically closed field k, under some mild hypotheses. The crucial point is the Stone-von Neumann Theorem (see Exercise 3), which says that there is only one irreducible representation of $\mathcal{G}(L)$ inducing the natural multiplication on the subgroup $\mathbb{C}^*$.

The *Heisenberg group* $\mathcal{H}(D)$ is an abstract version of the theta group $\mathcal{G}(L)$. It depends only on the type D of the polarization defined by L. A natural irreducible representation of $\mathcal{H}(D)$ is given by the vector space $k[\mathbb{Z}^g/D\mathbb{Z}^g]$ of k-valued functions on the finite group $\mathbb{Z}^g/D\mathbb{Z}^g$, called the *Schrödinger representation.* The Stone-von Neumann Theorem implies that the representations $H^0(L)$ and $k[\mathbb{Z}^g/D\mathbb{Z}^g]$ are isomorphic. So by Schur's lemma one knows $H^0(L)$ up to a nonzero constant. Hence it makes sense to call the elements of $k[\mathbb{Z}^g/D\mathbb{Z}^g]$ *finite theta functions.* They may be thought of as a replacement of the classical theta functions for abelian varieties over an arbitrary algebraically closed field.

Over the field of complex numbers one can do slightly better, since one can apply the theory of theta functions to determine the representation $H^0(L)$. We will see in Proposition 4.2 how the theta group acts on the basis of canonical theta functions of Theorem 3.2.7. As a consequence we obtain a direct proof (not using the Stone-von Neumann Theorem) for the irreducibility of the representation $H^0(L)$ (see Corollary 4.3). The effect of this is that the unknown constant appearing in Mumford's approach does not turn up. In fact, everything is normalized in such a way that the constant is 1. So instead of equations up to a constant we obtain equations here.

Strictly speaking, the theory of finite theta functions, that is, the Heisenberg group and its Schrödinger representation, are not necessary in our context. However psychologically it seems easier to work with finite theta functions, since one does not have to worry about the analytic structure of the theta functions. We will introduce this language in the second part of this chapter.

In Sections 1 to 4 we derive the basic properties of the theta group and its canonical representation. Section 5 contains a proof of the Isogeny Theorem, which describes the pullback of canonical theta functions via an isogeny in terms of the basis of Theorem 3.2.7. In Sections 6 and 7 we introduce the Heisenberg group and its Schrödinger representation. Section 8 contains a version of the Isogeny Theorem in terms of finite theta functions. Finally in Section 9 we compute for any symmetric line bundle the number of symmetric theta structures. These are isomorphisms between the theta and Heisenberg groups compatible with the action of (-1). Here we follow Birkenhake-Lange [2].

This chapter is essentially self-contained. Apart from Theorem 3.2.7 we only use some properties of symmetric line bundles from Section 4.6.

§ 1 Theta Groups

Let L be an arbitrary line bundle on an abelian variety $X = V/\Lambda$. In this section we want to introduce the theta group $\mathcal{G}(L)$ consisting of all automorphisms of L lying over translations of X. To get a good description of the elements of the theta group we show that $\mathcal{G}(L)$ is a quotient of a group

$\mathcal{G}(L)$ consisting of automorphisms of the line bundle $\pi^* L \simeq V \times \mathbb{C}$ on V, also called a theta group for L.

Suppose $x \in X$. A biholomorphic map $\varphi: L \to L$ is called an *automorphism of L over x*, if the diagram

$$\begin{array}{ccc} L & \xrightarrow{\varphi} & L \\ \downarrow & & \downarrow \\ X & \xrightarrow{t_x} & X \end{array}$$

commutes, and for every $y \in X$ the induced map $\varphi(y): L(y) \to L(x+y)$ on the fibres is $\mathbb{C}$-linear. Define $\mathcal{G}(L)$ to be the set of all automorphisms of L over points of X. We write the elements of $\mathcal{G}(L)$ in the form (φ, x), although φ determines x uniquely. The composition of automorphisms

$$(\varphi_1, x_1)(\varphi_2, x_2) = (\varphi_1 \varphi_2, x_1 + x_2)$$

defines a group structure on $\mathcal{G}(L)$. The group $\mathcal{G}(L)$ is called the *theta group of L*. Recall that $K(L)$ denotes the group of all $x \in X$ with $t_x^* L \simeq L$ over X.

(1.1) Proposition. *The sequence*

$$1 \longrightarrow \mathbb{C}^* \xrightarrow{i} \mathcal{G}(L) \xrightarrow{p} K(L) \longrightarrow 0$$

with $i(\alpha) = (\alpha, 0)$ and $p(\varphi, x) = x$ is exact. Moreover $\mathcal{G}(L)$ is a central extension of $K(L)$ by $\mathbb{C}^*$.

Proof. Suppose $(\varphi, x) \in \mathcal{G}(L)$. By definition $t_x^* L = X \times_X L$ is the fibre product of $t_x: X \to X$ with the bundle projection $L \to X$. According to the universal property of the fibre product there is a unique isomorphism $\widetilde{\varphi}: L \to t_x^* L$ of line bundles over X such that the following diagram commutes

$$\begin{array}{ccccc} L & & \xrightarrow{\varphi} & & L \\ \Big\downarrow & \searrow^{\widetilde{\varphi}} & & \nearrow & \Big\downarrow \\ & & t_x^* L & & \\ \Big\downarrow & \swarrow & & & \Big\downarrow \\ X & & \xrightarrow{t_x} & & X \end{array} \qquad (1)$$

In particular $t_x^* L \simeq L$, so $x \in K(L)$.

By definition of $K(L)$ there is an isomorphism $\widetilde{\varphi}: L \to t_x^* L$ over X for every $x \in K(L)$. The composition of $\widetilde{\varphi}$ with the natural projection $t_x^* L = X \times_X L \to L$ is an automorphism of L over x, so p is surjective. Since i and p are homomorphisms of groups, we have to show that $i(\mathbb{C}^*)$ is the kernel of p. But this is an immediate consequence of the fact that every automorphism of L over X is just multiplication by a nonzero constant. Obviously every such automorphism $(\alpha, 0)$ commutes with any automorphism of L over x. Hence $\mathcal{G}(L)$ is a central extension of $K(L)$ with $\mathbb{C}^*$. □

(1.2) Remark. Define $\widetilde{\mathcal{G}}(L)$ to be the set of linear isomorphisms $\widetilde{\varphi}: L \to t_x^* L$ over X. According to the proof above the map $\mathcal{G}(L) \to \widetilde{\mathcal{G}}(L)$, $\varphi \to \widetilde{\varphi}$ is a bijection. Hence the group structure of $\mathcal{G}(L)$ induces one of $\widetilde{\mathcal{G}}(L)$, namely

$$\widetilde{\varphi}_1 \cdot \widetilde{\varphi}_2 := (t_{x_2}^* \widetilde{\varphi}_1) \widetilde{\varphi}_2$$

for linear isomorphisms $\widetilde{\varphi}_\nu: L \to t_{x_\nu}^* L$, $\nu = 1, 2$. The group $\widetilde{\mathcal{G}}(L)$ is the theta group of L in the sense of Mumford, as defined in Mumford [1] p.289.

Let $\pi: V \to X$ denote the canonical projection. We use the fact that $\pi^* L$ is the trivial line bundle $V \times \mathbb{C}$ on V (see Lemma 2.1.1) in order to describe the elements of $\mathcal{G}(L)$. In fact, there is a group $G(L)$ of automorphisms of $V \times \mathbb{C}$ inducing the elements of $\mathcal{G}(L)$. It will turn out that $G(L)$ is a central extension of $\Lambda(L) = \pi^{\ 1}(K(L))$ with $\mathbb{C}^*$, appearing as the pullback of the sequence $1 \to \mathbb{C}^* \to \mathcal{G}(L) \to K(L) \to 0$ via the canonical projection $\pi: \Lambda(L) \to K(L)$.

Suppose $L = L(H, \chi)$. For $\alpha \in \mathbb{C}^*$ and $w \in \Lambda(L)$ define a holomorphic map $[\alpha, w]: V \times \mathbb{C} \to V \times \mathbb{C}$ by

$$[\alpha, w](v, t) = \big(v + w, \alpha e\big(\pi H(v, w)\big)t\big) .$$

Since the diagram

$$\begin{array}{ccc} V \times \mathbb{C} & \xrightarrow{[\alpha,w]} & V \times \mathbb{C} \\ \downarrow & & \downarrow \\ V & \xrightarrow{t_w} & V \end{array}$$

commutes and $[\alpha, w]$ restricts to vector space isomorphisms on the fibres of $V \times \mathbb{C} \to V$, it is a linear automorphism of the trivial line bundle on V over the translation t_w. Let $G(L)$ denote the set of all these automorphisms, i.e.

$$G(L) = \big\{[\alpha, w] \big| \alpha \in \mathbb{C}^*, w \in \Lambda(L)\big\} .$$

Composition of maps defines a group structure on $G(L)$, namely

$$[\alpha_1, w_1][\alpha_2, w_2] = \big[\alpha_1 \alpha_2 e\big(\pi H(w_2, w_1)\big), w_1 + w_2\big] . \tag{2}$$

Note that $[1, 0]$ is the unit in $G(L)$ and $[\alpha, w]^{-1} = [\alpha^{-1} e(\pi H(w, w)), -w]$. Moreover we have the following exact sequence

$$1 \longrightarrow \mathbb{C}^* \xrightarrow{\ j\ } G(L) \xrightarrow{\ q\ } \Lambda(L) \longrightarrow 0 \tag{3}$$

with $j(\alpha) = [\alpha, 0]$ and $q[\alpha, w] = w$. So (2) shows that $G(L)$ is a central extension of $\Lambda(L)$ with $\mathbb{C}^*$.

As usual let $a_L(\lambda, v) = \chi(\lambda)e(\pi H(v, \lambda) + \frac{\pi}{2}H(\lambda, \lambda))$ for $\lambda \in \Lambda$ and $v \in V$ denote the canonical factor of the line bundle L. The map

$$s_L: \Lambda \to G(L)\,,\; s_L(\lambda) = [a_L(\lambda, 0), \lambda]$$

is a section of $q: G(L) \to \Lambda(L)$ over Λ. Using the cocycle relation and Lemma 3.1.3 a), or just by an immediate computation, we see that s_L is an injective homomorphism of groups. We claim that $s_L(\Lambda)$ is contained in the center of $G(L)$. To see this note that by definition of $\Lambda(L)$ (see Section 2.4) $e(\pi H(\lambda, w) - \pi H(w, \lambda)) = e(2\pi i \operatorname{Im} H(\lambda, w)) = 1$ for all $w \in \Lambda(L)$ and $\lambda \in \Lambda$. Hence

$$[\alpha, w][a_L(\lambda, 0), \lambda] = [\alpha a_L(\lambda, 0)e(\pi H(\lambda, w)), \lambda + w] = [a_L(\lambda, 0), \lambda][\alpha, w]$$

for all $[\alpha, w] \in G(L)$ and $\lambda \in \Lambda$. This shows that (3) induces an exact sequence

$$1 \longrightarrow \mathbb{C}^* \longrightarrow G(L)/s_L(\Lambda) \longrightarrow \Lambda(L)/\Lambda \longrightarrow 0\,.$$

The main result of this section is the following

(1.3) Theorem. *There is a canonical isomorphism of exact sequences*

$$\begin{array}{ccccccccc} 1 & \longrightarrow & \mathbb{C}^* & \longrightarrow & G(L)/s_L(\Lambda) & \longrightarrow & \Lambda(L)/\Lambda & \longrightarrow & 0 \\ & & \| & & \downarrow{\scriptstyle\bar\sigma} & & \| & & \\ 1 & \longrightarrow & \mathbb{C}^* & \longrightarrow & \mathcal{G}(L) & \longrightarrow & K(L) & \longrightarrow & 0\,. \end{array}$$

Proof. As we saw in Section 2.2 we have $L \simeq V \times \mathbb{C}/\Lambda$, where Λ acts on $V \times \mathbb{C}$ via the canonical factor a_L of L. Since

$$\lambda(v, t) = \big(v + \lambda, a_L(\lambda, v)t\big) = \big[a_L(\lambda, 0), \lambda\big](v, t)$$

for all $\lambda \in \Lambda$ and $(v, t) \in V \times \mathbb{C}$, the action of Λ on $V \times \mathbb{C}$ coincides with the action of the subgroup $s_L(\Lambda)$ of $G(L)$ on $V \times \mathbb{C}$. So $L \simeq V \times \mathbb{C}/s_L(\Lambda)$.

Define a map $\sigma: G(L) \to \mathcal{G}(L)$ as follows. Every $[\alpha, w] \in G(L)$ defines an automorphism $\varphi_{\alpha,w}$ of L, since $s_L(\Lambda)$ is contained in the center of $G(L)$. It is obvious that $\varphi_{\alpha,w}$ is an automorphism over $\overline{w} \in \pi(\Lambda(L)) = K(L)$, so $\varphi_{\alpha,w} \in \mathcal{G}(L)$. Moreover the map $\sigma: G(L) \to \mathcal{G}(L)$, $[\alpha, w] \mapsto \varphi_{\alpha,w}$ is a homomorphism of groups and the following diagram commutes

$$\begin{array}{ccccccccc} 1 & \longrightarrow & \mathbb{C}^* & \longrightarrow & G(L) & \longrightarrow & \Lambda(L) & \longrightarrow & 0 \\ & & \| & & \downarrow{\scriptstyle\sigma} & & \downarrow{\scriptstyle\pi} & & \\ 1 & \longrightarrow & \mathbb{C}^* & \longrightarrow & \mathcal{G}(L) & \longrightarrow & K(L) & \longrightarrow & 0\,. \end{array} \tag{4}$$

By construction σ factorizes via $G(L)/s_L(\Lambda)$ and the assertion follows from the snake lemma. $\square$

The diagram (4) shows that the sequence (3) is the pullback of the sequence of Proposition 1.1 via $\pi\colon \Lambda(L) \to K(L)$. Since σ is surjective, the elements of $G(L)$ represent the elements of the theta group $\mathcal{G}(L)$. We call $G(L)$ also the *theta group of* L. Note that the theta group $G(L)$ depends only on the polarization H, whereas the canonical section s_L and thus the homomorphism $\sigma\colon G(L) \to \mathcal{G}(L)$ as well as the theta group $\mathcal{G}(L)$ depend on the the particular choice of the line bundle L within its algebraic equivalence class.

§ 2 Theta Groups under Homomorphisms

In this section we compare the theta groups of a line bundle L and its pullback via a surjective homomorphism of abelian varieties $f\colon Y \to X$. Again we will see that it is advantageous to work with $G(\cdot)$ instead of $\mathcal{G}(\cdot)$. Whereas $G(\cdot)$ is a functor for surjective homomorphisms, $\mathcal{G}(\cdot)$ is not. The group $\mathcal{G}(L)$ is only a quotient of a subgroup of $\mathcal{G}(f^*L)$. First we study the behaviour of the groups $K(L)$ and $\Lambda(L)$ under f.

Let $X = V/\Lambda$ and $Y = W/\Gamma$ be abelian varieties with canonical projections $\pi_X\colon V \to X$ and $\pi_Y\colon W \to Y$. Suppose $f\colon Y \to X$ is a surjective homomorphism with analytic representation $F\colon W \to V$ and $L = L(H, \chi)$ a line bundle on X. For any line bundle M on Y define $\Gamma(M) = \pi_Y^{-1}(K(M))$.

(2.1) Lemma. *$f^{-1}(K(L)) \subseteq K(f^*L)$ and $F^{-1}(\Lambda(L)) \subseteq \Gamma(f^*L)$ with equalities if and only if the kernel of f is connected.*

Proof. According to Corollary 2.4.6 c) the following diagram commutes

$$\begin{array}{ccc} Y & \xrightarrow{f} & X \\ {\scriptstyle \phi_{f^*L}}\big\downarrow & & \big\downarrow{\scriptstyle \phi_L} \\ \widehat{Y} & \xleftarrow{\hat{f}} & \widehat{X} \end{array} .$$

This implies $f^{-1}(K(L)) \subseteq K(f^*L)$, since $K(L) = \ker \phi_L$ and $K(f^*L) = \ker \phi_{f^*L}$. Furthermore we have equality if and only if $\widehat{f}$ is injective. According to Proposition 2.4.2 this is the case if and only if the kernel of f is connected. The second assertion is an immediate consequence. $\square$

According to the Stein factorization (see Section 1.2) every surjective homomorphism f factorizes into a surjective homomorphism with connected kernel and an isogeny. We will treat both cases separately.

(2.2) Proposition. *Let $f: Y \to X$ be a surjective homomorphism with connected kernel and analytic representation $\varrho_a(f) = F$. For any line bundle L on X*

a) the sequence

$$0 \longrightarrow \ker F \xrightarrow{\iota} G(f^*L) \xrightarrow{\widetilde{F}} G(L) \longrightarrow 0$$

is exact (here $\iota(w) = [1, w]$ and $\widetilde{F}[\alpha, w] = [\alpha, F(w)]$);

b) the sequence of a) induces an exact sequence

$$1 \longrightarrow \ker f \longrightarrow \mathcal{G}(f^*L) \longrightarrow \mathcal{G}(L) \longrightarrow 0 \, .$$

Proof. a): According to Lemma 2.1 we have $G(f^*L) = \mathbb{C}^* \times F^{-1}(\Lambda(L))$ as sets. Hence $\widetilde{F}$ is a well-defined surjective map. It suffices to show that $\widetilde{F}$ is a homomorphism. But this is an immediate computation using $c_1(f^*L) = F^*c_1(L)$.

b): Recall that $\mathcal{G}(f^*L) = G(f^*L)/s_{f^*L}(\Gamma)$ and $\mathcal{G}(L) = G(L)/s_L(\Lambda)$. Since $\ker f = \ker F/(\Gamma \cap \ker F)$, it suffices to show $\widetilde{F}(s_{f^*L}(\Gamma)) = s_L(\Lambda)$. But this is a consequence of $F(\Gamma) = \Lambda$. □

Consider now the case of an isogeny $f: Y \to X$. We may assume that $W = V$ and $F = \mathrm{id}_V$. Then we have

$$\Gamma \subset \Lambda \subset \Lambda(L) \subset \Gamma(f^*L) \, . \tag{1}$$

Arguing as in the proof of Proposition 2.2 this gives

(2.3) Lemma. *$G(L)$ is a subgroup of $G(f^*L)$.*

In particular $s_L(\Lambda) \subset G(L)$ is a subgroup of $G(f^*L)$ and the centralizer $\mathcal{Z}_{G(f^*L)}(s_L(\Lambda))$ of $s_L(\Lambda)$ in $G(f^*L)$ is defined.

(2.4) Proposition. *Let $f: Y \to X$ be an isogeny with $\varrho_a(f) = \mathrm{id}$. For any $L \in \mathrm{Pic}(X)$*

a) $G(L) = \mathcal{Z}_{G(f^*L)}(s_L(\Lambda))$.

b) $\mathcal{G}(L) = \mathcal{Z}_{G(f^*L)}(s_L(\Lambda))/s_L(\Lambda)$.

Proof. It suffices to prove a), since b) is an immediate consequence using Theorem 1.3. For any $[\alpha, w] \in G(f^*L)$ and $\lambda \in \Lambda$ we have

$$[a_L(\lambda, 0), \lambda][\alpha, w] = e\big(2\pi i \operatorname{Im} H(w, \lambda)\big)[\alpha, w][a_L(\lambda, 0), \lambda] \, .$$

But $e(2\pi i \operatorname{Im} H(w, \lambda)) = 1$ for all $\lambda \in \Lambda$ if and only if $w \in \Lambda(L)$, that is if and only if $[\alpha, w] \in G(L)$. □

§ 3 The Commutator Map

As above let $X = V/\Lambda$ be an abelian variety of dimension g and $L = L(H, \chi)$ a line bundle on X. The groups $\mathbb{C}^*$ and $K(L)$ are commutative, however the theta groups $\mathcal{G}(L)$ and $G(L)$ are not commutative in general. In this section we study the corresponding commutator maps.

The group $\mathcal{G}(L)$ is a central extension of abelian groups, so its commutator map induces a map

$$e^L: K(L) \times K(L) \longrightarrow \mathbb{C}^* .$$

Similarly the commutator map of $G(L)$ induces a map $\Lambda(L) \times \Lambda(L) \to \mathbb{C}^*$, also denoted by e^L. This makes sense, since by construction the commutator map of $G(L)$ is the pullback of the commutator map of $\mathcal{G}(L)$. Hence we have for all $[\alpha_\nu, w_\nu] \in G(L)$, $\nu = 1, 2$:

$$e^L(\overline{w}_1, \overline{w}_2) = e^L(w_1, w_2) = [\alpha_1, w_1][\alpha_2, w_2][\alpha_1, w_1]^{-1}[\alpha_2, w_2]^{-1} . \quad (1)$$

The map e^L can be expressed in terms of the first Chern class H of L.

(3.1) Proposition. *For all $w_1, w_2 \in \Lambda(L)$*

$$e^L(\overline{w}_1, \overline{w}_2) = e\big(-2\pi i \operatorname{Im} H(w_1, w_2)\big) .$$

Proof. Using (1) we get $e^L(\overline{w}_1, \overline{w}_2) = [e(\pi H(w_2, w_1) - \pi H(w_1, w_2)), 0] = [e(-2\pi i \operatorname{Im} H(w_1, w_2)), 0]$. □

As an immediate consequence we have for all $x_1, x_2, x \in K(L)$:

$$e^L(x_1 + x_2, x) = e^L(x_1, x)e^L(x_2, x)$$
$$e^L(x_1, x_2) = e^L(x_2, x_1)^{-1} \quad \text{and} \quad e^L(x, x) = 1 .$$

In other words e^L is a (multiplicative) alternating form on $K(L)$ (respectively $\Lambda(L)$) with values in $\mathbb{C}^*$. Another consequence of Proposition 3.1 and the Appell-Humbert Theorem is

(3.2) Corollary. *Let L_1 and L_2 be line bundles on X.*
a) $e^{L_1 \otimes L_2} = e^{L_1} e^{L_2}$ on $K(L_1) \cap K(L_2)$.
b) $e^{L_1} = e^{L_2}$, if L_1 and L_2 are algebraically equivalent.

Let $f: Y \to X$ be a surjective homomorphism of abelian varieties. The following proposition compares the forms e^{f^*L} and f^*e^L.

(3.3) Proposition. $e^{f^*L}(x, y) = e^L\big(f(x), f(y)\big)$ *for all $x, y \in f^{-1}(K(L))$.*

This follows from the fact that $c_1(f^*L) = f^*c_1(L)$. □

The results which we proved so far in this chapter are valid for arbitrary line bundles on abelian varieties. We will see now that for a nondegenerate line bundle L the theta group $\mathcal{G}(L)$ and the form e^L have some additional properties.

(3.4) Theorem. *For a line bundle L on X the following statements are equivalent:*
i) L is nondegenerate,
ii) $e^L: K(L) \times K(L) \to \mathbb{C}^*$ is nondegenerate,
iii) $\mathbb{C}^*$ is the center of $\mathcal{G}(L)$,
iv) there is a decomposition $K(L) = K_1 \oplus K_2$ with subgroups K_1 and K_2, isotropic with respect to e^L, such that the map $K_2 \to \widehat{K}_1 = \mathrm{Hom}(K_1, \mathbb{C}^*)$, $x \mapsto e^L(\cdot, x)$ is an isomorphism.

Proof. The equivalence ii) $\Longleftrightarrow$ iii) as well as the implication iv) $\Longrightarrow$ ii) are trivial. Moreover i) implies iv) by Lemma 3.1.4. Hence it suffices to show that ii) $\Longrightarrow$ i): Suppose L is degenerate, i.e. the group $K(L)$ is infinite (see Proposition 2.4.8). We have to show that e^L is degenerate. Consider the homomorphism $p: X \to X/K(L)_0$ of Section 3.3, where as usual $K(L)_0$ denotes the connected component of $K(L)$ containing zero.

We claim that we may assume that $L|K(L)_0$ is trivial. Since the canonical map $\mathrm{Pic}^0(X) \to \mathrm{Pic}^0(K(L)_0)$ is surjective (see Proposition 2.4.2), there is a $P \in \mathrm{Pic}^0(X)$ with $P|K(L)_0 = L|K(L)_0$. This means $L \otimes P^{-1}$ is trivial on $K(L)_0$. Since L and $L \otimes P^{-1}$ are algebraically equivalent, we may replace L by $L \otimes P^{-1}$.

By Lemma 3.3.2 there is a line bundle $\bar{L}$ on the abelian variety $X/K(L)_0$ with $p^*\bar{L} = L$. According to Proposition 3.3 we have $e^L = p^*e^{\bar{L}}$, since $p^{-1}(K(\bar{L})) = K(L)$ by Lemma 2.1. But $p^*e^{\bar{L}}$ is certainly degenerate, since $K(L)_0 \neq 0$. This completes the proof. □

Finally we reformulate Corollary 2.4.4 into terms of e^L and $K(L)$.

(3.5) Corollary. *For an isogeny $f: Y \to X$ of abelian varieties and a line bundle $L \in \mathrm{Pic}(Y)$ the following statements are equivalent:*
i) $L = f^*M$ for some $M \in \mathrm{Pic}(X)$,
ii) $\ker f$ is an isotropic subgroup of $K(L)$ with respect to e^L.

§ 4 The Canonical Representation of the Theta Group

Let $X = V/\Lambda$ be an abelian variety and $L = L(H, \chi)$ a line bundle on X. There is a natural action of the theta group $\mathcal{G}(L)$ on $H^0(L)$ to be introduced in this section. In order to avoid trivialities we assume that $h^0(L) > 0$ or equivalently L is semipositive and $L|K(L)_0$ is trivial (see Theorem 3.5.5).

Suppose s is a section of L and $(\varphi, x) \in \mathcal{G}(L)$. As the following commutative diagram shows

$$\begin{array}{ccc} L & \xrightarrow{\varphi} & L \\ {\scriptstyle s}\uparrow & & \uparrow{\scriptstyle \varphi s t_{-x}} \\ X & \xrightarrow{t_x} & X \end{array} ,$$

$\varphi s t_{-x}$ is also a section of L. The assignment $\big((\varphi, x), s\big) \mapsto \varphi s t_{-x}$ defines an action $\mathcal{G}(L) \times H^0(L) \to H^0(L)$ in a canonical way. The corresponding representation

$$\tilde{\varrho}: \mathcal{G}(L) \to \mathrm{GL}\big(H^0(L)\big) ,$$

is called the *canonical representation of the theta group* $\mathcal{G}(L)$. The subgroup $\mathbb{C}^*$ of $\mathcal{G}(L)$ acts by multiplication on $H^0(L)$. This implies that $\tilde{\varrho}$ induces a projective representation $\varrho: K(L) \to \mathrm{PGL}(H^0(L))$ such that the following diagram commutes

$$\begin{array}{ccccccccc} 1 & \longrightarrow & \mathbb{C}^* & \longrightarrow & \mathcal{G}(L) & \longrightarrow & K(L) & \longrightarrow & 0 \\ & & \| & & {\scriptstyle \tilde{\varrho}}\downarrow & & \downarrow{\scriptstyle \varrho} & & \\ 1 & \longrightarrow & \mathbb{C}^* & \longrightarrow & \mathrm{GL}\big(H^0(L)\big) & \longrightarrow & \mathrm{PGL}\big(H^0(L)\big) & \longrightarrow & 1 \end{array} .$$

In Section 1 we saw that there is a canonical surjective homomorphism $\sigma: G(L) \to \mathcal{G}(L)$. Composing σ with $\tilde{\varrho}$ we get a canonical representation of $G(L)$ also denoted by $\tilde{\varrho}$:

$$\tilde{\varrho}: G(L) \mapsto \mathrm{GL}\big(H^0(L)\big) .$$

Next we describe the representation $\tilde{\varrho}$ in terms of canonical theta functions. Consider $H^0(L)$ as the vector space of canonical theta functions. For elements $[\alpha, w] \in G(L)$ and $\vartheta \in H^0(L)$ denote by $[\alpha, w]\vartheta$ the image of ϑ under the action of $[\alpha, w]$. By definition of the action of $G(L)$ on $H^0(L)$ the following diagram commutes

$$\begin{array}{ccc} V \times \mathbb{C} & \xrightarrow{[\alpha,w]} & V \times \mathbb{C} \\ {\scriptstyle (\mathrm{id}_V, \vartheta)}\uparrow & & \uparrow{\scriptstyle (\mathrm{id}_V, [\alpha,w]\vartheta)} \\ V & \xrightarrow{t_w} & V \end{array} .$$

Hence
$$\begin{aligned}(\mathrm{id}_V,[\alpha,w]\vartheta)(v) &= \big([\alpha,w](\mathrm{id}_V,\vartheta)t_{-w}\big)(v)\\ &= [\alpha,w]\big(v-w,\vartheta(v-w)\big)\\ &= \big(v,\alpha e\big(\pi H(v-w,w)\big)\vartheta(v-w)\big)\end{aligned}$$

for all $v \in V$ and we obtain

$$[\alpha,w]\vartheta = \alpha e\big(\pi H(\cdot - w,w)\big)\vartheta(\cdot - w)\ . \tag{1}$$

By construction the function on the right hand side is of course a canonical theta function for L, but this can also be seen checking the functional equation. Similarly we get that the subgroups $\mathbb{C}^*$ and $s(\Lambda) = \{[a_L(\lambda,0),\lambda] | \lambda \in \Lambda\}$ of $G(L)$ act by multiplication respectively as identity on $H^0(L)$.

Next we will show that for studying the canonical representation it suffices to consider ample line bundles. Recall that by assumption L is positive semidefinite and $L|K(L)_0$ is trivial. Let $f: X \to \overline{X} = X/K(L)_0$ denote the natural homomorphism. By Lemma 3.3.2 there is an ample line bundle $\overline{L} = L(\overline{H},\chi)$ on $\overline{X}$ with $f^*\overline{L} \simeq L$. Pulling back canonical theta functions of $\overline{L}$ via f gives an isomorphism $f^*: H^0(\overline{L}) \to H^0(L)$. It induces an isomorphism $\mathrm{GL}(H^0(\overline{L})) \to \mathrm{GL}(H^0(L))$: $\varphi \mapsto f^*\varphi(f^*)^{-1}$. Using the homomorphism $\widetilde{F}: \mathcal{G}(L) \to \mathcal{G}(\overline{L})$ of Proposition 2.2 we have

(4.1) Lemma. *The following diagram commutes*

$$\begin{array}{ccc} G(L) & \xrightarrow{\tilde\varrho} & \mathrm{GL}(H^0(L)) \\ {\scriptstyle\widetilde{F}}\big\downarrow & & \big\uparrow \\ G(\overline{L}) & \xrightarrow{\tilde\varrho} & \mathrm{GL}(H^0(\overline{L})) \end{array} \ .$$

Proof. Denote by F the analytic representation of f. By definition of the maps it suffices to show that $[\alpha,w]f^*\vartheta = f^*((\widetilde{F}[\alpha,w])\vartheta)$ for all $[\alpha,w] \in G(L)$ and $\vartheta \in H^0(\overline{L})$. Using $f^*\vartheta = \vartheta(F(\cdot))$ and $F^*\overline{H} = H$ we get

$$\begin{aligned}[\alpha,w]f^*\vartheta &= \alpha e\big(\pi H(\cdot - w,w)\big)\vartheta\big(F(\cdot)-F(w)\big)\\ &= \alpha e\big(\pi\overline{H}(F(\cdot)-F(w),F(w))\big)\vartheta\big(F(\cdot)-F(w)\big)\\ &= [\alpha,F(w)]\vartheta(F(\cdot))\\ &= f^*\big((\widetilde{F}[\alpha,w])\vartheta\big)\ .\end{aligned}$$
□

Now let $L = L(H,\chi)$ be an ample line bundle on $X = V/\Lambda$. Choose a decomposition $\Lambda = \Lambda_1 \oplus \Lambda_2$ and a characteristic c for L. The decomposition induces decompositions $\Lambda(L) = \Lambda(L)_1 \oplus \Lambda(L)_2$ and $K(L) = K_1 \oplus K_2$ for L. According to Theorem 3.2.7 the canonical theta functions $\{\vartheta^c_{\bar{u}} \mid \bar{u} \in K_1\}$

form a basis for $H^0(L)$. The crucial point in this chapter is the following property of this basis with respect to the theta groups: the elements of $G(L)$ lying over $\Lambda(L)_1$ act as permutations whereas the elements over $\Lambda(L)_2$ act as dilatations on the functions $\vartheta^c_{\bar{u}}$. Roughly speaking the theta group acts on the basis $\{\vartheta^c_{\bar{u}} \mid \bar{u} \in K_1\}$ up to a constant. Denote by a_L the canonical factor of L extended to $V \times V$ (see Section 3.1).

(4.2) Proposition. *For all $\bar{u} \in K_1$ and $[\alpha, w] \in G(L)$ with $w = w_1 + w_2$, $w_\nu \in \Lambda(L)_\nu$*

$$[\alpha, w]\vartheta^c_{\bar{u}} = \alpha e^L(u - w_1, w_2) a_L(w, 0)^{-1} \vartheta^c_{\bar{u}-\bar{w}_1} \,.$$

In particular

$$[a_L(w_1, 0), w_1]\vartheta^c_{\bar{u}} = \vartheta^c_{\bar{u}-\bar{w}_1} \,, \tag{2}$$

$$[a_L(w_2, 0), w_2]\vartheta^c_{\bar{u}} = e^L(u, w_2)\vartheta^c_{\bar{u}} \,. \tag{3}$$

Proof. Using the definitions of $\vartheta^c_{\bar{u}}$ (see Section 3.2 equation (4)) and χ_0 (see Section 3.1) as well as Lemma 3.1.3 we get for all $w_1 \in \Lambda(L)_1$ and $v \in V$:

$$\begin{aligned}
[a_L(w_1, 0), w_1]\vartheta^c_{\bar{u}}(v) &= a_L(w_1, 0) e\big(\pi H(v - w_1, w_1)\big) \vartheta^c_{\bar{u}}(v - w_1) \\
&= a_L(w_1, v) e\big(-\pi H(w_1, w_1)\big) a_L(u, v - w_1)^{-1} \vartheta^c_{\bar{0}}(v + u - w_1) \\
&= a_L(-w_1, v)^{-1} a_L(u, v - w_1)^{-1} a_L(u - w_1, v) \vartheta^c_{\bar{u}-\bar{w}_1}(v) \\
&= \vartheta^c_{\bar{u}-\bar{w}_1}(v) \,.
\end{aligned}$$

Similarly we have for all $w_2 \in \Lambda(L)_2$ using in addition that $\vartheta^c_{\bar{0}}$ is a theta function with respect to the lattice $\Lambda_1 \oplus \Lambda(L)_2$ (see Corollary 3.2.5):

$$\begin{aligned}
&[a_L(w_2, 0), w_2]\vartheta^c_{\bar{u}}(v) = \\
&= a_L(w_2, 0) e\big(\pi H(v - w_2, w_2)\big) a_L(u, v - w_2)^{-1} a_L(-w_2, v + u) \vartheta^c_{\bar{0}}(v + u) \\
&= a_L(w_2, 0) e\big(\pi H(v - w_2, w_2)\big) a_L(u, v - w_2)^{-1} a_L(-w_2, v + u) a_L(u, v) \vartheta^c_{\bar{u}}(v) \\
&= e^L(u, w_2) \vartheta^c_{\bar{u}}(v) \,.
\end{aligned}$$

Combining both identities with

$$[\alpha, w] = \alpha a_L(w, 0)^{-1} e^L(w_1, w_2)^{-1} [a_L(w_1, 0), w_1][a_L(w_2, 0), w_2]$$

gives the assertion. □

A first consequence is

(4.3) Corollary. *For any line bundle L on X the canonical representation $\tilde{\varrho}\colon \mathcal{G}(L) \to \mathrm{GL}(H^0(L))$ (respectively $\tilde{\varrho}\colon G(L) \to \mathrm{GL}(H^0(L))$) is irreducible.*

Proof. By construction and Lemma 4.1 it suffices to prove the assertion only for an ample line bundle L and the theta group $G(L)$. Given $0 \neq \vartheta \in H^0(L)$ we have to show that $H^0(L) = <[\alpha, w]\vartheta \mid [\alpha, w] \in G(L)>$, the vector space spanned by all $[\alpha, w]\vartheta$. Let the notation be as above. First we claim that $<[\alpha, w]\vartheta|[\alpha, w] \in G(L)>$ contains one of the basis elements $\vartheta^c_{\overline{w}}$. To see this let $\vartheta' = \sum_{\overline{w} \in K_1} a_{\overline{w}} \vartheta^c_{\overline{w}}$ be an element of $<[\alpha, w]\vartheta|[\alpha, w] \in G(L)>$. Write $\vartheta' = a_1 \vartheta^c_{\overline{w}_1} + \cdots + a_n \vartheta^c_{\overline{w}_n}$ with $\overline{w}_i \neq \overline{w}_j$ for $i \neq j$ and $a_i \neq 0$. If $n \geq 2$, there is a $\bar{u} \in K_2$ such that $e^L(\overline{w}_1, \bar{u}) \neq e^L(\overline{w}_2, \bar{u})$, since $\overline{w}_1 \neq \overline{w}_2$ and e^L is nondegenerate. Using Proposition 4.2 we obtain

$$[e^L(\bar{u}, \overline{w}_1) a_L(u, 0), u]\vartheta' - \vartheta' = b_2 \vartheta^c_{\overline{w}_2} + \cdots + b_n \vartheta^c_{\overline{w}_n}$$

for some $b_\nu \in \mathbb{C}$ and $b_2 \neq 0$. Repeating this process, we get the assertion.

Hence there is a $\overline{w} \in \mathbb{C}$ with $\vartheta^c_{\overline{w}} \in <[\alpha, w]\vartheta|[\alpha, w] \in G(L)>$. Applying again Proposition 4.2, we see that $<[\alpha, w]\vartheta|[\alpha, w] \in G(L)>$ contains the whole basis $\{\vartheta^c_{\overline{w}} \mid \overline{w} \in K_1\}$ of $H^0(L)$. This completes the proof. □

(4.4) Remark. Using the isomorphism $\mathcal{G}(L) \to \widetilde{\mathcal{G}}(L)$ of Remark 1.2 the canonical representation induces a representation of the group $\widetilde{\mathcal{G}}(L)$. It is easy to see that it is given as follows:

$$\widetilde{\mathcal{G}}(L) \to \mathrm{GL}(H^0(L)), \quad (\widetilde{\varphi}, x)\sigma = t^*_{-x}(\widetilde{\varphi})(\sigma)) \ .$$

More generally, for an arbitrary line bundle L and any $p \geq 0$ the cohomology group $H^p(L)$ is a representation of $\widetilde{\mathcal{G}}(L)$. Namely $\widetilde{\mathcal{G}}(L)$ acts on $H^p(L)$ by

$$((\widetilde{\varphi}, x)\eta) \mapsto H^p(t^*_{-x})(H^p(\widetilde{\varphi})(\eta))$$

for $(\widetilde{\varphi}, x) \in \widetilde{\mathcal{G}}(L)$ and $\eta \in H^p(L)$. One can show (see Exercise 3) that the representation $\widetilde{\mathcal{G}}(L) \to \mathrm{GL}(H^p(L))$ is irreducible, if p is the number of negative eigenvalues of the hermitian form $H = c_1(L)$.

§ 5 The Isogeny Theorem

Let $f: Y = V/\Gamma \to X = V/\Lambda$ be an isogeny of abelian varieties and $L = L(H, \chi)$ an ample line bundle on X. We want to study the pullback homomorphism $f^*: H^0(L) \to H^0(f^*L)$ in terms of canonical theta functions. It turns out that f^* can be described explicitly using bases for $H^0(L)$ and $H^0(f^*L)$ as in Theorem 3.2.7 chosen in a compatible way.

As usual assume that the analytic representation of f is the identity on V. As we saw in Section 2 we have the following inclusions

$$\Gamma \subset \Lambda \subset \Lambda(L) \subset \Gamma(f^*L) \ .$$

Fix a decomposition $\Lambda = \Lambda_1 \oplus \Lambda_2$ for L. It induces decompositions of $\Lambda(L)$ for L, and Γ and $\Gamma(f^*L)$ for f^*L such that

$$\Gamma_\nu \subset \Lambda_\nu \subset \Lambda(L)_\nu \subset \Gamma(f^*L)_\nu$$

for $\nu = 1, 2$ (see Section 3.1). Such decompositions for L and f^*L are called to be *compatible*. Recall that we obtain induced decompositions

$$K(L) = K(L)_1 \oplus K(L)_2 \quad \text{and} \quad K(f^*L) = K(f^*L)_1 \oplus K(f^*L)_2$$

for L and f^*L with $K(L)_\nu = \Lambda(L)_\nu/\Lambda_\nu$ and $K(f^*L)_\nu = \Gamma(f^*L)_\nu/\Gamma_\nu$ for $\nu = 1, 2$.

Let $c \in V$ be a characteristic for L with respect to $\Lambda = \Lambda_1 \oplus \Lambda_2$. By construction c is also a characteristic for f^*L with respect to $\Gamma = \Gamma_1 \oplus \Gamma_2$ and the canonical factors a_L of L and a_{f^*L} of f^*L extended to $V \times V$ coincide:

$$a_L(v,w) = a_{f^*L}(v,w) = \chi_0(v) e\big(2\pi i \operatorname{Im} H(c,v) + \pi H(w,v) + \tfrac{\pi}{2} H(v,v)\big)$$

for all $v, w \in V$. According to Theorem 3.2.7 the decomposition for Λ respectively Γ and the characteristic c determine bases $\{\vartheta_x^L \mid x \in K(L)_1\}$ for $H^0(L)$ and $\{\vartheta_y^{f^*L} \mid y \in K(f^*L)_1\}$ for $H^0(f^*L)$. (Note that we changed notation from ϑ_x^c to ϑ_x^L and from ϑ_y^c to $\vartheta_y^{f^*L}$, since we are not interested in the particular characteristic but in the line bundle to which the theta function belongs). We want to describe the homomorphism $f^*: H^0(L) \to H^0(f^*L)$ in terms of these bases. Since by assumption the analytic representation $\varrho_a(f)$ is the identity on V, every canonical theta function for L is also a canonical theta function for f^*L. In other words, f^* is the canonical inclusion and we have

(5.1) Isogeny Theorem. $$f^*\vartheta_x^L = \sum_{\substack{y \in K(f^*L)_1 \\ f(y) = x}} \vartheta_y^{f^*L} \qquad \textit{for all } x \in K(L)_1.$$

Proof. First we consider the special case $x = 0$. We have to show that

$$f^*\vartheta_0^L = \sum_{\nu=1}^{s} \vartheta_{\overline{w}_\nu}^{f^*L},$$

where $\{w_1, \ldots, w_s\} \subset \Lambda_1 \subset \Gamma(f^*L)_1$ is a set of representatives of $\Lambda_1/\Gamma_1 = \{y \in K(f^*L)_1 \mid f(y) = 0\}$. According to Lemma 3.2.9 we may assume that the characteristic c is zero. Then we have with the notation as in Section 3.2

$$\vartheta_0^{f^*L}(v) = e\big(\tfrac{\pi}{2} B(v,v)\big) \sum_{\mu \in \Gamma_1} e\big(\pi(H-B)(v,\mu) - \tfrac{\pi}{2}(H-B)(\mu,\mu)\big) \quad \text{and}$$

$$\vartheta_0^L(v) = e\big(\tfrac{\pi}{2} B(v,v)\big) \sum_{\lambda \in \Lambda_1} e\big(\pi(H-B)(v,\lambda) - \tfrac{\pi}{2}(H-B)(\lambda,\lambda)\big)$$

for all $v \in V$. So we get

$$\begin{aligned}
\sum_{\nu=1}^{s} \vartheta_{w_\nu}^{f^*L}(v) &= \sum_{\nu=1}^{s} a_L(w_\nu, v)^{-1} \vartheta_0^{f^*L}(v + w_\nu) \\
&= \sum_{\nu=1}^{s} \sum_{\mu \in \Gamma_1} e\big(-\pi H(v, w_\nu) - \tfrac{\pi}{2} H(w_\nu, w_\nu)\big) \\
&\quad \cdot e\big(\tfrac{\pi}{2} B(v + w_\nu, v + w_\nu) + \pi(H - B)(v + w_\nu, \mu) - \tfrac{\pi}{2}(H - B)(\mu, \mu)\big) \\
&= e\big(\tfrac{\pi}{2} B(v, v)\big) \sum_{\nu=1}^{s} \sum_{\mu \in \Gamma_1} e\big(-\pi(H - B)(v, w_\nu) - \tfrac{\pi}{2}(H - B)(w_\nu, w_\nu)\big) \\
&\quad \cdot e\big(\pi(H - B)(v, \mu) + \pi(H - B)(w_\nu, \mu) - \tfrac{\pi}{2}(H - B)(\mu, \mu)\big) \\
&= e\big(\tfrac{\pi}{2} B(v, v)\big) \sum_{\nu=1}^{s} \sum_{\mu \in \Gamma_1} e\big(\pi(H - B)(v, \mu - w_\nu)\big) \\
&\quad \cdot e\big(-\tfrac{\pi}{2}(H - B)(\mu - w_\nu, \mu - w_\nu) + \tfrac{\pi}{2} H(w_\nu, \mu) - \tfrac{\pi}{2} H(\mu, w_\nu)\big) \\
&= e\big(\tfrac{\pi}{2} B(v, v)\big) \sum_{\lambda \in \Lambda_1} e\big(\pi(H - B)(v, \lambda) - \tfrac{\pi}{2}(H - B)(\lambda, \lambda)\big) = \vartheta_0^L(v) ,
\end{aligned}$$

where we used $e\big(\frac{\pi}{2} H(w_\nu, \mu) - \frac{\pi}{2} H(\mu, w_\nu)\big) = e\big(\pi i \operatorname{Im} H(w_\nu, \mu)\big) = 1$ and $\Lambda_1 = \{\mu - w_\nu \mid \mu \in \Gamma_1, \nu = 1, \ldots, s\}$. This proves the assertion for $x = 0$.

To prove the assertion for a general $x \in K(L)_1$ we use the action of the theta groups $G(L)$ and $G(f^*L)$. Let $\pi_X : V \to X$ and $\pi_Y : V \to Y$ be the canonical projections and $w \in \Lambda(L)_1$ a representative of x, i.e.

$$x = \pi_X(w) = f\pi_Y(w) .$$

Then we have using the first part of this proof and equation (2) in Section 4

$$\begin{aligned}
f^*\vartheta_x^L &= f^*\big([a_L(-w, 0), -w]\vartheta_0^L\big) \\
&= \sum_{\substack{y \in K(f^*L)_1 \\ f(y) = 0}} [a_{f^*L}(-w, 0), -w]\vartheta_y^{f^*L} \\
&= \sum_{\substack{y \in K(f^*L)_1 \\ f(y) = 0}} \vartheta_{y + \pi_Y(w)}^{f^*L} \\
&= \sum_{\substack{y \in K(f^*L)_1 \\ f(y) = x}} \vartheta_y^{f^*L} .
\end{aligned}$$

□

For later use we want to work out the hypotheses for the Isogeny Theorem: suppose we are given decompositions

$$K(L) = K(L)_1 \oplus K(L)_2 \tag{1}$$

$$K(f^*L) = K(f^*L)_1 \oplus K(f^*L)_2 \tag{2}$$

for L and f^*L. The following proposition gives a criterion for these decompositions to be induced by a pair of compatible decompositions of the underlying lattices.

(5.2) Proposition. *Let $f: Y = V/\Gamma \to X = V/\Lambda$ be an isogeny of abelian varieties and L an ample line bundle on X. The following statements are equivalent*

i) There are compatible decompositions of Λ for L and Γ for f^*L inducing the decompositions (1) and (2),

ii) $f(K(f^*L)_\nu) \cap K(L) = K(L)_\nu$ for $\nu = 1, 2$.

Hence it makes sense to call the decompositions (1) and (2) *compatible*, if condition ii) holds.

Proof. Without loss of generality we may assume that the analytic representation of f is the identity. Given compatible decompositions of Λ and Γ inducing (1) and (2), we have $\Gamma_\nu \subset \Lambda_\nu \subset \Lambda(L)_\nu \subset \Gamma(f^*L)_\nu$. Hence $\Gamma(f^*L)_\nu \cap \Lambda(L) = \Lambda(L)_\nu$ for $\nu = 1, 2$, which implies ii).

Conversely, suppose ii) is valid. By definition of a decomposition of $K(f^*L)$ for f^*L there is a decomposition $\Gamma(f^*L) = \Gamma(f^*L)_1 \oplus \Gamma(f^*L)_2$ inducing (2). Define $\Gamma_\nu = \Gamma(f^*L)_\nu \cap \Gamma$, $\Lambda_\nu = \Gamma(f^*L)_\nu \cap \Lambda$, and $\Lambda(L)_\nu = \Gamma(f^*L)_\nu \cap \Lambda(L)$ for $\nu = 1, 2$. This gives compatible decompositions of Λ and Γ (see Proposition 3.1.4). Now $\Lambda = \Lambda_1 \oplus \Lambda_2$ induces a decomposition $K(L) = \widetilde{K(L)}_1 \oplus \widetilde{K(L)}_2$. Applying the implication i) $\Longrightarrow$ ii) we get $\widetilde{K(L)}_\nu = f(K(f^*L)_\nu) \cap K(L)$ which is by assumption equal to $K(L)_\nu$. Hence $\Lambda = \Lambda_1 \oplus \Lambda_2$ induces the decomposition (1). □

Finally call characteristics c and c' for L and f^*L (with respect to the decompositions (1) and (2)) *compatible*, if $\varrho_a(f)(c') = c$. In these terms the *hypotheses of the Isogeny Theorem* are: $\{\vartheta_x^L \mid x \in K(L)_1\}$ and $\{\vartheta_y^{f^*L} \mid y \in K(f^*L)_1\}$ are bases for $H^0(L)$ and $H^0(f^*L)$ as in Theorem 3.2.7 with respect to compatible decompositions and compatible characteristics.

§ 6 Heisenberg Groups and Theta Structures

Let $L = L(H, \chi)$ be an ample line bundle on the abelian variety $X = V/\Lambda$ and $\vartheta_0, \ldots, \vartheta_N$ a basis of canonical theta functions for L. The associated rational map $\varphi_L: X \to \mathbb{P}_N$ is defined by $\varphi_L(\bar{v}) = (\vartheta_0(v) : \cdots : \vartheta_N(v))$.

Before we proceed, let us illustrate how one can use the theta group $G(L)$ and its canonical representation $\tilde{\varrho}$ to study the map φ_L. Recall that for any element $[\alpha, w] \in G(L)$ the automorphism $\tilde{\rho}[\alpha, w]: H^0(L) \to H^0(L)$ is defined by $\vartheta \mapsto [\alpha, w]\vartheta = \alpha e\big(\pi H(\cdot - w, w)\big)\vartheta(\cdot - w)$. The associated projective automorphism of $\mathbb{P}_N = P(H^0(L))$ depends only on $\overline{w} \in K(L)$. This defines the projective representation $\rho: K(L) \to \mathrm{PGL}_N(\mathbb{C})$ of Section 4. The following proposition shows that the map φ_L is equivariant with respect to the action of the group $K(L)$.

(6.1) Proposition. *For any $x \in K(L)$ the following diagram commutes*

$$\begin{array}{ccc} X & \xrightarrow{\varphi_L} & \mathbb{P}_N \\ {\scriptstyle t_{-x}}\downarrow & & \downarrow{\scriptstyle \rho(x)} \\ X & \xrightarrow{\varphi_L} & \mathbb{P}_N \end{array} .$$

Proof. For $v \in V$ and $\overline{w} \in K(L)$ we have

$$\begin{aligned} \varphi_L(\bar{v} - \overline{w}) &= \big(e\big(\pi H(v - w, w)\big)\vartheta_0(v - w) : \cdots : e\big(\pi H(v - w, w)\big)\vartheta_N(v - w)\big) \\ &= \big([1, w]\vartheta_0(v) : \cdots : [1, w]\vartheta_N(v)\big) \\ &= \rho(\overline{w})\big(\vartheta_0(v) : \cdots : \vartheta_N(v)\big) = \rho(\overline{w})\varphi_L(\bar{v}) . \end{aligned}$$

□

In particular the image $\varphi_L(X)$ is invariant under the action of the group $K(L)$. On the other hand Proposition 4.2 gives an explicit description of the automorphisms $\rho(x)$, $x \in K(L)$, in terms of a matrix with respect to a basis of $H^0(L)$. Thus we can use the action of $K(L)$ to get information about the variety $\varphi_L(X)$ in $\mathbb{P}_N$. For example, in many cases one can derive equations for $\varphi_L(X)$ in this way (see Sections 7.5 and 7.6). For these geometric applications the theory developed so far is completely sufficient. However the variety $\varphi_L(X)$ itself does not depend on the particular choice of the line bundle L within its algebraic equivalence class, whereas the formula of Proposition 4.2 describing the action of $K(L)$ on $\varphi_L(X)$ does. This suggests to look for a description of the theory of theta functions depending only on the polarization. In fact, this leads to the theory of Heisenberg groups, which we discuss now.

Let $H \in \mathrm{NS}(X)$ be a polarization of type $D = \mathrm{diag}(d_1, \ldots, d_g)$. Define as a set

$$\mathcal{H}(D) = \mathbb{C}^* \times K(D) ,$$

where $K(D) := \mathbb{Z}^g / D\mathbb{Z}^g \oplus \mathbb{Z}^g / D\mathbb{Z}^g$. The set $\mathcal{H}(D)$ admits a group structure. For this denote by $f_1, \ldots, f_{2g}$ the standard generators of $K(D)$. Define an alternating form $e^D: K(D)^2 \to \mathbb{C}^*$ by

$$e^D(f_\nu, f_\mu) = \begin{cases} e(-\frac{2\pi i}{d_\nu}) & \mu = g + \nu \\ e(\frac{2\pi i}{d_\nu}) \quad \text{if} & \nu = g + \mu \\ 1 & \text{otherwise} \end{cases}$$

and define for any $(\alpha, x_1, x_2), (\beta, y_1, y_2) \in \mathcal{H}(D)$

$$(\alpha, x_1, x_2)(\beta, y_1, y_2) = \big(\alpha\beta e^D(x_1, y_2), x_1 + y_1, x_2 + y_2\big) . \tag{1}$$

(6.2) Lemma. *Equation (1) defines a group structure on $\mathcal{H}(D)$ and*

$$1 \to \mathbb{C}^* \xrightarrow{i} \mathcal{H}(D) \xrightarrow{p} K(D) \to 0$$

is an exact sequence of groups. Here $i(\alpha) = (\alpha, 0, 0)$ and $p(\alpha, x_1, x_2) = (x_1, x_2)$.

The proof is an immediate computation which we omit. We call $\mathcal{H}(D)$ the *Heisenberg group (of type D)*. The reason for this name is the fact that the multiplication in $\mathcal{H}(D)$ is formally the same as in the usual Heisenberg group in Quantum Mechanics.

We will see that the theta group of every line bundle in $\operatorname{Pic}^H(X)$ is isomorphic to $\mathcal{H}(D)$. Let $L = L(H, \chi)$ be such a line bundle. An isomorphism

$$b: \mathcal{G}(L) \to \mathcal{H}(D)$$

restricting to the identity on $\mathbb{C}^*$ is called a *theta structure (for L)*. A theta structure b for L induces an isomorphism $\bar{b}: K(L) \to K(D)$ such that the following diagram commutes

$$\begin{array}{ccccccccc} 1 & \longrightarrow & \mathbb{C}^* & \longrightarrow & \mathcal{G}(L) & \longrightarrow & K(L) & \longrightarrow & 0 \\ & & \| & & \downarrow{\scriptstyle b} & & \downarrow{\scriptstyle \bar{b}} & & \\ 1 & \longrightarrow & \mathbb{C}^* & \longrightarrow & \mathcal{H}(D) & \longrightarrow & K(D) & \longrightarrow & 0 . \end{array} \tag{2}$$

(6.3) Lemma. *For any theta structure $b: \mathcal{G}(L) \to \mathcal{H}(D)$ the induced map $\bar{b}: K(L) \to K(D)$ is a symplectic isomorphism with respect to the forms e^L and e^D, i.e. $\bar{b}^* e^D = e^L$.*

Proof. The commutativity of the diagram (2) implies that the commutator map e^L of the upper row is the pullback via $\bar{b}$ of the commutator map of the lower row. But

$$(\alpha, x_1, x_2)(\beta, y_1, y_2)(\alpha, x_1, x_2)^{-1}(\beta, y_1, y_2)^{-1} = \big(e^D(x_1 + x_2, y_1 + y_2), 0, 0\big)$$

for all $(\alpha, x_1, x_2), (\beta, y_1, y_2) \in \mathcal{H}(D)$, which implies the assertion. □

(6.4) Remark. In Section 8.3 we define a *level D-structure* for (X, H) to be a symplectic basis of $K(L)$. Hence the symplectic isomorphism $\bar{b}$ is another description of a level D-structure. Thus we can say that a theta structure induces a level D-structure.

Our next aim is to determine all theta structures. We first show that every line bundle in $\mathrm{Pic}^H(X)$ admits theta structures. Fix a symplectic basis $\lambda_1, \dots, \lambda_g, \mu_1, \dots, \mu_g$ of Λ for H. It induces a symplectic isomorphism $\bar{b}: K(L) \to K(D)$

$$\bar{b}(\overline{\tfrac{1}{d_\nu}\lambda_\nu}) = f_\nu \quad \text{and} \quad \bar{b}(\overline{\tfrac{1}{d_\nu}\mu_\nu}) = f_{g+\nu}$$

for $\nu = 1, \dots, g$. Let $\Lambda(L) = \Lambda(L)_1 \oplus \Lambda(L)_2$ be the decomposition for L induced by the symplectic basis.

(6.5) Lemma. *Every characteristic c for L determines a theta structure $b_{\bar{c}}: \mathcal{G}(L) \to \mathcal{H}(D)$ inducing the symplectic isomorphism $\bar{b}$. The theta structure $b_{\bar{c}}$ depends only on the point $\bar{c} \in X$.*

Proof. Define a map $a_{\bar{c}}: \Lambda(L) \to \mathbb{C}^*$ by

$$a_{\bar{c}}(w) = \chi_0(w) e(2\pi i \operatorname{Im} H(c, w) + \tfrac{\pi}{2} H(w, w)) \ .$$

In fact, if $a_L: V \times V \to \mathbb{C}$ is the extension of the canonical factor of L determined by c, then $a_{\bar{c}}(w) = a_L(w, 0)$ for all $w \in \Lambda(L)$. We claim that the map $\tilde{b}_{\bar{c}}: G(L) \to \mathcal{H}(D)$,

$$\tilde{b}_{\bar{c}}([\alpha, w]) = (\alpha a_{\bar{c}}(w)^{-1}, \bar{b}(\overline{w}_1), \bar{b}(\overline{w}_2)) \tag{3}$$

is a homomorphism of groups. For the proof note first that by Lemma 3.1.1, Proposition 3.1 and Lemma 6.3 we have for all $w = w_1 + w_2$, $v = v_1 + v_2 \in \Lambda(L)$

$$\begin{aligned}
&a_{\bar{c}}(w + v)^{-1} e(\pi H(v, w)) = \\
&\quad = \chi_0(w + v)^{-1} e(-2\pi i \operatorname{Im} H(c, w + v) - \tfrac{\pi}{2} H(w + v, w + v) + \pi H(v, w)) \\
&\quad = a_{\bar{c}}(w)^{-1} a_{\bar{c}}(v)^{-1} e(-2\pi i \operatorname{Im} H(w_1, v_2)) \\
&\quad = a_{\bar{c}}(w)^{-1} a_{\bar{c}}(v)^{-1} e^D(\bar{b}(\overline{w}_1), \bar{b}(\overline{v}_2)) \ .
\end{aligned}$$

Hence

$$\begin{aligned}
\tilde{b}_{\bar{c}}([\alpha, w][\beta, v]) &= \tilde{b}_{\bar{c}}([\alpha\beta e(\pi H(v, w)), w + v]) \\
&= (\alpha\beta e(\pi H(v, w)) a_{\bar{c}}(w + v)^{-1}, \bar{b}(\overline{w}_1 + \overline{v}_1), \bar{b}(\overline{w}_2 + \overline{v}_2)) \\
&= (\alpha a_{\bar{c}}(w)^{-1}, \bar{b}(\overline{w}_1), \bar{b}(\overline{w}_2))(\beta a_{\bar{c}}(v)^{-1}, \bar{b}(\overline{v}_1), \bar{b}(\overline{v}_2)) \\
&= \tilde{b}_{\bar{c}}([\alpha, w]) \tilde{b}_{\bar{c}}([\beta, v]) \ .
\end{aligned}$$

The homomorphism $\tilde{b}_{\bar{c}}$ factorizes via $\sigma: G(L) \to \mathcal{G}(L) = G(L)/s_L(\Lambda)$, since $\tilde{b}_{\bar{c}}([a_L(\lambda, 0), \lambda]) = (1, 0, 0)$ for all $\lambda \in \Lambda$. Finally, $\tilde{b}_{\bar{c}}$ and thus the induced theta structure $b_{\bar{c}}$ depend only on $\bar{c} \in X$, since $a_{\bar{c}}$ does. □

Denote by $\mathrm{Aut}_{\mathbb{C}^*}\mathcal{H}(D)$ the group of all automorphisms of $\mathcal{H}(D)$ inducing the identity of $\mathbb{C}^*$. Obviously any two theta structures differ by an element of $\mathrm{Aut}_{\mathbb{C}^*}\mathcal{H}(D)$. In more sophisticated terms, the set of all theta structures for L is a principal homogenous space for the group $\mathrm{Aut}_{\mathbb{C}^*}\mathcal{H}(D)$.

In order to study $\mathrm{Aut}_{\mathbb{C}^*}\mathcal{H}(D)$, consider first the symplectic group $\mathrm{Sp}(D)$ consisting of all automorphisms of $K(D)$ preserving the alternating form e^D. With the same proof as for Lemma 6.3 one sees that any element of $\mathrm{Aut}_{\mathbb{C}^*}\mathcal{H}(D)$ induces a symplectic isomorphism of $K(D)$. This gives a homomorphism $p\colon \mathrm{Aut}_{\mathbb{C}^*}\mathcal{H}(D) \to \mathrm{Sp}(D)$. On the other hand, any $y \in K(D)$ defines an automorphism $\psi_y \in \mathrm{Aut}_{\mathbb{C}^*}\mathcal{H}(D)$, namely

$$\psi_y(\alpha, x_1, x_2) = (\alpha e^D(y, x_1 + x_2), x_1, x_2)$$

for all $(\alpha, x_1, x_2) \in \mathcal{H}(D)$. Since e^D is nondegenerate, the assignment $y \mapsto \psi_y$ is an injective homomorphism $\psi\colon K(D) \to \mathrm{Sp}(D)$.

(6.6) Lemma. *The sequence* $0 \longrightarrow K(D) \xrightarrow{\psi} \mathrm{Aut}_{\mathbb{C}^*}\mathcal{H}(D) \xrightarrow{p} \mathrm{Sp}(D) \longrightarrow 0$ *is exact.*

Proof. Any $\varphi \in \ker p$ is necessarily of the form

$$\varphi(\alpha, x_1, x_2) = (\alpha g(x_1 + x_2), x_1, x_2) .$$

The function $g\colon K(D) \to \mathbb{C}^*$ is linear, since φ is a homomorphism. So $g = e^D(y, \cdot)$ for some $y \in K(D)$ and the sequence is exact in the middle. It remains to show that p is surjective. Suppose $\sigma \in \mathrm{Sp}(D)$. Let $\bar{b}\colon K(L) \to K(D)$ be the symplectic isomorphism of Lemma 6.5. Then $\sigma\bar{b}\colon K(L) \to K(D)$ is also a symplectic isomorphism. Choose a symplectic basis $\lambda_1', \dots, \mu_g'$ of Λ for H such that $\sigma\bar{b}(\frac{1}{d_\nu}\lambda_\nu) = f_\nu$ and $\sigma\bar{b}(\frac{1}{d_\nu}\mu_\nu) = f_{g+\nu}$ for $\nu = 1, \dots, g$. Applying Lemma 6.5 there is a theta structure b' inducing $\sigma\bar{b}$. If b is a theta structure inducing $\bar{b}$, then $b'b^{-1} \in \mathrm{Aut}_{\mathbb{C}^*}\mathcal{H}(D)$ with $p(b'b^{-1}) = \sigma$. □

As a consequence we obtain

(6.7) Theorem. *Every ample line bundle L on X of type D admits exactly $h^0(L)^2 \cdot \#\,\mathrm{Sp}(D)$ theta structures.*

Given a symplectic isomorphism $\bar{b}\colon K(L) \to K(D)$, or equivalently a level D-structure for (X, H), we can describe explicitly all theta structures for L inducing $\bar{b}$. Suppose $c_0 \in V$ is a characteristic of L with respect to a basis $\lambda_1, \dots, \mu_g$ of Λ inducing the symplectic isomorphism $\bar{b}$ as in Lemma 6.5 (for the existence of such a basis see Section 8.3 or Exercise 8.3). Every $c \in c_0 + \Lambda(L)$ is a characteristic for L. As we saw above, every $\bar{c} \in \bar{c}_0 + K(L) = \pi(c_0 + \Lambda(L))$ determines a theta structure $b_{\bar{c}}\colon \mathcal{G}(L) \to \mathcal{H}(D)$ inducing $\bar{b}$.

(6.8) Lemma. $\psi_{\bar{b}(x)} b_{\bar{c}} = b_{\bar{c}+x}$ *for all* $x \in K(L)$.

Proof. It suffices to show that $\psi_{\bar{b}(x)} \tilde{b}_{\bar{c}} = \tilde{b}_{\bar{c}+x}$ where $\tilde{b}_{\bar{c}}$ and $\tilde{b}_{\bar{c}+x}$ are the homomorphisms $G(L) \to \mathcal{H}(D)$ defined in (3). Suppose $x = \pi(u)$ for some $u \in \Lambda(L)$. Using $\bar{b}^* e^D = e^L = e(-2\pi i \operatorname{Im} H)$ we have for all $[\alpha, w] \in G(L)$ with $w = w_1 + w_2$, $w_\nu \in \Lambda(L)_\nu$

$$\begin{aligned}
&\psi_{\bar{b}(x)} \tilde{b}_c([\alpha, w]) = \left(\alpha a_{\bar{c}}(w)^{-1} e^D\left(\bar{b}(x), \bar{b}(\overline{w})\right), \bar{b}(\overline{w}_1), \bar{b}(\overline{w}_2)\right) \\
&= \left(\alpha \chi_0(w)^{-1} e\left(-2\pi i \operatorname{Im} H(c, w) - \tfrac{\pi}{2} H(w, w) - 2\pi i \operatorname{Im} H(u, w)\right), \bar{b}(\overline{w}_1), \bar{b}(\overline{w}_2)\right) \\
&= \left(\alpha \chi_0(w)^{-1} e\left(-2\pi i \operatorname{Im} H(c+u, w) - \tfrac{\pi}{2} H(w, w)\right), \bar{b}(\overline{w}_1), \bar{b}(\overline{w}_2)\right) \\
&= \left(\alpha a_{\bar{c}+x}(w)^{-1}, \bar{b}(\overline{w}_1), \bar{b}(\overline{w}_2)\right) = \tilde{b}_{\bar{c}+x}\left([\alpha, w]\right) . \qquad \square
\end{aligned}$$

In particular the theta structures $b_{\bar{c}}$ and $b_{\bar{c}'}$ are different for $\bar{c} \neq \bar{c}'$. Using Lemma 6.5 we conclude that $\{b_{\bar{c}} \colon \mathcal{G}(L) \to \mathcal{H}(D) \mid \bar{c} \in \bar{c}_0 + K(L)\}$ is the set of all theta structures for L inducing the level D-structure $\bar{b}$. In other words we have

(6.9) Proposition. *The set of theta structures for* L *inducing a given symplectic isomorphism* $\bar{b} \colon K(L) \to K(D)$ *is a principal homogeneous space for the group* $K(L)$.

§ 7 The Schrödinger Representation

Let (X, H) be a polarized abelian variety of type $D = \operatorname{diag}(d_1, \ldots, d_g)$. In Section 4 we introduced the canonical representation $\tilde{\rho} \colon \mathcal{G}(L) \to \operatorname{GL}(H^0(L))$ of the theta group $\mathcal{G}(L)$ for every line bundle $L \in \operatorname{Pic}^H(X)$. Certainly this representation depends on the particular choice of L within its algebraic equivalence class. On the other hand the Heisenberg group $\mathcal{H}(D)$ of type D, discussed in the last section, is a model of the theta group of L depending only on H. In this section we introduce a representation of the Heisenberg group, the Schrödinger representation, which is a model of the canonical representation and also depends only on H.

Let $\mathbb{C}(\mathbb{Z}^g / D\mathbb{Z}^g)$ denote the $\mathbb{C}$-vector space of all complex valued functions on the finite group $\mathbb{Z}^g / D\mathbb{Z}^g$. The delta functions δ_x, $x \in \mathbb{Z}^g / D\mathbb{Z}^g$,

$$\delta_x(y) = \begin{cases} 1 & \text{if} \quad x = y \\ 0 & \phantom{\text{if}} \quad x \neq y \end{cases}$$

form a canonical basis of $\mathbb{C}(\mathbb{Z}^g / D\mathbb{Z}^g)$. Consider the action of the Heisenberg group $\mathcal{H}(D)$ on the vector space $\mathbb{C}(\mathbb{Z}^g / D\mathbb{Z}^g)$ defined by

$$(\alpha, x_1, x_2)\gamma = \alpha e^D(\,\cdot\,, x_2)\gamma(\,\cdot\, + x_1)$$

for all $\gamma \in \mathbb{C}(\mathbb{Z}^g/D\mathbb{Z}^g)$ and $(\alpha, x_1, x_2) \in \mathcal{H}(D)$. The corresponding representation $\tilde{\rho}: \mathcal{H}(D) \to \mathrm{GL}(\mathbb{C}(\mathbb{Z}^g/D\mathbb{Z}^g))$ is called the *Schrödinger representation of* $\mathcal{H}(D)$.

Let L be a line bundle in $\mathrm{Pic}^H(X)$. Fix a theta structure $b = b_{\bar{c}}: \mathcal{G}(L) \to \mathcal{H}(D)$ with induced symplectic isomorphism $\bar{b}: K(L) \to K(D)$. The isomorphism $\bar{b}$ determines a decomposition $K(L) = K_1 \oplus K_2$. Let $\{\vartheta_x^L \mid x \in K_1\}$ be the basis of $H^0(L)$ of Theorem 3.2.7 associated to $K(L) = K_1 \oplus K_2$ and a characteristic c of L over $\bar{c}$. Define an isomorphism

$$\beta = \beta_c: H^0(L) \to \mathbb{C}(\mathbb{Z}^g/D\mathbb{Z}^g)\,,\ \beta(\vartheta_x^L) = \delta_{\bar{b}(x)}\ .$$

According to Corollary 3.2.9 the bases of $H^0(L)$ corresponding to different representatives c and c' of $\bar{c}$ differ only by a multiplicative constant. So up to a constant the isomorphism β is uniquely determined by b. We call it the isomorphism *associated* to the theta structure b. The next proposition shows that the canonical representation of $\mathcal{G}(L)$ and the Schrödinger representation of $\mathcal{H}(D)$ are equivalent via the isomorphisms b and β.

(7.1) Proposition. *The following diagram commutes*

$$\begin{array}{ccc} \mathcal{G}(L) \times H^0(L) & \longrightarrow & H^0(L) \\ {\scriptstyle (b,\beta)}\downarrow & & \downarrow{\scriptstyle \beta} \\ \mathcal{H}(D) \times \mathbb{C}(\mathbb{Z}^g/D\mathbb{Z}^g) & \longrightarrow & \mathbb{C}(\mathbb{Z}^g/D\mathbb{Z}^g) \end{array}\ .$$

Proof. By construction the theta structure b lifts to the homomorphism $\tilde{b}_{\bar{c}}: G(L) \to \mathcal{H}(D)$, $[\alpha, w] \mapsto (\alpha a_L(w,0)^{-1}, \bar{b}(\overline{w}_1), \bar{b}(\overline{w}_2))$ (see the proof of Lemma 6.5). Using $\bar{b}^* e^D = e^L$ and Proposition 4.2 we have

$$\begin{aligned} \beta([\alpha, w]\vartheta_x^L) &= \beta(\alpha a_L(w,0)^{-1} e^L(x - \overline{w}_1, \overline{w}_2)\vartheta^L_{x-\overline{w}_1}) \\ &= \alpha a_L(w,0)^{-1} e^D(\,\cdot\,, \bar{b}(\overline{w}_2))\delta_{\bar{b}(x)}(\,\cdot\, + \bar{b}(\overline{w}_1)) \\ &= (\alpha a_L(w,0)^{-1}, \bar{b}(\overline{w}_1), \bar{b}(\overline{w}_2))\delta_{\bar{b}(x)} \\ &= \tilde{b}_{\bar{c}}[\alpha, w]\beta(\vartheta_x^L)\ . \end{aligned}$$

□

As a consequence of Proposition 7.1 and Corollary 4.3, the Schrödinger representation $\tilde{\rho}: \mathcal{H}(D) \to \mathrm{GL}(\mathbb{C}(\mathbb{Z}^g/D\mathbb{Z}^g))$ is irreducible and the center $\mathbb{C}^*$ of $\mathcal{H}(D)$ acts by multiplication. Hence $\tilde{\rho}$ descends to a projective representation $\rho: K(D) \to \mathrm{PGL}_N(\mathbb{C})$. Certainly ρ is equivalent to the representation $K(L) \to \mathrm{PGL}_N(\mathbb{C})$ of Proposition 6.1.

(7.2) Remarks. a) Since the isomorphism β identifies the functions on the finite set $K(D)$ with theta functions for L, it makes sense to call the elements of $\mathbb{C}(\mathbb{Z}^g/D\mathbb{Z}^g)$ *finite theta functions.*

b) Note that the classical theta functions for L (see Sections 3.2 and 8.5) are periodic with respect to Λ_1. Hence restricting classical theta functions to $\Lambda(L)_1$ gives a canonical map res: $H^0(L) \to \mathbb{C}(K_1)$. Combining this with the isomorphism $\mathbb{C}(K_1) \xrightarrow{\sim} \mathbb{C}(\mathbb{Z}^g/D\mathbb{Z}^g)$ induced by a theta structure b we obtain a map $H^0(L) \to \mathbb{C}(\mathbb{Z}^g/D\mathbb{Z}^g)$. But this map is not equivariant under the action of the theta group $\mathcal{G}(L)$ respectively the Heisenberg group $\mathcal{H}(D)$. In particular, this map is different from the isomorphism $\beta: H^0(L) \to \mathbb{C}(\mathbb{Z}^g/D\mathbb{Z}^g)$ associated to b.

Identifying $\mathbb{P}_N = P\big(H^0(L)\big)$ with $P\big(\mathbb{C}(\mathbb{Z}^g/D\mathbb{Z}^g)\big)$ via β we can work out the projective representation ρ in the case of an elliptic curve and an abelian surface:

(7.3) Example. Let X be an elliptic curve and H a polarization on X of type $D = (d)$. Then

$$\mathcal{H}(D) = \mathbb{C}^* \times \mathbb{Z}/d\mathbb{Z} \oplus \mathbb{Z}/d\mathbb{Z}\ ,$$

and

$$e^D\big((\bar{\nu}, \bar{\nu}'), (\bar{\mu}, \bar{\mu}')\big) = e\big(\tfrac{2\pi i}{d}(\nu'\mu - \nu\mu')\big)$$

for $\bar{\nu}, \bar{\nu}', \bar{\mu}, \bar{\mu}' \in \mathbb{Z}/d\mathbb{Z}$ with representatives $\nu, \nu', \mu, \mu' \in \mathbb{Z}$. The generators $(\bar{1}, 0)$ and $(0, \bar{1})$ of $\mathbb{Z}/d\mathbb{Z} \oplus \mathbb{Z}/d\mathbb{Z}$ are represented by $\sigma := (1, \bar{1}, 0)$ and $\tau := (1, 0, \bar{1})$ in $\mathcal{H}(D)$. By definition σ and τ act on the basis $\{\delta_{\bar{\nu}} \mid \bar{\nu} \in \mathbb{Z}/d\mathbb{Z}\}$ of $\mathbb{C}(\mathbb{Z}/d\mathbb{Z})$ as follows:

$$\sigma: \delta_{\bar{\nu}} \mapsto \delta_{\bar{\nu}-\bar{1}}\ , \ \tau: \delta_{\bar{\nu}} \mapsto \xi^{-\nu}\delta_{\bar{\nu}}$$

with $\xi = e(\frac{2\pi i}{d})$. Hence the automorphisms $\rho(\bar{1}, 0)$ and $\rho(0, \bar{1})$ of $\mathrm{PGL}_{d-1}(\mathbb{C})$ are represented by the matrices

$$\tilde{\rho}(\sigma) = \begin{pmatrix} 0 & 1 & & \\ & \ddots & \ddots & \\ & & \ddots & 1 \\ 1 & & & 0 \end{pmatrix} \quad \text{and} \quad \tilde{\rho}(\tau) = \begin{pmatrix} 1 & & & \\ & \xi^{-1} & & \\ & & \ddots & \\ & & & \xi^{-d+1} \end{pmatrix}.$$

(7.4) Example. Let X be an abelian surface and H a polarization on X of type $D = \mathrm{diag}(d_1, d_2)$. Thus we have

$$\mathcal{H}(D) = \mathbb{C}^* \times (\mathbb{Z}/d_1\mathbb{Z} \times \mathbb{Z}/d_2\mathbb{Z}) \oplus (\mathbb{Z}/d_1\mathbb{Z} \times \mathbb{Z}/d_2\mathbb{Z})$$

and

$$e^D\big((\bar{\nu}_1, \bar{\nu}_2, \bar{\nu}_1', \bar{\nu}_2',), (\bar{\mu}_1, \bar{\mu}_2, \bar{\mu}_1', \bar{\mu}_2')\big) = e\big(\tfrac{2\pi i}{d_1}(\nu_1'\mu_1 - \nu_1\mu_1')\big)e\big(\tfrac{2\pi i}{d_2}(\nu_2'\mu_2 - \nu_2\mu_2')\big)$$

for $\bar{\nu}_j, \bar{\nu}'_j, \bar{\mu}_j, \bar{\mu}'_j \in \mathbb{Z}/d_j\mathbb{Z}$ with representatives $\nu_j, \nu'_j, \mu_j, \mu'_j \in \mathbb{Z}$. The generators $(\bar{1},0,0,0)$, $(0,\bar{1},0,0)$, $(0,0,\bar{1},0)$ and $(0,0,0,\bar{1})$ of $(\mathbb{Z}^2/D\mathbb{Z}^2 \oplus \mathbb{Z}^2/D\mathbb{Z}^2)$ are represented by $\sigma_1 := (1,\bar{1},0,0,0)$, $\sigma_2 := (1,0,\bar{1},0,0)$, $\tau_1 := (1,0,0,\bar{1},0)$, and $\tau_2 := (1,0,0,0,\bar{1})$ in $\mathcal{H}(D)$. By definition these elements act on the basis $\{\delta_{(\bar{\nu}_1,\bar{\nu}_2)} \mid (\bar{\nu}_1,\bar{\nu}_2) \in \mathbb{Z}^2/D\mathbb{Z}^2\}$ of $\mathbb{C}(\mathbb{Z}^2/D\mathbb{Z}^2)$ as follows:

$$\begin{array}{ll}\sigma_1: \delta_{(\bar{\nu}_1,\bar{\nu}_2)} \mapsto \delta_{(\bar{\nu}_1-\bar{1},\bar{\nu}_2)} \quad, & \sigma_2: \delta_{(\bar{\nu}_1,\bar{\nu}_2)} \mapsto \delta_{(\bar{\nu}_1,\bar{\nu}_2-\bar{1})} \quad. \\ \tau_1: \delta_{(\bar{\nu}_1,\bar{\nu}_2)} \mapsto \xi_1^{-\nu_1}\delta_{(\bar{\nu}_1,\bar{\nu}_2)} \quad, & \tau_2: \delta_{(\bar{\nu}_1,\bar{\nu}_2)} \mapsto \xi_2^{-\nu_2}\delta_{(\bar{\nu}_1,\bar{\nu}_2)}\end{array}$$

with $\xi_j = e(\frac{2\pi i}{d_j})$ for $j = 1,2$.

§ 8 The Isogeny Theorem for Finite Theta Functions

In this section we translate the Isogeny Theorem 5.1 into terms of finite theta functions. As in Section 5 let $f: Y = V/\Gamma \to X = V/\Lambda$ be an isogeny of abelian varieties and $L = L(H,\chi)$ an ample line bundle on X. Suppose L is of type $D = \operatorname{diag}(d_1,\ldots,d_g)$ and f^*L is of type $D' = \operatorname{diag}(d'_1,\ldots,d'_g)$. Let

$$b = b_{\bar{c}}: \mathcal{G}(L) \to \mathcal{H}(D) \quad \text{and} \quad b' = b_{\bar{c}'}: \mathcal{G}(f^*L) \to \mathcal{H}(D')$$

be theta structures. The induced symplectic isomorphisms

$$\bar{b}: K(L) \to K(D) \quad \text{and} \quad \bar{b}': K(f^*L) \to K(D')$$

determine decompositions $K(L) = K(L)_1 \oplus K(L)_2$ and $K(f^*L) = K(f^*L)_1 \oplus K(f^*L)_2$. The theta structures b and b' are called *compatible*, if the following conditions are satisfied:

i) $f\big(K(f^*L)_\nu\big) \cap K(L) = K(L)_\nu$ for $\nu = 1,2$, i.e. the decompositions of $K(L)$ and $K(f^*L)$ are compatible,
ii) $\rho_a(f)(c') = c$, i.e. the characteristics c and c' are compatible.

There always exist compatible theta structures, as the construction in Section 5 shows. Note that i) and ii) are just the hypotheses of the Isogeny Theorem 5.1. For the rest of this section suppose b and b' are compatible. Let $\beta: H^0(L) \to \mathbb{C}(\mathbb{Z}^g/D\mathbb{Z}^g)$ and $\beta': H^0(f^*L) \to \mathbb{C}(\mathbb{Z}^g/D'\mathbb{Z}^g)$ be the associated isomorphisms as defined in Section 7. Our aim is to describe the composed map

$$F: \mathbb{C}(\mathbb{Z}^g/D\mathbb{Z}^g) \xrightarrow{\beta^{-1}} H^0(L) \xrightarrow{f^*} H^0(f^*L) \xrightarrow{\beta'} \mathbb{C}(\mathbb{Z}^g/D'\mathbb{Z}^g) \;.$$

We need some preliminaries: Consider the subgroups S and $(S^\perp)_1$ of $K(D')$:

$$S := D\mathbb{Z}^g/D'\mathbb{Z}^g \oplus D\mathbb{Z}^g/D'\mathbb{Z}^g$$

and $$(S^\perp)_1 := \big\{z \in K(D') \mid e^{D'}(z,S) = 1\big\} \cap \bar{b}'\big(K(f^*L)_1\big) \;.$$

Using the symplectic isomorphisms $\bar{b}$ and $\bar{b}'$ one can characterize the groups S and $(S^{\perp})_1$ as follows:

(8.1) Lemma. $S = \bar{b}'(\ker f)$ *and* $(S^{\perp})_1 = \bar{b}'\big(f^{-1}(K(L)_1) \cap K(f^*L)_1\big)$.

Proof. The first statement is obvious. Together with the fact that e^{f^*L} is the pullback of $e^{D'}$ via $\bar{b}'$ it implies

$$(\bar{b}')^{-1}(S^{\perp})_1 = \{y \in K(f^*L) \mid e^{f^*L}(y, \ker f) = 1\} \cap K(f^*L)_1 \ .$$

Suppose $y \in f^{-1}(K(L)_1) \cap K(f^*L)_1$. Using Proposition 3.3 we get that $e^{f^*L}(y, \ker f) = e^L(f(y), 0) = 1$ and thus $f^{-1}(K(L)_1) \cap K(f^*L)_1 \subset (\bar{b}')^{-1}(S^{\perp})_1$.

For the inverse inclusion let $\bar{v} = \pi_Y(v) \in (\bar{b}')^{-1}(S^{\perp})_1$. According to the compatibility condition i) it remains to show that $f(\bar{v}) \in K(L)$. But for all $\lambda \in \Lambda$ we have

$$\begin{aligned} e\big(-2\pi i \operatorname{Im} H(\rho_a(f)(v), \lambda)\big) &= e\big(-2\pi i \operatorname{Im} \rho_a(f)^* H(v, \rho_a(f)^{-1}(\lambda))\big) \\ &= e^{f^*L}\big(v, \rho_a(f)^{-1}(\lambda)\big) = 1 \ . \end{aligned}$$

This implies $\rho_a(f)(v) \in \Lambda(L)$ and hence $f(\bar{v}) = \pi_X \rho_a(f)(\lambda) \in K(L)$. □

According to Lemma 8.1 the composed map

$$\tilde{f}: (S^{\perp})_1 \xrightarrow{(\bar{b}')^{-1}} f^{-1}(K(L)_1) \xrightarrow{f} K(L)_1 \xrightarrow{\bar{b}} \mathbb{Z}^g/D\mathbb{Z}^g \ .$$

is well defined, and we are in position to formulate the Isogeny Theorem as follows:

(8.2) Isogeny Theorem for Finite Theta Functions. *Let $f: Y \to X$ be an isogeny, $L \in \operatorname{Pic}(X)$ ample of type D and assume f^*L is of type D'. For compatible theta structures $b: \mathcal{G}(L) \to \mathcal{H}(D)$ and $b': \mathcal{G}(f^*L) \to \mathcal{H}(D')$ let F and $\tilde{f}$ be defined as above. Then*

$$F(\gamma)(z) = \begin{cases} \gamma(\tilde{f}(z)) & \text{if } z \in (S^{\perp})_1 \\ 0 & z \notin (S^{\perp})_1 \end{cases}$$

for all $\gamma \in \mathbb{C}(\mathbb{Z}^g/D\mathbb{Z}^g)$ and $z \in \mathbb{Z}^g/D'\mathbb{Z}^g$.

Proof. It suffices to prove the assertion for the basis $\{\delta_x \mid x \in \mathbb{Z}^g/D\mathbb{Z}^g\}$ of $\mathbb{C}(\mathbb{Z}^g/D\mathbb{Z}^g)$. Since the theta structures b and b' are compatible we may apply the Isogeny Theorem 5.1.

$$F(\delta_x)(z) = (\beta' f^* \vartheta^L_{\bar{b}^{-1}(x)})(z)$$

$$= (\beta' \sum_{\substack{y \in K(f^*L)_1 \\ f(y)=\bar{b}^{-1}(x)}} \vartheta^{f^*L}_y)(z)$$

$$= \sum_{\substack{y \in K(f^*L)_1 \\ f(y)=\bar{b}^{-1}(x)}} \delta_{\bar{b}'(y)}(z)$$

$$= \begin{cases} 1 & \text{if } f((\bar{b}')^{-1}(z)) = \bar{b}^{-1}(x) \\ 0 & \text{otherwise} \end{cases}$$

$$= \begin{cases} \delta_x(\bar{b} f (\bar{b}')^{-1}(z)) & \text{if } z \in \bar{b}' f^{-1} \bar{b}^{-1}(\mathbb{Z}^g / D\mathbb{Z}^g) \\ 0 & \text{otherwise} \end{cases}$$

$$= \begin{cases} \delta_x \tilde{f}(z) & \text{if } z \in (S^{\perp})_1 \\ 0 & \text{otherwise} \end{cases} .$$

□

As an example, how to apply the Isogeny Theorem 8.2, let us consider the case of the multiplication $n_X: X \to X$ for $n \in \mathbb{Z}$, $n \neq 0$. Suppose $L = L(H, \chi)$ is an ample line bundle on X of type $D = \text{diag}(d_1, \ldots, d_g)$. Then $n_X^* L$ is of type $n^2 D = \text{diag}(n^2 d_1, \ldots, n^2 d_g)$. Choose compatible theta structures $b: \mathcal{G}(L) \to \mathcal{H}(D)$ and $b': \mathcal{G}(n_X^* L) \to \mathcal{H}(n^2 D)$ with associated isomorphisms $\beta: H^0(L) \to \mathbb{C}(\mathbb{Z}^g / D\mathbb{Z}^g)$ and $\beta': H^0(n_X^* L) \to \mathbb{C}(\mathbb{Z}^g / n^2 D\mathbb{Z}^g)$. Denote by (n) the composed map

$$(n): \mathbb{C}(\mathbb{Z}^g / D\mathbb{Z}^g) \xrightarrow{\beta^{-1}} H^0(L) \xrightarrow{n_X^*} H^0(n_X^* L) \xrightarrow{\beta'} \mathbb{C}(\mathbb{Z}^g / n^2 D\mathbb{Z}^g)$$

and identify $\mathbb{Z}^g / D\mathbb{Z}^g$ with its image $n^2 \mathbb{Z}^g / n^2 D\mathbb{Z}^g$ in $\mathbb{Z}^g / n^2 D\mathbb{Z}^g$. Then we have

(8.3) Proposition. *(Multiplication by n)*

$$((n)\gamma)(z) = \begin{cases} \gamma(nz) & \text{if } z \in n\mathbb{Z}^g / n^2 D\mathbb{Z}^g \\ 0 & \phantom{\text{if }} z \notin n\mathbb{Z}^g / n^2 D\mathbb{Z}^g \end{cases}$$

for all $\gamma \in \mathbb{C}(\mathbb{Z}^g / D\mathbb{Z}^g)$ *and* $z \in \mathbb{Z}^g / n^2 D\mathbb{Z}^g$.

Proof. We claim first that $(S^{\perp})_1 = n\mathbb{Z}^g / n^2 D\mathbb{Z}^g$. But this follows immediately from Lemma 8.1 together with

$$(\bar{b}')^{-1}(n\mathbb{Z}^g / n^2 D\mathbb{Z}^g \oplus \{0\}) = nK(L^{n^2})_1 = n_X^{-1} K(L)_1 \cap K(L^{n^2})_1 .$$

Now it is clear that the composed map

$$\tilde{n}_X: (S^{\perp})_1 \xrightarrow{\bar{b}'^{-1}} n_X^{-1}(K(L)_1) \xrightarrow{n_X} K(L)_1 \xrightarrow{\bar{b}} \mathbb{Z}^g / D\mathbb{Z}^g$$

is just multiplication by n and this completes the proof. □

§ 9 Symmetric Theta Structures

Let L be an ample line bundle on the abelian variety X. In Section 6 we saw that we can use the action of the Heisenberg group on $H^0(L)$ to study the geometry of the projective variety $\varphi_L(X)$. It turns out that for some applications it would be useful to have a bigger group acting on $H^0(L)$. For symmetric line bundles L there is such a group, since the involution $(-1)_X$ acts on L. This leads to the definition of the extended theta group $\mathcal{G}^e(L)$, a semidirect product of $\mathcal{G}(L)$ with $<(-1)_X>$. On the other hand, (-1) acts by inner automorphisms on the Heisenberg group in a canonical way. We call the corresponding semidirect product $\mathcal{H}^e(D)$ the extended Heisenberg group. A symmetric theta structure is a theta structure admitting an extension to an isomorphism $\mathcal{G}^e(L) \simeq \mathcal{H}^e(D)$. It turns out that not every symmetric line bundle admits symmetric theta structures. For a given symmetric line bundle we compute explicitly the number of symmetric theta structures.

Let $L = L(H, \chi)$ be a symmetric ample line bundle on the abelian variety $X = V/\Lambda$. For a moment consider $K(L)$ as a group of translations of X. Since $(-1)_X t_x (-1)_X = t_{-x}$ for every $x \in K(L)$, the group $<(-1)_X>$ acts on $K(L)$ by inner automorphisms. Hence we can form the semidirect product of $K(L)$ with $<(-1)_X>$

$$K^e(L) := K(L) \rtimes <(-1)_X> .$$

Recall that the theta group $\mathcal{G}(L)$ is a central extension of $K(L)$ by $\mathbb{C}^*$. Analogously we will define the extended theta group $\mathcal{G}^e(L)$ to be a central extension of $K^e(L)$ by $\mathbb{C}^*$. Consider first the involution $(-1) \times 1 \colon V \times \mathbb{C} \to V \times \mathbb{C}$. It acts on the theta group $G(L)$ by inner automorphisms as follows:

(9.1) Lemma. $\big((-1) \times 1\big)^{-1}[\alpha, w]\big((-1) \times 1\big) = [\alpha, -w]$ *for all* $[\alpha, w] \in G(L)$.

Proof. For any $(v, t) \in V \times \mathbb{C}$ we have

$$\begin{aligned}\big((-1) \times 1\big)^{-1}[\alpha, w]\big((-1) \times 1\big)(v, t) &= \big(v - w, \alpha e\big(\pi H(-v, w)\big)t\big) \\ &= [\alpha, -w](v, t) .\end{aligned}$$

□

Recall the canonical section $s_L(\Lambda) = \{[a_L(\lambda, 0), \lambda] \in G(L) \mid \lambda \in \Lambda\}$ of the theta group $G(L)$. Since L is symmetric, $a_L(\lambda, 0) = a_L(-\lambda, 0)$, and thus $(-1) \times 1$ commutes with the elements in $s_L(\Lambda)$ considered as automorphisms of $V \times \mathbb{C}$. Hence $(-1) \times 1$ descends to an automorphism $(-1)_L \colon L \to L$ over $(-1)_X$. In fact, $(-1)_L$ is just the normalized isomorphism of L over $(-1)_X$ of Lemma 4.6.3. According to Lemma 9.1 the involution $(-1)_L$ acts on $\mathcal{G}(L)$ by conjugation. Define the *extended theta groups* to be the semidirect products:

$$G^e(L) = G(L) \rtimes <(-1) \times 1>$$
$$\mathcal{G}^e(L) = \mathcal{G}(L) \rtimes <(-1)_L> .$$

Obviously we have $\mathcal{G}^e(L) = G^e(L)/s_L(\Lambda)$. The group $\mathcal{G}^e(L)$ fits into the following commutative diagram (of course there is a similar diagram for $G^e(L)$):

$$\begin{array}{ccccccccc}
 & & & & 1 & & 1 & & \\
 & & & & \downarrow & & \downarrow & & \\
1 & \longrightarrow & \mathbb{C}^* & \longrightarrow & \mathcal{G}(L) & \longrightarrow & K(L) & \longrightarrow & 0 \\
 & & \| & & \downarrow & & | & & \\
1 & \longrightarrow & \mathbb{C}^* & \longrightarrow & \mathcal{G}^e(L) & \longrightarrow & K^e(L) & \longrightarrow & 0 \\
 & & & & \downarrow & & \downarrow & & \\
 & & & & <(-1)_L> & \xrightarrow{\sim} & <(-1)_X> & & \\
 & & & & \downarrow & & \downarrow & & \\
 & & & & 1 & & 1 & &
\end{array} \qquad (1)$$

We may write the elements of $\mathcal{G}^e(L)$ as pairs $(g, (-1)_L^n)$ with $g \in \mathcal{G}(L)$ and $n \in \mathbb{Z}$. Then the multiplication in $\mathcal{G}^e(L)$ is

$$(g_1, (-1)_L^{n_1})(g_2, (-1)_L^{n_2}) = (g_1(-1)_L^{-n_1} g_2(-1)_L^{n_1}, (-1)_L^{n_1+n_2}) .$$

(9.2) Remark. The exact sequence $1 \to \mathcal{G}(L) \to \mathcal{G}^e(L) \to <(-1)_L> \to 1$ splits if and only if the group $K(L)$ is contained in X_2 and the semicharacter of L is trivial on $2\Lambda(L)$ (see Exercise 11).

Suppose H is of type $D = \operatorname{diag}(d_1, \ldots, d_g)$. In Section 6 we associated to H the Heisenberg group $\mathcal{H}(D)$. As we saw above, for every symmetric line bundle $L = L(H, \chi)$ there is the extended theta group. This suggests to introduce an extended Heisenberg group. Let ι denote the generator of $\mathbb{Z}/2\mathbb{Z}$. The action of $\mathbb{Z}/2\mathbb{Z}$ on $K(D) = \mathbb{Z}^g/D\mathbb{Z}^g \oplus \mathbb{Z}^g/D\mathbb{Z}^g$ given by

$$\iota\big((x_1, x_2)\big) = (-x_1, -x_2)$$

defines a semidirect product

$$K^e(D) = K(D) \rtimes \mathbb{Z}/2\mathbb{Z} .$$

The above action ι lifts to an action on the Heisenberg group $\mathcal{H}(D)$, namely

$$\iota\big((\alpha, x_1, x_2)\big) = (\alpha, -x_1, -x_2) .$$

The corresponding semidirect product

$$\mathcal{H}^e(D) = \mathcal{H}(D) \rtimes \mathbb{Z}/2\mathbb{Z}$$

is called the *extended Heisenberg group (of type D)*. Writing the elements of $\mathcal{H}^e(D)$ as pairs $(\tilde{x}, \iota^n)$ with $\tilde{x} \in \mathcal{H}(D)$ and $n \in \mathbb{Z}$ the multiplication in $\mathcal{H}^e(D)$ is:

$$(\tilde{x}, \iota^{n_1})(\tilde{y}, \iota^{n_2}) = (\tilde{x} \cdot \iota^{n_1}(\tilde{y}), \iota^{n_1+n_2}) . \tag{2}$$

The extended Heisenberg group fits into the following commutative diagram

$$\begin{array}{ccccccccc}
 & & & & 1 & & 1 & & \\
 & & & & \downarrow & & \downarrow & & \\
1 & \longrightarrow & \mathbb{C}^* & \longrightarrow & \mathcal{H}(D) & \longrightarrow & K(D) & \longrightarrow & 0 \\
 & & \| & & \downarrow & & \downarrow & & \\
1 & \longrightarrow & \mathbb{C}^* & \longrightarrow & \mathcal{H}^e(D) & \longrightarrow & K^e(D) & \longrightarrow & 0 \\
 & & & & \downarrow & & \downarrow & & \\
 & & & & \mathbb{Z}/2\mathbb{Z} & = & \mathbb{Z}/2\mathbb{Z} & & \\
 & & & & \downarrow & & \downarrow & & \\
 & & & & 0 & & 0 & &
\end{array} \tag{3}$$

Recall that a theta structure for L is an isomorphism between the groups $\mathcal{G}(L)$ and $\mathcal{H}(D)$ inducing the identity on $\mathbb{C}^*$. Similarly an isomorphism

$$b^e: \mathcal{G}^e(L) \to \mathcal{H}^e(D)$$

inducing the identity on $\mathbb{C}^*$ is called an *extended theta structure*. Comparing the diagrams (1) and (3) one concludes that every extended theta structure restricts to an ordinary theta structure for L. Conversely we will see that not every theta structure extends to an extended theta structure.

Call a theta structure $b: \mathcal{G}(L) \to \mathcal{H}(D)$ *symmetric*, if the diagram

$$\begin{array}{ccc}
\mathcal{G}(L) & \xrightarrow{(-1)_L} & \mathcal{G}(L) \\
{\scriptstyle b}\downarrow & & \downarrow{\scriptstyle b} \\
\mathcal{H}(D) & \xrightarrow{\iota} & \mathcal{H}(D)
\end{array}$$

commutes. Here $(-1)_L: \mathcal{G}(L) \to \mathcal{G}(L)$ means conjugation by $(-1)_L$. We have the following criterion for a theta structure to extend to an isomorphism $\mathcal{G}^e(L) \to \mathcal{H}^e(D)$.

(9.3) Proposition. *For a theta structure $b\colon \mathcal{G}(L) \to \mathcal{H}(D)$ the following conditions are equivalent:*

(i) there is an extended theta structure $b^e\colon \mathcal{G}^e(L) \to \mathcal{H}^e(D)$ such that $b^e|\mathcal{G}(L) = b$,

(ii) b is symmetric.

Moreover the extended theta structure b^e is uniquely determined by b.

Proof. We use the following universal property of semidirect products. For every homomorphism $p\colon \mathcal{G}^e(L) \to \mathbb{Z}/2\mathbb{Z}$ and every map $d\colon \mathcal{G}^e(L) \to \mathcal{H}(D)$ with

$$d(\tilde{g}_1\tilde{g}_2) = d(\tilde{g}_1) \cdot p(\tilde{g}_1)\big(d(\tilde{g}_2)\big) \tag{4}$$

for all $\tilde{g}_1, \tilde{g}_2 \in \mathcal{G}^e(L)$ there is a unique homomorphism $b^e\colon \mathcal{G}^e(L) \to \mathcal{H}^e(D)$ fitting into the following commutative diagram

$$\begin{array}{ccccccccc}
 & & & & \mathcal{G}^e(L) & & & & \\
 & & & \swarrow^{d} & \downarrow{\scriptstyle b^e} & \searrow^{p} & & & \\
1 & \longrightarrow & \mathcal{H}(D) & \longrightarrow & \mathcal{H}^e(D) & \longrightarrow & \mathbb{Z}/2\mathbb{Z} & \longrightarrow & 1\ .
\end{array}$$

Consider the diagram

$$\begin{array}{ccccccccc}
1 & \longrightarrow & \mathcal{G}(L) & \overset{q}{\underset{}{\leftrightarrows}} & \mathcal{G}^e(L) & \overset{p}{\longrightarrow} & <(-1)_L> & \longrightarrow & 1 \\
 & & {\scriptstyle b}\downarrow & & & & \Vert & & \\
1 & \longrightarrow & \mathcal{H}(D) & \longrightarrow & \mathcal{H}^e(D) & \longrightarrow & \mathbb{Z}/2\mathbb{Z} & \longrightarrow & 1
\end{array}$$

where $q\big(g, (-1)^n_L\big) = g$ and $p(g, (-1)^n_L) \doteq (-1)^n_L$ for all $\big(g, (-1)^n_L\big) \in \mathcal{G}^e(L)$. Since for all $\big(g_\nu, (-1)^{n_\nu}_L\big) \in \mathcal{G}^e(L)$, $\nu = 1, 2$

$$q\big(\big(g_1, (-1)^{n_1}_L\big)\big(g_2, (-1)^{n_2}_L\big)\big) = g_1(-1)^{-n_1}_L g_2(-1)^{n_1}_L\ ,$$

it is easy to check that the composed map $d := bq\colon \mathcal{G}^e(L) \to \mathcal{H}(D)$ satisfies (4) if and only if the theta structure b is symmetric. This implies the assertion. □

Given a symplectic isomorphism $\bar{b}\colon K(L) \to K(D)$ we want to determine the symmetric theta structures for L inducing $\bar{b}$. Choose a symplectic basis $\lambda_1, \ldots, \mu_g$ of Λ for H inducing $\bar{b}$ as in the construction before Lemma 6.5. According to Proposition 6.9 and Lemma 6.5 every theta structure inducing $\bar{b}$ is of the form $b_{\bar{c}}$ for a characteristic c of L. We want to work out what the symmetry of $b_{\bar{c}}$ means in terms of c.

(9.4) Proposition. *For a theta structure $b_{\bar{c}}: \mathcal{G}(L) \to \mathcal{H}(D)$ inducing the symplectic isomorphism $\bar{b}$ the following conditions are equivalent*
(i) $b_{\bar{c}}$ is symmetric,
(ii) $e^L(2\bar{c}, K(L)) = 1$,
(iii) $\bar{c} \in X_2$.

Proof. We use the notation of Section 6. By definition the theta structure $b_{\bar{c}}$ is symmetric if and only if the diagram

$$\begin{array}{ccc} G(L) & \xrightarrow{(-1)\times 1} & G(L) \\ {\scriptstyle \tilde{b}_{\bar{c}}}\downarrow & & \downarrow{\scriptstyle \tilde{b}_{\bar{c}}} \\ \mathcal{H}(D) & \xrightarrow{\iota} & \mathcal{H}(D) \end{array}$$

commutes. Note that $(-1) \times 1: G(L) \to G(L)$ means conjugation by the automorphism $(-1) \times 1$. By Lemma 9.1 we have for all $[\alpha, w] \in G(L)$ with $w = w_1 + w_2$, $w_\nu \in \Lambda(L)_\nu$

$$\tilde{b}_{\bar{c}}\big(((-1) \times 1)^{-1}[\alpha, w]((-1) \times 1)\big) = \big(\alpha a_L(-w, 0)^{-1}, -\bar{b}(\overline{w}_1), -\bar{b}(\overline{w}_2)\big)$$

and
$$\iota\tilde{b}_{\bar{c}}([\alpha, w]) = \big(\alpha a_L(w, 0)^{-1}, -\bar{b}(\overline{w}_1), -\bar{b}(\overline{w}_2)\big) .$$

Hence the diagram above commutes if and only if $a_L(-w, 0) = a_L(w, 0)$ for all $w \in \Lambda(L)$. But

$$\begin{aligned} a_L(-w, 0) &= \chi_0(-w)e\big(2\pi i \operatorname{Im} H(c, -w) + \tfrac{\pi}{2} H(-w, -w)\big) \\ &= e\big(-2\pi i \operatorname{Im} H(2c, w)\big)\chi_0(w)e\big(2\pi i \operatorname{Im} H(c, w) + \tfrac{\pi}{2} H(w, w)\big) \\ &= e^L(2\bar{c}, \overline{w})a_L(w, 0) , \end{aligned}$$

which implies the equivalence $(i) \iff (ii)$. As for $(ii) \iff (iii)$, we have only to note that e^L is nondegenerate (see Theorem 3.4), so $e^L(2\bar{c}, K(L)) = 1$ if and only if $2\bar{c} = 0$. □

We can use Propositions 9.3 and 9.4 to determine the number of extended theta structures for a given symmetric line bundle.

(9.5) Theorem. *Let H be a polarization of type $D = \operatorname{diag}(d_1, \ldots, d_g)$ with $d_1, \ldots, d_s$ odd and $d_{s+1}, \ldots, d_g$ even. There are 2^{2s} symmetric line bundles in $\operatorname{Pic}^H(X)$ each admitting exactly $2^{2(g-s)} \cdot \# \operatorname{Sp}(D)$ symmetric theta structures. The remaining $2^{2(g-s)}$ symmetric line bundles in $\operatorname{Pic}^H(X)$ do not admit any symmetric theta structure.*

In fact the proof gives more: for a given ample symmetric line bundle one can decide whether it admits extended theta structures or not (see Exercise 10).

In particular, the line bundle $L_0 \in \operatorname{Pic}^H(X)$ with characteristic zero admits $2^{2(g-s)} \cdot \# \operatorname{Sp}(D)$ extended theta structures.

Proof. Fix a symplectic basis $\lambda_1, \ldots, \lambda_g, \mu_1, \ldots, \mu_g$ of Λ for H and let $\bar{b}: K(L) \to K(D)$ be the induced symplectic isomorphism. Suppose $L \in \operatorname{Pic}^H(X)$ is symmetric and $c_0 \in \frac{1}{2}\Lambda(L)$ is a characteristic for L. Recall that $\{b_{\bar{c}}: \mathcal{G}(L) \to \mathcal{H}(D) \mid \bar{c} \in \bar{c}_0 + K(L)\}$ is the set of theta structures for L inducing b (see the remark before Proposition 6.9). According to Propositon 9.4 the theta structure $b_{\bar{c}}$ is symmetric if and only if $\bar{c} \in (\bar{c}_0 + K(L)) \cap X_2$. Since

$$K(L) \cap X_2 \simeq (\mathbb{Z}/2\mathbb{Z})^{2(g-s)}$$

the cardinality of $(\bar{c}_0 + K(L)) \cap X_2$ is either $2^{2(g-s)}$ or zero. Using Proposition 9.3 this gives that a symmetric line bundle in $\operatorname{Pic}^H(X)$ admits either $2^{2(g-s)} \# \operatorname{Sp}(D)$ extended theta structures or none.

In order to compute the number of line bundles admitting extended theta structures recall that the symmetric line bundles in $\operatorname{Pic}^H(X)$ are represented by the characteristics c out of

$$\tfrac{1}{2}\Lambda(L)/\Lambda(L) \simeq \tfrac{1}{2}K(L)/K(L) \ .$$

Hence the symmetric line bundles admitting symmetric theta structures are represented by the elements of

$$X_2/(K(L) \cap X_2) \simeq (\mathbb{Z}/2\mathbb{Z})^{2s} \ .$$

This gives the assertion. □

Exercises

(1) Let X be an abelian variety and L a line bundle on X defining a principal polarization. For any $x \in X_2$ denote by ϑ_x the basis of $H^0(t_x^* L)$ of Theorem 3.2.7. Use the canonical representation of $\mathcal{G}(L^4)$ to show that $\{2^* \vartheta_x \mid x \in X_2\}$ is a basis of $H^0(L^4)$.

(2) *(Theorem of Serre-Rosenlicht)* Let X be an abelian variety. Any extension of algebraic groups of X by $\mathbb{C}^*$ is of the form $0 \to \mathbb{C}^* \to \mathcal{G}(L) \to X \to 0$ for a uniquely determined line bundle $L \in \operatorname{Pic}^0(X)$. To be more precise, there is a canonical isomorphism $\operatorname{Ext}^1(X, \mathbb{C}^*) \simeq \operatorname{Pic}^0(X)$ (see Serre [2] Chapter 7).

(3) *(The Stone-von Neumann Theorem for Theta Groups)* Let L be a nondegenerate line bundle on an abelian variety.

a) The theta group $\mathcal{G}(L)$ admits a unique irreducible representation V such that the subgroup $\mathbb{C}^*$ acts as $\alpha \mapsto \alpha \, \mathrm{id}_V$. (Hint: let U be any irreducible representation, and K a maximal isotropic subgroup of $K(L)$ with respect to e^L. Show that any character $\chi \in \widehat{K}$ extends uniquely to a character of $p^{-1}(K)$ acting on $\mathbb{C}^*$ by the identity. Here $p: \mathcal{G}(L) \to K(L)$ denotes the natural projection. Then U splits into eigenspaces $U = \bigoplus_{\chi \in \widehat{K}} U_\chi$. Choose $0 \neq u_0 \in U_{\chi_0}$ for some $\chi_0 \in \widehat{K}$ and a section $\widetilde{K}$ of $\mathcal{G}(L) \to \mathcal{G}(L)/p^{-1}(K) \simeq \widehat{K}$. Show that $U = \bigoplus_{\varphi \in \widetilde{K}} \mathbb{C}\varphi(u_0)$ and the representation on the right hand side does not depend on the choice of u_0 and $\widetilde{K}$.)

b) Any representation U of $\mathcal{G}(L)$ such that $\mathbb{C}^*$ acts as in a) is a direct sum of r copies of the irreducible representation where $r = \dim U^{\widetilde{K}}$ with $\widetilde{K}$ as in the hint above and $U^{\widetilde{K}}$ the subspace of U of $\widetilde{K}$-invariants.

c) Use Proposition 2.2 and the methods of Section 3.3 to generalize a) and b) to an arbitrary line bundle on X.

(4) Let L be a nondegenerate line bundle of type $(d_1, \dots, d_g)$ on an abelian variety and n an integer.

a) Any irreducible representation of the theta group $\mathcal{G}(L)$, such that $\mathbb{C}^*$ acts as $\alpha \mapsto \alpha^n \, \mathrm{id}$, is of dimension $\geq \frac{d_1 \cdot \ldots \cdot d_g}{(n,d_1) \cdot \ldots \cdot (n,d_g)}$. (Hint: generalize the proof of Exercise 3 a).)

b) Let L be of type (d), $d > 1$, on an elliptic curve. Show that $\mathcal{G}(L)$ admits several nonisomorphic irreducible representations such that $\mathbb{C}^*$ acts as $\alpha \mapsto \alpha^d \, \mathrm{id}$. (Hint: use the d-th symmetric power of the canonical representation of $\mathcal{G}(L)$.)

(5) Let L be a line bundle on the abelian variety X. Let s be the number of negative eigenvalues of the hermitian form $c_1(L)$. Show that $H^s(L)$ is an irreducible representation of the theta group $\mathcal{G}(L)$ such that the subgroup $\mathbb{C}^*$ acts by multiplication. (Hint: use Exercise 2b).)

The next three exercises deal with the homomorphisms ϵ_n, δ_n and η_n between the theta groups $G(L)$ and $G(L^n)$. They were defined in Mumford [1] in terms of $\widetilde{\mathcal{G}}(L)$ and $\widetilde{\mathcal{G}}(L^n)$. Here L is again an ample line bundle on an abelian variety X.

(6) For any integer n define a map $\epsilon_n: G(L) \to G(L^n)$ by $[\alpha, w] \mapsto [\alpha^n, w]$.

a) Show that ϵ_n is a homomorphism extending the n-th power map on $\mathbb{C}^*$ and the identity on $\Lambda(L)$.

b) Suppose L is symmetric and $n \geq 1$. Show that ϵ_n fits into the following commutative diagram:

$$\begin{array}{ccc} G(L) \times H^0(L)^{\otimes n} & \longrightarrow & G(L^n) \times H^0(L^n) \\ \downarrow & & \downarrow \\ H^0(L)^{\otimes n} & \longrightarrow & H^0(L^n) \end{array} .$$

This induces an action of the theta group $\mathcal{G}(L)$ on $H^0(L^n)$ such that $\mathbb{C}^*$ acts by $\alpha \mapsto \alpha^n \operatorname{id}_{H^0(L^n)}$.

c) In terms of the groups $\widetilde{\mathcal{G}}(L)$ and $\widetilde{\mathcal{G}}(L^n)$ of Remark 1.2 the homomorphism ϵ_n translates to $\widetilde{\varphi} \to \widetilde{\varphi}^n$.

(7) For any integer n define a map $\delta_n: G(L) \to G(L)$ by $[\alpha, w] \mapsto [\alpha^{(n^2)}, nw]$.

a) δ_n is a homomorphism extending the n^2-power map on $\mathbb{C}^*$ and the multiplication by n on $\Lambda(L)$.

b) $\delta_n[\alpha, w] = [\alpha, w]^{\frac{n^2+n}{2}} [\alpha, -w]^{\frac{n^2-n}{2}}$.

c) Suppose L is symmetric. Show that δ_{-1} induces on $\mathcal{G}(L)$ the conjugation by the normalized isomorphism $(-1)_L: L \to L$ of Section 4.6.

d) Express δ_{-1} in terms of the group $\widetilde{\mathcal{G}}_L$ (see Mumford [1] p. 308).

(8) For any integer n define a map $\eta_n: G(L^n) \to G(L)$ by $[\alpha, w] \mapsto [\alpha^n, nw]$.

a) η_n is a homomorphism extending the n-th power map on $\mathbb{C}^*$ and the multiplication by n on $\Lambda(L)$.

b) $\delta_n = \eta_n \epsilon_n$.

c) Suppose L is a symmetric ample line bundle, then η_n descends to a homomorphism $\overline{\eta}_n: \mathcal{G}(L^n) \to \mathcal{G}(L)$.

d) Suppose L is symmetric. If $\varphi \in \mathcal{G}(L^2)$ is of order 2 and $x = p(\varphi)$ its image in $X_2 \subseteq K(L^2)$, then $\overline{\eta}_2(\varphi) = q_L(x)$ with q_L the quadratic form on X_2 associated to L defined in Section 4.7.

e) Express η_n in terms of the group $\widetilde{\mathcal{G}}(L)$ (see Mumford [1] p. 310).

(9) Give an example of a type $D = \operatorname{diag}(d_1, \ldots, d_g)$, such that $K(D)$ admits a maximal isotropic subgroup K with respect to the form e^D, which is not isomorphic to $\mathbb{Z}^g / D\mathbb{Z}^g$. Conclude that there is no maximal isotropic subgroup K' of $K(D)$ such that $K(D) = K \oplus K'$.

(10) Let L be an ample symmetric line bundle on an abelian variety X of dimension g. Show that L admits a symmetric theta structure if and only if the dimension of $H^0(L)_+$ or $H^0(L)_-$ is maximal. Here maximal means $h^0(L)_+$ or $h^0(L)_- = \frac{1}{2} h^0(L) + 2^{g-s-1}$ for some $0 \leq s \leq g$ (see Exercise 4.11). (Hint: use Exercise 4.11 and the proof of Theorem 9.5.)

(11) Let $L = L(H, \chi)$ be an ample symmetric line bundle on the abelian variety $X = V/\Lambda$. The extended theta group $\mathcal{G}^e(L)$ is isomorphic to the direct sum $\mathcal{G}(L) \oplus <(-1)_L>$ if and only if $K(L) \subset X_2$ and $q_L | K(L) \equiv 1$, with q_L the quadratic form on X_2 associated to L defined in Section 4.7. (Hint: $\mathcal{G}^e(L) \simeq \mathcal{G}(L) \oplus <(-1)_L>$ if and only if $(-1)_L$ acts on $\mathcal{G}(L)$ as the identity.)

(12) A symmetric nondegenerate line bundle $L = L(H, \chi)$ on the abelian variety X is called *totally symmetric*, if $\chi \equiv 1$. The following conditions are equivalent:

(i) L is totally symmetric.

(ii) L is the square of a symmetric line bundle.

(iii) $X_2 \subset K(L)$ and L is of characteristic 0 with respect to some decomposition.

(iv) Let $p: X \to K_X$ be the natural projection onto the Kummer variety $K_X = X/<(-1)_X>$. There is a line bundle $M \in \mathrm{Pic}(K_X)$ such that $L = p^*M$.

(13) Let $L = L(H,\chi)$ be an ample symmetric line bundle of type D on an abelian variety X, of characteristic zero with respect to some decomposition. The involution $(-1)_L$ on $H^0(L)$ induces an involution ι on $\mathbb{C}(\mathbb{Z}^g/D\mathbb{Z}^g)$ via the isomorphism $\beta: H^0(L) \to \mathbb{C}(\mathbb{Z}^g/D\mathbb{Z}^g)$ of Section 7.

a) (*Inverse Formula for Finite Theta Functions*): $\iota\delta_x = \delta_{-x}$ for any $x \in \mathbb{Z}^g/D\mathbb{Z}^g$ (Hint: use the Inverse Formula 4.6.4.)

a) The Schrödinger representation $\mathcal{H}(D) \to \mathrm{GL}(\mathbb{Z}^g/D\mathbb{Z}^g)$ extends to a representation $\mathcal{H}(D)^e \to \mathrm{GL}(\mathbb{Z}^g/D\mathbb{Z}^g)$.

(14) *(The Normalizer of the Heisenberg Group)* Let D be the type of an ample line bundle on an abelian variety of dimension g. Consider the Heisenberg group $\mathcal{H}(D)$ as subgroup of $\mathrm{GL}(\mathbb{C}(\mathbb{Z}^g/D\mathbb{Z}^g))$ via the Schrödinger representation. Let

$$\mathcal{N}(D) = \{\gamma \in \mathrm{GL}(\mathbb{C}(\mathbb{Z}^g/D\mathbb{Z}^g)) \mid \gamma^{-1}\sigma\gamma \in \mathcal{H}(D) \text{ for all } \sigma \in \mathcal{H}(D)\}$$

denote the normalizer of $\mathcal{H}(D)$ in $\mathrm{GL}(\mathbb{C}(\mathbb{Z}^g/D\mathbb{Z}^g))$

a) The map $\mathcal{N}(D) \to \mathrm{Aut}_{\mathbb{C}^*}\mathcal{H}(D)$, $\gamma \mapsto (\sigma \mapsto \gamma^{-1}\sigma\gamma)$ induces an isomorphism

$$\mathcal{N}(D)/\mathbb{C}^* \simeq \mathrm{Aut}_{\mathbb{C}^*}\mathcal{H}(D) .$$

b) Use Lemma 6.4 to conclude that the following diagram has exact rows

$$\begin{array}{ccccccccc} 0 & \longrightarrow & \mathcal{H}(D) & \longrightarrow & \mathcal{N}(D) & \longrightarrow & \mathrm{Sp}(D) & \longrightarrow & 0 \\ & & \downarrow & & \downarrow & & \| & & \\ 0 & \longrightarrow & K(D) & \longrightarrow & \mathcal{N}(D)/\mathbb{C}^* & \longrightarrow & \mathrm{Sp}(D) & \longrightarrow & 0 \end{array} .$$

Chapter 7

Equations for Abelian Varieties

In Chapter 4 we proved some criteria for a line bundle L on an abelian variety X to be very ample. The corresponding embedding $\varphi_L: X \to \mathbb{P}_N$ gives X the structure of a closed subvariety of $\mathbb{P}_N$. As such, X is the set of zeros of a homogeneous ideal I of polynomials in $N+1$ variables. Since the embedding φ_L is defined by means of a basis of theta functions of $H^0(L)$, the polynomials of I may be considered as relations among these theta functions. According to classical terminology they are called *theta relations*. The subject of this chapter is to find a set of theta relations which generates the ideal I, and thus describes the subvariety X of $\mathbb{P}_N$ completely in terms of equations.

The main tool is again the basis of canonical theta functions of Theorem 3.2.7. The Multiplication Formula (see Mumford [1]) describes the canonical map $\mu: H^0(L) \otimes H^0(L) \to H^0(L^2)$ in terms of this basis. By a slight change of the basis (due to Mumford [1]) the homomorphism μ may be diagonalized and this can be used to prove criteria for the multiplication map to be surjective. An immediate consequence are the theorems of Koizumi [1] and Ohbuchi [2] concerning the projective normality of X in $\mathbb{P}_N$. The next step is a theorem of Kempf [2], improving a slightly weaker result of Mumford [3], which says that under some additional hypotheses the homogeneous ideal I is generated already by the vector space I_2 of forms of degree 2 respectively by I_2 and I_3. Hence Riemann's Theta Relations, respectively the Cubic Theta Relations, turn out to be a system of generators of the ideal I.

In the first two more technical sections we prove the Multiplication Formula 1.3 and derive some consequences concerning the surjectivity of the multiplication map. Section 3 contains a proof of the Theorems of Koizumi and Ohbuchi on projective normality (Theorem 3.1). In the next section we prove Kempf's Theorem mentioned above. Using the special basis of canonical theta functions of Section 1, it is easy to deduce Riemann's Theta Relations in Section 5. Essentially the same method works for the proof of the Cubic Theta Relations, given in Section 6. Also here the Isogeny Theorem 6.5.1 is applied to express the multiplication map $\mu_3: H^0(L) \otimes H^0(L) \otimes H^0(L) \to H^0(L^3)$ in terms of the basis of canonical theta functions of Theorem 3.2.7, and again μ_3 can be diagonalized by a suitable change of the basis. The Cubic Theta Relations are a direct generalization of Hesse's cubic equation for plane elliptic curves (see Exercise 8).

As for the prerequisites, the first two and the last two sections require a good understanding of the results of Chapters 3 and 6 on canonical theta functions and the theta group. In the two middle sections some deeper results of algebraic geometry are applied, such as base change theorems, the Leray spectral sequence and relative Serre duality.

§ 1 The Multiplication Formula

Let L and L' be algebraically equivalent ample line bundles on the abelian variety $X = V/\Lambda$. The first step in the investigation of the rational map $\varphi_L: X \to \mathbb{P}_N$ is the study of the canonical multiplication map

$$\mu: H^0(L) \otimes H^0(L') \to H^0(L \otimes L') .$$

This is the theme of the present section. To be more precise, we will express μ explicitly in terms of the particular basis of canonical theta functions of Theorem 3.2.7. For this we will factorize μ suitably, so that we can apply the Isogeny Theorem.

Let $p_\nu: X \times X \to X$ denote the ν-th projection. The line bundle $M = p_1^*L \otimes p_2^*L'$ is ample on $X \times X$. By the Künneth formula the map

$$p_1^* \otimes p_2^*: H^0(L) \otimes H^0(L') \to H^0(M)$$

is an isomorphism. Let $\Delta: X \to X \times X$ be the diagonal map given by $\Delta(x) = (x, x)$; then the multiplication map μ factorizes as follows:

$$\mu: H^0(L) \otimes H^0(L') \xrightarrow{p_1^* \otimes p_2^*} H^0(M) \xrightarrow{\Delta^*} H^0(L \otimes L') .$$

The homomorphism Δ fits into the commutative diagram

$$\begin{array}{ccccc} & & X \times X & & \\ & {}^{s}\nearrow & & \searrow^{\alpha} & \\ X & & \xrightarrow{\quad\Delta\quad} & & X \times X \end{array}$$

with $s(x) = (x, 0)$ and $\alpha(x_1, x_2) = (x_1 + x_2, x_1 - x_2)$. Hence we obtain the following factorization of μ:

$$\mu: H^0(L) \otimes H^0(L') \xrightarrow{p_1^* \otimes p_2^*} H^0(M) \xrightarrow{\alpha^*} H^0(\alpha^* M) \xrightarrow{s^*} H^0(L \otimes L') .$$

We start by studying the homomorphism α:

(1.1) Lemma. *a) α is an isogeny with kernel $\{(x,x) \mid x \in X_2\}$.*
b) $\alpha^* M \simeq p_1^*(L \otimes L') \otimes p_2^*(L \otimes (-1)^* L')$.

Proof. The assertion a) is obvious. As for b), suppose $L = L(H,\chi)$ and $L' = L(H,\chi')$. According to the Appell-Humbert Theorem it suffices to compare the first Chern classes and the semicharacters. But for all $(v_1,v_2),(w_1,w_2) \in V \times V$ and $(\lambda_1,\lambda_2) \in \Lambda \times \Lambda$:

$$\begin{aligned}\alpha^*(p_1^*H + p_2^*H)\big((v_1,v_2),(w_1,w_2)\big) = \\ &= H(v_1+v_2, w_1+w_2) + H(v_1-v_2, w_1-w_2) \\ &= 2H(v_1,w_1) + 2H(v_2,w_2) \\ &= \big(p_1^*(2H) + p_2^*(2H)\big)\big((v_1,v_2),(w_1,w_2)\big)\ , \text{ and}\end{aligned}$$

$$\begin{aligned}\alpha^*(p_1^*\chi \cdot p_2^*\chi')(\lambda_1,\lambda_2) &= \chi(\lambda_1+\lambda_2)\chi'(\lambda_1-\lambda_2) \\ &= \chi(\lambda_1)\chi(\lambda_2)e\big(\pi i \operatorname{Im} H(\lambda_1,\lambda_2)\big)\chi'(\lambda_1)\chi'(-\lambda_2)e\big(\pi i \operatorname{Im} H(\lambda_1,-\lambda_2)\big) \\ &= \big(p_1^*(\chi\cdot\chi')p_2^*(\chi\cdot(-1)_\Lambda^*\chi')\big)(\lambda_1,\lambda_2)\ .\end{aligned}$$ □

In order to apply the Isogeny Theorem to the line bundle M and the isogeny α, we have to choose compatible decompositions and characteristics for M and α^*M. Since L and L' are algebraically equivalent, we have $K(L^\nu) = K(L'^\nu)$ for $\nu = 1,2$. Fix a decomposition $K(L^2) = K_1 \oplus K_2$ for L^2. Then

$$K(\alpha^*M) = (K_1 \times K_1) \oplus (K_2 \times K_2) \tag{1}$$

is a decomposition for α^*M, because $K(\alpha^*M) = K(L^2) \times K(L^2)$. According to Lemma 2.4.7 we have $2K(L^2) = K(L)$ and thus $2K_1 \oplus 2K_2$ is a decomposition of $K(L)$ for L. This implies that

$$K(M) = (2K_1 \times 2K_1) \oplus (2K_2 \times 2K_2) \tag{2}$$

is a decomposition for M. We claim that (1) and (2) are compatible decompositions. According to Proposition 6.5.2 this is a consequence of the following

(1.2) Lemma. $\alpha(K_\nu \times K_\nu) \cap K(M) = 2K_\nu \times 2K_\nu$ *for* $\nu = 1,2$.

Proof. It suffices to show that $2K_\nu \times 2K_\nu \subset \alpha(K_\nu \times K_\nu) \cap K(M)$, the converse inclusion being obvious. Suppose $u,v \in 2K_\nu$. There are $x,y \in K_\nu$ such that

$$2x = u+v \quad \text{and} \quad 2y = u-v\ .$$

This implies $x + y = u + z$ and $x - y = v + z'$ for some $z, z' \in K_\nu \cap X_2$. But $z = z'$, since

$$u + v = 2x = (x + y) + (x - y) = u + v + z + z' .$$

Thus $\alpha(x + z, y) = (u, v)$. □

Next choose characteristics for L and L' with respect to the underlying decompositions. This induces characteristics for $L \otimes L'$ and $L \otimes (-1)_X^* L'$ and for M and $\alpha^* M$, compatible in the sense of Section 6.5. Then Theorem 3.2.7 gives bases of canonical theta functions

$$\begin{aligned} \{\vartheta_x^L \mid x \in 2K_1\} &\quad \text{for } H^0(L) , \\ \{\vartheta_x^{L'} \mid x \in 2K_1\} &\quad \text{for } H^0(L') , \\ \{\vartheta_y^{L\otimes L'} \mid y \in K_1\} &\quad \text{for } H^0(L \otimes L') , \quad \text{and} \\ \{\vartheta_y^{L\otimes(-1)^* L'} \mid y \in K_1\} &\quad \text{for } H^0(L \otimes (-1)_X^* L') . \end{aligned}$$

Moreover by definition

$$\begin{aligned} &\{p_1^* \vartheta_{x_1}^L \otimes p_2^* \vartheta_{x_2}^{L'} \mid (x_1, x_2) \in 2K_1 \times 2K_1\}, \text{ and} \\ &\{p_1^* \vartheta_{y_1}^{L\otimes L'} \otimes p_2^* \vartheta_{y_2}^{L\otimes(-1)^* L'} \mid (y_1, y_2) \in K_1 \times K_1\} \end{aligned}$$

are the corresponding bases of $H^0(M)$ and $H^0(\alpha^* M)$. Finally define

$$Z_2 = K_1 \cap X_2 ,$$

the subgroup of 2-division points in K_1. It is of order 2^g, since $K_1 \simeq \mathbb{Z}/2d_1\mathbb{Z} \times \cdots \times \mathbb{Z}/2d_g\mathbb{Z}$. With this notation we can state

(1.3) Multiplication Formula. *For all* $(x_1, x_2) \in 2K_1 \times 2K_1$ *and* $(y_1, y_2) \in \alpha^{-1}(x_1, x_2) \cap K_1 \times K_1$

$$\mu(\vartheta_{x_1}^L \otimes \vartheta_{x_2}^{L'}) = \sum_{z\in Z_2} \vartheta_{y_2+z}^{L\otimes(-1)^* L'}(0) \cdot \vartheta_{y_1+z}^{L\otimes L'} .$$

Proof. This is a consequence of the Isogeny Theorem 6.5.1 and the fact that $\alpha^{-1}(x_1, x_2) \cap K_1 \times K_1 = \{(y_1 + z, y_2 + z) \mid z \in Z_2\}$:

$$\begin{aligned} \mu(\vartheta_{x_1}^L \otimes \vartheta_{x_2}^{L'}) &= s^* \alpha^* (p_1^* \vartheta_{x_1}^L \otimes p_2^* \vartheta_{x_2}^{L'}) \\ &= s^* \sum_{\substack{(z_1,z_2)\in \\ \alpha^{-1}(x_1,x_2)\cap K_1\times K_1}} p_1^* \vartheta_{z_1}^{L\otimes L'} \otimes p_2^* \vartheta_{z_2}^{L\otimes(-1)^* L'} \\ &= s^* \sum_{z\in Z_2} p_1^* \vartheta_{y_1+z}^{L\otimes L'} \otimes p_2^* \vartheta_{y_2+z}^{L\otimes(-1)^* L'} \\ &= \sum_{z\in Z_2} \vartheta_{y_2+z}^{L\otimes(-1)^* L'}(0) \cdot \vartheta_{y_1+z}^{L\otimes L'} . \end{aligned}$$

□

In order to check whether the multiplication map μ is surjective and to determine its kernel, the above version of the Multiplication Formula is not appropriate. With a slight additional assumption on L we can choose the bases in such a way that μ is given in diagonal form.

Assume that L is a square of an ample line bundle on X or equivalently that the group $K(L)$ contains all 2-division points of X. The crucial point of this is the fact that now Z_2 is already a subgroup of $2K_1$. So for any character $\rho \in \widehat{Z}_2 = \operatorname{Hom}(Z_2, \mathbb{C}_1)$ and $(x_1, x_2) \in 2K_1$

$$\theta_{(x_1,x_2),\rho} := \sum_{z\in Z_2} \rho(z)\vartheta^L_{x_1+z} \otimes \vartheta^{L'}_{x_2+z}$$

is a canonical theta function in $H^0(L) \otimes H^0(L')$. These functions generate the vector space $H^0(L) \otimes H^0(L')$, since

$$\sum_{\rho\in\widehat{Z}_2} \theta_{(x_1,x_2),\rho} = \sum_{z\in Z_2} \Big(\sum_{\rho\in\widehat{Z}_2} \rho(z)\Big) \vartheta^L_{x_1+z} \otimes \vartheta^{L'}_{x_2+z} = 2^g\, \vartheta^L_{x_1} \otimes \vartheta^{L'}_{x_2}$$

by the character relation. Similarly the canonical theta functions

$$\theta^{L\otimes L'}_{y,\rho} := \sum_{z\in Z_2} \rho(z)\vartheta^{L\otimes L'}_{y+z} \text{ , respectively}$$

$$\theta^{L\otimes(-1)^*L'}_{y,\rho} := \sum_{z\in Z_2} \rho(z)\vartheta^{L\otimes(-1)^*L'}_{y+z}$$

(for $y \in K_1$ and $\rho \in \widehat{Z}_2$) are generators of the vector space $H^0(L \otimes L')$, respectively $H^0(L \otimes (-1)^*L')$. In terms of these theta functions the Multiplication Formula reads:

(1.4) Corollary. *Suppose L is the square of an ample line bundle on X and L' is algebraically equivalent to L. Then for all $(x_1, x_2) \in 2K_1 \times 2K_1$, $(y_1, y_2) \in \alpha^{-1}(x_1, x_2) \cap K_1 \times K_1$ and $\rho \in \widehat{Z}_2$*

$$\mu(\theta_{(x_1,x_2),\rho}) = \theta^{L\otimes(-1)^*L'}_{y_2,\rho}(0) \cdot \theta^{L\otimes L'}_{y_1,\rho} \ .$$

Proof. Applying the Multiplication Formula 1.3 we have using $\alpha(y_1+z, y_2) = (x_1 + z, x_2 + z)$

$$\begin{aligned} \mu(\theta_{(x_1,x_2),\rho}) &= \sum_{z\in Z_2} \rho(z) \sum_{z'\in Z_2} \vartheta^{L\otimes(-1)^*L'}_{y_2+z'}(0)\vartheta^{L\otimes L'}_{y_1+z+z'} \\ &= \sum_{z'\in Z_2} \rho(z')\vartheta^{L\otimes(-1)^*L'}_{y_2+z'}(0) \sum_{z\in Z_2} \rho(z)\vartheta^{L\otimes L'}_{y_1+z} \\ &= \theta^{L\otimes(-1)^*L'}_{y_2,\rho}(0) \cdot \theta^{L\otimes L'}_{y_1,\rho} \ . \end{aligned}$$

□

It is easy to choose a basis of the vector space $H^0(L) \otimes H^0(L')$ out of the system of generators $\{\theta_{(x_1,x_2),\rho} \mid (x_1,x_2) \in 2K_1 \times 2K_1,\ \rho \in \widehat{Z}_2\}$. For any $z \in Z_2$ we have

$$\theta_{(x_1+z,x_2+z),\rho} = \rho(z)\theta_{(x_1,x_2),\rho}$$

for all $(x_1,x_2) \in 2K_1 \times 2K_1$ and $\rho \in \widehat{Z}_2$. Consequently the set of theta functions

$$\{\theta_{(x_1,x_2),\rho} \mid (x_1,x_2) \in (2K_1 \times 2K_1)/Z_2\ ,\ \rho \in \widehat{Z}_2\}$$

is already a system of generators of the vector space $H^0(L) \otimes H^0(L')$. But this is necessarily a basis, since $\#\big((2K_1 \times 2K_1)/Z_2\big) \times \widehat{Z}_2 = \#2K_1 \times 2K_1 = \dim(H^0(L) \otimes H^0(L'))$. Similarly on checks that

$$\{\theta^{L\otimes L'}_{y,\rho} \mid y \in K_1/Z_2,\ \rho \in \widehat{Z}_2\} \quad \text{and} \quad \{\theta^{L\otimes(-1)^*L'}_{y,\rho} \mid y \in K_1/Z_2,\ \rho \in \widehat{Z}_2\}$$

are bases for $H^0(L \otimes L')$ and $H^0(L \otimes (-1)^*L')$.

§ 2 Surjectivity of the Multiplication Map

In the last section we described explicitly the multiplication map μ in terms of canonical theta functions. We saw that for squares of ample line bundles μ can be diagonalized. Here we use this to deduce conditions for μ to be surjective.

Let L and L' be algebraically equivalent ample line bundles on the abelian variety $X = V/\Lambda$. We want to apply Corollary 1.4 to the multiplication map

$$\mu\colon H^0(L^2) \otimes H^0(L'^2) \to H^0(L^2 \otimes L'^2)\ .$$

For this fix a decomposition $K(L^4) = K(L'^4) = K_1 \oplus K_2$ and choose characteristics for L and L'. Write for abbreviation

$$M = L^2 \otimes L'^2 \quad \text{and} \quad M' = L^2 \otimes (-1)^*_X L'^2\ .$$

Then the canonical theta functions $\theta^M_{y_1,\rho} \in H^0(M)$, $\theta^{M'}_{y_2,\rho} \in H^0(M')$ and $\theta_{(x_1,x_2),\rho} \in H^0(L^2) \otimes H^0(L'^2)$, for $x_1, x_2 \in 2K_1$, $y_1, y_2 \in K_1$, $\rho \in \widehat{Z}_2 = (K_1 \cap X_2)\widehat{\ }$ are defined (for the definitions see the previous section) and Corollary 1.4 says:

$$\mu(\theta_{(x_1,x_2),\rho}) = \theta^{M'}_{y_2,\rho}(0) \cdot \theta^M_{y_1,\rho}$$

with $\alpha(y_1,y_2) = (x_1,x_2)$. As an immediate consequence we can state

(2.1) Lemma. *The following conditions are equivalent:*
i) $\mu: H^0(L^2) \otimes H^0(L'^2) \to H^0(L^2 \otimes L'^2)$ *is surjective,*
ii) for all $(y, \rho) \in K_1 \times \widehat{Z}_2$ *there is an* $x \in 2K_1$ *with* $\theta^{M'}_{y+x,\rho}(0) \neq 0$.

Proof. Suppose *ii)* holds. Then $\mu\big(\theta^{M'}_{y+x,\rho}(0)^{-1}\theta_{(2y+x,-x),\rho}\big) = \theta^{M}_{y,\rho}$, so μ is surjective. The converse implication is obvious. □

A first application of this lemma is

(2.2) Proposition. *There is a nonempty open set* $U \subset \mathrm{Pic}^0(X)$ *such that for all* $P \in U$ *the multiplication map*

$$\mu_P: H^0(L^2 \otimes P) \otimes H^0(L'^2) \to H^0(L^2 \otimes L'^2 \otimes P)$$

is surjective.

Proof. For any $(y, \rho) \in K_1 \times \widehat{Z}_2$ the set $W(y, \rho) := \mathrm{supp}\, \theta^{M'}_{y,\rho}$ is open and dense in V. Since $\pi: V \to X$ is a projection map and $\phi_{M'}: X \to \widehat{X} = \mathrm{Pic}^0(X)$ an isogeny, this implies that the set

$$\begin{aligned} U(y, \rho) &:= \phi_{M'}\big(\pi W(y, \rho)\big) \\ &= \big\{P \in \mathrm{Pic}^0(X) \mid t^*_{\bar{v}}M' = M' \otimes P \text{ for some } v \in W(y, \rho)\big\} \end{aligned}$$

is open and dense in $\mathrm{Pic}^0(X)$. On the other hand, for all $P = \phi_{M'}(\bar{v}) \in U(y, \rho)$ we have, using Corollary 3.2.9, $0 \neq \theta^{M'}_{y,\rho}(v) = \alpha\theta^{t^*_{\bar{v}}M'}_{y,\rho}(v - v) = \alpha\theta^{M'\otimes P}_{y,\rho}(0)$ for some nonzero constant α. Hence for any P out of the nonempty open set

$$U := \bigcap_{(y,\rho)\in K_1\times \widehat{Z}_2} U(y, \rho)$$

and any $(y, \rho) \in K_1 \times \widehat{Z}_2$ we have $\theta^{M'\otimes P}_{y,\rho}(0) \neq 0$ and the assertion follows from Lemma 2.1. □

For line bundles of characteristic zero we have

(2.3) Proposition. *For an ample line bundle* L_0 *on* X *of characteristic zero the multiplication map* $\mu: H^0(L_0^2) \otimes H^0(L_0^2) \to H^0(L_0^4)$ *is surjective if and only if* L_0 *has no base point in* $K(L_0^2)$.

Using Corollary 3.2.9 on can show that an analogous statement is true for an arbitrary ample line bundle (see Exercise 2), but we do not need this fact.

For the proof we need some preliminaries. With the notation as above, for any $(y_1, \rho) \in K_1 \times \widehat{Z}_2$ denote by

$$H(y_1,\rho) = <\theta^{L_0^4}_{y_1+x,\rho} \mid x \in 2K_1> ,$$

the subvector space of $H^0(L_0^4)$ spanned by the functions $\theta^{L_0^4}_{y_1+x,\rho}$, $x \in 2K_1$. For any $y \in K(L_0^4)$ the isogeny $2_X: X \to X$ induces an embedding $2_X^*: H^0(t_{2y}^* L_0) \hookrightarrow H^0(L_0^4)$, since $2_X^* t_{2y}^* L_0 \simeq L_0^4$. We will see that for a suitable choice of y and y_1 the vector spaces $2_X^* H^0(t_{2y}^* L_0)$ and $H(y_1,\rho)$ coincide. For this extend ρ to a character of K_1. According to Theorem 6.3.4 there is an $y_2 \in K_2$ with

$$\rho = e^{L_0^4}(y_2, \cdot\,) \,. \tag{1}$$

With this notation we have

(2.4) Lemma. $H(y_1,\rho) = \operatorname{Im}\bigl(H^0(t^*_{2(y_1+y_2)}L_0) \xrightarrow{2_X^*} H^0(L_0^4)\bigr)$.

Proof. We have to show that the canonical theta functions $\theta^{L_0^4}_{y_1+x,\rho}$, $x \in 2K_1$, descend via the homomorphism $\rho_a(2_X) = 2_V: V \to V$ to canonical theta functions for the line bundle $t^*_{2(y_1+y_2)}L_0$. For this it suffices to check that they satisfy the functional equation with respect to the factor $2_V^* a_{t^*_{2(y_1+y_2)}L_0}$ on the lattice $\frac{1}{2}\Lambda$.

Suppose $L_0 = L(H,\chi_0)$. Note first that for any $\lambda = \lambda_1 + \lambda_2 \in \Lambda$ (with respect to the underlying decomposition of Λ)

$$\begin{aligned} a_{L_0^4}(-\tfrac{1}{2}\lambda, 0)^{-1} &= \chi_0(-\tfrac{1}{2}\lambda)^{-4} e\bigl(-\tfrac{\pi}{2}(4H)(\tfrac{1}{2}\lambda, \tfrac{1}{2}\lambda)\bigr) \\ &= \chi_0(\lambda) e\bigl(\pi H(-\lambda,\lambda) + \tfrac{\pi}{2} H(\lambda,\lambda)\bigr) = a_{L_0}(\lambda, -\lambda) \,. \end{aligned}$$

Using the action of the theta group $G(L_0^4)$ (see equation 6.4(1) and Proposition 6.4.2) and the fact that $e^{L_0^4}(\,\cdot\,, \frac{1}{2}\lambda_2) \equiv 1$ on $2K_1$, we get

$$\begin{aligned} \theta^{L_0^4}_{y_1+x,\rho}(v + \tfrac{1}{2}\lambda) &= e\bigl(\pi(4H)(v + \tfrac{1}{2}\lambda, \tfrac{1}{2}\lambda)\bigr)[1, -\tfrac{1}{2}\lambda]\theta^{L_0^4}_{y_1+x,\rho}(v) \\ &= a_{L_0^4}(-\tfrac{1}{2}\lambda, 0)^{-1} e\bigl(\pi H(2v+\lambda,\lambda)\bigr) \cdot \\ &\quad \cdot \sum_{z\in Z_2} \rho(z) e^{L_0^4}(y_1 + x + z + \tfrac{1}{2}\lambda_1, -\tfrac{1}{2}\lambda_2) \vartheta^{L_0^4}_{y_1+x+z+\frac{1}{2}\lambda_1}(v) \\ &= a_{L_0}(\lambda,-\lambda) e\bigl(\pi H(2v+\lambda,\lambda)\bigr) e^{L_0^4}(y_1, -\tfrac{1}{2}\lambda_2)\rho(-\tfrac{1}{2}\lambda_1) \cdot \\ &\quad \cdot \sum_{z\in Z_2} \rho(z + \tfrac{1}{2}\lambda_1) \vartheta^{L_0^4}_{y_1+x+z+\frac{1}{2}\lambda_1}(v) \\ &= a_{L_0}(\lambda, 2v) e^{L_0^4}(y_1+y_2, -\tfrac{1}{2}\lambda)\theta^{L_0^4}_{y_1+x,\rho}(v) \\ &= 2_V^* a_{t^*_{2(y_1+y_2)}L_0}(\tfrac{1}{2}\lambda, v)\theta^{L_0^4}_{y_1+x,\rho}(v) \,. \end{aligned}$$

For the last equation we used Lemma 3.1.3 d). □

Now we are in position to prove Proposition 2.3.

Proof of Proposition 2.3. According to Lemmas 2.1 and 2.4 the map μ is surjective if and only if for any $(y_1, \rho) \in K_1 \times \widehat{Z}_2$ the point zero is not a base point of $t^*_{2(y_1+y_2)}L_0$. Here y_2 and ρ are related by (1). Since any $y_2 \in K_2$ leads to a $\rho \in \widehat{Z}_2$, this is the case if and only if for all $y \in K(L_0^4)$ zero is not a base point of $t^*_{2y}L_0$. According to Corollary 3.2.9 this is equivalent to saying that L_0 has no base point in $K(L_0^2)$. □

§ 3 Projective Normality

A projective variety $Y \subseteq \mathbb{P}_N$ is called *projectively normal* in $\mathbb{P}_N$ if its homogeneous coordinate ring is an integrally closed domain. According to Hartshorne [1] II Ex 5.14, a smooth variety Y is projectively normal in $\mathbb{P}_N$ if and only if the natural restriction map $H^0(\mathcal{O}_{\mathbb{P}_N}(n)) \to H^0(\mathcal{O}_Y(n))$ is surjective for every $n \geq 1$, or equivalently if the linear system of hypersurfaces of degree n in $\mathbb{P}_N$ restricts to a complete linear system on Y for every $n \geq 1$. Following Mumford [3] we call a line bundle M on Y *normally generated* if it is very ample and Y is projectively normal under the associated projective embedding. On the one hand there are canonical isomorphisms

$$H^0(\mathcal{O}_{\mathbb{P}_N}(n)) \simeq S^n H^0(\mathcal{O}_{\mathbb{P}_N}(1)) \simeq S^n H^0(M)$$

for all $n \geq 1$. On the other hand by Mumford [3] p.38, the surjectivity of the canonical maps $S^n H^0(M) \to H^0(M^n)$ for all $n \geq 2$ implies that M is very ample. Hence an ample line bundle M on a smooth projective variety is normally generated if and only if the canonical map $S^n H^0(M) \to H^0(M^n)$ is surjective for every $n \geq 2$.

In Section 4.5 we saw that for every ample line bundle L on an abelian variety X the line bundles L^n are very ample for $n \geq 3$ and gave a criterion for the case $n = 2$. The aim of this section is to investigate which of the corresponding embeddings $\varphi_{L^n}: X \hookrightarrow \mathbb{P}_N$ are projectively normal. The main result is

(3.1) Theorem. *Let L be an ample line bundle on an abelian variety X.*
a) L^n is normally generated for any $n \geq 3$.
b) Suppose L is of characteristic c, then L^2 is normally generated if and only if no point of $t^*_{\bar{c}}K(L^2)$ is a base point of L.

Note that this gives a second proof of the Theorem of Lefschetz 4.5.1, since a normally generated line bundle is automatically very ample. We need the following characterization for a line bundle to be normally generated.

(3.2) Lemma. *For an ample line bundle M on X the following conditions are equivalent*

i) M *is normally generated,*

ii) $H^0(M^m) \otimes H^0(M) \to H^0(M^{m+1})$ *is surjective for every* $m \geq 1$.

Proof. The assertion follows by induction on m using the commutative diagram

$$\begin{array}{ccccc} & \nearrow & S^m H^0(M) \otimes H^0(M) & \longrightarrow & H^0(M^m) \otimes H^0(M) \\ H^0(M)^{\otimes(m+1)} & & \downarrow & & \downarrow \\ & \searrow & S^{m+1} H^0(M) & \longrightarrow & H^0(M^{m+1}) \end{array} . \quad \square$$

Denote for any line bundles M and M' and any subvector spaces $V \subseteq H^0(M)$ and $V' \subseteq H^0(M')$ by $V \cdot V'$ the image of $V \otimes V'$ under the multiplication map:

$$V \cdot V' = \operatorname{Im}\{V \otimes V' \hookrightarrow H^0(M) \otimes H^0(M') \xrightarrow{\mu} H^0(M \otimes M')\} .$$

With this notation we have

(3.3) Lemma. *Let M and M' be ample line bundles on X. For every nonempty open subset U of* $\operatorname{Pic}^0(X)$

$$\sum_{P \in U} H^0(M \otimes P) \cdot H^0(M' \otimes P^{-1}) = H^0(M \otimes M') .$$

Proof. Step I: Denote $P_x = \phi_{M'}(x)$ for all $x \in K(M \otimes M')$. We claim that

$$\sum_{x \in K(M \otimes M')} H^0(M \otimes P_x) \cdot H^0(M' \otimes P_x^{-1}) = H^0(M \otimes M') .$$

According to Corollary 6.4.3 and Remark 6.1.2 the vector space $H^0(M \otimes M')$ is an irreducible representation of the theta group $\widetilde{\mathcal{G}}(M \otimes M')$ consisting of all isomorphisms $\varphi_x: M \otimes M' \to t_x^*(M \otimes M')$. Hence it suffices to show that $\sum_{x \in K(M \otimes M')} H^0(M \otimes P_x) \cdot H^0(M' \otimes P_x^{-1})$ is invariant under the action of $\widetilde{\mathcal{G}}(M \otimes M')$.

By definition of P_x we have $P_x^{-1} = P_{-x}$. So $t_x^*(M' \otimes P_x^{-1}) \simeq t_x^* M' \otimes M' \otimes t_x^* M'^{-1} \simeq M'$. On the other hand, by choice of x we have $t_x^*(M \otimes M') \simeq M \otimes M'$ and thus $t_x^*(M \otimes P_x) \simeq t_x^*(M \otimes P_x \otimes M' \otimes P_x^{-1}) \otimes M'^{-1} \simeq M$. Choose isomorphisms $\psi: M \to t_x^*(M \otimes P_x)$ and $\psi': M' \to t_x^*(M' \otimes P_x^{-1})$. Then

$$\varphi_x := \psi \otimes \psi': M \otimes M' \to t_x^*(M \otimes M')$$

is an element of $\widetilde{\mathcal{G}}(M \otimes M')$. Conversely every element of $\widetilde{\mathcal{G}}(M \otimes M')$ is of this form. By Remark 6.4.4 the isomorphism φ_x acts on $H^0(M \otimes M')$ by $t_{-x}^*\varphi_x$. Hence we have for all $\sigma \in H^0(M)$ and $\sigma' \in H^0(M')$:

$$t_{-x}^*\varphi_x(\sigma \otimes \sigma') = t_{-x}^*\psi\sigma \otimes t_{-x}^*\psi'\sigma' \in H^0(M \otimes P_x) \otimes H^0(M' \otimes P_x^{-1}) .$$

This implies that the subspace $\sum_{x\in K(M\otimes M')} H^0(M \otimes P_x) \cdot H^0(M' \otimes P_x^{-1})$ is invariant under the action of $\widetilde{\mathcal{G}}(M \otimes M')$.

Step II: Let now U be a nonempty open subset of $\mathrm{Pic}^0(X)$. We claim that there are finitely many points $P_1, \dots, P_n \in U$ with

$$\sum_{\nu=1}^{n} H^0(M \otimes P_\nu) \cdot H^0(M' \otimes P_\nu^{-1}) = H^0(M \otimes M') .$$

Let p_X and $p_{\widehat{X}}$ denote the natural projections of $X \times \widehat{X}$ and $\mathcal{P}$ the Poincaré bundle on $X \times \widehat{X}$. Define

$$\mathcal{M} = p_{\widehat{X}*}(p_X^* M \otimes \mathcal{P}) \quad \text{and} \quad \mathcal{M}' = p_{\widehat{X}*}(p_X^* M' \otimes \mathcal{P}^{-1}) .$$

The fibres $\mathcal{M}(P) = H^0(M \otimes P)$ and $\mathcal{M}'(P) = H^0(M' \otimes P^{-1})$ over $P \in \widehat{X}$ are all of the same dimension according to Riemann-Roch. So Grauert's Theorem (see Hartshorne [1] III 12.9) implies that $\mathcal{M}$ and $\mathcal{M}'$ are vector bundles on $\widehat{X}$. Denote by

$$\varphi: \mathcal{M} \otimes_{\mathcal{O}_{\widehat{X}}} M' \to H^0(M \otimes M') \otimes_{\mathbb{C}} \mathcal{O}_{\widehat{X}}$$

the composition of the generalized multiplication map $p_{\widehat{X}*}(p_X^* M \otimes \mathcal{P}) \otimes p_{\widehat{X}*}(p_X^* M' \otimes \mathcal{P}^{-1}) \to p_{\widehat{X}*} p_X^*(M \otimes M')$ and the base change homomorphism $p_{\widehat{X}*} p_X^*(M \otimes M') \to H^0(M \otimes M') \otimes \mathcal{O}_{\widehat{X}}$. Furthermore denote by q_ν the ν-th projection of $(\widehat{X})^n$ for $\nu = 1, \dots, n$. Consider

$$\phi := \sum_{\nu=1}^{n} q_\nu^*\varphi: \bigoplus_{\nu=1}^{n} q_\nu^*(\mathcal{M} \otimes_{\mathcal{O}_{\widehat{X}}} \mathcal{M}') \to H^0(M \otimes M') \otimes_{\mathbb{C}} \mathcal{O}_{(\widehat{X})^n} .$$

Over every point $Q = (Q_1, \dots, Q_n) \in (\widehat{X})^n$ the induced map $\phi(Q)$ is given by

$$\begin{aligned} \phi(Q)\colon \bigoplus_{\nu=1}^{n} q_\nu^*(\mathcal{M} \otimes_{\mathcal{O}_{\widehat{X}}} \mathcal{M}')(Q) &= \\ = \bigoplus_{\nu=1}^{n} H^0(M \otimes Q_\nu) \otimes H^0(M' \otimes Q_\nu^{-1}) &\to H^0(M \otimes M') . \end{aligned}$$

By semicontinuity (see Hartshorne [1] III 12.8) the set W of points $Q \in (\widehat{X})^n$, for which $\phi(Q)$ is surjective, is open. According to Step I the set W is nonempty for $n = \#K(M \otimes M')$. Hence for any nonempty open set $U \subset \widehat{X}$

the intersection $(\overset{n}{\underset{\nu=1}{\times}} U) \cap W$ is nonempty and every point $(P_1, \ldots, P_n) \in (\overset{n}{\underset{\nu=1}{\times}} U) \cap W$ satisfies the assertion. □

Using this and Proposition 2.2 we deduce

(3.4) Proposition. *For any algebraically equivalent ample line bundles L and L' on X the multiplication map $H^0(L^m) \otimes L^0(L'^n) \to H^0(L^m \otimes L'^n)$ is surjective for all $m \geq 3$ and $n \geq 2$.*

Proof. By Proposition 2.2 there is a nonempty open subset $U \subset \mathrm{Pic}^0(X)$ such that the multiplication $\mu_P \colon H^0(L^2 \otimes P) \otimes H^0(L'^2) \to H^0(L^2 \otimes L'^2 \otimes P)$ is surjective for all $P \in U$. Applying Lemma 3.3 twice, we see that the composed map

$$H^0(L^m) \otimes H^0(L'^2) = \big[\sum_{P \in U} H^0(L^{m-2} \otimes P^{-1}) \cdot H^0(L^2 \otimes P)\big] \otimes H^0(L'^2)$$
$$\to \sum_{P \in U} H^0(L^{m-2} \otimes P^{-1}) \cdot H^0(L^2 \otimes L'^2 \otimes P) = H^0(L^m \otimes L'^2)$$

is surjective. Induction on n yields the assertion. □

From this and Proposition 2.3 Theorem 3.1 is an easy consequence.

Proof of Theorem 3.1. According to Lemma 3.2 we have to check the surjectivity of $H^0(L^{nm}) \otimes H^0(L^n) \to H^0(L^{n(m+1)})$ for every $m \geq 1$. For $n \geq 3$ and m arbitrary, and for $n = 2$ and $m \geq 2$ this is a special case of Proposition 3.4. It remains to show that $H^0(L^2) \otimes H^0(L^2) \to H^0(L^4)$ is surjective if and only if no point of $t_{\bar{c}}^* K(L^2)$ is a base point of L. For this note that $L = t_{\bar{c}}^* L_0$, where L_0 is of characteristic zero. Then the diagram

$$\begin{array}{ccc} X & \searrow^{\varphi_{L^2}} & \\ {\scriptstyle t_{\bar{c}}}\downarrow & & \mathbb{P}_N \\ X & \nearrow_{\varphi_{L_0^2}} & \end{array}$$

commutes up to an automorphism of $\mathbb{P}_N$ (see Lemma 4.6.1). It shows that L^2 is normally generated if and only if L_0^2 is normally generated. Now the assertion follows from Proposition 2.3. □

§ 4 The Ideal of an Abelian Variety in $\mathbb{P}_N$

Let M be a very ample line bundle on the abelian variety X with associated embedding $\varphi_M : X \hookrightarrow \mathbb{P}_N$. If $\mathcal{I}$ denotes the ideal sheaf of the variety $\varphi_M(X)$, the exact sequence $0 \longrightarrow \mathcal{I} \longrightarrow \mathcal{O}_{\mathbb{P}_N} \longrightarrow \mathcal{O}_{\varphi_M(X)} \longrightarrow 0$ induces the following diagram

$$\begin{array}{ccccccc} 0 & \longrightarrow & H^0(\mathcal{I}(m)) & \longrightarrow & H^0(\mathcal{O}_{\mathbb{P}_N}(m)) & \longrightarrow & H^0(\mathcal{O}_{\varphi_M(X)}(m)) \\ & & \| & & \| & & \| \\ 0 & \longrightarrow & I_m & \longrightarrow & S^m H^0(M) & \longrightarrow & H^0(M^m) \end{array}$$

for every $m \geq 0$. Here I_m denotes the vector space of forms of degree m vanishing on $\varphi_M(X) \subset \mathbb{P}_N$. Note that $I_0 = I_1 = 0$, since $\varphi_M(X)$ is nondegenerate in $\mathbb{P}_N$, and define

$$I(M) = \bigoplus_{m \geq 2} I_m \ .$$

It is a graded ideal in the symmetric algebra $SH^0(M) = \bigoplus_{m \geq 0} S^m H^0(M)$, namely the kernel of the canonical map $SH^0(M) \to \bigoplus_{m \geq 0} H^0(M^m)$. The aim of this chapter is to determine an explicit set of generators for the ideal $I(M)$ under some additional assumptions on M. In this section we want to find out for which integers k the vector space $\bigoplus_{m=2}^{k} I_m$ generates the graded ideal $I(M)$. If this is known, a basis of the vector space $\bigoplus_{m=2}^{k} I_m$ will be a set of generators of $I(M)$.

The ideal $I(M)$ is called *(homogeneously) generated by forms of degree* $\leq k$ if the canonical map

$$I_k \otimes S^{m-k} H^0(M) \to I_m$$

is surjective for every $m \geq k$. The main result of this section is the following theorem, due to Kempf [2]

(4.1) Theorem. *Suppose L is an ample line bundle on X and L^n is normally generated, then*

a) the ideal $I(L^n)$ is generated by forms of degree 2 whenever $n \geq 4$,
b) the ideal $I(L^3)$ is generated by forms of degree 2 and 3,
c) the ideal $I(L^2)$ is generated by forms of degree 2, 3 and 4.

In the statements a) and b) the assumption that L^n is normally generated is automatically fulfilled by Theorem 3.1.

For the proof we need some preliminaries. Suppose M_1, M_2 and M_3 are ample line bundles on X. Define $\mathcal{R}(M_1, M_2)$ to be the kernel of the multiplication map

$$\mathcal{R}(M_1, M_2) = \ker\{H^0(M_1) \otimes H^0(M_2) \to H^0(M_1 \otimes M_2)\} \ .$$

The commutative diagram

$$\begin{array}{ccccccc} 0 \longrightarrow & H^0(M_1)\otimes\mathcal{R}(M_2,M_3) & \longrightarrow & H^0(M_1)\otimes H^0(M_2)\otimes H^0(M_3) & \longrightarrow & H^0(M_1)\otimes H^0(M_2\otimes M_3) & \\ & & & \downarrow & & \downarrow & (1) \\ 0 \longrightarrow & \mathcal{R}(M_1\otimes M_2,M_3) & \longrightarrow & H^0(M_1\otimes M_2)\otimes H^0(M_3) & \longrightarrow & H^0(M_1\otimes M_2\otimes M_3) & \end{array}$$

induces a canonical map $H^0(M_1) \otimes \mathcal{R}(M_2, M_3) \to \mathcal{R}(M_1 \otimes M_2, M_3)$. In particular we have

$$H^0(M_1) \cdot \mathcal{R}(M_2, M_3) \subseteq \mathcal{R}(M_1 \otimes M_2, M_3) \ . \tag{2}$$

The following lemma gives a criterion for $I(M)$ to be generated by forms of degree $\leq k$.

(4.2) Lemma. *Suppose $M \in \operatorname{Pic}(X)$ is normally generated and there is a $k \geq 1$ such that the canonical map $H^0(M) \otimes \mathcal{R}(M^m, M) \to \mathcal{R}(M^{m+1}, M)$ is surjective for every $m \geq k$. Then the ideal $I(M)$ is generated by forms of degree $\leq k+1$.*

Proof. For $m \geq 1$ denote by $\mathcal{R}_m(M)$ the kernel of the canonical map $H^0(M)^m := H^0(M)^{\otimes m} \to H^0(M^m)$. Consider the canonical map

$$\phi = \sum_\nu \phi_\nu : \bigoplus_\nu \left(\mathcal{R}_{k+1}(M) \otimes H^0(M)^{m-k}\right) \to \mathcal{R}_{m+1}(M) \ , \tag{3}$$

where the sums have to be taken over all $\nu = (\nu_1, \ldots, \nu_{k+1})$ with $1 \leq \nu_1 < \cdots < \nu_{k+1} \leq m+1$ and the map ϕ_ν is given by

$$\begin{aligned} &a_1 \otimes \cdots \otimes a_{k+1} \otimes b_1 \otimes \cdots \otimes b_{m-k} \mapsto \\ &\quad b_1 \otimes \cdots \otimes b_{\nu_1 - 1} \otimes a_1 \otimes b_{\nu_1} \otimes \cdots \otimes b_{\nu_{k+1}-1} \otimes a_{k+1} \otimes b_{\nu_{k+1}} \otimes \cdots \otimes b_{m-k} \ . \end{aligned}$$

Roughly speaking, a_i is inserted in the ν_i-th place. The map (3) is just the desymmetrization of the map

$$I_{k+1} \otimes S^{m-k} H^0(M) \to I_{m+1} \ . \tag{4}$$

Hence it suffices to show that (3) is surjective.

Consider the commutative diagram

$$\begin{array}{c} \bigoplus_\nu (\mathcal{R}_{k+1}(M) \otimes H^0(M)^{m-k}) \\ \swarrow^{\alpha_0 = \phi} \qquad \downarrow^{\alpha_1} \qquad \searrow^{\alpha_{m-k}} \\ H^0(M)^{m+1} \xrightarrow{\beta_0} H^0(M^{k+1}) \otimes H^0(M)^{m-k} \xrightarrow{\beta_1} \cdots \xrightarrow{\beta_{m-k-1}} H^0(M^m) \otimes H^0(M) \xrightarrow{\beta_{m-k}} H^0(M^{m+1}) \ . \end{array}$$

Here for $1 \leq i \leq m-k$ the β_i's are the canonical maps

$$\beta_i \colon H^0(M^{k+i}) \otimes H^0(M)^{m-k-i+1} \to H^0(M^{k-i+1}) \otimes H^0(M)^{m-k-i}$$

and $\alpha_i = \beta_{i-1} \circ \alpha_{i-1}$. We have to show that $\operatorname{Im} \alpha_0$ contains $\mathcal{R}_{m+1}(M) = \ker(\beta_{m-k} \circ \cdots \circ \beta_0)$. By assumption M is normally generated, so β_i is surjective for every $i = 0, \ldots, m-k$. Hence it suffices to show that $\ker \beta_i \subseteq \operatorname{Im} \alpha_i$ for every $i = 0, \ldots, m-k$.

This is true for $i = 0$ by definition of the maps. For $i = 1, \ldots, m-k$ we have $\ker \beta_i = \mathcal{R}(M^{k+i}, M) \otimes H^0(M)^{m-k-i}$. By restriction to a suitable direct summand of $\bigoplus_\nu (\mathcal{R}_{k+1}(M) \otimes H^0(M)^{m-k})$ and omission of some tensor factors $H^0(M)$ we see that it suffices to show that the canonical map

$$\tilde{\alpha}_i \colon H^0(M)^i \otimes \mathcal{R}_{k+1}(M) \to \mathcal{R}(M^{k+i}, M)$$

is surjective for $i = 1, \ldots, k-i$. But $\tilde{\alpha}_i$ factorizes canonically as follows

$$\begin{array}{ccc}
H^0(M)^i \otimes \mathcal{R}_{k+1}(M) & \xrightarrow{\tilde{\alpha}_i} & \mathcal{R}(M^{k+i}, M) \\
{\scriptstyle 1\otimes\gamma} \searrow & & \nearrow {\scriptstyle \delta_i} \\
& H^0(M)^i \otimes \mathcal{R}(M^k, M) & .
\end{array}$$

The map δ_i is surjective according to the assumption. To show the surjectivity of $\gamma \colon \mathcal{R}_{k+1}(M) \to \mathcal{R}(M^k, M)$ consider the diagram

$$\begin{array}{ccccccccc}
 & & & & & & 0 & & \\
 & & & & & & \downarrow & & \\
 & & & & 0 & & \mathcal{R}(M^k, M) & & \\
 & & & & \downarrow & & \downarrow & & \\
0 & \longrightarrow & \ker\sigma & \longrightarrow & H^0(M)^k \otimes H^0(M) & \xrightarrow{\sigma} & H^0(M^k) \otimes H^0(M) & \longrightarrow & 0 \\
 & & \downarrow{\scriptstyle \psi} & & \| & & \downarrow & & \\
0 & \longrightarrow & \mathcal{R}_{k+1}(M) & \longrightarrow & H^0(M)^{k+1} & \longrightarrow & H^0(M^{k+1}) & \longrightarrow & 0 \\
 & & \downarrow{\scriptstyle \pi} & & \downarrow & & & & \\
 & & \operatorname{coker}\psi & & 0 & & & & \\
 & & \downarrow & & & & & & \\
 & & 0 & & & & & & .
\end{array}$$

According to the snake lemma $\mathcal{R}(M^k, M)$ is canonically isomorphic to $\operatorname{coker} \psi$, and under this canonical isomorphism π identifies with γ. This completes the proof. $\square$

The main step in the proof of Theorem 4.1 is the following proposition. In the proof we follow Kempf [2].

(4.3) Proposition. *Let L be an ample line bundle on X and n_1, n_2, n_3 integers with $n_1 \geq 1$, $n_2, n_3 \geq 2$ such that $n_2 + n_3 \geq 5$. Then*

$$\sum_{P \in \mathrm{Pic}^0(X)} H^0(L^{n_1} \otimes P) \cdot \mathcal{R}(L^{n_2} \otimes P^{-1}, L^{n_3}) = \mathcal{R}(L^{n_1+n_2}, L^{n_3}) \,.$$

Proof. Step I: According to (1) we have for every $P \in \mathrm{Pic}^0(X)$ the following commutative diagram

$$\begin{array}{ccc}
0 & & 0 \\
\downarrow & & \downarrow \\
H^0(L^{n_1} \otimes P) \otimes \mathcal{R}(L^{n_2} \otimes P^{-1}, L^{n_3}) & \xrightarrow{\varphi} & \mathcal{R}(L^{n_1+n_2}, L^{n_3}) \\
\downarrow & & \downarrow \\
H^0(L^{n_1} \otimes P) \otimes H^0(L^{n_2} \otimes P^{-1}) \otimes H^0(L^{n_3}) & \longrightarrow & H^0(L^{n_1+n_2}) \otimes H^0(L^{n_3}) \\
\downarrow & & \downarrow \\
H^0(L^{n_1} \otimes P) \otimes H^0(L^{n_2+n_3} \otimes P^{-1}) & \longrightarrow & H^0(L^{n_1+n_2+n_3}) \\
\downarrow & & \downarrow \\
0 & & 0 \,.
\end{array}$$

The exactness of the columns follows from Proposition 3.4 and the assumption on n_1, n_2, n_3.

It suffices to show that every linear form on $H^0(L^{n_1+n_2}) \otimes H^0(L^{n_3})$ inducing zero on $\mathrm{im}\,\varphi = H^0(L^{n_1} \otimes P) \cdot \mathcal{R}(L^{n_2} \otimes P^{-1}, L^{n_3})$ for every $P \in \mathrm{Pic}^0(X)$, induces zero on $\mathcal{R}(L^{n_1+n_2}, L^{n_3})$. According to the exactness of the columns in the diagram this is equivalent to the condition that every linear form on $H^0(L^{n_1+n_2}) \otimes H^0(L^{n_3})$ inducing a linear form on $H^0(L^{n_1} \otimes P) \otimes H^0(L^{n_2+n_3} \otimes P^{-1})$ for every $P \in \mathrm{Pic}^0(X)$ descends to a linear form on $H^0(L^{n_1+n_2+n_3})$. We want to describe these families of linear forms as sections of a vector bundle on $\widehat{X}$:

Let $\mathcal{P}$ denote the Poincaré bundle on $X \times \widehat{X}$ and p_X, $p_{\widehat{X}}$ the natural projections of $X \times \widehat{X}$. The families of linear forms on the vector spaces $H^0(L^{n_1} \otimes P) \otimes H^0(L^{n_2+n_3} \otimes P^{-1})$, $P \in \mathrm{Pic}^0(X)$, (and in particular those which come from linear forms on $H^0(L^{n_1+n_2}) \otimes H^0(L^{n_3})$) can be considered as global sections of the sheaf

$$\mathcal{F} := \left(p_{\widehat{X}*}(p_X^* L^{n_1} \otimes \mathcal{P})\right)^* \otimes \left(p_{\widehat{X}*}(p_X^* L^{n_2+n_3} \otimes \mathcal{P}^{-1})\right)^*$$

on $\widehat{X}$. To see this, note that by Grauert's Theorem (see Hartshorne [1], III 12.9) the sheaf $\mathcal{F}$ is a vector bundle on $\widehat{X}$ with fibres

$$\begin{aligned}\mathcal{F}(P) &= H^0(L^{n_1} \otimes P)^* \otimes H^0(L^{n_2+n_3} \otimes P^{-1})^* \\ &= \operatorname{Hom}\bigl(H^0(L^{n_1} \otimes P) \otimes H^0(L^{n_2+n_3} \otimes P^{-1}), \mathbb{C}\bigr) .\end{aligned}$$

As in the proof of Lemma 3.3 there is a natural map

$$p_{\widehat{X}*}(p_X^* L^{n_1} \otimes \mathcal{P}) \otimes p_{\widehat{X}*}(p_X^* L^{n_2+n_3} \otimes \mathcal{P}^{-1}) \to H^0(L^{n_1+n_2+n_3}) \otimes \mathcal{O}_{\widehat{X}} .$$

Dualizing and taking sections we get a map

$$H^0(L^{n_1+n_2+n_3})^* \to H^0(\mathcal{F}) . \tag{5}$$

This is the map associating to a linear form on $H^0(L^{n_1+n_2+n_3})$ a family of linear forms on the vector spaces $H^0(L^{n_1} \otimes P) \otimes H^0(L^{n_2+n_3} \otimes P^{-1})$ (parametrized by $P \in \operatorname{Pic}^0(X)$). Hence it suffices to show that the map (5) is an isomorphism.

Step II: According to Lemma 3.3

$$H^0(L^{n_1+n_2+n_3}) = \sum_{P \in \operatorname{Pic}^0(X)} H^0(L^{n_1} \otimes P) \cdot H^0(L^{n_2+n_3} \otimes P^{-1}) .$$

Consequently every linear form on $H^0(L^{n_1+n_2+n_3})$ comes from at most one family of linear forms on the vector spaces $H^0(L^{n_1} \otimes P) \otimes H^0(L^{n_2+n_3} \otimes P^{-1})$. This implies that the map (5) is injective and it suffices to show that $h^0(L^{n_1+n_2+n_3}) = h^0(\mathcal{F})$.

Step III: The idea to compute $h^0(\mathcal{F})$ is to replace the vector bundle $\mathcal{F}$ on $\widehat{X}$ by a line bundle M on $X \times X \times \widehat{X}$ with $h^0(\mathcal{F}) = h^{2g}(M)$. Consider the line bundle

$$M := (q_1, q_3)^*(p_X^* L^{-n_1} \otimes \mathcal{P}^{-1}) \otimes (q_2, q_3)^*(p_X^* L^{-n_2-n_3} \otimes \mathcal{P}) ,$$

where q_ν denotes the ν-th projection of $X \times X \times \widehat{X}$. The line bundles L^{-n_1} and $L^{-n_2-n_3}$ are of index g, so by Grauert's Theorem and Künneth's Formula $R^i q_{3*} M = 0$ for all $i \neq 2g$. Hence the Leray spectral sequence (see Godement [1] I 4.6.2) gives

$$H^{2g}(M) = H^0(R^{2g} q_{3*} M) .$$

On the other hand, using relative Serre duality (see Kleimann [1] Cor. 24 and I 1.5) and the generalized Künneth Formula (see Grothendieck [2] III, 6.7.8)

$$\begin{aligned}R^{2g} q_{3*} M &= (q_{3*} M^{-1})^* \\ &= \bigl[q_{3*}\bigl((q_1, q_3)^*(p_X^* L^{n_1} \otimes \mathcal{P}) \otimes (q_2, q_3)^*(p_X^* L^{n_2+n_3} \otimes \mathcal{P}^{-1})\bigr)\bigr]^* \\ &= \bigl[p_{\widehat{X}*}(p_X^* L^{n_1} \otimes \mathcal{P}) \otimes p_{\widehat{X}*}(p_X^* L^{n_2+n_3} \otimes \mathcal{P}^{-1})\bigr]^* = \mathcal{F} ,\end{aligned}$$

so $h^0(\mathcal{F}) = h^{2g}(M)$.
Step IV: We claim that $h^{2g}(M) = h^0(L^{n_1+n_2+n_3})$. First, for all $i \leq g$, using the fact that $\big((-1)_X \times \mathrm{id}_{\widehat{X}}\big)^* \mathcal{P} = \mathcal{P}^{-1}$, the projection formula and flat base change (see Hartshorne [1] III 9.3), we have

$$\begin{aligned} R^i(q_1, q_2)_* M &= R^i(q_1, q_2)_* \big[(q_1, q_2)^*(p_1^* L^{-n_1} \otimes p_2^* L^{-n_2-n_3}) \otimes (q_2 - q_1, q_3)^* \mathcal{P}\big] \\ &= p_1^* L^{-n_1} \otimes p_2^* L^{-n_2-n_3} \otimes R^i(q_1, q_2)_*(q_2 - q_1, q_3)^* \mathcal{P} \\ &= p_1^* L^{-n_1} \otimes p_2^* L^{-n_2-n_3} \otimes (p_2 - p_1)^* R^i p_{X^*} \mathcal{P} \ . \end{aligned}$$

Here p_ν denotes the ν-th projection of $X \times X$.

Next we claim that $R^i p_{X^*} \mathcal{P} = 0$ for $i < g$ and $R^g p_{X^*} \mathcal{P}$ is the sheaf supported in zero with fibre $\mathbb{C}$. For the proof note that $R^i p_{X^*} \mathcal{P}$ is supported in zero, since $H^i(P) = 0$ for all $P \in \mathrm{Pic}^0(X)$, $P \neq \mathcal{O}_X$. Using the Leray spectral sequence and Godement [1] I, 4.6.1 we get

$$H^0(R^i p_{X^*} \mathcal{P}) = H^i(P) = \begin{cases} \mathbb{C} & \text{if } i = g \\ 0 & \text{if } i < g \end{cases} ,$$

where for the second equation we used Exercise 3.7. This implies the claim.

It follows that $(p_2 - p_1)^* R^g p_{X^*} \mathcal{P} = \mathcal{O}_\Delta$ where $\Delta \subset X \times X$ is the diagonal, and thus

$$R^i(q_1, q_2)_* M = \begin{cases} p_1^* L^{-n_1} \otimes p_2^* L^{-n_2-n_3} \otimes \mathcal{O}_\Delta & \text{if } i = g \\ 0 & \text{if } i < g \end{cases} .$$

Consequently $H^j\big(R^i(q_1, q_2)_* M\big) \neq 0$ if and only if $i = j = g$. Applying again the Leray spectral sequence and Godement [1] I, 4.6.2 we obtain

$$\begin{aligned} h^{2g}(M) &= h^g\big(R^g(q_1, q_2)_* M\big) = h^g(p_1^* L^{-n_1} \otimes p_2^* L^{-n_2-n_3} \otimes \mathcal{O}_\Delta) \\ &= h^g(L^{-n_1-n_2-n_3}) = h^0(L^{n_1+n_2+n_3}) \ . \end{aligned}$$

Step V: Summing up we have $h^0(\mathcal{F}) \overset{III}{=} h^{2g}(M) \overset{IV}{=} h^0(L^{n_1+n_2+n_3})$. According to Step II this completes the proof. □

Proof of Theorem 4.1. According to Lemma 4.2 it suffices to show that for an ample line bundle L on X such that L^n is normally generated:

$$H^0(L^n)\mathcal{R}(L^{nm}, L^n) = \mathcal{R}(L^{n(m+1)}, L^n) \tag{6}$$

for all integers n and m with

a) $n \geq 4$ and $m \geq 1$,
b) $n = 3$ and $m \geq 2$,
c) $n = 2$ and $m \geq 3$.

According to Propositions 4.3 and 3.4 and (2) we have

$$\begin{aligned}
\mathcal{R}(L^{n(m+1)}, L^n) &= \sum_{P \in \mathrm{Pic}^0(X)} H^0(L^{n(m+1)-n_2} \otimes P) \cdot \mathcal{R}(L^{n_2} \otimes P^{-1}, L^n) \\
&\qquad \text{(for } n \geq 2 \text{ and some } n_2 \geq 2 \text{ with } n_2 + n \geq 5 \text{ and} \\
&\qquad n(m+1) - n_2 \geq 1) \\
&= \sum_{P \in \mathrm{Pic}^0(X)} H^0(L^n) H^0(L^{nm-n_2} \otimes P) \mathcal{R}(L^{n_2} \otimes P^{-1}, L^n) \\
&\qquad \text{(for } n \geq 3 \text{ and } nm - n_2 \geq 2 \quad \text{or} \quad n \geq 2 \text{ and } nm - n_2 \geq 3) \\
&\subset H^0(L^n) \cdot \mathcal{R}(L^{nm}, L^n) \subset \mathcal{R}(L^{n(m+1)}, L^n) \,.
\end{aligned}$$

Hence we have equality and it remains to verify the restrictions on n, n_2 and m in the three cases a), b), and c). Summing up these restrictions we have to make sure that

i) $n \geq 2$, $n_2 \geq 2$ and $n_2 + n \geq 5$, and
ii) $n \geq 3$ and $nm \geq n_2 + 2$ or $n \geq 2$ and $nm \geq n_2 + 3$.

For $n \geq 4$ let $n_2 = 2$, so that i) and ii) hold if and only if $m \geq 1$. This gives a). In case $n = 3$ let $n_2 = 2$, then i) and ii) hold if and only if $m \geq 2$, which gives b). Finally, for $n = 2$ let $n_2 = 3$. Then i) and ii) hold only for $m \geq 3$, which gives c). This completes the proof. □

§ 5 Riemann's Theta Relations

Let $X = V/\Lambda$ be an abelian variety and M a very ample line bundle on X with associated embedding $\varphi_M \colon X \hookrightarrow \mathbb{P}_N$. Choosing a basis of canonical theta functions $\vartheta_0, \ldots, \vartheta_N$ of the vector space $H^0(M)$, the map φ_M is given by $\varphi_M(\bar{v}) = \big(\vartheta_0(v) : \cdots : \vartheta_N(v)\big)$. From this point of view equations for the variety $\varphi_M(X)$ are just relations among the theta functions. For applications, for example on moduli problems, one would like to have that the coefficients of such equations are values at the point zero of V of certain theta functions determined by M. Classically such equations are called *theta relations*. In this section we derive a system of quadratic equations, namely *Riemann's Theta Relations*. Under the assumption that M is an n-th power, n even ≥ 4, of an ample line bundle on X they describe the variety $\varphi_M(X)$ in $\mathbb{P}_N$ completely.

Let L be an ample line bundle on X and $M = L^n$ with n even ≥ 4. In the last section we saw that the space of quadratic equations for $\varphi_M(X)$ is $I_2 = \ker\big\{S^2 H^0(M) \to H^0(M^2)\big\}$. According to Theorem 4.1 the vector space I_2 generates the homogeneous ideal $I(M)$ of all equations of $\varphi_M(X)$ in $\mathbb{P}_N$. In order to determine generators for I_2 we first desymmetrize:

$$\begin{array}{ccccccccc}
0 & \longrightarrow & \mathcal{R}(M,M) & \longrightarrow & H^0(M)\otimes H^0(M) & \stackrel{\mu}{\longrightarrow} & H^0(M^2) & \longrightarrow & 0 \\
 & & \downarrow & & \downarrow & & \| & & \\
0 & \longrightarrow & I_2 & \longrightarrow & S^2H^0(M) & \longrightarrow & H^0(M^2) & \longrightarrow & 0 .
\end{array}$$

The canonical map $\mathcal{R}(M,M) \to I_2$ is surjective. So it suffices to determine generators for the kernel $\mathcal{R}(M,M)$ of the multiplication map $\mu: H^0(M) \otimes H^0(M) \to H^0(M^2)$. For this we will use Corollary 1.4.

Recall the notation: Fix a decomposition $K(M^2) = K_1 \oplus K_2$ for M^2 and define $Z_2 = K_1 \cap X_2$. In Section 1 we defined for $y \in K_1$ and $\rho \in \widehat{Z}_2$ canonical theta functions generating $H^0(M^2)$:

$$\theta_{y,\rho} = \theta_{y,\rho}^{M\otimes(-1)^*M} = \sum_{z\in Z_2} \rho(z)\vartheta_{y+z}^{M\otimes(-1)^*M} .$$

With this notation we get as a consequence of Corollary 1.4:

(5.1) Riemann's Theta Relations. *The functions*

$$\theta_{y_1,\rho}(0) \sum_{z\in Z_2} \rho(z)\vartheta^M_{y+y_2+z} \otimes \vartheta^M_{y-y_2+z} - \theta_{y_2,\rho}(0) \sum_{z\in Z_2} \rho(z)\vartheta^M_{y+y_1+z} \otimes \vartheta^M_{y-y_1+z}$$

with $y, y_1, y_2 \in K_1$, $y \equiv y_1 \equiv y_2 \pmod{2K_1}$ *and* $\rho \in \widehat{Z}_2$, *generate the vector space* $\mathcal{R}(M,M)$.

Proof. As we saw in Section 1, the functions

$$\theta_{(x_1,x_2),\rho} = \sum_{z\in Z_2} \rho(z)\vartheta^M_{x_1+z} \otimes \vartheta^M_{x_2+z} ,$$

$x_1, x_2 \in 2K_1$, $\rho \in \widehat{Z}_2$, generate the vector space $H^0(M) \otimes H^0(M)$, and Corollary 1.4 stated

$$\mu(\theta_{(x_1,x_2),\rho}) = \theta_{y_1,\rho}^{M\otimes(-1)^*M}(0) \cdot \theta_{y,\rho}^{M^2}$$

for $(y, y_1) \in \alpha^{-1}(x_1, x_2) \cap (K_1 \times K_1)$. The vector space $H^0(M) \otimes H^0(M)$ decomposes into the direct sum of subvector spaces

$$C(y,\rho) := <\theta_{(y+y_1,y-y_1),\rho} \in H^0(M)^{\otimes 2} \mid y_1 \in y + 2K_1> ,$$

namely
$$H^0(M) \otimes H^0(M) = \bigoplus_{(y,\rho)\in(K_1/Z_2)\times\widehat{Z}_2} C(y,\rho)$$

(for this see also the remark after Corollary 1.4). The multiplication map μ restricts to linear forms

$$\mu|C(y,\rho): C(y,\rho) \to \mathbb{C}\cdot\theta_{y,\rho}^{M^2} .$$

Now the assertion follows from the following elementary observation: given a nonzero linear form l on a finite dimensional $\mathbb{C}$-vector space W with basis $w_1, \dots, w_n$, then $\{l(w_\nu)w_\mu - l(w_\mu)w_\nu \mid 1 \leq \nu < \mu \leq n\}$ generates the kernel of l. □

For any $y \in 2K_1$ let X_y denote the coordinate function of $\mathbb{P}_N = P(H^0(M))$ corresponding to the basis element ϑ_y^M of $H^0(M)$. Combining everything we obtain

(5.2) Riemann's Equations. *Assume M is an even n-th power, $n \geq 4$, of an ample line bundle on X. The equations*

$$\theta_{y_1,\rho}(0) \sum_{z \in Z_2} \rho(z) X_{y+y_2+z} X_{y-y_2+z} = \theta_{y_2,\rho}(0) \sum_{z \in Z_2} \rho(z) X_{y+y_1+z} X_{y-y_1+z} ,$$

for $y, y_1, y_2 \in K_1$ with $y \equiv y_1 \equiv y_2 \pmod{2K_1}$ and $\rho \in \widehat{Z}_2$, generate the homogeneous ideal $I(M)$ of the variety $\varphi_M(X)$ in $\mathbb{P}_N$.

§ 6 Cubic Theta Relations

Suppose now M is the third power of an ample line bundle on the abelian variety X. According to Theorem 4.1 the vector spaces I_2 and I_3 of quadratic and cubic equations generate the homogeneous ideal $I(M)$ of the variety $\varphi_M(X)$ in $\mathbb{P}_n$. Here the cubic equations are indispensable, as the example of an elliptic curve in $\mathbb{P}_3$ shows. However, the cubic equations describe the variety $\varphi_M(X)$ in $\mathbb{P}_N$ completely, since $I_2 \cdot H^0(M) \subseteq I_3$. In analogy to the last section, where we considered I_2, the aim of this section is to determine a system of generators for I_3. Here we follow Birkenhake-Lange [1].

Let L be an ample line bundle on X and $M = L^3$. Without loss of generality we may assume that M is symmetric (see Lemma 4.6.1). According to Theorem 3.1 the canonical map $S^3 H^0(M) \to H^0(M^3)$ is surjective. By definition cubic theta relations are elements of its kernel I_3. In order to compute I_3, we first desymmetrize

$$\begin{array}{ccccccccc} 0 & \longrightarrow & \mathcal{R}_3(M) & \longrightarrow & H^0(M)^{\otimes 3} & \xrightarrow{\mu_3} & H^0(M^3) & \longrightarrow & 0 \\ & & \downarrow & & \downarrow & & \| & & \\ 0 & \longrightarrow & I_3 & \longrightarrow & S^3 H^0(M) & \longrightarrow & H^0(M^3) & \longrightarrow & 0 \end{array} .$$

Since the canonical map $\mathcal{R}_3(M) \to I_3$ is surjective, it suffices to compute the kernel $\mathcal{R}_3(M)$ of the multiplication map μ_3. For this we first study the map μ_3. In analogy to the procedure of Section 1 we factorize μ_3.

The line bundle $N = p_1^*M \otimes p_2^*M \otimes p_3^*M$ is ample on $X \times X \times X$. According to the Künneth formula the map

$$p_1^* \otimes p_2^* \otimes p_3^* \colon H^0(M)^{\otimes 3} \to H^0(N)$$

is an isomorphism. Denoting by $\Delta\colon X \to X \times X \times X$, $\Delta(x) = (x, x, x)$ the diagonal map, the multiplication μ_3 factorizes as follows

$$\mu_3 \colon H^0(M)^{\otimes 3} \xrightarrow{p_1^* \otimes p_2^* \otimes p_3^*} H^0(N) \xrightarrow{\Delta^*} H^0(M^3) \,.$$

The homomorphism Δ fits into the commutative diagram

$$\begin{array}{ccc} & X \times X \times X & \\ {}^{s}\nearrow & & \searrow^{\alpha} \\ X & \xrightarrow{\quad\Delta\quad} & X \times X \times X \end{array}$$

with $s(x) := (0, 0, x)$ and

$$\alpha(x_1, x_2, x_3) := (x_1 + x_2 + x_3, x_1 - x_2 + x_3, -2x_1 + x_3) \,.$$

We obtain the following factorization of μ_3:

$$\mu_3 \colon H^0(M)^{\otimes 3} \xrightarrow{p_1^* \otimes p_2^* \otimes p_3^*} H^0(N) \xrightarrow{\alpha^*} H^0(\alpha^* N) \xrightarrow{s^*} H^0(M^3) \,.$$

We start by studying the homomorphism α:

(6.1) Lemma. *a) α is an isogeny with kernel $\{(x, 3x, 2x) \mid x \in X_6\} \simeq X_6$,*
b) $\alpha^* N \simeq p_1^* M^6 \otimes p_2^* M^2 \otimes p_3^* M^3$.

Proof. As for a): suppose $\alpha(x_1, x_2, x_3) = 0$. Adding all 3 equations we get $3x_3 = 0$. Since $x_3 = 2x_1$ by the third, and thus $x_2 = x_1 + x_3 = 3x_1$ by the second equation, we obtain $(x_1, x_2, x_3) = (x_1, 3x_1, 2x_1)$ with $x_1 \in X_6$. Conversely every element of this form is in the kernel of α. In particular $\ker \alpha$ is finite and α is an isogeny.
As for b): according to the Appell-Humbert Theorem it suffices to compare the hermitian forms and the semicharacters. Writing $M = L(H, \chi)$ we have for all $v_\nu, w_\nu \in X$, $\nu = 1, 2, 3$:

$$\begin{aligned} \alpha^*(p_1^*H + p_2^*H + p_3^*H)&\big((v_1, v_2, v_3), (w_1, w_2, w_3)\big) = \\ &= H(v_1 + v_2 + v_3, w_1 + w_2 + w_3) + H(v_1 - v_2 + v_3, w_1 - w_2 + w_3) + \\ &\qquad + H(-2v_1 + v_3, -2w_1 + w_3) \\ &= 6H(v_1, w_1) + 2H(v_2, w_2) + 3H(v_3, w_3) \\ &= (p_1^*6H + p_2^*2H + p_3^*3H)\big((v_1, v_2, v_3), (w_1, w_2, w_3)\big) \,. \end{aligned}$$

On the other hand $\chi(\Lambda) \subseteq \{\pm 1\}$ the line bundle M being symmetric, so

$$\begin{aligned}\alpha^*(p_1^*\chi p_2^*\chi p_3^*\chi)(\lambda_1,\lambda_2,\lambda_3) &= \chi(\lambda_1+\lambda_2+\lambda_3)\chi(\lambda_1-\lambda_2+\lambda_3)\chi(-2\lambda_1+\lambda_3)\\ &= \chi(\lambda_3) = \chi(\lambda_1)^6\chi(\lambda_2)^2\chi(\lambda_3)^3\\ &= (p_1^*\chi^6 p_2^*\chi^2 p_3^*\chi^3)(\lambda_1,\lambda_2,\lambda_3)\end{aligned}$$

for all $\lambda_1, \lambda_2, \lambda_3 \in \Lambda$. □

In order to apply the Isogeny Theorem 6.5.1, we have to choose compatible decompositions and characteristics for N and $\alpha^* N$. For this fix a decomposition

$$K(M^6) = K_1 \oplus K_2$$

for M^6. It induces decompositions $K(M) = 6K_1 \oplus 6K_2$ for M, $K(M^2) = 3K_1 \oplus 3K_2$ for M^2 and $K(M^3) = 2K_1 \oplus 2K_2$ for M^3, and hence decompositions for $\alpha^* N$ and N:

$$K(\alpha^*N) = (K_1 \times 3K_1 \times 2K_1) \oplus (K_2 \times 3K_2 \times 2K_2)\ , \quad \text{and} \qquad (1)$$

$$K(N) = (6K_1 \times 6K_1 \times 6K_1) \oplus (6K_2 \times 6K_2 \times 6K_2)\ . \qquad (2)$$

We have to check whether (1) and (2) are compatible (in the sense of Section 6.5). By Proposition 6.5.2 the compatibility is a consequence of

(6.2) Lemma. *$\alpha(K_\nu \times 3K_\nu \times 2K_\nu) \cap K(N) = 6K_\nu \times 6K_\nu \times 6K_\nu$ for $\nu = 1, 2$.*

Proof. It suffices to show that $6K_\nu \times 6K_\nu \times 6K_\nu \subset \alpha(K_\nu \times 3K_\nu \times 2K_\nu) \cap K(N)$, the converse inclusion being obvious. Suppose $x_1, x_2, x_3 \in 6K_\nu$. There are $y_1 \in K_\nu$, $y_2 \in 3K_\nu$ and $y_3 \in 2K_\nu$ such that

$$\begin{aligned} x_1 + x_2 - 2x_3 &= 6y_1\\ x_1 - x_2 &= 2y_2\\ x_1 + x_2 + x_3 &= 3y_3\ . \end{aligned}$$

Then

$$\begin{aligned} 6(y_1+y_2+y_3) &= x_1+x_2-2x_3+3(x_1-x_2)+2(x_1+x_2+x_3) = 6x_1\\ 6(y_1-y_2+y_3) &= x_1+x_2-2x_3-3(x_1-x_2)+2(x_1+x_2+x_3) = 6x_2\\ 3(-2y_1+y_3) &= -x_1-x_2+2x_3+x_1+x_2+x_3 = 3x_3\ . \end{aligned}$$

Hence there are 6-division points z and z' and a 3-division point z'' (all these points in K_1) such that

$$\begin{aligned} y_1 + y_2 + y_3 + z &= x_1\\ y_1 - y_2 + y_3 + z' &= x_2\\ -2y_1 + y_3 + z'' &= x_3\ . \end{aligned}$$

But $2y_2 = x_1 - x_2 = 2y_2 + z - z'$ and $3y_3 = x_1 + x_2 + x_3 = 3y_3 + z + z' + z''$ implying $z = z'$ and $z'' = -2z$. So $\alpha(y_1 + z, y_2, y_3) = (x_1, x_2, x_3)$, which implies the assertion. □

Now choose a characteristic for M. It induces characteristics for M^ν and thus for N and $\alpha^* N$ which are compatible in the sense of Section 6.5. By Theorem 3.2.7 this gives bases $\{\vartheta_x^{M^\nu} \mid x \in \frac{6}{\nu} K_1\}$ for $H^0(M^\nu)$, $\nu = 1, 2, 3$ and 6. Moreover by definition

$$\{p_1^* \vartheta_{x_1}^M \otimes p_2^* \vartheta_{x_2}^M \otimes p_3^* \vartheta_{x_3}^M \mid x_1, x_2, x_3 \in 6K_1\} \quad \text{and}$$

$$\{p_1^* \vartheta_{y_1}^{M^6} \otimes p_2^* \vartheta_{y_2}^{M^2} \otimes p_3^* \vartheta_{y_3}^{M^3} \mid y_1 \in K_1,\ y_2 \in 3K_1,\ y_3 \in 2K_1\}$$

are the corresponding bases of $H^0(N)$ and $H^0(\alpha^* N)$. Denote by Z_6 the subgroup of 6-division points in K_1

$$Z_6 = K_1 \cap X_6 .$$

It is a group of order 6^g, since $K_1 \simeq \mathbb{Z}/6d_1\mathbb{Z} \oplus \cdots \oplus \mathbb{Z}/6d_g\mathbb{Z}$. With this notation

(6.3) Proposition. *For all $x_1, x_2, x_3 \in 6K_1$ and $(y_1, y_2, y_3) \in K_1 \times 3K_1 \times 2K_1$ with $\alpha(y_1, y_2, y_3) = (x_1, x_2, x_3)$*

$$\mu_3(\vartheta_{x_1}^M \otimes \vartheta_{x_2}^M \otimes \vartheta_{x_3}^M) = \sum_{z \in Z_6} \vartheta_{y_1+z}^{M^6}(0) \vartheta_{y_2+3z}^{M^2}(0) \cdot \vartheta_{y_3+2z}^{M^3} .$$

Note that for all $z \in Z_6$ the 2-division point $3z$ is contained in $3K_1 \subset K(M^2)$ and $2z$ is contained in $2K_1 \subset K(M^3)$, so the right hand side of the equation above makes sense.

Proof. Applying the Isogeny Theorem 6.5.1 and Lemma 6.1 a) we get

$$\begin{aligned}
\mu_3(\vartheta_{x_1}^M \otimes \vartheta_{x_2}^M \otimes \vartheta_{x_3}^M) &= s^* \alpha^* (p_1^* \vartheta_{x_1}^M \otimes p_2^* \vartheta_{x_2}^M \otimes p_3^* \vartheta_{x_3}^M) \\
&= s^* \sum_{\substack{(z_1, z_2, z_3) \in \\ \alpha^{-1}(x_1, x_2, x_3) \cap K_1 \times 3K_1 \times 2K_1}} p_1^* \vartheta_{z_1}^{M^6} \otimes p_2^* \vartheta_{z_2}^{M^2} \otimes p_3^* \vartheta_{z_3}^{M^3} \\
&= s^* \sum_{z \in Z_6} p_1^* \vartheta_{y_1+z}^{M^6} \otimes p_2^* \vartheta_{y_2+3z}^{M^2} \otimes p_3^* \vartheta_{y_3+2z}^{M^3} \\
&= \sum_{z \in Z_6} \vartheta_{y_1+z}^{M^6}(0) \vartheta_{y_2+3z}^{M^2}(0) \cdot \vartheta_{y_3+2z}^{M^3} .
\end{aligned}$$

□

The description of μ_3 in Proposition 6.3 is not appropriate in order to compute the kernel $\mathcal{R}_3(M)$. As in Section 1 we change the bases in such a way that μ_3 appears in diagonal form.

Denote by $\widehat{Z}_6$ the character group of Z_6. For every $x_1, x_2, x_3 \in 6K_1$ and $\rho \in \widehat{Z}_6$,

$$\theta_{(x_1,x_2,x_3),\rho} := \sum_{z\in Z_6} \rho(z)\vartheta^M_{x_1+2z} \otimes \vartheta^M_{x_2+2z} \otimes \vartheta^M_{x_3+2z}$$

is a canonical theta function in $H^0(M)^{\otimes 3}$. To see this, note that $z \in Z_6 \subset K(M^2)$, since $M = L^3$. So $2z \in K(M) \cap Z_6 \subset 6K_1$, which gives the assertion. These functions generate the vector space $H^0(M)^{\otimes 3}$, because

$$\begin{aligned}\sum_{\rho\in\widehat{Z}_6} \theta_{(x_1,x_2,x_3),\rho} &= \sum_{z\in Z_6} \Big(\sum_{\rho\in\widehat{Z}_6} \rho(z)\Big) \vartheta^M_{x_1+2z} \otimes \vartheta^M_{x_2+2z} \otimes \vartheta^M_{x_3+2z} \\ &= 6^g \vartheta^M_{x_1} \otimes \vartheta^M_{x_2} \otimes \vartheta^M_{x_3}\end{aligned}$$

by the character relation. Similarly the canonical theta functions

$$\begin{aligned}\theta^{M^3}_{y,\rho} &:= \sum_{z\in Z_6} \rho(z)\vartheta^{M^3}_{y+2z}\ , \text{ and} \\ \theta_{(y_1,y_2),\rho} &:= \sum_{z\in Z_6} \rho(z)\vartheta^{M^6}_{y_1-z} \otimes \vartheta^{M^2}_{y_2-3z}\end{aligned}$$

for $y \in 2K_1$, $y_1 \in K_1$, $y_2 \in 3K_1$, and $\rho \in \widehat{Z}_6$ are generators of the vector spaces $H^0(M^3)$ and $H^0(M^6) \otimes H^0(M^2)$. With respect to these new generators the multiplication map μ_3 is given as follows:

(6.4) Corollary. *For all $x_1, x_2, x_3 \in 6K_1$, and $\rho \in \widehat{Z}_6$*

$$\mu_3(\theta_{(x_1,x_2,x_3),\rho}) = \theta_{(y_1,y_2),\rho}(0) \cdot \theta^{M^3}_{y_3,\rho}\ ,$$

where $(y_1, y_2, y_3) \in \alpha^{-1}(x_1, x_2, x_3) \cap K_1 \times 3K_1 \times 2K_1$.

Proof. Applying Proposition 6.3 and the fact that $\alpha(y_1 - z, y_2 - 3z, y_3) = (x_1 + 2z, x_2 + 2z, x_3 + 2z)$ for all $z \in Z_6$ we obtain

$$\begin{aligned}\mu_3(\theta_{(x_1,x_2,x_3),\rho}) &= \sum_{z\in Z_6} \rho(z)\mu_3(\vartheta^M_{x_1+2z} \otimes \vartheta^M_{x_2+2z} \otimes \vartheta^M_{x_3+2z}) \\ &= \sum_{z\in Z_6} \sum_{z'\in Z_6} \rho(z-z')\rho(z')\vartheta^{M^6}_{y_1-z+z'}(0)\vartheta^{M^2}_{y_2-3z+3z'}(0) \cdot \vartheta^{M^3}_{y_3+2z'} \\ &= \sum_{z\in Z_6} \rho(z)\vartheta^{M^6}_{y_1-z}(0)\vartheta^{M^2}_{y_2-3z}(0) \cdot \sum_{z'\in Z_6} \rho(z')\vartheta^{M^3}_{y_3+2z'} \\ &= \theta_{(y_1,y_2),\rho}(0) \cdot \theta^{M^3}_{y_3,\rho}\ .\end{aligned}$$

□

With a proof completely analogous to that of Proposition 5.1 we conclude

(6.5) Cubic Theta Relations. *The functions*

$$\theta_{(y_1,y_2),\rho}(0) \sum_{z\in Z_6} \rho(z)\vartheta^M_{y_1'+y_2'+y_3+2z} \otimes \vartheta^M_{y_1'-y_2'+y_3+2z} \otimes \vartheta^M_{-2y_1'+y_3+2z} -$$
$$-\theta_{(y_1',y_2'),\rho}(0) \sum_{z\in Z_6} \rho(z)\vartheta^M_{y_1+y_2+y_3+2z} \otimes \vartheta^M_{y_1-y_2+y_3+2z} \otimes \vartheta^M_{-2y_1+y_3+2z}$$

with $y_1, y_1' \in K_1$, $y_2, y_2' \in 3K_1$, $y_3 \in 2K_1$ *so that* $\alpha(y_1,y_2,y_3), \alpha(y_1',y_2',y_3) \in 6K_1 \times 6K_1 \times 6K_1$ *and* $\rho \in \widehat{Z}_6$, *generate the vector space* $\mathcal{R}_3(M)$.

For any $y \in 6K_1$ denote by X_y the coordinate function of $\mathbb{P}_N = P(H^0(M))$ corresponding to the basis element ϑ^M_y of $H^0(M)$. Combining everything we obtain

(6.6) Cubic Equations. *Let* M *be an ample symmetric line bundle on the abelian variety* X *and assume* M *is a third power. The equations*

$$\theta_{(y_1,y_2),\rho}(0) \sum_{z\in Z_6} \rho(z) X_{y_1'+y_2'+y_3+2z} X_{y_1'-y_2'+y_3+2z} X_{-2y_1'+y_3+2z}$$
$$= \theta_{(y_1',y_2'),\rho}(0) \sum_{z\in Z_6} \rho(z) X_{y_1+y_2+y_3+2z} X_{y_1-y_2+y_3+2z} X_{-2y_1+y_3+2z}$$

with $y_1, y_1' \in K_1$, $y_2, y_2' \in 3K_1$, $y_3 \in 2K_1$ *so that* $\alpha(y_1,y_2,y_3), \alpha(y_1',y_2',y_3) \in 6K_1 \times 6K_1 \times 6K_1$ *and* $\rho \in \widehat{Z}_6$, *generate the vector space of cubic equations for the variety* $\varphi_M(X)$ *in* $\mathbb{P}_N$.

Certainly many of these cubic equations are redundant: on the one hand those equations coming from the kernel of the map $\mathcal{R}_3(M) \to I_3$ are trivial. On the other hand some equations appear several times. As an example let us consider the case of an elliptic curve X embedded into $\mathbb{P}_2$ by an ample symmetric line bundle M of degree 3. The map φ_M embeds X as a plane cubic $X \xrightarrow{\sim} E \subset \mathbb{P}_2$. Now (6.6) gives us lots of equations for E. However one can single out the unique nontrivial one, namely Hesse's equation:

$$X^3 + Y^3 + Z^3 = 3\lambda XYZ \ ,$$

(see Exercise 8). Thus one may consider the cubic equations as a generalization of Hesse's equation.

Exercises

(1) *(Addition Formula)* Let L and L' be algebraically equivalent ample line bundles on the abelian variety $X = V/\Lambda$ and $\alpha: X \times X \to X \times X$, $\alpha(x_1, x_2) = (x_1 + x_2, x_1 - x_2)$. Use the notation of Section 1 and apply the Isogeny Theorem 6.5.1 to show

$$\vartheta_{x_1}^{L}(v_1+v_2)\vartheta_{x_2}^{L'}(v_1-v_2) = \sum_{\substack{(z_1,z_2)\in K_1\times K_1 \\ \alpha(z_1,z_2)=(x_1,x_2)}} \vartheta_{z_1}^{L\otimes L'}(v_1)\vartheta_{z_2}^{L\otimes(-1)^*L'}(v_2)$$

for all $v_1, v_2 \in V$ and $x_1, x_2 \in 2K_1$. In particular, if L defines a principal polarization of X and $L = L'$ is of characteristic zero, then

$$\vartheta_0^L(v_1+v_2)\vartheta_0^L(v_1-v_2) = \sum_{z\in K_1} \vartheta_z^{L^2}(v_1)\vartheta_z^{L^2}(v_2)$$

for all $v_1, v_2 \in V$. (These are special cases of Schröter's Formula, see Krazer [1] p. 89, Formel (L).)

(2) Let L be an ample line bundle of characteristic c with respect to some decomposition on an abelian variety X. Show that the multiplication map $H^0(L^2) \times H^0(L^2) \to H^0(L^4)$ is surjective if and only if L has no base point in $K(L^2) - \bar{c}$. (Hint: see Proposition 2.3)

(3) Let L be a symmetric line bundle on an abelian surface X defining a principal polarization and consider the map $\varphi_{L^2}: X \to \mathbb{P}_3$. Its image, the Kummer surface associated to X, is a quartic in $\mathbb{P}_3$ which is invariant under the Heisenberg group, with 16 double points (see Section 10.3).

a) Use Example 6.7.4 to show that for a suitable choice of coordinates of $\mathbb{P}_3$ the Kummer surface $\varphi_{L^2}(X)$ is given by an equation

$$\begin{aligned} Q_4(X_0,X_1,X_2,X_3) = {}& \lambda_0(X_0^4+X_1^4+X_2^4+X_3^4)+ \\ &+\lambda_1(X_0^2X_1^2+X_2^2X_3^2)+\lambda_2(X_0^2X_2^2+X_1^2X_3^2)+\lambda_3(X_0^2X_3^2+X_1^2X_2^2)+ \\ &+\lambda_4X_0X_1X_2X_3 = 0 \end{aligned}$$

with $(\lambda_0 : \lambda_1 : \lambda_2 : \lambda_3 : \lambda_4) \in \mathbb{P}_4$.

b) Show that a general member of the family of quartics in a) is smooth. As soon as a surface in this family is singular at a general point of $\mathbb{P}_3$, it admits 16 singular points.

c) A quartic surface in $\mathbb{P}_3$ admits a singular point if and only if the discriminant $D(\lambda_0, \dots, \lambda_4)$ of its polynomial vanishes. The discriminant is

$$\begin{aligned}D(\lambda_0,\dots,\lambda_4) = \\ &\lambda_0(2\lambda_0+\lambda_1)(2\lambda_0-\lambda_1)(2\lambda_0+\lambda_2)(2\lambda_0-\lambda_2)(2\lambda_0+\lambda_3)(2\lambda_0-\lambda_3) \\ &(4\lambda_0+2\lambda_1+2\lambda_2+2\lambda_3+\lambda_4)(4\lambda_0+2\lambda_1+2\lambda_2+2\lambda_3-\lambda_4) \\ &(4\lambda_0+2\lambda_1-2\lambda_2-2\lambda_3+\lambda_4)(4\lambda_0+2\lambda_1-2\lambda_2-2\lambda_3-\lambda_4) \\ &(4\lambda_0-2\lambda_1+2\lambda_2-2\lambda_3+\lambda_4)(4\lambda_0-2\lambda_1+2\lambda_2-2\lambda_3-\lambda_4) \\ &(4\lambda_0-2\lambda_1-2\lambda_2+2\lambda_3+\lambda_4)(4\lambda_0-2\lambda_1-2\lambda_2+2\lambda_3-\lambda_4) \\ &[\lambda_0(16\lambda_0^2-4\lambda_1^2-4\lambda_2^2-4\lambda_3^2+\lambda_4^2)+4\lambda_1\lambda_2\lambda_3]\end{aligned}$$

d) Show that the Kummer surfaces in the above family form an open set in the hypersurface of $\mathbb{P}_4$ given by the equation

$$\lambda_0(16\lambda_0^2-4\lambda_1^2-4\lambda_2^2-4\lambda_3^2+\lambda_4^2)+4\lambda_1\lambda_2\lambda_3 = 0\,.$$

(For a different proof of this fact see Jessop [1] p.99.)

(4) Let L be a very ample line bundle of type $(1,d)$ on an abelian surface X. Show that L is normally generated for d odd ≥ 7 or d even ≥ 14 (see Lazarsfeld [1]. Hint: Use Exercise 6.5).

(5) Let L be an ample line bundle of type $(1,5)$ on an abelian surface X and $I(L) = \bigoplus_{m\geq 2} I_m$ its graded ideal.

a) L is not normally generated.

b) $I_2 = I_3 = I_4 = 0$.

c) $\dim I_5 = 3$. The coordinates of $\mathbb{P}_4$ can be chosen in such a way that $Y = \prod_{i\in\mathbf{Z}/5\mathbf{Z}} X_i$, $Q' = \sum_{i\in\mathbf{Z}/5\mathbf{Z}} X_i^3X_{i+2}X_{i+3}$ and $R' = \sum_{i\in\mathbf{Z}/5\mathbf{Z}} X_i^2X_{i+1}^2X_{i+3}$ form a basis for I_5. The zero set of Y, Q' and R' consists of $\varphi_L(X)$ together with 25 disjoint lines.

d) $\dim I_6 = 30$ and the graded ideal $I(L)$ is generated by I_5 and I_6. (See Manolache [2], the idea is to use the Horrocks-Mumford bundle, a stable rank 2 vector bundle on $\mathbb{P}_4$ closely related to $\varphi_L(X) \subset \mathbb{P}_4$.)

(6) Let X be an abelian surface admitting an irreducible principal polarization L_0. Assume L_1 is a second principal polarization on X with $(L_0\cdot L_1) = 3$ (for the existence of such surfaces see Exercise 5.14). According to Theorem 10.1.6 the line bundle $L = L_0 \otimes L_1$ yields an embedding $\varphi\colon X \to \mathbb{P}_4$. Let C_i be the unique curve in the linear system $|L_i|$, $i = 0,1$. We may assume that C_0 and C_1 are symmetric. The fixed locus of $(-1)_X$, considered as an involution on $\mathbb{P}_4$, consists of a plane S and a line (see Corollary 4.6.6).

a) The plane S contains exactly 10 2-division points of X.

b) Consider the automorphism $\sigma = -\phi_{L_1}^{-1}\phi_{L_0}\colon X \to X$. Show that for any 2-division point $x \in X_2$ the divisor $t_x^*C_0 + t_{\sigma(x)}^*C_1$ is contained in the linear system $|L|$. Denote by H_x the corresponding hyperplane of $\mathbb{P}_4$. Either $S \subset H_x$ or $S \cap H_x$ is a line in $\mathbb{P}_4$. Show that there are exactly 10 hyperplanes H_x such that $S \cap H_x$ is a line.

c) Show that the 10 points of a) and the 10 lines of b) form a *Desargues configuration*, i.e. the usual 10_3-configuration of the theorem of Desargues in the projective plane. (See Comessatti [2] and Lange [1].)

(7) Let L be an ample line bundle on an abelian variety X and $M = L^n$ for some $n \geq 3$. As in Section 4 denote by $I(M)$ the graded ideal in the symmetric algebra $S = SH^0(M)$ associated to M. Consider a minimal graded free resolution

$$0 \longrightarrow F_r \xrightarrow{\varphi_r} F_{r-1} \longrightarrow \cdots \longrightarrow F_1 \xrightarrow{\varphi_1} I(M) \longrightarrow 0 \ .$$

Here each F_i is a direct sum $F_i = \bigoplus_j S(-b_{ij})$. The S-submodule $\ker \varphi_i$ of F_i is called the i-th *syzygy-module* of $I(M)$. Show: if $n \geq 2i+4$ for some $i \geq 1$, then the i-th syzygy-module of $I(M)$ is generated by linear forms (see Kempf [2]).

(8) Let X be an elliptic curve and $M \in \operatorname{Pic}(X)$ the line bundle of type (3) and of characteristic zero with respect to the decomposition induced by some decomposition $K(M^6) = K_1 \oplus K_2$ for M^6. Then $K_1 \simeq \mathbb{Z}/18\mathbb{Z}$. Denote its elements with $\{0, \ldots, 17\}$. Use the notation of Section 6. In particular X_0, X_6, X_{12} denote coordinate functions of $\mathbb{P}_2$. Deduce from the Cubic Equations 6.6 that

$$X_0^3 + X_6^3 + X_{12}^3 = 3\frac{\theta_{(0,0),1}(0)}{\theta_{(0,6),1}(0)} X_0 X_6 X_{12}$$

is the equation of the elliptic curve $\varphi_M(X)$ in $\mathbb{P}_2$ (see Birkenhake-Lange [2]).

(9) *(Classical Form of Riemann's Theta Relations)* For $Z \in \mathfrak{H}_g$ and $c \in \mathbb{R}^{2g}$ consider the classical Riemann theta function with characteristic c, $\vartheta[c] = \vartheta[c](\cdot, Z)\colon \mathbb{C}^g \to \mathbb{C}$ of Section 8.5. Show that for all $v_1, v_2, v_3, v_4 \in \mathbb{C}^g$ and $a, b_1, b_2 \in \frac{1}{2}\mathbb{Z}^{2g}$

$$\begin{aligned} &2^g \vartheta[a](v_1)\vartheta[a+b_1](v_2)\vartheta[a+b_2](v_3)\vartheta[a-b_1-b_2](v_4) = \\ &= \sum_{\bar{c} \in \frac{1}{2}\mathbb{Z}^{2g}/\mathbb{Z}^{2g}} e(4\pi i E(a,c))\vartheta[c](v_1+v_2+v_3+v_4)\vartheta[c+b_1](v_1+v_2-v_3-v_4)\cdot \\ &\qquad \cdot\vartheta[c+b_2](v_1-v_2+v_3-v_4)\vartheta[c-b_1-b_2](v_1-v_2-v_3+v_4) \ . \end{aligned}$$

Here E denotes the alternating form on $\mathbb{R}^{2g}$ with matrix $\begin{pmatrix} 0 & \mathbb{1}_g \\ -\mathbb{1}_g & 0 \end{pmatrix}$ (see Krazer [1] p. 307 or Krazer-Wirtinger [1] p. 673).

(10) Use the notation of Exercise 9 and denote by A the subgroup $\frac{1}{2}\mathbb{Z}^g \oplus \{0\}^g/\mathbb{Z}^{2g}$ of $\frac{1}{2}\mathbb{Z}^{2g}/\mathbb{Z}^{2g}$. Show that for all $v_1, v_2, v_3, v_4 \in \mathbb{C}^g$ and $b_1, b_2, c, d \in \frac{1}{2}\mathbb{Z}^{2g}$

$$\begin{aligned} &\sum_{\bar{a} \in A} e(4\pi i E(a,d))\vartheta[a+c](v_1)\vartheta[a+c+b_1](v_2)\cdot \\ &\qquad \vartheta[a+c+b_2](v_3)\vartheta[a+c-b_1-b_2](v_4) = \\ &= \sum_{\bar{a} \in A} e(4\pi i E(a,c+d))\vartheta[a+d](v_1+v_2+v_3+v_4)\cdot \\ &\qquad \cdot\vartheta[a+d+b_1](v_1+v_2-v_3-v_4)\vartheta[a+d+b_2](v_1-v_2+v_3-v_4)\cdot \\ &\qquad \cdot\vartheta[a+d-b_1-b_2](v_1-v_2-v_3+v_4) \end{aligned}$$

(see Krazer [1] p. 311. For a more general version of this formula see Krazer [1] p. 309 Satz XXXIX).

(11) Let $(X = V/\Lambda, H)$ be a principally polarized abelian variety of dimension 3 and $L \in \operatorname{Pic}^H(X)$ a line bundle of characteristic zero. For $\bar{c} \in X_2$ denote by ϑ^c the canonical theta function on V generating $H^0(t_{\bar{c}}^* L)$ as in Theorem 3.2.7. Choose a fundamental system $A \subset X_2$ (see Exercise 4.16) and let $\bar{c}$ be the sum of all points $\bar{a} \in A$ with $\operatorname{mult}_a(\vartheta^0) \equiv 1 \pmod 2$. Moreover let $e^H: X_2 \times X_2 \to \mathbb{C}^*$ be the alternating form and $q_L: X_2 \to \mathbb{C}^*$ the quadratic form of Section 4.7. Show that for all $u, v, w \in \mathbb{C}^3$ and $\bar{b}_0, \bar{b}_1, \bar{b}_2 \in X_2$

$$q_L(\bar{c})\vartheta^c(0)\vartheta^{c+b_1}(u+v)\vartheta^{c+b_2}(u+w)\vartheta^{c-b_1-b_2}(-v-w) =$$

$$= \sum_{\bar{a}\in A} q_L(\bar{a}+\bar{b}_0)e^H(\bar{a}+\bar{b}_0,\bar{c})\vartheta^{a+b_0}(u)\vartheta^{a+b_0+b_1}(v)\vartheta^{a+b_0+b_2}(w)\cdot$$
$$\cdot\vartheta^{a+b_0-b_1-b_2}(-u-v-w)$$

(see Krazer-Wirtinger [1] p. 675).

(12) Let $(X = V/\Lambda, H)$ be a principally polarized abelian surface. Let the notation be as in Exercise 7. Show that for all $u, v, w, \in \mathbb{C}^2$, $\bar{b}, \bar{b}_1, \bar{b}_2 \in X_2$ and every $\bar{a}_0 \in A$

$$2q_L(a_0+b)\vartheta^{a_0+b}(u)\vartheta^{a_0+b+b_1}(v)\vartheta^{a_0+b+b_2}(w)\vartheta^{a_0+b-b_1-b_2}(-u-v-w) =$$
$$= \sum_{\bar{a}\in A} q_L(\bar{a}+\bar{b})e^H(\bar{a}_0,\bar{a})\vartheta^{a+b}(u)\vartheta^{a+b+b_1}(v)\vartheta^{a+b+b_2}(w)\cdot$$
$$\cdot\vartheta^{a+b-b_1-b_2}(-u-v-w)$$

(See Krazer [1] p. 339 and p. 355 for a generalization to arbitrary dimensions.)

Chapter 8

Moduli

In this chapter we construct several moduli spaces of polarized abelian varieties with additional structure. We would like to stress in advance that we take a slightly naive point of view of the notion of "moduli space": a *moduli space for a set* of abelian varieties with some additional structure means a complex analytic space or a complex manifold whose points are in some natural one to one correspondence with the elements of the set. We disregard uniqueness and functorial properties of these spaces.

The starting point is the Siegel upper half space $\mathfrak{H}_g$ of complex symmetric $(g \times g)$-matrices with positive definite imaginary part. It parametrizes the set of polarized abelian varieties of a given type $D = \operatorname{diag}(d_1, \dots, d_g)$ with a symplectic basis. The corresponding symplectic group G_D acts on $\mathfrak{H}_g$ in a natural way and the quotient $\mathcal{A}_D = \mathfrak{H}_g / G_D$ is a moduli space for polarized abelian varieties of type D (Theorem 2.6). We also introduce several level structures. For example, a level n-structure for a polarized abelian variety is a certain isomorphism between the group of n-division points and the group $(\mathbb{Z}/n\mathbb{Z})^{2g}$. The corresponding moduli space is the quotient of $\mathfrak{H}_g$ by the principal congruence subgroup of level n.

The last part of the chapter is devoted to Theorem 10.1, due to Igusa [1], which provides an analytic embedding of the moduli space of polarized abelian varieties $\mathcal{A}_D(D)_0$ with orthogonal level D-structure into projective space. It is not difficult to conclude from this that $\mathcal{A}_D(D)_0$ is a quasi-projective algebraic variety (see Remark 10.4). Since $\mathcal{A}_D(D)_0$ is a finite covering of the moduli space $\mathcal{A}_D$, this implies that also the other moduli spaces $\mathcal{A}_D$, $\mathcal{A}_D(D)$, etc. are algebraic. Let us outline the approach to Igusa's Theorem: we construct first a universal family of abelian varieties $\mathfrak{X}_D \to \mathfrak{H}_g$ parametrizing all polarized abelian varieties of type D with symplectic basis. Since the classical factor of automorphy is holomorphic in $Z \in \mathfrak{H}_g$, it extends to a factor on $\mathfrak{X}_D$ and thus defines a line bundle $\mathcal{L}$ on $\mathfrak{X}_D$. Composing the zero section s_0 of $\mathfrak{X}_D \to \mathfrak{H}_g$ with the map $\mathfrak{X}_D \to \mathbb{P}_N$ associated to a certain sublinear system of $|\mathcal{L}|$, we get a rational map $\psi_D \colon \mathfrak{H}_g \to \mathbb{P}_N$. Now the classical Theta Transformation Formula 6.1 implies that ψ_D factorizes via the quotient $\mathcal{A}_D(D)_0$ of $\mathfrak{H}_g$ to give a holomorphic map $\bar{\psi}_D \colon \mathcal{A}_D(D)_0 \to \mathbb{P}_N$, which turns out to be an embedding.

In the first two sections we construct the moduli spaces of polarized abelian varieties. A specialization of this leads to the moduli spaces with level structures in Section 3. Sections 4 to 6 are devoted to the proof of the Theta Transformation Formula. In Sections 7 and 8 we construct the universal family of polarized abelian varieties and study the action of the symplectic group on it. Sections 9 and 10 contain the proof of Igusa's Theorem.

Finally a word about period matrices: in almost every book they appear in a different form. The following two requirements seemed natural to us: (1) Period matrices should be $(g \times 2g)$-matrices rather that $(2g \times g)$-matrices. (2) A symplectic matrix $\begin{pmatrix} \alpha & \beta \\ \gamma & \delta \end{pmatrix} \in \mathrm{Sp}_{2g}(\mathbb{R})$ should act on the Siegel upper half space $\mathfrak{H}_g$ by $Z \mapsto (\alpha Z + \beta)(\gamma Z + \delta)^{-1}$. Under these two conditions the period matrices are necessarily of the form (Z, D) with $Z \in \mathfrak{H}_g$ and D the type of the polarization. There is a slight disadvantage: there is a transposition coming up. Namely, if R is the rational representation of an isomorphism of polarized abelian varieties, the corresponding action on $\mathfrak{H}_g$ is given by the matrix ${}^tR$.

§ 1 The Siegel Upper Half Space

In this section we introduce the Siegel upper half space $\mathfrak{H}_g$ and show that it parametrizes the set of polarized abelian varieties of a given type D with symplectic basis. Moreover we work out what it means for two points of $\mathfrak{H}_g$ that the associated polarized abelian varieties are isomorphic.

Suppose $X = V/\Lambda$ is an abelian variety of dimension g and $H \in \mathrm{NS}(X)$ a hermitian form on V defining a polarization of type $D = \mathrm{diag}(d_1, \ldots, d_g)$. Let $\lambda_1, \ldots, \lambda_g, \mu_1, \ldots, \mu_g$ denote a symplectic basis of Λ for H. By definition, the alternating form $\mathrm{Im}\, H$ is given by the matrix $\begin{pmatrix} 0 & D \\ -D & 0 \end{pmatrix}$ with respect to this basis. Define $e_\nu = \frac{1}{d_\nu}\mu_\nu$ for $\nu = 1, \ldots, g$. According to Lemma 3.2.1 the vectors $e_1, \ldots, e_g$ form a $\mathbb{C}$-basis for V. With respect to these bases the period matrix of X is of the form

$$\Pi = (Z, D)$$

for some $Z \in M_g(\mathbb{C})$. The matrix Z has the following properties

(1.1) Proposition. *a)* ${}^tZ = Z$ *and* $\mathrm{Im}\, Z > 0$.
b) $(\mathrm{Im}\, Z)^{-1}$ *is the matrix of the hermitian form* H *with respect to the basis* $e_1, \ldots, e_g$.

Proof. The assertions of a) are just the Riemann Relations with $\Pi_1 = Z$ and $\Pi_2 = D$ as stated at the very end of Section 4.2. By Lemma 4.2.3 the matrix of H is $2i\left(\overline{\Pi}\left(\begin{smallmatrix} 0 & D \\ -D & 0 \end{smallmatrix}\right)^{-1} {}^t\Pi\right)^{-1} = (\operatorname{Im} Z)^{-1}$. □

Define a *polarized abelian variety of type D with symplectic basis* to be a triplet

$$(X, H, \{\lambda_1, \dots, \lambda_g, \mu_1, \dots, \mu_g\})$$

with $X = V/\Lambda$ an abelian variety, H a polarization of type D on X, and $\{\lambda_1, \dots, \lambda_g, \mu_1, \dots, \mu_g\}$ a symplectic basis of Λ for H. The set

$$\mathfrak{H}_g := \{Z \in M_g(\mathbb{C}) \mid {}^tZ = Z, \operatorname{Im} Z > 0\}$$

is called the *Siegel upper half space*. It is a $\frac{1}{2}g(g+1)$-dimensional open submanifold of the vector space of symmetric matrices in $M_g(\mathbb{C})$.

We have seen that a polarized abelian variety of type D with symplectic basis determines a point Z in $\mathfrak{H}_g$. Conversely, given a type D, any $Z \in \mathfrak{H}_g$ determines a polarized abelian variety with symplectic basis as follows. Since $\Lambda_Z := (Z, D)\mathbb{Z}^{2g}$ is a lattice in $V = \mathbb{C}^g$, the quotient

$$X_Z := \mathbb{C}^g/\Lambda_Z$$

is a complex torus. Define a hermitian form H_Z by the matrix $(\operatorname{Im} Z)^{-1}$ with respect to the standard basis of $\mathbb{C}^g$.

We claim that H_Z is a polarization of type D on X. To see this note that it is positive definite. Moreover, consider the $\mathbb{R}$-linear isomorphism $\mathbb{R}^{2g} \to \mathbb{C}^g$ defined by the matrix (Z, D). Let $\lambda_1, \dots, \lambda_g, \mu_1, \dots, \mu_g$ denote the images of the standard basis of $\mathbb{R}^{2g}$ in $\mathbb{C}^g$. They are just the columns of the matrix (Z, D) with respect to the standard basis of $\mathbb{C}^g$. By definition $\lambda_1, \dots, \lambda_g, \mu_1, \dots, \mu_g$ is a basis of Λ_Z. With respect to this basis $\operatorname{Im} H_Z | \Lambda_Z \times \Lambda_Z$ is given by the matrix

$$\operatorname{Im}\left({}^t(Z, D)(\operatorname{Im} Z)^{-1}\overline{(Z, D)}\right) = \begin{pmatrix} 0 & D \\ -D & 0 \end{pmatrix} . \tag{1}$$

This completes the proof of the claim.

So to every $Z \in \mathfrak{H}_g$ one can associate a polarized abelian variety of type D with symplectic basis in a natural way. Summing up we get: the assignment $Z \mapsto (X_Z, H_Z, \{\text{columns of } (Z, D)\})$ gives a bijection between the Siegel upper half space $\mathfrak{H}_g$ and the set of (isomorphism classes of) polarized abelian varieties of type D with symplectic basis. According to our loose notion of moduli spaces this can be expressed as follows.

(1.2) Proposition. *Given a type D, the Siegel upper half space $\mathfrak{H}_g$ is a moduli space for polarized abelian varieties of type D with symplectic basis.*

Our aim is to construct an analytic moduli space for the polarized abelian varieties of type D. For this we have to analyze, which points in $\mathfrak{H}_g$ determine isomorphic polarized abelian varieties.

Suppose $Z, Z' \in \mathfrak{H}_g$ and that there is an isomorphism of polarized abelian varieties $\varphi: (X_{Z'}, H_{Z'}) \to (X_Z, H_Z)$. Let A and R denote the matrices of the analytic and rational representation of φ with respect to the standard basis of $\mathbb{C}^g$ and the symplectic bases of $\Lambda_{Z'}$ and Λ_Z determined by Z' and Z. According to equation 1.2(1) the matrices A and R are related by $A(Z', D) = (Z, D)R$. Define

$$N = \begin{pmatrix} \mathbb{1} & \\ & D \end{pmatrix} R \begin{pmatrix} \mathbb{1} & \\ & D \end{pmatrix}^{-1} = {}^t\!\begin{pmatrix} \alpha & \beta \\ \gamma & \delta \end{pmatrix} \tag{2}$$

with $(g \times g)$-blocks $\alpha, \beta, \gamma, \delta$. Then this equation is equivalent to

$$AZ' = Z\,{}^t\alpha + {}^t\beta \quad \text{and} \quad A = Z\,{}^t\gamma + {}^t\delta \,. \tag{3}$$

Since φ is an isomorphism, the matrix ${}^tA = (\gamma Z + \delta)$ is invertible. Thus we can express Z' in terms of Z and $N = {}^t\!\begin{pmatrix} \alpha & \beta \\ \gamma & \delta \end{pmatrix}$ as follows

$$Z' = {}^tZ' = {}^t(Z\,{}^t\alpha + {}^t\beta)\,{}^tA^{-1} = (\alpha Z + \beta)(\gamma Z + \delta)^{-1} \,.$$

Moreover $\varphi^* H_Z = H_{Z'}$ means ${}^tR \begin{pmatrix} 0 & D \\ -D & 0 \end{pmatrix} R = \begin{pmatrix} 0 & D \\ -D & 0 \end{pmatrix}$. In terms of N this translates to

$${}^tN \begin{pmatrix} 0 & \mathbb{1} \\ -\mathbb{1} & 0 \end{pmatrix} N = \begin{pmatrix} 0 & \mathbb{1} \\ -\mathbb{1} & 0 \end{pmatrix} \,. \tag{4}$$

Recall that for any commutative ring $\mathcal{R}$ with 1 the *symplectic group* $\mathrm{Sp}_{2g}(\mathcal{R})$ is the group of matrices in $M_{2g}(\mathcal{R})$ satisfying (4). Hence $N \in \mathrm{Sp}_{2g}(\mathbb{Q})$. Moreover for

$$\Lambda_D := \begin{pmatrix} \mathbb{1} & \\ & D \end{pmatrix} \mathbb{Z}^{2g}$$

equation (2) means $N\Lambda_D \subseteq \Lambda_D$, since $R \in M_{2g}(\mathbb{Z})$.

Noting that $\mathrm{Sp}_{2g}(\mathbb{Q})$ is invariant under transposition, we have seen that the matrix $M := {}^tN$ is an element of the group

$$G_D := \left\{ M \in \mathrm{Sp}_{2g}(\mathbb{Q}) \mid {}^tM\Lambda_D \subseteq \Lambda_D \right\} .$$

For $M = \begin{pmatrix} \alpha & \beta \\ \gamma & \delta \end{pmatrix} \in G_D$ and $Z \in \mathfrak{H}_g$ define

$$M(Z) = (\alpha Z + \beta)(\gamma Z + \delta)^{-1} \,.$$

Summing up we have proved the implication (i) $\Longrightarrow$ (ii) of the following Proposition.

(1.3) Proposition. *For $Z, Z' \in \mathfrak{H}_g$ the following statements are equivalent*
(i) the polarized abelian varieties (X_Z, H_Z) and $(X_{Z'}, H_{Z'})$ of type D are isomorphic.
(ii) $Z' = M(Z)$ for some $M \in G_D$

Proof. Suppose we are in case (ii). From the arguments of above it is easy to see that the matrix $\begin{pmatrix} \mathbb{1} & \\ & D \end{pmatrix}^{-1} {}^tM \begin{pmatrix} \mathbb{1} & \\ & D \end{pmatrix}$ is the rational representation of an isomorphism $(X_{Z'}, H_{Z'}) \to (X_Z, H_Z)$ with respect to the symplectic bases determined by Z and Z'. □

(1.4) Remark. For later use we observe that for any $Z \in \mathfrak{H}_g$ and $M = \begin{pmatrix} \alpha & \beta \\ \gamma & \delta \end{pmatrix} \in G_D$ the isomorphism $X_{M(Z)} \to X_Z$ of Proposition 1.3 is given by the equation

$$A\big(M(Z), \mathbb{1}\big) = (Z, \mathbb{1}){}^tM \ . \tag{6}$$

Here $A = {}^t(\gamma Z + \delta)$ is the matrix of the corresponding analytic representation and $\begin{pmatrix} \mathbb{1} & \\ & D \end{pmatrix}^{-1} {}^tM \begin{pmatrix} \mathbb{1} & \\ & D \end{pmatrix}$ is the matrix of the rational representation with respect to the chosen bases.

§ 2 The Analytic Moduli Space

In the last section we saw that the Siegel upper half space parametrizes the set of polarized abelian varieties of a given type with symplectic basis, and that two polarized abelian varieties (X_Z, H_Z) and $(X_{Z'}, H_{Z'})$ of type D are isomorphic if and only if $Z' = (\alpha Z + \beta)(\gamma Z + \delta)^{-1}$ for some $\begin{pmatrix} \alpha & \beta \\ \gamma & \delta \end{pmatrix} \in G_D$. In this section we want to show that the last equation is induced by an action of the group $\mathrm{Sp}_{2g}(\mathbb{R})$ on $\mathfrak{H}_g$. We need the following characterization for a matrix to be symplectic. Denote again by $\mathcal{R}$ a commutative ring with 1.

(2.1) Lemma. *a) The group $\mathrm{Sp}_{2g}(\mathcal{R})$ is closed under transposition.*
b) For a matrix $M = \begin{pmatrix} \alpha & \beta \\ \gamma & \delta \end{pmatrix} \in \mathrm{Sp}_{2g}(\mathcal{R})$ the following conditions are equivalent

(i) $M \in \mathrm{Sp}_{2g}(\mathcal{R})$,
(ii) ${}^t\alpha\gamma$ and ${}^t\beta\delta$ are symmetric and ${}^t\alpha\delta - {}^t\gamma\beta = \mathbb{1}_g$,
(iii) $\alpha\,{}^t\beta$ and $\gamma\,{}^t\delta$ are symmetric and $\alpha\,{}^t\delta - \beta\,{}^t\gamma = \mathbb{1}_g$.

Proof. For the proof of a) apply definition (4) to $M^{-1} = \begin{pmatrix} 0 & -\mathbb{1} \\ \mathbb{1} & 0 \end{pmatrix} {}^tM \begin{pmatrix} 0 & \mathbb{1} \\ -\mathbb{1} & 0 \end{pmatrix}$. b) follows directly from the definition and a). □

(2.2) Proposition. *The group* $\mathrm{Sp}_{2g}(\mathbb{R})$ *acts biholomorphically on* $\mathfrak{H}_g$ *by*

$$Z \mapsto M(Z) = (\alpha Z + \beta)\,(\gamma Z + \delta)^{-1}$$

for all $M = \begin{pmatrix} \alpha & \beta \\ \gamma & \delta \end{pmatrix} \in \mathrm{Sp}_{2g}(\mathbb{R})$.

Proof. First we claim that the matrix $(\gamma Z + \delta)$ is invertible. To see this apply Lemma 2.1 (ii) to get

$$^t\overline{(\gamma Z + \delta)}\,(\alpha Z + \beta) - {}^t\overline{(\alpha Z + \beta)}\,(\gamma Z + \delta) = Z - \overline{Z} = 2i\,\mathrm{Im}\,Z. \qquad (1)$$

Suppose $(\gamma Z + \delta)v = 0$ for some $v \in \mathbb{C}^g$. Then (1) implies ${}^t\bar{v}(\mathrm{Im}\,Z)v = 0$ and thus $v = 0$, since $\mathrm{Im}\,Z > 0$. This proves the first assertion, and thus $M(Z)$ is well defined. Similarly one obtains

$$^t(\gamma Z + \delta)\,\big(M(Z) - {}^tM(Z)\big)\,(\gamma Z + \delta) = Z - {}^tZ = 0\,.$$

This implies that $M(Z)$ is symmetric. Now by (1) and the symmetry of $M(Z)$

$$^t\overline{(\gamma Z + \delta)}\;\mathrm{Im}\,M(Z)\,(\gamma Z + \delta) = \mathrm{Im}\,Z > 0$$

and thus $M(Z) \in \mathfrak{H}_g$. It remains to show that $M_1(M_2(Z)) = (M_1M_2)(Z)$ for all $M_1, M_2 \in G_D$ and $Z \in \mathfrak{H}_g$. But this is an immediate computation, which we omit here. □

We need some properties of the action of $\mathrm{Sp}_{2g}(\mathbb{R})$ on $\mathfrak{H}_g$. First note

(2.3) Proposition. *a) The group* $\mathrm{Sp}_{2g}(\mathbb{R})$ *acts transitively on* $\mathfrak{H}_g$.
b) The stabilizer of $i\mathbb{1}_g \in \mathfrak{H}_g$ *is the compact group*

$$\mathrm{Sp}_{2g}(\mathbb{R}) \cap O_{2g}(\mathbb{R}) = \left\{ \begin{pmatrix} \alpha & \beta \\ -\beta & \alpha \end{pmatrix} \in M_{2g}(\mathbb{R}) \;\middle|\; {\alpha{}^t\beta = \beta{}^t\alpha \atop \alpha{}^t\alpha + \beta{}^t\beta = \mathbb{1}_g} \right\}.$$

Proof. a) Suppose $Z = X + iY \in \mathfrak{H}_g$. Since Y is positive definite and symmetric, there is an $\alpha \in \mathrm{GL}_g(\mathbb{R})$ with $Y = \alpha{}^t\alpha$. The matrix $N = \begin{pmatrix} \alpha & X{}^t\alpha^{-1} \\ 0 & {}^t\alpha^{-1} \end{pmatrix}$ is obviously symplectic and $N(i\mathbb{1}_g) = Z$. Finally, b) follows by an immediate computation, since ${}^tM^{-1} = \begin{pmatrix} \delta & -\gamma \\ -\beta & \alpha \end{pmatrix}$ for any symplectic matrix $M = \begin{pmatrix} \alpha & \beta \\ \gamma & \delta \end{pmatrix}$. □

As a consequence of Proposition 2.3 the map

$$h\colon \mathrm{Sp}_{2g}(\mathbb{R}) \to \mathfrak{H}_g,\, M \mapsto M(i\mathbb{1}_g)$$

is surjective and all its fibres are of the form $M(\mathrm{Sp}_{2g}(\mathbb{R}) \cap O_{2g}(\mathbb{R}))$ for some $M \in \mathrm{Sp}_{2g}(\mathbb{R})$. In particular, the fibres are compact in $\mathrm{Sp}_{2g}(\mathbb{R})$. Moreover we have

(2.4) Proposition. *The map $h\colon \mathrm{Sp}_{2g}(\mathbb{R}) \to \mathfrak{H}_g$ is proper, that is $h^{-1}(K)$ is compact for any compact set $K \subset \mathfrak{H}_g$.*

For the proof we only note that the Iwasawa decomposition gives a diffeomorphism $\mathrm{Sp}_{2g}(\mathbb{R}) \xrightarrow{\sim} N \times \Delta \times O$ with $O = O_{2g}(\mathbb{R}) \cap \mathrm{Sp}_{2g}(\mathbb{R})$ (see Exercise 8). Under this diffeomorphism the map h corresponds to the projection $N \times \Delta \times O \to N \times \Delta$ which obviously is proper, since O is compact. □

For the convenience of the reader we also give a second direct proof not using the Iwasawa decomposition: it suffices to show that any sequence $(M_n)_{n\in\mathbb{N}}$ of matrices in $\mathrm{Sp}_{2g}(\mathbb{R})$, for which the sequence $(h(M_n))_{n\in\mathbb{N}}$ converges in $\mathfrak{H}_g$, admits a convergent subsequence. For the proof note that we can write $h(M_n) = X_n + i\alpha_n {}^t\alpha_n$, since the imaginary part of $h(M_n)$ is symmetric and positive definite. Then $N_n = \begin{pmatrix} \alpha_n & X_n {}^t\alpha_n^{-1} \\ 0 & {}^t\alpha_n^{-1} \end{pmatrix} \in \mathrm{Sp}_{2g}(\mathbb{R})$ and $h(M_n) = h(N_n)$ for any $n \in \mathbb{N}$. It follows that there are matrices P_n in the stabilizer $\mathrm{Sp}_{2g}(\mathbb{R}) \cap O_{2g}(\mathbb{R})$ of $\mathbb{1}_g$, such that $M_n = N_n P_n$ for all $n \in \mathbb{N}$. In fact, the sequence $(P_n)_{n\in\mathbb{N}}$ admits a convergent subsequence, the stabilizer being compact. So without loss of generality we may assume that $(P_n)_{n\in\mathbb{N}}$ is convergent.

By assumption the sequences (X_n) and $(\alpha_n {}^t\alpha_n)_{n\in\mathbb{N}}$ are convergent. It follows that the sequence $(\alpha_n)_{n\in\mathbb{N}}$ is bounded and thus admits a convergent subsequence. Again without loss of generality, we may assume that $(\alpha_n)_{n\in\mathbb{N}}$ converges to a nonsingular real matrix. This implies that also the sequence $(\alpha_n^{-1})_{n\in\mathbb{N}}$ converges.

Summing up, the sequence $(N_n)_{n\in\mathbb{N}}$ and thus also the sequence $(M_n)_{n\in\mathbb{N}} = (N_n P_n)_{n\in\mathbb{N}}$ converges. □

More important than the action of $\mathrm{Sp}_{2g}(\mathbb{R})$ on $\mathfrak{H}_g$ are the induced actions of certain subgroups. For discrete subgroups we have:

(2.5) Proposition. *Any discrete subgroup $G \subseteq \mathrm{Sp}_{2g}(\mathbb{R})$ acts properly and discontinuously on $\mathfrak{H}_g$.*

Proof. According to the definition of a proper and discontinuous action (see Appendix A) we have to show that for all compact subsets $K_1, K_2 \subseteq \mathfrak{H}_g$ there are only finitely many $M \in G$ with $M(K_1) \cap K_2 \neq \emptyset$. From the definition of the map h it follows that $M(K_1) \cap K_2 \neq \emptyset$ if and only if

$$M \in h^{-1}(K_2)\big(h^{-1}(K_1)\big)^{-1} = \big\{M_2 M_1^{-1} \mid M_\nu \in h^{-1}(K_\nu), \nu = 1, 2\big\} .$$

Hence it suffices to show that $h^{-1}(K_2)(h^{-1}(K_1))^{-1}$ is compact in $\mathrm{Sp}_{2g}(\mathbb{R})$. But $h^{-1}(K_\nu)$, $\nu = 1,2$, is compact, since the map h is proper. Thus $h^{-1}(K_2)(h^{-1}(K_1))^{-1}$ is compact as the image of the compact set $h^{-1}(K_1) \times h^{-1}(K_2)$ under the continuous map $(M_1, M_2) \mapsto M_2 M_1^{-1}$. □

We apply this results to construct a moduli space for polarized abelian varieties of type D. The group G_D defined in Section 1 is a discrete subgroup of $\mathrm{Sp}_{2g}(\mathbb{R})$. Hence by Proposition 2.5 it acts properly and discontinuously on $\mathfrak{H}_g$. According to Theorem A.6 the quotient

$$\mathcal{A}_D := \mathfrak{H}_g / G_D$$

with its natural quotient structure is a normal complex analytic space of dimension $\frac{1}{2}g(g+1)$. Applying Propositions 1.2 and 1.3 we see that the elements of $\mathcal{A}_D$ are in one to one correspondence to the isomorphism classes of polarized abelian varieties of type D. Thus we have proved

(2.6) Theorem. *The normal complex analytic space $\mathcal{A}_D = \mathfrak{H}_g / G_D$ is a moduli space for polarized abelian varieties of type D.*

In Remark 10.4 we will see that $\mathcal{A}_D$ admits the structure of an algebraic variety.

There is another approach to the moduli space for polarized abelian varieties of type D, which for some purposes is more convenient: for any commutative ring $\mathcal{R}$ with 1 of characteristic 0 define the group

$$\mathrm{Sp}_{2g}^D(\mathcal{R}) = \left\{ R \in M_{2g}(\mathcal{R}) \mid R \begin{pmatrix} 0 & D \\ -D & 0 \end{pmatrix} {}^t R = \begin{pmatrix} 0 & D \\ -D & 0 \end{pmatrix} \right\} .$$

The map $\sigma_D \colon \mathrm{Sp}_{2g}^D(\mathbb{R}) \to \mathrm{Sp}_{2g}(\mathbb{R})$,

$$\sigma_D(R) = \begin{pmatrix} \mathbb{1} & 0 \\ 0 & D \end{pmatrix}^{-1} R \begin{pmatrix} \mathbb{1} & 0 \\ 0 & D \end{pmatrix}$$

is an isomorphism of groups, since $\mathrm{Sp}_{2g}(\mathbb{R})$ is invariant under transposition. We observe that in general $\mathrm{Sp}_{2g}^D(\mathbb{R})$ is not invariant under transposition. The action of $\mathrm{Sp}_{2g}(\mathbb{R})$ on $\mathfrak{H}_g$ induces an action of $\mathrm{Sp}_{2g}^D(\mathbb{R})$ on $\mathfrak{H}_g$ via σ_D, namely

$$R\langle Z \rangle := (aZ + bD)\,(D^{-1}cZ + D^{-1}dD)^{-1}$$

for all $R = \begin{pmatrix} a & b \\ c & d \end{pmatrix} \in \mathrm{Sp}_{2g}^D(\mathbb{R})$ and $Z \in \mathfrak{H}_g$. Note that $\sigma_D(\mathrm{Sp}_{2g}^D(\mathbb{Z}))$ is just the group G_D defined in Section 1. For abbreviation write $\Gamma_D = \mathrm{Sp}_{2g}^D(\mathbb{Z})$, then

$$\widetilde{\mathcal{A}}_D := \mathfrak{H}_g / \Gamma_D$$

is a normal complex analytic space and the identity on $\mathfrak{H}_g$ induces an isomorphism $\widetilde{\mathcal{A}}_D \xrightarrow{\sim} \mathcal{A}_D$. Hence Theorem 2.6 implies

(2.7) Corollary. *The normal complex analytic space $\widetilde{\mathcal{A}}_D = \mathfrak{H}_g/\Gamma_D$ is a moduli space for polarized abelian varieties of type D.*

Of course one can also give a direct proof of this Corollary, reasoning analogously as in Sections 1 and 2. Notice that for a principal polarization the isomorphism $\sigma_{\mathbb{1}}$ is the identity on $\mathrm{Sp}_{2g}(\mathbb{R})$. So $\Gamma_{\mathbb{1}} = G_{\mathbb{1}}$ and $\widetilde{\mathcal{A}}_{\mathbb{1}} = \mathcal{A}_{\mathbb{1}}$. For later use let us point out the following interpretation of the elements of Γ_D.

(2.8) Remark. According to Corollary 2.7 for $Z \in \mathfrak{H}_g$ and $R \in \Gamma_D$ the polarized abelian varieties (X_Z, H_Z) and $(X_{R\langle Z\rangle}, H_{R\langle Z\rangle})$ are isomorphic. In fact, ${}^tR$ is the rational representation of the corresponding isomorphism $X_{R\langle Z\rangle} \to X_Z$ with respect to the symplectic bases determined by Z and $R\langle Z\rangle$ (see Remark 1.4). □

For some purposes the second approach is more convenient, since the group Γ_D acting on $\mathfrak{H}_g$ has integer coefficients. On the other hand, sometimes the first approach has advantages the action of G_D on $\mathfrak{H}_g$ being easier and more familiar. It depends on the particular problem which point of view we will take subsequently.

§ 3 Level Structures

In the last section we constructed the moduli spaces of polarized abelian varieties of type D as the quotient of the Siegel upper half space $\mathfrak{H}_g$ modulo the discrete subgroup Γ_D. On the other hand, according to Proposition 1.2 the Siegel upper half space itself may be considered as the moduli space of polarized abelian varieties with symplectic basis. A symplectic basis can not be defined in algebraic terms. A *level structure* on a polarized abelian variety is roughly speaking a replacement of the notion of a symplectic basis or only some properties of it. The corresponding moduli spaces are quotients of $\mathfrak{H}_g$ by suitable subgroups of Γ_D (respectively G_D) and hence are situated between $\mathfrak{H}_g$ and $\widetilde{\mathcal{A}}_D$ (respectively $\mathcal{A}_D$).

Moduli spaces for polarized abelian varieties with level structures have various applications in arithmetic and geometry. In this section we present the most important examples. For other examples see Section 9 as well as Exercises 4 to 7.

Level D-Structure. Let $(X = V/\Lambda, H)$ be a polarized abelian variety of type $D = \mathrm{diag}(d_1, \ldots, d_g)$. Recall the finite group $K(H) = \Lambda(H)/\Lambda$ and the (multiplicative) alternating form $e^H: K(H) \times K(H) \to \mathbb{C}^*$, $e^H(\bar{v}, \bar{w}) = e(-2\pi i \operatorname{Im} H(v,w))$ (see Proposition 6.3.1). In Section 6.6 we introduced the group $K(D) = (\mathbb{Z}^g/D\mathbb{Z}^g)^2$ and the (multiplicative) alternating form $e^D: K(D) \times K(D) \to \mathbb{C}^*$. A *level D-structure* on (X, H) is by definition a symplectic isomorphism $\bar{b}: K(H) \to K(D)$. Given a symplectic isomorphism $\bar{b}$, there is a symplectic basis $\lambda_1, \ldots, \lambda_g, \mu_1, \ldots, \mu_g$ of Λ for H such that $\bar{b}(\frac{1}{d_i}\lambda_i) = f_i$ and $\bar{b}(\frac{1}{d_i}\mu_i) = f_{g+i}$ for $1 \le i \le g$ (see Exercise 3). Here $f_1, \ldots, f_{2g}$ denote the standard generators of $K(D)$.

Every $Z \in \mathfrak{H}_g$ determines a polarized abelian variety of type D with level D-structure:

$$Z \mapsto \left(X_Z, H_Z, \{\overline{\tfrac{1}{d_1}\lambda_1}, \ldots, \overline{\tfrac{1}{d_g}\lambda_g}, \overline{\tfrac{1}{d_1}\mu_1}, \ldots, \overline{\tfrac{1}{d_g}\mu_g}, \}\right),$$

where $(X_Z, H_Z, \{\lambda_1, \ldots \mu_g\})$ is the polarized abelian variety of type D with symplectic basis of Proposition 1.2. Conversely it is clear by what we said above that every polarized abelian variety with level D-structure is isomorphic to one of these, and we have to analyze when two of them are isomorphic. Suppose $Z, Z' \in \mathfrak{H}_g$ and

$$\varphi: (X_{Z'}, H_{Z'}, \{\overline{\tfrac{1}{d_1}\lambda'_1}, \ldots, \overline{\tfrac{1}{d_g}\mu'_g}\}) \to (X_Z, H_Z, \{\overline{\tfrac{1}{d_1}\lambda_1}, \ldots, \overline{\tfrac{1}{d_g}\mu_g}\})$$

is an isomorphism, that is i): φ is an isomorphism of polarized abelian varieties $\varphi: (X_{Z'}, H_{Z'}) \to (X_Z, H_Z)$, and ii): it satisfies $\varphi(\overline{\frac{1}{d_\nu}\lambda'_\nu}) = \overline{\frac{1}{d_\nu}\lambda_\nu}$ and $\varphi(\overline{\frac{1}{d_\nu}\mu'_\nu}) = \overline{\frac{1}{d_\nu}\mu_\nu}$ for $1 \le \nu \le g$. Denote by A and ${}^tR$ the matrices of the analytic and rational representations of φ with respect to the chosen bases. Then i) is equivalent to $A(Z', D) = (Z, D)\,{}^tR$ and $R \in \Gamma_D$ according to Remark 2.8. In particular $Z' = R\langle Z\rangle$. In terms of matrices condition ii) reads

$$A(Z'D^{-1}, \mathbb{1}_g) \equiv (ZD^{-1}, \mathbb{1}_g) \pmod{\Lambda_Z = (Z, D)\mathbb{Z}^{2g}}\,.$$

In other words

$$(Z, D)({}^tR - \mathbb{1}_{2g}) = A(Z', D) - (Z, D) \in (Z, D) M_{2g}(\mathbb{Z}) \begin{pmatrix} D & 0 \\ 0 & D \end{pmatrix}.$$

This means
$$R - \mathbb{1}_{2g} \in \begin{pmatrix} D & 0 \\ 0 & D \end{pmatrix} M_{2g}(\mathbb{Z})\,.$$

Summing up we have shown that Z and $Z' \in \mathfrak{H}_g$ determine isomorphic polarized abelian varieties of type D with level D-structure if and only if $Z' = R\langle Z\rangle$, where R is an element of the group

$$\Gamma_D(D) := \left\{ \begin{pmatrix} a & b \\ c & d \end{pmatrix} \in \Gamma_D \mid a - \mathbb{1}_g \equiv b \equiv c \equiv d - \mathbb{1}_g \equiv 0 \ (\mathrm{mod}\ D) \right\}.$$

Here we write $\quad a \equiv 0 \ (\mathrm{mod}\ D) \quad \text{if } a \in D \cdot M_g(\mathbb{Z}).$

The group $\Gamma_D(D)$ is a *congruence subgroup* of Γ_D, i.e. a subgroup containing some principal congruence subgroup which is defined in the following subsection (see p. 222). In fact, $\Gamma_D(D)$ contains $\Gamma_D(d_g)$. In particular, $\Gamma_D(D)$ is of finite index in Γ_D. Moreover note that the subgroup $\Gamma_D(D)$ is normal in Γ_D. As a subgroup of Γ_D also $\Gamma_D(D)$ acts properly and discontinuously on $\mathfrak{H}_g$ and we obtain

(3.1) Theorem. *The normal complex analytic space $\widetilde{\mathcal{A}}_D(D) := \mathfrak{H}_g/\Gamma_D(D)$ is a moduli space for polarized abelian varieties of type $D = \operatorname{diag}(d_1, \dots d_g)$ with level D-structure. The embedding $\Gamma_D(D) \hookrightarrow \Gamma_D$ induces a holomorphic map $\widetilde{\mathcal{A}}_D(D) \to \widetilde{\mathcal{A}}_D$ of finite degree.*

Generalized Level n-Structure. A *level n-structure*, $n > 1$, on a principally polarized abelian variety (X, H) is by definition a level $(n\mathbb{1}_g)$-structure on the polarized abelian variety (X, nH). We want to generalize the notion of a level n-structure to a polarized abelian variety (X, H) of arbitrary type D. The problem is that the hermitian form H in general does not induce a nondegenerate multiplicative alternating form on the n-division points X_n.

Let $(X = V/\Lambda, H)$ be a polarized abelian variety of type D. A symplectic basis $\lambda_1, \dots, \lambda_g, \mu_1, \dots, \mu_g$ of Λ for H determines a basis for the group of n-division points X_n in X, namely $\overline{\frac{1}{n}\lambda_1}, \dots, \overline{\frac{1}{n}\lambda_g}, \overline{\frac{1}{n}\mu_1}, \dots, \overline{\frac{1}{n}\mu_g}$. A *generalized level n-structure* for (X, H) is by definition a basis of X_n coming from a symplectic basis in this way.

Every $Z \in \mathfrak{H}_g$ determines a polarized abelian variety of type D with level n-structure in the following way:

$$Z \mapsto \left(X_Z, H_Z, \{\overline{\tfrac{1}{n}\lambda_1}, \dots, \overline{\tfrac{1}{n}\lambda_g}, \overline{\tfrac{1}{n}\mu_1}, \dots, \overline{\tfrac{1}{n}\mu_g}\}\right) ,$$

where $(X_Z, H_Z, \{\lambda_1, \dots, \lambda_g, \mu_1, \dots, \mu_g\})$ is the polarized abelian variety of type D with symplectic basis of Proposition 1.2. Conversely it is clear by definition that every polarized abelian variety with generalized level n-structure is isomorphic to one of these, and again we have to analyze when two of them are isomorphic. Suppose $Z, Z' \in \mathfrak{H}_g$ and

$$\varphi : (X_{Z'}, H_{Z'}, \{\overline{\tfrac{1}{n}\lambda'_1}, \dots, \overline{\tfrac{1}{n}\mu'_g}\}) \to (X_Z, H_Z, \{\overline{\tfrac{1}{n}\lambda_1}, \dots, \overline{\tfrac{1}{n}\mu_g}\})$$

is an isomorphism, that is an isomorphism of polarized abelian varieties $\varphi : (X_{Z'}, H_{Z'}) \to (X_Z, H_Z)$ with $\varphi(\overline{\frac{1}{n}\lambda'_\nu}) = \overline{\frac{1}{n}\lambda_\nu}$ and $\varphi(\overline{\frac{1}{n}\mu'_\nu}) = \overline{\frac{1}{n}\mu_\nu}$ for $1 \le \nu \le g$. Denote by A and ${}^tR$ the matrices of the analytic and rational representations of φ with respect to the chosen bases. Then $A(Z', D) = (Z, D)\,{}^tR$ and $R \in \Gamma_D$ by Remark 2.8. In terms of matrices the condition on the bases translates to $(\frac{1}{n}Z, \frac{1}{n}D)\,{}^tR = A(\frac{1}{n}Z', \frac{1}{n}D) \equiv (\frac{1}{n}Z, \frac{1}{n}D) \pmod{\Lambda_Z}$. This is equivalent to $R \equiv \mathbb{1}_{2g} \pmod n$.

Conversely, given $Z \in \mathfrak{H}_g$, any $R \in \Gamma_D$ with the extra condition $R \equiv \mathbb{1}_{2g}$ (mod n) gives an isomorphism of polarized abelian varieties with level n-structure associated to Z and $Z' = R\langle Z \rangle$ by reversing the above arguments.

For any $n > 1$ the *principal congruence subgroup* $\Gamma_D(n)$ of Γ_D is defined to be

$$\Gamma_D(n) = \left\{R \in \Gamma_D \mid R \equiv \mathbb{1}_{2g} \pmod n\right\} .$$

Note that $\Gamma_D(n)$ is a normal subgroup of Γ_D. As a subgroup of Γ_D also $\Gamma_D(n)$ acts properly and discontinuously on $\mathfrak{H}_g$ and we obtain

(3.2) Theorem. *The normal complex analytic space $\widetilde{\mathcal{A}}_D(n) := \mathfrak{H}_g/\Gamma_D(n)$ is a moduli space for polarized abelian varieties of type D with generalized level n-structure. The embedding $\Gamma_D(n) \hookrightarrow \Gamma_D$ induces a holomorphic map $\widetilde{\mathcal{A}}_D(n) \to \widetilde{\mathcal{A}}_D$ of finite degree.*

Proof. It remains to show that $[\Gamma_D : \Gamma_D(n)] < \infty$. But the quotient $\Gamma_D/\Gamma_D(n)$ is a subgroup of the finite group $\mathrm{Sl}_{2g}(\mathbb{Z}/n\mathbb{Z})$. □

Decomposition of the Lattice. Let $(X = V/\Lambda, H)$ be a polarized abelian variety of type D. Recall from Section 3.1 that a *decomposition (for H)* is a decomposition $\Lambda = \Lambda_1 \oplus \Lambda_2$ with isotropic subgroups Λ_1 and Λ_2 of Λ for H. Any symplectic basis $\lambda_1, \dots, \lambda_g, \mu_1, \dots, \mu_g$ of Λ for H determines such a decomposition: Λ_1 is the group spanned by $\lambda_1, \dots, \lambda_g$ and Λ_2 the group spanned by $\mu_1, \dots, \mu_g$. Thus every $Z \in \mathfrak{H}_g$ determines a polarized abelian variety of type D with a decomposition, namely

$$Z \mapsto (X_Z, H_Z, \Lambda_Z = \Lambda_1 \oplus \Lambda_2) ,$$

where $\Lambda_1 := Z\mathbb{Z}^g$ and $\Lambda_2 := D\mathbb{Z}^g$. Conversely, it follows from the proof of the elementary divisor theorem that every polarized abelian variety of type D with such a decomposition is of this form. Again we have to analyze when two of these triplets are isomorphic. Suppose $Z, Z' \in \mathfrak{H}_g$ and

$$\varphi\colon \left(X_{Z'}, H_{Z'}, \Lambda_{Z'} = \Lambda_1' \oplus \Lambda_2'\right) \to \left(X_Z, H_Z, \Lambda_Z = \Lambda_1 \oplus \Lambda_2\right)$$

is an isomorphism, i.e. $\varphi(X_{Z'}, H_{Z'}) \to (X_Z, H_Z)$ is an isomorphism such that $\varrho_r(\varphi)(\Lambda_\nu') = \Lambda_\nu$ for $\nu = 1, 2$. Let the notation be as above. The first condition is again $A(Z', D) = (Z, D)\,{}^tR$. Write $R = \left(\begin{smallmatrix} a & b \\ c & d \end{smallmatrix}\right)$, then $\varrho_r(\varphi)(\Lambda_1') = (Z\,{}^ta + D\,{}^tb)\mathbb{Z}^g$ and $\varrho_r(\varphi)(\Lambda_2') = (Z\,{}^tc + D\,{}^td)\mathbb{Z}^g$. So the second condition on φ gives

$$(Z\,{}^ta + D\,{}^tb)\mathbb{Z}^g = Z\mathbb{Z}^g \quad \text{and} \quad (Z\,{}^tc + D\,{}^td)\mathbb{Z}^g = D\mathbb{Z}^g .$$

Using $\operatorname{Im} Z > 0$ it is easy to see that this is equivalent to $b = c = 0$. Hence denoting $\Delta_D := \left\{ \left(\begin{smallmatrix} a & b \\ c & d \end{smallmatrix} \right) \in \Gamma_D \mid b = c = 0 \right\}$, we obtain

(3.3) Proposition. *The normal complex analytic space $\widetilde{\mathcal{A}}_D^\Delta := \mathfrak{H}_g/\Delta_D$ is a moduli space for polarized abelian varieties of type D with a decomposition.*

The embedding $\Delta_D \hookrightarrow \Gamma_D$ yields holomorphic maps $\mathfrak{H}_g \xrightarrow{\pi_1} \widetilde{\mathcal{A}}_D^\Delta \xrightarrow{\pi_2} \widetilde{\mathcal{A}}_D$ and π_1 and π_2 both have infinite fibres.

§ 4 The Theta Transformation Formula, Preliminary Version

Theta functions with respect to a lattice $\Lambda_Z = (Z, D)\mathbb{Z}^{2g}$ in $\mathbb{C}^g$ (for some $Z \in \mathfrak{H}_g$) are holomorphic functions on $\mathbb{C}^g$ with a certain functional behaviour with respect to translations by elements of Λ_Z. Varying Z within $\mathfrak{H}_g$ one may ask how the corresponding theta functions are related. We will see that the action of the symplectic group on $\mathfrak{H}_g$ induces such a relation, namely the theta transformation formula.

Let $D = \operatorname{diag}(d_1, \ldots, d_g)$ be a type. Suppose $Z \in \mathfrak{H}_g$, $M = \left(\begin{smallmatrix} \alpha & \beta \\ \gamma & \delta \end{smallmatrix} \right) \in G_D$, and $Z' = M(Z)$. Let $(X_{Z'}, H_{Z'})$ and (X_Z, H_Z) be the polarized abelian varieties of type D corresponding to Z' and Z. According to Proposition 1.3 the matrix M induces an isomorphism $\varphi: (X_{Z'}, H_{Z'}) \xrightarrow{\sim} (X_Z, H_Z)$ given by the equation $A(Z', \mathbb{1}) = (Z, \mathbb{1})\, {}^tM$ (see Remark 1.4). Here $A = {}^t(\gamma Z + \delta)$ is the matrix of the analytic representation $\varrho_a(\varphi)$ with respect to the standard basis of $\mathbb{C}^g$.

Recall the decompositions $\Lambda_Z = Z\mathbb{Z}^g \oplus D\mathbb{Z}^g$ and $\Lambda_{Z'} = Z'\mathbb{Z}^g \oplus D\mathbb{Z}^g$ for H_Z and $H_{Z'}$, and let $L = L(H_Z, \chi)$ denote the line bundle with characteristic $c \in \mathbb{C}^g$ on X_Z with respect to the decomposition of Λ_Z. We want to compute the characteristic of $L' = \varphi^* L$ in terms of c and M. For any $S = (s_{ij}) \in M_g(\mathbb{R})$ denote by $(S)_0$ the vector of diagonal elements

$$(S)_0 = {}^t(s_{11}, \ldots, s_{gg}) \in \mathbb{R}^g \, .$$

With this notation we have

(4.1) Lemma. *a) The line bundle $\varphi^* L$ is of characteristic*

$$M[c] := A^{-1}c + \tfrac{1}{2}(Z', \mathbb{1}) \begin{pmatrix} (\gamma\, {}^t\delta)_0 \\ (\alpha\, {}^t\beta)_0 \end{pmatrix}$$

with respect to the decomposition $\Lambda_{Z'} = Z'\mathbb{Z}^g \oplus D\mathbb{Z}^g$.

b) If $c = Zc^1 + c^2$ *with* $c^1, c^2 \in \mathbb{R}^g$ *and* $M[c] = M(Z)M[c]^1 + M[c]^2$*, then*

$$M[c]^1 = \delta c^1 - \gamma c^2 + \tfrac{1}{2}(\gamma\,{}^t\delta)_0$$
$$M[c]^2 = -\beta c^1 + \alpha c^2 + \tfrac{1}{2}(\alpha\,{}^t\beta)_0$$

Proof. a) According to Lemma 2.3.4 the semicharacter of L' is $\varrho_r(\varphi)^*\chi$. Suppose $\mu = Z'\mu^1 + \mu^2 \in \Lambda_{Z'}$ and $\lambda = \varrho_r(\varphi)\mu = Z\lambda^1 + \lambda^2 \in \Lambda_Z$. In terms of matrices this reads $\binom{\lambda^1}{\lambda^2} = {}^tM\binom{\mu^1}{\mu^2}$. Since $\begin{pmatrix} 0 & \mathbb{1} \\ -\mathbb{1} & 0 \end{pmatrix}$ is the matrix of $\operatorname{Im} H_Z$ (respectively $\operatorname{Im} H_{Z'}$) with respect to the $\mathbb{R}$-basis of $\mathbb{C}^g$ given by the columns of $(Z, \mathbb{1})$ (respectively $(Z', \mathbb{1})$), we have by Lemma 3.1.2

$$\begin{aligned}\varrho_r(\varphi)^*\chi(\mu) = \chi(\lambda) &= e(\pi i \operatorname{Im} H_Z(Z\lambda^1, \lambda^2) + 2\pi i \operatorname{Im} H_Z(c, \lambda)) \\ &= e(\pi i\,{}^t\lambda^1\lambda^2 + 2\pi i\,{}^t\tbinom{c^1}{c^2}\begin{pmatrix} 0 & \mathbb{1} \\ -\mathbb{1} & 0 \end{pmatrix}\tbinom{\lambda^1}{\lambda^2}) \\ &= e(\pi i\,{}^t\mu^1(\alpha\,{}^t\delta - \beta\,{}^t\gamma)\mu^2 + \pi i\,{}^t\mu^2\gamma\,{}^t\delta\mu^2 - \pi i\,{}^t\mu^1\alpha\,{}^t\beta\mu^1 \\ &\quad + 2\pi i\,{}^t\tbinom{c^1}{c^2}M^{-1}\begin{pmatrix} 0 & \mathbb{1} \\ -\mathbb{1} & 0 \end{pmatrix}\tbinom{\mu^1}{\mu^2}) .\end{aligned}$$

For the last equation we used that ${}^t\mu^1\beta\,{}^t\gamma\mu^2$ and ${}^t\mu^1\alpha\,{}^t\beta\mu^1$ are integers and that $\begin{pmatrix} 0 & \mathbb{1} \\ -\mathbb{1} & 0 \end{pmatrix}{}^tM = M^{-1}\begin{pmatrix} 0 & \mathbb{1} \\ -\mathbb{1} & 0 \end{pmatrix}$. Note that for any $l \in \mathbb{Z}^g$ and any symmetric matrix $S = (s_{ij}) \in M_g(\mathbb{Z})$

$${}^tlSl = \sum_{i=1}^{g} s_{ii}l_i^2 + 2\sum_{1\le i<j\le g} s_{ij}l_il_j \equiv {}^t(S)_0 l \pmod 2 .$$

Hence using Lemma 2.1 we get

$$\begin{aligned}\varrho_r(\varphi)^*\chi(\mu) &= e\big(\pi i\,{}^t\mu^1\mu^2 + 2\pi i\,{}^t[{}^tM^{-1}\tbinom{c^1}{c^2} + \tfrac{1}{2}\tbinom{(\gamma\,{}^t\delta)_0}{(\alpha\,{}^t\beta)_0}]\begin{pmatrix} 0 & \mathbb{1} \\ -\mathbb{1} & 0 \end{pmatrix}\tbinom{\mu^1}{\mu^2}\big) \\ &= e(\pi i \operatorname{Im} H_{Z'}(Z'\mu^1, \mu^2) + 2\pi i \operatorname{Im} H_{Z'}(M[c], \mu)) ,\end{aligned}$$

since $M[c] = A^{-1}c + \frac{1}{2}(Z', \mathbb{1})\big(\binom{(\gamma\,{}^t\delta)_0}{(\alpha\,{}^t\beta)_0}\big) = (Z', \mathbb{1})({}^tM^{-1}\binom{c^1}{c^2} + \frac{1}{2}\big(\binom{(\gamma\,{}^t\delta)_0}{(\alpha\,{}^t\beta)_0}\big))$.

b)
$$\begin{aligned}M[c] &= A^{-1}c + \tfrac{1}{2}(M(Z), \mathbb{1})\tbinom{(\gamma\,{}^t\delta)_0}{(\alpha\,{}^t\beta)_0} \\ &= (M(Z), \mathbb{1})({}^tM^{-1}\tbinom{c^1}{c^2} + \tfrac{1}{2}\big(\tbinom{(\gamma\,{}^t\delta)_0}{(\alpha\,{}^t\beta)_0}\big)) \\ &= M(Z)(\delta c^1 - \gamma c^2 + \tfrac{1}{2}(\gamma\,{}^t\delta)_0) + (-\beta c^1 + \alpha c^2 + \tfrac{1}{2}(\alpha\,{}^t\beta)_0).\end{aligned}$$

This implies the assertion. □

From now on suppose $D = \mathbb{1}$, i.e. the polarizations are principal. So $H^0(X_Z, L)$ is a one-dimensional vector space. According to Theorem 3.2.7 the decomposition of $\Lambda_Z = Z\mathbb{Z}^g \oplus \mathbb{Z}^g$ together with the characteristic c of L

determines a basis of $H^0(X_Z, L)$, i.e. the canonical theta function $\vartheta^c = \vartheta^c_Z$ defined in 3.2(2). Here the index Z indicates that the theta function belongs to the lattice Λ_Z. Similarly the canonical theta function $\vartheta^{M[c]}_{Z'}$ determined by $\Lambda_{Z'} = Z'\mathbb{Z}^g \oplus \mathbb{Z}^g$ and the characteristic $M[c]$ of $\varphi^* L$ is a basis of $H^0(X_{Z'}, \varphi^* L)$. Certainly the map $A^*\colon H^0(X_Z, L) \to H^0(X_{Z'}, \varphi^* L)$ with $A = \varrho_a(\varphi)$ is an isomorphism. Hence there is a constant $C(Z, M, c) \in \mathbb{C}^*$ depending only on Z, M and c such that

$$A^* \vartheta^c_Z = C(Z, M, c) \vartheta^{M[c]}_{Z'} \ . \tag{1}$$

This equation may be considered as a preliminary version of the *theta transformation formula*. For its final version (see Section 6) we have only to compute the constant $C(Z, M, c)$. As a first step we show

(4.2) Lemma. $C(Z, M, c) = C(Z, M, 0) e(\pi i \operatorname{Im} H_{Z'}(M[0], A^{-1}c))$.

Proof. Denote by L_0 the line bundle on X_Z of characteristic 0 algebraically equivalent to L, and by $\vartheta^0_Z \in H^0(X_Z, L_0)$ the corresponding theta function. According to Lemma 4.1 the line bundle $\varphi^* L_0$ on $X_{Z'}$ is of characteristic $M[0]$ and we denote by $\vartheta^{M[0]}_{Z'}$ the associated basis of $H^0(X_{Z'}, \varphi^* L_0)$. According to equation 3.2(5) the canonical theta functions ϑ^0_Z, ϑ^c_Z, $\vartheta^{M[0]}_{Z'}$ and $\vartheta^{M[c]}_{Z'}$ are related as follows

$$\vartheta^c_Z = \tau \cdot t^*_c \vartheta^0_Z \quad \text{and} \quad \vartheta^{M[0]}_{Z'} = \tau' \cdot t^*_{-A^{-1}c} \vartheta^{M[c]}_{Z'}$$

with $\tau = e(-\pi H_Z(\cdot, c) - \frac{\pi}{2} H_Z(c, c))$ and $\tau' = e(\pi i \operatorname{Im} H_{Z'}(M[0], A^{-1}c) + \pi H_{Z'}(\cdot, A^{-1}c) - \frac{\pi}{2} H_{Z'}(A^{-1}c, A^{-1}c))$. This gives using (1) and Lemma 4.1

$$\begin{aligned} A^* \vartheta^c_Z &= A^*(\tau \cdot t^*_c \vartheta^0_Z) = A^*\tau \cdot t^*_{A^{-1}c} A^* \vartheta^0_Z \\ &= A^*\tau \cdot t^*_{A^{-1}c} \big(C(Z, M, 0) \vartheta^{M[0]}_{Z'}\big) \\ &= C(Z, M, 0) A^*\tau \cdot t^*_{A^{-1}c} \tau' \cdot \vartheta^{M[c]}_{Z'} \end{aligned}$$

So
$$\begin{aligned} C(Z, M, c) &= C(Z, M, 0) A^*\tau \cdot t^*_{A^{-1}c} \tau' \\ &= e\big(\pi i \operatorname{Im} H_{Z'}(M[0], A^{-1}c)\big)\, C(Z, M, 0) \ . \end{aligned}$$

The second equation follows from an immediate computation using $A^* H_Z = H_{Z'}$. □

§ 5 Classical Theta Functions

In Section 3.2 we introduced apart from canonical theta functions also classical theta functions. Sometimes it turns out to be more convenient to work with the latter ones. For example, we will see that they depend holomorphically on $Z \in \mathcal{H}_g$. Moreover, they are helpful for computing the norm of the corresponding canonical theta function with respect to the hermitian metric introduced in Section 3.4. We do this at the end of this section. This norm is the main ingredient for the final version of the theta transformation formula. We start by introducing the relation between the canonical theta functions $\vartheta^c = \vartheta^c_Z$ and the classical Riemann theta functions with real characteristics.

Let $(X_Z = \mathbb{C}^g/\Lambda_Z, H = H_Z)$ denote the principally polarized abelian variety corresponding to $Z \in \mathcal{H}_g$. Then $\Lambda_Z = \Lambda_1 \oplus \Lambda_2$ with $\Lambda_1 = Z\mathbb{Z}^g$ and $\Lambda_2 = \mathbb{Z}^g$ is a decomposition for H. It induces a decomposition $\mathbb{C}^g = V_1 \oplus V_2$ for H with real vector spaces $V_1 = Z\mathbb{R}^g$ and $V_2 = \mathbb{R}^g$ and we can write every $v \in \mathbb{C}^g$ uniquely as

$$v = Zv^1 + v^2$$

with $v^1, v^2 \in \mathbb{R}^g$. Denote by B the $\mathbb{C}$-bilinear extension of the symmetric bilinear form $H|V_2 \times V_2$ (see Section 3.2).

(5.1) Lemma. *For all $v, w \in \mathbb{C}^g$ we have*
a) $B(v,w) = {}^t v (\operatorname{Im} Z)^{-1} w$,
b) $(H - B)(v,w) = -2i\,{}^t v w^1$.

Proof. a) follows immediately from the definition of B and Proposition 1.1. Using a) we get

$$(H-B)(v,w) = {}^t\overline{v}(\operatorname{Im} Z)^{-1}(\overline{w}-w) = {}^t\overline{v}(\operatorname{Im} Z)^{-1}(\overline{Z}-Z)w^1 = -2i\,{}^t\overline{v}w^1 \ . \ \square$$

Let $L = L(H,\chi)$ be the line bundle on X_Z with characteristic $c = Zc^1 + c^2$ with $c^1, c^2 \in \mathbb{R}^g$. Moreover, as in the last section, let $\vartheta^c = \vartheta^c_Z$ denote the canonical theta function for L generating $H^0(X_Z, L)$. The classical *Riemann theta function* with the (real) characteristic $\left[\begin{smallmatrix} c^1 \\ c^2 \end{smallmatrix}\right]$ is defined by

$$\vartheta\left[\begin{smallmatrix} c^1 \\ c^2 \end{smallmatrix}\right](v,Z) = \sum_{l \in \mathbb{Z}^g} e(\pi i\,{}^t(l+c^1)\,Z(l+c^1) + 2\pi i\,{}^t(v+c^2)(l+c^1)) \qquad (1)$$

for all $v \in \mathbb{C}^g$. Note that this notation coincides with the classical notation (see Krazer [1] p. 30). The functions ϑ^c_Z and $\vartheta\left[\begin{smallmatrix} c^1 \\ c^2 \end{smallmatrix}\right]$ are related as follows:

(5.2) Lemma. $\vartheta_Z^{Zc^1+c^2} = e(\frac{\pi}{2}B(\,\cdot\,,\cdot\,) - \pi i\,{}^t c^1 c^2)\,\vartheta\left[\begin{smallmatrix} c^1 \\ c^2 \end{smallmatrix}\right](\cdot, Z)$

This shows in particular that in the principally polarized case our definition of characteristics of theta functions coincides with the classical one. Hence our notion of characteristics of nondegenerate line bundles of arbitrary type is a natural generalization of this.

Proof. Write $c = Zc^1 + c^2$. By the definition of ϑ_Z^c (see 3.2(2)) we have for all $v \in \mathbb{C}^g$

$$\begin{aligned}
\vartheta_Z^{Zc^1+c^2}(v) &= e\big(-\pi H(v,c) - \tfrac{\pi}{2}H(c,c) + \tfrac{\pi}{2}B(v+c,v+c)\big) \\
&\quad \cdot \sum_{\lambda\in\Lambda_1} e\big(\pi(H-B)(v+c,\lambda) - \tfrac{\pi}{2}(H-B)(\lambda,\lambda)\big) \\
&= e\big(\tfrac{\pi}{2}B(v,v)\big) \sum_{\lambda\in\Lambda_1} e\big(-\tfrac{\pi}{2}(H-B)(\lambda,\lambda) - \tfrac{\pi}{2}(H-B)(c,c)\big) \\
&\quad \cdot e\big(-\pi(H-B)(c,\lambda) - \pi(H-B)(v,\lambda+c)\big) \\
&\qquad\qquad\qquad\qquad \text{(where we replaced } \lambda \text{ by } -\lambda) \\
&= e\big(\tfrac{\pi}{2}B(v,v)\big) \sum_{\lambda^1\in\mathbb{Z}^g} e\big(\pi i\,{}^t\lambda^1 Z\lambda^1 + \pi i\,{}^t(Zc^1+c^2)c^1\big) \\
&\quad \cdot e\big(2\pi i\,{}^t(Zc^1+c^2)\lambda^1 + 2\pi i\,{}^tv(\lambda^1+c^1)\big) \quad \text{(by Lemma 5.1.b))} \\
&= e\big(\tfrac{\pi}{2}B(v,v) - \pi i\,{}^tc^1c^2\big) \sum_{\lambda^1\in\mathbb{Z}^g} e\big(\pi i\,{}^t(\lambda^1+c^1)Z(\lambda^1+c^1) \\
&\quad + 2\pi i\,{}^t(v+c^2)(\lambda^1+c^1)\big)\,. \qquad \square
\end{aligned}$$

Using Lemma 5.2 one can translate all properties of canonical theta functions discussed in Chapter 3 into terms of $\vartheta\left[\begin{smallmatrix}c^1\\c^2\end{smallmatrix}\right]$. For later use we compile some of these here:

(5.3) Remarks. Fix $Z \in \mathfrak{H}_g$.

a) For every $c^1, c^2 \in \mathbb{R}^g$ the function $\vartheta\left[\begin{smallmatrix}c^1\\c^2\end{smallmatrix}\right](\cdot, Z)$ is a theta function with respect to the lattice $Z\mathbb{Z}^g \oplus \mathbb{Z}^g$ with functional equation

$$\begin{aligned}
&\vartheta\left[\begin{smallmatrix}c^1\\c^2\end{smallmatrix}\right](v + Z\lambda^1 + \lambda^2, Z) = \\
&\qquad = e\big(2\pi i({}^tc^1\lambda^2 - {}^tc^2\lambda^1) - \pi i\,{}^t\lambda^1 Z\lambda^1 - 2\pi i\,{}^tv\lambda^1\big)\vartheta\left[\begin{smallmatrix}c^1\\c^2\end{smallmatrix}\right](v,Z)
\end{aligned}$$

for all $v \in \mathbb{C}^g$ and $\lambda^1, \lambda^2 \in \mathbb{Z}^g$.

b) For all $c^1, c^2 \in \mathbb{R}^g$ the lattice $Z\mathbb{Z}^g \oplus \mathbb{Z}^g$ is a maximal lattice in $\mathbb{C}^g$ for which $\vartheta\left[\begin{smallmatrix}c^1\\c^2\end{smallmatrix}\right](\cdot, Z)$ satisfies a functional equation.

To see this, suppose Λ is a lattice in $\mathbb{C}^g$ containing $Z\mathbb{Z}^g \oplus \mathbb{Z}^g$ and for every $\lambda \in \Lambda$ there is a holomorphic function $f_\lambda\colon \mathbb{C}^g \to \mathbb{C}^*$ such that $\vartheta\left[\begin{smallmatrix}c^1\\c^2\end{smallmatrix}\right](v+\lambda, Z) = f_\lambda(v)\vartheta\left[\begin{smallmatrix}c^1\\c^2\end{smallmatrix}\right](v,Z)$ for all $v \in \mathbb{C}^g$. The family $\{f_\lambda\}$ is necessarily a factor of automorphy with respect to the lattice Λ. This means that there is a line bundle M on $\mathbb{C}^g/\Lambda$ such that $L \simeq p^*M$, where $p\colon X_Z \to \mathbb{C}^g/\Lambda$ denotes the isogeny induced by the inclusion $Z\mathbb{Z}^g \oplus \mathbb{Z}^g \subset \Lambda$. But L defines a principal polarization. So $\deg p = 1$ by Corollary 3.6.6 and hence $\Lambda = Z\mathbb{Z}^g \oplus \mathbb{Z}^g$.

c) For all $c^1, c^2 \in \mathbb{R}^g$

$$\vartheta\begin{bmatrix} c^1+l^1 \\ c^2+l^2 \end{bmatrix} = \vartheta\begin{bmatrix} c^1 \\ c^2 \end{bmatrix} \Longleftrightarrow l^1, l^2 \in \mathbb{Z}^g .$$

This reflects the fact that in the principally polarized case the characteristic $c \in \mathbb{C}^g$ of a line bundle on X_Z is uniquely determined modulo the lattice $Z\mathbb{Z}^g \oplus \mathbb{Z}^g$.

d) Define $e\begin{bmatrix} 0 \\ 0 \end{bmatrix}\colon \Lambda_Z \times \mathbb{C}^g \to \mathbb{C}^*$ by

$$e\begin{bmatrix} 0 \\ 0 \end{bmatrix}(Z\lambda^1 + \lambda^2, v) = e(-\pi i\, {}^t\lambda^1 Z \lambda^1 - 2\pi i\, {}^t v \lambda^1)$$

for all $\lambda^1, \lambda^2 \in \mathbb{Z}^g$ and $v \in \mathbb{C}^g$. The notation $\begin{bmatrix} 0 \\ 0 \end{bmatrix}$ indicates that $e\begin{bmatrix} 0 \\ 0 \end{bmatrix}$ is the classical factor of automorphy for the line bundle of characteristic zero in $\operatorname{Pic}^{H_Z}(X_Z)$ with respect to the decomposition $\Lambda_Z = Z\mathbb{Z}^g \oplus \mathbb{Z}^g$.

Let $D = \operatorname{diag}(d_1, \ldots, d_g)$ be a type. Then $\vartheta\begin{bmatrix} c^1 \\ c^2 \end{bmatrix}(\cdot, Z)$ is a theta function with factor $e\begin{bmatrix} 0 \\ 0 \end{bmatrix}$ with respect to the lattice $Z\mathbb{Z}^g \oplus D\mathbb{Z}^g$ if and only if $\binom{c^1}{c^2} \in D^{-1}\mathbb{Z}^g \oplus \mathbb{Z}^g$. According to Theorem 3.2.7 the functions $\vartheta\begin{bmatrix} c_0 \\ 0 \end{bmatrix}(\cdot, Z), \ldots, \vartheta\begin{bmatrix} c_N \\ 0 \end{bmatrix}(\cdot, Z)$, where $c_0, \ldots, c_N$ is a set of representatives of $D^{-1}\mathbb{Z}^g/\mathbb{Z}^g$, form a basis of the vector space of classical theta functions for the line bundle on $X_{(Z,D)} = \mathbb{C}^g/(Z\mathbb{Z}^g \oplus D\mathbb{Z}^g)$ determined by the factor $e\begin{bmatrix} 0 \\ 0 \end{bmatrix}$.

e) Corollary 3.2.9 translates to

$$\vartheta\begin{bmatrix} c^1 \\ c^2 \end{bmatrix}(v, Z) = e\big(\pi i\, {}^t c^1 Z c^1 + 2\pi i\, {}^t c^1 (v + c^2) \vartheta\begin{bmatrix} 0 \\ 0 \end{bmatrix}(v + Zc^1 + c^2, Z)\big)$$

for all $c^1, c^2 \in \mathbb{R}^g$ and $v \in \mathbb{C}^g$.

(5.4) Proposition. *The Riemann theta function $\vartheta\begin{bmatrix} c^1 \\ c^2 \end{bmatrix}$ is holomorphic on the complex manifold $\mathbb{C}^g \times \mathfrak{H}_g$ for every $c^1, c^2 \in \mathbb{R}^g$.*

Proof. It suffices to show that the series (1) converges absolutely and uniformly on every compact set in $\mathbb{C}^g \times \mathfrak{H}_g$. Suppose $K \subset \mathbb{C}^g \times \mathfrak{H}_g$ is compact. There is a point (v_0, Z_0) in K such that $\operatorname{Im} Z \geq \operatorname{Im} Z_0$ for all $(v, Z) \in K$. Then for every $(v, Z) \in K$

$$\sum_{l \in \mathbb{Z}^g} \left| e(\pi i\, {}^t(l+c^1)\, Z(l+c^1) + 2\pi i\, {}^t(v+c^2)(l+c^1)) \right|$$
$$\leq \sum_{l \in \mathbb{Z}^g} e(-\pi\, {}^t(l+c^1) \operatorname{Im} Z_0 (l+c^1) - 2\pi\, {}^t(\operatorname{Im} v)(l+c^1)) .$$

According to Lemmas 3.2.4 and 5.2 the series of $\vartheta\begin{bmatrix} c^1 \\ c^2 \end{bmatrix}(v, Z_0)$ converges absolutely and uniformly on compact sets in $\mathbb{C}^g$. This means that the series on the right hand side converges uniformly on $p_{\mathbb{C}^g}(K) \subset \mathbb{C}^g$, which completes the proof. □

Since $\vartheta\left[\begin{smallmatrix} c^1 \\ c^2 \end{smallmatrix}\right]$ is holomorphic on $\mathbb{C}^g \times \mathfrak{H}_g$, we can differentiate the series (1) term by term. In this way it is easy to derive the well known differential equation for $\vartheta\left[\begin{smallmatrix} c^1 \\ c^2 \end{smallmatrix}\right]$. In fact, for every $v = {}^t(v_1, \ldots, v_g) \in \mathbb{C}^g$ and $Z = (z_{jk}) \in \mathfrak{H}_g$

$$\frac{\partial^2 \vartheta\left[\begin{smallmatrix} c^1 \\ c^2 \end{smallmatrix}\right]}{\partial v_j\, \partial v_k}(v, Z) = \begin{cases} 2\pi i \dfrac{\partial \vartheta\left[\begin{smallmatrix} c^1 \\ c^2 \end{smallmatrix}\right]}{\partial z_{jk}}(v, Z) & \text{if } j \neq k \\ 4\pi i \dfrac{\partial \vartheta\left[\begin{smallmatrix} c^1 \\ c^2 \end{smallmatrix}\right]}{\partial z_{jj}}(v, Z) & \phantom{\text{if }} j = k \end{cases}$$

This gives

(5.5) Proposition. *For every symmetric matrix $(s_{jk}) \in M_g(\mathbb{C})$ and all vectors $c^1, c^2 \in \mathbb{R}^g$*

$$\sum_{j,k=1}^{g} s_{jk} \frac{\partial^2 \vartheta\left[\begin{smallmatrix} c^1 \\ c^2 \end{smallmatrix}\right]}{\partial v_j\, \partial v_k} = 4\pi i \sum_{1 \leq j \leq k \leq g} s_{jk} \frac{\partial \vartheta\left[\begin{smallmatrix} c^1 \\ c^2 \end{smallmatrix}\right]}{\partial z_{jk}}.$$

For a suitable choice of (s_{jk}) and certain restrictions for the values of Z this equation appears in physics. It describes the conduction of heat. Therefore the differential equation of Proposition 5.5 is called the *heat equation.*

In Section 3.4 we introduced the hermitian metric $e(-\pi H(\cdot,\cdot))$ on the line bundle $L = L(H, \chi)$. In particular, for all canonical theta functions $\vartheta_1, \vartheta_2 \in H^0(L)$ the function $\langle \vartheta_1, \vartheta_2 \rangle : \mathbb{C}^g \to \mathbb{C}$, $v \mapsto \vartheta_1(v)\overline{\vartheta_2(v)}e(-\pi H(v,v))$ is $\mathcal{C}^\infty$ on $\mathbb{C}^g$ and periodic with respect to Λ_Z (see Section 3.4). Denote by dv the volume element of $\mathbb{C}^g$ (respectively X_Z) corresponding to the symplectic basis of Λ_Z given by the columns of the period matrix $(Z, \mathbb{1})$, i.e.

$$dv = dv_1 \wedge \cdots \wedge dv_{2g}$$

where $v_1, \ldots, v_{2g}$ are the real coordinate functions with respect to this basis. Then

$$(\vartheta_1, \vartheta_2) := \int_{\mathbb{C}^g/\Lambda_Z} \langle \vartheta_1, \vartheta_2 \rangle dv$$

defines a hermitian inner product on the vector space of canonical theta functions $H^0(L)$. We want to compute the norm $\|\vartheta_Z^c\|$ of ϑ_Z^c with respect to this inner product.

(5.6) Proposition. $\|\vartheta_Z^c\| = (\det 2 \operatorname{Im} Z)^{-\frac{1}{4}}$.

Proof. From equation 3.2(3) we conclude $\langle \vartheta_Z^c, \vartheta_Z^c \rangle(v) = \langle \vartheta_Z^0, \vartheta_Z^0 \rangle(v + c)$ for all $v \in \mathbb{C}^g$. Hence it suffices to prove the assertion for $c = 0$. Using Lemmas 5.1 and 5.2 we get

$$\begin{aligned}\langle\vartheta^0,\vartheta^0\rangle(v) &= \vartheta\left[\begin{smallmatrix}0\\0\end{smallmatrix}\right](v,Z)\overline{\vartheta\left[\begin{smallmatrix}0\\0\end{smallmatrix}\right](v,Z)}\cdot e\big(-\tfrac{\pi}{2}(H-B)(v,v)-\tfrac{\pi}{2}\overline{(H-B)(v,v)}\big)\\ &= \sum_{l\in\mathbb{Z}^g} e(\pi i\,{}^t lZl+2\pi i\,{}^t vl)\\ &\qquad\cdot\sum_{m\in\mathbb{Z}^g} e(-\pi i\,{}^t m\bar{Z}m-2\pi i\,{}^t\bar{v}m)\cdot e(-2\pi\,{}^t v^1\operatorname{Im}Zv^1)\\ &= \sum_{l,m\in\mathbb{Z}^g} e\big(\pi i\,{}^t lZl-\pi i\,{}^t m\bar{Z}m+2\pi i\,{}^t v^1(Zl-\bar{Z}m)\big)\\ &\qquad\cdot e\big(-2\pi\,{}^t v^1\operatorname{Im}Zv^1\big)\cdot e\big(2\pi i\,{}^t v^2(l-m)\big)\ .\end{aligned}$$

Since this series converges absolutely and uniformely on every compact set in $\mathbb{C}^g$, we get, writing $dv^1=dv_1\wedge\cdots\wedge dv_g$ and $dv^2=dv_{g+1}\wedge\cdots\wedge dv_{2g}$,

$$\begin{aligned}\|\vartheta^0\|^2 &= \sum_{l,m\in\mathbb{Z}^g}\int_{\mathbb{R}^g/\mathbb{Z}^g} e\big(\pi i\,{}^t lZl-\pi i\,{}^t m\bar{Z}m+2\pi i\,{}^t v^1(Zl-\bar{Z}m)\big)\\ &\qquad\cdot e\big(-2\pi\,{}^t v^1\operatorname{Im}Zv^1\big)dv^1\int_{\mathbb{R}^g/\mathbb{Z}^g} e\big(2\pi i\,{}^t v^2(l-m)\big)dv^2\ .\end{aligned}$$

But $\displaystyle\int_{\mathbb{R}^g/\mathbb{Z}^g} e\big(2\pi i\,{}^t v^2(l-m)\big)dv^2=\begin{cases}1 & \text{if } l=m\\ 0 & \text{if } l\neq m\end{cases}$ such that

$$\begin{aligned}\|\vartheta^0\|^2 &= \sum_{l\in\mathbb{Z}^g}\int_{\mathbb{R}^g/\mathbb{Z}^g} e\big(\pi i\,{}^t l(Z-\bar{Z})l+2\pi i\,{}^t v^1(Z-\bar{Z})l-2\pi\,{}^t v^1\operatorname{Im}Zv^1\big)dv^1\\ &= \sum_{l\in\mathbb{Z}^g}\int_{\mathbb{R}^g/\mathbb{Z}^g} e\big(-{}^t(l+v^1)(2\pi\operatorname{Im}Z)(l+v^1)\big)dv^1\\ &= \int_{\mathbb{R}^g} e\big(-{}^t v^1(2\pi\operatorname{Im}Z)v^1\big)dv^1=(\det 2\operatorname{Im}Z)^{-\frac{1}{2}}\ .\end{aligned}$$

The last equation can be easily deduced from the well known formula $\int_{-\infty}^{+\infty}e(-x^2)=\pi^{\frac{1}{2}}$. □

§ 6 The Theta Transformation Formula, Final Version

In Section 4 we proved a preliminary version of the theta transformation formula: a relation between canonical theta functions. To get the following final version, we translate it into terms of classical theta functions and compute the factor $C(Z,M,0)$.

(6.1) Theta Transformation Formula. *For all* $(v,Z)\in\mathbb{C}^g\times\mathfrak{H}_g$, $c=Zc^1+c^2$ *with* $c^1,c^2\in\mathbb{R}^g$ *and* $M=\left(\begin{smallmatrix}\alpha&\beta\\\gamma&\delta\end{smallmatrix}\right)\in G_{\mathbb{1}}=\operatorname{Sp}_{2g}(\mathbb{Z})$

$$\vartheta\left[\begin{smallmatrix}M[c]^1\\M[c]^2\end{smallmatrix}\right]\big({}^t(\gamma Z+\delta)^{-1}v,M(Z)\big)=$$

$$=\kappa(M)\det(\gamma Z+\delta)^{\frac{1}{2}}e\big(\pi ik(M,c^1,c^2)+\pi i\,{}^t v(\gamma Z+\delta)^{-1}\gamma v\big)\vartheta\left[\begin{smallmatrix}c^1\\c^2\end{smallmatrix}\right](v,Z)\ ,$$

where $k(M,c^1,c^2)={}^t(\delta c^1-\gamma c^2)(-\beta c^1+\alpha c^2+(\alpha\,{}^t\beta)_0)-{}^t c^1c^2$ *and* $\kappa(M)\in\mathbb{C}_1$ *is a constant with the same sign ambiguity as* $\det(\gamma Z+\delta)^{\frac{1}{2}}$.

Proof of 6.1. Step I: Translation of equation 4(1) into terms of classical theta functions.

With the notation of Section 4 let ϑ_Z^c and $\vartheta_{M(Z)}^{M[c]}$ be the canonical theta functions corresponding to $\vartheta\left[\begin{smallmatrix}c^1\\c^2\end{smallmatrix}\right](\cdot,Z)$ and $\vartheta\left[\begin{smallmatrix}M[c]^1\\M[c]^2\end{smallmatrix}\right](\cdot,M(Z))$. According to Lemma 5.2 and Lemma 5.1 a)

$$\vartheta_Z^c(v)=e(\tfrac{\pi}{2}\,{}^tv(\operatorname{Im}Z)^{-1}v-\pi i\,{}^tc^1c^2)\vartheta\left[\begin{smallmatrix}c^1\\c^2\end{smallmatrix}\right](v,Z)$$
$$\vartheta_{M(Z)}^{M[c]}(v)=e(\tfrac{\pi}{2}\,{}^tv(\operatorname{Im}M(Z))^{-1}v-\pi i\,{}^tM[c]^1M[c]^2)\vartheta\left[\begin{smallmatrix}M[c]^1\\M[c]^2\end{smallmatrix}\right](v,M(Z))$$

for all $v\in\mathbb{C}^g$. Since $A={}^t(\gamma Z+\delta)$ (by Remark 1.4), equation 4(1) gives $\vartheta_{M(Z)}^{M[c]}({}^t(\gamma Z+\delta)^{-1}v)=C(Z,M,c)^{-1}\,\vartheta_Z^c(v)$. Replacing now canonical theta functions by the classical ones, we get using $A^*H_Z=H_{M(Z)}$, which reads in terms of matrices $(\operatorname{Im}M(Z))^{-1}={}^tA(\operatorname{Im}Z)^{-1}\overline{A}$,

$$\begin{aligned}
&\vartheta\left[\begin{smallmatrix}M[c]^1\\M[c]^2\end{smallmatrix}\right]({}^t(\gamma Z+\delta)^{-1}v,M(Z))\\
&\quad=C(Z,M,c)^{-1}e(\pi i\,{}^tM[c]^1M[c]^2-\pi i\,{}^tc^1c^2)\\
&\qquad\cdot e(-\tfrac{\pi}{2}\,{}^tv\,{}^tA^{-1}(\operatorname{Im}M(Z))^{-1}A^{-1}v+\tfrac{\pi}{2}\,{}^tv(\operatorname{Im}Z)^{-1}v)\vartheta\left[\begin{smallmatrix}c^1\\c^2\end{smallmatrix}\right](v,Z)\\
&\quad=C(Z,M,c)^{-1}e(\pi i\,{}^tM[c]^1M[c]^2-\pi i\,{}^tc^1c^2)\\
&\qquad\cdot e(-\tfrac{\pi}{2}\,{}^tv(\operatorname{Im}Z)^{-1}(\overline{A}A^{-1}-\mathbb{1})v)\vartheta\left[\begin{smallmatrix}c^1\\c^2\end{smallmatrix}\right](v,Z)\\
&\quad=C(Z,M,c)^{-1}\\
&\qquad\cdot e(\pi i\,{}^tM[c]^1M[c]^2-\pi i\,{}^tc^1c^2+\pi i\,{}^tv(\gamma Z+\delta)^{-1}\gamma v)\vartheta\left[\begin{smallmatrix}c^1\\c^2\end{smallmatrix}\right](v,Z)
\end{aligned}$$

For the last equation we used that $(\operatorname{Im}Z)^{-1}(\overline{A}A^{-1}-\mathbb{1})=-2i\,{}^t\gamma\,{}^t(\gamma Z+\delta)^{-1}=-2i(\gamma Z+\delta)^{-1}\gamma$ by Lemma 2.1 iii). Write $\binom{d^1}{d^2}={}^tM^{-1}\binom{c^1}{c^2}$. So $d^1=\delta c^1-\gamma c^2$ and $d^2=-\beta c^1+\alpha c^2$. Then by Lemma 4.1 b) we have $M[c]^1=d^1+\frac{1}{2}(\gamma\,{}^t\delta)_0$ and $M[c]^2=d^2+\frac{1}{2}(\alpha\,{}^t\beta)_0$. Together with Lemma 4.2 this implies

$$\begin{aligned}
&C(Z,M,c)^{-1}e(\pi i\,{}^tM[c]^1M[c]^2-\pi i\,{}^tc^1c^2)\\
&\quad=C(Z,M,0)^{-1}e\Big(-\tfrac{1}{2}\pi i\,{}^t\begin{pmatrix}(\gamma\,{}^t\delta)_0\\(\alpha\,{}^t\beta)_0\end{pmatrix}\begin{pmatrix}0&\mathbb{1}\\-\mathbb{1}&0\end{pmatrix}\begin{pmatrix}d^1\\d^2\end{pmatrix}\Big)\\
&\qquad\cdot e(\pi i\,{}^t(d^1+\tfrac{1}{2}(\gamma\,{}^t\delta)_0)(d^2+\tfrac{1}{2}(\alpha\,{}^t\beta)_0)-\pi i\,{}^tc^1c^2)\\
&\quad=C(Z,M,0)^{-1}e\big(\tfrac{1}{4}\pi i\,{}^t(\gamma\,{}^t\delta)_0(\alpha\,{}^t\beta)_0+\pi i\,{}^td^1(d^2+(\alpha\,{}^t\beta)_0)-\pi i\,{}^tc^1c^2\big)\,.
\end{aligned}$$

So finally we obtain with $k(M,c^1,c^2)$ as above

$$\begin{aligned}
&\vartheta\left[\begin{smallmatrix}M[c]^1\\M[c]^2\end{smallmatrix}\right]({}^t(\gamma Z+\delta)^{-1}v,M(Z))\\
&\quad=C(Z,M,0)^{-1}e\big(\tfrac{1}{4}\pi i\,{}^t(\gamma\,{}^t\delta)_0(\alpha\,{}^t\beta)_0\big) \qquad (1)\\
&\qquad\cdot e(\pi ik(M,c^1,c^2)+\pi i\,{}^tv(\gamma Z+\delta)^{-1}\gamma v)\vartheta\left[\begin{smallmatrix}c^1\\c^2\end{smallmatrix}\right](v,Z)\,.
\end{aligned}$$

Step II: The factor $C(Z,M,0)^{-1}e(\frac{1}{4}\pi i\,{}^t(\gamma\,{}^t\delta)_0(\alpha\,{}^t\beta)_0)$.
According to equation 4(1) we have

$$|C(Z,M,0)^{-1}|^2 = \|A^*\vartheta_Z^0\|^{-2}\cdot\|\vartheta_{M(Z)}^{M[0]}\|^2 .$$

But $\|A^*\vartheta_Z^0\| = \|\vartheta_Z^0\|$, since the corresponding change of variables of $\mathbb{R}^{2g}$ is given by the matrix ${}^tM$, which has determinant 1. Applying Proposition 5.6 we get

$$\begin{aligned}|C(Z,M,0)^{-1}|^2 &= (\det 2\operatorname{Im} Z)^{\frac{1}{2}}(\det 2\operatorname{Im} M(Z))^{-\frac{1}{2}}\\ &= (\det 2\operatorname{Im} Z)^{\frac{1}{2}}(\det 2\overline{A}^{-1}\operatorname{Im} Z\,{}^tA^{-1})^{-\frac{1}{2}}\\ &= |\det A| = |\det(\gamma Z+\delta)| .\end{aligned}$$

Now equation (1) implies that $C(Z,M,0)^{-1}$ depends holomorphically on Z such that $C(Z,M,0)^{-1}$ and $(\det(\gamma Z+\delta))^{\frac{1}{2}}$ differ only by a constant. This gives

$$C(Z,M,0)^{-1}e(\tfrac{1}{4}\pi i\,{}^t(\gamma\,{}^t\delta)_0(\alpha\,{}^t\beta)_0) = \kappa(M)\det(\gamma Z+\delta)^{\frac{1}{2}} ,$$

where the factor $\kappa(M)\in\mathbb{C}_1$ depends only on M and the chosen root of $\det(\gamma Z+\delta)^{\frac{1}{2}}$. □

As for the constant $\kappa(M)$, one can say more (see Exercise 9). In particular $\kappa(M)$ is an 8-th root of unity for every $M\in \mathrm{Sp}_{2g}(\mathbb{Z})$.

§ 7 The Universal Family

Fix a type $D=\operatorname{diag}(d_1,\dots,d_g)$. In Proposition 1.2 we saw that the Siegel upper half space $\mathfrak{H}_g$ is a moduli space for polarized abelian varieties of type D with symplectic basis. In fact, we associated to every $Z\in\mathfrak{H}_g$ the triplet $(X_Z,H_Z,\{\text{columns of }(Z,D)\})$. In this section we construct a holomorphic family over $\mathfrak{H}_g$ parametrizing these triplets.

We need some notation. For any $Z\in\mathfrak{H}_g$ consider the isomorphism of $\mathbb{R}$-vector spaces

$$j_Z\colon \mathbb{R}^{2g}\to\mathbb{C}^g,\quad x\mapsto (Z,\mathbb{1})x .$$

If Λ_D denotes the lattice

$$\Lambda_D = \left(\begin{smallmatrix}\mathbb{1} & 0\\ 0 & D\end{smallmatrix}\right)\mathbb{Z}^{2g}$$

in $\mathbb{R}^{2g}$, then $j_Z(\Lambda_D)$ is just the lattice $\Lambda_Z=(Z,D)\mathbb{Z}^{2g}$ in $\mathbb{C}^g$ determined by Z. In other words, if $f_1,\dots,f_{2g}$ denotes the standard basis of $\mathbb{R}^{2g}$ and $\lambda_1,\dots,\lambda_g,\mu_1,\dots,\mu_g$ the symplectic basis of Λ_Z associated to Z, then we have

$$\lambda_\nu = j_Z(f_\nu) \quad \text{and} \quad \mu_\nu = j_Z(d_\nu f_{g+\nu})$$

for $1 \leq \nu \leq g$. The lattice Λ_D acts freely and properly discontinuously on $\mathbb{C}^g \times \mathfrak{H}_g$ by

$$l(v, Z) = (v + j_Z(l), Z)$$

for all $l \in \Lambda_D$ and $(v, Z) \in \mathbb{C}^g \times \mathfrak{H}_g$. According to Corollary A.7 the quotient

$$\mathfrak{X}_D := (\mathbb{C}^g \times \mathfrak{H}_g)/\Lambda_D$$

is a complex manifold. Let $p: \mathfrak{X}_D \to \mathfrak{H}_g$ denote the natural projection map. For every $Z \in \mathfrak{H}_g$ the fibre $p^{-1}(Z) = \mathbb{C}^g / j_Z(\Lambda_D) = X_Z$ is the abelian variety associated to Z. Define a map $e_{\Lambda_D}: \Lambda_D \times (\mathbb{C}^g \times \mathfrak{H}_g) \to \mathbb{C}^*$ by

$$e_{\Lambda_D}(l, (v, Z)) = e(-\pi i {}^t l^1 Z l^1 - 2\pi i {}^t v l^1) \ ,$$

where for any $l \in \mathbb{R}^{2g}$ we denote by $l^1 \in \mathbb{R}^g$ the vector of the first g components of l. The map e_{Λ_D} is a cocycle in $Z^1(\Lambda_D, H^0(\mathcal{O}^*_{\mathbb{C}^g \times \mathfrak{H}_g}))$, since it is holomorphic in (v, Z) and satisfies the cocycle relation $e_{\Lambda_D}(l + m, (v, Z)) = e_{\Lambda_D}(l, (v + j_Z(m), Z)) e_{\Lambda_D}(m, (v, Z))$ for all $l, m \in \Lambda_D$, $(v, Z) \in \mathbb{C}^g \times \mathfrak{H}_g$. According to Proposition B.1 the cocycle e_D defines a holomorphic line bundle $\mathcal{L}$ on $\mathfrak{X}_D$.

(7.1) Lemma. *For any $Z \in \mathfrak{H}_g$ the restriction of the line bundle $\mathcal{L}$ on $\mathfrak{X}_D$ to the fibre X_Z defines the polarization H_Z. To be more precise,*

$$\mathcal{L}|X_Z \simeq L(H_Z, \chi_0) \ ,$$

the line bundle of characteristic 0 with respect to the decomposition $\Lambda_Z = Z\mathbb{Z}^g \oplus D\mathbb{Z}^g$ for H_Z.

Proof. It suffices to show that the restriction of e_{Λ_D} to $\Lambda_D \times \mathbb{C}^g \times \{Z\}$ is a factor for $L(H_Z, \chi_0)$. But $e_{\Lambda_D}(\cdot, (\cdot, Z)) = e\left[\begin{smallmatrix}0\\0\end{smallmatrix}\right]$, the classical factor of automorphy for $L(H_Z, \chi_0)$ (see Remark 5.3 d)). □

Define holomorphic maps $\lambda_\nu, \mu_\nu: \mathfrak{H}_g \to \mathbb{C}^g$ by

$$\lambda_\nu(Z) = j_Z(f_\nu) \quad \text{and} \quad \mu_\nu(Z) = j_Z(d_\nu f_{g+\nu})$$

for $1 \leq \nu \leq g$. For every $Z \in \mathfrak{H}_g$ the set $\{\lambda_1(Z), \ldots, \mu_g(Z)\}$ is a symplectic basis of Λ_Z for H_Z. Summarizing we can say that

$$(p: \mathfrak{X}_D \to \mathfrak{H}_g, \mathcal{L}, \{\lambda_1, \ldots, \mu_g\})$$

is a holomorphic family parametrizing the set of polarized abelian varieties of type D with symplectic basis.

Every sublinear system of $|\mathcal{L}|$ induces a rational map of $\mathfrak{X}_D$ into some projective space. We want to single out a sublinear system $|U_D|$ with the property that the restriction $|U_D||X_Z$ coincides with the complete linear system $|\mathcal{L}|X_Z| = |L(H_Z, \chi_0)|$ for every fibre $X_Z = p^{-1}(Z)$. For this consider $H^0(\mathcal{L})$ as the vector space of holomorphic functions $f\colon \mathbb{C}^g \times \mathfrak{H}_g \to \mathbb{C}$ satisfying the functional equation $f(v + j_Z(l), Z) = e_{\Lambda_D}(l,(v,Z))\, f(v,Z)$ for all $l \in \Lambda_D$ and $(v,Z) \in \mathbb{C}^g \times \mathfrak{H}_g$ (see Appendix B). Proposition 5.4 suggests that the classical Riemann theta function might be a global section of $\mathcal{L}$.

Define an alternating form J on $\mathbb{R}^{2g}$ by the matrix $\left(\begin{smallmatrix} 0 & \mathbb{1} \\ -\mathbb{1} & 0 \end{smallmatrix}\right)$ with respect to the standard basis $f_1, \ldots, f_{2g}$. We observe that for every $Z \in \mathfrak{H}_g$ the isomorphism j_Z is defined in such a way that

$$J = j_Z^*(\operatorname{Im} H_Z)\,. \tag{1}$$

Define
$$\Lambda_D^\perp = \left(\begin{smallmatrix} D^{-1} & 0 \\ 0 & \mathbb{1} \end{smallmatrix}\right) \mathbb{Z}^{2g}\,. \tag{2}$$

So $\Lambda_D^\perp$ is the orthogonal complement of the lattice Λ_D with respect to the form $e(2\pi i J)$.

(7.2) Lemma. $\vartheta\left[\begin{smallmatrix} c^1 \\ c^2 \end{smallmatrix}\right] \in H^0(\mathcal{L})$ *for every* $\left(\begin{smallmatrix} c^1 \\ c^2 \end{smallmatrix}\right) \in \Lambda_D^\perp$.

For the proof it suffices to check the functional behaviour of $\vartheta\left[\begin{smallmatrix} c^1 \\ c^2 \end{smallmatrix}\right]$ with respect to Λ_D, which is immediate. This follows also from Remark 5.3 d). □

Define U_D to be the $\mathbb{C}$-vector space spanned by the theta functions $\vartheta\left[\begin{smallmatrix} c^1 \\ c^2 \end{smallmatrix}\right]$, with $\left(\begin{smallmatrix} c^1 \\ c^2 \end{smallmatrix}\right) \in \Lambda_D^\perp$. Since the characteristic of $\vartheta\left[\begin{smallmatrix} c^1 \\ c^2 \end{smallmatrix}\right]$ is determined modulo $\mathbb{Z}^{2g}$, U_D is a subvector space of $H^0(\mathcal{L})$ of finite dimension. Denote by $\{c_0, \ldots, c_N\}$ a set of representatives of $D^{-1}\mathbb{Z}^g/\mathbb{Z}^g$. Then we have

(7.3) Proposition. *a)* $\{\vartheta\left[\begin{smallmatrix} c_\nu \\ 0 \end{smallmatrix}\right], 0 \le \nu \le N\}$ *is a basis for* U_D.
b) For every $Z \in \mathfrak{H}_g$ *the restriction of the linear system* $|U_D|$ *to the fibre* X_Z *coincides with the complete linear system* $|L(H_Z, \chi_0)|$.

Proof. a) and b) follow from Remarks 5.3 c) and d) respectively. □

Denote by $\varphi_D\colon \mathfrak{X}_D \to \mathbb{P}_N$ the rational map associated to the linear system $|U_D|$. By definition $\varphi_D(\bar{v}, Z) = (\vartheta\left[\begin{smallmatrix} c_0 \\ 0 \end{smallmatrix}\right](v,Z) : \ldots : \vartheta\left[\begin{smallmatrix} c_N \\ 0 \end{smallmatrix}\right](v,Z))$ for every $(\bar{v}, Z) \in \mathfrak{X}_D$. According to Proposition 7.3 b) for every $Z \in \mathfrak{H}_g$ the restriction of φ_D to the fibre X_Z is just the rational map $\varphi_{L(H_Z,\chi_0)}\colon X_Z \to \mathbb{P}_N$ associated to the line bundle $L(H_Z, \chi_0)$.

Let $s_0 : \mathfrak{H}_g \to \mathfrak{X}_D$ denote the zero section and define a rational map ψ_D as the composition $\psi_D = \varphi_D s_0$. Then we have the following commutative diagram

$$\begin{array}{ccc} \mathfrak{X}_D & & \\ {\scriptstyle p}\downarrow \uparrow{\scriptstyle s_0} & \searrow^{\varphi_D} & \\ \mathfrak{H}_g & \xrightarrow{\psi_D} & \mathbb{P}_N \end{array} .$$

By definition ψ_D is given by means of *theta null values*:

$$\psi_D(Z) = (\vartheta\left[\begin{smallmatrix} c_0 \\ 0 \end{smallmatrix}\right](0, Z) : \ldots : \vartheta\left[\begin{smallmatrix} c_N \\ 0 \end{smallmatrix}\right](0, Z)).$$

Recall the action $(M, Z) \mapsto M(Z)$ of $G_D \subset \mathrm{Sp}_{2g}(\mathbb{R})$ on $\mathfrak{H}_g$ and the moduli space $\mathcal{A}_D = \mathfrak{H}_g / G_D$. In general ψ_D does not factorize via $\mathfrak{H}_g \to \mathcal{A}_D$. However we will see that for a suitable subgroup G of G_D the map ψ_D factorizes via $\mathfrak{H}_g \to \mathfrak{H}_g / G$ and induces an embedding $\mathfrak{H}_g / G \hookrightarrow \mathbb{P}_N$.

§ 8 The Action of the Symplectic Group

In Section 2 we introduced an action of the symplectic group on the Siegel upper half space $\mathfrak{H}_g$, namely $Z \mapsto M(Z) = (\alpha Z + \beta)(\gamma Z + \delta)^{-1}$ for $M = \left(\begin{smallmatrix} \alpha & \beta \\ \gamma & \delta \end{smallmatrix}\right) \in \mathrm{Sp}_{2g}(\mathbb{R})$. We will see next that for suitable subgroups this action extends to an action on the manifold $\mathfrak{X}_D$.

Recall the group $G_D = \{M \in \mathrm{Sp}_{2g}(\mathbb{R}) \mid {}^tM \Lambda_D = \Lambda_D\}$. According to Proposition 1.3 for every $Z \in \mathfrak{H}_g$ and $M = \left(\begin{smallmatrix} \alpha & \beta \\ \gamma & \delta \end{smallmatrix}\right) \in G_D$ there is an isomorphism $X_{M(Z)} \to X_Z$ given by the equation ${}^t(\gamma Z + \delta)(M(Z), \mathbb{1}) = (Z, \mathbb{1}){}^tM$ (see Remark 1.4). Denote $M_Z = {}^t(\gamma Z + \delta)^{-1}$ which is the analytic representation of the inverse map $X_Z \to X_{M(Z)}$. Then the following diagram commutes

$$\begin{array}{ccc} \mathbb{R}^{2g} & \xrightarrow{j_Z} & \mathbb{C}^g \\ {\scriptstyle {}^tM^{-1}}\downarrow & & \downarrow{\scriptstyle M_Z} \\ \mathbb{R}^{2g} & \xrightarrow{j_{M(Z)}} & \mathbb{C}^g \end{array} . \tag{1}$$

This immediately implies that

$$M(v, Z) := (M_Z v, M(Z)) \tag{2}$$

defines an action of G_D on $\mathbb{C}^g \times \mathfrak{H}_g$. Actually, (2) gives an action of the whole symplectic group $\mathrm{Sp}_{2g}(\mathbb{R})$, but we do not need this fact.

(8.1) Lemma. *The action of the group G_D on $\mathbb{C}^g \times \mathfrak{H}_g$ descends to an action on the family of abelian varieties $p: \mathfrak{X}_D \to \mathfrak{H}_g$.*

Here an *action on the family of abelian varieties* $p: \mathfrak{X}_D \to \mathfrak{H}_g$ means an action $\tau: G_D \times \mathfrak{X}_D \to \mathfrak{X}_D$ of G_D on the variety $\mathfrak{X}_D$ in such a way that the restriction $\tau|\{M\} \times p^{-1}(Z)$ is an isomorphism of the abelian varieties X_Z and $X_{M(Z)}$ for every $M \in G_D$ and $Z \in \mathfrak{H}_g$.

Proof. For all $Z \in \mathfrak{H}_g$ and $M \in G_D$ we have, applying diagram (1),

$$M(j_Z(\Lambda_D), Z) = (j_{M(Z)}({}^tM^{-1}\Lambda_D), M(Z)) = (j_{M(Z)}(\Lambda_D), M(Z)) ,$$

since $M^{-1} \in G_D$. This implies the assertion. □

Any subgroup G of G_D acts properly and discontinuously on $\mathfrak{H}_g$ and thus on $\mathfrak{X}_D$. Hence $\mathfrak{X}_D/G$ is a normal complex analytic space and p induces a holomorphic map

$$\bar{p}: \mathfrak{X}_D/G \to \mathfrak{H}_g/G .$$

Since the action of the whole group G_D on $\mathfrak{H}_g$ has fixed points, the holomorphic map $\bar{p}: \mathfrak{X}_D/G_D \to \mathfrak{H}_g/G_D$ is not a family of abelian varieties. In fact, the fibre of $\bar{p}$ over a fixed point $Z \in \mathfrak{H}_g$ is just the quotient of X_Z modulo the isotropy subgroup $(G_D)_Z$ of G_D in Z. We will see that there exist suitable subgroups of G_D such that the corresponding quotient is a family of abelian varieties.

Recall the group

$$\Gamma_D(D) = \{\left(\begin{smallmatrix} a & b \\ c & d \end{smallmatrix}\right) \in \Gamma_D \mid a - \mathbb{1}_g \equiv b \equiv c \equiv d - \mathbb{1}_g \equiv 0 \pmod D\}$$

of Section 3, representing isomorphisms of polarized abelian varieties with level D-structure. Denote by $G_D(D)$ the image of $\Gamma_D(D)$ under the isomorphism $\sigma_D: \Gamma_D \to G_D$, $R \mapsto \left(\begin{smallmatrix} \mathbb{1} & 0 \\ 0 & D \end{smallmatrix}\right)^{-1} R \left(\begin{smallmatrix} \mathbb{1} & 0 \\ 0 & D \end{smallmatrix}\right)$ (see Section 2). We have

$$G_D(D) = \{\left(\begin{smallmatrix} \mathbb{1}+D\tilde{a} & D\tilde{b}D \\ \tilde{c} & \mathbb{1}+\tilde{d}D \end{smallmatrix}\right) \in G_D \mid \tilde{a}, \tilde{b}, \tilde{c}, \tilde{d} \in M_g(\mathbb{Z})\} . \tag{4}$$

Note that $G_D(D)$ is a subgroup of $M_{2g}(\mathbb{Z}) \cap G_D \subseteq G_{\mathbb{1}} = \mathrm{Sp}_{2g}(\mathbb{Z})$. This observation will turn out to be important in the sequel. According to Theorem 3.1 the quotient $\mathcal{A}_D(D) := \mathfrak{H}_g/G_D(D)$ is a moduli space of polarized abelian varieties with level D-structure.

(8.2) Proposition. *Suppose* $D = \operatorname{diag}(d_1, \dots, d_g)$ *with* $d_1 \geq 3$. *Then*

$$\overline{p}: \mathfrak{X}_D / G_D(D) \to \mathcal{A}_D(D)$$

is a family of abelian varieties.

Proof. It suffices to show that the action of $G_D(D)$ on $\mathfrak{H}_g$ is fixed point free. Suppose $M(Z) = Z$ for some $M \in G_D(D)$ and $Z \in \mathfrak{H}_g$. According to Theorem 3.1 the corresponding automorphism φ_M of X_Z restricts to the identity on the group $K(H_Z)$. But $K(H_Z)$ contains $(X_Z)_{d_1}$, the group of d_1-division points in X_Z. Since $d_1 \geq 3$, Corollary 5.1.10 implies $\varphi_M = \mathrm{id}_{X_Z}$. □

§ 9 Orthogonal Level D-Structures

Recall the lattice $\Lambda_D^{\perp} = \left(\begin{smallmatrix} D & 0 \\ 0 & \mathbb{1} \end{smallmatrix}\right)^{-1} \mathbb{Z}^{2g}$. By definition, the group G_D, and thus also $G_D(D)$ acts from the right on $\Lambda_D^{\perp}$ by $(M, \left(\begin{smallmatrix} c^1 \\ c^2 \end{smallmatrix}\right)) \mapsto {}^tM \left(\begin{smallmatrix} c^1 \\ c^2 \end{smallmatrix}\right)$. Consider the quadratic form

$$Q: \Lambda_D^{\perp} \to \mathbb{C}_1 \, , \; Q\left(\begin{smallmatrix} c^1 \\ c^2 \end{smallmatrix}\right) = e(\pi i {}^tc^1 c^2) \; .$$

We define $G_D(D)_0$ to be the subgroup of $G_D(D)$ preserving Q. Roughly speaking, the quotient $\mathcal{A}_D(D)_0 := \mathfrak{H}_g / G_D(D)_0$ is the space of polarized abelian varieties with *orthogonal level* D*-structure*. In this section we show that the meromorphic map $\psi_D: \mathfrak{H}_g \to \mathbb{P}_N$, introduced in Section 7, factorizes via $\mathfrak{H}_g \to \mathcal{A}_D(D)_0$. We need the following characterization of $G_D(D)_0$:

(9.1) Lemma. *a) For any* $M = \left(\begin{smallmatrix} \alpha & \beta \\ \gamma & \delta \end{smallmatrix}\right) \in G_D(D)$ *the following conditions are equivalent*

i) $M \in G_D(D)_0$,
ii) $(D^{-1}\alpha {}^t\beta D^{-1})_0 \equiv (\gamma {}^t\delta)_0 \equiv 0 \pmod 2$,
iii) $(D^{-1} {}^t\delta \beta D^{-1})_0 \equiv ({}^t\gamma \alpha)_0 \equiv 0 \pmod 2$.

b) $G_D(D)_0$ *is of finite index in* G_D.

Proof. a) For every $l^1, l^2 \in \mathbb{Z}^g$ we have

$$\begin{aligned} & Q({}^tM \left(\begin{smallmatrix} D^{-1} l^1 \\ l^2 \end{smallmatrix}\right)) Q\left(\begin{smallmatrix} D^{-1} l^1 \\ l^2 \end{smallmatrix}\right)^{-1} \\ &= e(\pi i {}^tl^1 D^{-1}(\alpha {}^t\delta + \beta {}^t\gamma - \mathbb{1}) l^2 + \pi i {}^tl^1 D^{-1} \alpha {}^t\beta D^{-1} l^1 + \pi i {}^tl^2 \gamma {}^t\delta l^2) \\ &= e(\pi i {}^tl^1 D^{-1} \alpha {}^t\beta D^{-1} l^1 + \pi i {}^tl^2 \gamma {}^t\delta l^2) \\ & \qquad \text{(by Lemma 2.1 (iii), since } {}^tl^1 D^{-1} \beta {}^t\gamma l^2 \in \mathbb{Z} \text{ by equation 8(4))} \\ &= e(\pi i {}^t(D^{-1}\alpha {}^t\beta D^{-1})_0 l^1 + \pi i {}^t(\gamma {}^t\delta)_0 l^2) \; . \end{aligned}$$

For the last equation we used on the one hand that $D^{-1}\alpha\,{}^t\beta D^{-1} \in M_g(\mathbb{Z})$ according to 8(4), and on the other hand, as in the proof of Lemma 4.1, that ${}^t l S l \equiv {}^t(S)_0 l \pmod 2$ for any $l \in \mathbb{Z}^g$ and symmetric $S \in M_g(\mathbb{Z})$. This implies i)$\Longleftrightarrow$ii). The equivalence i)$\Longleftrightarrow$iii) can be proven analogously replacing only M by M^{-1}.

b) Recall that $D = \operatorname{diag}(d_1, \ldots, d_g)$ with $d_\nu | d_{\nu+1}$ for $1 \le \nu \le g$. Since the principal congruence subgroup $\Gamma(2d_g)$ is of finite index in Γ_D, it suffices to show that its image under the isomorphism $\sigma_D\colon \Gamma_D \to G_D$ is contained in $G_D(D)_0$. But

$$\sigma_D(\Gamma(2d_g)) = \left\{ \begin{pmatrix} \mathbb{1}+2d_g a & 2d_g bD \\ 2d_g D^{-1}c & \mathbb{1}+2d_g D^{-1}dD \end{pmatrix} \in G_D \mid a,b,c,d \in M_g(\mathbb{Z}) \right\}$$

and one easily checks using a) that this is a subgroup of $G_D(D)_0$. □

(9.2) Lemma. *There is a meromorphic map $\overline{\psi}_D\colon \mathcal{A}_D(D)_0 \to \mathbb{P}_N$ such that the following diagram commutes*

$$\begin{array}{ccc} & \mathfrak{H}_g & \\ \swarrow & & \searrow {\scriptstyle \psi_D} \\ \mathcal{A}_D(D)_0 & \xrightarrow{\ \overline{\psi}_D\ } & \mathbb{P}_N \end{array} .$$

Proof. According to the definition of ψ_D, it suffices to show that for every $M \in G_D(D)_0$ there is a holomorphic $\tau_M\colon \mathfrak{H}_g \to \mathbb{C}^*$ such that

$$\vartheta\begin{bmatrix} D^{-1}l \\ 0 \end{bmatrix}(0, M(Z)) = \tau_M(Z)\,\vartheta\begin{bmatrix} D^{-1}l \\ 0 \end{bmatrix}(0, Z)$$

for all $Z \in \mathfrak{H}_g$ and $l \in \mathbb{Z}^g$.

The essential observation is that we can apply the Theta Transformation Formula 6.1, since $G_D(D)_0$ is a subgroup of $G_{\mathbb{1}}$. It gives for every $M = \begin{pmatrix} \alpha & \beta \\ \gamma & \delta \end{pmatrix} \in G_D(D)_0$ and every characteristic $c = ZD^{-1}l$

$$\begin{aligned} &\vartheta\begin{bmatrix} \delta D^{-1}l + \frac{1}{2}(\gamma\,{}^t\delta)_0 \\ -\beta D^{-1}l + \frac{1}{2}(\alpha\,{}^t\beta)_0 \end{bmatrix}(0, M(Z)) \\ &\qquad = \kappa(M)\det(\gamma Z + \delta)^{\frac{1}{2}} e(\pi i k(M, D^{-1}l, 0))\,\vartheta\begin{bmatrix} D^{-1}l \\ 0 \end{bmatrix}(0, Z) . \end{aligned}$$

Hence, according to Remark 5.3 c), it suffices to show that

i) $\delta D^{-1}l + \frac{1}{2}(\gamma\,{}^t\delta)_0 \equiv D^{-1}l \pmod{\mathbb{Z}}$,
ii) $-\beta D^{-1}l + \frac{1}{2}(\alpha\,{}^t\beta)_0 \equiv 0 \pmod{\mathbb{Z}}$, and
iii) $k(M, D^{-1}l, 0) = {}^t l D^{-1}\,{}^t\delta(-\beta D^{-1}l + (\alpha\,{}^t\beta)_0) \equiv 0 \pmod 2$.

But this is an immediate computation using the special form of the elements of $G_D(D)_0$ given in equation 8(4) and Lemma 9.1 a). □

(9.3) Remark. A slight modification of the proof above shows that also the map $\varphi_D\colon \mathfrak{X}_D \to \mathbb{P}_N$ factorizes via $\mathfrak{X}_D \to \mathfrak{X}_D / G_D(D)_0$.

§ 10 The Embedding of $\mathcal{A}_D(D)_0$ into Projective Space

Let $D = \operatorname{diag}(d_1, \ldots, d_g)$ be a type with $d_g > 0$. In this section we show that under some additional hypotheses the meromorphic map $\overline{\psi}_D$ of Lemma 9.2 is an embedding. The main result is

(10.1) Theorem. *If $d_1 \geq 4$ and $2|d_1$ or $3|d_1$, then $\overline{\psi}_D\colon \mathcal{A}_D(D)_0 \hookrightarrow \mathbb{P}_N$ is an analytic embedding.*

By definition of $\overline{\psi}_D$ the Theorem can also be interpreted by saying that the theta null values $\{\vartheta\left[\begin{smallmatrix} c_\nu \\ 0 \end{smallmatrix}\right](0,Z) \mid \bar{c}_\nu \in D^{-1}\mathbb{Z}^g/\mathbb{Z}^g\}$ embed the moduli space. The most important case is $D = \operatorname{diag}(4,\ldots,4)$, i.e. the type of the fourth power of a principal polarization L_Z. In this case the theta functions $2^*\vartheta\left[\begin{smallmatrix} c^1 \\ c^2 \end{smallmatrix}\right](\cdot,Z)$ with $\bar{c^1},\bar{c^2} \in \frac{1}{2}\mathbb{Z}^g/\mathbb{Z}^g$ are a basis for $H^0(X_Z, L_Z^4)$ (see Exercise 6.1). Hence Theorem 10.1 implies that the theta null values $\{\vartheta\left[\begin{smallmatrix} c_\nu \\ 0 \end{smallmatrix}\right](0,Z) \mid \bar{c^1},\bar{c^2} \in \frac{1}{2}\mathbb{Z}^g/\mathbb{Z}^g\}$ give an analytic embedding of the moduli space $\mathcal{A}_{4\cdot\mathbb{1}_g}(4\cdot\mathbb{1}_g)_0$ in $\mathbb{P}_{2^{2g}-1}$.

The Theorem is a consequence of the following two propositions.

(10.2) Proposition. *If $d_1 \geq 3$ with $2|d_1$ or $3|d_1$, then $\overline{\psi}_D\colon \mathcal{A}_D(D)_0 \to \mathbb{P}_N$ is an injective holomorphic map.*

Proof. According to Proposition 4.1.6 the map ψ_D and hence also $\overline{\psi}_D$ is holomorphic. Suppose $Z, Z' \in \mathfrak{H}_g$ with $\psi_D(Z) = \psi_D(Z')$. We have to show that there is an $M \in G_D(D)_0$ with $Z' = M(Z)$.
Step I: There is an $M \in G_D$ with $Z' = M(Z)$.
By definition of ψ_D and Remark 5.3 c) there is a constant $\tau \in \mathbb{C}^*$ such that

$$\vartheta\left[\begin{smallmatrix} c^1 \\ c^2 \end{smallmatrix}\right](0,Z') = \tau\,\vartheta\left[\begin{smallmatrix} c^1 \\ c^2 \end{smallmatrix}\right](0,Z)$$

for all $\binom{c^1}{c^2} \in \Lambda_D^\perp$. By the assumption on d_1 we can apply Riemann's Equations 7.5.2 respectively the Cubic Equations 7.6.6 together with Theorem 7.3.1 to conclude that these theta null values determine the abelian varieties $X_{Z'}$ and X_Z. To be more precise, the images of the abelian varieties $X_{Z'}$ and X_Z in $\mathbb{P}_N$ under the embeddings associated to the linear systems $|L(H_{Z'}, \chi_0')|$ and $|L(H_Z, \chi_0)|$ coincide. In other words, there is an isomorphism $f\colon X_{Z'} \to X_Z$ such that the following diagram commutes

$$\begin{array}{ccc} X_{Z'} & \xrightarrow{f} & X_Z \\ & {\scriptstyle \varphi_{L(H_{Z'},\chi_0')}} \searrow \quad \swarrow {\scriptstyle \varphi_{L(H_Z,\chi_0)}} & \\ & \mathbb{P}_N & . \end{array}$$

Hence by Proposition 1.3 there is an $M \in G_D$ with $Z' = M(Z)$ and $f: X_{M(Z)} \to X_Z$ is the isomorphism defined by the equation $M_Z^{-1}(M(Z), \mathbb{1}) = (Z, \mathbb{1})\, {}^tM$ (see Section 8). In particular $M_Z: \mathbb{C}^g \to \mathbb{C}^g$ is the analytic representation of f^{-1} and $f^*\varphi_{L(H_Z,\chi_0)} = \varphi_{L(H_{M(Z)},\chi_0')}$ translates to

$$\vartheta\left[\begin{smallmatrix}c^1\\c^2\end{smallmatrix}\right](M_Z v, M(Z)) = \tau(v)\,\vartheta\left[\begin{smallmatrix}c^1\\c^2\end{smallmatrix}\right](v, Z) \tag{1}$$

for all $v \in \mathbb{C}^g$ and $\binom{c^1}{c^2} \in \Lambda_D^\perp$, with some holomorphic function $\tau: \mathbb{C}^g \to \mathbb{C}^*$, not depending on $\binom{c^1}{c^2}$.

Step II: $\vartheta\left[\begin{smallmatrix}c^1\\c^2\end{smallmatrix}\right](v, Z) = e(\pi i\, {}^tc^1 c^2 - \pi i\, {}^td^1 d^2)\,\vartheta\left[\begin{smallmatrix}d^1\\d^2\end{smallmatrix}\right](v, Z)$ *for all* $\binom{c^1}{c^2} \in \Lambda_D^\perp$ *with* $\binom{d^1}{d^2} := {}^tM^{-1}\binom{c^1}{c^2}$.

Note that $j_Z(\Lambda_D^\perp) = \{v \in \mathbb{C}^g \mid \operatorname{Im} H_Z(v, j_Z(\Lambda_D)) \subset \mathbb{Z}\} = \Lambda(H_Z)$ (see Section 2.4). Hence $Zc^1 + c^2 \in \Lambda(H_Z)$ and $\vartheta_Z^{Zc^1+c^2}$ is a canonical theta function for the line bundle $L(H_Z, \chi_0)$ of type D and characteristic 0 on $X_Z = \mathbb{C}^g/(Z\mathbb{Z}^g \oplus D\mathbb{Z}^g)$. By Lemmas 5.1 and 5.2 we have $\vartheta_Z^{Zc^1+c^2}(v) = e(\frac{\pi}{2}\, {}^tv(\operatorname{Im} Z)^{-1}v - \pi i\, {}^tc^1 c^2)\vartheta\left[\begin{smallmatrix}c^1\\c^2\end{smallmatrix}\right](v, Z)$, and (1) translates to

$$\vartheta_{M(Z)}^{M(Z)c^1+c^2}(M_Z v) = \tilde{\tau}(v)\,\vartheta_Z^{Zc^1+c^2}(v) \tag{2}$$

for all $v \in \mathbb{C}^g$ and $\binom{c^1}{c^2} \in \Lambda_D^\perp$ and some holomorphic $\tilde{\tau}: \mathbb{C}^g \to \mathbb{C}^*$. But the canonical factors of automorphy $a_{L(H_Z,\chi_0)}$ of $\vartheta_Z^{Zc^1+c^2}$ and $a_{L(H_{M(Z)},\chi_0')}$ of $\vartheta_{M(Z)}^{M(Z)c^1+c^2}$ are related by $M_Z^* a_{L(H_{M(Z)},\chi_0')} = a_{L(H_Z,\chi_0)}$, since M_Z is the analytic representation of the isomorphism f^{-1} of Step I and $f^{-1*}L(H_{M(Z)}, \chi_0') = L(H_Z, \chi_0)$. This implies that $\tilde{\tau}$ is periodic with respect to the lattice Λ_Z and thus constant.

Using equation 3.2(3) twice and the equations $H_Z = M_Z^* H_{M(Z)}$ and $M_Z j_Z\binom{c^1}{c^2} = j_{M(Z)}({}^tM^{-1}\binom{c^1}{c^2}) = j_{M(Z)}\binom{d^1}{d^2}$, we deduce from (2)

$$\begin{aligned}
&\vartheta_Z^{Zc^1+c^2}(v) = \\
&= e\left(-\pi H_Z(v, j_Z\tbinom{c^1}{c^2}) - \tfrac{\pi}{2} H_Z(j_Z\tbinom{c^1}{c^2}, j_Z\tbinom{c^1}{c^2})\right)\vartheta_Z^0(v + j_Z\tbinom{c^1}{c^2}) \\
&= \tilde{\tau}(0)^{-1} e(-\pi M_Z^* H_{M(Z)}(v, j_Z\tbinom{c^1}{c^2}) - \tfrac{\pi}{2} M_Z^* H_{M(Z)}(j_Z\tbinom{c^1}{c^2}, j_Z\tbinom{c^1}{c^2})) \\
&\quad \cdot \vartheta_{M(Z)}^0(M_Z v + M_Z j_Z\tbinom{c^1}{c^2}) \\
&= \tilde{\tau}(0)^{-1}\vartheta_{M(Z)}^{M(Z)d^1+d^2}(M_Z v) = \vartheta_Z^{Zd^1+d^2}(v)\,.
\end{aligned}$$

For the last equation note that $\binom{d^1}{d^2} \in \Lambda_D^\perp$. Translating this back into terms of classical theta functions we obtain the assertion.

Step III: $M \in G_D(D)_0$.

According to Step II the theta functions $\vartheta\left[\begin{smallmatrix}c^1\\c^2\end{smallmatrix}\right](\cdot,Z)$ and $\vartheta\left[\begin{smallmatrix}d^1\\d^2\end{smallmatrix}\right](\cdot,Z)$ differ only by a constant for all $\binom{c^1}{c^2} \in \Lambda_D^\perp$. Hence their factors of automorphy with respect to the lattice $Z\mathbb{Z}^g \oplus \mathbb{Z}^g$ coincide. This gives, see Remark 5.3 c),

$$ {}^tM^{-1}\binom{c^1}{c^2} = \binom{d^1}{d^2} \equiv \binom{c^1}{c^2} \pmod{\mathbb{Z}} \tag{3}$$

for all $\binom{c^1}{c^2} \in \Lambda_D^\perp = \begin{pmatrix} D^{-1} & 0\\ 0 & \mathbb{1}\end{pmatrix}\mathbb{Z}^{2g}$, or equivalently

$$(\mathbb{1}_{2g} - {}^tM^{-1}) \in M_{2g}(\mathbb{Z}) \cdot \begin{pmatrix} D & 0\\ 0 & \mathbb{1}\end{pmatrix}.$$

Using the fact that this relation holds also for ${}^tM$ instead of ${}^tM^{-1}$, we derive that M is of the form

$$M = \begin{pmatrix} \alpha & \beta\\ \gamma & \delta\end{pmatrix} = \begin{pmatrix} \mathbb{1}+Da & bD\\ c & \mathbb{1}+dD\end{pmatrix}$$

for some $a,b,c,d \in M_g(\mathbb{Z})$. On the other hand combining (3) with the assertion of Step II gives

$$\vartheta\left[\begin{smallmatrix}c^1\\c^2\end{smallmatrix}\right](v,Z) = e(\pi i\, {}^tc^1c^2 - \pi i\, {}^td^1d^2)\vartheta\left[\begin{smallmatrix}c^1\\c^2\end{smallmatrix}\right](v,Z)\,.$$

Inserting $\binom{d^1}{d^2} = \binom{\delta c^1 - \gamma c^2}{-\beta c^1 + \alpha c^2}$, this implies

$${}^tc^1c^2 - {}^td^1d^2 = {}^tc^1(\mathbb{1} - {}^t\delta\alpha - {}^t\beta\gamma)c^2 + {}^tc^1\,{}^t\delta\beta c^1 + {}^tc^2\,{}^t\gamma\alpha c^2 \equiv 0 \pmod 2\,.$$

Since M is symplectic, $\mathbb{1} - {}^t\delta\alpha + {}^t\beta\gamma = 0$ by Lemma 2.1 b). We may replace c^1 by $D^{-1}l^1$ and c^2 by l^2 for some $l^1, l^2 \in \mathbb{Z}^g$. Then the above equation reads

$${}^tl^1D^{-1}\,{}^t\delta\beta D^{-1}l^1 + {}^tl^2\,{}^t\gamma\alpha l^2 \equiv {}^t(D^{-1}\,{}^t\delta\beta D^{-1})_0 l^1 + {}^t({}^t\gamma\alpha)_0 l^2 \equiv 0 \pmod 2$$

for all $l^1, l^2 \in \mathbb{Z}^g$. So according to Lemma 9.1 and equation 8(4) it remains to show that $\beta = D\tilde{b}D$ for some $\tilde{b} \in M_g(\mathbb{Z})$. Using $\beta = bD$ and $\delta = \mathbb{1} + dD$ this follows from the subsequent computation

$$D^{-1}b \equiv D^{-1}b + {}^tdb \equiv D^{-1}\,{}^t(\mathbb{1}+dD)b = D^{-1}\,{}^t\delta\beta D^{-1} \equiv 0 \pmod{\mathbb{Z}} \quad \square$$

(10.3) Proposition. *For $d_1 \geq 4$ the differential $d\psi_{D,Z}$ is injective at any point $Z \in \mathcal{A}_D(D)_0$.*

Proof. According to the proof of Proposition 8.2 the projection map $\mathfrak{H}_g \to \mathcal{A}_D(D)_0$ is unramified. Hence it suffices to show that $d\psi_{D,Z}$ is injective for all $Z \in \mathfrak{H}_g$.

Suppose $Z = (z_{jk}) \in \mathfrak{H}_g$. There is a $c \in \{c_0,\dots,c_N\}$, the set of representatives of $\Lambda_D^\perp$ in $\Lambda_D^\perp/\Lambda_D$, such that $\vartheta\left[\begin{smallmatrix}c\\0\end{smallmatrix}\right](0,Z) \neq 0$ and we have to prove

$$\operatorname{rk}\left(\frac{\partial\,\vartheta\left[\begin{smallmatrix}c_\nu\\0\end{smallmatrix}\right]/\vartheta\left[\begin{smallmatrix}c\\0\end{smallmatrix}\right]}{\partial z_{11}},\frac{\partial\,\vartheta\left[\begin{smallmatrix}c_\nu\\0\end{smallmatrix}\right]/\vartheta\left[\begin{smallmatrix}c\\0\end{smallmatrix}\right]}{\partial z_{12}},\dots,\frac{\partial\,\vartheta\left[\begin{smallmatrix}c_\nu\\0\end{smallmatrix}\right]/\vartheta\left[\begin{smallmatrix}c\\0\end{smallmatrix}\right]}{\partial z_{gg}}\right)_{0\le\nu\le N}(0,Z)=\tfrac12 g(g+1)\,.$$

For this it suffices to show that

$$\operatorname{rk}\left(\vartheta\left[\begin{smallmatrix}c_\nu\\0\end{smallmatrix}\right],\frac{\partial\,\vartheta\left[\begin{smallmatrix}c_\nu\\0\end{smallmatrix}\right]}{\partial z_{11}},\frac{\partial\,\vartheta\left[\begin{smallmatrix}c_\nu\\0\end{smallmatrix}\right]}{\partial z_{12}},\dots,\frac{\partial\,\vartheta\left[\begin{smallmatrix}c_\nu\\0\end{smallmatrix}\right]}{\partial z_{gg}}\right)_{0\le\nu\le N}(0,Z)=\tfrac12 g(g+1)+1\,,\qquad(5)$$

since this rank is less or equal to $\operatorname{rk}\left(\frac{\partial\,\vartheta\left[\begin{smallmatrix}c_\nu\\0\end{smallmatrix}\right]/\vartheta\left[\begin{smallmatrix}c\\0\end{smallmatrix}\right]}{\partial z_{jk}}\right)+1$.

Assume (5) does not hold. In other words, assume that the columns of the matrix in (5) are linearly dependent. Since the functions $\vartheta\left[\begin{smallmatrix}c_0\\0\end{smallmatrix}\right](\cdot,Z),\dots,$ $\vartheta\left[\begin{smallmatrix}c_N\\0\end{smallmatrix}\right](\cdot,Z)$ span the vector space $H^0(L(H_Z,\chi_0))$, this means that for all (classical theta functions) $\vartheta=\vartheta(\cdot,Z)\in H^0(L(H_Z,\chi_0))$

$$s\,\vartheta(0)=\sum_{1\le j\le k\le g} s_{jk}\frac{\partial\vartheta}{\partial z_{jk}}(0)\qquad(6)$$

for some constants $s_{jk},s\in\mathbb{C}$, not all zero and not depending on ϑ. Defining $s_{kj}=s_{jk}$ for $j<k$ we can apply the heat equation (Proposition 5.5) for the matrix $S=(s_{jk})$ to get

$$4\pi i\,s\,\vartheta(0)=\sum_{j,k=1}^{g} s_{jk}\frac{\partial^2\vartheta}{\partial v_j\partial v_k}(0)\qquad(7)$$

for all $\vartheta\in H^0(L(H_Z,\chi_0))$.

According to Lemma 2.5.6 the line bundle $L(H_Z,\chi_0)$ is of the form M^{d_1} for some ample line bundle M on X_Z. Choose a classical theta function $\theta\neq0$ in $H^0(M)$ and points a_1,a_2 in its divisor (θ). Since $d_1\ge4$ there are $a_3,\dots,a_{d_1}\notin(\theta)$ with $\sum_{\nu=1}^{d_1}a_\nu=0$. According to Lemma 4.1.5 the function $\vartheta_0:=\prod_{\nu=1}^{d_1}\theta(\cdot+a_\nu)$ is a theta function for $L(H_Z,\chi_0)$. Equation (7) reads for $\vartheta=\vartheta_0$:

$$0=4\pi i\,s\,\vartheta_0(0)=2\theta(a_3)\cdot\ldots\cdot\theta(a_{d_1})\sum_{j,k=1}^{g}s_{jk}\frac{\partial\theta}{\partial v_j}(a_1)\frac{\partial\theta}{\partial v_k}(a_2)\,,$$

and equivalently

$${}^t\!\left(\frac{\partial\theta}{\partial v_1}(a_1),\dots\frac{\partial\theta}{\partial v_g}(a_1)\right)S\left(\frac{\partial\theta}{\partial v_1}(a_2),\dots\frac{\partial\theta}{\partial v_g}(a_2)\right)=0\,.\qquad(8)$$

Note that $(\frac{\partial\theta}{\partial v_1}(a_\nu):\ldots:\frac{\partial\theta}{\partial v_g}(a_\nu))\in\mathbb{P}_{g-1}$ is just the image of the point $a_\nu\in X_Z$ under the Gauss map for the divisor (θ). But by Proposition 4.4.1 the image of the Gauss map spans $\mathbb{P}_{g-1}$. Now varying a_1 and a_2 within (θ) equation (8) implies $s_{jk}=0$ for all $1\le j,k\le g$. Inserting this into

(6) we conclude that either $s = 0$ or 0 is a base point of the linear system $|L(H_Z, \chi_0)|$, a contradiction in both cases. □

(10.4) Remark. Let for a moment X denote the image of the analytic embedding $\psi_D: \mathcal{A}_D(D)_0 \to \mathbb{P}_N$ of Theorem 10.1 and $\overline{X}$ its closure in $\mathbb{P}_N$ with respect to the euclidean topology. Using the theory of modular forms it is not difficult to show that $\overline{X}$ coincides with the closure of X in the Zariski topology (see Igusa [1] Theorem V.8). So $\overline{X}$ is a projective algebraic variety by Chow's Theorem A.3. Moreover, according to Igusa [1] Remark p. 224, the variety X is Zariski open in $\overline{X}$, hence

The moduli space $\mathfrak{H}_g/G_D(D)_0$ is a quasi-projective algebraic variety of dimension $\frac{1}{2}g(g+1)$ over the complex numbers.

Using the Theorem of Grauert and Remmert A.5 and the fact that the quotient of an algebraic variety under a good action of a finite group is again algebraic this implies that all the moduli spaces $\mathcal{A}_D$, $\mathcal{A}_D(n)$ and $\mathcal{A}_D(D)$ (respectively $\widetilde{\mathcal{A}}_D$, $\widetilde{\mathcal{A}}_D(n)$ and $\widetilde{\mathcal{A}}_D(D)$) are algebraic varieties of dimension $\frac{1}{2}g(g+1)$ over $\mathbb{C}$. We omit the details.

Exercises

(1) a) Show that there are at most countably many proper analytic subvarieties A_i of the moduli space $\mathcal{A}_D$ of polarized abelian varieties of type D such that every $(X, L) \in \mathcal{A}_D - \bigcup_i A_i$ has endomorphism ring $\mathbb{Z}$. (Hint: consider for any $\begin{pmatrix} a & b \\ c & d \end{pmatrix} \in M_{2g}(\mathbb{Z})$ the equation $ZD^{-1}(cZ + dD) = aZ + bD$ in $\mathfrak{H}_g$.)

b) Deduce that $\mathrm{NS}(X) = \mathbb{Z}$ for every $(X, L) \in \mathcal{A}_D - \bigcup_i A_i$.

(2) Suppose $D = \mathrm{diag}(d_1, \ldots, d_g)$ is the type of a polarization and $M = \begin{pmatrix} \alpha & \beta \\ \gamma & \delta \end{pmatrix} \in \mathrm{Sp}_{2g}(\mathbb{Q})$. Show that M is contained in the symplectic group G_D if and only if $\alpha, D\gamma, \beta D^{-1}, D\delta D^{-1} \in M_g(\mathbb{Z})$.

In particular, if $g = 1$ and $D = (d)$, then $G_D = \mathrm{Sp}_2(\mathbb{Q}) \cap \begin{pmatrix} \mathbb{Z} & d\mathbb{Z} \\ \frac{1}{d}\mathbb{Z} & \mathbb{Z} \end{pmatrix}$, and if $g = 2$ and $D = \begin{pmatrix} 1 & 0 \\ 0 & d \end{pmatrix}$, then $G_D = \mathrm{Sp}_4(\mathbb{Q}) \cap \begin{pmatrix} \mathbb{Z} & \mathbb{Z} & \mathbb{Z} & d\mathbb{Z} \\ \mathbb{Z} & \mathbb{Z} & \mathbb{Z} & d\mathbb{Z} \\ \mathbb{Z} & \mathbb{Z} & \mathbb{Z} & d\mathbb{Z} \\ \frac{1}{d}\mathbb{Z} & \frac{1}{d}\mathbb{Z} & \frac{1}{d}\mathbb{Z} & \mathbb{Z} \end{pmatrix}$.

(3) Let $D = \mathrm{diag}(d_1, \ldots, d_g)$ be a type of an ample line bundle on an abelian variety. As in Section 6.6 let $\mathrm{Sp}(D)$ denote the group of automorphisms of $K(D) = (\mathbb{Z}^g/D\mathbb{Z}^g)^2$ preserving the alternating form e^D.

a) Show that the canonical map $\Gamma_D = \mathrm{Sp}_{2g}^D(\mathbb{Z}) \to \mathrm{Sp}(D)$ is surjective.

b) Let $(X = V/\Lambda, H)$ be a polarized abelian variety of type D. Use a) to show that any symplectic basis of $K(H) \subset X$ is of the form $\frac{1}{d_1}\lambda_1, \ldots, \frac{1}{d_g}\lambda_g, \frac{1}{d_1}\mu_1, \ldots, \frac{1}{d_g}\mu_g$ for some symplectic basis $\lambda_1, \ldots, \lambda_g, \mu_1 \ldots, \mu_g$ of Λ for H.

(See Brasch [1].)

(4) *(Moduli Space of Polarized Abelian Varieties with Isogeny of Type D)* Let (X, L) be a polarized abelian variety of type D. An isogeny $p: (X, L) \to (Y, M)$ onto a principally polarized abelian variety (Y, M) is called *of type D*, if $\ker p \simeq \mathbb{Z}^g/D\mathbb{Z}^g$. The triplet (X, L, p) is called *polarized abelian variety with isogeny of type D*. Two such triplets $(X_i, L_i, p_i: (X_i, L_i) \to (Y_i, M_i))$, $i = 1, 2$, are isomorphic, if there are isomorphisms $\varphi: (X_1, L_1) \to (X_2, L_2)$ and $\psi: (Y_1, M_1) \to (Y_2, M_2)$ such that $p_2\varphi = \psi p_1$.

a) Show that the normal complex analytic space $\mathcal{A}_D^0 = \mathfrak{H}_g/G_D \cap G_{\mathbb{1}_g}$ is a moduli space for polarized abelian varieties with isogeny of type D.

b) There is a canonical Galois covering $\mathcal{A}_D(D) \to \mathcal{A}_D^0$ of finite degree.

(In the special case $g = 2$ and $D = \begin{pmatrix} 1 & 0 \\ 0 & d \end{pmatrix}$ an isogeny of type D is sometimes called a *root*. So $\mathcal{A}_{\begin{pmatrix} 1 & \\ & d \end{pmatrix}}$ is the moduli space of rooted abelian surfaces of type $(1, d)$.)

(5) *(Moduli Space of Elliptic Curves with Cyclic Subgroup)* For a given integer $n \geq 1$ consider the set of elliptic curves E with a cyclic subgroup K of order n. Two such pairs (E_i, K_i), $i = 1, 2$ are called isomorphic, if there is an isomorphism $\varphi: E_1 \to E_2$ with $\varphi(K_1) = K_2$.

a) Show that the space $\mathcal{A}_{(n)}^0 = \mathfrak{H}_2/G_{(n)} \cap G_{\mathbb{1}}$ of Exercise 3 is a moduli space for elliptic curves with cyclic subgroup of order n.

b) $G_{(n)} \cap G_{\mathbb{1}} = \Gamma^0(n) := \left\{ \begin{pmatrix} a & b \\ c & d \end{pmatrix} \in \mathrm{Sp}_2(\mathbb{Z}) \mid b \equiv 0 \pmod{n} \right\}$.

(6) *(Moduli Space of Elliptic Curves with n-Division Point)* For a given integer $n \geq 1$ consider the set of elliptic curves E with a point $x \in E$ of order n. Two such pairs (E_i, x_i), $i = 1, 2$ are called isomorphic, if there is an isomorphism $\varphi: E_1 \to E_2$ with $\varphi(x_1) = x_2$. Denote

$$\Gamma^{1,0}(n) = \left\{ \begin{pmatrix} a & b \\ c & d \end{pmatrix} \in \mathrm{Sp}_2(\mathbb{Z}) \mid a \equiv 1\,,\, b \equiv 0 \pmod{n} \right\}.$$

Show that $\mathcal{A}^{1,0}(n) = \mathfrak{H}_1/\Gamma^{1,0}(n)$ is a moduli space for elliptic curves with n-division point.

(7) For a prime number p consider the set of triplets (X, L, A_p) with (X, L) a principally polarized abelian surface and A_p a subgroup of order p^2 of the group of p-division points X_p, nonisotropic for the alternating form e^{L^p} on X_p. Two such triplets (X, L, A_p) and (X', L', A'_p) are called isomorphic, if there is an isomorphism $\varphi: (X, L) \to (X', L')$ with $\varphi(A_p) = A'_p$. Denote

$$\Gamma^0_{\mathbb{1}_2}(p) = \mathrm{Sp}_4(\mathbb{Z}) \cap \begin{pmatrix} \mathbb{Z} & p\mathbb{Z} & \mathbb{Z} & p\mathbb{Z} \\ \mathbb{Z} & \mathbb{Z} & \mathbb{Z} & \mathbb{Z} \\ \mathbb{Z} & p\mathbb{Z} & \mathbb{Z} & p\mathbb{Z} \\ \mathbb{Z} & \mathbb{Z} & \mathbb{Z} & \mathbb{Z} \end{pmatrix}$$

a) Show that the normal complex analytic space $\mathcal{A}^0_{\mathbb{1}_2}(p) = \mathfrak{H}_2/\Gamma^0_{\mathbb{1}_2}(p)$ is a moduli space for triplets (X, L, A_p) as above.

b) Show that there is a canonical isomorphism of $\mathcal{A}^0_{\mathbb{1}_2}(p)$ with the moduli space $\mathcal{A}_{\left(\begin{smallmatrix} 1 & \\ & p^2 \end{smallmatrix}\right)}$ of abelian surfaces with polarization of type $(1, p^2)$.

(8) *(Iwasawa Decomposition of* $\mathrm{Sp}_{2g}(\mathbb{R})$*)* Consider the group $O = O_{2g}(\mathbb{R}) \cap \mathrm{Sp}_{2g}(\mathbb{R})$ and the sets $\Delta = \left\{\left(\begin{smallmatrix} C & 0 \\ 0 & C^{-1} \end{smallmatrix}\right) \in \mathrm{Sl}_{2g}(\mathbb{R}) \mid C = \mathrm{diag}(c_1, \ldots, c_g)\right\}$ and

$$N = \left\{ \left(\begin{smallmatrix} A & 0 \\ 0 & {}^tA^{-1} \end{smallmatrix}\right) \left(\begin{smallmatrix} \mathbb{1}_g & B \\ 0 & \mathbb{1}_g \end{smallmatrix}\right) \in \mathrm{Sl}_{2g}(\mathbb{R}) \;\middle|\; \begin{array}{c} A \text{ unipotent upper triangular} \\ B \text{ symmetric} \end{array} \right\} .$$

a) Show that Δ and N are subgroups of $\mathrm{Sp}_{2g}(\mathbb{R})$.

b) The canonical map $N \times \Delta \times O \to \mathrm{Sp}_{2g}(\mathbb{R})$ is a diffeomorphism. In particular, any symplectic matrix M can be written uniquely as a product $M = M_1 M_2 M_3$ with $M_1 \in N$, $M_2 \in \Delta$ and $M_3 \in O$. This is called the *Iwasawa decomposition* of M.

(9) For every $M = \left(\begin{smallmatrix} \alpha & \beta \\ \gamma & \delta \end{smallmatrix}\right) \in \mathrm{Sp}_{2g}(\mathbb{Z})$ let $\kappa(M)$ be the constant in the Theta Transformation Formula 6.1.

a) Show that $\kappa^2: \mathrm{Sp}_{2g}(\mathbb{Z}) \to \mathbb{C}^*_1$ is a homomorphism of groups.

b) The matrices $\left(\begin{smallmatrix} 0 & \mathbb{1}_g \\ -\mathbb{1}_g & 0 \end{smallmatrix}\right)$, $\left(\begin{smallmatrix} \mathbb{1}_g & \beta \\ 0 & \mathbb{1}_g \end{smallmatrix}\right)$ and $\left(\begin{smallmatrix} \alpha & 0 \\ 0 & {}^t\alpha^{-1} \end{smallmatrix}\right)$, with $\beta \in M_g(\mathbb{Z})$ symmetric and $\alpha \in \mathrm{GL}_g(\mathbb{Z})$, generate the subgroup $\mathrm{Sp}_{2g}(\mathbb{Z})$.

c) Show that $\kappa^2\left(\begin{smallmatrix} 0 & \mathbb{1}_g \\ -\mathbb{1}_g & 0 \end{smallmatrix}\right) = (-i)^g$, $\kappa^2\left(\begin{smallmatrix} \mathbb{1}_g & \beta \\ 0 & \mathbb{1}_g \end{smallmatrix}\right) = 1$, $\kappa^2\left(\begin{smallmatrix} \alpha & 0 \\ 0 & {}^t\alpha^{-1} \end{smallmatrix}\right) = \det \alpha = \pm 1$.

(10) For any $Z = X + iY \in \mathfrak{H}_g$ consider the $\mathbb{R}$-linear isomorphism $j_Z: \mathbb{R}^{2g} \to \mathbb{C}^g$, $x \mapsto (Z, \mathbb{1})x$ of Section 8. The complex structure on $\mathbb{C}^g$ induces a complex structure on $\mathbb{R}^{2g}$ given by a matrix $J \in M_{2g}(\mathbb{R})$, i.e. $ij_Z = j_Z J$. Show that $J = \left(\begin{smallmatrix} Y^{-1}X & Y^{-1} \\ -Y-XY^{-1}X & -XY^{-1} \end{smallmatrix}\right)$.

Chapter 9

Moduli Spaces of Abelian Varieties with Endomorphism Structure

In Chapter 5 we saw that the endomorphism algebra of a simple abelian variety is a skew field F of finite dimension over $\mathbb{Q}$ admitting a positive anti-involution $'$, the Rosati involution. Moreover, in Section 5.5 we classified all such pairs $(F, ')$. In this chapter we study the converse question: which of the pairs $(F, ')$ actually occur as the endomorphism algebra of a polarized abelian variety? To be more precise, for every pair $(F, ')$ we construct families of polarized abelian varieties (X, H) together with an *endomorphism structure*. Roughly speaking, this is an embedding $F \hookrightarrow \mathrm{End}_{\mathbb{Q}}(X)$. The parameter spaces of these families are certain complex manifolds $\mathcal{H}$, which are generalizations of Siegel's upper half space. The families themselves are parametrized by pairs $(\mathcal{M}, T)$, where $\mathcal{M}$ is a certain $\mathbb{Z}$-submodule of the left F-vector space F^m and T a nondegenerate skew-hermitian form on F^m. Moreover we analyze, when two elements of $\mathcal{H}$ represent isomorphic polarized abelian varieties with endomorphism structure. It turns out that, given $(\mathcal{M}, T)$, there is a group $G(\mathcal{M}, T)$ acting properly and discontinuously on $\mathcal{H}$, such that the quotient $\mathcal{A}(\mathcal{M}, T) = \mathcal{H}/G(\mathcal{M}, T)$ is a moduli space (in the sense of Chapter 8) for the polarized abelian varieties with endomorphism structure associated to $(\mathcal{M}, T)$. Finally we show that any polarized abelian variety with endomorphism structure is represented in one of these moduli spaces and that for a general element of $\mathcal{A}(\mathcal{M}, T)$ the endomorphism algebra equals the skew field F, except for some particular cases. This answers the question we started with.

All results of this chapter are due to Shimura [1]. Actually Shimura proves more: he shows that there is a canonical compactification $\overline{\mathcal{A}(\mathcal{M}, T)}$ of $\mathcal{A}(\mathcal{M}, T)$, which is a projective algebraic variety. This implies that the moduli spaces $\mathcal{A}(\mathcal{M}, T)$ are algebraic varieties. We do not prove this here, since our main interest lies in providing a list of all polarized abelian varieties with an endomorphism algebra containing one of the skew fields F of Section 5.5.

The basic idea, which is already very old and goes back at least to Scorza [1] p. 24 and Lefschetz [1] p. 397, is to write the period matrices in a form exhibiting the presence of endomorphisms of a given type. To be more precise, Shimura proceeds as follows: suppose $X = \mathbb{C}^g/\Lambda$ is an abelian variety with $F \subseteq \mathrm{End}_{\mathbb{Q}}(X)$. Then $\Lambda \otimes \mathbb{Q}$ is an F-left vector space. Hence there is an $\mathbb{R}$-linear isomorphism

$$(F \otimes \mathbb{R})^m \simeq \Lambda \otimes \mathbb{R} = \mathbb{C}^g \qquad (*)$$

for some $m \geq 1$. The preimage of Λ under this isomorphism is a $\mathbb{Z}$-module $\mathcal{M}$ of F^m and the polarization is essentially given by the reduced trace of F over $\mathbb{Q}$. So start with a suitable $\mathbb{Z}$-submodule $\mathcal{M}$ of F^m. An $\mathbb{R}$-isomorphism $(*)$ defines a complex torus X with $F \subset \mathrm{End}_{\mathbb{Q}}(X)$. Finally, define a polarization on X via the reduced trace map.

Our approach is slightly different from Shimura's original construction. We want the proof in this book to be a direct generalization of the construction of the moduli space of polarized abelian varieties in Chapter 8. We choose a suitable $\mathbb{R}$-vector space basis of $(F \otimes \mathbb{R})^m$, such that the isomorphisms $(*)$ can be defined by $(g \times 2g)$- matrices, which generalize the period matrices $(Z, \mathbb{1})$ of Chapter 8 (see Example 2.4). The isomorphism also allows to introduce the polarization by defining a hermitian form in terms of a matrix. It generalizes directly the matrix $(\mathrm{Im}\, Z)^{-1}$ defining the hermitian form in Chapter 8.

In Section 2 we construct the moduli spaces of polarized abelian varieties with real multiplication. This is the easiest case, which should serve as a model for the construction of the moduli space for the other 3 types. In the latter cases the choice of the basis of $(F \otimes \mathbb{R})^m$ is complicated by the fact that the completions of F themselves are already matrix algebras. We need a particular vector space isomorphism $\tilde{}: M_d(\mathbb{C})^n \to \mathbb{C}^{d^2 n}$ and some of its properties. This is the contents of Section 3. We proceed to the construction of the families of polarized abelian varieties with totally indefinite quaternion multiplication, with totally definite quaternion multiplication, and with complex multiplication in Sections 4, 5 and 6 respectively. In Section 7 we analyze, when 2 elements of the same family are isomorphic, and study the corresponding group action. It is analogous to the action of the symplectic group on the Siegel upper half space. In Section 8 we construct the moduli spaces $\mathcal{A}(\mathcal{M}, T)$ for the last three types simultaneously. Finally in Section 9, we follow Shimura [1] to show that the endomorphism algebra of a general member of the families is F itself, except in some special cases.

The methods of proof are essentially elementary, however rather complicated, due to the fact that the matrices are of considerable size. A good understanding of the *Kronecker product of matrices* is required, which we recall here: Let $A = (a_{ij}) \in M(m \times n, \mathbb{C})$ and $B \in M(r \times s, \mathbb{C})$. Then

$$A \otimes B := \begin{pmatrix} a_{11}B & \cdots & a_{1n}B \\ \vdots & & \vdots \\ a_{m1}B & \cdots & a_{mn}B \end{pmatrix} \in M(mr \times ns, \mathbb{C}) \ .$$

A generalization of the usual notion for a diagonal matrix will be useful: if $A_1, \ldots, A_r \in M(m \times n, \mathbb{C})$, we write

$$\operatorname{diag}(A_1, \ldots, A_r) = \begin{pmatrix} A_1 & & \\ & \ddots & \\ & & A_r \end{pmatrix}$$

although this is not a diagonal matrix in the usual sense.

§1 Abelian Varieties with Endomorphism Structure

In this section we introduce the terminology in preparation for the construction of families of abelian varieties with endomorphism structure.

As in Section 5.5, let $(F, \ ')$ denote a skew field of finite dimension over $\mathbb{Q}$ with positive anti-involution $'$. Fix a representation $\varrho: F \to M_g(\mathbb{C})$. A *polarized abelian variety with endomorphism structure* $(F, \ ', \varrho)$ is by definition a triplet (X, H, ι) with an abelian variety $X = \mathbb{C}^g/\Lambda$, a positive definite hermitian form H on $\mathbb{C}^g$ defining a polarization on X, and an embedding $\iota: F \hookrightarrow \operatorname{End}_{\mathbb{Q}}(X) \subseteq M_g(\mathbb{C})$ (here we consider $\operatorname{End}_{\mathbb{Q}}(X)$ as a subspace of $M_g(\mathbb{C})$ via the analytic representation), such that

a) ι and ϱ are equivalent representations, and
b) the Rosati involution on $\operatorname{End}_{\mathbb{Q}}(X)$ with respect to H extends the anti-involution $'$ on F via ι.

Suppose (X, H, ι) and $(\widetilde{X}, \widetilde{H}, \tilde{\iota})$ are polarized abelian varieties with endomorphism structure $(F, \ ', \varrho)$. An *isomorphism* $f: (\widetilde{X}, \widetilde{H}, \tilde{\iota}) \to (X, H, \iota)$ is by definition an isomorphism of polarized abelian varieties $f: (\widetilde{X}, \widetilde{H}) \to (X, H)$ such that the diagram

$$\begin{array}{ccc} \widetilde{X} & \xrightarrow{f} & X \\ {\scriptstyle \tilde{\iota}(a)}\downarrow & & \downarrow{\scriptstyle \iota(a)} \\ \widetilde{X} & \xrightarrow{f} & X \end{array}$$

commutes for all $a \in F$.

The representation $\varrho: F \to M_g(\mathbb{C})$ cannot be arbitrary in order to ensure the existence of a polarized abelian variety of type $(F, \ ', \varrho)$. In fact, suppose (X, H, ι) is of type $(F, \ ', \varrho)$. Then the analytic representation $\rho_a: \operatorname{End}_{\mathbb{Q}}(X) \to M_g(\mathbb{C})$ restricts via ι to the representation ρ. According to Proposition 1.2.3 the representation $\varrho_a \oplus \overline{\varrho_a}$ is equivalent (over $\mathbb{C}$) to the extended rational representation $\varrho_r \otimes 1_{\mathbb{C}}: \mathrm{End}_{\mathbb{Q}}(X) \otimes \mathbb{C} \to M_{2g}(\mathbb{C})$. This implies that there is a rational representation $\psi: F \to M_{2g}(\mathbb{Q})$, equivalent to $\varrho \oplus \overline{\varrho}$ over $\mathbb{C}$. The following lemma says that such a ψ and thus ϱ is of a special form.

(1.1) Lemma. *Let $\sigma_1, \ldots, \sigma_e$ denote the irreducible $\mathbb{C}$-representations of $F \otimes_{\mathbb{Q}} \mathbb{C}$. For any rational representation $\psi: F \to M_{2g}(\mathbb{Q})$ there is an integer $m \geq 1$ such that*

$$\psi \otimes 1_{\mathbb{C}} \simeq m \sum_{\nu=1}^{e} \sigma_\nu \ .$$

Proof. The group $\mathrm{Aut}(\mathbb{C}/\mathbb{Q})$ acts on the set $\{\sigma_1, \ldots, \sigma_e\}$ transitively. On the other hand $\psi \otimes 1_{\mathbb{C}}$ is invariant under the action of $\mathrm{Aut}(\mathbb{C}/\mathbb{Q})$, since ψ is a rational representation. This implies the assertion. □

In Theorems 5.5.3 and 5.5.6 we classified the skew fields of finite dimension over $\mathbb{Q}$ with positive anti-involution $(F, \ ')$. We obtained the four types:

I: F = totally real number field
II: F = totally indefinite quaternion algebra
III: F = totally definite quaternion algebra
IV: $(F, \ ')$ of the second kind.

We will see that in every case one can find a representation ϱ of F such that $(F, \ ', \varrho)$ is an endomorphism structure of a polarized abelian variety. To each such endomorphism structure we construct families of polarized abelian varieties.

§ 2 Abelian Varieties with Real Multiplication

Let F be a totally real number field of degree e over $\mathbb{Q}$. In this section we construct a set of families of polarized abelian varieties of dimension g with endomorphism structure (F, id_F, ρ) for every representation $\rho: F \to M_g(\mathbb{C})$ satisfying the condition of Lemma 1.1. Moreover we show that conversely every abelian variety with endomorphism structure (F, id_F, ρ) is contained in one of these families. Finally we construct the corresponding moduli spaces.

We say that an abelian variety X *admits real multiplication* by F, if there is an embedding $F \hookrightarrow \mathrm{End}_{\mathbb{Q}}(X)$. Proposition 5.5.7 gives the following condition for g:

$$g = em$$

for some integer $m \geq 1$. The e different embeddings $F \hookrightarrow \mathbb{R}$ determine an isomorphism of $F \otimes_{\mathbb{Q}} \mathbb{R}$ with $\mathbb{R}^e$. Identifying both sides, we may write any $a \in F \otimes_{\mathbb{Q}} \mathbb{R}$ in the form

$$a = \begin{pmatrix} a^1 \\ \vdots \\ a^e \end{pmatrix}.$$

Similarly we write the elements of $(F \otimes_{\mathbb{Q}} \mathbb{R})^{2m} = (\mathbb{R}^{2m})^e = \mathbb{R}^{2g}$ as

$$\underline{a} = \begin{pmatrix} \underline{a}^1 \\ \vdots \\ \underline{a}^e \end{pmatrix} \quad \text{with} \quad \underline{a}^\nu = \begin{pmatrix} a_1^\nu \\ \vdots \\ a_{2m}^\nu \end{pmatrix} \in \mathbb{R}^{2m} \quad \text{for} \quad 1 \leq \nu \leq e \,.$$

Define a representation $\varrho\colon F \to M_g(\mathbb{C})$ by

$$\varrho(a) = \operatorname{diag}(a^1 \mathbb{1}_m, \ldots, a^e \mathbb{1}_m) = \operatorname{diag}(a^1, \ldots, a^e) \otimes \mathbb{1}_m \,.$$

Obviously every representation $F \to M_g(\mathbb{C})$, satisfying the conditions of Lemma 1.1, is equivalent to ϱ. So it suffices to consider the representation ϱ.

Our first aim is to associate to every element Z of the e-fold product $\mathfrak{H}_m^e$ of the Siegel upper half space $\mathfrak{H}_m$ a polarized abelian variety (X_Z, H_Z, ι_Z) with endomorphism structure $(F, \mathrm{id}_F, \varrho)$: fix a free $\mathbb{Z}$-submodule $\mathcal{M}$ of F^{2m} of rank $2g = 2em$ such that

$$\operatorname{tr}_{F/\mathbb{Q}}\left({}^t\underline{a} \begin{pmatrix} 0 & \mathbb{1}_m \\ -\mathbb{1}_m & 0 \end{pmatrix} \underline{b}\right) \in \mathbb{Z} \tag{1}$$

for all $\underline{a}, \underline{b} \in \mathcal{M} \subset (F \otimes_{\mathbb{Q}} \mathbb{R})^{2m}$. Note that such modules exist: for example, if $\mathfrak{o}$ denotes the ring of integers in F, then the module $\mathcal{M} = \mathfrak{o}^{2m}$ satisfies condition (1).

For every $Z = (Z^1, \ldots, Z^e) \in \mathfrak{H}_m^e$ define a map

$$J_Z\colon (F \otimes_{\mathbb{Q}} \mathbb{R})^{2m} = \mathbb{R}^{2g} \to \mathbb{C}^g \,, \quad \underline{a} \mapsto \operatorname{diag}((Z^1, \mathbb{1}_m), \ldots, (Z^e, \mathbb{1}_m))\underline{a} \,.$$

This is an isomorphism of $\mathbb{R}$-vector spaces, since the columns of the defining matrix are linearly independent over $\mathbb{R}$. Hence $J_Z(\mathcal{M})$ is a lattice in $\mathbb{C}^g$ and the quotient

$$X_Z := \mathbb{C}^g / J_Z(\mathcal{M})$$

is a complex torus. Define a hermitian form H_Z on $\mathbb{C}^g$ by

$$H_Z(x, y) = {}^t x \operatorname{diag}(\operatorname{Im} Z^1, \ldots, \operatorname{Im} Z^e)^{-1} \bar{y} \,.$$

By definition H_Z is a positive definite hermitian form on $\mathbb{C}^g$. Moreover H_Z defines a polarization on X_Z. In order to see this, we have to check that $\operatorname{Im} H_Z$ is integer valued on the lattice $J_Z(\mathcal{M})$. But by definition of $\mathcal{M}$

$$\begin{aligned}\operatorname{Im} H_Z(J_Z(\underline{a}), J_Z(\underline{b})) &= \sum_{\nu=1}^{e} {}^t\underline{a}^\nu \operatorname{Im}({}^t(Z^\nu, \mathbb{1})(\operatorname{Im} Z^\nu)^{-1}\overline{(Z^\nu, \mathbb{1})})\underline{b}^\nu \\ &= \sum_{\nu=1}^{e} {}^t\underline{a}^\nu \begin{pmatrix} 0 & \mathbb{1}_m \\ -\mathbb{1}_m & 0 \end{pmatrix} \underline{b}^\nu \\ &= \operatorname{tr}_{F/\mathbb{Q}}({}^t\underline{a} \begin{pmatrix} 0 & \mathbb{1}_m \\ -\mathbb{1}_m & 0 \end{pmatrix} \underline{b}) \in \mathbb{Z}\end{aligned}$$

for all $\underline{a}, \underline{b} \in \mathcal{M}$.

Finally we have $\varrho(a)\, J_Z(\underline{b}) = J_Z(a \cdot \underline{b})$ for all $a \in F$ and $\underline{b} \in F^{2m}$. This follows from an immediate matrix computation, using the fact that the multiplication $F \times (F \otimes_{\mathbb{Q}} \mathbb{R})^{2m} \to (F \otimes_{\mathbb{Q}} \mathbb{R})^{2m}$ translates to

$$a \cdot \underline{b} = \operatorname{diag}(a^1 \mathbb{1}_{2m}, \ldots, a^e \mathbb{1}_{2m}) \cdot \underline{b}\ .$$

Since $\mathcal{M} \otimes \mathbb{Q} = F^{2m}$, this yields $\varrho(na)\, J_Z(\mathcal{M}) \subseteq J_Z(\mathcal{M})$ for some integer $n > 0$, and we obtain

$$\varrho(F) \subset \operatorname{End}_{\mathbb{Q}}(X_Z)\ .$$

Combining everything and setting $\iota_Z = \varrho$ we get

(2.1) Proposition. *For every $Z \in \mathfrak{H}_m^e$ the triplet (X_Z, H_Z, ι_Z) is a polarized abelian variety with endomorphism structure $(F, \mathrm{id}_F, \varrho)$.*

For the proof it remains to show that the Rosati involution with respect to H_Z restricts to the identity on F. But an immediate matrix computation shows that $H_Z(\iota_Z(a)x, y) = H_Z(x, \iota_Z(a)y)$ for all $x, y \in \mathbb{C}^g$ and $a \in F$. This implies the assertion, since the Rosati involution is the adjoint operator for H (see Proposition 5.1.1). □

In order to construct moduli spaces out of this, we have to analyze when $(X_{Z'}, H_{Z'}, \iota_{Z'}) \simeq (X_Z, H_Z, \iota_Z)$ for points $Z, Z' \in \mathfrak{H}_m^e$. The action of $\mathrm{Sp}_{2m}(\mathbb{R})$ on $\mathfrak{H}_m$ of Proposition 8.2.2 induces an action of the group

$$G := \mathop{\times}_{i=1}^{e} \mathrm{Sp}_{2m}(\mathbb{R})$$

on $\mathfrak{H}_m^e$: for $Z = (Z^1, \ldots, Z^e) \in \mathfrak{H}_m^e$ and $M = (M^1, \ldots, M^e) \in G$ define

$$M(Z) = (M^1(Z^1), \ldots, M^e(Z^e))$$

with $M^\nu(Z^\nu) = (\alpha^\nu Z^\nu + \beta^\nu)(\gamma^\nu Z^\nu + \delta^\nu)^{-1}$, where $M^\nu = \begin{pmatrix} \alpha^\nu & \beta^\nu \\ \gamma^\nu & \delta^\nu \end{pmatrix}$ with $(m \times m)$-matrices α^ν, β^ν, γ^ν and δ^ν. For any $\mathbb{Z}$-submodule $\mathcal{M}$ of F^{2m} as above define the subgroup $G(\mathcal{M})$ of G by

$$G(\mathcal{M}) = \{M \in G \mid \operatorname{diag}({}^tM^1, \ldots, {}^tM^e)\mathcal{M} \subseteq \mathcal{M}\}\ .$$

Here we consider $\mathcal{M}$ as a submodule of $(F \otimes_{\mathbb{Q}} \mathbb{R})^{2m} = \mathbb{R}^{2g}$.

(2.2) Proposition. *Let Z and Z' be two points in $\mathfrak{H}_m^e$. The polarized abelian varieties $(X_{Z'}, H_{Z'}, \iota_{Z'})$ and (X_Z, H_Z, ι_Z) with endomorphism structure $(F, {}', \varrho)$ are isomorphic if and only if there is an $M \in G(\mathcal{M})$ such that $Z' = M(Z)$.*

Proof. Suppose $f: (X_{Z'}, H_{Z'}, \iota_{Z'}) \to (X_Z, H_Z, \iota_Z)$ is an isomorphism. Let $A = \varrho_a(f) \in M_g(\mathbb{C})$ be the analytic representation of f. By definition of an isomorphism of polarized abelian varieties with endomorphism structure $(F, {}', \varrho)$ the matrix A satisfies

$$AJ_{Z'}(\mathcal{M}) \subset J_Z(\mathcal{M}) \qquad (2)$$
$$A^* H_Z = H_{Z'} \qquad (3)$$
$$\varrho(a)A = A\varrho(a) \qquad (4)$$

for all $a \in F$. Analyzing equation (4), we get that A is of the form

$$A = \operatorname{diag}(A^1, \ldots, A^e)$$

with $A^\nu \in M_m(\mathbb{C})$ for $\nu = 1, \ldots, e$. Using (4) and the fact that $\varrho(a)J_Z(\underline{b}) = J_Z(a \cdot \underline{b})$ for all $a \in F$, $\underline{b} \in (F \otimes_{\mathbb{Q}} \mathbb{R})^{2m}$, equation (2) implies that the composed map

$$J_Z^{-1} A J_{Z'}: F^{2m} \to F^{2m}$$

is an isomorphism of F-vector spaces. Hence there is a matrix $M \in \mathrm{GL}_{2m}(F)$ such that $J_Z^{-1} A J_{Z'}(\underline{a}) = {}^tM\underline{a}$ for all $\underline{a} \in F^{2m}$. If we consider as usual M as a matrix in $M_{2m}(F \otimes_{\mathbb{Q}} \mathbb{R}) = M_{2m}(\mathbb{R}^e) = M_{2m}(\mathbb{R})^e$ and write $M = (M^1, \ldots, M^e)$, then this means

$$AJ_{Z'} = J_Z \operatorname{diag}({}^tM^1, \ldots, {}^tM^e) . \qquad (5)$$

Inserting the special form of the matrix A and the definition of J_Z and $J_{Z'}$, equation (5) gives for all ν

$$A^\nu (Z'^\nu, \mathbb{1}_m) = (Z^\nu, \mathbb{1}_m)\, {}^tM^\nu .$$

Setting as usual $M^\nu = \begin{pmatrix} \alpha^\nu & \beta^\nu \\ \gamma^\nu & \delta^\nu \end{pmatrix}$ this is equivalent to

$$Z'^\nu = M^\nu(Z^\nu) = (\alpha^\nu Z^\nu + \beta^\nu)(\gamma^\nu Z^\nu + \delta^\nu)^{-1} .$$

It remains to show that $M = (M^1, \ldots, M^e) \in G(\mathcal{M})$: by (3) we get using (5)

$$\operatorname{Im} J_{Z'}^* H_{Z'} = \operatorname{Im} J_{Z'}^* A^* H_Z = \operatorname{diag}({}^tM^1, \ldots, {}^tM^e)^* \operatorname{Im} J_Z^* H_Z. \qquad (6)$$

In the proof of Proposition 2.1 we saw that

$$\operatorname{Im} H_Z(J_Z(\underline{a}), J_Z(\underline{b})) = {}^t\underline{a} \operatorname{diag}\left(\begin{pmatrix} 0 & \mathbb{1}_m \\ -\mathbb{1}_m & 0 \end{pmatrix}, \ldots, \begin{pmatrix} 0 & \mathbb{1}_m \\ -\mathbb{1}_m & 0 \end{pmatrix}\right) \underline{b}$$

for all $\underline{a}, \underline{b} \in (F \otimes_{\mathbb{Q}} \mathbb{R})^{2m} = \mathbb{R}^{2me}$ and similarly for Z'. Inserting this into (6), we get $\begin{pmatrix} 0 & \mathbb{1}_m \\ -\mathbb{1}_m & 0 \end{pmatrix} = M^\nu \begin{pmatrix} 0 & \mathbb{1}_m \\ -\mathbb{1}_m & 0 \end{pmatrix} {}^tM^\nu$ for all ν. Since the symplectic group is invariant under transposition, this means $M \in G$. As $\operatorname{diag}({}^tM^1, \ldots, {}^tM^e)\mathcal{M} \subseteq \mathcal{M}$ by (2) and the definition of M, we even obtain $M \in G(\mathcal{M})$. The converse implication of the proposition follows easily by reading the above arguments upside down. □

The group $G(\mathcal{M})$ is discrete in G, since $\mathcal{M}$ is a lattice in some real vector space on which G acts. According to Proposition 8.2.5 the group $G(\mathcal{M})$ acts properly and discontinuously on $\mathfrak{H}_m^e$. Hence by Theorem A.6 the quotient

$$\mathcal{A}(\mathcal{M}) := \mathfrak{H}_m^e / G(\mathcal{M})$$

is a normal complex analytic space. It is a moduli space in the sense of Chapter 8. We call it the *moduli space of polarized abelian varieties with endomorphism structure* $(F, {}', \varrho)$ associated to the F-module $\mathcal{M}$. Since $G(\mathcal{M})$ acts properly and discontinuously on $\mathfrak{H}_m^e$, we have

$$\dim \mathcal{A}(\mathcal{M}) = \dim \mathfrak{H}_m^e = \tfrac{e}{2} m(m+1) .$$

Note that the choice of $\mathcal{M}$ fixes the type of the polarization. Changing $\mathcal{M}$ means also that the type of the polarization changes. In this way we obtain infinitely many moduli spaces for polarized abelian varieties with endomorphism structure $(F, {}', \varrho)$. Every type of polarization occurs at least once. The following proposition shows that conversely every triplet (X, H, ι) is contained in one of these.

(2.3) Proposition. *Every polarized abelian variety (X, H, ι) of dimension g with endomorphism structure $(F, \mathrm{id}_F, \varrho)$ is contained in some moduli space $\mathcal{A}(\mathcal{M})$ as above.*

Proof. As we saw above we may assume $\iota = \varrho$. Suppose $X = \mathbb{C}^g/\Lambda$. We may choose the basis of $\mathbb{C}^g$ in such a way that the matrix of H is diagonal. By assumption $\Lambda \otimes \mathbb{Q}$ is an F-vector space of dimension $2m = \frac{2}{e} g$ via ϱ. Choose an isomorphism of F-vector spaces

$$\eta \colon F^{2m} \to \Lambda \otimes \mathbb{Q} . \tag{7}$$

For any $\underline{b}, \underline{c} \in F^{2m}$ define a $\mathbb{Q}$-linear map $F \to \mathbb{Q}$ by $a \mapsto (\eta^* \operatorname{Im} H)(\varrho(a)\underline{b}, \underline{c})$. Since the trace $\operatorname{tr}_{F/\mathbb{Q}}$ is a nondegenerate bilinear form, there is an element $T(\underline{b}, \underline{c}) \in F$ such that

$$(\eta^* \operatorname{Im} H)(\varrho(a)\underline{b}, \underline{c}) = \operatorname{tr}_{F/\mathbb{Q}}(aT(\underline{b}, \underline{c})) .$$

Since $\operatorname{Im} H$ is an alternating form and by Proposition 5.1.1, the map $T: F^{2m} \times F^{2m} \to F$, $(\underline{b}, \underline{c}) \mapsto T(\underline{b}, \underline{c})$ is an alternating F-bilinear form. As usual we may choose the basis of F^{2m} in such a way that

$$T(\underline{b}, \underline{c}) = {}^t\underline{b} \begin{pmatrix} 0 & \mathbb{1}_m \\ -\mathbb{1}_m & 0 \end{pmatrix} \underline{c}$$

for all $\underline{b}, \underline{c} \in F^{2m}$. Let $U \in M(g \times 2g, \mathbb{C})$ be the matrix of the $\mathbb{R}$-vector space isomorphism $\eta \otimes 1_{\mathbb{R}}: (F \otimes_{\mathbb{Q}} \mathbb{R})^{2m} = \mathbb{R}^{2g} \to \mathbb{C}^g$. Since η is an F-vector space isomorphism, we get $\varrho(a)(\eta \otimes 1_{\mathbb{R}})(\underline{b}) = (\eta \otimes 1_{\mathbb{R}})(a \cdot \underline{b})$ for all $a \in F$ and $\underline{b} \in (F \otimes_{\mathbb{Q}} \mathbb{R})^{2m}$. From this it is easy to see that the matrix U is of the form $U = \operatorname{diag}((U^1, V^1), \ldots, (U^e, V^e))$ with $U^\nu, V^\nu \in \mathrm{GL}_m(\mathbb{C})$. Define $Z^\nu = (V^\nu)^{-1} U^\nu$, $1 \le \nu \le e$. Then, changing the coordinates of $\mathbb{C}^g$ by the matrix $\operatorname{diag}((V^1)^{-1}, \ldots, (V^e)^{-1})$, the map $\eta \otimes 1_{\mathbb{R}}$ is given by the matrix

$$\operatorname{diag}((Z^1, \mathbb{1}), \ldots, (Z^e, \mathbb{1}))$$

with respect to the new bases. Finally define a free $\mathbb{Z}$-submodule $\mathcal{M}$ of rank $2g$ of F^{2m} by

$$\mathcal{M} = \eta^{-1}(\Lambda) .$$

We claim that $Z = (Z^1, \ldots, Z^e) \in \mathfrak{H}_m^e$ and that the triplet (X, H, ϱ) is isomorphic to $(X_Z = \mathbb{C}^g / J_Z(\mathcal{M}), H_Z, \iota_Z)$. For the proof note first that

$$((\eta \otimes 1_{\mathbb{R}})^* \operatorname{Im} H)(\underline{a}, \underline{b}) = \operatorname{tr}_{F \otimes \mathbb{R}/\mathbb{R}}\left({}^t\underline{a} \begin{pmatrix} 0 & \mathbb{1}_m \\ -\mathbb{1}_m & 0 \end{pmatrix} \underline{b}\right) = \sum_{\nu=1}^{e} {}^t\underline{a}^\nu \begin{pmatrix} 0 & \mathbb{1}_m \\ -\mathbb{1}_m & 0 \end{pmatrix} \underline{b}^\nu \tag{8}$$

for all $\underline{a}, \underline{b} \in (F \otimes_{\mathbb{Q}} \mathbb{R})^{2m}$. On the other hand, with respect to the new bases of $\mathbb{C}^g$ the matrix of H is of the form $\operatorname{diag}(H^1, \ldots, H^e)$ with positive definite hermitian $(m \times m)$-matrices H^ν, and we have

$$(\eta \otimes 1_{\mathbb{R}})^* \operatorname{Im} H(\underline{a}, \underline{b}) = \sum_{\nu=1}^{e} {}^t\underline{a}^\nu \operatorname{Im}\left({}^t(Z^\nu, \mathbb{1}) H^\nu \overline{(Z^\nu, \mathbb{1})}\right) \underline{b}^\nu$$

for all $\underline{a}, \underline{b} \in (F \otimes_{\mathbb{Q}} \mathbb{R})^{2m}$. Comparing this with (8) gives

$$\operatorname{Im} \begin{pmatrix} {}^tZ^\nu H^\nu \overline{Z^\nu} & {}^tZ^\nu H^\nu \\ H^\nu \overline{Z^\nu} & H^\nu \end{pmatrix} = \begin{pmatrix} 0 & \mathbb{1}_m \\ -\mathbb{1}_m & 0 \end{pmatrix}$$

for $1 \le \nu \le e$. This yields $Z^\nu \in \mathfrak{H}_m$ and $H^\nu = (\operatorname{Im} Z^\nu)^{-1}$ for $1 \le \nu \le e$. So $\eta \otimes 1_{\mathbb{R}} = J_Z$. Since by definition $\iota_Z = \varrho$, we obtain $(X, H, \iota) \simeq (X_Z, H_Z, \iota_Z)$. □

(2.4) Example. Consider the special case $e = 1$, $m = g$. Then F is the field of rational numbers and ϱ the natural representation of $\mathbb{Q}$ in $M_g(\mathbb{C})$. Here the endomorphism structure is not an additional information. This reflects the fact that the parameter space is the Siegel upper half space which appeared already in the construction of the moduli spaces of polarized abelian varieties in Section 8.1. Moreover J_Z coincides with the isomorphism $j_Z\colon \mathbb{R}^{2g} \to \mathbb{C}^g$ as defined in Section 8.7. Thus for $\mathcal{M} = \mathbb{Z}^g \times d_1\mathbb{Z} \times \cdots \times d_g\mathbb{Z}$ Theorem 8.2.6 is a special case of the above construction.

(2.5) Example. Consider the special case $e = g$, $m = 1$. Then F is a totally real number field of degree g over $\mathbb{Q}$, and $\varrho\colon F \to M_g(\mathbb{C})$ is given by $a \mapsto \operatorname{diag}(a^1, \ldots, a^g)$, where as usual $a^1, \ldots, a^g$ denote the images of a under the different embeddings $F \hookrightarrow \mathbb{R}$. Let $\mathcal{M}$ denote a fractional ideal in the field F. The most important case is $\mathcal{M} = \mathfrak{o}$, the ring of integers in F. For any $z = (z^1, \ldots, z^g) \in \mathfrak{H}_1^g$ the isomorphism $J_Z = \mathbb{R}^{2g} \to \mathbb{C}^g$ is given by the matrix $\operatorname{diag}((z^1, 1), \ldots, (z^g, 1))$. Hence

$$X_Z = \mathbb{C}^g/\mathcal{M}z + \mathcal{M} \ ,$$

where $\mathcal{M}z + \mathcal{M} = \{ {}^t(\lambda^1 z^1 + \mu^1, \ldots, \lambda^g z^g + \mu^g) \in \mathbb{C}^g \mid \lambda, \mu \in \mathcal{M}\}$ according to our notation. The corresponding moduli spaces $\mathcal{A}(\mathcal{M})$ are called *Hilbert modular varieties.*

§ 3 Some Notation

In the following three sections we construct families of abelian varieties with endomorphism structure of the remaining three types. The construction in each case is essentially the same as for type I in the last section. However it is complicated by the fact that $F \otimes_{\mathbb{Q}} \mathbb{R}$ is a direct sum of matrix algebras rather than a direct sum of copies of $\mathbb{R}$, as in the real multiplication case. So we have to deal with matrix algebras over rings of matrices. In order to describe isomorphisms of these $\mathbb{R}$-algebras with $\mathbb{C}^g$ in terms of matrices, we have to choose suitable bases. This is the aim of this section. Note that for type II and III we only need the special case $d = 2$.

Consider the direct sum $M_d(\mathbb{C})^{em}$ of em matrix algebras of degree d for some positive integers d, m and e. The elements of $M_d(\mathbb{C})^{em}$ are vectors of length em with entries in $M_d(\mathbb{C})$. We write them in the form

$$\underline{a} = \begin{pmatrix} \underline{a}^1 \\ \vdots \\ \underline{a}^e \end{pmatrix}, \quad \text{where} \quad \underline{a}^\nu = \begin{pmatrix} a_1^\nu \\ \vdots \\ a_m^\nu \end{pmatrix} \in M_d(\mathbb{C})^m \ ,$$

$$\text{and} \quad a_\mu^\nu = \begin{pmatrix} (a_\mu^\nu)_{11} & \cdots & (a_\mu^\nu)_{1d} \\ \vdots & & \\ (a_\mu^\nu)_{d1} & \cdots & (a_\mu^\nu)_{dd} \end{pmatrix} \in M_d(\mathbb{C}) \ .$$

The $\mathbb{C}$-vector space $M_d(\mathbb{C})^{em}$ is of dimension d^2em. Define a $\mathbb{C}$-vector space isomorphism

$$\tilde{\ }: M_d(\mathbb{C})^{em} \to \mathbb{C}^{d^2em}, \quad \underline{a} \mapsto \underline{\tilde{a}} = \begin{pmatrix} \underline{\tilde{a}}^1 \\ \vdots \\ \underline{\tilde{a}}^e \end{pmatrix}, \tag{1}$$

$$\text{where} \quad \underline{\tilde{a}}^\nu = \begin{pmatrix} (\underline{a}^\nu)_{11} \\ (\underline{a}^\nu)_{12} \\ \vdots \\ (\underline{a}^\nu)_{dd} \end{pmatrix} \in \mathbb{C}^{d^2m} \quad \text{and} \quad (\underline{a}^\nu)_{jk} = \begin{pmatrix} (a_1^\nu)_{jk} \\ \vdots \\ (a_m^\nu)_{jk} \end{pmatrix} \in \mathbb{C}^m .$$

Accordingly, for a submodule $\mathcal{M}$ of $M_d(\mathbb{C})^{em}$ over $\mathbb{Z}$ we denote by $\mathcal{M}^\sim$ its image in $\mathbb{C}^{d^2em}$. If in particular $e = 1$, we omit the upper index in the notation of above.

The map $': M_d(\mathbb{C})^m \to M_d(\mathbb{C})^m$,

$$\underline{a} = \begin{pmatrix} a_1 \\ \vdots \\ a_m \end{pmatrix} \mapsto \underline{a}' = \begin{pmatrix} {}^t\bar{a}_1 \\ \vdots \\ {}^t\bar{a}_m \end{pmatrix}$$

is a canonical anti-involution on the $\mathbb{C}$-algebra $M_d(\mathbb{C})^m$. Consider $M_d(\mathbb{C})^m$ as a left-module over the ring $M_d(\mathbb{C})$. Let T be an $(m \times m)$-matrix with entries in $M_d(\mathbb{C})$, i.e. $T \in M_m(M_d(\mathbb{C}))$. The matrix T defines a map on $M_d(\mathbb{C})^m \times M_d(\mathbb{C})^m$ with values in $M_d(\mathbb{C})$ by $(\underline{a}, \underline{b}) \mapsto {}^t\underline{a}T\underline{b}'$, where ${}^t\underline{a} = (a_1, \ldots, a_m)$. This leads to a map $M_d(\mathbb{C})^m \times M_d(\mathbb{C})^m \to \mathbb{C}$,

$$(\underline{a}, \underline{b}) \mapsto \operatorname{tr}_{M_d(\mathbb{C})/\mathbb{C}}({}^t\underline{a}T\underline{b}') . \tag{2}$$

Certainly this map is $\mathbb{C}$-linear in the first and $\mathbb{C}$-antilinear in the second component. In order to translate (2) into terms of the vector space $\mathbb{C}^{d^2m}$ via the isomorphism (1), we write the matrix $T \in M_m(M_d(\mathbb{C}))$ in the following form

$$T = (T_{\lambda\mu})_{\lambda,\mu=1,\ldots,m} \quad \text{with} \quad T_{\lambda\mu} = \big((T_{\lambda\mu})_{jk}\big)_{j,k=1,\ldots,d} \in M_d(\mathbb{C})$$

and associate to T the matrix $\widetilde{T} \in M_d(M_m(\mathbb{C})) = M_{dm}(\mathbb{C})$ defined by

$$\widetilde{T} = (\widetilde{T}_{jk})_{j,k=1,\ldots,d} \quad \text{with} \quad \widetilde{T}_{jk} = \big((T_{\lambda\mu})_{jk}\big)_{\lambda,\mu=1,\ldots,m} \in M_m(\mathbb{C}) .$$

For example, in the special case $d = m = 2$, the matrices T and $\widetilde{T}$ are

$$T = \begin{pmatrix} (T_{11})_{11} & (T_{11})_{12} & (T_{12})_{11} & (T_{12})_{12} \\ (T_{11})_{21} & (T_{11})_{22} & (T_{12})_{21} & (T_{12})_{22} \\ (T_{21})_{11} & (T_{21})_{12} & (T_{22})_{11} & (T_{22})_{12} \\ (T_{21})_{21} & (T_{21})_{22} & (T_{22})_{21} & (T_{22})_{22} \end{pmatrix}, \quad \text{and}$$

$$\widetilde{T} = \begin{pmatrix} (T_{11})_{11} & (T_{12})_{11} & (T_{11})_{12} & (T_{12})_{12} \\ (T_{21})_{11} & (T_{22})_{11} & (T_{21})_{12} & (T_{22})_{12} \\ (T_{11})_{21} & (T_{12})_{21} & (T_{11})_{22} & (T_{12})_{22} \\ (T_{21})_{21} & (T_{22})_{21} & (T_{21})_{22} & (T_{22})_{22} \end{pmatrix} .$$

It is easy to see that $\widetilde{T}$ can be obtained from T by conjugation with a permutation matrix, namely

(3.1) Lemma. *Define the matrix $P \in M_{dm}(\mathbb{C})$ by the Kronecker product $P = (\mathbb{1}_m \otimes e_1, \dots, \mathbb{1}_m \otimes e_d)$, where $e_1, \dots, e_d$ denotes the standard basis of $\mathbb{C}^d$, then*

$$\widetilde{T} = {}^tPTP$$

for every $T \in M_m(M_d(\mathbb{C}))$. In particular, the map

$$\tilde{}: M_m(M_d(\mathbb{C})) \to M_d(M_m(\mathbb{C})), \quad T \mapsto \widetilde{T}$$

is an isomorphism of $\mathbb{C}$-algebras.

Applying the isomorphism $T \mapsto \widetilde{T}$, we can determine the matrix of the sesquilinear form (2) with respect to the standard basis of $\mathbb{C}^{d^2m}$.

(3.2) Lemma. *For all $\underline{a}, \underline{b} \in M_d(\mathbb{C})^m$ and $T \in M_m(M_d(\mathbb{C}))$*

$$\operatorname{tr}_{M_d(\mathbb{C})/\mathbb{C}}({}^t\underline{a}T\underline{b}') = {}^t\tilde{\underline{a}}(\mathbb{1}_d \otimes \widetilde{T})\bar{\tilde{\underline{b}}} \ .$$

Proof.
$$\operatorname{tr}_{M_d(\mathbb{C})/\mathbb{C}}({}^t\underline{a}T\underline{b}') = \sum_{\lambda,\mu=1}^{m} \operatorname{tr}_{M_d(\mathbb{C})/\mathbb{C}}(a_\lambda T_{\lambda\mu}{}^t\bar{b}_\mu)$$

$$= \sum_{\lambda,\mu=1}^{m} \sum_{j,k=1}^{d} \sum_{p=1}^{d} (a_\lambda)_{pj}(T_{\lambda\mu})_{jk}(\bar{b}_\mu)_{pk} = \sum_{p=1}^{d} \sum_{j,k=1}^{d} {}^t(\underline{a})_{pj}\widetilde{T}_{jk}(\bar{\underline{b}})_{pk}$$

$$= \sum_{p=1}^{d} {}^t\begin{pmatrix} (\underline{a})_{p1} \\ \vdots \\ (\underline{a})_{pd} \end{pmatrix} \widetilde{T} \begin{pmatrix} \bar{\underline{b}}_{p1} \\ \vdots \\ \bar{\underline{b}}_{pd} \end{pmatrix} = {}^t\tilde{\underline{a}} \begin{pmatrix} \widetilde{T} & & \\ & \ddots & \\ & & \widetilde{T} \end{pmatrix} \bar{\tilde{\underline{b}}} \ . \qquad \square$$

Finally consider the homomorphism $\varphi_T: M_d(\mathbb{C})^m \to M_d(\mathbb{C})^m$ of left $M_d(\mathbb{C})$-modules defined by

$$\varphi_T(\underline{a}) = ({}^tT'\underline{a}')' \ ,$$

where $T' = (T'_{\lambda\mu})$ and ^{t} means transposition of the $(m \times m)$-matrix T' (but not of its entries $T'_{\lambda\mu} \in M_d(\mathbb{C})$). Via the isomorphism (1) the map φ_T induces an $\mathbb{R}$-vector space homomorphism $\mathbb{C}^{d^2m} \to \mathbb{C}^{d^2m}$. With a computation similar as in the previous proof one can compute the corresponding matrix:

(3.3) Lemma. *The following diagram commutes*

$$\begin{array}{ccc} M_d(\mathbb{C})^m & \xrightarrow{\sim} & \mathbb{C}^{d^2m} \\ {\scriptstyle\varphi_T}\downarrow & & \downarrow{\scriptstyle\mathbb{1}_d \otimes {}^t\widetilde{T}} \\ M_d(\mathbb{C})^m & \xrightarrow{\sim} & \mathbb{C}^{d^2m} \end{array} .$$

(Here ^{t} means transposition of the $(dm \times dm)$-matrix $\widetilde{T}$.)

§ 4 Families of Abelian Varieties with Totally Indefinite Quaternion Multiplication

Let F be a totally indefinite quaternion algebra over a totally real number field K with $[K:\mathbb{Q}] = e$ and $'$ a positive anti-involution on F. We want to construct families of g-dimensional abelian varieties X with $F \hookrightarrow \mathrm{End}_{\mathbb{Q}}(X)$.

We say that an abelian variety X *admits totally indefinite quaternion multiplication* by F if there is an embedding $F \hookrightarrow \mathrm{End}_{\mathbb{Q}}(X)$ with F as above. Proposition 5.5.7 gives the following condition for g

$$g = 2em$$

for some integer $m \geq 1$. The e different embeddings $K \hookrightarrow \mathbb{R}$ induce an isomorphism of $F \otimes_{\mathbb{Q}} \mathbb{R}$ with $M_2(\mathbb{R})^e$ and of $(F \otimes_{\mathbb{Q}} \mathbb{R})^m$ with $M_2(\mathbb{R})^{em}$. We identify the elements of $F \otimes_{\mathbb{Q}} \mathbb{R}$ (respectively $(F \otimes_{\mathbb{Q}} \mathbb{R})^m$) with their images in $M_2(\mathbb{R})^e$ (respectively $M_2(\mathbb{R})^{em}$). According to Lemma 5.5.1 we may assume that the anti-involution $'$ extends to matrix transposition on every factor $M_2(\mathbb{R})$. We use the notation of Section 3 with $d = 2$. In particular, the isomorphism $\tilde{}$ restricts to the isomorphism

$$(F \otimes_{\mathbb{Q}} \mathbb{R})^m = M_2(\mathbb{R})^{em} \to \mathbb{R}^{4em}, \quad \underline{a} \mapsto \tilde{\underline{a}} \, .$$

Define a representation $\varrho \colon F \to M_g(\mathbb{C})$ by

$$\varrho(a) = \mathrm{diag}(a^1 \otimes \mathbb{1}_m, \ldots, a^e \otimes \mathbb{1}_m) \, .$$

Obviously every representation of $F \mapsto M_g(\mathbb{C})$, satisfying the conditions of Lemma 1.1, is equivalent to ϱ.

Our first aim is to associate to every element Z of the e-fold product $\mathfrak{H}_m^e$ of the Siegel upper half space $\mathfrak{H}_m$ a polarized abelian variety (X_Z, H_Z, ι_Z) with endomorphism structure $(F, {}', \varrho)$: fix a pair $(\mathcal{M}, T)$ with $\mathcal{M}$ a free $\mathbb{Z}$-submodule of F^m of rank $4em$ and T a nondegenerate $(m \times m)$-matrix over F with ${}^tT' = -T$ such that

$$\mathrm{tr}_{F/\mathbb{Q}}({}^t\underline{a}T\underline{b}') \in \mathbb{Z}$$

for all $\underline{a}, \underline{b} \in \mathcal{M} \subset F^m$. It is easy to see that such pairs $(\mathcal{M}, T)$ exist.

Via the embedding $F \hookrightarrow F \otimes_{\mathbb{Q}} \mathbb{R} = M_2(\mathbb{R})^e$ we may consider T as an element of $M_m(M_2(\mathbb{R}))^e$ and we write T in the form $(T^1, \ldots, T^e)$ with $T^\nu \in M_m(M_2(\mathbb{R})) = M_{2m}(\mathbb{R})$. With T also the matrices T^ν are nondegenerate, whereas the property ${}^tT' = -T$ implies that the T^ν's are alternating, considered as elements of $M_{2m}(\mathbb{R})$. Applying the isomorphism $\tilde{}\colon M_m(M_2(\mathbb{R})) \to M_2(M_m(\mathbb{R}))$ of Lemma 3.1, we obtain nondegenerate alternating matrices $\widetilde{T}^\nu \in M_2(M_m(\mathbb{R})) = M_{2m}(\mathbb{R})$. Hence there are matrices $W^\nu \in \mathrm{GL}_{2m}(\mathbb{R})$ such that

$$\widetilde{T}^\nu = {}^tW^\nu \begin{pmatrix} 0 & \mathbb{1}_m \\ -\mathbb{1}_m & 0 \end{pmatrix} W^\nu$$

for $\nu = 1, \dots, e$.

For any $Z = (Z^1, \dots, Z^e) \in \mathfrak{H}_m^e$ define an $\mathbb{R}$-vector space homomorphism

$$J_Z \colon \mathbb{R}^{4em} \to \mathbb{C}^{2em} = \mathbb{C}^g$$

by the matrix

$$\operatorname{diag}(J_Z^1, \dots, J_Z^e), \quad \text{where} \quad J_Z^\nu = \begin{pmatrix} (Z^\nu, \mathbb{1})W^\nu & 0 \\ 0 & (Z^\nu, \mathbb{1})W^\nu \end{pmatrix},$$

with respect to the standard bases. In fact, J_Z is an isomorphism, since the columns of the defining matrix are linearly independent over $\mathbb{R}$. Hence $J_Z(\mathcal{M}^{\smallsmile})$ is a lattice in $\mathbb{C}^g$ and the quotient

$$X_Z := \mathbb{C}^g / J_Z(\mathcal{M}^{\smallsmile})$$

is a complex torus. Moreover, define a map $H_Z \colon \mathbb{C}^g \times \mathbb{C}^g \to \mathbb{C}^g$ by

$$H_Z(x, y) = {}^tx \operatorname{diag}(H^1, \dots, H^e)\, \bar{y}, \quad \text{where} \quad H^\nu = \begin{pmatrix} \operatorname{Im} Z^\nu & 0 \\ 0 & \operatorname{Im} Z^\nu \end{pmatrix}^{-1}.$$

By definition H_Z is a positive definite hermitian form. We claim that H_Z defines a polarization on X_Z. For this we have only to check that its imaginary part is integral on the lattice $J_Z(\mathcal{M}^{\smallsmile})$. This follows from the definition of $\mathcal{M}$ together with the following

(4.1) Lemma. *For all $\underline{a}, \underline{b} \in (F \otimes_{\mathbb{Q}} \mathbb{R})^m$*

$$\operatorname{Im} H_Z(J_Z(\underline{\tilde{a}}), J_Z(\underline{\tilde{b}})) = \operatorname{tr}_{F \otimes_{\mathbb{Q}} \mathbb{R}/\mathbb{R}}({}^t\underline{a} T \underline{b}') .$$

Proof. Obviously it suffices to prove the assertion in the case $e = 1$. Correspondingly we omit the upper indices. Then we have applying Lemma 3.2

$$\begin{aligned}
&\operatorname{Im} H_Z(J_Z(\underline{\tilde{a}}), J_Z(\underline{\tilde{b}})) = \\
&= {}^t\underline{\tilde{a}} \operatorname{Im} \begin{pmatrix} {}^tW\, {}^t(Z,\mathbb{1})(\operatorname{Im} Z)^{-1}\overline{(Z,\mathbb{1})}W & \\ & {}^tW\, {}^t(Z,\mathbb{1})(\operatorname{Im} Z)^{-1}\overline{(Z,\mathbb{1})}W \end{pmatrix} \underline{\tilde{b}} \\
&= {}^t\underline{\tilde{a}} \begin{pmatrix} {}^tW & \\ & {}^tW \end{pmatrix} \begin{pmatrix} \begin{pmatrix} 0 & \mathbb{1}_m \\ -\mathbb{1}_m & 0 \end{pmatrix} & \\ & \begin{pmatrix} 0 & \mathbb{1}_m \\ -\mathbb{1}_m & 0 \end{pmatrix} \end{pmatrix} \begin{pmatrix} W & \\ & W \end{pmatrix} \underline{\tilde{b}} \\
&= {}^t\underline{\tilde{a}} \begin{pmatrix} \widetilde{T} & \\ & \widetilde{T} \end{pmatrix} \underline{\tilde{b}} = \operatorname{tr}_{M_2(\mathbb{R})/\mathbb{R}}({}^t\underline{a} T \underline{b}') = \operatorname{tr}_{F \otimes_{\mathbb{Q}} \mathbb{R}/\mathbb{R}}({}^t\underline{a} T \underline{b}') . \qquad \square
\end{aligned}$$

Finally, we claim that for all $a \in F$ and $\underline{b} \in (F \otimes_{\mathbb{Q}} \mathbb{R})^m$

$$\varrho(a)\, J_Z(\underline{\tilde{b}}) = J_Z(a \cdot \underline{\tilde{b}}) .$$

This follows from an immediate matrix computation, using the fact that the multiplication $F \times (F \otimes_{\mathbb{Q}} \mathbb{R})^m \to (F \otimes_{\mathbb{Q}} \mathbb{R})^m$ translates to

$$a \cdot \tilde{\underline{b}} := (a \cdot \underline{b})^{\tilde{}} = \operatorname{diag}(a^1 \otimes \mathbb{1}_{2m}, \ldots, a^e \otimes \mathbb{1}_{2m})\tilde{\underline{b}} \ .$$

Since $\mathcal{M} \otimes \mathbb{Q} = F^m$, this yields that for any $a \in F$ there is an integer $n > 0$ such that $\varrho(na)\, J_Z(\mathcal{M}^{\tilde{}}) \subseteq J_Z(\mathcal{M}^{\tilde{}})$. So we obtain $\varrho(F) \subset \operatorname{End}_{\mathbb{Q}}(X_Z)$. Combining everything and setting $\iota_Z = \varrho$, we have

(4.2) Proposition. *Given a pair* $(\mathcal{M}, T)$ *as above, for every* $Z \in \mathfrak{H}_m^e$ *the triplet* (X_Z, H_Z, ι_Z) *is a polarized abelian variety with endomorphism structure* $(F, {}' , \varrho)$.

For the proof it remains to show that the Rosati involution with respect to H_Z restricts to the anti-involution $'$ on F. But an immediate matrix computation shows that $H_Z(\iota_Z(a)x, y) = H_Z(x, \iota_Z(a')y)$ for all $a \in F$ and $x, y \in \mathbb{C}^g$. This implies the assertion, since the Rosati involution is the adjoint operator for H_Z (see Proposition 5.1.1). □

We have to analyze, when two points $Z, Z' \in \mathfrak{H}_m^e$ lead to isomorphic polarized abelian varieties with endomorphism structure $(F, {}' , \varrho)$. In Section 8 we will see that there is a group $G(\mathcal{M}, T)$, acting on $\mathfrak{H}_m^e$, such that for $Z, Z' \in \mathfrak{H}_m^e$ there is an isomorphism $(X_{Z'}, H_{Z'}, \iota_{Z'}) \simeq (X_Z, H_Z, \iota_Z)$ if and only if there is an $M \in G(\mathcal{M}, T)$ such that $Z' = M(Z)$. Since moreover $G(\mathcal{M})$ acts properly and discontinuously on $\mathfrak{H}_m^e$, the quotient

$$\mathcal{A}(\mathcal{M}, T) := \mathfrak{H}_m^e / G(\mathcal{M}, T)$$

is a normal complex analytic space of dimension $\frac{e}{2}m(m+1)$. It is a moduli space in the sense of Chapter 8. We call it the *moduli space of polarized abelian varieties with endomorphism structure* $(F, {}' , \varrho)$ *associated* to the pair $(\mathcal{M}, T)$.

(4.3) Remark. According to the proof of Proposition 4.2 the map $\mathfrak{H}_m^e \to \mathcal{A}(\mathcal{M}, T)$ depends on the particular choice of the matrices W^ν, $\nu = 1, \ldots, e$. In Section 8 we will see that maps $\mathfrak{H}_m^e \to \mathcal{A}(\mathcal{M}, T)$ corresponding to different choices of W^ν, $\nu = 1, \ldots, e$, differ only by an automorphism of $\mathfrak{H}_m^e$. Hence the moduli space $\mathcal{A}(\mathcal{M}, T)$ does not depend on W^ν, $\nu = 1, \ldots, e$, which justifies our notation.

Note that the choice of $(\mathcal{M}, T)$ fixes the type of the polarization. Every type $(d_1, \ldots, d_g)$ occurs at least once. The following proposition shows that, conversely, every triplet (X, H, ι) with totally indefinite quaternion multiplication by F is contained in one of these moduli spaces.

(4.4) Proposition. *Every (X, H, ι) with endomorphism structure $(F, {}', \varrho)$ is contained in $\mathcal{A}(\mathcal{M}, T)$ for some $(\mathcal{M}, T)$ as above.*

We omit the proof, since it is very similar to the proof of Proposition 2.3.

(4.5) Example. For the convenience of the reader let us specialize the construction above to the case of an abelian surface, that is $e = m = 1$. Here F is an indefinite quaternion algebra over $\mathbb{Q}$ and the representation $\varrho: F \to M_2(\mathbb{C})$ the natural one. Choose a pair $(\mathcal{M}, T)$ as above. In particular $\mathcal{M} \subset F \subset M_2(\mathbb{R})$ and T is an element of F with $T' = -T$ and thus $T = \left(\begin{smallmatrix} 0 & t \\ -t & 0 \end{smallmatrix}\right)$, considered as an element of $M_2(\mathbb{R})$. For $z \in \mathfrak{H}_1$ the isomorphism $J_z: \mathbb{R}^4 \to \mathbb{C}^2$ is given by the matrix $\left(\begin{smallmatrix} z & t & 0 & 0 \\ 0 & 0 & z & t \end{smallmatrix}\right)$ with respect to the standard bases. Since moreover $J_z(\mathcal{M}\check{}) = \mathcal{M}\left(\begin{smallmatrix} z \\ t \end{smallmatrix}\right)$, where we consider $\mathcal{M}$ as a submodule of $(F \otimes_{\mathbb{Q}} \mathbb{R}) = M_2(\mathbb{R})$,

$$X_z = \mathbb{C}^2/\mathcal{M}\left(\begin{smallmatrix} z \\ t \end{smallmatrix}\right) \quad \text{and} \quad H_z(x, y) = {}^t x \begin{pmatrix} \operatorname{Im} z & 0 \\ 0 & \operatorname{Im} z \end{pmatrix}^{-1} \bar{y}$$

for all $x, y \in \mathbb{C}^2$.

§ 5 Families of Abelian Varieties with Totally Definite Quaternion Multiplication

Let F be a totally definite quaternion algebra over a totally real number field K of degree e over $\mathbb{Q}$ and $'$ a positive anti-involution on F. We want to construct families of g-dimensional abelian varieties with $F \hookrightarrow \operatorname{End}_{\mathbb{Q}}(X)$.

We say that an abelian variety X *admits totally definite quaternion multiplication* by F, if there is an embedding $F \hookrightarrow \operatorname{End}_{\mathbb{Q}}(X)$ with F as above. As in the previous section Proposition 5.5.7 gives the following condition for g:

$$g = 2em$$

for some integer $m \geq 1$. The e different embeddings $K \hookrightarrow \mathbb{R}$ induce an isomorphism of $F \otimes_{\mathbb{Q}} \mathbb{R}$ with $\mathbb{H}^e$. It is convenient to consider $\mathbb{H}$ as a subalgebra of $M_2(\mathbb{C})$ via the representation $\mathbb{H} = \mathbb{C} \oplus j\mathbb{C} \to M_2(\mathbb{C})$,

$$\alpha + j\beta \mapsto \begin{pmatrix} \alpha & \beta \\ -\bar{\beta} & \bar{\alpha} \end{pmatrix}. \tag{1}$$

Thus we get an embedding $F \otimes_{\mathbb{Q}} \mathbb{R} \hookrightarrow M_2(\mathbb{C})^e$ and similarly $(F \otimes_{\mathbb{Q}} \mathbb{R})^m \hookrightarrow M_2(\mathbb{C})^{em}$. According to Lemma 5.5.1 we may assume that the anti-involution $'$ extends to the usual quaternion conjugation, i.e. $x' = {}^t\bar{x}$ on every factor $M_2(\mathbb{C})$. We identify the elements of $F \otimes_{\mathbb{Q}} \mathbb{R}$ (respectively $(F \otimes_{\mathbb{Q}} \mathbb{R})^m$) with their images in $M_2(\mathbb{C})^e$ (respectively $M_2(\mathbb{C})^{em}$). Moreover we use the isomorphism

$$M_2(\mathbb{C})^{em} \to \mathbb{C}^{4em}, \quad \underline{a} \mapsto \tilde{\underline{a}}$$

of Section 3 (with $d = 2$). Note that by (1)

$$(\underline{a}^\nu)_{21} = -\overline{(\underline{a}^\nu)_{12}} \quad \text{and} \quad (\underline{a}^\nu)_{22} = \overline{(\underline{a}^\nu)_{11}}$$

for all $\underline{a} \in (F \otimes_{\mathbb{Q}} \mathbb{R})^m \subset M_2(\mathbb{C})^{em}$ and $\nu = 1, \ldots, e$. Define a representation $\varrho \colon F \to M_g(\mathbb{C})$ by

$$\varrho(a) = \operatorname{diag}(a^1 \otimes \mathbb{1}_m, \ldots, a^e \otimes \mathbb{1}_m) \, .$$

Every representation $F \to M_g(\mathbb{C})$, satisfying the conditions of Lemma 1.1, is equivalent to ϱ.

Consider the complex manifold

$$\mathcal{H}_m := \{Z \in M_m(\mathbb{C}) \mid {}^tZ = -Z, \ \mathbb{1}_m - {}^t\overline{Z}Z > 0\}.$$

For all $Z \in \mathcal{H}_m$ the matrix $\mathbb{1}_m - {}^t\overline{Z}Z$ defines a positive definite hermitian form on $\mathbb{C}^m$. Our first aim is to associate to every member of the e-fold product $\mathcal{H}_m^e$ a polarized abelian variety with endomorphism structure $(F, {}', \varrho)$: fix a pair $(\mathcal{M}, T)$ with $\mathcal{M}$ a free $\mathbb{Z}$-submodule of F^m of rank $4em$, and T a nondegenerate matrix in $M_m(F)$ with ${}^tT' = -T$ such that

$$\operatorname{tr}_{F/\mathbb{Q}}({}^t\underline{a}T\underline{b}') \in \mathbb{Z}$$

for all $\underline{a}, \underline{b} \in \mathcal{M}$. It is easy to see that such pairs $(\mathcal{M}, T)$ exist.

If we consider $M_m(F)$ as a subspace of $M_m(F \otimes_{\mathbb{Q}} \mathbb{R}) = M_m(\mathbb{H})^e \subset M_m(M_2(\mathbb{C}))^e$, then T is of the form $(T^1, \ldots, T^e)$ with nondegenerate matrices T^ν contained in the image of the map $M_m(\mathbb{H}) \to M_m(M_2(\mathbb{C}))$. The condition ${}^tT' = -T$ implies that T^ν is skew-hermitian for $\nu = 1, \ldots, e$. Any nondegenerate skew-hermitian matrix in $M_m(\mathbb{H})$ is equivalent to $i\mathbb{1}_m$ over $\mathbb{H}$ (see Scharlau [1] Theorem 10.3.7). Since $i\mathbb{1}_m$ maps to $\mathbb{1}_m \otimes \begin{pmatrix} i & \\ & -i \end{pmatrix} = \operatorname{diag}(i, -i, \ldots, i, -i)$ in $M_m(M_2(\mathbb{C}))$, it follows that there are nonsingular matrices W^ν in the image of $M_m(\mathbb{H}) \to M_m(M_2(\mathbb{C}))$ such that $T^\nu = {}^tW^\nu \left(\mathbb{1}_d \otimes \begin{pmatrix} i & \\ & -i \end{pmatrix}\right) \overline{W^\nu}$ for $\nu = 1, \ldots, e$. Applying the isomorphism $\tilde{}$ of Lemma 3.1 this implies

$$\widetilde{T^\nu} = {}^t\widetilde{W^\nu} \begin{pmatrix} i\mathbb{1}_m & \\ & -i\mathbb{1}_m \end{pmatrix} \overline{\widetilde{W^\nu}}.$$

Moreover T^ν and W^ν being in the image of $M_m(\mathbb{H}) \to M_m(M_2(\mathbb{C}))$ means

$$\widetilde{T}^\nu = \begin{pmatrix} \widetilde{T}^\nu_{11} & \widetilde{T}^\nu_{12} \\ -\overline{\widetilde{T}^\nu_{12}} & \overline{\widetilde{T}^\nu_{11}} \end{pmatrix} \quad \text{and} \quad \widetilde{W}^\nu = \begin{pmatrix} \widetilde{W}^\nu_{11} & \widetilde{W}^\nu_{12} \\ -\overline{\widetilde{W}^\nu_{12}} & \overline{\widetilde{W}^\nu_{11}} \end{pmatrix}$$

with $\widetilde{T}^\nu_{jk}, \widetilde{W}^\nu_{jk} \in M_m(\mathbb{C})$.

For every $Z = (Z^1, \ldots, Z^e) \in \mathcal{H}^e_m$ define the $\mathbb{C}$-vector space homomorphism

$$J_Z\colon \mathbb{C}^{4em} \to \mathbb{C}^{2em} = \mathbb{C}^g$$

by the matrix

$$\operatorname{diag}(J^1_Z, \ldots, J^e_Z) \quad \text{with} \quad J^\nu_Z = \begin{pmatrix} ({}^tZ^\nu, \mathbb{1}_m)\widetilde{W}^\nu & 0 \\ 0 & ({}^t\overline{Z^\nu}, \mathbb{1}_m)\overline{\widetilde{W}^\nu} \end{pmatrix},$$

with respect to the standard bases.

(5.1) Lemma. *J_Z restricted to the subspace $((F \otimes_{\mathbb{Q}} \mathbb{R})^m)^{\tilde{}}$ of $\mathbb{C}^{4em}$ is an isomorphism of $\mathbb{R}$-vector spaces.*

Proof. It suffices to prove the assertion for $e = 1$. Correspondingly we omit the upper index. Moreover we may assume $\widetilde{W} = \mathbb{1}$, the matrix $\widetilde{W}$ being nonsingular. Consider the composed map

$$\mathbb{C}^{2m} \xhookrightarrow{\psi} \mathbb{C}^{4m} \xrightarrow{J_Z} \mathbb{C}^{2m} \xrightarrow{\varphi} \mathbb{C}^{2m},$$

where $$\psi\begin{pmatrix} x \\ y \end{pmatrix} := \begin{pmatrix} x \\ y \\ -\bar{y} \\ \bar{x} \end{pmatrix} \quad \text{and} \quad \varphi\begin{pmatrix} x \\ y \end{pmatrix} := \begin{pmatrix} x \\ \bar{y} \end{pmatrix}$$

for all $x, y \in \mathbb{C}^m$. We have to show that $J_Z\psi$ is an isomorphism of $\mathbb{R}$-vector spaces. Since φ is an $\mathbb{R}$-vector space isomorphism, it suffices to show that $\varphi J_Z \psi$ is an isomorphism of $\mathbb{R}$-vector spaces. But $\varphi J_Z \psi$ is even a $\mathbb{C}$-vector space isomorphism. In fact, it is given by the matrix $\begin{pmatrix} {}^tZ & \mathbb{1}_m \\ \mathbb{1}_m & -{}^t\overline{Z} \end{pmatrix}$, which is nonsingular, since

$$\begin{pmatrix} {}^tZ & \mathbb{1}_m \\ \mathbb{1}_m & -{}^t\overline{Z} \end{pmatrix} \begin{pmatrix} -\overline{Z} & \mathbb{1}_m \\ \mathbb{1}_m & Z \end{pmatrix} = \begin{pmatrix} \mathbb{1}_m - {}^tZ\overline{Z} & 0 \\ 0 & \mathbb{1}_m - {}^t\overline{Z}Z \end{pmatrix}$$

and $\mathbb{1}_m - {}^tZ\overline{Z}$ and $\mathbb{1}_m - {}^t\overline{Z}Z$ are positive definite by assumption. □

It follows that $J_Z(\mathcal{M}^{\tilde{}})$ is a lattice in $\mathbb{C}^g$ and the quotient

$$X_Z := \mathbb{C}^g / J_Z(\mathcal{M}^{\tilde{}})$$

is a complex torus. Define a map $H_Z\colon \mathbb{C}^g \times \mathbb{C}^g \to \mathbb{C}$ by

$$H_Z(x,y) = 2\,{}^t x\,\mathrm{diag}(H^1,\ldots,H^e)\,\bar{y}\;,\quad H^\nu = \begin{pmatrix} (\mathbb{1}_m - {}^tZ^\nu Z^\nu)^{-1} & \\ & (\mathbb{1}_m - Z^\nu\,{}^tZ^\nu)^{-1} \end{pmatrix}.$$

By definition, for every $Z \in \mathcal{H}_m$ the matrix $(\mathbb{1}_m - {}^tZZ)^{-1} = (\mathbb{1}_m - Z\,{}^tZ)^{-1}$ is positive definite and hermitian. So H_Z is a positive definite hermitian form on $\mathbb{C}^g$. We claim that H_Z defines a polarization on X_Z. For this we have to show that its imaginary part $\mathrm{Im}\,H_Z$ takes only integer values on the lattice $J_Z(\mathcal{M}\tilde{\ })$. But this is a consequence of the following

(5.2) Lemma. *For all $a, b \in (F \otimes_{\mathbb{Q}} \mathbb{R})^m$*

$$\mathrm{Im}\,H_Z(J_Z(\tilde{a}), J_Z(\tilde{b})) = \mathrm{tr}_{F\otimes\mathbb{R}/\mathbb{R}}({}^t a T b')\;.$$

Proof. It suffices to prove the assertion in the case $e = 1$. Correspondingly we omit the upper indices.

$\mathrm{Im}\,H_Z(J_Z(\tilde{a}), J_Z(\tilde{b})) =$

$$= 2\,\mathrm{Im}\;{}^t\!\begin{pmatrix} (a)_{11} \\ (a)_{12} \\ -(a)_{12} \\ (a)_{11} \end{pmatrix} \begin{pmatrix} {}^t\widetilde{W}\binom{Z}{\mathbb{1}}(\mathbb{1} - {}^tZZ)^{-1}({}^tZ, \mathbb{1})\overline{\widetilde{W}} & 0 \\ 0 & {}^t\widetilde{W}\binom{Z}{\mathbb{1}}(\mathbb{1} - Z\,{}^tZ)^{-1}({}^tZ, \mathbb{1})\overline{\widetilde{W}} \end{pmatrix} \overline{\begin{pmatrix} (b)_{11} \\ (b)_{12} \\ -(b)_{12} \\ (b)_{11} \end{pmatrix}}$$

$$\overset{(*)}{=} 2\,\mathrm{Im}\;{}^t\!\begin{pmatrix} (a)_{11} \\ (a)_{12} \end{pmatrix} {}^t\widetilde{W}\binom{Z}{\mathbb{1}}(\mathbb{1} - {}^tZZ)^{-1}({}^tZ, \mathbb{1})\overline{\widetilde{W}}\,\overline{\begin{pmatrix} (b)_{11} \\ (b)_{12} \end{pmatrix}}$$

$$-2\,\mathrm{Im}\;{}^t\!\begin{pmatrix} (a)_{11} \\ (a)_{12} \end{pmatrix} {}^t\widetilde{W}\,\overline{\binom{\mathbb{1}}{{}^tZ}(\mathbb{1} - Z\,{}^tZ)^{-1}(\mathbb{1}, Z)}\,\overline{\widetilde{W}}\,\overline{\begin{pmatrix} (b)_{11} \\ (b)_{12} \end{pmatrix}}$$

$$\overset{(**)}{=} 2\,\mathrm{Im}\;{}^t\!\begin{pmatrix} (a)_{11} \\ (a)_{12} \end{pmatrix} {}^t\widetilde{W}\,\overline{\begin{pmatrix} -\mathbb{1} & \\ & \mathbb{1} \end{pmatrix}}\,\overline{\widetilde{W}}\,\overline{\begin{pmatrix} (b)_{11} \\ (b)_{12} \end{pmatrix}}$$

$$= 2\,\mathrm{Im}\;{}^t\!\begin{pmatrix} (a)_{11} \\ (a)_{12} \end{pmatrix} (i\widetilde{T}) \begin{pmatrix} (b)_{11} \\ (b)_{12} \end{pmatrix}$$

by definition of $\widetilde{W}$. For equation $(*)$ we used that, according to the special form of the matrix $\widetilde{W}$, we have $\widetilde{W} = \begin{pmatrix} 0 & \mathbb{1} \\ -\mathbb{1} & 0 \end{pmatrix} \overline{\widetilde{W}} \begin{pmatrix} 0 & -\mathbb{1} \\ \mathbb{1} & 0 \end{pmatrix}$. For equation $(**)$ we used that $Z(1 - {}^tZZ)^{-1} = (1 - {}^tZZ)^{-1}Z$ for all $Z \in \mathcal{H}_m$. On the other hand, the trace of $\mathbb{H}$ over $\mathbb{R}$ translates to $\mathrm{tr}_{M_2(\mathbb{C})/\mathbb{C}}$ under the representation (1) and we have

$\mathrm{tr}_{\mathbb{H}/\mathbb{R}}({}^t aTb') = \mathrm{tr}_{M_2(\mathbb{C})/\mathbb{C}}({}^t aTb') =$

$$= {}^t\!\begin{pmatrix} (a)_{11} \\ (a)_{12} \\ -(a)_{12} \\ (a)_{11} \end{pmatrix} \begin{pmatrix} \widetilde{T} & 0 \\ 0 & \widetilde{T} \end{pmatrix} \begin{pmatrix} (b)_{11} \\ (b)_{12} \\ -(b)_{12} \\ (b)_{11} \end{pmatrix} \qquad \text{(by Lemma 3.2)}$$

$$= {}^t\begin{pmatrix} \underline{(a)}_{11} \\ \underline{(a)}_{12} \end{pmatrix} \widetilde{T} \begin{pmatrix} \overline{\underline{(b)}_{11}} \\ \overline{\underline{(b)}_{12}} \end{pmatrix} + \overline{{}^t\begin{pmatrix} \underline{(a)}_{11} \\ \underline{(a)}_{12} \end{pmatrix} \begin{pmatrix} 0 & \mathbb{1} \\ -\mathbb{1} & 0 \end{pmatrix} \overline{\widetilde{T}} \begin{pmatrix} 0 & -\mathbb{1} \\ \mathbb{1} & 0 \end{pmatrix} \begin{pmatrix} \overline{\underline{(b)}_{11}} \\ \overline{\underline{(b)}_{12}} \end{pmatrix}}$$

$$\overset{(*)}{=} 2\,\mathrm{Re}\; {}^t\begin{pmatrix} \underline{(a)}_{11} \\ \underline{(a)}_{12} \end{pmatrix} \widetilde{T} \begin{pmatrix} \overline{\underline{(b)}_{11}} \\ \overline{\underline{(b)}_{12}} \end{pmatrix}$$

$$= 2\,\mathrm{Im}\; {}^t\begin{pmatrix} \underline{(a)}_{11} \\ \underline{(a)}_{12} \end{pmatrix} (i\widetilde{T}) \begin{pmatrix} \overline{\underline{(b)}_{11}} \\ \overline{\underline{(b)}_{12}} \end{pmatrix} .$$

For equation $(*)$ we used that $\widetilde{T}$ satifies $\widetilde{T} = \left(\begin{smallmatrix} 0 & \mathbb{1} \\ -\mathbb{1} & 0 \end{smallmatrix}\right) \overline{\widetilde{T}} \left(\begin{smallmatrix} 0 & -\mathbb{1} \\ \mathbb{1} & 0 \end{smallmatrix}\right)$ by the special form of $\widetilde{T}$. □

Finally, we claim that for all $a \in F$ and $\underline{b} \in (F \otimes_{\mathbb{Q}} \mathbb{R})^m$

$$\varrho(a)\, J_Z(\tilde{\underline{b}}) = J_Z(a \cdot \tilde{\underline{b}}) \; .$$

This follows from an immediate matrix computation, using the fact that the multiplication $F \times (F \otimes_{\mathbb{Q}} \mathbb{R})^m \to (F \otimes_{\mathbb{Q}} \mathbb{R})^m$ translates to

$$a \cdot \tilde{\underline{b}} := (a \cdot \underline{b})\tilde{} = \mathrm{diag}(a^1 \otimes \mathbb{1}_{2m}, \dots, a^e \otimes \mathbb{1}_{2m})\, \tilde{\underline{b}} \; .$$

Since $\mathcal{M} \otimes \mathbb{Q} = F^m$, this yields that for any $a \in F$ there is an integer $n > 0$ such that $\varrho(na)\, J_Z(\mathcal{M}\tilde{}) \subseteq J_Z(\mathcal{M}\tilde{})$. So we obtain $\varrho(F) \subset \mathrm{End}_{\mathbb{Q}}(X_Z)$. Combining everything and setting $\iota_Z = \varrho$, we have

(5.3) Proposition. *Given a pair $(\mathcal{M}, T)$ as above, for every $Z \in \mathcal{H}_m^e$ the triplet (X_Z, H_Z, ι_Z) is a polarized abelian variety with endomorphism structure $(F, {}', \varrho)$.*

For the proof it remains to show that the Rosati involution with respect to H_Z restricts to the anti-involution $'$ on F. But for all $a \in F$ and $\underline{b}, \underline{c} \in (F \otimes_{\mathbb{Q}} \mathbb{R})^m$ we have

$$\begin{aligned} \mathrm{Im}\, H_Z(\iota_Z(a)\, J_Z(\tilde{\underline{b}}), J_Z(\tilde{\underline{c}})) &= \mathrm{tr}_{F\otimes_{\mathbb{Q}}\mathbb{R}/\mathbb{R}}(a\, {}^t\underline{b} T \underline{c}') \\ &= \mathrm{tr}_{F\otimes_{\mathbb{Q}}\mathbb{R}/\mathbb{R}}({}^t\underline{b} T (a'\underline{c})') \\ &= \mathrm{Im}\, H_Z(J_Z(\tilde{\underline{b}}), \iota_Z(a')\, J_Z(\tilde{\underline{c}})) \end{aligned}$$

where we used Lemma 5.2 and the fact that ${}^t(a\underline{b}) = a\,{}^t\underline{b}$. This implies the assertion, since the Rosati involution is the adjoint operator for $\mathrm{Im}\, H_Z$ (see Proposition 5.1.1). □

We have to analyze, when two points $Z, Z' \in \mathcal{H}_m^e$ lead to isomorphic polarized abelian varieties with endomorphism structure $(F, {}', \varrho)$. In Section 8 we will see that there is a group $G(\mathcal{M}, T)$, acting on $\mathcal{H}_m^e$, such that for $Z, Z' \in \mathcal{H}_m^e$ there is an isomorphism $(X_{Z'}, H_{Z'}, \iota_{Z'}) \simeq (X_Z, H_Z, \iota_Z)$ if and only if there is an $M \in G(\mathcal{M}, T)$ such that $Z' = M(Z)$. Since moreover $G(\mathcal{M}, T)$ acts properly and discontinuously on $\mathcal{H}_m^e$, the quotient

$$\mathcal{A}(\mathcal{M}, T) := \mathcal{H}_m^e / G(\mathcal{M}, T)$$

is a normal complex analytic space of dimension $\frac{e}{2}m(m-1)$. It is a moduli space in the sense of Chapter 8. We call it the *moduli space of polarized abelian varieties with endomorphism structure* $(F, {}', \varrho)$ *associated* to the pair $(\mathcal{M}, T)$. Note that Remark 4.3 also applies here. Note moreover that the choice of $(\mathcal{M}, T)$ fixes the type of the polarization. Every type $(d_1, \ldots, d_g)$ occurs at least once. The following proposition shows that, conversely, every triplet (X, H, ι) with totally indefinite quaternion multiplication by F is contained in one of these moduli spaces.

(5.4) Proposition. *Every* (X, H, ι) *of type* $(F, {}', \varrho)$ *is contained in* $\mathcal{A}(\mathcal{M}, T)$ *for some* $(\mathcal{M}, T)$ *as above.*

We omit the proof, since it is very similar to that of Proposition 2.3.

(5.5) Example. Consider the case $e = m = 1$ and the definite quaternion algebra $F = \mathbb{Q} + i\mathbb{Q} + j\mathbb{Q} + k\mathbb{Q}$ over $\mathbb{Q}$. Let $'$ denote the usual quaternion conjugation. Then $(\mathcal{M}, T)$ with $\mathcal{M} = \mathbb{Z} + i\mathbb{Z} + j\mathbb{Z} + k\mathbb{Z}$ and $T = i$ is a pair as above. The space $\mathcal{H}_1$ consists only of the complex number 0. The corresponding map $J_0 \colon \mathbb{C}^4 \to \mathbb{C}^2$ is given by the matrix $\left(\begin{smallmatrix} 0 & 1 & 0 & 0 \\ 0 & 0 & 0 & 1 \end{smallmatrix}\right)$ with respect to the standard bases. Hence

$$X_0 = \mathbb{C}^2 / \{ \left(\begin{smallmatrix} \lambda \\ \mu \end{smallmatrix}\right) \mid \lambda, \mu \in \mathbb{Z} + i\mathbb{Z} \} = (\mathbb{C}/\mathbb{Z} + i\mathbb{Z})^2 \ .$$

One can show (see Exercise 1 and Corollary 10.6.3) that any abelian surface with endomorphism structure a definite quaternion algebra over $\mathbb{Q}$ is the product of an elliptic curve with complex multiplication with itself.

(5.6) Example. Consider the case $e = 1$ and $m = 2$ and suppose again $F = \mathbb{Q} + i\mathbb{Q} + j\mathbb{Q} + k\mathbb{Q}$. Then $(\mathcal{M}, T)$ with $\mathcal{M} = (\mathbb{Z} + i\mathbb{Z} + j\mathbb{Z} + k\mathbb{Z})^2$ and $T = i\mathbb{1}_2$ is a pair as above. For the space $\mathcal{H}_2$ we have

$$\mathcal{H}_2 = \{ \left(\begin{smallmatrix} 0 & z \\ -z & 0 \end{smallmatrix}\right) \mid |z| < 1 \} \quad (\simeq \mathfrak{H}_1) \ .$$

The map $J_Z \colon \mathbb{C}^8 \to \mathbb{C}^4$ corresponding to $Z = \left(\begin{smallmatrix} 0 & z \\ -z & 0 \end{smallmatrix}\right) \in \mathcal{H}_2$ is given by the matrix $\mathbb{1}_2 \otimes \left(\begin{smallmatrix} 0 & -z & 1 & 0 \\ z & 0 & 0 & 1 \end{smallmatrix}\right)$ with respect to the standard bases. Hence

$$X_Z = \mathbb{C}^4 / \{ \begin{pmatrix} -\lambda_1 z + \mu_1 \\ \lambda_2 z + \mu_2 \\ \bar{\mu}_2 z + \bar{\lambda}_2 \\ -\bar{\mu}_1 z + \bar{\lambda}_1 \end{pmatrix} \mid \lambda_\mu, \mu_\nu \in \mathbb{Z} + i\mathbb{Z} \} \simeq Y_1 \times Y_2$$

with $Y_1 = \mathbb{C}^2/\{\binom{\lambda z+\mu}{\bar\mu z+\bar\lambda}\}$ and $Y_2 = \mathbb{C}^2/\{\binom{-\lambda z+\mu}{-\bar\mu z+\bar\lambda}\}$. It is easy to see that Y_1 and Y_2 are isomorphic abelian surfaces. More generally for $m = 2$ and e arbitrary one can work out an explicit condition on T for X to be a nonsimple abelian variety (see Exercise 2).

§ 6 Families of Abelian Varieties with Complex Multiplication

Let F be a skew field of degree d^2 over its center K which is a totally complex quadratic extension of a totally real number field K_0 with $[K_0 : \mathbb{Q}] = e_0$, and let $'$ be a positive anti-involution on F extending complex conjugation on K. We want to construct families of g-dimensional abelian varieties X with $F \hookrightarrow \mathrm{End}_{\mathbb{Q}}(X)$.

We say that an abelian variety X *admits complex multiplication* by F, if there is an embedding $F \hookrightarrow \mathrm{End}_{\mathbb{Q}}(X)$ with F as above. Note that this notion does not coincide with the more restrictive notion of complex multiplication used in number theory (see Example 6.6). Our definition is closer to the classical terminology (see Krazer-Wirtinger [1]). Proposition 5.5.7 gives the following condition for g

$$g = d^2 e_0 m$$

for some integer $m \geq 1$. Choose an isomorphism $K_0 \otimes_{\mathbb{Q}} \mathbb{R} \to \mathbb{R}^{e_0}$ and extend this to isomorphisms $F \otimes_{\mathbb{Q}} \mathbb{R} \simeq M_d(\mathbb{C})^{e_0}$ and $(F \otimes_{\mathbb{Q}} \mathbb{R})^m \simeq M_d(\mathbb{C})^{e_0 m}$. We identify the elements of $F \otimes_{\mathbb{Q}} \mathbb{R}$ (respectively $(F \otimes_{\mathbb{Q}} \mathbb{R})^m$) with their images in $M_d(\mathbb{C})^{e_0}$ (respectively $M_d(\mathbb{C})^{e_0 m}$). According to Lemma 5.5.1 we may assume that the anti-involution $'$ extends to $x' = {}^t\bar{x}$ on every factor $M_d(\mathbb{C})$. Recall the isomorphism $\tilde{\ }: M_d(\mathbb{C})^{e_0 m} \to \mathbb{C}^{d^2 e_0 m}$, $\underline{a} \to \underline{\tilde{a}}$ of Section 3. Define an $\mathbb{R}$-linear embedding

$$\tilde{\tilde{\ }}: \mathbb{C}^{d^2 e_0 m} \longrightarrow \mathbb{C}^{2d^2 e_0 m}, \ \underline{\tilde{a}} = \begin{pmatrix} \underline{\tilde{a}}^1 \\ \vdots \\ \underline{\tilde{a}}^{e_0} \end{pmatrix} \mapsto \underline{\tilde{\tilde{a}}} = \begin{pmatrix} \underline{\tilde{a}}^1 \\ \bar{\underline{\tilde{a}}}^1 \\ \vdots \\ \underline{\tilde{a}}^{e_0} \\ \bar{\underline{\tilde{a}}}^{e_0} \end{pmatrix}. \tag{1}$$

For abbreviation we denote also the composed map $\tilde{\tilde{\ }} \circ \tilde{\ }: (F \otimes_{\mathbb{Q}} \mathbb{R})^m = M_d(\mathbb{C})^{e_0 m} \to \mathbb{C}^{2d^2 e_0 m}$ by $\tilde{\tilde{\ }}$. Fix integers $r_1, s_1, \ldots, r_{e_0}, s_{e_0} \geq 0$ with $r_\nu + s_\nu = dm$ for $\nu = 1, \ldots, e_0$, and define a representation $\varrho: F \to M_g(\mathbb{C})$ by

$$\varrho(a) = \mathrm{diag}(a^1 \otimes \mathbb{1}_{s_1}, \bar{a}^1 \otimes \mathbb{1}_{r_1}, \ldots, a^{e_0} \otimes \mathbb{1}_{s_{e_0}}, \bar{a}^{e_0} \otimes \mathbb{1}_{r_{e_0}}) . \tag{2}$$

The equation $r_\nu + s_\nu = dm$ for $\nu = 1, \dots, e_0$ implies that the representation ϱ fulfills the condition of Lemma 1.1, which is necessary for the existence of polarized abelian varieties of type $(F, {}', \varrho)$. Conversely every representation $F \to M_g(\mathbb{C})$ satisfying the condition of Lemma 1.1, is equivalent to a representation (2) for some integers r_ν, s_ν as above.

For integers $r, s \geq 1$ consider the complex manifold

$$\mathcal{H}_{r,s} := \left\{ Z \in M(r \times s, \mathbb{C}) \mid \mathbb{1}_s - {}^t\overline{Z}Z > 0 \right\} .$$

In case $rs = 0$ let $\mathcal{H}_{r,s}$ be the space consisting of a single point, which we denote by 0. Our first aim is to associate to every element of the product $\mathcal{H}_{r_1,s_1} \times \cdots \times \mathcal{H}_{r_{e_0},s_{e_0}}$ a polarized abelian variety with endomorphism structure $(F, {}', \varrho)$.

(6.1) Remark. By definition, the matrix $\mathbb{1}_s - {}^t\overline{Z}Z$ defines a positive definite hermitian form on $\mathbb{C}^s$ for every $Z \in \mathcal{H}_{r,s}$. Note that also the $(r \times r)$-matrix $\mathbb{1}_r - Z{}^t\overline{Z}$ is positive definite and hermitian, which means that conjugation and transposition of matrices in $M(r \times s, \mathbb{C})$ restricts to a bijective map from $\mathcal{H}_{r,s}$ to $\mathcal{H}_{s,r}$. To see this, it suffices to note that $\mathbb{1}_r - Z{}^t\overline{Z}$ is the inverse of the positive definite matrix $\mathbb{1}_r + Z(\mathbb{1}_s - {}^t\overline{Z}Z)^{-1}{}^t\overline{Z}$.

Fix a pair $(\mathcal{M}, T)$ with $\mathcal{M}$ a free $\mathbb{Z}$-submodule of F^m of rank $2d^2e_0m$, and T a nondegenerate matrix in $M_m(F)$ with ${}^tT' = -T$ and signature $((r_1, s_1), \dots, (r_{e_0}, s_{e_0}))$ (to be defined below), such that

$$\operatorname{tr}_{F/\mathbb{Q}}({}^t\underline{a}T\underline{b}') \in \mathbb{Z}$$

for all $\underline{a}, \underline{b} \in \mathcal{M}$. It remains to define the signature: considering $M_m(F)$ as a subspace of $M_m(F \otimes_\mathbb{Q} \mathbb{R}) = M_m(M_d(\mathbb{C}))^{e_0}$, the matrix T is of the form $(T^1, \dots, T^{e_0})$ with nondegenerate matrices $T^\nu \in M_m(M_d(\mathbb{C})) = M_{dm}(\mathbb{C})$. Moreover the condition ${}^tT' = -T$ implies that T^ν is skew-hermitian considered as an element of $M_{dm}(\mathbb{C})$. We define T to be of *signature* $((r_1, s_1), \dots, (r_{e_0}, s_{e_0}))$, if its skew-hermitian component T^ν is of signature (r_ν, s_ν) for $\nu = 1, \dots, e_0$. (Recall that for any nondegenerate skew-hermitian matrix $S \in M_{dm}(\mathbb{C})$ there are uniquely determined integers r, s with $r + s = dm$ and a matrix $V \in \mathrm{GL}_{dm}(\mathbb{C})$ such that $S = {}^tV \operatorname{diag}(i\mathbb{1}_r, -i\mathbb{1}_s)\overline{V}$. The pair (r, s) is called the *signature* of S). It is easy to see that such pairs $(\mathcal{M}, T)$ exist.

According to Lemma 3.1 the matrix $\widetilde{T}^\nu$, the image of T^ν under the isomorphism of Section 3, is also nondegenerate and skew-hermitian with signature (r_ν, s_ν). Hence there are matrices $W^\nu \in \mathrm{GL}_{dm}(\mathbb{C})$ such that

$$\widetilde{T}^\nu = {}^tW^\nu \begin{pmatrix} i\mathbb{1}_{r_\nu} & 0 \\ 0 & -i\mathbb{1}_{s_\nu} \end{pmatrix} \overline{W}^\nu$$

for $\nu = 1, \ldots, e_0$.

For every $Z = (Z^1, \ldots, Z^{e_0}) \in \mathcal{H}_{r_1,s_1} \times \cdots \times \mathcal{H}_{r_{e_0},s_{e_0}}$ define a $\mathbb{C}$-vector space homomorphism

$$J_Z \colon \mathbb{C}^{2d^2 e_0 m} \to \mathbb{C}^{d^2 e_0 m} = \mathbb{C}^g$$

by the matrix

$$\operatorname{diag}(J_Z^1, \ldots, J_Z^{e_0}), \quad \text{where} \quad J_Z^\nu = \begin{pmatrix} \mathbb{1}_d \otimes ({}^tZ^\nu, \mathbb{1}_{s_\nu}) W^\nu & 0 \\ 0 & \mathbb{1}_d \otimes (\mathbb{1}_{r_\nu}, Z^\nu)\overline{W^\nu} \end{pmatrix}$$

with respect to the standard bases. In case $r_\nu s_\nu = 0$ for some $\nu = 1, \ldots, e_0$ the matrix J^ν has to be interpreted appropriately, see Examples 6.6 and 6.7.

(6.2) Lemma. *J_Z restricted to the subspace $((F \otimes_{\mathbb{Q}} \mathbb{R})^m)^{\tilde{}}$ of $\mathbb{C}^{2d^2 e_0 m}$ is an isomorphism of $\mathbb{R}$-vector spaces.*

Proof. It suffices to prove the assertion for $e_0 = 1$. Correspondingly we omit the upper index. Moreover we may assume $W = \mathbb{1}_{dm}$, since it is nonsingular. Consider the composed map

$$\mathbb{C}^{d^2m} \overset{\tilde{}}{\hookrightarrow} \mathbb{C}^{2d^2m} \xrightarrow{J_Z} \mathbb{C}^{d^2m} = \mathbb{C}^{ds+dr} \xrightarrow{\varphi} \mathbb{C}^{d^2m}$$

where $\tilde{}$ is the embedding (1) (for $e_0 = 1$), and $\varphi\binom{x}{y} := \binom{x}{\bar{y}}$ for $x \in \mathbb{C}^{ds}$ and $y \in \mathbb{C}^{dr}$. Since φ is an $\mathbb{R}$-vector space isomorphism, it suffices to show that $\varphi J_Z \tilde{}$ is an isomorphism of $\mathbb{R}$-vector spaces. But $\varphi J_Z \tilde{}$ is even an isomorphism of $\mathbb{C}$-vector spaces. In fact, it is given by the matrix $\begin{pmatrix} \mathbb{1}_d \otimes ({}^tZ, \mathbb{1}_s) \\ \mathbb{1}_d \otimes (\mathbb{1}_r, \overline{Z}) \end{pmatrix}$, and hence it suffices to show that the matrix $\begin{pmatrix} {}^tZ & \mathbb{1}_s \\ \mathbb{1}_r & \overline{Z} \end{pmatrix}$ is nonsingular. But

$$\begin{pmatrix} {}^tZ & \mathbb{1}_s \\ \mathbb{1}_r & \overline{Z} \end{pmatrix} \begin{pmatrix} -\overline{Z} & \mathbb{1}_r \\ \mathbb{1}_s & -{}^tZ \end{pmatrix} = \begin{pmatrix} \mathbb{1}_s - {}^tZ\overline{Z} & 0 \\ 0 & \mathbb{1}_r - \overline{Z}{}^tZ \end{pmatrix},$$

and the matrices $\mathbb{1}_r - \overline{Z}{}^tZ$, $\mathbb{1}_s - {}^tZ\overline{Z}$ are nonsingular by definition of $\mathcal{H}_{r,s}$ and Remark 6.1. □

It follows that $J_Z(\mathcal{M}^{\tilde{}})$ is a lattice in $\mathbb{C}^g$ and the quotient

$$X_Z := \mathbb{C}^g / J_Z(\mathcal{M}^{\tilde{}})$$

is a complex torus. Define a map $H_Z \colon \mathbb{C}^g \times \mathbb{C}^g \to \mathbb{C}$ by

$$H_Z(x, y) = 2\,{}^tx \operatorname{diag}(H^1, \ldots, H^{e_0})\, \bar{y}\ ,$$

with $$H^\nu = \begin{pmatrix} \mathbb{1}_d \otimes (\mathbb{1}_{s_\nu} - {}^t\overline{Z^\nu} Z^\nu)^{-1} & 0 \\ 0 & \mathbb{1}_d \otimes (\mathbb{1}_{r_\nu} - \overline{Z^\nu}{}^tZ^\nu)^{-1} \end{pmatrix}.$$

By definition H_Z is a positive definite hermitian form. We claim that H_Z defines a polarization on X_Z. For this we have to show that $\operatorname{Im} H_Z$ is integral on the lattice $J_Z(\mathcal{M}^{\tilde{}})$. But this is a consequence of

(6.3) Lemma. *For all $\underline{a}, \underline{b} \in (F \otimes_{\mathbb{Q}} \mathbb{R})^m$*

$$\operatorname{Im} H_Z(J_Z(\tilde{\underline{a}}), J_Z(\tilde{\tilde{\underline{b}}})) = \operatorname{tr}_{F\otimes\mathbb{R}/\mathbb{R}}({}^t\underline{a}T\underline{b}') \ .$$

We omit the proof, since it is completely analogous to the proof of Lemma 5.2. □

Finally we claim that for all $a \in F$ and $\underline{b} \in (F \otimes_{\mathbb{Q}} \mathbb{R})^m$

$$\varrho(a)\, J_Z(\tilde{\tilde{\underline{b}}}) = J_Z(a \cdot \tilde{\tilde{\underline{b}}}) \ .$$

This is an immediate matrix computation, using the fact that the multiplication $F \times (F \otimes_{\mathbb{Q}} \mathbb{R})^m \to (F \otimes_{\mathbb{Q}} \mathbb{R})^m$ translates to

$$a \cdot \tilde{\tilde{\underline{b}}} := (a \cdot \tilde{\underline{b}})^{\tilde{}} = (\operatorname{diag}(a^1, \bar{a}^1, \ldots, a^{e_0}, \bar{a}^{e_0}) \otimes \mathbb{1}_{dm})\tilde{\tilde{\underline{b}}} \ .$$

Since $\mathcal{M} \otimes \mathbb{Q} = F^m$, for any $a \in F$ there is an $n > 0$ such that $\varrho(na)\, J_Z(\mathcal{M}^{\tilde{}}) \subseteq J_Z(\mathcal{M}^{\tilde{}})$. We obtain $\varrho(F) \subset \operatorname{End}_{\mathbb{Q}}(X_Z)$. Combining everything and setting $\iota_Z = \varrho$, we have

(6.4) Proposition. *Let $(\mathcal{M}, T)$ be as above and $Z \in \mathcal{H}_{r_1,s_1} \times \cdots \times \mathcal{H}_{r_{e_0},s_{e_0}}$. The triplet (X_Z, H_Z, ι_Z) is a polarized abelian variety with endomorphism structure $(F, {}', \varrho)$.*

For the proof it remains to show that the Rosati involution with respect to H_Z restricts to the anti-involution $'$ on F. But this follows with the same argument as in the proof of Proposition 5.3. □

We have to analyze, when two points $Z, Z' \in \mathcal{H}_{r_1,s_1} \times \cdots \times \mathcal{H}_{r_{e_0},s_{e_0}}$ lead to isomorphic polarized abelian varieties with endomorphism structure $(F, {}', \varrho)$. In Section 8 we will see that there is a group $G(\mathcal{M}, T)$, acting on $\mathcal{H}_{r_1,s_1} \times \cdots \times \mathcal{H}_{r_{e_0},s_{e_0}}$, such that there is an isomorphism $(X_{Z'}, H_{Z'}, \iota_{Z'}) \simeq (X_Z, H_Z, \iota_Z)$ if and only if there is an $M \in G(\mathcal{M}, T)$ with $Z' = M(Z)$. Since moreover $G(\mathcal{M}, T)$ acts properly and discontinuously on $\mathcal{H}_{r_1,s_1} \times \cdots \times \mathcal{H}_{r_{e_0},s_{e_0}}$, the quotient

$$\mathcal{A}(\mathcal{M}, T) := (\mathcal{H}_{r_1,s_1} \times \cdots \times \mathcal{H}_{r_{e_0},s_{e_0}})/G(\mathcal{M}, T)$$

is a normal complex analytic space of dimension $\sum_{\nu=1}^{e_0} r_\nu s_\nu$. It is a moduli space in the sense of Chapter 8. We call it the *moduli space of polarized abelian varieties with endomorphism structure* $(F, {}', \varrho)$ associated to the pair $(\mathcal{M}, T)$. Note that Remark 4.3 also applies here. Note moreover that the choice of $(\mathcal{M}, T)$ fixes the type of the polarization. Every type $(d_1, \ldots, d_g)$ occurs at least once. The following proposition shows that, conversely, every triplet (X, H, ι) with complex multiplication is contained in one of these moduli spaces.

(6.5) Proposition. *Every (X, H, ι) with endomorphism structure $(F, {}', \varrho)$ is contained in $\mathcal{A}(\mathcal{M}, T)$ for some $(\mathcal{M}, T)$ as above.*

We omit the proof, since it is very similar to the proof of Proposition 2.3.

(6.6) Example. For the convenience of the reader we work out the most important case, namely the case $d = m = 1$ and $e_0 = g$. Usually in number theory the name *complex multiplication* is reserved for polarized abelian varieties with endomorphism structure of this type.

We have $F = K$, a totally complex quadratic extension of a totally real number field K_0 of degree g over $\mathbb{Q}$. Choose nonnegative integers r_ν, s_ν with $r_\nu + s_\nu = 1$ for $\nu = 1, \ldots, g$, and fix a pair $(\mathcal{M}, t)$ with $\mathcal{M}$ a free $\mathbb{Z}$-submodule of K of rank $2g$, and a purely imaginary element $t \in K$, which is of signature $((r_1, s_1), \ldots, (r_g, s_g))$, and such that $\operatorname{tr}_{K|\mathbb{Q}}(atb') \in \mathbb{Z}$ for all $a, b \in \mathcal{M}$. By definition of the signature there are uniquely determined positive real numbers w_ν such that $t^\nu = (-1)^{s_\nu} i w_\nu^2$ for $\nu = 1, \ldots, g$.

The space $\mathcal{H}_{r_1,s_1} \times \cdots \times \mathcal{H}_{r_g,s_g}$ consists of a single point which we denote by 0. The corresponding map $J_0 \colon \mathbb{C}^{2g} \to \mathbb{C}^g$ is given by the matrix

$$\operatorname{diag}(J^1, \ldots, J^g), \quad \text{where} \quad J^\nu = \begin{cases} (w_\nu, 0) & \text{if} \quad r_\nu = 1 \\ (0, w_\nu) & \phantom{\text{if}} \quad r_\nu = 0 \end{cases},$$

with respect to the standard bases. Hence

$$X_0 = \mathbb{C}^g / J_0(\mathcal{M}^{\tilde{}}) \quad \text{with}$$

$$J_0(\mathcal{M}^{\tilde{}}) = \left\{ \begin{pmatrix} \hat{\lambda}^1 w_1 \\ \vdots \\ \hat{\lambda}^g w_g \end{pmatrix} \;\middle|\; \lambda \in \mathcal{M}\,, \quad \hat{\lambda}^\nu = \begin{cases} \lambda^\nu & \text{if} \quad r_\nu = 1 \\ \bar{\lambda}^\nu & \phantom{\text{if}} \quad r_\nu = 0 \end{cases} \right\}.$$

Note that for any signature there exists a pair $(\mathcal{M}, t)$ as above and the space $\mathcal{A}(\mathcal{M}, t)$ is not empty. Since there are 2^g possible signatures, we get in this way 2^g nonisomorphic polarized abelian varieties with endomorphism structure $(F, {}', \varrho)$. (Here $\varrho(a) = \operatorname{diag}(a^1 \mathbb{1}_{r_1}, \bar{a}^1 \mathbb{1}_{s_1}, \ldots, a^g \mathbb{1}_{r_g}, \bar{a}^g \mathbb{1}_{s_g})$ for $a \in F$ as defined above.) One can show that these represent exactly the 2^g isogeny classes of polarized abelian varieties, which can be associated to K for a given $\mathcal{M}$ (see Exercise 6).

(6.7) Example. Consider the case $d = e_0 = 1$, $r_1 = m = g$, $s_1 = 0$ and suppose $F = K = \mathbb{Q}(i)$. Then $(\mathcal{M}, T)$ with $\mathcal{M} = (\mathbb{Z} + i\mathbb{Z})^g$ and $T = i\mathbb{1}_g$ is a pair as above. The space $\mathcal{H}_{g,0}$ again consists of a single point denoted by 0, and the corresponding map $J_0 \colon \mathbb{C}^{2g} \to \mathbb{C}^g$ is given by the matrix $(\mathbb{1}_g, 0)$ with respect to the standard bases. We obtain the abelian variety

$$X_0 = (\mathbb{C}/\mathbb{Z} + i\mathbb{Z})^g .$$

So X_0 is the g-fold product of the elliptic curve $\mathbb{C}/(\mathbb{Z}+i\mathbb{Z})$ with itself. Moreover the polarization H_0 on X_0 is given by the matrix $\mathbb{1}_g$ with respect to the standard bases which implies that H_0 is the canonical principal polarization of the product of elliptic curves.

This example generalizes as follows (see Exercise 3): whenever we have $\sum_{\nu=1}^{e_0} r_\nu s_\nu = 0$ (d, m and e_0 arbitrary) then X_0 is an d^2m-fold product of an e_0-dimensional abelian variety Y with $K \subseteq \operatorname{End}(Y)$.

(6.8) Example. Consider the case $d = e_0 = 1$, $m = 2$ and $r_1 = s_1 = 1$, and suppose $F = K = \mathbb{Q}(i)$. Then $(\mathcal{M}, T)$ with $\mathcal{M} = (\mathbb{Z} + i\mathbb{Z})^2$ and $T = \left(\begin{smallmatrix} i & 0 \\ 0 & -i \end{smallmatrix}\right)$ is a pair as above. The space $\mathcal{H}_{1,1}$ is the unit disc $\{z \in \mathbb{C} \mid |z| < 1\}$ in $\mathbb{C}$. For any $z \in \mathcal{H}_{1,1}$ the homomorphism $J_z \colon \mathbb{C}^4 \to \mathbb{C}^2$ is given by the matrix $\left(\begin{smallmatrix} z & 1 & 0 & 0 \\ 0 & 0 & 1 & z \end{smallmatrix}\right)$ such that

$$X_z = \mathbb{C}^2 / \left\{ \left(\begin{smallmatrix} \lambda z + \mu \\ \bar{\mu} z + \bar{\lambda} \end{smallmatrix}\right) \mid \lambda, \mu \in \mathbb{Z} + i\mathbb{Z} \right\} .$$

Note that this is the abelian surface Y_1 appearing already in Example 5.6. We will see in Exercise 10.11 that X_z is isogenous to a product of elliptic curves, but splits only for special $z \in \mathcal{H}_{1,1}$. One can show (see Exercises 4 and 5): whenever $r_\nu = s_\nu = 1$ for $\nu = 1, \ldots, e_0$, then $\operatorname{End}_{\mathbb{Q}}(X_Z)$ is bigger than F.

§ 7 Group Actions on $\mathcal{H}_{r,s}$ and $\mathcal{H}_m$

In the last sections we constructed families of abelian varieties with prescribed endomorphism algebra structure of the possible four types. Our aim is to construct the corresponding moduli spaces. For this we have to analyze, when two polarized abelian varieties with the same endomorphism structure are isomorphic. The isomorphism classes can be described by a group action on the corresponding spaces. In the first two cases the basic ingredient is the usual action of the symplectic group on the Siegel upper half space (see Section 8.2). In this section we introduce and study the corresponding group actions on $\mathcal{H}_m$ and $\mathcal{H}_{r,s}$.

Recall the complex manifolds

$$\mathcal{H}_m = \{Z \in M_m(\mathbb{C}) \mid {}^tZ = -Z,\ \mathbb{1}_m - {}^t\overline{Z}Z > 0\} \quad \text{and}$$
$$\mathcal{H}_{r,s} = \{Z \in M(r \times s, \mathbb{C}) \mid \mathbb{1}_s - {}^t\overline{Z}Z > 0\}\ .$$

for integers $m > 0$ and $r, s \geq 0$. (If $rs = 0$, the space $\mathcal{H}_{r,s}$ consists of a single point which we denote by 0.) Define $U_{r,s}$ to be the unitary group associated to the hermitian form $\begin{pmatrix} \mathbb{1}_r & 0 \\ 0 & -\mathbb{1}_s \end{pmatrix}$, i.e.

$$U_{r,s} = \left\{M \in M_{r+s}(\mathbb{C}) \mid {}^tM \begin{pmatrix} \mathbb{1}_r & 0 \\ 0 & -\mathbb{1}_s \end{pmatrix} \overline{M} = \begin{pmatrix} \mathbb{1}_r & 0 \\ 0 & -\mathbb{1}_s \end{pmatrix}\right\}\ .$$

We will write the elements of $U_{r,s}$ in the form $M = \begin{pmatrix} \alpha & \beta \\ \gamma & \delta \end{pmatrix}$ with $\alpha \in M_r(\mathbb{C})$, $\beta, {}^t\gamma \in M(r \times s, \mathbb{C})$ and $\delta \in M_s(\mathbb{C})$. Moreover, define the subgroup $\widetilde{U}_{m,m}$ of $U_{m,m}$ as follows

$$\widetilde{U}_{m,m} = \left\{M = \begin{pmatrix} \alpha & \beta \\ -\bar{\beta} & \bar{\alpha} \end{pmatrix} \in M_{2m}(\mathbb{C}) \mid {}^tM \begin{pmatrix} \mathbb{1}_m & 0 \\ 0 & -\mathbb{1}_m \end{pmatrix} \overline{M} = \begin{pmatrix} \mathbb{1}_m & 0 \\ 0 & -\mathbb{1}_m \end{pmatrix}\right\}\ .$$

In more sophisticated terms, $\widetilde{U}_{m,m}$ is the image of the group

$$U_m(\mathbb{H}) := \{N \in M_m(\mathbb{H}) \mid {}^tN(i\mathbb{1}_m)N' = i\mathbb{1}_m\}$$

under the composed map $M_m(\mathbb{H}) \to M_m(M_2(\mathbb{C})) \xrightarrow{\sim} M_{2m}(\mathbb{C})$, where the second map denotes the isomorphism " $\tilde{}$ "of Lemma 3.1. The elements of $U_{r,s}$ and $\widetilde{U}_{m,m}$ have similar properties as the symplectic matrices:

(7.1) Lemma. *a) The groups $U_{r,s}$ and $\widetilde{U}_{m,m}$ are closed under transposition.*

b) For a matrix $M = \begin{pmatrix} \alpha & \beta \\ -\bar{\beta} & \bar{\alpha} \end{pmatrix} \in M_{2m}(\mathbb{C})$ we have: $M \in \widetilde{U}_{m,m}$ if and only if ${}^t\alpha\bar{\alpha} - {}^t\bar{\beta}\beta = \mathbb{1}_m$, and ${}^t\alpha\bar{\beta} + {}^t\bar{\beta}\alpha = 0$.

c) For a matrix $M = \begin{pmatrix} \alpha & \beta \\ \gamma & \delta \end{pmatrix} \in M_{r+s}(\mathbb{C})$ we have: $M \in U_{r,s}$ if and only if ${}^t\alpha\bar{\alpha} - {}^t\gamma\bar{\gamma} = \mathbb{1}_r$, ${}^t\alpha\bar{\beta} = {}^t\gamma\bar{\delta}$ and ${}^t\delta\bar{\delta} - {}^t\beta\bar{\beta} = \mathbb{1}_s$.

Proof. For a) apply the defining equation to $M^{-1} = \begin{pmatrix} \mathbb{1}_m & \\ & -\mathbb{1}_m \end{pmatrix} {}^t\overline{M} \begin{pmatrix} \mathbb{1}_m & \\ & -\mathbb{1}_m \end{pmatrix}$ respectively $M^{-1} = \begin{pmatrix} \mathbb{1}_r & \\ & -\mathbb{1}_s \end{pmatrix} {}^t\overline{M} \begin{pmatrix} \mathbb{1}_r & \\ & -\mathbb{1}_s \end{pmatrix}$. b) and c) are direct consequences of the definition. □

(7.2) Proposition. *The group $\widetilde{U}_{m,m}$, respectively $U_{r,s}$, acts biholomorphically on $\mathcal{H}_m$, respectively $\mathcal{H}_{r,s}$, by*

$$Z \mapsto M(Z) := (\alpha Z + \beta)(\gamma Z + \delta)^{-1}$$

for all $Z \in \mathcal{H}_m$, $M = \begin{pmatrix} \alpha & \beta \\ \gamma & \delta \end{pmatrix} \in \widetilde{U}_{m,m}$, *respectively all* $Z \in \mathcal{H}_{r,s}$, $M = \begin{pmatrix} \alpha & \beta \\ \gamma & \delta \end{pmatrix} \in U_{r,s}$.

Note that the action in both cases is formally the same as the action of the symplectic group $\mathrm{Sp}_{2m}(\mathbb{R})$ on the Siegel upper half space $\mathfrak{H}_m$ (see Proposition 8.2.2). But in the case of the group $\widetilde{U}_{m,m}$ we have $\gamma = -\bar{\beta}$ and $\delta = \bar{\alpha}$, whereas in the case of the group $U_{r,s}$ we have $(\alpha Z + \beta) \in M(r \times s, \mathbb{C})$ and $(\gamma Z + \delta) \in M_s(\mathbb{C})$. If $rs = 0$, the action has to be interpreted appropriately.

Proof. First we claim that $\det(\gamma Z + \delta) \neq 0$ for any $M = \begin{pmatrix} \alpha & \beta \\ \gamma & \delta \end{pmatrix} \in U_{r,s}$. Using Lemma 7.1 c) we get for any $Z \in \mathcal{H}_{r,s}$:

$$ {}^t(\gamma Z + \delta)\overline{(\gamma Z + \delta)} = 1\!\!1_s - {}^tZ\overline{Z} + {}^t(\alpha Z + \beta)\overline{(\alpha Z + \beta)} \,. \tag{1} $$

This implies the assertion, since ${}^t(\alpha Z + \beta)\overline{(\alpha Z + \beta)}$ is positive semidefinite and $1\!\!1_s - {}^tZ\overline{Z}$ is positive definite by definition. Hence $M(Z)$ is well-defined. In order to see that $M(Z) \in \mathcal{H}_{r,s}$, use (1) to get

$$ \begin{aligned} &{}^t(\gamma Z + \delta)\left(1\!\!1_s - {}^tM(Z)\overline{M(Z)}\right)\overline{(\gamma Z + \delta)} \\ &\quad = {}^t(\gamma Z + \delta)\overline{(\gamma Z + \delta)} - {}^t(\alpha Z + \beta)\overline{(\alpha Z + \beta)} = 1\!\!1_s - {}^tZ\overline{Z} > 0 \,. \end{aligned} $$

Suppose now $M = \begin{pmatrix} \alpha & \beta \\ -\bar{\beta} & \bar{\alpha} \end{pmatrix} \in \widetilde{U}_{m,m}$ and $Z \in \mathcal{H}_m$. The proof of above applied for $r = s = m$ gives that $M(Z)$ is well defined and $1\!\!1_m - {}^tM(Z)\overline{M(Z)} > 0$. In order to see that $M(Z) \in \mathcal{H}_m$, it remains to show that ${}^tM(Z) = -M(Z)$. But using Lemma 7.1 b)

$$ \begin{aligned} &{}^t(-\bar{\beta}Z + \bar{\alpha})(M(Z) + {}^tM(Z))(-\bar{\beta}Z + \bar{\alpha}) \\ &\quad = -\,{}^tZ({}^t\bar{\beta}\alpha + {}^t\alpha\bar{\beta})Z + {}^tZ({}^t\alpha\bar{\alpha} - {}^t\bar{\beta}\beta) + ({}^t\bar{\alpha}\alpha - {}^t\beta\bar{\beta})Z + {}^t\bar{\alpha}\beta + {}^t\beta\bar{\alpha} \\ &\quad = {}^tZ + Z = 0, \end{aligned} $$

whence $M(Z) \in \mathcal{H}_m$. Finally the same computation as in the case of the Siegel upper half space shows that $M_1(M_2(Z)) = (M_1M_2)(Z)$ for all $M_1, M_2 \in U_{r,s}$ (respectively $\widetilde{U}_{m,m}$) and $Z \in \mathcal{H}_{r,s}$ (respectively $\mathcal{H}_m$). □

We need some properties of the actions of $U_{r,s}$ and $\widetilde{U}_{m,m}$ on $\mathcal{H}_{r,s}$ respectively $\mathcal{H}_m$.

(7.3) Proposition. *a) The groups* $U_{r,s}$ *and* $\widetilde{U}_{m,m}$ *act transitively on* $\mathcal{H}_{r,s}$ *respectively* $\mathcal{H}_m$.

b) The stabilizer of $0 \in \mathcal{H}_{r,s}$*, respectively* $\mathcal{H}_m$*, is the compact group* $U_{r,s} \cap U_{r+s}(\mathbb{C}) = \left\{ \begin{pmatrix} \alpha & 0 \\ 0 & \delta \end{pmatrix} \in M_{r+s}(\mathbb{C}) \mid {}^t\alpha\bar{\alpha} = \mathbb{1}_r,\ {}^t\delta\bar{\delta} = \mathbb{1}_s \right\}$*, respectively* $\widetilde{U}_{m,m} \cap U_{2m}(\mathbb{C}) = \left\{ \begin{pmatrix} \alpha & 0 \\ 0 & \bar{\alpha} \end{pmatrix} \in M_{2m}(\mathbb{C}) \mid {}^t\alpha\bar{\alpha} = \mathbb{1}_m \right\}$.

Proof. a) Suppose $Z \in \mathcal{H}_{r,s}$. Since $(\mathbb{1}_r - \bar{Z}\,{}^tZ)^{-1}$ and $(\mathbb{1}_s - {}^tZ\bar{Z})^{-1}$ are positive definite and hermitian (see Remark 6.1), there are matrices $\alpha \in M_r(\mathbb{C})$ and $\delta \in M_s(\mathbb{C})$ such that $(\mathbb{1}_r - \bar{Z}\,{}^tZ)^{-1} = \bar{\alpha}\,{}^t\alpha$ and $(\mathbb{1}_s - {}^tZ\bar{Z})^{-1} = \bar{\delta}\,{}^t\delta$. Then the matrix $M := \begin{pmatrix} \alpha & Z\delta \\ {}^t\bar{Z}\alpha & \delta \end{pmatrix}$ is obviously an element of $U_{r,s}$ and satisfies $M(0) = Z$. In particular, in case $r = s = m$ and $Z \in \mathcal{H}_m$ the matrix $M = \begin{pmatrix} \alpha & Z\bar{\alpha} \\ -\bar{Z}\alpha & \bar{\alpha} \end{pmatrix}$, with α as above, is contained in $\widetilde{U}_{m,m}$ and satisfies $M(0) = Z$.

b) follows with an immediate computation using Lemma 7.1. □

As a consequence the maps $h\colon U_{r,s} \to \mathcal{H}_{r,s}$ and $h\colon \widetilde{U}_{m,m} \to \mathcal{H}_m$ defined by $h(M) = M(0)$ are surjective with compact fibres. Using this, a slight modification of the proof of Proposition 8.2.4, which states that the map $\mathrm{Sp}_{2g}(\mathbb{R}) \to \mathcal{H}_g$, $M \mapsto M(i\mathbb{1})$ is proper, gives that also the maps h here are proper. As a consequence we obtain

(7.4) Proposition. *Any discrete subgroup of* $U_{r,s}$*, respectively* $\widetilde{U}_{m,m}$ *acts properly and discontinuously on* $\mathcal{H}_{r,s}$*, respectively* $\mathcal{H}_m$*.*

The proof is the same as the proof of Proposition 8.2.5. □

§ 8 Shimura Varieties

Let $(F, {}', \varrho)$ be a type of an endomorphism structure and $(\mathcal{M}, T)$ a pair as in Sections 4, 5 or 6 respectively. We want to analyze, when two polarized abelian varieties with endomorphism structure $(F, {}', \varrho)$ out of the same family are isomorphic. The proof follows the same lines as the proof of Proposition 2.2. There are however some technical difficulties. In order to do this simultaneously for the three cases II, III, and IV, define

$$\mathcal{H} = \begin{cases} \mathfrak{H}_m \times \cdots \times \mathfrak{H}_m & \\ \mathcal{H}_m \times \cdots \times \mathcal{H}_m & \text{for type} \\ \mathcal{H}_{r_1,s_1} \times \cdots \times \mathcal{H}_{r_{e_0},s_{e_0}} & \end{cases} \quad \begin{matrix} \mathrm{II} \\ \mathrm{III} \\ \mathrm{IV} \end{matrix}$$

and

$$G = \begin{cases} \mathrm{Sp}_{2m}(\mathbb{R}) \times \cdots \times \mathrm{Sp}_{2m}(\mathbb{R}) & \\ \widetilde{U}_{m,m} \times \cdots \times \widetilde{U}_{m,m} & \text{for type} \\ U_{r_1,s_1} \times \cdots \times U_{r_{e_0},s_{e_0}} & \end{cases} \quad \begin{matrix} \mathrm{II} \\ \mathrm{III} \\ \mathrm{IV} \end{matrix}$$

where the number of factors is $e = e_0$ in the cases II and III. Here e, e_0, m and $r_1, s_1, \ldots, r_{e_0}, s_{e_0}$ have the same meaning as in Sections 4, 5, and 6 respectively. The actions of $\mathrm{Sp}_{2m}(\mathbb{R})$ on $\mathfrak{H}_m$, of $\widetilde{U}_{m,m}$ on $\mathcal{H}_m$, and of U_{r_ν,s_ν} on $\mathcal{H}_{r_\nu,s_\nu}$, $\nu = 1, \ldots, e_0$, (see Propositions 8.2.2 and 7.2) induce an action of G on $\mathcal{H}$: For $Z = (Z^1, \ldots, Z^{e_0}) \in \mathcal{H}$ and $M = (M^1, \ldots, M^{e_0}) \in G$ define

$$M(Z) = (M^1(Z^1), \ldots, M^{e_0}(Z^{e_0}))$$

with
$$M^\nu(Z^\nu) = (\alpha^\nu Z^\nu + \beta^\nu)(\gamma^\nu Z^\nu + \beta^\nu)^{-1},$$

where $M^\nu = \begin{pmatrix} \alpha^\nu & \beta^\nu \\ \gamma^\nu & \delta^\nu \end{pmatrix}$ with $(m \times m)$-matrices $\alpha^\nu, \beta^\nu, \gamma^\nu$ and δ^ν for type II and III and an $(r_\nu \times r_\nu)$-matrix α^ν and an $(s_\nu \times s_\nu)$-matrix δ^ν for type IV. We will see that there is a certain discrete subgroup $G(\mathcal{M}, T)$ of G with the property that two triplets $(X_{Z'}, H_{Z'}, \iota_{Z'})$ and (X_Z, H_Z, ι_Z) are isomorphic if and only if there is an $M \in G(\mathcal{M}, T)$ such that $Z' = M(Z)$.

Write as usual $T = (T^1, \ldots, T^{e_0})$ with $T^\nu \in M_m(M_2(\mathbb{R}))$ or $M_m(\mathbb{H}) \subset M_m(M_2(\mathbb{C}))$ or $M_m(M_d(\mathbb{C}))$ respectively. Let $\widetilde{T}^\nu$ be the image of T^ν under the isomorphism of Lemma 3.1. The matrices $\widetilde{T}^\nu$ are nondegenerate and alternating (in case II) respectively skew-hermitian (in cases III and IV). Hence we have in case

$$\text{II:}\quad \widetilde{T}^\nu = {}^t\widetilde{W}^\nu \begin{pmatrix} 0 & \mathbb{1}_m \\ -\mathbb{1}_m & 0 \end{pmatrix} \widetilde{W}^\nu \text{ with } \widetilde{W}^\nu \in \mathrm{GL}_{2m}(\mathbb{R})$$

$$\text{III:}\quad \widetilde{T}^\nu = {}^t\widetilde{W}^\nu \begin{pmatrix} i\mathbb{1}_m & 0 \\ 0 & -i\mathbb{1}_m \end{pmatrix} \overline{\widetilde{W}^\nu} \text{ with } \widetilde{W}^\nu = \begin{pmatrix} \widetilde{W}^\nu_{11} & \widetilde{W}^\nu_{12} \\ -\overline{\widetilde{W}^\nu_{12}} & \overline{\widetilde{W}^\nu_{11}} \end{pmatrix} \in \mathrm{GL}_{2m}(\mathbb{C}) \quad (1)$$

$$\text{IV:}\quad \widetilde{T}^\nu = {}^t\widetilde{W}^\nu \begin{pmatrix} i\mathbb{1}_{r_\nu} & 0 \\ 0 & -i\mathbb{1}_{s_\nu} \end{pmatrix} \overline{\widetilde{W}^\nu} \text{ with } \widetilde{W}^\nu \in \mathrm{GL}_{r_\nu+s_\nu}(\mathbb{C})$$

for all ν. Recall moreover that $\mathcal{M}^\sim$ denotes the image of the $\mathbb{Z}$-module $\mathcal{M}$ under the isomorphism $^\sim$ of Section 3. For $\widetilde{W} = (\widetilde{W}^1, \ldots, \widetilde{W}^{e_0})$ define the group $G(\mathcal{M}, T) = G_{\widetilde{W}}(\mathcal{M}, T)$ by

$$G(\mathcal{M}, T) = \{M = (M^1, \ldots, M^{e_0}) \in G \mid$$
$$\mathrm{diag}\big(\mathbb{1}_d \otimes (\widetilde{W}^1\, {}^tM^1 \widetilde{W}^{1\,-1}), \ldots, \mathbb{1}_d \otimes (\widetilde{W}^{e_0}\, {}^tM^{e_0} \widetilde{W}^{e_0\,-1})\big)\mathcal{M}^\sim \subset \mathcal{M}^\sim\},$$

where as usual $d = 2$ in the cases II and III and $d = \frac{1}{m}(r_\nu + s_\nu)$ in case IV. The group $G(\mathcal{M}, T)$ is discrete in G, since $\mathcal{M}^\sim$ is a lattice in a real vector space on which G acts. According to Proposition 8.2.5 and Proposition 7.4 in the previous section the group $G(\mathcal{M}, T)$ acts properly and discontinuously on $\mathcal{H}$. Hence by Theorem A.6 the quotient $\mathcal{H}/G(\mathcal{M}, T)$ is a normal complex analytic space.

(8.1) Remark. Suppose $\widetilde{W}_1 = (\widetilde{W}_1^1, \ldots, \widetilde{W}_1^{e_0})$ and $\widetilde{W}_2 = (\widetilde{W}_2^1, \ldots, \widetilde{W}_2^{e_0})$ are matrices as in (1). Then there is an element $N = (N^1, \ldots, N^{e_0}) \in G$

such that $N^\nu = \widetilde{W}_1^{\nu\,-1}\widetilde{W}_2^\nu$ for $\nu = 1, \ldots, e_0$ and we have $G_{\widetilde{W}_2}(\mathcal{M}, T) = {}^\iota N G_{\widetilde{W}_1}(\mathcal{M}, T)\, {}^\iota N^{-1}$. Since G is invariant under transposition (see Lemma 8.2.1 and Lemma 7.1 above), this means that the groups $G_{\widetilde{W}_1}(\mathcal{M}, T)$ and $G_{\widetilde{W}_2}(\mathcal{M}, T)$ are conjugate in G, which justifies our notation. In particular the analytic structure on $\mathcal{H}/G(\mathcal{M}, T)$ does not depend on the choice of $\widetilde{W}$.

Fix $\widetilde{W} = (\widetilde{W}^1, \ldots, \widetilde{W}^{e_0})$ satisfying (1). Then

(8.2) Proposition. *Let Z and Z' be two points of $\mathcal{H}$. The polarized abelian varieties $(X_{Z'}, H_{Z'}, \iota_{Z'})$ and (X_Z, H_Z, ι_Z) with endomorphism structure $(F, \;'\;, \varrho)$ are isomorphic if and only if there is an $M \in G(\mathcal{M}, T)$ such that $Z' = M(Z)$.*

Proof. Suppose $f\colon (X_{Z'}, H_{Z'}, \iota_{Z'}) \to (X_Z, H_Z, \iota_Z)$ is an isomorphism. Let $A = \varrho_a(f) \in M_g(\mathbb{C})$ be the analytic representation of f. By definition of an isomorphism of polarized abelian varieties with endomorphism structure $(F, \;'\;, \varrho)$ the matrix A satisfies

$$\begin{aligned} A\, J_{Z'}(\mathcal{M}\tilde{}\,) &\subset J_Z(\mathcal{M}\tilde{}\,) \quad \text{for type II and III} \\ A J_{Z'}(\mathcal{M}\tilde{\tilde{}}\,) &\subset J_Z(\mathcal{M}\tilde{\tilde{}}\,) \quad \text{for type IV} \end{aligned} \tag{2}$$

$$A^*\, H_Z = H_{Z'} \tag{3}$$

$$\varrho(a)\, A = A\, \varrho(a) \tag{4}$$

for all $a \in F$. Analyzing equation (4), we obtain that A is of the form

$$A = \begin{cases} \operatorname{diag}(\mathbb{1}_2 \otimes A^1, \ldots, \mathbb{1}_2 \otimes A^{e_0}) & \text{for type II and III} \\ \operatorname{diag}(\mathbb{1}_d \otimes A_1^1, \mathbb{1}_d \otimes A_2^1, \ldots, \mathbb{1}_d \otimes A_1^{e_0}, \mathbb{1}_d \otimes A_2^{e_0}) & \text{for type IV,} \end{cases}$$

with $A^\nu \in M_m(\mathbb{C})$, $A_1^\nu \in M_{s_\nu}(\mathbb{C})$ and $A_2^\nu \in M_{r_\nu}(\mathbb{C})$. The homomorphism $J_Z\colon \mathbb{C}^{2g} \to \mathbb{C}^g$ restricts to an $\mathbb{R}$-linear isomorphism from a $2g$-dimensional real subvector space $(F \otimes \mathbb{R})^m$ of $\mathbb{C}^{2g}$ onto $\mathbb{C}^g$. Let J_Z^{-1} denote the inverse of this isomorphism. Using (4) and the fact that $\varrho(a) J_Z(\tilde{\underline{b}}) = J_Z(a \cdot \tilde{\underline{b}})$, respectively $\varrho(a) J_Z(\tilde{\tilde{\underline{b}}}) = J_Z(a \cdot \tilde{\tilde{\underline{b}}})$, for all $a \in F$ and $\underline{b} \in (F \otimes_{\mathbb{Q}} \mathbb{R})^m$, equation (2) implies that the composed map

$$\tilde{}^{-1} J_Z^{-1} A J_{Z'}\tilde{}\,\colon F^m \to F^m \qquad \text{for type II and III}$$

$$\tilde{\tilde{}}^{-1} J_Z^{-1} A J_{Z'}\tilde{\tilde{}}\,\colon F^m \to F^m \qquad \text{for type IV}$$

is an isomorphism of left F-vector spaces. Hence it is of the form

$$\underline{a} \mapsto \varphi_R(\underline{a}) = ({}^t R' \underline{a}')'$$

for some $R \in M_m(F)$. In particular ${}^t R' \mathcal{M}' \subset \mathcal{M}'$ by (2) and

$$A\,J_{Z'}{}^{\sim} = J_Z\,\widetilde{\varphi}_R \quad \text{respectively} \quad A\,J_{Z'}{}^{\approx} = J_Z\,\widetilde{\widetilde{\varphi}}_R\,.$$

In order to compute the composed maps $\tilde{\varphi}_R$ and $\widetilde{\widetilde{\varphi}}_R$, we consider as usual R as a matrix with entries in $F \otimes_{\mathbb{Q}} \mathbb{R} = M_2(\mathbb{R})^{e_0}$, (or $\mathbb{H}^{e_0} \subset M_2(\mathbb{C})^{e_0}$ or $M_d(\mathbb{C})^{e_0}$ respectively) and write $R = (R^1, \dots, R^{e_0})$. Now we can apply Lemma 3.3 to each component R^ν to get that

$$A\,J_{Z'} = J_Z \hat{R} \tag{5}$$

with

$${}^t\hat{R} = \begin{cases} \operatorname{diag}(\mathbb{1}_2 \otimes \widetilde{R}^1, \dots, \mathbb{1}_2 \otimes \widetilde{R}^{e_0}) & \text{for type II and III} \\ \operatorname{diag}(\mathbb{1}_d \otimes \widetilde{R}^1, \mathbb{1}_d \otimes \overline{\widetilde{R}}{}^1, \dots, \mathbb{1}_d \otimes \widetilde{R}^{e_0}, \mathbb{1}_d \otimes \overline{\widetilde{R}}{}^{e_0}) & \text{for type IV.} \end{cases}$$

Inserting the special form of the matrix A and the definition of J_Z, equation (5) gives for all ν

$$A^\nu({}^tZ'^\nu, \mathbb{1}_m) = ({}^tZ^\nu, \mathbb{1}_m)\widetilde{W}^\nu\,{}^t\widetilde{R}^\nu \widetilde{W}^{\nu\,-1}$$

for type II and III, respectively

$$A_1^\nu({}^tZ'^\nu, \mathbb{1}_{s_\nu}) = ({}^tZ^\nu, \mathbb{1}_{s_\nu})\widetilde{W}^\nu\,{}^t\widetilde{R}^\nu \widetilde{W}^{\nu\,-1}, \quad \text{and}$$

$$A_2^\nu(\mathbb{1}_{r_\nu}, Z'^\nu) = (\mathbb{1}_{r_\nu}, Z^\nu)\overline{\widetilde{W}^\nu\,{}^t\widetilde{R}^\nu \widetilde{W}^{\nu\,-1}}$$

for type IV. Setting

$$M^\nu = {}^t\widetilde{W}^{\nu\,-1}\widetilde{R}^\nu\,{}^t\widetilde{W}^\nu = \begin{pmatrix} \alpha^\nu & \beta^\nu \\ \gamma^\nu & \delta^\nu \end{pmatrix},$$

this yields $\qquad Z'^\nu = M^\nu(Z^\nu) = (\alpha^\nu Z^\nu + \beta^\nu)(\gamma^\nu Z^\nu + \delta^\nu)^{-1}$

in any case. (Note that in case IV the second equation is equivalent to $Z'^\nu = (Z^\nu\,{}^t\bar{\beta}^\nu + {}^t\bar{\alpha}^\nu)^{-1}(Z^\nu\,{}^t\bar{\delta}^\nu + {}^t\bar{\gamma}^\nu)$. But using Lemma 7.1 c) one easily sees that $(Z^\nu\,{}^t\bar{\beta}^\nu + {}^t\bar{\alpha}^\nu)^{-1}(Z^\nu\,{}^t\bar{\delta}^\nu + {}^t\bar{\gamma}^\nu) = M^\nu(Z^\nu)$.)

It remains to show that $M = (M^1, \dots, M^{e_0})$ is in $G_{\widetilde{W}}(\mathcal{M}, T)$. But by (3) we get using (5)

$$\operatorname{Im} J_{Z'}^* H_{Z'} = \operatorname{Im} J_{Z'}^* A^* H_Z = \hat{R}^* \operatorname{Im} J_Z^* H_Z. \tag{6}$$

In the proofs of the Lemmas 4.1, 5.2 and 6.3 respectively, we saw that

$$\operatorname{Im} H_Z(J_Z(\underline{\tilde{a}}), J_Z(\underline{\tilde{b}})) = {}^t\underline{\tilde{a}}\,\operatorname{diag}(\mathbb{1}_2 \otimes \widetilde{T}^1, \dots, \mathbb{1}_2 \otimes \widetilde{T}^e)\underline{\bar{\tilde{b}}}$$

for type II and III, whereas in type IV

$$\operatorname{Im} H_Z(J_Z(\underline{\tilde{a}}), J_Z(\underline{\tilde{b}})) = {}^t\underline{\tilde{a}}\,\operatorname{diag}\bigl(\mathbb{1}_d \otimes \widetilde{T}^1, \mathbb{1}_d \otimes \overline{\widetilde{T}}{}^1, \dots, \mathbb{1}_d \otimes \widetilde{T}^{e_0}, \mathbb{1}_d \otimes \overline{\widetilde{T}}{}^{e_0}\bigr)\underline{\bar{\tilde{b}}}$$

for all $a, b \in (F \otimes_{\mathbb{Q}} \mathbb{R})^m$. Inserting this into (6), we get $\widetilde{T}^\nu = \widetilde{R}^\nu \widetilde{T}^\nu \, {}^t\overline{\widetilde{R}^\nu}$ for all ν. In terms of M^ν this reads

$$ {}^t\overline{\widetilde{W}^{\nu}}^{-1} \widetilde{T}^\nu \widetilde{W}^{\nu -1} = M^\nu ({}^t\overline{\widetilde{W}^{\nu}}^{-1} \widetilde{T}^\nu \widetilde{W}^{\nu -1}) \, {}^t\overline{M^\nu} \, . $$

By (1) this means $M \in G$. On the other hand by construction ${}^t\!R' \mathcal{M}' \subset \mathcal{M}'$ and equivalently $\hat{R}\mathcal{M}^{\check{}} \subset \mathcal{M}^{\check{}}$ for type II and III, and $\hat{R}\mathcal{M}^{\tilde{\check{}}} \subset \mathcal{M}^{\tilde{\check{}}}$ for type IV. By definition of M this gives in any case

$$ \operatorname{diag}(\mathbb{1}_d \otimes (\overline{\widetilde{W}^1}^{-1} \, {}^t\overline{M^1} \widetilde{W}^1), \ldots, \mathbb{1}_d \otimes (\overline{\widetilde{W}^{e_0}}^{-1} \, {}^t\overline{M^{e_0}} \widetilde{W}^{e_0})) \mathcal{M}^{\check{}} \subset \mathcal{M}^{\check{}} \, . $$

This completes the proof. □

Proposition 8.2 justifies our terminology to call the normal complex spaces $\mathcal{A}(\mathcal{M}, T)$ moduli spaces of polarized abelian varieties with endomorphism structure $(F, {}' , \varrho)$ associated to the pair $(\mathcal{M}, T)$. We call these spaces *(analytic) Shimura varieties.* Note that in number theory the terminology is slightly different. There the name Shimura variety is reserved for the canonical compactifications $\overline{\mathcal{A}(\mathcal{M}, T)}$ mentioned in the introduction to this chapter and their extension to schemes over $\mathbb{Z}$.

§ 9 The Endomorphism Algebra of a General Member

Consider the moduli space $\mathcal{A}(\mathcal{M}, T)$ of polarized abelian varieties with endomorphism structure $(F, {}' , \varrho)$ associated to a pair $(\mathcal{M}, T)$. Let (X, H, ι) be a member of $\mathcal{A}(\mathcal{M}, T)$. By definition ι is an embedding $F \hookrightarrow \operatorname{End}_{\mathbb{Q}}(X)$. As we saw in some examples, the endomorphism algebra $\operatorname{End}_{\mathbb{Q}}(X)$ may be bigger than $\iota(F)$. It is the aim of this section to show that for a general member of $\mathcal{A}(\mathcal{M}, T)$ we have $\iota(F) = \operatorname{End}_{\mathbb{Q}}(X)$, except in some special cases. Combining this with the results of Section 5.5, we obtain a complete list of the skew fields with positive anti-involution occuring as endomorphism algebras of simple abelian varieties.

First recall some definitions: as usual K denotes the center of F with $[F : K] = d^2$ and $[K : \mathbb{Q}] = e$, and $K_0 \subset K$ the fixed field of the anti-involution $'$ with $[K_0 : \mathbb{Q}] = e_0$. If g denotes the dimension of the abelian variety X, the integer m is defined by $m = \frac{g}{e}$ for type I, $m = \frac{g}{2e}$ for types II and III and $m = \frac{g}{d^2 e_0}$ for type IV. Moreover in case IV the tuplet $((r_1, s_1), \ldots, (r_{e_0}, s_{e_0}))$ is the signature of the skew hermitian form T. The main result of this section is:

(9.1) Theorem. *For a general member (X, H, ι) of the space $\mathcal{A}(\mathcal{M}, T)$ associated to $(F, {}', \varrho)$ we have*

$$\mathrm{End}_{\mathbb{Q}}(X) = \iota(F)$$

except in the following cases:

a) $(F, {}', \varrho)$ *is of type* III *and* $m \leq 2$.

b) $(F, {}', \varrho)$ *is of type* IV *and* $\sum_{\nu=1}^{e_0} r_\nu s_\nu = 0$.

c) $(F, {}', \varrho)$ *is of type* IV *and* $r_\nu = s_\nu = 1$ *for* $\nu = 1, \ldots, e_0$.

Here a general member means a member of $\mathcal{A}(\mathcal{M}, T)$ outside an union of countably many proper analytic subspaces, which can be given explicitly. For the exceptional cases see Exercises 1 to 5. The proof of Theorem 9.1 is a direct consequence of the following three propositions.

Fix matrices $\widetilde{W}^\nu$, $\nu = 1, \ldots, e_0$, associated to T as in 8(1) in the cases II, III and IV. In case I let $\widetilde{W}^\nu = -\mathbb{1}_{2m}$ for all ν. For a matrix $R \in M_{2m}(F)$ in case I, respectively $M_m(F)$ otherwise, with ${}^tR'\mathcal{M} \subset \mathcal{M}$ define

$$M^\nu = {}^t\widetilde{W}^{\nu^{-1}} \widetilde{R}^\nu \, {}^t\widetilde{W}^\nu = \begin{pmatrix} \alpha^\nu & \beta^\nu \\ \gamma^\nu & \delta^\nu \end{pmatrix}$$

for all ν. Then the equation

$$Z(\gamma^\nu Z + \delta^\nu) = \alpha^\nu Z + \beta^\nu$$

defines an analytic subspace of the space $\mathfrak{H}_m$ for type I and II, $\mathcal{H}_m$ for type III and $\mathcal{H}_{r_\nu, s_\nu}$ for type IV respectively, which we denote by $S^\nu(R)$. With this notation

(9.2) Proposition. *If R is not an element of the center of $M_{2m}(F)$ respectively $M_m(F)$, then for every ν the set $S^\nu(R)$ is a proper analytic subspace of $\mathfrak{H}_m$, $\mathcal{H}_m$ or $\mathcal{H}_{r_\nu, s_\nu}$ respectively, except in the cases a) and b) of Theorem 9.1.*

Proof. Suppose $S^\nu(R)$ is the whole space for some ν. This means $\gamma^\nu = \beta^\nu = 0$ and

$$Z\delta^\nu = \alpha^\nu Z \tag{1}$$

for all Z. Since $\mathfrak{H}_m$, $\mathcal{H}_m$ respectively $\mathcal{H}_{r_\nu, s_\nu}$ is an open submanifold of $\{Z \in M_m(\mathbb{C}) \mid Z \text{ symmetric}\}$, $\{Z \in M_m(\mathbb{C}) \mid Z \text{ alternating}\}$ respectively $M(r_\nu \times s_\nu, \mathbb{C})$, equation (1) is valid for all Z in these spaces. Analyzing (1) in each case separately, one immediately checks that $M^\nu = r\mathbb{1}$ for some $r \in \mathbb{R}$ for type I, II, III and $r \in \mathbb{C}$ for type IV. (Note that this conclusion is not valid in the cases a) and b)). This implies that $r\mathbb{1} = \widetilde{R}^\nu = R^\nu$ and thus R^ν is contained in the center of $M_{2m}(\mathbb{R})$ or $M_m(\mathbb{R})$ or $M_{dm}(\mathbb{C})$ respectively.

Hence its preimage R is also an element of the center of $M_{2m}(F)$ respectively $M_m(F)$. □

The vector space $M_{2m}(F)$, respectively $M_m(F)$, consists of countably many elements. This fact together with Proposition 9.2 implies that

$$S := \bigcup_R S^1(R) \times \cdots \times S^{e_0}(R)$$

is a proper subset of $\mathcal{H}$. (Here R runs through the set of matrices in $M_{2m}(F)$, respectively $M_m(F)$, not in the center and with ${}^tR'\mathcal{M}' \subset \mathcal{M}'$). Recall that to any $Z \in \mathcal{H}$ we associated the polarized abelian variety (X_Z, H_Z, ι_Z) with endomorphism structure $(F, {}', \varrho)$.

(9.3) Proposition. *If $Z \in \mathcal{H} - S$, then $\iota_Z(K)$ is the centralizer of $\iota_Z(F)$ in* $\mathrm{End}_{\mathbb{Q}}(X_Z)$.

Proof. Suppose A is an element of the centralizer of $\iota_Z(F)$ in $\mathrm{End}_{\mathbb{Q}}(X_Z)$. We have to show that $A = \iota_Z(r)$ for some $r \in K$. As usual, we consider A as a matrix in $M_g(\mathbb{C})$. By construction of X_Z we have $A\,J_Z(\mathcal{M}\check{}\,) \subset J_Z(\mathcal{M}\check{}\,)$ respectively $AJ_Z(\mathcal{M}^{\tilde{}}) \subset J_Z(\mathcal{M}^{\tilde{}})$. As in the proofs of Proposition 2.2 and Theorem 8.2 there is an $R \in M_{2m}(F)$ respectively $M_m(F)$ with ${}^tR'\mathcal{M} \subset \mathcal{M}$ and

$$A\,J_Z = J_Z\hat{R} \tag{2}$$

with

$${}^t\hat{R} = \begin{cases} \mathrm{diag}(\mathbb{1}_d \otimes \widetilde{R}^1, \ldots, \mathbb{1}_d \otimes \widetilde{R}^e) & \text{for type I, II and III} \\ \mathrm{diag}(\mathbb{1}_d \otimes \widetilde{R}^1, \mathbb{1}_d \otimes \overline{\widetilde{R}^1}, \ldots, \mathbb{1}_d \otimes \widetilde{R}^{e_0}, \mathbb{1}_d \otimes \overline{\widetilde{R}^{e_0}}) & \text{for type IV.} \end{cases}$$

On the other hand, since $A\iota_Z(a) = \iota_Z(a)A$ for all $a \in F$ by assumption, the matrix A is necessarily of the form

$$A = \begin{cases} \mathrm{diag}(\mathbb{1}_d \otimes A^1, \ldots, \mathbb{1}_d \otimes A^e) & \text{for type I, II and III} \\ \mathrm{diag}\left(\mathbb{1}_d \otimes A_1^1, \mathbb{1}_d \otimes A_2^1, \ldots, \mathbb{1}_d \otimes A_1^{e_0}, \mathbb{1}_d \otimes A_2^{e_0}\right) & \text{for type IV.} \end{cases}$$

Inserting the special form of the matrices A, $\hat{R}$ and J_Z, equation (2) gives $A^\nu({}^tZ^\nu, \mathbb{1}) = ({}^tZ^\nu, \mathbb{1})\,{}^tM^\nu$ with $M^\nu = {}^t\widetilde{W}^{\nu\,-1}\widetilde{R}^\nu\,{}^t\widetilde{W}^\nu$. Writing $M^\nu = \left(\begin{smallmatrix} \alpha^\nu & \beta^\nu \\ \gamma^\nu & \delta^\nu \end{smallmatrix}\right)$ as usual this means

$$Z^\nu(\gamma^\nu Z^\nu + \delta^\nu) = (\alpha^\nu Z^\nu + \beta^\nu)$$

and thus $Z \in S(R)$. By assumption Z is an element of $\mathcal{H} - S$. This implies that R lies in the center of $M_{2m}(F)$ (respectively $M_m(F)$). In other words $R = r\mathbb{1}$ for some $r \in K$ and thus $\hat{R} = \mathrm{diag}(r^1, \ldots, r^{e_0}) \otimes \mathbb{1}_{2m}$ respectively

$\hat{R} = \operatorname{diag}(r^1, \dots, r^{e_0}) \otimes \mathbb{1}_{dm}$. Now $J_Z \hat{R} = \iota_Z(r) J_Z$ and comparing this with (2) gives $A = \iota_Z(r)$. □

(9.4) Proposition. *Let (X, H_X, ι_X) be a polarized abelian variety with endomorphism structure $(F, {}', \varrho)$. If $\iota_X(K)$ is the centralizer of $\iota_X(F)$ in $\operatorname{End}_{\mathbb{Q}}(X)$, then*

$$\operatorname{End}_{\mathbb{Q}}(X) = \iota_X(F)$$

except in the case c) of Theorem 9.1.

Proof. For simplicity we identify F with its image in $\operatorname{End}_{\mathbb{Q}}(X)$ via ι_X. Let $\widetilde{K}$ denote the center of $\operatorname{End}_{\mathbb{Q}}(X)$. By assumption $\widetilde{K}$ is a subfield of the center K of F. This implies that $\operatorname{End}_{\mathbb{Q}}(X)$ is a simple algebra. According to Poincaré's Complete Reducibility Theorem 5.3.7 there is a simple abelian variety Y such that X is isogenous to Y^n for some integer n, and $\operatorname{End}_{\mathbb{Q}}(X)$ is isomorphic to $M_n(\operatorname{End}_{\mathbb{Q}}(Y))$. Necessarily $\widetilde{K}$ is the center of $\operatorname{End}_{\mathbb{Q}}(Y)$. Fix an isogeny

$$f: Y^n \to X ,$$

and denote by H_Y the polarization on Y induced by H_X via f and the diagonal map $\Delta: Y \to Y^n$. By definition ι_X is the analytic representation restricted to $F \subset \operatorname{End}_{\mathbb{Q}}(X)$. Via the analytic represenation of the composed map $f\Delta$, the analytic representation of $\operatorname{End}_{\mathbb{Q}}(X)$ induces the analytic representation ι_Y of $\operatorname{End}_{\mathbb{Q}}(Y)$. The triplet (Y, H_Y, ι_Y) is a polarized abelian variety with endomorphism structure $(\operatorname{End}_{\mathbb{Q}}(Y), {}', \iota_Y)$, where $'$ denotes the Rosati involution with respect to H_Y. The idea of the proof is to analyze the type of $(\operatorname{End}_{\mathbb{Q}}(Y), {}', \iota_Y)$ and to compare it with the type $(F, {}', \varrho)$ of (X, H_X, ι_X).

We may consider F as a subalgebra of $M_n(\operatorname{End}_{\mathbb{Q}}(Y))$ via f. Then by assumption, K is the centralizer of F in $M_n(\operatorname{End}_{\mathbb{Q}}(Y))$. Applying the double centralizer theorem (see Jacobson [1] II Theorem 4.10), F is the centralizer of K in $M_n(\operatorname{End}_{\mathbb{Q}}(Y))$. Moreover Jacobson [1] II Theorem 4.11 and a little additional argument yield

$$[F : \widetilde{K}]\,[K : \widetilde{K}] = [M_n(\operatorname{End}_{\mathbb{Q}}(Y)) : \widetilde{K}] \tag{3}$$

and

$$\operatorname{End}_{\mathbb{Q}}(Y) \otimes_{\widetilde{K}} K \simeq M_p(F) \tag{4}$$

for some integer $p \geq 1$. By (3) we have $d^2[K : \widetilde{K}]^2 = n^2[\operatorname{End}_{\mathbb{Q}}(Y) : \widetilde{K}]$. On the other hand (4) gives $[\operatorname{End}_{\mathbb{Q}}(Y) : \widetilde{K}] = p^2 d^2$. Combining both we get

$$[K : \widetilde{K}] = np \quad \text{and} \quad e' := [\widetilde{K} : \mathbb{Q}] = \tfrac{e}{np} \, . \tag{5}$$

Let $\mathcal{H}_X$ respectively $\mathcal{H}_Y$ denote the space $\mathcal{H}$ as in Sections 2 and 8, of which (X, H_X, ι_X) respectively (Y, H_Y, ι_Y) is a member. Intuitively it is clear that the assignment $(X, H_X, \iota_X) \mapsto (Y, H_Y, \iota_Y)$ gives a holomorphic map $\psi: U \to \mathcal{H}_Y$ for some neighbourhood U of (X, H_X, ι_X) in $\mathcal{H}_X$. For a precise proof of this statement we would have to introduce the notion of families of polarized abelian varieties with endomorphism structure and use the functional properties of the corresponding moduli spaces. We omit the details. However, if we have this holomorphic map, we can proceed as follows: the fibres of ψ are 0-dimensional, since an abelian variety admits only countably many isogenies, whence

$$\dim \mathcal{H}_X \leq \dim \mathcal{H}_Y . \tag{6}$$

In order to compute the dimensions of $\mathcal{H}_X$ and $\mathcal{H}_Y$, let d', e_0', m', r_ν' and s_ν' be integers associated to $(\mathrm{End}_{\mathbb{Q}}(Y), \ ' \ , \iota_Y)$ with the same meaning as d, e_0, m, r_ν and s_ν with respect to $(F, \ ' \ , \varrho)$. Suppose first F is of type I, II or III. Then (4) implies

$$\mathrm{End}_{\mathbb{Q}}(Y) \otimes_{\widetilde{K}} \mathbb{R} \simeq \begin{cases} M_p(\mathbb{R}) & \\ M_{2p}(\mathbb{R}) & \text{if } F \text{ is of type} \\ M_p(\mathbb{H}) & \end{cases} \begin{matrix} \mathrm{I} \\ \mathrm{II} \\ \mathrm{III}. \end{matrix}$$

Note that $(\mathrm{End}_{\mathbb{Q}}(Y), \ ' \ , e_Y)$ can not be of type IV, since $\widetilde{K} \subset K$ is totally real. Using this, (5), and $\dim X = n \dim Y$ we end up with the following possibilities:

F	p	$\mathrm{End}_{\mathbb{Q}}(Y)$	e'	m'	$\dim \mathcal{H}_X$	$\dim \mathcal{H}_Y$
type I	1	type I	$\frac{e}{n}$	m	$\frac{e}{2}m(m+1)$	$\frac{1}{n}\frac{e}{2}m(m+1)$
type I	2	type II	$\frac{e}{2n}$	m	$\frac{e}{2}m(m+1)$	$\frac{1}{2n}\frac{e}{2}m(m+1)$
type II	1	type II	$\frac{e}{n}$	m	$\frac{e}{2}m(m+1)$	$\frac{1}{n}\frac{e}{2}m(m+1)$
type III	1	type III	$\frac{e}{n}$	m	$\frac{e}{2}m(m-1)$	$\frac{1}{n}\frac{e}{2}m(m-1)$

Using (6) we get $n = 1$ in any of the cases (the second case does not occur). This implies $\mathrm{End}_{\mathbb{Q}}(X) \simeq \mathrm{End}_{\mathbb{Q}}(Y)$ and thus $K \simeq \widetilde{K}$. Hence $p = 1$ and equation (4) gives the assertion of the theorem in the cases I, II and III.

Suppose next that F is of type IV and $\widetilde{K}$ is totally real. Then (4) implies $\mathrm{End}_{\mathbb{Q}}(Y) \otimes_{\widetilde{K}} \mathbb{C} \simeq M_{dp}(\mathbb{C})$ and thus

$$d' = dp .$$

Moreover by (5) $\qquad np \equiv 0 \pmod 2 ,$

K being not real. Finally, we must have $r_\nu = s_\nu = \frac{dm}{2}$ and thus $\dim \mathcal{H}_X = \frac{1}{4} e_0 d^2 m^2$, since $\widetilde{K}$ is totally real such that $\mathrm{End}_{\mathbb{Q}}(Y)$ can only be of type I, II ,III. Here we end up with the following possibilities:

$\operatorname{End}_{\mathbb{Q}}(Y)$	d	p	e'	m'	$\dim \mathcal{H}_X$	$\dim \mathcal{H}_Y$
type I	1	1	$\frac{2e_0}{n}$	$\frac{m}{2}$	$\frac{1}{4}e_0 m^2$	$\frac{1}{n}e_0\frac{m}{2}(\frac{m}{2}+1)$
type II	1	2	$\frac{e_0}{n}$	$\frac{m}{2}$	$\frac{1}{4}e_0 m^2$	$\frac{1}{4n}e_0 m(\frac{m}{2}+1)$
type II	2	1	$\frac{2e_0}{n}$	m	$e_0 m^2$	$\frac{1}{n}e_0 m(m+1)$
type III	1	2	$\frac{e_0}{n}$	$\frac{m}{2}$	$\frac{1}{4}e_0 m^2$	$\frac{1}{4n}e_0 m(\frac{m}{2}-1)$
type III	2	1	$\frac{2e_0}{n}$	m	$e_0 m^2$	$\frac{1}{n}e_0 m(m-1)$

Using (6) we get a contradiction in any of these cases, since necessarily $n \geq 2$ and we excluded case c).

Finally, suppose F is of type IV and $\widetilde{K}$ is not totally real. According to the classification of Section 5.5 the field $\widetilde{K}$ is a totally complex quadratic extension of a totally real number field $\widetilde{K}_0$ and we have the following diagram

$$\begin{array}{ccccccc}
 & & K_0 & \overset{2}{\text{———}} & K & \overset{d^2}{\text{———}} & F \\
 & {\scriptstyle e_0}\diagup & \Big| {\scriptstyle np} & & \Big| & & \\
\mathbb{Q} & \overset{e_0' = \frac{e_0}{np}}{\text{———}} & \widetilde{K}_0 & \overset{2}{\text{———}} & \widetilde{K} & \overset{d'^2 = d^2 p^2}{\text{———}} & \operatorname{End}_{\mathbb{Q}}(Y)
\end{array}$$

Let $\sigma_1, \ldots, \sigma_{e_0}, \bar{\sigma}_1, \ldots, \bar{\sigma}_{e_0} : K \hookrightarrow \mathbb{C}$ be the different embeddings of K in $\mathbb{C}$. Restricting $\sigma_1, \ldots, \sigma_{e_0}$ to $\widetilde{K}$ we obtain e_0' different embeddings $\sigma_1', \ldots, \sigma_{e_0'}' : \widetilde{K} \hookrightarrow \mathbb{C}$. Reordering the σ_ν's suitably, we may assume that

$$\sigma_{knp+\nu} | \widetilde{K} = \sigma_{k+1}' \tag{7}$$

for $k = 0, \ldots, e_0' - 1$ and $\nu = 1, \ldots, np$. By definition of the representation ι_Y we have

$$\iota_X(a) = \mathbb{1}_n \otimes \iota_Y(a) \tag{8}$$

for all $a \in \widetilde{K}$ and by definition of the integers $r_1, s_1, \ldots, r_{e_0}, s_{e_0}$ and (7)

$$\begin{aligned}
\iota_X(a) &= \big(\operatorname{diag}(\sigma_1 \mathbb{1}_{s_1}, \bar{\sigma}_1 \mathbb{1}_{r_1}, \ldots, \sigma_{e_0} \mathbb{1}_{s_{e_0}}, \bar{\sigma}_{e_0} \mathbb{1}_{r_{e_0}}) \otimes \mathbb{1}_d\big)(a) \\
&= \big(\operatorname{diag}(\sigma_1' \mathbb{1}_{s_1}, \bar{\sigma}_1' \mathbb{1}_{r_1}, \ldots, \sigma_1' \mathbb{1}_{s_{np}}, \bar{\sigma}_1' \mathbb{1}_{r_{np}}, \sigma_2' \mathbb{1}_{s_{np+1}}, \ldots, \bar{\sigma}_{e_0'}' \mathbb{1}_{r_{e_0'}}) \otimes \mathbb{1}_d\big)(a)
\end{aligned}$$

for all $a \in \widetilde{K}$. On the other hand we have

$$\begin{aligned}
\mathbb{1}_n \otimes \iota_Y(a) &= \big(\mathbb{1}_n \otimes \big(\operatorname{diag}(\sigma_1' \mathbb{1}_{s_1'}, \bar{\sigma}_1' \mathbb{1}_{r_1'}, \ldots, \sigma_{e_0'}' \mathbb{1}_{s_{e_0'}'}, \bar{\sigma}_{e_0'}' \mathbb{1}_{r_{e_0'}'}) \otimes \mathbb{1}_{d'}\big)(a) \\
&= \big(\mathbb{1}_n \otimes \operatorname{diag}(\sigma_1' \mathbb{1}_{ps_1'}, \bar{\sigma}_1' \mathbb{1}_{pr_1'}, \ldots, \sigma_{e_0'}' \mathbb{1}_{ps_{e_0'}'}, \bar{\sigma}_{e_0'}' \mathbb{1}_{pr_{e_0'}'}) \otimes \mathbb{1}_d\big)(a) \\
&= \big(\operatorname{diag}(\sigma_1' \mathbb{1}_{ps_1'}, \bar{\sigma}_1' \mathbb{1}_{pr_1'}, \ldots, \bar{\sigma}_{e_0'}' \mathbb{1}_{pr_{e_0'}'}, \ldots, \sigma_1' \mathbb{1}_{ps_1'}, \ldots, \bar{\sigma}_{e_0'}' \mathbb{1}_{pr_{e_0'}'}) \otimes \mathbb{1}_d\big)(a)
\end{aligned}$$

for all $\widetilde{K}$. Comparing both identities with (8) we obtain $n = p = 1$, which together with equation (4) gives the assertion. □

Exercises

In the following exercises let the integers d, e, e_0, g, m, r_ν and s_ν, the algebra F and its subfields K and K_0 have the same meaning as in the Sections 2, 4, 5 and 6 respectively. For Exercises 1 to 5 see Shimura [1].

(1) Let (X, H, ι) be a polarized abelian variety with endomorphism structure $(F, {}', \varrho)$ with F a totally definite quaternion algebra over K and with $m = 1$. Show that there is an abelian variety Y of dimension e such that X is isogenous to $Y \times Y$ and $\mathrm{End}_{\mathbb{Q}}(Y)$ is a maximal subfield of F. Conclude that $F \neq \mathrm{End}_{\mathbb{Q}}(X)$. In particular there is no abelian surface X such that $\mathrm{End}_{\mathbb{Q}}(X)$ is a totally definite quaternion algebra over $\mathbb{Q}$.

(2) Let $(F, {}', \varrho)$ be a type of an endomorphism structure of a polarized abelian variety with F a totally definite quaternion algebra over K and with $m = 2$. Moreover, let $(\mathcal{M}, T)$ be a pair as in Section 5, and assume $(X, H, \iota) \in \mathcal{A}(\mathcal{M}, T)$. Denote by N the reduced norm of $M_2(F)$ over K. Show that
  a) if $N(T)$ is not the square of a totally positive element in K, then for a general $(X, {}', \varrho)$ in $\mathcal{A}(\mathcal{M}, T)$ we have $\mathrm{End}_{\mathbb{Q}}(X) = F$.
  b) if $N(T)$ is the square of a totally positive element in K, then there is an abelian variety Y of dimension $2e$ such that X is isogenous to $Y \times Y$ and $\mathrm{End}_{\mathbb{Q}}(Y)$ contains a totally indefinite quaternion algebra over K (not necessarily a skew field).

(3) Let (X, H, ι) be a polarized abelian variety with endomorphism structure $(F, {}', \varrho)$ with $(F, {}')$ a skew field of the second kind and $\sum_{\nu=1}^{e_0} r_\nu s_\nu = 0$. Show that X is isogenous to a d^2m-fold product of an e_0-dimensional abelian variety Y, where $\mathrm{End}_{\mathbb{Q}}(Y)$ contains the center K of F.

(4) Let (X, H, ι) be a polarized abelian variety with endomorphism structure $(F, {}', \varrho)$ with $(F, {}')$ a skew field of the second kind, $d = 1$, and $r_\nu = s_\nu = 1$ for $\nu = 1, \ldots, e_0$. Show that $\mathrm{End}_{\mathbb{Q}}(X)$ contains a totally indefinite quaternion algebra $\widetilde{F}$ over K_0 with $F = K \subset \widetilde{F}$. Conclude from this and Exercise 3 that there is no abelian surface such that $\mathrm{End}_{\mathbb{Q}}(X)$ is an imaginary quadratic number field.

(5) Let (X, H, ι) be a polarized abelian variety with endomorphism structure $(F, {}', \varrho)$ with $(F, {}')$ a skew field of the second kind, $d = 2$, and $r_\nu = s_\nu = 1$ for $\nu = 1, \ldots, e_0$. Show that there is an abelian variety Y of dimension $2e_0$ such that X is isogenous to $Y \times Y$ and $\mathrm{End}_{\mathbb{Q}}(Y)$ contains a totally indefinite quaternion algebra over K_0.

(6) Define an equivalence relation on the set of polarized abelian varieties with endomorphism structure $(F, {}', \varrho)$ as follows: (X_1, H_1, ι_1) and (X_2, H_2, ι_2) are equivalent if there is an isogeny $f: X_1 \to X_2$ such that $f \circ \iota_1(a) = \iota_2(a) \circ f$ for all $a \in F$.

Let K be a totally complex quadratic extension of a totally real number field K_0 of degree g. Show that for a given $\mathbb{Z}$-submodule $\mathcal{M}$ as above there are exactly 2^g equivalence classes of polarized abelian varieties with endomorphism structure K.

(7) Give proofs for Propositions 4.4, 5.4 and 6.5. (Hint: generalize the proof of Proposition 2.3.)

Chapter 10

Abelian Surfaces

So far we have studied abelian varieties of arbitrary dimension. In low dimensions one can prove, of course, some more specific results. Since there are many books on elliptic curves, we do not say anything about them, but refer to Hulek [1] and the literature quoted there. This chapter deals with the next interesting case, namely abelian varieties of dimension two, called *abelian surfaces*.

Let X be an abelian surface and L a polarization on X of type (d_1, d_2). If $d_1 \geq 3$, the associated map $\varphi_L: X \to \mathbb{P}_N$ is an embedding according to the Theorem of Lefschetz 4.5.1. Theorems 4.3.1, 4.5.5, and 4.8.2 give a complete answer to the question as to whether φ_L is an embedding, if $d_1 = 2$. Moreover, if $2|d_1$, $d_1 \geq 4$ respectively $3|d_1$, Riemann's Theta Relations 7.5.1 and the Cubic Theta Relations 7.6.5 give a set of equations for the image $\varphi_L(X)$ in $\mathbb{P}_N$. Here we want to investigate the map $\varphi_L: X \to \mathbb{P}_N$ in the remaining case $d_1 = 1$. If $d_2 \geq 5$, Reider's Theorem 4.1 gives an explicit criterion for φ_L to be an embedding. It implies that in general L is very ample. Moreover, we give equations for the image $\varphi_L(X)$ in the cases of a polarization of type $(2,2)$ and $(1,4)$.

After some preliminaries in Section 1 we study Kummer surfaces in Sections 2 and 3. A *Kummer surface* K in $\mathbb{P}_3$ is by definition the image of the map φ_{L^2} associated to an irreducible principal polarization L. We will see that there are 16 distinguished planes in $\mathbb{P}_3$, called *singular planes*, which together with the 16 singular points of K form a 16_6-configuration. This implies in particular that K admits a *Rosenhain tetrahedron*, i.e. a tetrahedron in $\mathbb{P}_3$ with singular planes of K as faces and singular points of K as vertices. It will be applied to derive an equation for K in $\mathbb{P}_3$. In Section 4 we prove Reider's Theorem. The idea is to investigate a vector bundle on X, which arises naturally out of the assumption that two points p and q on X have the same image under the map φ_L. In Section 5 we study polarizations of type $(1,4)$. We will see that generically φ_L is birational with image a singular octic in $\mathbb{P}_3$. We derive an equation for it, again using a Rosenhain tetrahedron. Finally in Section 6 we prove Ruppert's criterion for an abelian surface to split into a product of elliptic curves.

In this chapter we use some easy results on algebraic surfaces such as the adjunction formula and the Hodge Index Theorem. In Section 4 we need Chern characters and Serre duality. Finally in Section 6 we apply some facts about the Plücker quadric.

§ 1 Preliminaries

Let (X, L) be a polarized abelian surface. In this chapter we study some geometrical properties of the associated rational map $\varphi_L : X \to \mathbb{P}_N$ and its image. First we recall some general results.

For any line bundle L on the abelian surface X the Riemann-Roch Theorem states

$$\chi(L) = \tfrac{1}{2}(L^2) \ .$$

By Theorem 3.4.5 and Proposition 4.5.2 the line bundle L is ample if and only if $h^i(L) = 0$ for $i = 1, 2$ and $(L^2) > 0$. Suppose L is an ample line bundle of type (d_1, d_2). Then the Riemann-Roch Theorem implies

$$h^0(L) = \tfrac{1}{2}(L^2) = d_1 d_2 \ .$$

So the line bundle L defines a rational map $\varphi_L : X \to \mathbb{P}_{d_1 d_2 - 1}$.

Effective divisors on a surface can be interpreted as curves. We will use here both terms. For any curve C on X the *arithmetic genus* $p_a(C)$ is defined as

$$p_a(C) = 1 - \chi(\mathcal{O}_C) \ .$$

The *adjunction formula* (see Hartshorne [1] Exercise V.1.3) says

$$2p_a(C) - 2 = (C^2) \ . \tag{1}$$

Here the intersection number of two curves is defined to be the intersection number of the corresponding line bundles. Hence $p_a(C)$ depends only on the line bundle $\mathcal{O}_X(C)$ and not on C itself. So for any curve C in the linear system of the ample line bundle L we obtain

$$p_a(C) = d_1 d_2 + 1 \tag{2}$$

According to the Theorem of Lefschetz 4.5.1, whenever $d_1 \geq 3$, the line bundle L is very ample, i.e. the map $\varphi_L : X \to \mathbb{P}_{d_1 d_2 - 1}$ is an embedding. Riemann's theta relations and the cubic theta relations give equations for the image of X in $\mathbb{P}_{d_1 d_2 - 1}$ in most cases. In this chapter we want to study the remaining cases, namely $d_1 = 1$ or 2.

Suppose first $d_1 = 2$. According to Proposition 4.1.6 the linear system $|L|$ is base point free. So φ_L is a morphism. By Lemma 2.5.6 there is an ample line bundle M on X with $L = M^2$. If the polarization M on X splits, i.e. $(X, M) \simeq (E_1 \times E_2, p_1^* M_1 \otimes p_2^* M_2)$ with elliptic curves E_1 and E_2 and a principal polarization M_1 on E_1, then $\varphi_L : X \to \mathbb{P}_{2d_2-1}$ is the composition of $\varphi_{M_1^2} \times \varphi_{M_2^2} : E_1 \times E_2 \to \mathbb{P}_1 \times \mathbb{P}_{d_2-1}$ and the Segre embedding $\mathbb{P}_1 \times \mathbb{P}_{d_2-1} \hookrightarrow \mathbb{P}_{2d_1-1}$. Suppose now the polarization M does not split. If $d_2 > 2$, then φ_L is an embedding according to Theorem 4.5.5. Finally if $d_2 = 2$, then by Theorem 4.8.1 the map $\varphi_L : X \to K \subseteq \mathbb{P}_3$ is of degree two onto its image K, which is a Kummer surface in $\mathbb{P}_3$. We study Kummer surfaces in more detail in the next two sections.

Suppose now that L is of type $(1, d)$. Then the Decomposition Theorem 4.3.1 reads

(1.1) Lemma. *L has a fixed component if and only if there are elliptic curves E_1 and E_2 such that $(X, L) \simeq (E_1 \times E_2, p_1^* L_1 \otimes p_2^* L_2)$ with line bundles L_1 of type (1) on E_1 and L_2 of type (d) on E_2.*

From now on we assume that L has no fixed component. Then we have

(1.2) Lemma. *a) If $d \geq 3$, the line bundle L has no base point.*
b) If $d = 2$, the line bundle L has exactly four base points.

Proof. Suppose L has a base point. The group $K(L)$ acts on the base locus of L by translations. Hence L has at least $d^2 = \# K(L)$ base points. On the other hand there are at most $(L^2) = 2d$ base points, implying $d \leq 2$.

If $d = 2$, then φ_L maps the abelian surface X to $\mathbb{P}_1$. Since $(L^2) > 0$, the map $\varphi_L : X \to \mathbb{P}_1$ is not a fibration, so L has a base point. By what we have said above the base locus of L consists exactly of four points. □

According to the Theorem of Bertini (see Griffiths-Harris [1] p. 137) a general member of $|L|$ is singular at most in the base locus of L. This gives immediately

(1.3) Proposition. *If $d \geq 2$, the general member of the linear system $|L|$ is smooth.*

Proof. For $d \geq 3$ the assertion follows from the previous lemma. Let L be of type $(1, 2)$ and $D \in |L|$ be singular in one of the four base points of L. Then $(D^2) > 4$, a contradiction. □

(1.4) Example. Suppose L is a symmetric line bundle on X of type $(1, 2)$. We claim that the four base points of L are 4-division points.

For a proof let $b \in X$ be a base point. Since L is symmetric, the point $-b$ is also a base point of L. But $K(L)$ acts on the set of base points, whence $2b \in K(L) \subset X_2$, which implies the assertion.

Let $p: \widetilde{X} \to X$ denote the blow-up of the four base points. The self-intersection number of the strict transform $\widetilde{D}$ of $D \in |L|$ is zero. Hence by (2) and Proposition 1.3 the composition $\varphi_L \circ p: \widetilde{X} \to \mathbb{P}_1$ is a family of curves of genus three over $\mathbb{P}_1$. For the possible degenerations see Exercise 1.

(1.5) Example. Suppose L is of type (1,3). We claim that the associated morphism $\varphi_L: X \to \mathbb{P}_2$ is surjective. Otherwise $\varphi_L(X)$ would be a nondegenerate curve in $\mathbb{P}_2$. This curve would necessarily intersect a general line in at least 2 points. So the general member of the linear system $|L|$ would be reducible, contradicting Theorem 4.3.5.

A general curve C in the linear system $|L|$ is smooth of genus 4 and the restricted linear system $|L||C$ is base point free (see (2), Lemma 1.2 and Proposition 1.3). Since $\deg L|C = (L^2) = 6$, the restriction $\varphi_L|C: C \to \mathbb{P}_1$ is a morphism of degree 6. By Hurwitz's formula it is ramified in 18 points. This implies that the ramification locus of $\varphi_L: X \to \mathbb{P}_2$ is a plane curve of degree 18.

Finally we want to give examples for very ample line bundles L of type $(1, d)$ with $d \geq 5$. Let X be an abelian surface admitting two principal polarizations L_0 and L_1 with $(L_0 \cdot L_1) = n$ for some $n \geq 3$. The following theorem is due to Comessatti (see Lange [1]).

(1.6) Theorem. *If L_0 is an irreducible principal polarization, then the product $L = L_0 \otimes L_1$ is very ample and induces an embedding $X \hookrightarrow \mathbb{P}_{n+1}$.*

For the existence of such surfaces see Exercise 5.14. In particular this gives examples of abelian surfaces in $\mathbb{P}_4$. If $d = n + 2$ is a squarefree integer, then L is of type $(1, d)$.

Proof. Note first that $h^0(L) = \frac{1}{2}(L^2) = \frac{1}{2}(L_0^2) + (L_0 \cdot L_1) + \frac{1}{2}(L_1^2) = n + 2$. Denote by C_0 the unique curve in the linear system $|L_0|$. By Corollary 11.8.2 the curve C_0 is smooth of genus 2. According to Corollary 11.1.6 it suffices to show that for any $x \in X$ the restriction of the linear system $|L|$ to the translate $t_x^* C_0$ is very ample on the curve $t_x^* C_0$. But the exact sequence

$$0 \longrightarrow L(-t_x^* C_0) \longrightarrow L \longrightarrow L|t_x^* C_0 \longrightarrow 0$$

induces

$$H^0(L) \xrightarrow{r} H^0(L|t_x^* C_0) \longrightarrow H^1\big(L(-t_x^* C_0)\big) .$$

The line bundle $L(-t_x^*C_0)$ is algebraically equivalent to L_1 and therefore ample. So $H^1\big(L(-t_x^*C_0)\big) = 0$, whence the map r is surjective, i.e. the restriction $|L|\big|t_x^*C_0$ is a complete linear system. Since it is of degree $(L \cdot L_0) = (L_0 + L_1) \cdot L_0 = 2 + n \geq 5$ on a curve of genus 2, it is very ample. □

§ 2 The 16_6-Configuration of the Kummer Surface

In this section let $X = V/\Lambda$ be an abelian surface with an irreducible principal polarization. Denote by K the *Kummer surface* associated to X, i.e. the quotient

$$K = X/<(-1)_X> .$$

Suppose L is a symmetric line bundle on X defining the principal polarization. According to Section 4.8 the map $\varphi = \varphi_{L^2}: X \to \mathbb{P}_3$ factors via an embedding $\psi: K \hookrightarrow \mathbb{P}_3$. We identify K with its image under ψ. The Kummer surface has exactly 16 singularities, namely the images of the 2-division points of X. We will see that the 16 symmetric line bundles algebraically equivalent to L determine 16 planes in $\mathbb{P}_3$ touching K along a conic. The main object of this section is to show that the 16 singular points and the 16 planes form a 16_6-configuration and to study the underlying geometry.

According to the Riemann-Roch Theorem there is exactly one curve D on X such that $L = \mathcal{O}_X(D)$. In particular D is a symmetric divisor.

(2.1) Lemma. *D contains exactly 6 two-division points, i.e.* $\#D \cap X_2 = 6$.

Proof. Obviously D contains at least those $x \in X_2$ for which the multiplicity $\operatorname{mult}_x(D)$ of D in x is odd. So by Proposition 4.7.5 we have $\#D \cap X_2 \geq \#X_2^-(D) \geq 6$.

On the other hand φ restricts to a double covering $\varphi: D \to C = \varphi(D)$, ramified exactly in the 2-division points contained in D. According to Corollary 11.8.2 the curve D is smooth of genus 2. Hence by the Riemann-Hurwitz formula $\#D \cap X_2 \leq 6$. □

Next we study curves on the Kummer surface K. If C is a curve on K, then φ^*C is a symmetric divisor on X. Conversely every symmetric effective divisor on X yields a curve on K (not necessarily reduced). This suggests to study symmetric line bundles on X.

(2.2) Lemma. *For $x \in X$ and a positive integer n the line bundle $t_x^*L^n$ is symmetric if and only if $x \in X_{2n}$.*

Proof. By definition $t_x^*L^n$ is symmetric if and only if $t_{2x}^*L^n \simeq (-1)^*L^n \simeq L^n$ or equivalently $2x \in K(L^n) = X_n$. □

According to Lemma 4.7.1 any linear system $|t_x^*L^n|$ with $x \in X_{2n}$ contains symmetric divisors. If D is any symmetric divisor in $|t_x^*L^n|$ denote by $\varphi(D)$ its image in K. To be more precise, if $D = \sum n_i D_i$ with reduced symmetric curves D_i, then $\varphi(D) = \sum n_i \varphi(D_i)$. Note that $\varphi(D)$ represents the class $\frac{1}{2}\varphi_*\{D\}$ in the homology group $H_2(K, \mathbb{Z})$.

(2.3) Proposition. *a) If D is a symmetric divisor in $|t_x^*L^n|$ and $C = \varphi(D)$, then C is of degree $2n$ in $\mathbb{P}_3$.*

b) The map $D \mapsto C = \varphi(D)$ induces a bijection between the sets of

1) complete intersections of K with a surface of degree n, and

2) even symmetric divisors in $|L^{2n}|$.

Recall that a symmetric divisor on X is called *even* if the corresponding theta function is an even function (see Section 4.7).

Proof. a) $\deg C = \left(C \cdot \mathcal{O}_{\mathbb{P}_3}(1)\right) = \frac{1}{2}(D \cdot L^2) = (L^n \cdot L) = 2n.$

b) Suppose $F \in \mathbb{C}[X_0, \dots, X_3]$ is a quartic homogeneous polynomial defining K in $\mathbb{P}_3$. The complete intersection curve of K with a surface of degree n in $\mathbb{P}_n$ is defined by an element of the degree n part V_n of the graded algebra $\mathbb{C}[X_0, \dots, X_3]/F \cdot \mathbb{C}[X_0 \dots, X_3]$.

Recall that $H^0(L^{2n})_+$ is the vector space of even theta functions corresponding to the even symmetric divsors in $|L^{2n}|$. Let $\vartheta_0, \dots, \vartheta_3$ be a basis of $H^0(L^2) = H^0(L^2)_+$ corresponding to the coordinates $X_0, \dots, X_3$ of $\mathbb{P}_3 = P(H^0(L^2))$. The assignment $g \mapsto g(\vartheta_0, \dots, \vartheta_3)$ defines an embedding $V_n \hookrightarrow H^0(L^{2n})_+$. But $\dim V_n = \binom{n+3}{3} - \binom{n-1}{3} = 2n^2 + 2$. Since by Corollary 4.6.6 also $h^0(L^{2n})_+ = 2n^2 + 2$, the map $V_n \to H^0(L^{2n})_+$ is an isomorphism. This implies the assertion. □

(2.4) Corollary. *Let D be a symmetric divisor in $|t_x^*L^n|$ and $C = \varphi(D)$. Then $2C$ is the complete intersection of K with a surface S of degree n. Geometrically this means that S touches K along C.*

Proof. By Lemma 2.2 we have $x \in X_{2n} = K(L^{2n})$. This implies that $2D$ is an even symmetric divisor in $|L^{2n}|$. □

For any $x \in X_2$ denote by D_x the unique divisor in the linear system $|t_x^*L|$. According to Proposition 2.3 the curve $C_x = \varphi(D_x)$ is a conic and by the previous corollary $2C_x$ is a complete intersection of the Kummer surface K with a plane in $\mathbb{P}_3$. We denote this plane by P_x. The 16 planes P_x, $x \in X_2$ are called *singular planes* of K in $\mathbb{P}_3$. The double covering $\varphi\colon X \to K$ maps the 16 two-division points to the 16 singular points on K in $\mathbb{P}_3$. We denote both by thc same symbol.

As a first step for the investigation of the configuration of the singular planes and the singular points on K we work out, when a singular point is contained in a singular plane. For this let us fix some notation: suppose $L = L(H, \chi)$. Choose a symplectic basis $\lambda_1, \lambda_2, \mu_1, \mu_2$ of Λ for L. It induces a basis $e_\nu := \frac{1}{2}\lambda_\nu$, $f_\nu := \frac{1}{2}\mu_\nu$, $\nu = 1, 2$ of X_2. Let $L_0 = L(H, \chi_0)$ denote the line bundle with characteristic zero with respect to $\lambda_1, \lambda_2, \mu_1, \mu_2$. Note that we chose L arbitrarily among the symmetric line bundles in $\mathrm{Pic}^H(X)$. Without loss of generality (see Lemma 4.6.1) we may assume that

$$L = t^*_{e_1+e_2+f_1+f_2} L_0 \ . \tag{1}$$

Finally it turns out to be useful to identify X_2, considered as the group of singular points of K, with the $\mathbb{F}_2$-vector space $\mathbb{F}_2^4$ in the following way

$$\epsilon' e_1 + \epsilon'' f_1 + \eta' e_2 + \eta'' f_2 \leftrightarrow \binom{\epsilon}{\eta} \quad \text{with} \quad \epsilon = \binom{\epsilon'}{\epsilon''} \quad \text{and} \quad \eta = \binom{\eta'}{\eta''} \ .$$

With this notation we have:

(2.5) Proposition. *For singular points $x_1 = \binom{\epsilon_1}{\eta_1}$ and $x_2 = \binom{\epsilon_2}{\eta_2}$ the following conditions are equivalent:*

i) x_1 is contained in the singular plane P_{x_2},

ii) either $\epsilon_1 = \epsilon_2$ and $\eta_1 \neq \eta_2$ or $\epsilon_1 \neq \epsilon_2$ and $\eta_1 = \eta_2$.

Proof. Let $D = |L_0|$. Then i) is equivalent to $x_1 \in D_{x_2} = t^*_{x_2+e_1+e_2+f_1+f_2}D$. Arguing as in the proof of Lemma 2.1 this means that $x_1 + x_2 + e_1 + e_2 + f_1 + f_2 \in X_2^-(D)$. Note that $x_i = \bar{v}_i$ with $v_i = \frac{1}{2}(\epsilon_i'\lambda_1 + \epsilon_i''\mu_1 + \eta_i'\lambda_2 + \eta_i''\mu_2)$. So, according to Proposition 4.7.2 and the definition of χ_0 (in Section 3.1), condition i) is equivalent to

$$\begin{aligned} -1 &= \chi_0(\lambda_1 + \lambda_2 + \mu_1 + \mu_2 + 2v_1 + 2v_2) \\ &= e\left(\pi i \, {}^t\!\begin{pmatrix} 1+\epsilon_1'+\epsilon_2' \\ 1+\eta_1'+\eta_2' \end{pmatrix} \begin{pmatrix} 1+\epsilon_1''+\epsilon_2'' \\ 1+\eta_1''+\eta_2'' \end{pmatrix}\right) . \end{aligned}$$

Now the assertion is a consequence of the following observation: for $\binom{\epsilon}{\eta} \in \mathbb{F}_2^4$ we have ${}^t\!\begin{pmatrix} 1+\epsilon' \\ 1+\eta' \end{pmatrix} \begin{pmatrix} 1+\epsilon'' \\ 1+\eta'' \end{pmatrix} \equiv 1 \pmod 2$ if and only if either $\epsilon = \binom{0}{0} \neq \eta$ or $\epsilon \neq \binom{0}{0} = \eta$. □

From Proposition 2.5 the following corollaries are immediate consequences.

(2.6) Corollary. *The* 16 *singular planes and the* 16 *singular points of K form a 16_6-configuration, i.e.*

a) any singular plane contains exactly 6 *singular points, and*

b) any singular point is contained in exactly 6 *singular planes.*

To be more precise, using Proposition 2.5, one can figure out explicitly which singular points lie in a given singular plane, respectively which singular planes pass through a given singular point.

(2.7) Example. Denote for a moment the elements of $\mathbb{F}_2^2$ by $0 = \binom{0}{0}$, $\alpha = \binom{1}{0}$, $\beta = \binom{0}{1}$, $\gamma = \binom{1}{1}$. The plane $P_{\binom{0}{0}}$ contains the singular points

$$\binom{0}{\alpha}, \binom{0}{\beta}, \binom{0}{\gamma}, \quad \binom{\alpha}{0}, \binom{\beta}{0}, \binom{\gamma}{0},$$

whereas the singular point $\binom{0}{0}$ is contained in the singular planes

$$P_{\binom{0}{\alpha}}, P_{\binom{0}{\beta}}, P_{\binom{0}{\gamma}}, \quad P_{\binom{\alpha}{0}}, P_{\binom{\beta}{0}}, P_{\binom{\gamma}{0}}.$$

(2.8) Corollary. *Any two different singular planes have exactly two singular points in common.*

To be more precise, suppose $\mathbb{F}_2^2 = \{\epsilon_1, \epsilon_2, \epsilon_3, \epsilon_4\} = \{\eta_1, \eta_2, \eta_3, \eta_4\}$. We have to distinguish three cases:

a) $P_{\binom{\epsilon_1}{\eta}} \cap P_{\binom{\epsilon_2}{\eta}}$ contains the singular points $\binom{\epsilon_3}{\eta}$ and $\binom{\epsilon_4}{\eta}$,

b) $P_{\binom{\epsilon}{\eta_1}} \cap P_{\binom{\epsilon}{\eta_2}}$ contains the singular points $\binom{\epsilon}{\eta_3}$ and $\binom{\epsilon}{\eta_4}$, and

c) $P_{\binom{\epsilon_1}{\eta_1}} \cap P_{\binom{\epsilon_2}{\eta_2}}$ contains the singular points $\binom{\epsilon_1}{\eta_2}$ and $\binom{\epsilon_2}{\eta_1}$.

Three different singular planes intersect in one point. This point may be singular or not. The following corollary works out both cases.

(2.9) Corollary. *Let $P_i = P_{\binom{\epsilon_i}{\eta_i}}$, $i = 1, 2, 3$, be pairwise different singular planes of K.*

a) P_1, P_2 and P_3 intersect in a singular point if and only if either

$$\begin{array}{lll} \epsilon_i \neq \epsilon_j \text{ for all } i \neq j & \text{and} & \eta_k = \eta_l \text{ for some } k \neq l, \quad \text{or} \\ \eta_i \neq \eta_j \text{ for all } i \neq j & \text{and} & \epsilon_k = \epsilon_l \text{ for some } k \neq l. \end{array}$$

b) P_1, P_2 and P_3 do not intersect in a singular point if and only if either

$$\begin{array}{llll} \epsilon_i \neq \epsilon_j & \text{and} & \eta_i \neq \eta_j & \text{for } i \neq j, \quad \text{or} \\ \epsilon_i = \epsilon_j & \text{and} & \eta_i = \eta_k & \text{where } \{i, j, k\} = \{1, 2, 3\}. \end{array}$$

(2.10) Remark. Corollary 2.9 can also be expressed as follows: pairwise different singular planes $P_{x_1}, P_{x_2}, P_{x_3}$ intersect in a singular point z if and only if the singular points x_1, x_2, x_3 span the singular plane P_z. This statement is due to the fact that the notation reflects the *self-duality of the Kummer surface.* We do not study this property of K (see Hudson [1]).

Finally we introduce Rosenhain and Göpel tetrahedra. They can be applied to derive theta relations (see Exercise 7.10). In the next section we will use a Rosenhain tetrahedron to deduce an equation for the Kummer surface K. A *Rosenhain tetrahedron* for K is by definition a tetrahedron in $\mathbb{P}_3$ with singular planes of K as faces and singular points of K as vertices.

(2.11) Proposition. *The singular planes*

a) $P_{\binom{\epsilon_1}{\eta}}, P_{\binom{\epsilon_2}{\eta}}, P_{\binom{\epsilon_3}{\eta'}}, P_{\binom{\epsilon_4}{\eta'}}$ *where* $\epsilon_j \neq \epsilon_j$ *for* $i \neq j$, *and*

b) $P_{\binom{\epsilon}{\eta_1}}, P_{\binom{\epsilon}{\eta_2}}, P_{\binom{\epsilon'}{\eta_3}}, P_{\binom{\epsilon'}{\eta_4}}$ *where* $\eta_i \neq \eta_j$ *for* $i \neq j$,

form Rosenhain tetrahedra for K. Conversely every Rosenhain tetrahedron is one of these.

Proof. According to Corollary 2.9 a) the quadruples of pairwise different singular planes such that any three of them intersect in a singular point are those of a) and b) as well as of

c) $P_{\binom{\epsilon_1}{\eta}}, P_{\binom{\epsilon_2}{\eta}}, P_{\binom{\epsilon_3}{\eta}}, P_{\binom{\epsilon_4}{\eta'}}$ where $\eta \neq \eta'$ and $\epsilon_i \neq \epsilon_j$ for $i \neq j$, and

d) $P_{\binom{\epsilon}{\eta_1}}, P_{\binom{\epsilon}{\eta_2}}, P_{\binom{\epsilon}{\eta_3}}, P_{\binom{\epsilon'}{\eta_4}}$ where $\epsilon \neq \epsilon'$ and $\eta_i \neq \eta_j$ for $i \neq j$.

But the planes of c) and d) do not form tetrahedra, since all four planes intersect in a point, namely in $\binom{\epsilon_4}{\eta}$ in case c) and, respectively, in $\binom{\epsilon}{\eta_4}$ in case d). On the other hand, in the cases a) and b) the four planes have no common point, so they form tetrahedra. □

(2.12) Corollary. *Singular planes P_{x_1}, P_{x_2}, P_{x_3} and P_{x_4} form a Rosenhain tetrahedron if and only if x_1, x_2 and x_3 span a singular plane and $x_4 = x_1 + x_2 + x_3$.*

(2.13) Corollary. *There are exactly* 80 *Rosenhain tetrahedra for K.*

Proof. By Remark 2.10 any three singular points out of the 6 singular points of a given singular plane P determine a Rosenhain tetrahedron. Hence there are $\binom{6}{3} = 20$ Rosenhain tetrahedra with P as a face. Counting plane-tetrahedra incidences shows that there are exactly $\frac{1}{4} \cdot 16 \cdot 20 = 80$ Rosenhain tetrahedra. □

A *Göpel tetrahedron* for K is by definition a tetrahedron in $\mathbb{P}_3$ with singular planes of K as faces such that the vertices are not singular points.

(2.14) Proposition. *The singular planes*

$$\begin{array}{lllll} a) & P_{\binom{\epsilon_1}{\eta_1}}, & P_{\binom{\epsilon_2}{\eta_2}}, & P_{\binom{\epsilon_3}{\eta_3}}, & P_{\binom{\epsilon_4}{\eta_4}}, \quad \text{and} \\ b) & P_{\binom{\epsilon_1}{\eta_1}}, & P_{\binom{\epsilon_1}{\eta_2}}, & P_{\binom{\epsilon_2}{\eta_2}}, & P_{\binom{\epsilon_2}{\eta_1}}, \end{array}$$

where $\epsilon_i \neq \epsilon_j$ and $\eta_i \neq \eta_j$ for $i \neq j$, form Göpel tetrahedra for K. Conversely every Göpel tetrahedron is one of these.

Proof. According to Corollary 2.9 b) three planes do not intersect in a singular point if and only if they are either of the form $P_{\binom{\epsilon_1}{\eta_1}}, P_{\binom{\epsilon_2}{\eta_2}}, P_{\binom{\epsilon_3}{\eta_3}}$ or $P_{\binom{\epsilon_1}{\eta_1}}, P_{\binom{\epsilon_1}{\eta_2}}, P_{\binom{\epsilon_2}{\eta_2}}$ with $\epsilon_i \neq \epsilon_j$ and $\eta_i \neq \eta_j$. Moreover, it is immediate to check that there is exactly one singular plane, namely $P_{\binom{\epsilon_4}{\eta_4}}$ respectively $P_{\binom{\epsilon_2}{\eta_1}}$, completing the planes to a Göpel tetrahedron. □

(2.15) Corollary. *Singular planes $P_{x_1}, P_{x_2}, P_{x_3}$ and P_{x_4} form a Göpel tetrahedron for K if and only if x_1, x_2, x_3 do not span a singular plane and $x_4 = x_1 + x_2 + x_3$.*

Finally we have for the number of Göpel tetrahedra

(2.16) Corollary. *There are exactly* 60 *Göpel tetrahedra for K.*

Proof. Any three given singular planes P_{x_1}, P_{x_2} and P_{x_3} intersect either in a singular point or not. According to Corollaries 2.12 and 2.15 there is a unique singular plane, namely $P_{x_1+x_2+x_3}$, such that the four planes form a Rosenhain (respectively Göpel) tetrahedron. This implies that the union of the set of Rosenhain tetrahedra with the set of Göpel tetrahedra contains $\frac{1}{4}\binom{16}{3} = 140$ elements. Now the assertion follows from the fact that there are 80 Rosenhain tetrahedra. □

§ 3 An Equation for the Kummer Surface

In the last section we studied some geometrical properties of the Kummer surface K in $\mathbb{P}_3$ associated to an abelian surface $X = V/\Lambda$ with an irreducible principal polarization. In this section we want to deduce an equation for K using a Rosenhain tetrahedron and the theory of theta groups.

Let H be an irreducible principal polarization on X with symplectic basis $\lambda_1, \lambda_2, \mu_1, \mu_2$ of Λ and $L_0 = L(H, \chi_0)$ the associated line bundle of characteristic zero. Suppose e_1, e_2, f_1, f_2 is the induced basis of $X_2 = K(L_0^2)$. Let K_1 (respectively K_2) denote the isotropic subgroups of X_2 spanned by

e_1 and e_2 (respectively f_1 and f_2). Then $X_2 = K(L_0^2) = K_1 \oplus K_2$ is the decomposition for L_0^2 associated to the symplectic basis of Λ in the sense of Section 3.1. Using the identification $X_2 \leftrightarrow \mathbb{F}_2^4$ of Section 2, we have with $e := e_1 + e_2 + f_1 + f_2$

$$e = \begin{pmatrix} 1 \\ 1 \\ 1 \\ 1 \end{pmatrix}, \; e_1 + e = \begin{pmatrix} 0 \\ 1 \\ 1 \\ 1 \end{pmatrix}, \; f_1 + e = \begin{pmatrix} 1 \\ 0 \\ 1 \\ 1 \end{pmatrix}, \; e_1 + f_1 + e = \begin{pmatrix} 0 \\ 0 \\ 1 \\ 1 \end{pmatrix}.$$

As in the last section define $L = t_e^* L_0$. According to Proposition 2.11 the singular planes $P_e, P_{e_1+e}, P_{f_1+e}$ and $P_{e_1+f_1+e}$ form a Rosenhain tetrahedron for K. For any $x \in X_2$ denote by ϑ^x the basis of the 1-dimensional vector space (of canonical theta functions) $H^0(t_x^* L_0)$ as in Theorem 3.2.7. The squares $(\vartheta^x)^2$ are canonical theta functions for the line bundle L_0^2. Since $t_x^* L = t_{x+e}^* L_0$, the divisors corresponding to ϑ^0, ϑ^{e_1}, ϑ^{f_1} and $\vartheta^{e_1+f_1}$ define the singular planes $P_e, P_{e_1+e}, P_{f_1+e}$ and $P_{e_1+f_1+e}$. The fact that these planes form a tetrahedron implies that $(\vartheta^0)^2$, $(\vartheta^{e_1})^2$, $(\vartheta^{f_1})^2$ and $i(\vartheta^{e_1+f_1})^2$ form a basis for $H^0(L_0^2)$. Denote by Y_0, Y_1, Y_2, Y_3 the homogeneous coordinates of $\mathbb{P}_3$ corresponding to this basis. In these terms the property of P_e, P_{e_1+e}, P_{f_1+e} and $P_{e_1+f_1+e}$ being a Rosenhain tetrahedron means

a) the Kummer surface K is singular in the coordinate points,
b) the coordinate planes touch K in smooth conics.

We want to use these properties to deduce an equation for K. The map $\varphi_{L_0^2}: X \to K \subset \mathbb{P}_3$ is $K(L_0^2)$-equivariant. In order to determine an equation for K in $\mathbb{P}_3$, we have to compute the action of the group $K(L_0^2)$ on the chosen coordinates. It turns out to be sufficient to know the action of the subgroup generated by e_1 and f_1. The automorphisms e_1 and f_1 of $\mathbb{P}_3$ act on the homogeneous coordinates $Y_0, \ldots, Y_3$ as follows.

(3.1) Lemma. $$e_1: \begin{pmatrix} Y_0 \\ Y_1 \\ Y_2 \\ Y_3 \end{pmatrix} \mapsto \begin{pmatrix} Y_1 \\ Y_0 \\ -Y_3 \\ -Y_2 \end{pmatrix}, \qquad f_1: \begin{pmatrix} Y_0 \\ Y_1 \\ Y_2 \\ Y_3 \end{pmatrix} \mapsto \begin{pmatrix} Y_2 \\ Y_3 \\ Y_0 \\ Y_1 \end{pmatrix}.$$

Proof. By definition the action of $K(L_0^2)$ on $\mathbb{P}_3$ lifts to the canonical representation $H^0(L_0^2)$ of the theta group $\mathcal{G}(L_0^2)$. Proposition 6.4.2 gives an explicit description of the canonical representation in terms of the basis $\{\vartheta_x^{L_0^2} \mid x \in K_1\}$ of Theorem 3.2.7. In Step I we express the basis of $H^0(L_0^2)$ corresponding to the coordinates $Y_0, \ldots, Y_3$ in terms of this basis.

Step I: For all $x = x_1 + x_2$ with $x_1 \in \{0, e_1\}$, $x_2 \in \{0, f_1\}$ and with $c_0 = c_{e_1} = c_{f_1} = 1$ and $c_{e_1+f_1} = i$ we have

$$(\vartheta^x)^2 = c_x \sum_{y \in K_1} e^{L_0^2}(x_2, y) \vartheta_y^{L_0^2}(0) \vartheta_{x_1+y}^{L_0^2}.$$

Consider the multiplication map $\mu: H^0(L_0) \otimes H^0(L_0) \to H^0(L_0^2)$. The bases of $H^0(L_0)$ and $H^0(L_0^2)$ are chosen in such a way that we may apply the Multiplication Formula 7.1.3 to obtain

$$(\vartheta^0)^2 = \mu(\vartheta^0 \otimes \vartheta^0) = \sum_{y \in K_1 \cap X_2} \vartheta_{y+z}^{L_0^2}(0)\vartheta_{y+z}^{L_0^2} = \sum_{y \in K_1} \vartheta_y^{L_0^2}(0)\vartheta_y^{L_0^2}$$

for any $z \in K_1$. Applying Corollary 3.2.9 and the definition of $\vartheta_{e_1}^{L_0^2}$ (see equation 3.2(4)) we get in the case $x = e_1$:

$$\begin{aligned}
\left(\vartheta^{e_1}(v)\right)^2 &= e\left(-\pi(2H)(v, \tfrac{1}{2}\lambda_1) - \tfrac{\pi}{2}(2H)(\tfrac{1}{2}\lambda_1, \tfrac{1}{2}\lambda_1)\right)\left(\vartheta^0(v + \tfrac{1}{2}\lambda_1)\right)^2 \\
&= a_{L_0^2}(\tfrac{1}{2}\lambda_1, v)^{-1}\left(\vartheta^0(v + \tfrac{1}{2}\lambda_1)\right)^2 \\
&= \sum_{y \in K_1} \vartheta_y^{L_0^2}(0) a_{L_0^2}(\tfrac{1}{2}\lambda_1, v)^{-1} \vartheta_y^{L_0^2}(v + \tfrac{1}{2}\lambda_1) \\
&= \sum_{y \in K_1} \vartheta_y^{L_0^2}(0)\vartheta_{e_1+y}^{L_0^2}(v) \quad .
\end{aligned}$$

Since $e^{L_0^2}(0,y) = e^{L_0^2}(e_1,y) = 1$ for all $y \in K_1$, this gives the assertion in the case $x \in \{0, e_1\}$. With a similar computation, using Corollary 3.2.5 or Exercise 3.3, one obtains the assertion in the remaining cases.
Step II: In terms of the chosen coordinates of $\mathbb{P}_3$ Lemma 3.1 can be interpreted as

$$Y_0 \leftrightarrow \sum_{y \in K_1} \vartheta_y^{L_0^2}(0)\vartheta_y^{L_0^2}\,, \qquad Y_1 \leftrightarrow \sum_{y \in K_1} \vartheta_y^{L_0^2}(0)\vartheta_{e_1+y}^{L_0^2}\,,$$
$$Y_2 \leftrightarrow \sum_{y \in K_1} e^{L_0^2}(f_1,y)\vartheta_y^{L_0^2}(0)\vartheta_y^{L_0^2}\,, Y_3 \leftrightarrow -\sum_{y \in K_1} e^{L_0^2}(f_1,y)\vartheta_y^{L_0^2}(0)\vartheta_{e_1+y}^{L_0^2}\,.$$

The elements $[a_{L_0^2}(\frac{1}{2}\lambda_1, 0), \frac{1}{2}\lambda_1]$ and $[a_{L_0^2}(\frac{1}{2}\mu_1, 0), \frac{1}{2}\mu_1]$ are liftings of e_1 and f_1 to the theta group $G(L_0^2)$. Now the assertion is a direct application of Proposition 6.4.2 respectively equations 6.4(2) and 6.4(3). □

Let $Q \in \mathbb{C}[Y_0, Y_1, Y_2, Y_3]$ be a quartic defining the Kummer surface K, i.e. $K = \{x \in \mathbb{P}_3 \mid Q(x) = 0\}$. According to Proposition 6.6.1 the Kummer surface K is invariant under the action of the theta group. In terms of Q this means that there is a character $\kappa: K(L_0^2) \to \mathbb{C}^*$ such that

$$\sigma^* Q = \kappa(\sigma) Q \tag{1}$$

for all $\sigma \in K(L_0^2)$. On the other hand by property a) of above we have in particular

$$Q(1,0,0,0) = Q(0,1,0,0) = Q(0,0,1,0) = Q(0,0,0,1) = 0\,. \tag{2}$$

Moreover by b) there are quadratic forms F_0, F_1, F_2 and F_3 such that

$$\begin{aligned} Q(0,Y_1,Y_2,Y_3) &= F_0(Y_1,Y_2,Y_3)^2, \quad Q(Y_0,0,Y_2,Y_3) = F_1(Y_0,Y_2,Y_3)^2, \\ Q(Y_0,Y_1,0,Y_3) &= F_2(Y_0,Y_1,Y_3)^2, \quad Q(Y_0,Y_1,Y_2,0) = F_3(Y_0,Y_1,Y_2)^2. \end{aligned} \tag{3}$$

By (2) we have $F_0(1,0,0) = F_0(0,1,0) = F_0(0,0,1) = 0$, so

$$F_0(Y_1,Y_2,Y_3) = \lambda_1 Y_2 Y_3 + \lambda_2 Y_1 Y_3 + \lambda_3 Y_1 Y_2 \tag{4}$$

for some $\lambda_1, \lambda_2, \lambda_3 \in \mathbb{C}$. Using (1) and (3) we get

$$F_1(Y_0,Y_2,Y_3)^2 = e_1^* Q(0,Y_0,-Y_3,-Y_2) = \kappa(e_1) F_0(Y_0,-Y_3,-Y_2)^2 \; .$$

Similarly we obtain

$$\begin{aligned} F_2(Y_0,Y_1,Y_3)^2 &= \kappa(f_1) F_0(Y_3,Y_0,Y_1)^2 \quad \text{and} \\ F_3(Y_0,Y_1,Y_2)^2 &= \kappa(e_1 f_1) F_0(-Y_2,Y_1,Y_0)^2 \; . \end{aligned}$$

Define a polynomial p by

$$\begin{aligned} p(Y_0,\ldots,Y_3) = Q(Y_0,Y_1,Y_2,Y_3) &- F_0(Y_1,Y_2,Y_3)^2 - F_1(Y_0,Y_2,Y_3)^2 \\ &- F_2(Y_0,Y_1,Y_3) - F_3(Y_0,Y_1,Y_2)^2 \; . \end{aligned}$$

Then, using (4) we can compute $Q - p$ as follows:

$$\begin{aligned} Q - p &= F_0(Y_1,Y_2,Y_3)^2 + \kappa(e_1) F_0(Y_0,-Y_3,-Y_2)^2 + \kappa(f_1) F_0(Y_3,Y_0,Y_1)^2 \\ &\quad + \kappa(e_1 f_1) F_0(-Y_2,Y_1,Y_0)^2 \\ &= \lambda_1^2 \big(1+\kappa(e_1)\big)\big(\kappa(f_1) Y_0^2 Y_1^2 + Y_2^2 Y_3^2\big) \\ &\quad + \lambda_2^2 \big(1+\kappa(f_1)\big)\big(\kappa(e_1) Y_0^2 Y_2^2 + Y_1^2 Y_3^2\big) \\ &\quad + \lambda_3^2 \big(1+\kappa(e_1 f_1)\big)\big(\kappa(e_1) Y_0^2 Y_3^2 + Y_1^2 Y_2^2\big) \\ &\quad - 2\lambda_1\lambda_2 \big(\kappa(f_1) Y_0 Y_1 + Y_2 Y_3\big)\big(\kappa(e_1) Y_0 Y_2 - Y_1 Y_3\big) \\ &\quad + 2\lambda_1\lambda_3 \big(\kappa(e_1 f_1) Y_0 Y_1 - Y_2 Y_3\big)\big(\kappa(e_1) Y_0 Y_3 - Y_1 Y_2\big) \\ &\quad + 2\lambda_2\lambda_3 \big(\kappa(e_1 f_1) Y_0 Y_2 + Y_1 Y_3\big)\big(\kappa(f_1) Y_0 Y_3 + Y_1 Y_2\big) \quad . \end{aligned}$$

This together with (3) gives

$$\begin{aligned} p(0,Y_1,Y_2,Y_3) &= F_0(Y_1,Y_2,Y_3)^2 - \lambda_1^2(1+\kappa(e_1)) Y_2^2 Y_3^2 \\ &\qquad - \lambda_2^2(1+\kappa(f_1)) Y_1^2 Y_3^2 - \lambda_3^2(1+\kappa(e_1 f_1)) Y_1^2 Y_2^2 \\ &\qquad - 2\lambda_1\lambda_2 Y_1 Y_2 Y_3^2 - 2\lambda_1\lambda_3 Y_1 Y_2^2 Y_3 - 2\lambda_2\lambda_3 Y_1^2 Y_2 Y_3 \\ &= -\lambda_1^2 \kappa(e_1) Y_2^2 Y_3^2 - \lambda_2^2 \kappa(f_1) Y_1^2 Y_3^2 - \lambda_3^2 \kappa(e_1 f_1) Y_1^2 Y_2^2 \; . \end{aligned}$$

Similarly we obtain

$$p(Y_0, 0, Y_2, Y_3) = -\lambda_1^2 Y_2^2 Y_3^2 - \lambda_2^2 \kappa(e_1 f_1) Y_0^2 Y_2^2 - \lambda_3^2 \kappa(f_1) Y_0^2 Y_3^2 \quad \text{and}$$
$$p(Y_0, Y_1, 0, Y_3) = -\lambda_1^2 \kappa(e_1 f_1) Y_0^2 Y_1^2 - \lambda_2^2 Y_1^2 Y_3^2 - \lambda_3^2 \kappa(e_1) Y_0^2 Y_3^2 .$$

Setting $Y_1 = 0$ and $Y_0 = 0$ in $p(0, Y_1, Y_2, Y_3)$ and $p(Y_0, 0, Y_2, Y_3)$ respectively, we deduce $\kappa(e_1) = 1$. Similarly one gets $\kappa(f_1) = 1$, and thus p is necessarily of the form

$$\begin{aligned} p(Y_0, Y_1, Y_2, Y_3) = & -\lambda_1^2 (Y_0^2 Y_1^2 + Y_2^2 Y_3^2) - \lambda_2^2 (Y_0^2 Y_2^2 + Y_1^2 Y_3^2) \\ & -\lambda_3^2 (Y_0^2 Y_3^2 + Y_1^2 Y_2^2) + \lambda_0^2 Y_0 Y_1 Y_2 Y_3 \end{aligned}$$

for some $\lambda_0 \in \mathbb{C}$. This proves

(3.2) Proposition. *The coordinates of $\mathbb{P}_3$ can be chosen in such a way that the Kummer surface K associated to the abelian surface X with irreducible principal polarization L is given by the equation*

$$\begin{aligned} & \lambda_1^2 (Y_0^2 Y_1^2 + Y_2^2 Y_3^2) + \lambda_2^2 (Y_0^2 Y_2^2 + Y_1^2 Y_3^2) + \lambda_3^2 (Y_0^2 Y_3^2 + Y_1^2 Y_2^2) \\ & + 2\lambda_1 \lambda_2 (Y_0 Y_1 + Y_2 Y_3)(-Y_0 Y_2 + Y_1 Y_3) \\ & + 2\lambda_1 \lambda_3 (Y_0 Y_1 - Y_2 Y_3)(Y_0 Y_3 - Y_1 Y_2) \\ & + 2\lambda_2 \lambda_3 (Y_0 Y_2 + Y_1 Y_3)(Y_0 Y_3 + Y_1 Y_2) \\ & + \lambda_0^2 Y_0 Y_1 Y_2 Y_3 = 0 \end{aligned} \tag{5}$$

for some $(\lambda_0 : \lambda_1 : \lambda_2 : \lambda_3) \in \mathbb{P}_3$.

(3.3) Remark. The moduli space $\mathcal{A}_{(1,1)}$ of principally polarized abelian surfaces is of dimension three. Hence the family of Kummer surfaces is also of dimension three. Since the family of quartics above is parametrized by $\mathbb{P}_3$, this implies that for a general $(\lambda_0 : \cdots : \lambda_3)$ equation (5) defines a Kummer surface. For another form of the equations for the Kummer surfaces see Exercise 7.3.

§ 4 Reider's Theorem

Suppose L is an ample line bundle of type $(1, d)$ with $d \geq 5$ on an abelian surface X. The object of this section is to give a criterion for the induced map $\varphi_L : X \to \mathbb{P}_{d-1}$ to be an embedding. In order to avoid trivialities, suppose (X, L) does not split, i.e. is not isomorphic to a polarized product of elliptic curves $(E_1 \times E_2, p_1^* L_1 \otimes p_2^* L_2)$. The main result of this section is

(4.1) Reider's Theorem. *Suppose L is an ample line bundle of type $(1, d)$ with $d \geq 5$ on X. Then the morphism $\varphi_L : X \to \mathbb{P}_{d-1}$ is an embedding if and only if there is no elliptic curve C on X with $(C \cdot L) = 2$.*

A general abelian surface X with polarization L of type $(1,d)$ does not contain an elliptic curve, since its Néron Severi group $\mathrm{NS}(X)$ is isomorphic to $\mathbb{Z}$ (see Exercise 8.1). So generically the map φ_L is an embedding.

On the other hand, special abelian surfaces X may contain infinitely many elliptic curves. For applications, in particular to moduli problems, it is more useful to have a criterion involving only one curve: according to Remark 3.1.5 there is a cyclic isogeny of abelian varieties $q: X \to Y$ of degree d and a line bundle M on Y with $L = q^*M$. The line bundle M defines a principal polarization on Y. According to Corollary 11.8.2 the unique curve D_Y in $|M|$ is either smooth of genus 2 or a union of two elliptic curves intersecting exactly in one point. Define D_X by the cartesian diagram

$$\begin{array}{ccc} D_X & \hookrightarrow & X \\ {\scriptstyle q}\downarrow & & \downarrow{\scriptstyle q} \\ D_Y & \hookrightarrow & Y \end{array}$$

and recall that an *elliptic involution* on a curve is an involution whose quotient is an elliptic curve.

The following criterion was given by Ramanan [1] in the case of a smooth curve D_Y and by Hulek-Lange [1] in the remaining case. Here we deduce it as a consequence of Reider's Theorem.

(4.2) Corollary. *For $d \geq 5$ the morphism φ_L is an embedding if and only if the curves D_X and D_Y do not admit elliptic involutions compatible with the covering q.*

Proof. It suffices to show that the following two conditions are equivalent:

i) there is an elliptic curve $C \subset X$ with $(C \cdot L) = 2$, and

ii) the curves D_X and D_Y admit elliptic involutions compatible with q.

i) $\Longrightarrow$ ii) Let $C \subset X$ be an elliptic curve with $(C \cdot L) = 2$. We may assume that C is an abelian subvariety of X. Then the quotient X/C is also an elliptic curve. Denote by f the composition $f: D_X \hookrightarrow X \to X/C$.

We claim that f is a finite map of degree 2. If D_X is irreducible this is clear, since $\deg f = (C \cdot L) = 2$. Otherwise $D_X = F_1 + F_2$ where F_1 and F_2 are elliptic curves with $(F_1 \cdot F_2) = d$ and $(C \cdot F_1) + (C \cdot F_2) = (C \cdot L) = 2$. Assume $(C \cdot F_1) = 0$. Then $C \equiv F_1$ and thus $2 = (C \cdot F_2) = (F_1 \cdot F_2) = d \geq 5$, a contradiction. By symmetry this gives $(C \cdot F_1) = (C \cdot F_2) = 1$, which implies the assertion.

Define $C' = q(C)$. Since by Proposition 4.9.5 the abelian surface Y does not contain any rational curve, C' and thus also Y/C' are elliptic curves. Let f' denote the composition $f': D_Y \hookrightarrow Y \to Y/C'$. Then we have the following commutative diagram

$$\begin{array}{ccc} D_X & \xrightarrow{f} & X/C \\ {\scriptstyle q}\downarrow & & \downarrow{\scriptstyle \bar{q}} \\ D_Y & \xrightarrow{f'} & Y/C' \end{array} .$$

Let $d' = \deg(q|C: C \to C')$. The composed map $\bar{q}f$ is finite of degree $\frac{2d}{d'}$. Since $q: D_X \to D_Y$ is of degree d, this implies that f' is finite of degree $\frac{2}{d'}$. But f' cannot be of degree 1, since Y/C' is an elliptic curve and either D_Y is smooth of genus 2 or reducible. So f and f' provide a pair of elliptic involutions compatible with the covering q.
ii) $\Longrightarrow$ i) Suppose ι_X and ι_Y are elliptic involutions on D_X and D_Y compatible with q, i.e. the following diagram commutes

$$\begin{array}{ccc} D_X & \xrightarrow{\iota_X} & D_X \\ {\scriptstyle q}\downarrow & & \downarrow{\scriptstyle q} \\ D_Y & \xrightarrow{\iota_Y} & D_Y \end{array} .$$

We may assume that $0 \in D_Y$ and $\iota_Y(0) = 0$. This implies that $\ker(q: X \to Y)$ is contained in D_X and ι_X acts trivially on it.

We first claim that ι_X extends to an involution $\tilde{\iota}_X$ on X. Since D_X generates X, every $x \in X$ is of the form $x = \sum_{i=1}^n n_i p_i$ with $p_i \in D_X$ and $n_i \in \mathbb{Z}$. Define $\tilde{\iota}_X(x) = \sum_{i=1}^n n_i \iota_X(p_i)$. We have to show that this definition is independent of the chosen representation of x. It suffices to show that for any representation of the origin $0 = \sum n_i p_i$ we have $\sum n_i \iota_X(p_i) = 0$. According to the Universal Property of the Jacobian 11.4.1 the involution ι_Y on D_Y extends to an involution $\tilde{\iota}_Y$ on Y. Note that this also holds for reducible D_Y. So

$$q(\sum n_i \iota_X(p_i)) = \sum n_i \iota_Y(q(p_i)) = \tilde{\iota}_Y(q(\sum n_i p_i)) = 0 .$$

This implies $\sum n_i \iota_X(p_i) =: p \in \ker q$. But $p = \iota_X(p) = \sum n_i p_i = 0$, which completes the proof of the claim.

Denote $F = \operatorname{im}(1_X + \tilde{\iota}_X)$. It is easy to see that this is an elliptic curve. Moreover, by definition the composed map $f: D_X \hookrightarrow X \to F$ is the double covering associated to the involution ι_X. The kernel $\ker(1_X + \tilde{\iota}_X)$ is a disjoint union of translates of an elliptic curve C, say $\ker(1_X + \tilde{\iota}_X) = \bigcup_{i=1}^n t_{x_i}^* C$. Since $n(C \cdot L) = \deg f = 2$, we obtain $(C \cdot L) = 1$ or 2. If $(C \cdot L) = 1$, the polarized abelian variety (X, L) splits according to Lemma 4.6, a contradiction. So $(C \cdot L) = 2$. □

We go back to Theorem 4.1. It suffices to show that φ_L is an embedding provided there is no elliptic curve C on X with $(C \cdot L) = 2$. The converse implication is trivial, since a line bundle of degree 2 on an elliptic curve is never very ample. The idea of the proof is as follows: assuming there are points p and q on X (possibly infinitely near) with $\varphi_L(p) = \varphi_L(q)$, one constructs a vector bundle F on X from which a contradiction can be derived.

The proof consists of several propositions and lemmas. First we recall some definitions and easy statements on vector bundles on (abelian) surfaces: a vector bundle F of rank 2 on X is called *(semi-)stable* (with respect to a polarization H of X) if

$$(c_1(G) \cdot H) \underset{(=)}{<} \tfrac{1}{2}(c_1(F) \cdot H) \tag{1}$$

for every coherent subsheaf G of F of rank 1. Any coherent subsheaf of F is contained in a unique coherent subsheaf of F with a torsion free quotient sheaf. On the other hand, any coherent subsheaf of F with torsion free quotient is a line bundle. Hence in the definition of (semi-)stability if suffices to require the inequality (1) for all line bundles contained in F. Finally, recall that any coherent torsion free sheaf of rank 1 on X is of the form $I_Z \otimes M$, where M is a line bundle on X and I_Z the ideal sheaf of a zero dimensional subscheme Z of X.

Denote as usual by $\mathcal{E}nd\, F = F^* \otimes F$ the vector bundle of endomorphisms of F.

(4.3) Lemma. $h^0(\mathcal{E}nd\, F) = 1$ *for any stable vector bundle* F *of rank* 2 *on* X.

Proof. It suffices to show that any nonzero endomorphism $f\colon F \to F$ is multiplication by a constant. Suppose first that f is of rank 1. Consider the following exact sequences of coherent sheaves

$$\begin{array}{ccccccccc} 0 & \longrightarrow & \ker f & \longrightarrow & F & \longrightarrow & \operatorname{im} f & \longrightarrow & 0 \ , \\ 0 & \longrightarrow & \operatorname{im} f & \longrightarrow & F & \longrightarrow & \operatorname{coker} f & \longrightarrow & 0 \ . \end{array}$$

The stability of F implies

$$\tfrac{1}{2}\big(c_1(F)\cdot H\big) < \big(c_1(F)\cdot H\big) - \big(c_1(\ker f)\cdot H\big) = \big(c_1(\operatorname{im} f)\cdot H\big) < \tfrac{1}{2}\big(c_1(F)\cdot H\big) \ ,$$

a contradiction. Hence f is of rank 2. But then $\det f$ is a nonzero endomorphism of the line bundle $\det F$ and as such multiplication by a nonzero constant. This implies that f is an isomorphism. It follows that the $\mathbb{C}$-algebra of endomorphisms of F is a skew field of finite dimension over $\mathbb{C}$ and thus isomorphic to $\mathbb{C}$ itself. □

The main tool for the proof of Theorem 4.1 is the following result

(4.4) Bogomolov's Inequality. $\big(c_1(F)^2\big) - 4c_2(F) \leq 0$ *for any semistable vector bundle* F *of rank* 2 *on* X.

Proof. Step I: Suppose first F is stable with respect to a polarization H. According to Lemma 4.3 we have $h^2(\mathcal{E}nd\, F) = h^0(\mathcal{E}nd\, F) = 1$. Moreover, the trace yields a splitting of the natural inclusion $\mathcal{O}_X \hookrightarrow \mathcal{E}nd\, F$, $\alpha \mapsto \alpha\, \mathrm{id}_F$. This implies $h^1(\mathcal{E}nd\, F) \geq h^1(\mathcal{O}_X) = 2$. So

$$\chi(\mathcal{E}nd\, F) = \sum_{\nu=0}^{2} (-1)^\nu h^\nu(\mathcal{E}nd\, F) \leq 2 - 2 = 0 \tag{2}$$

On the other hand the Riemann-Roch Theorem states

$$\chi(\mathcal{E}nd\, F) = ch(\mathcal{E}nd\, F)_2 \tag{3}$$

(see Exercise 3.9). Using the calculus of Chern classes we obtain

$$ch(\mathcal{E}nd\, F)_2 = \big(c_1(F)^2\big) - 4c_2(F).$$

Combining this with (2) and (3) gives the assertion in the stable case.
Step II: Suppose F is semistable, but not stable with respect to H. Then there is an exact sequence

$$0 \longrightarrow G_1 \longrightarrow F \longrightarrow I_Z \otimes G_2 \longrightarrow 0$$

with line bundles G_1 and G_2 on X such that $(c_1(G_1) \cdot H) = (c_1(G_2) \cdot H)$, and an ideal sheaf I_Z of a zero dimensional subscheme Z of X. Tensoring with G_1^{-1}, we obtain the exact sequence

$$0 \longrightarrow \mathcal{O}_X \longrightarrow F \otimes G_1^{-1} \longrightarrow I_Z \otimes G_2 \otimes G_1^{-1} \longrightarrow 0$$

with $\big(c_1(G_2 \otimes G_1^{-1}) \cdot H\big) = 0$. On the other hand, comparing the Chern classes of F and $F \otimes G_1^{-1}$ we get

$$\begin{aligned}\big(c_1(F)^2\big) - 4c_2(F) &= \big(c_1(F \otimes G_1^{-1})^2\big) - 4c_2(F \otimes G_1^{-1})\\ &= \big(c_1(G_2 \otimes G_1^{-1})^2\big) - 4 \deg Z \leq 0\,.\end{aligned}$$

For the last inequality we used that

$$\big(c_1(G_2 \otimes G_1^{-1})^2\big) \cdot (H^2) \leq \big(c_1(G_2 \otimes G_1^{-1}) \cdot H\big)^2 = 0$$

by the Hodge index theorem (see Hartshorne [1] Ex. V.1.9). □

Recall that L is an ample line bundle of type $(1, d)$ with $d \geq 5$ without fixed components on X. According to Lemma 1.2 it is generated by global sections. Assume now that the linear system $|L|$ fails to separate points p and $q \in X$ possibly infinitely near (see Hartshorne [1] p. 392), i.e.

$$h^0(I_{p+q} \otimes L) = h^0(L) - 1 = d - 1 \tag{4}$$

Here I_{p+q} denotes the ideal of the zero dimensional subscheme $p + q$ on X. Then we can construct a vector bundle F of rank 2 on X as follows: the exact sequence $0 \longrightarrow I_{p+q} \otimes L \longrightarrow L \longrightarrow L|p+q \longrightarrow 0$ gives the cohomology sequence

$$0 \to H^0(I_{p+q} \otimes L) \to H^0(L) \to H^0(L|p+q) \to H^1(I_{p+q} \otimes L) \to 0 .$$

Together with (4) we deduce $h^1(I_{p+q} \otimes L) = 1$. According to Serre duality this gives

$$\mathrm{Ext}^1(I_{p+q} \otimes L, \mathcal{O}_X)^* \simeq H^1(I_{p+q} \otimes L) \simeq \mathbb{C} .$$

Hence there exists a nontrivial extension

$$0 \longrightarrow \mathcal{O}_X \longrightarrow F \longrightarrow I_{p+q} \otimes L \longrightarrow 0 \tag{5}$$

uniquely determined up to multiplication by a constant. Since $\mathrm{Ext}^1(L, \mathcal{O}_X) \simeq H^1(L)^* = 0$, the extension is not in the image of the canonical map $\mathrm{Ext}^1(L, \mathcal{O}_X) \to \mathrm{Ext}^1(I_{p+q} \otimes L, \mathcal{O}_X)$, implying that F is a vector bundle. For the proof of this fact we refer to Griffiths-Harris [1] p. 727. With this notation we have

(4.5) Proposition. *If F is not semistable with respect to $H = c_1(L)$, then there exists a curve C on X with $(C^2) = 0$ and $(C \cdot L) = 1$ or 2.*

Proof. The assumption on F implies that there is a diagram

$$\begin{array}{ccccccccc}
 & & & & 0 & & & & \\
 & & & & \downarrow & & & & \\
 & & & & G_1 & & & & \\
 & & & & \downarrow & & & & \\
0 & \longrightarrow & \mathcal{O}_X & \longrightarrow & F & \longrightarrow & I_{p+q} \otimes L & \longrightarrow & 0 \\
 & & & \overset{\sigma}{\searrow} & \downarrow & & & & \\
 & & & & I_Z \otimes G_2 & & & & \\
 & & & & \downarrow & & & & \\
 & & & & 0 & & & &
\end{array}$$

where Z is a zero dimensional subscheme of X and G_1 and G_2 are line bundles on X with $\big(c_1(G_1) \cdot H\big) > \frac{1}{2}\big(c_1(F) \cdot H\big)$ or equivalently

$$(G_1 \cdot L) > (G_2 \cdot L) . \tag{6}$$

We claim that there is an effective divisor C on X with $G_2 = \mathcal{O}_X(C)$ and $G_1 = L(-C)$. For the proof it suffices to show that the composed map

$\sigma: \mathcal{O}_X \to F \to I_Z \otimes G_2$ is nonzero. Otherwise there would be a nonzero homomorphism $I_{p+q} \otimes L \to I_Z \otimes G_2$ and thus, taking double duals, a nonzero homomorphism $L \to G_2$. This would imply that $G_2 \otimes L^{-1} = G_1^{-1}$ is effective, contradicting $2(G_1 \cdot L) > \big(c_1(F) \cdot L\big) = (L^2) > 0$.

Next we claim:

$$0 \le \big(C \cdot L(-C)\big) \le 2 \,. \tag{7}$$

The right hand inequality follows from $2 = c_2(F) = \big(C \cdot L(-C)\big) + \deg Z$. For the left hand inequality consider the diagram above restricted to an irreducible component C_i of C. Since the section σ vanishes on C_i we obtain an injective homomorphism of sheaves $0 \to \mathcal{O}_{C_i} \to L(-C) \otimes \mathcal{O}_{C_i}$ implying that $\big(C_i \cdot L(-C)\big) = \deg L(-C)|C_i \ge 0$. This proves (7).

The next point to observe is $(C^2) = 0$.

In fact, we have $\big((L(-C) \otimes \mathcal{O}_X(C))^2\big) = (L^2) = 2d \ge 10$. Using the right hand inequality of (7), this gives

$$\big(L(-C)^2\big) + (C^2) \ge 6 \,.$$

On the other hand, by (6)

$$\begin{aligned}\big(L(-C)^2\big) - (C^2) &= \Big(\big(L(-C) \otimes \mathcal{O}_X(-C)\big) \cdot \big(L(-C) \otimes \mathcal{O}_X(C)\big)\Big) \\ &= \big(L(-C) \cdot L\big) - (C \cdot L) = (G_1 \cdot L) - (G_2 \cdot L) > 0 \,.\end{aligned}$$

Combining both inequalities gives $\big(L(-C)^2\big) > 3$. So by the Hodge index theorem (see Hartshorne [1] Ex. V.1.9) and (7) we obtain

$$3(C^2) < (C^2)\big(L(-C)^2\big) \le \big(C \cdot L(-C)\big)^2 \le 4 \,.$$

Since (C^2) is an even nonnegative number, this is only possible for $(C^2) = 0$.

Finally we have $0 < (C \cdot L) = \big(C \cdot (L(-C) \otimes \mathcal{O}_X(C))\big) = \big(C \cdot L(-C)\big) \le 2$, which is the last assertion. □

The next lemma shows that for the proof of Theorem 4.1 one has only to take into account the case $(C \cdot L) = 2$ in the previous proposition.

(4.6) Lemma. *For an ample line bundle L of type $(1, d)$ on X the following conditions are equivalent:*

i) There is a curve C on X with $(C^2) = 0$ and $(C \cdot L) = 1$.

ii) The polarized abelian variety (X, L) is isomorphic to a polarized product of elliptic curves $(E_1 \times E_2, p_1^* L_1 \otimes p_2^* L_2)$.

Proof. ii) $\Longrightarrow$ i) Without loss of generality we may assume that L_1 is of type (1) on E_1 and L_2 is type (d) on E_2. Suppose there is an isomorphism $\varphi: (X, L) \to (E_1 \times E_2, p_1^* L_1 \otimes p_2^* L_2)$. Then $C := \varphi^{-1} E_1$ satisfies $(C^2) = 0$ by the adjunction formula (see 1(1)) and moreover $(C\cdot L) = (E_1\cdot p_1^*L_1\otimes p_2^*L_2) = (E_1\cdot E_2) + d(E_1^2) = 1$.

i) $\Longrightarrow$ ii) Note first that necessarily C is an elliptic curve by the adjunction formula 1(1) and Proposition 4.9.5. According to the Nakai-Moishezon Criterion (Corollary 4.3.3) the line bundle $L(-(d-1)C)$ is ample. Hence $h^0(L(-(d-1)C)) = \frac{1}{2}(L(-(d-1)C)^2) = 1$ and $h^0(L(-(d-1)C)|C) = \deg L(-(d-1)C)|C = 1$. The exact sequence

$$0\to H^0(L(-dC))\to H^0(L(-(d-1)C))\to H^0(L(-(d-1)C)|C)\to H^1(L(-dC))\to 0$$

yields $h^0(L(-dC)) = h^1(L(-dC)) = 1$ or 0. On the other hand, $\chi(L(-dC)) = \frac{1}{2}(L(-dC)^2) = 0$. So, twisting eventually $L(-dC)$ by a line bundle of $\mathrm{Pic}^0(X)$, we may assume that $h^0(L(-dC)) = 1$ (see Theorem 3.5.5).

Finally, let $E_1 = C$ and E_2 the unique curve in $|L(-dC)|$. Since then $(E_1^2) = (E_2^2) = 0$ and $(E_1\cdot E_2) = 1$, the curves E_1 and E_2 are elliptic and the map $\varphi: E_1\times E_2 \to X$, $(p,q)\mapsto p-q$ is an isomorphism of abelian varieties. Moreover, for $L_1 = \mathcal{O}_{E_1}(0)$ and $L_2 = \mathcal{O}_{E_2}(d\cdot 0)$ we have $(X,L)\simeq (E_1\times E_2, p_1^*L_1\otimes p_2^*L_2)$ as polarized abelian varieties. □

Proof of Theorem 4.1. If φ_L is not an embedding, then there exists the extension (5). But $c_1(F)^2 - 4c_2(F) = 2d-8 > 0$. So F is not semistable by Proposition 4.4, and Proposition 4.5 and Lemma 4.6 predict the existence of a curve C on X with $(C^2) = 0$ and $(C\cdot L) = 2$. Finally, C is elliptic by the adjunction formula 1(1) and Proposition 4.9.5. □

§ 5 Polarizations of Type (1, 4)

Let L be an ample line bundle of type $(1,4)$ on an abelian surface X and $\varphi_L: X\to \mathbb{P}_3$ the associated map. By a theorem of Lefschetz any smooth surface in $\mathbb{P}_3$ is simply connected, so φ_L cannot be an embedding. In this section we will see that nevertheless generically φ_L is birational onto its image. We will derive an equation for $\varphi_L(X)\subseteq \mathbb{P}_3$ and study its singularities. Again, in order to avoid trivialities, we assume that the linear system $|L|$ has no fixed components.

For simplicity suppose L is of characteristic zero with respect to some decomposition for L. According to Example 6.7.4 and Exercise 6.13 we may choose coordinates $X_0,\dots,X_3$ of $\mathbb{P}_3$ in such a way that the elements σ, τ and ι of the extended Heisenberg group $\mathcal{H}^e(1,4)$ of type $\left(\begin{smallmatrix}1&0\\0&4\end{smallmatrix}\right)$ act on $\mathbb{P}_3$ by

$$\sigma: X_j\mapsto X_{j-1}\ ,\qquad \tau: X_j\mapsto i^{-j}X_j\ ,\qquad \iota: X_j\mapsto X_{-j}$$

(here the indices of the coordinates are considered as elements of $\mathbb{Z}/4\mathbb{Z}$ and $i = e(\frac{\pi i}{2})$). We denote the corresponding elements of $K^e(L)$ by the same letter. It turns out to be convenient to change the coordinates. Define new coordinates by

$$\begin{aligned} Z_0 &= X_0 + X_2 & \qquad Z_2 &= X_3 + X_1 \\ Z_1 &= X_0 - X_2 & \qquad Z_3 &= X_3 - X_1 \quad . \end{aligned}$$

On these coordinates σ, τ and ι act as follows

$$\sigma\colon \left\{ \begin{matrix} Z_0 \\ Z_1 \\ Z_2 \\ Z_3 \end{matrix} \right. \mapsto \begin{matrix} Z_2 \\ Z_3 \\ Z_0 \\ -Z_1 \end{matrix} \ , \qquad \tau\colon \left\{ \begin{matrix} Z_0 \\ Z_1 \\ Z_2 \\ Z_3 \end{matrix} \right. \mapsto \begin{matrix} Z_1 \\ Z_0 \\ iZ_3 \\ iZ_2 \end{matrix} \ , \qquad \iota\colon \left\{ \begin{matrix} Z_0 \\ Z_1 \\ Z_2 \\ Z_3 \end{matrix} \right. \mapsto \begin{matrix} Z_0 \\ Z_1 \\ Z_2 \\ -Z_3 \end{matrix} \ .$$

Denote the coordinate points with respect to $Z_0, \ldots, Z_3$ by

$$p_0 = (1:0:0:0) \ , \quad \ldots \ , \quad p_3 = (0:0:0:1) \ ,$$

and for $\nu = 0, \ldots, 3$ let H_ν be the coordinate plane $\{Z_\nu = 0\}$.

(5.1) Lemma. *a) The variety $\varphi_L(X)$ is a surface of degree 8, 4 or 2 in $\mathbb{P}_3$.*
b) The coordinate points $p_0, \ldots, p_3$ are of multiplicity 4 (2 or 1, respectively) in $\varphi_L(X)$ if $\deg \varphi_L(X) = 8$ (4 or 2, respectively).

Proof. a) follows from the fact that $(L^2) = 8$ and Theorem 4.3.5.
b) According to Corollary 4.6.6 we have $h^0(L)_- = 1$ and $h^0(L)_+ = 3$. The point p_3 (respectively the plane H_3) corresponds to the eigenspace $H^0(L)_-$ (respectively $H^0(L)_+$). On the other hand, the set X_2^- (respectively X_2^+) of 2-division points $x \in X_2$, where the normalized isomorphism $(-1)_L\colon L \to L$ acts on the fibre $L(x)$ by multiplication with -1 (respectively $+1$), is of order 4 (respectively 12) (see Remark 4.7.8). Since φ_L is $K^e(L)$-equivariant, it maps the 4 points in X_2^- to p_3. But $\sigma(p_3) = p_1$, $\tau(p_3) = p_2$ and $\sigma\tau(p_3) = p_0$. So the preimage of any p_ν consists of at least 4 points. Now the assertion follows from a) noting that any coordinate line contains exactly 2 of the points p_ν. □

We will use the action of $K^e(L)$ to determine an equation for $\varphi_L(X)$ in $\mathbb{P}_3$. Let $Q \in \mathbb{C}[Z_0, \ldots, Z_3]$ denote a homogenous polynomial with zero set $\varphi_L(X)$. There is a linear character $\kappa\colon K^e(L) \to \mathbb{C}^*$ such that

$$\alpha^* Q = \kappa(\alpha) \cdot Q \ . \tag{1}$$

The group $K^e(L)$ contains the four reflections

$$\begin{aligned} \iota\tau\sigma^2\tau\colon Z_0 &\mapsto -Z_0 & \quad \iota\tau^2\colon Z_2 &\mapsto -Z_2 \\ \iota\sigma^2\colon Z_1 &\mapsto -Z_1 & \quad \iota\colon Z_3 &\mapsto -Z_3 \ . \end{aligned}$$

The quotient of $K^e(L)$ by its commutator subgroup is isomorphic to $(\mathbb{Z}/2\mathbb{Z})^3$ so that the linear characters of $K^e(L)$ take only values in $\{\pm 1\}$. Hence $\kappa(\tau\sigma^2\tau) = \kappa(\sigma^2) = \kappa(\tau^2) = 1$, and for all four reflections the character κ takes the same value, namely $\kappa(\iota)$.

We claim that $\kappa(\iota) = +1$. Otherwise (1) would imply that Q consists only of monomials of the form $aZ_0^j Z_1^k Z_2^l Z_3^m$ with $a \in \mathbb{C}$ and positive odd integers j, k, l, m such that $j + k + l + m = \deg \varphi_L(X)$. But this would contradict Lemma 5.1 b).

Hence $\kappa(\iota) = +1$, and Q is actually a polynomial in the squares $Z_0^2, \ldots, Z_3^2$. So there is a polynomial $\widetilde{Q}$ over $\mathbb{C}$ with

$$Q(Z_0, \ldots, Z_3) = \widetilde{Q}(Z_0^2, \ldots, Z_3^2). \tag{2}$$

Denote by K the surface in $\mathbb{P}_3 = \mathbb{P}(Y_0, \ldots, Y_3)$ defined by the equation $\widetilde{Q}(Y_0, \ldots, Y_3) = 0$. Then (2) means geometrically:

(5.2) Lemma. *The map $\mathbb{P}(Z_0, \ldots, Z_3) \to \mathbb{P}(Y_0, \ldots, Y_3)$, $Y_\nu = Z_\nu^2$ induces a covering $\bar{p}: \varphi_L(X) \to K$ which is $8:1$ outside the coordinate planes.*

As a consequence we obtain

(5.3) Corollary. *$\varphi_L(X)$ is of degree 8 or 4, whereas K is of degree 4 or 2 in $\mathbb{P}_3$.*

Proof. Suppose $\varphi_L(X)$ is of degree 2 in $\mathbb{P}_3$. Then K would be a plane in $\mathbb{P}_3$ containing the four coordinate vertices, a contradiction. □

Our next aim is to show that K is a Kummer surface. Consider the groups $K(L)_2 = K(L) \cap X_2$ of 2-division points of $K(L)$. The group $K(L)_2$ is isotropic with respect to the alternating form e^L, so by Corollary 6.3.5 the line bundle L descends via the isogeny $p: X \to Y := X/K(L)_2$ to a line bundle N on Y.

(5.4) Proposition. *The following diagram commutes*

$$\begin{array}{ccccc} X & \xrightarrow{\varphi_L} & \varphi_L(X) & \subset \mathbb{P}_3 \\ {\scriptstyle p}\downarrow & & \downarrow{\scriptstyle \bar{p}} & \\ Y & \xrightarrow{\varphi_{N^2}} & K & \subset \mathbb{P}_3 & . \end{array}$$

In particular, N defines a principal polarization on Y.

Proof. The composed map $\varphi_{N^2}p: X \to \mathbb{P}_3$ is given by the linear system $\operatorname{Im}(p^*: H^0(N^2) \to H^0(L^2))$ which is the subspace $H^0(L^2)^{K(L)_2}$ of sections invariant under the action of $K(L)_2$. On the other hand $Z_0^2, \dots, Z_3^2$ can be considered as elements of $H^0(L^2)$ and the map $\bar{p}\varphi_L$ is just defined by these sections. Since $K(L)_2 \simeq <2\sigma, 2\tau>$, it is easy to check that the subvector space $<Z_0^2, \dots, Z_3^2>$ of $H^0(L^2)$ is invariant under the action of $K(L)_2$. Hence $<Z_0^2, \dots, Z_3^2> = H^0(L^2)^{K(L)_2}$. This gives the assertion. □

According to Proposition 5.4 the map φ_L is birational if and only if φ_{N^2} is of degree 2. By Theorem 4.8.2 this is the case if and only if N defines an irreducible principal polarization, or in other words, if (Y, N) is not isomorphic to a product of elliptic curves. Since the space of étale 4-fold coverings of products of two elliptic curves is 2-dimensional, we conclude that for a general abelian surface X with an ample line bundle L of type $(1, 4)$ the map φ_L is birational and thus $\varphi_L(X)$ is an octic. There is also a criterion for the birationality of φ_L analogous to Corollary 4.2. For this we refer to Birkenhake-Lange-v. Straten [1] and Exercise 4.

From now on we assume that φ_L is birational. According to Proposition 5.4 the surface K is then a Kummer surface. This fact turns out to be the essential tool for studying some geometrical properties of $\varphi_L(X)$ and deriving an equation.

Note first that in the notation we do not distinguish between the coordinate planes respectively the coordinate points of $\mathbb{P}(Y_0, \dots, Y_3)$ and $\mathbb{P}(Z_0, \dots, Z_3)$. The involution ι acts as identity on the curve $\varphi_L(X) \cap H_3$. Since moreover the map φ_{N^2} restricted to $\varphi_{N^2}^{-1}(K \cap H_3)$ is of degree 2, the morphism φ_L restricts also to a double covering $\varphi_L^{-1}(\varphi_L(X) \cap H_3) \to \varphi_L(X) \cap H_3$. Applying the automorphisms σ, τ and $\sigma\tau$ this shows:

(5.5) Lemma. *The octic $\varphi_L(X)$ has double curves along the coordinate planes H_ν for $\nu = 0, \dots, 3$.*

As a consequence we obtain

(5.6) Lemma. *The coordinate planes $H_0, \dots, H_3$ form a Rosenhain tetrahedron for K.*

Proof. Lemma 5.5 together with Proposition 5.4 shows that the coordinate planes touch K along a conic, i.e. the planes $H_0, \dots, H_3$ are singular planes for K (in the sense of Section 2). Moreover by Lemma 5.1 b) the vertices $p_0, \dots, p_3$ of the tetrahedron $H_0, \dots, H_3$ are singular points. □

Now we can proceed as in Section 3 to compute the polynomial $\widetilde{Q}$ describing the Kummer surface K. The elements σ and τ of above act on the coordinates $Y_0, \dots, Y_3$ of $\mathbb{P}_3$ as follows

$$\sigma\colon \begin{cases} Y_0 \\ Y_1 \\ Y_2 \\ Y_3 \end{cases} \mapsto \begin{matrix} Y_2 \\ Y_3 \\ Y_0 \\ Y_1 \end{matrix}, \qquad \tau\colon \begin{cases} Y_0 \\ Y_1 \\ Y_2 \\ Y_3 \end{cases} \mapsto \begin{matrix} Y_1 \\ Y_0 \\ -Y_3 \\ -Y_2 \end{matrix}.$$

But this is exactly the action of Lemma 3.1 with $\sigma = f_1$ and $\tau = e_1$. So $\widetilde{Q}$ is just the polynomial (5) in Proposition 3.2. Inserting $Y_\nu = Z_\nu^2$ we obtain

(5.7) Proposition. *If L is an ample symmetric line bundle of type $(1,4)$ on the abelian surface X such that the induced map φ_L is birational, then the coordinates of $\mathbb{P}_3$ can be chosen in such a way that the surface $\varphi_L(X)$ is given by the equation*

$$\begin{aligned} \widetilde{Q}(Y_0,\dots,Y_3) = {} & \lambda_1^2(Z_0^4Z_1^4+Z_2^4Z_3^4)+\lambda_2^2(Z_0^4Z_2^4+Z_1^4Z_3^4)+\lambda_3^2(Z_0^4Z_3^4+Z_1^4Z_2^4) \\ & + 2\lambda_1\lambda_2(Z_0^2Z_1^2+Z_2^2Z_3^2)(-Z_0^2Z_2^2+Z_1^2Z_3^2) \\ & + 2\lambda_1\lambda_3(Z_0^2Z_1^2-Z_2^2Z_3^2)(Z_0^2Z_3^2-Z_1^2Z_2^2) \\ & + 2\lambda_2\lambda_3(Z_0^2Z_2^2+Z_1^2Z_3^2)(Z_0^2Z_3^2+Z_1^2Z_2^2) \\ & + \lambda_0^2Z_0^2Z_1^2Z_2^2Z_3^2 = 0 \end{aligned}$$

for some $(\lambda_0 : \dots : \lambda_3) \in \mathbb{P}_3$.

In Exercise 4 an equation for $\varphi_L(X)$ is given in the case that the map φ_L is of degree 2 onto its image.

Finally consider the map $\bar{p}\colon \varphi_L(X) \to K$ restricted to a coordinate plane, say H_3. According to Section 3 the curve $K \cap H_3$ is a conic passing through 6 of the 16 singular points of K. Three of these 6 points are the coordinate points p_0, p_1, p_2. Denote the other three points by q_0, q_1, q_2. The preimages of p_0, p_1, p_2 are the coordinate points of p_0, p_1, p_2 in $\mathbb{P}_3 = \mathbb{P}(Z_0, \dots, Z_3)$, whereas $\bar{p}$ is étale over q_0, q_1 and q_2. Thus the corresponding preimages consist of four points for each q_ν, say $q_{\nu 1}, q_{\nu 2}, q_{\nu 3}$, and $q_{\nu 4}$. These 12 points are exactly the pinch points of the surface $\varphi_L(X)$ in the plane H_3.

The covering $\bar{p}|\varphi_L(X) \cap H_3\colon \varphi_L(X) \cap H_3 \to K \cap H_3$ looks as follows:

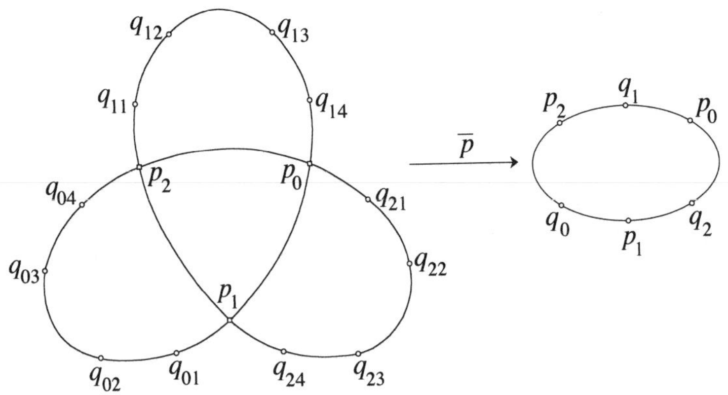

§ 6 Products of Elliptic Curves

In this section we present Ruppert's method to check whether an abelian surface is isomorphic or isogenous to a product of elliptic curves (see Ruppert [1]).

Let $X = \mathbb{C}^2/\Lambda$ be an abelian surface. We associate to X an alternating form $\alpha\colon \Lambda \times \Lambda \to \mathbb{C}$ defined by

$$\alpha(u,v) = \det(u,v) = {}^t u \begin{pmatrix} 0 & 1 \\ -1 & 0 \end{pmatrix} v \ .$$

The form α is called *hyperbolic* if there is a decomposition

$$\Lambda = \Lambda_1 \oplus \Lambda_2$$

into submodules Λ_1 and Λ_2 which are isotropic with respect to α. In other words, α is hyperbolic if and only if there is a basis $\lambda_1, \lambda_2, \mu_1, \mu_2$ of Λ such that $\alpha(\lambda_1, \lambda_2) = \alpha(\mu_1, \mu_2) = 0$. By abuse of notation we denote the extension of α to $\Lambda \otimes \mathbb{Q}$ also by α. The form $\alpha\colon (\Lambda \otimes \mathbb{Q})^2 \to \mathbb{C}$ is called *hyperbolic over* $\mathbb{Q}$ if there is a decomposition

$$\Lambda \otimes \mathbb{Q} = V_1 \oplus V_2$$

into subvector spaces V_1 and V_2 which are isotropic with respect to α. Note that both notions are independent of the choice of the coordinates of $\mathbb{C}^2$, since a coordinate transformation in $\mathbb{C}^2$ changes α by a multiplicative constant.

(6.1) Proposition. *For an abelian surface $X = \mathbb{C}^2/\Lambda$ the following conditions are equivalent:*

i) X is isomorphic (respectively isogenous) to a product of elliptic curves.

ii) The form α is hyperbolic (respectively hyperbolic over $\mathbb{Q}$).

Proof. i) $\Longrightarrow$ ii) Suppose X is isomorphic to a product of elliptic curves. Then there are bases of $\mathbb{C}^2$ and Λ such that the period matrix of X with respect to these bases is

$$\Pi = \begin{pmatrix} z_1 & 1 & 0 & 0 \\ 0 & 0 & z_2 & 1 \end{pmatrix} .$$

Obviously the columns $\lambda_1, \lambda_2, \mu_1, \mu_2$ of Π satisfy $\alpha(\lambda_1, \lambda_2) = \alpha(\mu_1, \mu_2) = 0$, so α is hyperbolic.

ii) $\Longrightarrow$ i) Suppose $\lambda_1, \lambda_2, \mu_1, \mu_2$ is a $\mathbb{Z}$-basis of Λ with $\alpha(\lambda_1, \lambda_2) = \alpha(\mu_1, \mu_2) = 0$. Then there are constants $z_1, z_2 \in \mathbb{C}^*$ such that $\lambda_1 = z_1 \lambda_2$ and $\mu_1 = z_2 \mu_2$ and

$$(\lambda_1, \lambda_2, \mu_1, \mu_2) = (z_1\lambda_2, \lambda_2, z_2\mu_2, \mu_2) = (\lambda_2, \mu_2) \begin{pmatrix} z_1 & 1 & 0 & 0 \\ 0 & 0 & z_2 & 1 \end{pmatrix} ,$$

so X is isomorphic to a product of elliptic curves. The assertion that X is isogenous to a product of elliptic curves if and only if α is hyperbolic over $\mathbb{Q}$ follows by the same proof. □

Consequently we have to find a criterion for the form α to be hyperbolic. For this we introduce the rank of α: The image $\alpha(\Lambda \times \Lambda)$ is a free $\mathbb{Z}$-submodule of $\mathbb{C}$ of some rank r. We call r the *rank of* α.

If $y_1, \ldots, y_r$ is a basis of the $\mathbb{Z}$-module $\alpha(\Lambda \times \Lambda)$, we can decompose α as

$$\alpha(u,v) = \alpha_1(u,v)y_1 + \cdots + \alpha_r(u,v)y_r$$

with alternating forms $\alpha_\nu : \Lambda \times \Lambda \to \mathbb{Z}$. Note that the forms $\alpha_1, \ldots, \alpha_r$ are necessarily linearly independent over $\mathbb{Z}$. Moreover it is clear that α is hyperbolic (over $\mathbb{Q}$) if and only if $\alpha_1, \ldots, \alpha_r$ are hyperbolic (over $\mathbb{Q}$) all with the same decomposition of Λ (respectively $\Lambda \otimes \mathbb{Q}$).

We claim that the rank r of a hyperbolic form α can only be 2, 3 or 4. To see this, note first that $r \geq 2$, since α is nondegenerate. On the other hand, if $\lambda_1, \lambda_2, \mu_1, \mu_2$ is a $\mathbb{Z}$-basis of Λ with $\alpha(\lambda_1, \lambda_2) = \alpha(\mu_1, \mu_2) = 0$, then $\operatorname{Im} \alpha = \mathbb{Z}\alpha(\lambda_1, \mu_1) + \mathbb{Z}\alpha(\lambda_1, \mu_2) + \mathbb{Z}\alpha(\lambda_2, \mu_1) + \mathbb{Z}\alpha(\lambda_2, \mu_2)$, so $r \leq 4$.

Hence it suffices to consider the cases $r = 2, 3$ and 4. Suppose first $r = 2$.

(6.2) Proposition. *Any alternating form* $\alpha : \Lambda \times \Lambda \to \mathbb{C}$ *of rank* 2 *is hyperbolic.*

Proof. Let y_1, y_2 be a basis of $\alpha(\Lambda \times \Lambda)$ and $\alpha = \alpha_1 y_1 + \alpha_2 y_2$. The alternating forms α_1 and α_2 are given by integral (4×4)-matrices $A = (a_{ij})$ and $B = (b_{ij})$ with respect to some basis of Λ. It suffices to transform A and B simultaneously so that $a_{12} = a_{34} = b_{12} = b_{34} = 0$. This can be done for A using the elementary divisor theorem (see Bourbaki [1] Alg. IX. 5.1 Th. 1) and for B by a refinement of the proof of the elementary divisor theorem using only transformations preserving A. We leave the details as an exercise to the reader or refer to Ruppert [1]. □

An immediate consequence is the following theorem of Shioda and Mitani (see Shioda-Mitani [1]).

(6.3) Corollary. *An abelian surface which is isogenous to a product of isogenous elliptic curves with complex multiplication is isomorphic to a product of elliptic curves.*

Of course, the predicted isomorphism does not respect the polarizations. For a generalization of the corollary to abelian varieties of arbitrary dimensions see Exercise 5.

Proof. If $E_1 = \mathbb{C}/(z_1,1)\mathbb{Z}^2$ and $E_2 = \mathbb{C}/(z_2,1)\mathbb{Z}^2$ are isogenous elliptic curves with complex multiplication, then z_1 and z_2 are elements of the same imaginary quadratic field $\mathbb{Q}(\sqrt{-m})$. Every abelian surface X isogenous to $E_1 \times E_2$ admits a period matrix of the form

$$\begin{pmatrix} z_1 & 1 & 0 & 0 \\ 0 & 0 & z_2 & 1 \end{pmatrix} R$$

with some $R \in M_4(\mathbb{Q})$. This implies that the alternating form α associated to X takes values in $\mathbb{Q}(\sqrt{-m})$. Consequently its rank is 2. Now Propositions 6.1 and 6.2 imply the assertion. □

It remains to consider forms α of rank 3 and 4. For this we first recall some properties of the Plücker quadric (see Griffiths-Harris [1] p. 211). Let k be a field. If X_{ij}, $0 \le i < j \le 3$ denote the coordinates of $\mathbb{P}_5 = P(\bigwedge^2 k^4)$, the Plücker quadric Q is defined by the equation

$$X_{01}X_{23} - X_{02}X_{13} + X_{03}X_{12} = 0$$

The 2-dimensional subvector spaces of the vector space k^4 correspond bijectively to the points of the Plücker quadric: for any 2-dimensional subvector space $U = ku + kv$ with ${}^t u = (u_0, \dots, u_3)$, ${}^t v = (v_0, \dots, v_3) \in k^4$ the elements $p_{ij} := u_i v_j - u_j v_i$ are the Plücker coordinates of U and

$$p(U) := (p_{01} : p_{02} : p_{03} : p_{12} : p_{13} : p_{23})$$

is the corresponding point on the Plücker quadric Q. If V is another 2-dimensional subvector space of k^4, then

$$k^4 = U \oplus V \iff \begin{cases} \textit{the line joining } p(U) \textit{ and } p(V) \textit{ is not} \\ \textit{contained in the Plücker quadric.} \end{cases}$$

Finally we associate to any alternating form $A: k^4 \times k^4 \to k$, $A(u,v) := \sum_{i,j=0}^{3} a_{ij} u_i v_j$ (with $a_{ij} = -a_{ji} \in k$) the hyperplane defined by

$$H(A) = \Big\{ \sum_{0 \le i < j \le 3} a_{ij} X_{ij} = 0 \Big\}$$

of $\mathbb{P}_5 = P(\Lambda^2 k^4)$. The form A can also be written as $A(u,v) = \sum_{i<j} a_{ij} p_{ij}$. This implies that a 2-dimensional subvector space U of k^4 is isotropic with respect to A if and only if its corresponding point $p(U)$ in Q lies on $H(A)$.

Let $\alpha = \alpha_1 y_1 + \cdots + \alpha_r y_r$ be the alternating form associated to the abelian surface $X = \mathbb{C}^2/\Lambda$ as above. Since the forms $\alpha_1, \dots, \alpha_r$ are linearly independent over $\mathbb{Z}$, the intersection $E(\alpha) := H(\alpha_1) \cap \ldots \cap H(\alpha_r)$ is a $(5-r)$-plane in $\mathbb{P}_5$.

(6.4) Proposition. *a) α is hyperbolic if and only if there are different rational points q_1 and q_2 in $Q \cap E(\alpha)$ such that the following conditions hold*

1) The line joining q_1 and q_2 is not contained in Q.

2) For every prime p statement 1) holds modulo p.

b) α is hyperbolic over $\mathbb{Q}$ if and only if there are different rational points q_1 and q_2 in $Q \cap E(\alpha)$ such that 1) holds.

Proof. a) The form α is hyperbolic if there are $\mathbb{Z}$-submodules Λ_1 and Λ_2 of rank 2 of Λ such that $\Lambda = \Lambda_1 \oplus \Lambda_2$, and Λ_1 and Λ_2 are isotropic with respect to α_ν for $\nu = 1, \ldots, r$. But $\Lambda = \Lambda_1 \oplus \Lambda_2$ if and only if $\Lambda \otimes \mathbb{Q} = (\Lambda_1 \otimes \mathbb{Q}) \oplus (\Lambda_2 \otimes \mathbb{Q})$ and $\Lambda \otimes \mathbb{F}_p = (\Lambda_1 \otimes \mathbb{F}_p) \oplus (\Lambda_2 \otimes \mathbb{F}_p)$ for all primes p. By what we have said above, this means that the line joining $p(\Lambda_1 \otimes \mathbb{Q})$ and $p(\Lambda_2 \otimes \mathbb{Q})$ respectively $p(\Lambda_1 \otimes \mathbb{F}_p)$ and $p(\Lambda_2 \otimes \mathbb{F}_p)$ is not contained in the Plücker quadric Q over $\mathbb{Q}$ respectively $\mathbb{F}_p$ for all primes p.

On the other hand, Λ_1 and Λ_2 are isotropic with respect to α_ν for $\nu = 1, \ldots, r$ if and only if $p(\Lambda_1)$ and $p(\Lambda_2)$ are contained in $E(\alpha)$. This completes the proof of a). Assertion b) follows similarly. □

Note that Proposition 6.4 holds also for $r = 2$ and Proposition 6.2 can also be deduced from this.

It is easy to work out the cases $r = 3$ and $r = 4$:

Case $r = 3$: Here $Q \cap E(\alpha)$ is a quadric in $\mathbb{P}_2$ given by a symmetric matrix $F \in M_3(\mathbb{Z})$.

a) If $\operatorname{rank} F = 0$ or 1, then α is not hyperbolic

b) If $\operatorname{rank} F = 2$, then either the singular point of $Q \cap E(\alpha)$ is the unique rational point, so α is not hyperbolic, or α is hyperbolic if and only of the two lines of the quadric are different modulo every prime p.

c) If $\operatorname{rank} F = 3$, then α is hyperbolic if and only if for every prime p dividing $\det F$ there is a nonsingular rational point on $Q \cap E(\alpha) \pmod{p}$.

Case $r = 4$: Here $Q \cap E(\alpha)$ is a quadric in $\mathbb{P}_1$. So α is hyperbolic if and only if this quadric consists of two different rational points, which are also different modulo p for every prime p.

The hyperbolicity of α over $\mathbb{Q}$ can be worked out similarly.

Ruppert's method to check whether an abelian surface is isomorphic or isogenous to a product of elliptic curves is effective and is easy to apply. For examples see Exercises 11 and 12.

Exercises

(1) Let L be an ample line bundle of type $(1,2)$ on an abelian surface X. Any curve $D \in |L|$ is of one of the following types
 a) D smooth of genus 3, admitting an elliptic involution.
 b) D irreducible of genus 2 with one double point, admitting an elliptic involution.
 c) $D = E_1 + E_2$ with elliptic curves E_1 and E_2 and $(E_1 \cdot E_2) = 2$.
 d) $D = E_0 + E_1 + E_2$ with elliptic curves E_i, such that $(E_0 \cdot E_1) = (E_0 \cdot E_2) = 1$ and $(E_1 \cdot E_2) = 0$.

The linear system $|L|$ always contains singular curves. In case d) we have $(X, L) \simeq (E_0 \times E_1, p_1^* \mathcal{O}_{E_0}(0) \otimes p_2^* \mathcal{O}_{E_1}(2 \cdot 0))$.

(2) Show that any polarized abelian surface of type $(1,d)$ contains a curve of genus 2, not necessarily smooth and irreducible.

(3) *(Generalization of the* 16_6*-Configuration)* Let X be an abelian variety of dimension g and L a symmetric line bundle on X defining an irreducible principal polariztion. According to Theorem 4.8.1 the map $\varphi = \varphi_{L^2}: X \to K \subset \mathbb{P}_{2^g-1}$ is of degree 2. Its image K is called *Kummer variety* of X. The singular points of K are exactly the images of the 2-division points $x \in X_2$. For $x \in X_2$ denote by D_x the unique divisor in the linear system $|t_x^* L|$. Then $2D_x \in |L^2|$ and thus corresponds to a uniquely determined hyperplane P_x in $\mathbb{P}_{2^g-1}$, called *singular hyperplane*. Show that the 2^{2g} singular points of K and the 2^{2g} singular hyperplanes form a $(2^{2g})_{2^{g-1}(2^g-1)}$-configuration, i.e. any singular hyperplane contains exactly $2^{g-1}(2^g-1)$ singular points and any singular point lies in exactly $2^{g-1}(2^g-1)$ singular hyperplanes.

(4) Suppose L is an ample line bundle of type $(1,4)$ on an abelian surface X. Let the abelian surface Y and the curves D_X on X and D_Y on Y be as in Corollary 4.2.
 a) If D_X and D_Y do not admit elliptic involutions, compatible with the covering q, then $\varphi_L: X \to \mathbb{P}_3$ is birational onto its image.
 b) If D_X and D_Y admit elliptic involutions, compatible with the covering q, then $\varphi_L: X \to \mathbb{P}_3$ is a double covering of a singular quartic $\overline{X}$, which is birational to an elliptic scroll. Moreover, the coordinates of $\mathbb{P}_3$ can be chosen in such a way that $\overline{X}$ is given by the equation

$$\lambda_1(Y_0^2 Y_1^2 + Y_2^2 Y_3^2) - \lambda_2(Y_0^2 Y_2^2 - Y_1^2 Y_3^2) = 0$$

 for some $(\lambda_1 : \lambda_2) \in \mathbb{P}_1 - \{(1:0),(0:1),(1:i),(1:-i)\}$. The surface $\overline{X}$ is singular exactly along the coordinate lines $\{Y_0 = Y_3 = 0\}$ and $\{Y_1 = Y_2 = 0\}$.

(5) Generalize Corollary 6.3 to abelian varieties of arbitrary dimension: If X is an abelian variety of dimension g, isogenous to a product $\times_{i=1}^{g} E$ with E an elliptic curve with complex multiplication, then X is isomorphic to a product of elliptic curves (see Schoen [1]).

(6) Give an example of elliptic curves E, E_1, E_2 such that $E \times E_1 \simeq E \times E_2$, but E_1 not isomorphic to E_2.

(7) Let X be a complex torus of dimension 2 with period matrix $(iY, \mathbb{1}_2)$, where $Y = (y_{ij}) \in M_2(\mathbb{R})$.

a) Use Exercise 2.4 to show that the Picard number of X is

$$\rho(x) = 4 - \dim_{\mathbb{Q}} \mathbb{Q}(y_{11}, y_{12}, y_{21}, y_{22}) + \begin{cases} 1 & \text{if} \quad \det Y \in \mathbb{Q} \\ 0 & \phantom{\text{if}} \quad \det Y \notin \mathbb{Q}. \end{cases}$$

b) The matrices Y in the following table give examples of complex tori X realizing all possible values for the Picard number $\rho = \rho(X)$ and the algebraic dimension $a = a(X)$. Here p, q, r denote pairwise different prime numbers.

ρ \ a	0	1	2
0	$\begin{pmatrix} \sqrt{p} & \sqrt{q} \\ \sqrt{r} & 1 \end{pmatrix}$	impossible	impossible
1	$\frac{1}{\sqrt{\sqrt{p}-\sqrt{qr}}} \begin{pmatrix} \sqrt{p} & \sqrt{q} \\ \sqrt{r} & 1 \end{pmatrix}$	$\begin{pmatrix} \sqrt{p} & \sqrt{q} \\ 0 & 1 \end{pmatrix}$	$\begin{pmatrix} \sqrt{p} & 1 \\ 1 & \sqrt{q} \end{pmatrix}$
2	$\begin{pmatrix} 1 & -\sqrt[3]{p} \\ \sqrt[3]{p} & 1 \end{pmatrix}$	$\begin{pmatrix} \sqrt[3]{p} & 1 \\ 0 & \sqrt[3]{p} \end{pmatrix}$	$\begin{pmatrix} \sqrt{p} & 0 \\ 0 & 1 \end{pmatrix}$
3	$\begin{pmatrix} 1 & -\sqrt{p} \\ \sqrt{p} & 1 \end{pmatrix}$	$\begin{pmatrix} 1 & \sqrt{p} \\ 0 & 1 \end{pmatrix}$	$\begin{pmatrix} \sqrt{p} & 1 \\ 1 & \sqrt{p} \end{pmatrix}$
4	impossible	impossible	$\begin{pmatrix} 1 & 0 \\ 0 & 1 \end{pmatrix}$

(Hint: for the computation of $a(X)$ use Exercises 2.4 and 4.9 a). For the restrictions in the first line use Exercise 4.9 b), for the restrictions in the last line show that $\rho(X)$ being maximal implies that X is abelian (see Exercise 5.10).)

(8) Let X be a complex torus of dimension 2 with $a(X) = 0$. Show that any line bundle on X, not algebraically equivalent zero, is nondegenerate of index 1. For examples see Exercise 7.

(9) Let X be a complex torus of dimension 2 with $\rho(X) = a(X) = 1$. Show that up to translation X contains exactly one elliptic curve. In particular, Poincaré's Reducibility Theorem (Exercise 3.8) does not hold for X.

(10) Let X be a polarized abelian surface of type $(1,n)$ with period matrix $\left(Z, \begin{smallmatrix} 1 & 0 \\ 0 & n \end{smallmatrix}\right)$. The surface X contains an elliptic curve if and only if there exist integers $a, b, c, d, e, f \in \mathbb{Z}$ satisfying
1) $na + nez_{11} + (f - nb)z_{12} - dz_{22} + c\det Z = 0$ and
2) $ac + de - bf = 0$.
(Hint: use Exercise 2.4)

(11) Consider for $z \in \mathbb{C}$ with $|z| < 1$ the abelian surface X with period matrix $\left(\begin{smallmatrix} 1 & i & z & iz \\ z & -iz & 1 & -i \end{smallmatrix}\right)$. Use the method of Section 6 to show that X is isogenous to a product of elliptic curves. Moreover, X is isomorphic to a product of elliptic curves if and only if iz, $1 - z^2$, $i + iz^2$ are linearly dependent over $\mathbb{Q}$.

(12) Let C be a smooth projective curve of genus 2 with nontrivial reduced automorphism group. According to Bolza [1] the curve C is isomorphic to one of the 6 types of curves in the list of Section 11.7. Use the method of Section 6 to show that
a) if C is of type I, its Jacobian J is isogenous to a product of elliptic curves. In general J is not isomorphic to a product of elliptic curves, for example, if $1, z, z', zz'$ are linearly independent over $\mathbb{Q}$.
b) if C is of type II, its Jacobian is isomorphic to a product of elliptic curves if and only if z is contained in some imaginary quadratic field.
c) if C is of type III, IV or V, its Jacobian is isomorphic to a product of elliptic curves.
d) if C is of type VI, its Jacobian is a simple abelian surface.

(13) *(Torelli Problem for Complex Tori of Dimension 2)* Let X be a complex torus of dimension 2 and ω a nonvanishing holomorphic 2-form on X. The map $\pi_X \colon H_2(X, \mathbb{Z}) \to \mathbb{C}$, $\gamma \mapsto \int_\gamma \omega$ is called the *period of X (with respect to 2-forms)*. Since ω is unique up to a nonzero constant, so is the period π_X. Given another complex torus Y of dimension 2 with period π_Y, we say that an isomorphism $\varphi \colon H_2(X, \mathbb{Z}) \to H_2(Y, \mathbb{Z})$ *preserves the periods*, if $\pi_X \varphi = c\pi_Y$ for some $c \in \mathbb{C}$. Assume that there exists an isomorphism $\varphi \colon H_2(X, \mathbb{Z}) \to H_2(Y, \mathbb{Z})$ preserving the intersection form and the periods. Show that either $Y \simeq X$ or $Y \simeq \widehat{X}$.

In particular, for self-dual complex tori of dimension 2 the intersection form and the period determine the complex torus. This is called Torelli's Theorem for self-dual 2-dimensional complex tori (see Shioda [1]).

Chapter 11

Jacobian Varieties

To every smooth projective curve C one can associate in a natural way a principally polarized abelian variety, its *Jacobian* $J(C)$. As we mentioned already in the introduction, the theory of abelian varieties originated with the investigation of Jacobians. They are not only the most important, but also the best-known examples of abelian varieties. Much more can be said about them than about a general principally polarized abelian variety. In fact, presenting the theory of Jacobian varieties in a satisfactory way would require a whole volume for itself. This chapter contains, apart from the basic definitions and constructions, only some selected topics on Jacobian varieties. We focus on results and in particular on proofs, which on the one hand apply the results of the earlier chapters and on the other hand are not yet contained in other books.

The main topics are: a) We show that the Kummer variety of any Jacobian admits a 4-dimensional family of trisecants. A more precise version of this is Fay's Trisecant Identity, proved by Fay [1] with analytic methods. We give an algebraic proof, due to Raina [1], which only uses elementary methods of algebraic geometry. b) Ran [1] gave an improvement of a criterion of Matsusaka [2] for an abelian variety to be the Jacobian of a curve. The proof, which we give here, is due to Collino [1]. Matsusaka's original criterion will be a direct consequence of a result in Chapter 12 (see Remark 12.2.5). c) It is in general difficult, to compute the period matrix of the Jacobian of a curve which is given in terms of equations. Under the supervision of F. Klein Bolza [1] developed in his thesis a method for doing this, provided that the curve admits a large group of automorphisms. We explain the method and give an example.

Section 1 contains the basic definitions of the Jacobian $J(C)$ of a curve C. Moreover we show that the Abel-Jacobi map $C \to J(C)$ is an embedding. Without proof we state the Abel-Jacobi Theorem and the Torelli Theorem, proofs of these theorems are contained in almost every book on compact Riemann surfaces or algebraic curves. In Section 2 we study the theta divisor of a Jacobian variety. We prove Poincaré's Formula and Riemann's Theorem and state without proof Riemann's Singularity Theorem. As a consequence we can compute the number of even and odd theta characteristics, as well as the dimension of the singularity locus of the theta divisor. In Sections 3 and 4 we construct the Poincaré bundles of a curve C out of the Poincaré bundle for $J(C)$ (defined in Section 2.5) and derive the universal property of the Jacobian. In Section 5 we show that the endomorphism ring of the Jacobian $J(C)$ may be interpreted as the ring of equivalence classes of correspondences on C. To the best of our knowledge this result is due to Hurwitz [1]. Formula 5.2, expressing the rational trace of an endomorphism of $J(C)$ in terms of an associated correspondence of C, is due to Weil [1]. In Section 6 we prove a criterion, due to Matsusaka [2], for an algebraic 1-cycle on an abelian variety to be numerically equivalent to zero. It will be applied in the final chapter. Finally, Sections 7–10 contain the selected topics mentioned above.

We use some basic results of the theory of algebraic curves, such as the Riemann-Roch Theorem, Plücker's Formulas, and the description of finite coverings. By g^r_d we denote a linear system of dimension r and degree d on a curve C. Finally, we use Bertini's Theorem and the inequality for the dimension of the nonempty intersection of two closed subvarieties of a smooth projective variety.

§ 1 Definition of the Jacobian Variety

Let C be a smooth projective curve of genus g over the field of complex numbers. In this section we introduce the Jacobian variety and the Abel-Jacobi map of C.

Recall the g-dimensional $\mathbb{C}$-vector space $H^0(\omega_C)$ of holomorphic 1-forms on C. The homology group $H_1(C,\mathbb{Z})$ is a free abelian group of rank $2g$. For convenience we use the same letter for (topological) 1-cycles on C and their corresponding classes in $H_1(C,\mathbb{Z})$. By Stokes' theorem any element $\gamma \in H_1(C,\mathbb{Z})$ yields in a canonical way a linear form on the vector space $H^0(\omega_C)$, which we also denote by γ:

$$\gamma\colon H^0(\omega_C) \to \mathbb{C}\,,\ \omega \mapsto \int_\gamma \omega\ .$$

(1.1) Lemma. *The canonical map* $H_1(C,\mathbb{Z}) \to H^0(\omega_C)^* = \operatorname{Hom}(H^0(\omega_C),\mathbb{C})$ *is injective.*

Proof. By the universal coefficient theorem the canonical map $H_1(C,\mathbb{Z}) \hookrightarrow H_1(C,\mathbb{C}) = H^1_{\mathrm{DR}}(C)^*\,,\ \gamma \mapsto \{\omega \mapsto \int_\gamma \omega\}$ is injective. Recall the Hodge decomposition $H^1_{\mathrm{DR}}(C) = H^0(\omega_C) \oplus \overline{H^0(\omega_C)}$. Clearly the canonical map in question is the composition

$$H_1(C,\mathbb{Z}) \longrightarrow H^1_{\mathrm{DR}}(C)^* = H^0(\omega_C)^* \oplus \overline{H^0(\omega_C)}^* \xrightarrow{p} H^0(\omega_C)^*$$

where p denotes the projection. Since the image of any $\gamma \in H_1(C,\mathbb{Z})$ in $H^0(\omega_C)^* \oplus \overline{H^0(\omega_C)}^*$ is invariant under complex conjugation, it is necessarily of the form $l + \bar{l}$ with $l \in H^0(\omega_C)^*$. This implies the assertion. □

It follows that $H_1(C,\mathbb{Z})$ is a lattice in $H^0(\omega_C)^*$ and the quotient

$$J(C) := H^0(\omega_C)^*/H_1(C,\mathbb{Z})$$

is a complex torus of dimension g, called the *Jacobian variety* or simply the *Jacobian* of C. Note that $J(C) = 0$ for $g = 0$. In the sequel we often assume $g \geq 1$ in order to avoid trivialities.

In order to describe $J(C)$ in terms of period matrices, choose bases $\lambda_1, \dots, \lambda_{2g}$ of $H_1(C,\mathbb{Z})$ and $\omega_1, \dots, \omega_g$ of $H^0(\omega_C)$. Let $l_1, \dots, l_g$ denote the basis of $H^0(\omega_C)^*$ dual to $\omega_1, \dots, \omega_g$, i.e. $l_i(\omega_j) = \delta_{ij}$. Considering λ_i as a linear form on $H^0(\omega_C)$ as above, we have $\lambda_i = \sum_{j=1}^g (\int_{\lambda_i} \omega_j) l_j$ for $i = 1, \dots, 2g$. Hence

$$\Pi = \begin{pmatrix} \int_{\lambda_1} \omega_1 & \cdots & \cdots & \int_{\lambda_{2g}} \omega_1 \\ \vdots & & & \vdots \\ \int_{\lambda_1} \omega_g & \cdots & \cdots & \int_{\lambda_{2g}} \omega_g \end{pmatrix}$$

is a period matrix for $J(C)$ with respect to these bases.

The complex torus $J(C)$ turns out to be an abelian variety. In fact, there is a canonical principal polarization on $J(C)$, which we introduce now: fix a homology basis $\lambda_1, \dots, \lambda_{2g}$ of $H_1(C,\mathbb{Z})$ with intersection matrix $\left(\begin{smallmatrix} 0 & -\mathbb{1} \\ \mathbb{1} & 0 \end{smallmatrix}\right)$ as indicated in the following picture.

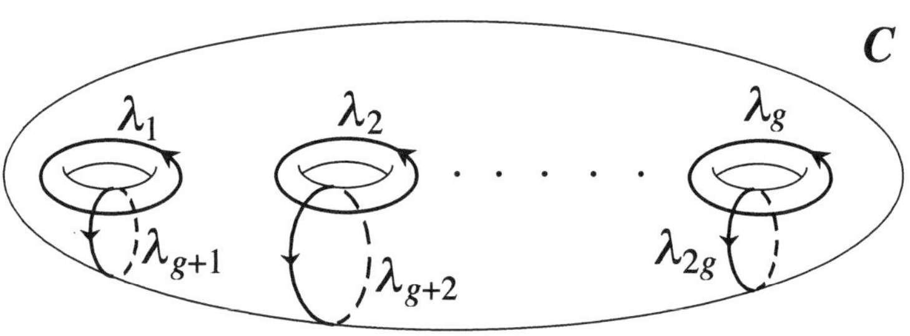

By what we have said above, $\lambda_1, \dots, \lambda_{2g}$ is a basis of $H^0(\omega_C)^*$ considered as an $\mathbb{R}$-vector space. Denote by E the alternating form on $H^0(\omega_C)^*$ with matrix $\left(\begin{smallmatrix} 0 & \mathbb{1} \\ -\mathbb{1} & 0 \end{smallmatrix}\right)$ with respect to the basis $\lambda_1, \dots, \lambda_{2g}$ of $H^0(\omega_C)^*$ and define $H\colon H^0(\omega_C)^* \times H^0(\omega_C)^* \to \mathbb{C}$ by $H(u,v) = E(iu,v) + iE(u,v)$.

(1.2) Proposition. *The hermitian form H defines a principal polarization on $J(C)$.*

The polarization H is called the *canonical polarization of* $J(C)$. Any divisor Θ on $J(C)$ such that the line bundle $\mathcal{O}_{J(C)}(\Theta)$ defines the canonical polarization is called a *theta divisor* of the Jacobian $J(C)$. We often write $(J(C), \Theta)$ for the canonically polarized Jacobian. Let us say one word about the relation between the intersection product on C and the canonical polarization on $J(C)$: the Riemann Relations 4.2.1 imply (see the proof of Proposition 1.2) that if A is the matrix of the intersection product on C with respect to some basis of $H_1(C, \mathbb{Z})$, then A^{-1} is the matrix of the alternating form defining the canonical polarization with respect to the same basis. In particular, the cycles $\lambda_1, \ldots, \lambda_{2g}$ form a symplectic basis in the sense of Section 3.1 of the lattice $H_1(C, \mathbb{Z})$ in $H^0(\omega_C)^*$ for H, since $\left(\begin{smallmatrix} 0 & -\mathbb{1} \\ \mathbb{1} & 0 \end{smallmatrix}\right)^{-1} = \left(\begin{smallmatrix} 0 & \mathbb{1} \\ -\mathbb{1} & 0 \end{smallmatrix}\right)$.

Proof. By definition of E is suffices to show that H is a positive definite hermitian form. According to Theorem 4.2.1 this is the case if and only if the period matrix Π of $J(C)$ (with respect to the chosen bases) satisfies the Riemann Relations:

$$\Pi \begin{pmatrix} 0 & -\mathbb{1} \\ \mathbb{1} & 0 \end{pmatrix} {}^t\Pi = 0 \quad \text{and} \quad i\Pi \begin{pmatrix} 0 & -\mathbb{1} \\ \mathbb{1} & 0 \end{pmatrix} {}^t\overline{\Pi} > 0\,.$$

We will check only the inequality, the proof of the equality being very similar. Recall that Π is the matrix of the $\mathbb{C}$-linear map $H^0(\omega_C)^* \to H^1_{\mathrm{DR}}(C)^*$ with respect to the bases $l_1, \ldots, l_g$ of $H^0(\omega_C)^*$, dual to $\omega_1, \ldots, \omega_g$, and $\lambda_1, \ldots, \lambda_{2g}$ of $H^1_{\mathrm{DR}}(C)^* = H_1(C, \mathbb{C})$. Let $\varphi_1, \ldots, \varphi_{2g}$ denote the basis of $H^1_{\mathrm{DR}}(C)$, dual to $\lambda_1, \ldots, \lambda_{2g}$. The dual map $H^1_{\mathrm{DR}}(C) \to H^0(\omega_C)$ is given by ${}^t\Pi$ with respect to the bases $\varphi_1, \ldots, \varphi_{2g}$ and $\omega_1, \ldots, \omega_g$, so

$$\omega_s = \sum_{t=1}^{2g} \Big(\int_{\lambda_t} \omega_s\Big)\varphi_t \quad \text{for } s = 1, \ldots, g.$$

Recall the well known fact (see Griffiths-Harris [1] p.59) that the intersection product in $H_1(C, \mathbb{Z})$ is Poincaré dual to the cup product in $H^1_{\mathrm{DR}}(C)$. Denoting by $P\colon H_1(C, \mathbb{C}) \to H^1_{\mathrm{DR}}(C)$ the Poincaré duality isomorphism, this means $(\lambda_i \cdot \lambda_j) = \int_C P(\lambda_i) \wedge P(\lambda_j)$ for $i, j = 1, \ldots, 2g$. The bases $\varphi_1, \ldots, \varphi_{2g}$ and $P(\lambda_1), \ldots, P(\lambda_{2g})$ of $H^1_{\mathrm{DR}}(C)$ are related by $P(\lambda_j) = \sum_{i=1}^{2g} \left(\begin{smallmatrix} 0 & -\mathbb{1} \\ \mathbb{1} & 0 \end{smallmatrix}\right)_{ij} \varphi_i$ for $i = 1, \ldots, 2g$. To see this write $P(\lambda_j) = \sum_{i=1}^{2g} a_{ij}\varphi_i$. Then using the definition of the Poincaré duality

$$a_{ij} = \sum_{\nu=1}^{2g} a_{\nu j} \int_{\lambda_i} \varphi_\nu = \int_{\lambda_i} P(\lambda_j) = \int_C P(\lambda_i) \wedge P(\lambda_j) = (\lambda_i \cdot \lambda_j) = \begin{pmatrix} 0 & -\mathbb{1} \\ \mathbb{1} & 0 \end{pmatrix}_{ij}\,.$$

Now an immediate matrix computation gives

$$\int_C \varphi_s \wedge \varphi_t = \begin{pmatrix} 0 & -\mathbb{1} \\ \mathbb{1} & 0 \end{pmatrix}_{st} \tag{1}$$

for $s, t = 1, \ldots, 2g$. Using this we get

$$\begin{aligned} i \int_C \omega_\mu \wedge \overline{\omega}_\nu &= i \sum_{s,t=1}^{2g} \Big(\int_{\lambda_s} \omega_\mu\Big)\Big(\int_{\lambda_t} \overline{\omega}_\nu\Big) \int_C \varphi_s \wedge \varphi_t \\ &= i \sum_{s,t=1}^{2g} \Pi_{\mu s} \begin{pmatrix} 0 & -\mathbb{1} \\ \mathbb{1} & 0 \end{pmatrix}_{st} \overline{\Pi}_{\nu t} = i\Big(\Pi \begin{pmatrix} 0 & -\mathbb{1} \\ \mathbb{1} & 0 \end{pmatrix} {}^t\overline{\Pi}\Big)_{\mu\nu} \end{aligned}$$

for $\mu, \nu = 1, \ldots, g$. So for any holomorphic 1-form $\omega = \sum \alpha_\mu \omega_\mu$

$$i \int \omega \wedge \overline{\omega} = i(\alpha_1, \ldots, \alpha_g) \Pi \begin{pmatrix} 0 & -\mathbb{1} \\ \mathbb{1} & 0 \end{pmatrix} {}^t\overline{\Pi} \begin{pmatrix} \overline{\alpha}_1 \\ \vdots \\ \overline{\alpha}_g \end{pmatrix}.$$

Since $i \int \omega \wedge \overline{\omega} > 0$ for every $\omega \neq 0$, this implies that $i\Pi \begin{pmatrix} 0 & -\mathbb{1} \\ \mathbb{1} & 0 \end{pmatrix} {}^t\overline{\Pi}$ is positive definite. □

There is another approach to define the Jacobian variety of C namely as the group $\mathrm{Pic}^0(C)$ of line bundles of degree zero on C. The Abel-Jacobi Theorem states that there is a canonical isomorphism between $\mathrm{Pic}^0(C)$ and $J(C)$.

Recall that $\mathrm{Pic}^0(C)$ is the quotient of the group $\mathrm{Div}^0(C)$ of divisors of degree zero on C modulo the subgroup of principal divisors. Define a canonical map

$$\mathrm{Div}^0(C) \to J(C) = H^0(\omega_C)^*/H_1(C, \mathbb{Z})$$

as follows: any divisor $D \in \mathrm{Div}^0(C)$ can be written as a finite sum $D = \sum_{\nu=1}^N (p_\nu - q_\nu)$ for some points $p_\nu, q_\nu \in C$. The class of the linear form $\omega \mapsto \sum_{\nu=1}^N \int_{q_\nu}^{p_\nu} \omega$ in $H^0(\omega_C)^*/H_1(C, \mathbb{Z})$ depends only on the divisor D, but not on its special representation as a sum of differences of points. So

$$D \mapsto \{\omega \mapsto \sum_{\nu=1}^N \int_{q_\nu}^{p_\nu} \omega\} \pmod{H_1(C, \mathbb{Z})}$$

gives a well defined map $\mathrm{Div}^0(C) \to J(C)$. Obviously this is a homomorphism of groups. It is called the *Abel-Jacobi map* of C.

Jacobi's Inversion Theorem states that the Abel-Jacobi map is surjective. On the other hand, by a theorem of Abel its kernel is the subgroup of principal divisors in $\mathrm{Div}^0(C)$. Combining both theorems we obtain the

(1.3) Abel-Jacobi Theorem. *The Abel-Jacobi map induces a canonical isomorphism* $\mathrm{Pic}^0(C) \xrightarrow{\sim} J(C)$.

Via this isomorphism $\mathrm{Pic}^0(C)$ inherits the structure of a principally polarized abelian variety. For the proof of Theorem 1.3 we refer to the standard books on Riemann surfaces or algebraic curves. In the sequel we identify $J(C) = \mathrm{Pic}^0(C)$ via the canonical isomorphism. We work with both interpretations without further notice: sometimes we consider elements of $J(C)$ as points and write $+$ for the group law and sometimes as line bundles on C and write $\otimes$ for the group law. The respective meaning will be clear from the context.

For any $n \in \mathbb{Z}$ denote by $\mathrm{Div}^n(C)$ the set of divisors of degree n on C. It is a principal homogeneous space for the group $\mathrm{Div}^0(C)$. This suggests that one can define more generally an Abel-Jacobi map from $\mathrm{Div}^n(C)$ to $J(C)$. Certainly, for $n \neq 0$ this map will not be canonical, but depends on the choice of a divisor in $\mathrm{Div}^n(C)$. To be more precise, fix a divisor $D_n \in \mathrm{Div}^n(C)$ and define

$$\mathrm{Div}^n(C) \to J(C)\,,\; D \mapsto \mathcal{O}_C(D - D_n)\,. \tag{2}$$

The most important case is $D_n = nc$ for some point $c \in C$. In this case the map $\mathrm{Div}^n(C) \to J(C)$ can also be written as

$$D = \sum r_\nu p_\nu \mapsto \{\omega \mapsto \sum r_\nu \int_c^{p_\nu} \omega\} \pmod{H_1(C, \mathbb{Z})}\,.$$

For $n \geq 1$ let $C^{(n)}$ denote the n-fold symmetric product of C. Recall that $C^{(n)}$ is the quotient of the cartesian product C^n by the natural action of the symmetric group of degree n. As such it is a smooth projective variety of dimension n (see Griffiths-Harris [1] p. 236). The elements of $C^{(n)}$ can be considered as effective divisors of degree n on C. In this way $C^{(n)}$ is a subset of $\mathrm{Div}^n(C)$ and we denote the restriction to $C^{(n)}$ of the map (2) by $\alpha_{D_n}: C^{(n)} \to J(C)$. The map α_{D_n} is also called *Abel-Jacobi map.*

Let $\mathrm{Pic}^n(C)$ denote the set of line bundles of degree n on C. It is a principal homogeneous space for the group $\mathrm{Pic}^0(C)$: given a line bundle L_n of degree n on C, the map

$$\alpha_{L_n}: \mathrm{Pic}^n(C) \to J(C)\,,\; L \mapsto L \otimes L_n^{-1}$$

is bijective. Finally, consider the canonical map $\rho: C^{(n)} \to \mathrm{Pic}^n(C)$ sending an effective divisor D in $C^{(n)}$ to its class $\mathcal{O}_C(D)$ in $\mathrm{Pic}^n(C)$. These maps fit into the following commutative diagram

$$\begin{array}{ccc} & \overset{\rho}{\nearrow} & \mathrm{Pic}^n(C) \\ C^{(n)} & & \big\downarrow \alpha_{\mathcal{O}(D_n)} \\ & \underset{\alpha_{D_n}}{\searrow} & J(C) \end{array}\,.$$

For any line bundle $L \in \mathrm{Pic}^n(C)$ the fibre $\rho^{-1}(L)$ is by definition the complete linear system $|L|$ on C. So for any $M \in \mathrm{Pic}^0(C)$ the fibre $\alpha_{D_n}^{-1}(M)$ is the complete linear system $|M \otimes \mathcal{O}_C(D_n)|$.

Suppose now $g \geq 1$ and fix a point $c \in C$. The Abel-Jacobi map $\alpha = \alpha_c : C \to J = J(C)$ is of special interest. In order to show that α is an embedding, we first study its differential $d\alpha$. Recall that $d\alpha$ is a holomorphic map from the tangent bundle $\mathcal{T}_C$ of C to the tangent bundle $\mathcal{T}_J$ of J. According to Lemma 1.4.2 the tangent bundle of J is trivial: $\mathcal{T}_J = J \times \mathbb{C}^g$. The projectivization of the composed map $\mathcal{T}_C \to \mathcal{T}_J \simeq J \times \mathbb{C}^g \to \mathbb{C}^g$ is a priori a rational map $C \to \mathbb{P}^{g-1}$ called the *projectivized differential of* α.

(1.4) Proposition. *The projectivized differential of the Abel-Jacobi map* $\alpha: C \to J$ *is the canonical map* $\varphi_{\omega_C}: C \to \mathbb{P}_{g-1}$.

Proof. For $x \in J$ consider the tangent space $T_x J$ at the point x. The canonical isomorphisms $T_x J = T_0 J = H^0(\omega_C)^*$ yield an isomorphism $\mathcal{T}_J = J \times H^0(\omega_C)^*$. Choose a basis $\omega_1, \ldots, \omega_g$ of $H^0(\omega_C)$ and identify $H^0(\omega_C)^* = \mathbb{C}^g$. Then the Abel-Jacobi map $\alpha: C \to \mathbb{C}^g / H_1(C, \mathbb{Z})$ is given by $\alpha(p) = {}^t(\int_c^p \omega_1, \ldots, \int_c^p \omega_g) \pmod{H_1(C, \mathbb{Z})}$. Hence by the fundamental theorem of calculus, the projectivization of the composed map $\mathcal{T}_C \xrightarrow{d\alpha} \mathcal{T}_J \simeq J \times \mathbb{C}^g \to \mathbb{C}^g$ is given by $p \mapsto (\omega_1(p) : \cdots : \omega_g(p))$. But this is just the canonical map. □

(1.5) Corollary. *For any* $g \geq 1$ *the Abel-Jacobi map* $\alpha: C \to J(C)$ *is an embedding.*

Proof. The map α is injective, since for every line bundle L of degree 1 on a curve of genus ≥ 1 we have $h^0(L) \leq 1$. From Proposition 1.4 we conclude that the differential of α is injective at every point $p \in C$, the canonical line bundle ω_C on C being base point free. □

As a second corollary we get a statement, which we applied already in the proof of Theorem 10.1.6.

(1.6) Corollary. *Suppose* C *is a curve of genus 2. Let* x *and* y *be different points on* $J = J(C)$ *and* $t \in T_x J$ *a tangent vector.*

a) There is a point $z \in J$ *such that the translated curve* $t_z^* \alpha(C)$ *passes through* x *and* y.

b) There is a point $z \in J$ *such that the translated curve* $t_z^* \alpha(C)$ *passes through* x *and* t *is tangential to* $t_z^* \alpha(C)$ *at* x.

In fact, one can be more precise (see Exercise 7): there are exactly two translates of $\alpha(C)$ passing through x and y.

Proof. As for a): let ι denote the hyperelliptic involution of C. The map $C^{(2)} \to J$, $(p,q) \mapsto \mathcal{O}_J(p - \iota q)$ is surjective. Hence there are points p and q on C such that $x - y = \mathcal{O}_J(p - \iota q)$. Then the point $z = \alpha(p) - x$ satisfies the assertion: $x = \alpha(p) - z$ and $y = \alpha(\iota q) - z$.
As for b): according to Proposition 1.4 the projectivized differential of $\alpha: C \to J$ is a double covering $C \to \mathbb{P}_1$. Hence there is a point $p \in C$ such that t is tangent to C in $\alpha(p)$. Then the point $z = \alpha(p) - x$ satisfies the assertion. □

Finally, we state for the sake of completeness

(1.7) Torelli's Theorem. *Suppose C and C' are compact Riemann surfaces of genus g. If their Jacobians $(J(C), \Theta)$ and $(J(C'), \Theta')$ are isomorphic as polarized abelian varieties, then C is isomorphic to C'.*

Proofs of Torelli's Theorem can be found in almost every book on compact Riemann surfaces or algebraic curves. We do not repeat it here, but see Exercise 8.

§ 2 The Theta Divisor

Let (J, Θ) be the Jacobian of a smooth projective curve C of genus $g \geq 1$. In this section we study the geometry of the theta divisor Θ. In particular we will see that some properties of Θ reflect geometrical properties of the curve C itself.

As we saw in the last section, the varieties $\operatorname{Pic}^n(C)$ are principal homogenous spaces for $J = \operatorname{Pic}^0(C)$. We will see next that there is an intrinsic way of defining a theta divisor in $\operatorname{Pic}^{g-1}(C)$. Recall the canonical map $\rho: C^{(n)} \to \operatorname{Pic}^n(C)$ for $n \geq 1$. Its image

$$W_n := \rho(C^{(n)})$$

is the subset of $\operatorname{Pic}^n(C)$ of line bundles with nonempty linear system. For $n \geq g$ we have $W_n = \operatorname{Pic}^n(C)$ as an immediate consequence of the Riemann-Roch Theorem for curves. For $1 \leq n \leq g$ it is well known (see Griffith-Harris [1] p. 338) that $h^0(\mathcal{O}_C(D)) = 1$ for a general divisor $D \in C^{(n)}$. This means that $\rho: C^{(n)} \to \operatorname{Pic}^n(C)$ is birational onto its image W_n. Since moreover ρ is a proper morphism, W_n is an irreducible closed subvariety of $\operatorname{Pic}^n(C)$ of dimension n. In particular, W_{g-1} is a divisor in $\operatorname{Pic}^{g-1}(C)$.

We want to study the relation between the varieties W_n and the theta divisor Θ: fix a point $c \in C$ and denote by $\widetilde{W}_n$ the image of W_n in J under the bijection $\alpha_{\mathcal{O}(nc)}: \operatorname{Pic}^n(C) \to J$, i.e.

$$\widetilde{W}_n = \alpha_{\mathcal{O}_C(nc)}(W_n) = \alpha_{nc}(C^{(n)}) \ .$$

The next theorem compares the fundamental classes of Θ and $\widetilde{W}_n$. For this recall that the *fundamental class* $[Y]$ of an n-dimensional subvariety Y of a smooth projective variety X of dimension g is by definition the element in $H^{2g-2n}(X, \mathbb{Z})$, Poincaré dual to the homology class $\{Y\}$ of Y in $H_{2n}(X, \mathbb{Z})$.

(2.1) Poincaré's Formula. $[\widetilde{W}_n] = \frac{1}{(g-n)!} \overset{g-n}{\bigwedge} [\Theta]$ *for any* $1 \leq n \leq g$.

Notice that the formula also holds for $n = 0$, if we define $\widetilde{W}_0$ to be a point. By definition $[\Theta] = c_1(\Theta)$, the first Chern class of Θ (see Griffiths-Harris [1] p. 141). Thus $\bigwedge^g [\Theta]$ equals the intersection number (Θ^g) times the class of a point (see Section 3.6). So for $n = 0$ the formula is equivalent to $(\Theta^g) = g!$ which is a consequence of Riemann-Roch.

Proof. We will prove the Poincaré dual of the above formula:

$$\{\widetilde{W}_n\} = \frac{1}{(g-n)!} \{\Theta\}^{g-n} \tag{1}$$

using the Pontryagin product $\star$.

First we claim $\{\widetilde{W}_n\} = \frac{1}{n!} \star_{i=1}^n \{\widetilde{W}_1\}$ for all $1 \leq n \leq g$. The proof proceeds by induction on n. For $n = 1$ there is nothing to show. So suppose the formula is valid for some $n \geq 1$. Denote by $p: C^{(n)} \times C \to C^{(n+1)}$ the natural map and by μ the addition map. Then the commutativity of the diagram

$$\begin{array}{ccc} C^{(n)} \times C & \xrightarrow{\alpha_{nc} \times \alpha_c} & J \times J \\ {\scriptstyle p}\downarrow & & \downarrow{\scriptstyle \mu} \\ C^{(n+1)} & \xrightarrow{\alpha_{(n+1)c}} & J \end{array}$$

and the induction hypothesis imply

$$\begin{aligned} \{\widetilde{W}_{n+1}\} &= \tfrac{1}{n+1} (\alpha_{(n+1)c})_* p_* \{C^{(n)} \times C\} \\ &= \tfrac{1}{n+1} \mu_* (\alpha_{nc} \times \alpha_c)_* \{C^{(n)} \times C\} \\ &= \tfrac{1}{n+1} \mu_* (\alpha_{nc*} \{C^{(n)}\} \times \alpha_{c*} \{C\}) \\ &= \tfrac{1}{n+1} \{\widetilde{W}_n\} \star \{\widetilde{W}_1\}) = \tfrac{1}{(n+1)!} \star_{i=1}^n \{\widetilde{W}_1\} \end{aligned}$$

This completes the proof of the claim.

Now suppose $\lambda_1, \dots, \lambda_{2g}$ is a symplectic basis of $H_1(C, \mathbb{Z}) = H_1(J, \mathbb{Z})$ with corresponding real coordinate functions $x_1, \dots, x_{2g}$ of $H^0(\omega_C)^*$. By construction the bases $dx_1, \dots, dx_{2g}$ of $H^1_{\mathrm{DR}}(J)$ and $\lambda_1, \dots, \lambda_{2g}$ of $H_1(J, \mathbb{Z})$ are dual to each other. So under the identification $H_1(C, \mathbb{Z}) = H_1(J, \mathbb{Z})$ also the basis $\alpha_c^* dx_1, \dots, \alpha_c^* dx_{2g}$ of $H^1_{\mathrm{DR}}(C)$ is dual to $\lambda_1, \dots, \lambda_{2g}$. So arguing as in the proof of Proposition 1.2 (see equation 1(1)) we obtain

$$\int_{\widetilde{W}_1} dx_i \wedge dx_j = \int_C \alpha_c^* dx_i \wedge \alpha_c^* dx_j = (\lambda_i \cdot \lambda_j) = -\delta_{g+i,j} = \int_{-\sum_\nu \lambda_\nu \star \lambda_{g+\nu}} dx_i \wedge dx_j \,.$$

For the last equation we used the fact that the basis $\{dx_i \wedge dx_j \mid i < j\}$ of $H^2(J, \mathbb{Z})$ is dual to the basis $\{\lambda_i \star \lambda_j \mid i < j\}$ of $H_2(J, \mathbb{Z})$ according to Lemma 4.10.1. This implies $\{\widetilde{W}_1\} = -\sum_{\nu=1}^g \lambda_\nu \star \lambda_{g+\nu}$. Hence using Theorem 4.10.4

$$\{W_n\} = \tfrac{1}{n!} \star_{i=1}^n \Big(-\sum_{\nu=1}^{g} \lambda_\nu \star \lambda_{g+\nu}\Big) = \tfrac{1}{(g-n)!}\{\Theta\}^{g-n} \,. \qquad \square$$

We want to work out Poincaré's Formula in the cases $n = 1$ and $g - 1$. According to Corollary 1.5 the Abel-Jacobi map $\alpha\colon C \to J$ is an embedding. Identifying C with its image $\widetilde{W}_1 = \alpha(C)$, Poincaré's formula for $n = 1$ gives

(2.2) Corollary. $(C \cdot \Theta) = g$.

Proof. According to Poincaré's Formula we have

$$[C] \wedge [\Theta] = \tfrac{1}{(g-1)!} \bigwedge^{g-1} [\Theta] \wedge [\Theta] = \tfrac{1}{(g-1)!} \bigwedge^{g} [\Theta] \,.$$

So $(C \cdot \Theta) = \frac{1}{(g-1)!}(\Theta^g) = g$ by Riemann-Roch and the definition of the intersection number. □

In case $n = g - 1$ we get:

(2.3) Corollary. *There is an $\eta \in \mathrm{Pic}^{g-1}(C)$ such that $W_{g-1} = \alpha_\eta^* \Theta$.*

Proof. By Poincaré's Formula $[\widetilde{W}_{g-1}] = [\Theta]$, so the first Chern classes of the corresponding line bundles coincide. Hence according to Corollary 2.5.4 there is an $x \in J = \mathrm{Pic}^0(C)$ such that $\widetilde{W}_{g-1} = t_x^* \Theta$. This implies $W_{g-1} = \alpha^*_{\mathcal{O}((g-1)c)} \widetilde{W}_{g-1} = \alpha_\eta^* \Theta$ with $\eta = \mathcal{O}((g-1)c) \otimes x^{-1}$. □

There is no canonical way to distinguish a theta divisor Θ within its algebraic equivalence class. On the other hand, the divisor W_{g-1} in $\mathrm{Pic}^{g-1}(C)$ is intrinsic. So the line bundle η in Corollary 2.3 depends on the choice of Θ. If Θ is one of the 2^{2g} symmetric theta divisors (see Lemma 4.6.2), then we can say more about the line bundle η. For this recall that a *theta characteristic* on C is a line bundle κ on C with $\kappa^2 = \omega_C$.

(2.4) Riemann's Theorem. *For any symmetric theta divisor Θ there is a theta characteristic κ on C such that*

$$W_{g-1} = \alpha_\kappa^* \Theta \ .$$

In classical terminology Riemann's Theorem reads:

$$W_{g-1} - \kappa = \Theta \ .$$

In view of Riemann's Theorem, it makes sense to call W_{g-1} the *canonical theta divisor* of C. Note that in particular the theta divisor Θ corresponding to a symplectic basis is symmetric. To be more precise, let $\lambda_1, \dots, \lambda_{2g}$ be a symplectic basis for the canonical polarization and L_0 the corresponding line bundle of characteristic zero on J. Then the unique divisor Θ in the linear system $|L_0|$ is symmetric. In fact, Θ may be considered as the zero divisor of the classical Riemann theta function. In this case the theta characteristic κ in the theorem is called *Riemann's constant.*

Proof. We have to show that $\kappa^2 = \omega_C$. For this consider the involution ι on $\mathrm{Pic}^{g-1}(C)$, sending a line bundle L to $\omega_C \otimes L^{-1}$. Since by Riemann-Roch and Serre duality $h^0(L) = h^0(\omega_C \otimes L^{-1})$ for every $L \in \mathrm{Pic}^{g-1}(C)$, the divisor W_{g-1} is invariant under ι, i.e. $\iota^* W_{g-1} \simeq W_{g-1}$. Using this and the fact that Θ is symmetric, we have

$$\iota^* \alpha_\kappa^* (-1)^* \Theta \simeq \iota^* \alpha_\kappa^* \Theta \simeq \iota^* W_{g-1} \simeq W_{g-1} \simeq \alpha_\kappa^* \Theta \ .$$

Now $(-1)\alpha_\kappa \iota = \alpha_{\omega_C \otimes \kappa^{-1}}$ yields $\alpha_\kappa^* \Theta \simeq \alpha_{\omega_C \otimes \kappa^{-1}}^* \Theta$. This implies the assertion, since Θ defines a principal polarization. □

Consider again the canonical map $\rho: C^{(g-1)} \to W_{g-1} \subset \mathrm{Pic}^{g-1}(C)$ defined by $D \mapsto \mathcal{O}_C(D)$. It blows down the whole linear system $|D| \subset C^{(g-1)}$ to the point $\mathcal{O}_C(D)$ of W_{g-1}. This suggests that for positive dimensional linear systems the corresponding point is a singular point of W_{g-1}. In fact, this is the contents of

(2.5) Riemann's Singularity Theorem. *For every $L \in \mathrm{Pic}^{(g-1)}(C)$*

$$\mathrm{mult}_L \, W_{g-1} = h^0(L) \ .$$

For the proof we refer to Arbarello et al. [1], we do not repeat it here. However, we want to give some applications.

Theta Characteristics

Riemann's Theorem 2.4 implies in particular that there is a theta characteristic κ on C. We claim that the map $\alpha_\kappa\colon \mathrm{Pic}^{g-1}(C) \to J$ induces a bijection between the set of theta characteristics on C and the set of 2-division points J_2 of J. In fact, $x = \alpha_\kappa(\eta) = \eta \otimes \kappa^{-1}$ is a 2-division point if and only if $\eta^2 = \omega_C$, i.e. η is a theta characteristic. This implies that the curve C admits exactly 2^{2g} theta characteristics.

A theta characteristic η is called *even* (respectively *odd*) if $h^0(\eta) \equiv 0 \pmod 2$ (respectively $h^0(\eta) \equiv 1 \pmod 2$).

(2.6) Proposition. *Let Θ be a symmetric theta divisor on J and κ the theta characteristic with $W_{g-1} = \alpha_\kappa^*\Theta$. Then κ is even (respectively odd) if and only if Θ is even (respectively odd).*

In particular Riemann's constant is an even theta characteristic, since the classical Riemann theta function with characteristic $\begin{bmatrix}0\\0\end{bmatrix}$ is even.

Proof. Recall from Section 4.7 that Θ is even (respectively odd) if $\mathrm{mult}_0\,\Theta \equiv 0 \pmod 2$ (respectively $\mathrm{mult}_0\,\Theta \equiv 1 \pmod 2$). By Riemann's Theorem α_κ induces a biholomorphic map $W_{g-1} \to \Theta$. So Riemann's Singularity Theorem gives $h^0(\kappa) = \mathrm{mult}_{\alpha_\kappa(\kappa)}(\Theta) = \mathrm{mult}_0(\Theta)$ which implies the assertion. □

Fix an even symmetric theta divisor Θ and let κ be the corresponding theta characteristic. As in the proof of Proposition 2.6 one has $h^0(\eta) = \mathrm{mult}_{\alpha_\kappa(\eta)}(\Theta)$ for every theta characteristic η. So Corollary 4.7.7 implies

(2.7) Proposition. *The curve C admits exactly $2^{g-1}(2^g+1)$ (respectively $2^{g-1}(2^g-1)$) even (respectively odd) theta characteristics.*

The Singularity Locus of Θ

The singularity locus $\mathrm{sing}\,\Theta$ of a theta divisor Θ on the Jacobian $J(C)$ is a closed algebraic subset. Using Riemann's Singularity Theorem one can compute its dimension. For this we assume $g \geq 4$, the cases of smaller genus being trivial.

(2.9) Proposition. $$\dim \mathrm{sing}\,\Theta = \begin{cases} g-4 & \text{if } C \text{ is not hyperelliptic} \\ g-3 & \text{if } C \text{ is hyperelliptic.} \end{cases}$$

Proof. According to Riemann's Theorem we have to compute $\dim \mathrm{sing}\, W_{g-1}$. By Riemann's Singularity Theorem a line bundle L is contained in $\mathrm{sing}\, W_{g-1}$ if and only if $h^0(L) \geq 2$.

Suppose first that C is hyperelliptic. Let ι denote the hyperelliptic involution on C and $L_0 = \mathcal{O}_C(p + \iota p)$ the unique line bundle on C with $h^0(L_0) = \deg L_0 = 2$. Consider the map

$$\phi: C^{(g-3)} \to \operatorname{sing} W_{g-1}, \quad p_1 + \cdots + p_{g-3} \mapsto \mathcal{O}_C(p_1 + \cdots + p_{g-3}) \otimes L_0 .$$

It suffices to show that ϕ is birational and surjective. For this recall the *geometric version of the Riemann-Roch Theorem* for (not necessarily hyperelliptic) curves (see Arbarello et al. [1] p. 12):

If $\varphi_{\omega_C}: C \to \mathbb{P}_{g-1}$ is the canonical map and D an effective divisor on C, denote by $\operatorname{span} \varphi_\omega(D)$ *the intersection of the hyperplanes in $\mathbb{P}_{g-1}$ containing the zero-dimensional scheme $\varphi_\omega(D)$. Then*

$$h^0(\mathcal{O}_C(D)) = \deg D - \dim \operatorname{span} \varphi_{\omega_C}(D) . \tag{1}$$

In order to show that ϕ is surjective, suppose $\mathcal{O}_C(D) \in \operatorname{sing} W_{g-1}$. Then $2 \leq h^0(\mathcal{O}_C(D)) = g-1-\dim \operatorname{span} \varphi_{\omega_C}(D)$ implies $\dim \operatorname{span} \varphi_{\omega_C}(D) \leq g-3$. Using the well known facts that $\varphi_{\omega_C}(C)$ is the rational normal curve in $\mathbb{P}_{g-1}$ (see Hartshorne [1] p. 343) and that any $g-1$ pairwise different points on it span a $\mathbb{P}_{g-2}$, we conclude that D is of the form $D = p_1 + \cdots + p_{g-3} + p + \iota p$.

It remains to show that ϕ is generically injective. But, if $p_1, \ldots, p_{g-3}$ are points in C, no two of which correspond to each other under the involution ι, then $h^0(\mathcal{O}_C(p_1 + \cdots + p_{g-3}) \otimes L_0) = 2$, again by (1). This implies that $p_1 + \ldots + p_{g-3}$ is uniquely determined by $\mathcal{O}_C(p_1 + \cdots + p_{g-3}) \otimes L_0$.

Finally, suppose that C is not hyperelliptic. Then the canonical map $\varphi_{\omega_C}: C \to \mathbb{P}_{g-1}$ is an embedding. Consider the natural map $\rho: C^{(g-1)} \to W_{g-1} \subset \operatorname{Pic}^{g-1}(C)$.

We claim $\dim \rho^{-1}(\operatorname{sing} W_{g-1}) \leq g-3$. For the proof fix a general divisor $D = p_1 + \cdots + p_{g-3}$. According to the General Position Theorem (see Arbarello et al. [1] p. 109) $\dim \operatorname{span} \varphi_{\omega_C}(D) = g-4$ and $C \cap \operatorname{span} \varphi_{\omega_C}(D) = \varphi_{\omega_C}(D)$ and moreover the linear projection with center $\operatorname{span} \varphi_{\omega_C}(D)$ leads to a birational morphism $p: C \to \overline{C} \subset \mathbb{P}_2$. Since $g \geq 4$ and $\deg(\overline{C}) = \deg C - g + 3 = g+1$, the curve $\overline{C}$ is not smooth by Plückers's formula for plane curves. Let q be a singular point of $\overline{C}$ and consider the linear projection $\mathbb{P}_2 \to \mathbb{P}_1$ with center q. Let ν denote the multiplicity of the singular point q of $\overline{C}$ and $p^{-1}(q) = \{q_1, \ldots, q_\nu\}$ counted with multiplicities. The composed map $C \to \mathbb{P}_1$ is given by a linear system $g^1_{g+1-\nu}$, the corresponding line bundle of which is $\omega_C(-D - \sum_{i=1}^{\nu} q_i)$. By Riemann-Roch and Serre duality this implies

$$2 \leq h^0(\omega_C(-D - \sum_{i=1}^{\nu} q_i)) \leq h^0(\omega_C(-D - q_1 - q_2)) = h^0(\mathcal{O}_C(D + q_1 + q_2)) .$$

Thus $D + q_1 + q_2 \in \rho^{-1}(\operatorname{sing} W_{g-1})$ by Riemann's Singularity Theorem.

Let F_D denote the surface of divisors in $C^{(g-1)}$ containing D. Since $\overline{C}$ contains only finitely many singularities, the argument above shows that F_D intersects $\rho^{-1}(\operatorname{sing} W_{g-1})$ at most in finitely many points. So we get

$$\begin{aligned} 0 &= \dim(\rho^{-1}(\operatorname{sing} W_{g-1}) \cap F_D) \\ &\geq \dim \rho^{-1}(\operatorname{sing} W_{g-1}) + \dim F_D - \dim C^{(g-1)} \\ &= \dim \rho^{-1}(\operatorname{sing} W_{g-1}) - g + 3 , \end{aligned}$$

which completes the proof of the claim.

On the other hand, varying the divisor D within an open dense subset of $C^{(g-3)}$, the construction of above shows that the dimension of a component of $\rho^{-1}(\operatorname{sing} W_{g-1})$ is $\geq g-3$. By what we have said above, this gives $\dim \rho^{-1}(\operatorname{sing} W_{g-1}) = g-3$.

The fibre of ρ over $L \in W_{g-1}$ is just the linear system $|L|$, and $h^0(L) \geq 2$ for all $L \in \operatorname{sing} W_{g-1}$. Moreover it is easy to see using (1) that every component of $\operatorname{sing} W_{g-1}$ contains a line bundle L with $h^0(L) = 2$. So the general fibre of ρ over any component is of dimension one and thus $\dim \operatorname{sing} W_{g-1} = g-4$. □

Note that the proof shows that $\operatorname{sing} W_{g-1}$ is irreducible for hyperelliptic C. On the other hand, together with the existence theorem of Brill-Noether theory (see Arbarello et al. [1] p. 206), the proof gives that $\operatorname{sing} W_{g-1}$ is equidimensional for non-hyperelliptic C.

§ 3 The Poincaré Bundles for a Curve C

Let C be a smooth curve of genus g and $J = J(C)$ its Jacobian. For any integer n we will construct a universal family of line bundles of degree n on C, the Poincaré bundle of degree n for C.

Fix a point $c \in C$ and let $\alpha_c : C \hookrightarrow J$ be the corresponding Abel-Jacobi map.

(3.1) Lemma. *The restriction* $\alpha_c^* : \widehat{J} = \operatorname{Pic}^0(J) \to \operatorname{Pic}^0(C)$ *is an isomorphism.*

We will see in Proposition 3.5 below that $(\alpha_c^*)^{-1} = -\phi_\Theta$.

Proof. The exponential sequences of J and C induce the following commutative diagram

$$\begin{array}{ccccccc} H^1(J,\mathbb{Z}) & \longrightarrow & H^1(\mathcal{O}_J) & \longrightarrow & \mathrm{Pic}^0(J) & \longrightarrow & 0\\ \downarrow{\scriptstyle\beta} && \downarrow{\scriptstyle\gamma} && \downarrow{\scriptstyle\alpha_c^*} && \\ H^1(C,\mathbb{Z}) & \longrightarrow & H^1(\mathcal{O}_C) & \longrightarrow & \mathrm{Pic}^0(C) & \longrightarrow & 0 \quad . \end{array}$$

It suffices to show that the restriction maps β and γ are isomorphisms. As for β : we have $H^1(J,\mathbb{Z}) = \mathrm{Hom}(H_1(J,\mathbb{Z}),\mathbb{Z})$, as we saw in Section 1.3. Moreover $H^1(C,\mathbb{Z}) = \mathrm{Hom}(H_1(C,\mathbb{Z}),\mathbb{Z})$ and β is the transposed map of the isomorphism $\alpha_{c*}: H_1(C,\mathbb{Z}) \to H_1(J,\mathbb{Z})$. As for γ: the functoriality of the Hodge duality means that the following diagram is commutative

$$\begin{array}{ccc} H^1(\mathcal{O}_J) & = & H^0(\Omega_J)^* \\ {\scriptstyle\gamma}\downarrow & & \uparrow{\scriptstyle r^*} \\ H^1(\mathcal{O}_C) & = & H^0(\Omega_C)^* \quad . \end{array}$$

Here $r: H^0(\Omega_J) \to H^0(\Omega_C)$ denotes the restriction map. By definition of the Jacobian r is an isomorphism, and so is γ. □

Lemma 3.1 allows us to use the Poincaré bundle for the Jacobian J in order to construct Poincaré bundles for the curve C itself. Recall that a *Poincaré bundle of degree n for C* (normalized with respect to the point $c \in C$) is a line bundle $\mathcal{P}_C^n$ on $C \times \mathrm{Pic}^n(C)$ satisfying

i) $\mathcal{P}_C^n | C \times \{L\} \simeq L$ for every $L \in \mathrm{Pic}^n(C)$, and
ii) $\mathcal{P}_C^n | \{c\} \times \mathrm{Pic}^n(C)$ is trivial.

(3.2) Proposition. *For every $n \in \mathbb{Z}$ there exists a Poincaré bundle $\mathcal{P}_C^n$ for C, uniquely determined by the choice of the point $c \in C$.*

Proof. For any $L \in Pic^n(C)$ consider the commutative diagram

$$\begin{array}{ccccc} C \times \mathrm{Pic}^n(C) & \xrightarrow{\mathrm{id}\times\alpha_{\mathcal{O}(nc)}} & C \times \mathrm{Pic}^0(C) & \xrightarrow{\alpha_c \times \alpha_c^{*-1}} & J \times \widehat{J} \\ \uparrow & & \uparrow & & \uparrow \\ C \times \{L\} & \longrightarrow & C \times \{\alpha_{\mathcal{O}(nc)}(L)\} & \longrightarrow & J \times \{\alpha_c^{*-1}(\alpha_{\mathcal{O}(nc)}(L))\} , \end{array}$$

Here the vertical maps are the natural embeddings and the lower horizontal maps are the restrictions of the upper ones. Denote by γ the composed map $\gamma = (\alpha_c \times \alpha_c^{*-1})(\mathrm{id} \times \alpha_{\mathcal{O}_C(nc)}): C \times \mathrm{Pic}^n(C) \to J \times \widehat{J}$. If $\mathcal{P}$ is the Poincaré bundle on $J \times \widehat{J}$, define

$$\mathcal{P}_C^n = \gamma^* \mathcal{P} \ .$$

The commutativity of the diagram above and the corresponding property of $\mathcal{P}$, i.e. $\mathcal{P}|J \times \{M\} \simeq M$ for all $M \in \widehat{J}$, implies that $\mathcal{P}_C^n$ satisfies condition i).

With a similar diagram as above, and using the facts that $\alpha_c: C \to J$ maps c to $0 \in J$ and that $\mathcal{P}|\{0\} \times \widehat{J}$ is trivial, one verifies condition ii). Finally, the uniqueness statement follows from the Seesaw Theorem A.8. □

The Poincaré bundle $\mathcal{P}^n_C$ satisfies the following universal property:

(3.3) Proposition. *For any normal algebraic variety T and any line bundle $\mathcal{L}$ on $C \times T$ with*

i) $\mathcal{L}|C \times \{t\} \in \operatorname{Pic}^n(C)$ for every $t \in T$, and

ii) $\mathcal{L}|\{c\} \times T$ is trivial,

there is a unique morphism $\psi: T \to \operatorname{Pic}^n(C)$ such that $\mathcal{L} \simeq (\mathrm{id} \times \psi)^* \mathcal{P}^n_C$.

Notice that the underlying set-theoretical map of ψ is $t \mapsto \mathcal{L}|C \times \{t\}$. Moreover, if T is irreducible, one can show that it suffices to assume condition i) for one point $t_0 \in T$. We omit the proof of Proposition 3.3, since it is completely analogous to that of Proposition 2.5.2.

As a consequence we show that $(\alpha_c^*)^{-1} = -\phi_\Theta$. For this we need a technical lemma. We prove it in a slightly more general form, which we need in the next chapter. Fix a line bundle $\kappa \in \operatorname{Pic}^{g-1}(C)$ (not necessarily a theta characteristic) and define a theta divisor Θ on J by $W_{g-1} = \alpha_\kappa^* \Theta$ $(= \Theta + \kappa)$.

(3.4) Lemma. $\alpha_c^* \mathcal{O}_J(t_x^*(-1)^*\Theta) = x^{-1} \otimes \kappa \otimes \mathcal{O}_C(c)$ *for all* $x \in J = \operatorname{Pic}^0(C)$.

Proof. Step I: Let U denote the subset of J consisting of all points x such that i) the curve $\alpha_c(C)$ intersects the divisor $t_x^*(-1)^*\Theta$ in g pairwise different points (see Corollary 2.2) and ii) $h^0(x^{-1} \otimes \kappa \otimes \mathcal{O}_C(c)) = 1$. According to the Moving Lemma 5.4.1 and by semicontinuity U is open and dense in J. We claim that the assertion holds for every $x \in U$.

Suppose $x \in U$. There are pairwise different points $p_1, \dots, p_g$ of C, such that $\alpha_c^*(t_x^*(-1)^*\Theta) = p_1 + \dots + p_g$. We have to show that $\mathcal{O}_C(p_1 + \dots + p_g) = x^{-1} \otimes \kappa \otimes \mathcal{O}_C(c)$.

By assumption, we have $\alpha_c(p_i) \in t_x^*(-1)^*\Theta$ for $i = 1, \dots, g$, or equivalently $(x^{-1} \otimes \kappa \otimes \mathcal{O}_C(c)) \otimes \mathcal{O}_C(-p_i) = \alpha_c(p_i)^{-1} \otimes x^{-1} \otimes \kappa \in \alpha_\kappa^* \Theta = W_{g-1}$ for $i = 1, \dots, g$. Since $h^0(x^{-1} \otimes \kappa \otimes \mathcal{O}_C(c)) = 1$ and the points p_i are pairwise different, this implies the claim.

Step II: The proof of the lemma is completed by applying the Seesaw Principle to a globalized version of the assertion: denoting by p_C and p_J the natural projections of $C \times J$ and by $\mu: J \times J \to J$ the addition map, we claim

$$(\alpha_c \times \mathrm{id})^* \mu^* \mathcal{O}_J((-1)^*\Theta) \otimes p_J^* \mathcal{O}_J(-(-1)^*\Theta) \simeq \mathcal{P}_C^{0\,-1} \otimes p_C^*(\kappa \otimes \mathcal{O}_C(c)) \,. \quad (1)$$

But this follows from Step I and Corollary A.9, restricting both sides to $C \times \{x\}$ with $x \in U$ and $\{c\} \times J$. Restricting (1) to $C \times \{x\}$ for any $x \in J$ gives the assertion. □

(3.5) Proposition. $(\alpha_c^*)^{-1} = -\phi_\Theta$.

Proof. By Lemma 3.4 we have for all $x \in J = \mathrm{Pic}^0(X)$

$$\alpha_c^* \phi_\Theta(x) = \alpha_c^* \phi_{(-1)^*\Theta}(x) = \alpha_c^* \mathcal{O}_J(t_x^*(-1)^*\Theta - (-1)^*\Theta)$$
$$= x^{-1} \otimes \kappa \otimes \mathcal{O}_C(c) \otimes (\kappa \otimes \mathcal{O}_C(c))^{-1} = x^{-1} \ . \ \square$$

§ 4 The Universal Property

The Jacobian J of a curve C of genus g admits a universal property: maps from C into abelian varieties factorize via the Jacobian. In this section we prove this fact and deduce some consequences.

(4.1) Universal Property of the Jacobian. *Suppose X is an abelian variety and $\varphi: C \to X$ is a rational map. Then there exists a unique homomorphism $\widetilde{\varphi}: J \to X$ such that for every $c \in C$ the following diagram is commutative*

$$\begin{array}{ccc} C & \xrightarrow{\varphi} & X \\ {\scriptstyle \alpha_c}\downarrow & & \downarrow{\scriptstyle t_{-\varphi(c)}} \\ J & \xrightarrow{\widetilde{\varphi}} & X \end{array} .$$

Proof. According to Theorem 4.9.4 the map φ is everywhere defined. Consider the morphism $(t_{-\varphi(c)}\varphi)^{(g)}: C^{(g)} \to X$ defined by $(t_{-\varphi(c)}\varphi)^{(g)}(p_1 + \cdots + p_g) = \varphi(p_1) + \cdots + \varphi(p_g) - g\varphi(c)$. Since $\alpha_{gc}: C^{(g)} \to J$ is birational, there is a rational map $\widetilde{\varphi}_c: J \to X$ such that

$$(t_{-\varphi(c)}\varphi)^{(g)} = \widetilde{\varphi}_c \alpha_{gc} \tag{1}$$

on an open dense set of $C^{(g)}$. Again by Theorem 4.9.4 the map $\widetilde{\varphi}_c$ is a morphism, so (1) holds everywhere. Now $\widetilde{\varphi}_c(0) = (t_{-\varphi(c)}\varphi)^{(g)}(gc) = 0$ implies that $\widetilde{\varphi}_c$ is a homomorphism (see Proposition 1.2.1). Moreover the diagram commutes, since $\alpha_c(p) = \alpha_{gc}(p + (g-1)c)$ and $\varphi(p) - \varphi(c) = (t_{-\varphi(c)}\varphi)^{(g)}(p + (g-1)c)$ for all $p \in C$. The uniqueness of $\widetilde{\varphi}_c$ follows from the fact that $\alpha_c(C)$ generates J as a group.

It remains to show that $\widetilde{\varphi}_c = \widetilde{\varphi}_{c'}$ for all $c, c' \in C$. But

$$(\widetilde{\varphi}_c - \widetilde{\varphi}_{c'})\alpha_c(p) = \widetilde{\varphi}_c \alpha_c(p) - \widetilde{\varphi}_{c'}(\alpha_{c'}(p) - \alpha_{c'}(c))$$
$$= t_{-\varphi(c)}\varphi(p) - t_{-\varphi(c')}\varphi(p) + t_{-\varphi(c')}\varphi(c) = 0$$

for all $p \in C$. Since $\alpha_c(C)$ generates J, this implies the assertion. $\square$

The dual of the homomorphism $\widetilde{\varphi}$ is

(4.2) Corollary. $\widehat{\widehat{\varphi}} = -\phi_\Theta \varphi^*$

Proof. Under the identifications $\widehat{X} = \operatorname{Pic}^0(X)$ and $\widehat{J} = \operatorname{Pic}^0(J)$ we have $\widehat{\widehat{\varphi}} = \widetilde{\varphi}^*$. Moreover $t^*_{-\varphi(c)}\colon \operatorname{Pic}^0(X) \to \operatorname{Pic}^0(X)$ is the identity according to Lemma 2.4.7 a). So using Proposition 3.5 and the Universal Property of the Jacobian we obtain

$$\varphi^* = (t_{-\varphi(c)}\varphi)^* = (\widetilde{\varphi}\alpha_c)^* = \alpha_c^* \widehat{\widehat{\varphi}} = -\phi_\Theta^{-1} \widehat{\widehat{\varphi}} \,. \qquad \square$$

Let $f\colon C \to C'$ be a finite morphism of smooth projective curves. Denote by J' the Jacobian of C' and consider the composed morphism $\alpha_{f(c)} f\colon C \to J'$. According to the Universal Property 4.1 there is a unique homomorphism N_f fitting into the following commutative diagram

$$\begin{array}{ccc} C & \xrightarrow{f} & C' \\ {\scriptstyle \alpha_c}\downarrow & & \downarrow{\scriptstyle \alpha_{f(c)}} \\ J & \xrightarrow[N_f]{} & J' \end{array} .$$

By definition N_f is just the map $\mathcal{O}_C(\sum r_i p_i) \mapsto \mathcal{O}_{C'}(\sum r_i f(p_i))$, classically called the *norm map* of f. Denote by Θ' a theta divisor on J'. Dualizing the equation $\alpha_{f(c)} f = N_f \alpha_c$ and applying Corollary 4.2 gives

$$\widehat{N}_f \phi_{\Theta'} = \phi_\Theta f^* \,. \tag{2}$$

So the investigation of N_f is equivalent to the investigation of f^*.

(4.3) Proposition. *The homomorphism $f^*\colon J' \to J$ is not injective if and only if f factorizes via a cyclic étale covering f' of degree $n \geq 2$:*

$$\begin{array}{ccc} C & \xrightarrow{f} & C' \\ & {\scriptstyle f''}\searrow \quad \nearrow{\scriptstyle f'} & \\ & C'' & \end{array} .$$

Proof. Suppose first that f factorizes via a cyclic étale covering f' of degree $n \geq 2$. It suffices to show that the homomorphism $f'^*\colon J' \to J'' = J(C'')$ is not injective. To see this, recall that f' is given as follows: there exists a line bundle L on C' of order n in $\operatorname{Pic}^0(C')$ such that C'' is the inverse image of the unit section of $L^n = C' \times \mathbb{C}$ under the n-th power map $L \to L^n$ and $f'\colon C'' \to C'$ is the restriction of $L \to L^n$ to C''. Denote by $p\colon L \to C'$ the natural projection. Since the tautological line bundle p^*L is trivial, so is $f'^*L = p^*L|C''$, and thus f'^* is not injective.

Conversely, suppose f^* is not injective. Choose a nontrivial line bundle $L \in \ker f^* \subset \operatorname{Pic}^0(C')$. Necessarily L is of finite order, say $n \geq 2$, since

$$L^{\deg f} = N_f f^* L = N_f \mathcal{O}_C = \mathcal{O}_{C'} \ .$$

Then the cyclic étale covering $f': C'' \to C'$ associated to L is of degree n. Consider the pullback diagram

$$\begin{array}{ccc} C \times_{C'} C'' & \xrightarrow{q} & C'' \\ {\scriptstyle p}\downarrow & & \downarrow{\scriptstyle f'} \\ C & \xrightarrow[f]{} & C' \ . \end{array}$$

The étale covering p is given by the trivial line bundle $f^*L = \mathcal{O}_C$. Hence $C \times_{C'} C''$ is the disjoint union of n copies of C. In particular there exists a section $s: C \to C \times_{C'} C''$ and f factorizes as $f = f'qs$. □

From the proof of Proposition 4.3 one easily deduces that for the cyclic étale covering $f': C'' \to C'$ the kernel $\ker\{f^*: J' \to J''\}$ is generated by the line bundle L defining f'. If $(f'')^*: J'' \to J$ is not injective, one can apply the proposition again and factorize f''. Repeating this process we obtain

(4.4) Corollary. *For any finite morphism $f: C \to C'$ of smooth projective curves C and C' there is a factorization*

$$\begin{array}{ccc} C & \xrightarrow{f} & C' \\ & {\scriptstyle g}\searrow \quad \nearrow{\scriptstyle f_e} & \\ & C_e & \end{array}$$

with f_e étale, $\ker f^* = \ker f_e^*$, and $g^*: J(C_e) \to J$ injective.

The *difference map* $\delta: C \times C \to J$ is defined by $\delta(x, y) = \alpha_c(x) - \alpha_c(y)$. Considering J as $\operatorname{Pic}^0(C)$ as usual, we have $\delta(x, y) = \mathcal{O}_C(x - y)$. In particular δ is independent of the choice of c and vanishes on the diagonal Δ of $C \times C$.

(4.5) Proposition. *Let X be an abelian variety and $\varphi: C \times C \to X$ a rational map with $\varphi(\Delta) = 0$. Then there exists a unique homomorphism $\widetilde{\varphi}: J \to X$ such that the following diagram is commutative*

$$\begin{array}{ccc} C \times C & \xrightarrow{\varphi} & X \\ & {\scriptstyle \delta}\searrow \quad \nearrow{\scriptstyle \widetilde{\varphi}} & \\ & J & . \end{array}$$

Proof. According to Theorem 4.9.4 the map φ is everywhere defined. Since $\varphi(c, c) = 0$, Corollary 4.9.3 provides unique morphisms $\varphi_i: C \to X$, $i = 1, 2$

with $\varphi_i(c) = 0$ and $\varphi(x,y) = \varphi_1(x) + \varphi_2(y)$ for all $x, y \in C$. Now $\varphi(\Delta) = 0$ implies $\varphi_2 = -\varphi_1$. By the Universal Property of the Jacobian there is a unique homomorphism $\widetilde{\varphi}: J \to X$ such that $\varphi_1 = \widetilde{\varphi}\alpha_c$ and we have for all $x, y \in C$

$$\widetilde{\varphi}\delta(x,y) = \widetilde{\varphi}\alpha_c(x) - \widetilde{\varphi}\alpha_c(y) = \varphi_1(x) + \varphi_2(y) = \varphi(x,y) . \qquad \square$$

(4.6) Remark. The Universal Property of the Jacobian as well as Proposition 4.5 mean that the Jacobian J is the *Albanese variety* of the curve C (see Ueno [1] p. 101 and Lang [1] p. 45. See also Exercise 2.).

Finally we show that every abelian variety is a quotient of a Jacobian. This has been very important in the development of the theory of abelian varieties (see Weil [1]).

(4.7) Proposition. *For any abelian variety X there is a smooth projective curve C whose Jacobian $J(C)$ admits a surjective homomorphism*

$$J(C) \to X .$$

Proof. An essential ingredient of the proof is the following fact:

Suppose $Y \subseteq \mathbb{P}_N$ is a smooth irreducible projective variety of dimension ≥ 2 and $f: Z \to Y$ is a finite morphism of a smooth irreducible variety Z onto Y. Then $f^{-1}(Y \cap H)$ is connected for every hyperplane H in $\mathbb{P}_N$.

For the proof note that $f^{-1}(Y \cap H)$ is the support of an ample divisor, since f is finite. So the statement follows from Hartshorne [1], III 7.9.

Without loss of generality we may assume that $g = \dim X \geq 2$. Choose a projective embedding $X \hookrightarrow \mathbb{P}_N$. According to Bertini's Theorem (see Hartshorne [1] II 8.18) there is a linear subspace $\mathbb{P}_{N-g-1}$ of $\mathbb{P}_N$ such that $C = X \cap \mathbb{P}_{N-g-1}$ is a smooth irreducible curve. Translating, if necessary, we may assume $0 \in C$. By the Universal Property of the Jacobian the embedding $C \hookrightarrow X$ factorizes via a homomorphism $\widetilde{\varphi}: J(C) \to X$. Assume that $\widetilde{\varphi}$ is not surjective and denote by X_1 the abelian subvariety $\operatorname{im} \widetilde{\varphi}$. Moreover denote by X_2 the complementary abelian subvariety of X_1 in X as defined in Section 5.3. By Corollary 5.3.6 the addition map $\mu: X_1 \times X_2 \to X$ is an isogeny. Since $C \subset X_1$ and $X_1 \cap X_2$ is finite, so is $C \cap X_2$. Let $f: X_1 \times X_2 \to X$ be the composition of $1_{X_1} \times 2_{X_2}: X_1 \times X_2 \to X_1 \times X_2$ with μ. Then $f^{-1}(C)$ is not connected. But this contradicts the above fact applied $g+1$ times. $\square$

§ 5 Correspondences

In the theory of algebraic curves a correspondence between two curves C_1 and C_2 is defined to be a divisor D on the product $C_1 \times C_2$. We will see that to any such correspondence one can associate a homomorphism between the corresponding Jacobians in a canonical way. This homomorphism depends only on the line bundle $L = \mathcal{O}_{C_1 \times C_2}(D)$. So for our purposes it is more convenient to define: a *correspondence* between two smooth projective curves C_1 and C_2 is a line bundle L on $C_1 \times C_2$.

Let J_1 and J_2 denote the Jacobians of C_1 and C_2 respectively. For any correspondence L between C_1 and C_2 and any point $p \in C_1$ define

$$L(p) = L|\{p\} \times C_2 ,$$

considered as a line bundle on C_2. Define the homomorphism $\gamma_L : J_1 \to J_2$ by

$$\gamma_L \mathcal{O}_{C_1}(\sum_{i=1}^{n} r_i p_i) = L(p_1)^{r_1} \otimes \cdots \otimes L(p_n)^{r_n} .$$

Note that γ_L is well defined, since it is the homomorphism $J_1 \to J_2$ induced by the morphism $C_1 \to J_2$, $p \mapsto L(p) \otimes L(c)^{-1}$ according to the Universal Property 4.1.

Two correspondences L and L' between C_1 and C_2 are called *equivalent*, if there are line bundles L_i on C_i, $i = 1, 2$, such that

$$L' = L \otimes q_1^* L_1 \otimes q_2^* L_2$$

where q_1 and q_2 denote the canonical projections of $C_1 \times C_2$. This defines an equivalence relation on the set of all correpondences between C_1 and C_2. Denote by $\operatorname{Corr}(C_1, C_2)$ the $\mathbb{Z}$-module of equivalence classes of correspondences between C_1 and C_2.

(5.1) Theorem. *The assignment $L \mapsto \gamma_L$ induces a canonical isomorphism of abelian groups*

$$\operatorname{Corr}(C_1, C_2) \to \operatorname{Hom}(J_1, J_2) .$$

Proof. First note that $\gamma_L = \gamma_{L'}$ for equivalent correspondences L and L', since $(q_1^* L_1)(p) = \mathcal{O}_{C_2}$ and $(q_2^* L_2)(p) = L_2$ for all $p \in C_1$. So the map $\operatorname{Corr}(C_1, C_2) \to \operatorname{Hom}(J_1, J_2)$ is well defined. It is obviously a homomorphism of abelian groups. In order to show that it is bijective, fix points $c_i \in C_i$ and denote by $\alpha_i : C_i \to J_i$ the corresponding embeddings. For $\gamma \in \operatorname{Hom}(J_1, J_2)$ define a correspondence L_γ by

$$L_\gamma = (\gamma\alpha_1 \times \alpha_2)^* \mu^* \mathcal{O}_{J_2}(\Theta_2)^{-1} \qquad (1)$$

where $\mu: J_2 \times J_2 \to J_2$ is the addition map and Θ_2 is some theta divisor on J_2.

First we claim $\gamma_{L_\gamma} = \gamma$ for every $\gamma \in \operatorname{Hom}(J_1, J_2)$. Since the curve $\alpha_1(C_1)$ generates J_1, this follows from the fact that for all $p \in C_1$

$$\begin{aligned}\gamma_{L_\gamma}\alpha_1(p) &= \gamma_{L_\gamma}\mathcal{O}_{J_1}(p - c_1) = L_\gamma(p) \otimes L_\gamma(c_1)^{-1} \\ &= \alpha_2^* t_{\gamma\alpha_1(p)}^* \mathcal{O}(\Theta_2)^{-1} \otimes \alpha_2^* \mathcal{O}(\Theta_2) = -\alpha_2^* \phi_{\Theta_2} \gamma\alpha_1(p) = \gamma\alpha_1(p) \,,\end{aligned}$$

where we used Proposition 3.5. So $\operatorname{Corr}(C_1, C_2) \to \operatorname{Hom}(J_1, J_2)$ is surjective. In order to show that it is injective, note that $\gamma_L = 0$ means $\gamma_L\alpha_1(p) = L(p) \otimes L(c_1)^{-1} = \mathcal{O}_{C_2}$ for all $p \in C_1$. By the Seesaw Theorem A.8 this implies $L = q_1^* L_1 \otimes q_2^* L(c_1)$ for some line bundle L_1 on C_1. □

Suppose now $C_1 = C_2 = C$ is a curve of genus g with Jacobian variety J. The ring structure of $\operatorname{Hom}(J, J) = \operatorname{End}(J)$ induces a ring structure on $\operatorname{Corr}(C, C)$, which is easy to work out (see Exercise 14). For every $\gamma \in \operatorname{End}(J)$ we can use the associated correspondence L_γ (see (1)) to compute the trace of the rational representation $\operatorname{Tr}_r(\gamma)$. For this define the *bidegree* (d_1, d_2) of a correspondence L on $C \times C$ by

$$d_1 = d_1(L) = \deg L|C \times \{p\} \quad \text{and} \quad d_2 = d_2(L) = \deg L|\{p\} \times C \,.$$

Certainly this definition is independent of the point $p \in C$. Denoting by Δ the diagonal on $C \times C$ we have

(5.2) Proposition. $\operatorname{Tr}_r(\gamma) = d_1(L_\gamma) + d_2(L_\gamma) - (\Delta \cdot L_\gamma)$ *for every* $\gamma \in \operatorname{End}(J)$.

Since the right hand side of the formula is constant on the equivalence classes of correspondences, the proposition implies slightly more generally

$$\operatorname{Tr}_r(\gamma_L) = d_1(L) + d_2(L) - (\Delta \cdot L)$$

for all line bundles $L \in \operatorname{Pic}(C \times C)$.

Proof. Let $\alpha: C \to J$ denote the embedding of C with base point $c \in C$. Since $L_\gamma|C \times \{p\} = \alpha^*\gamma^*\mathcal{O}_J(t_{\alpha(p)}^*\Theta)^{-1}$ and $L_\gamma|\{p\} \times C = \alpha^*\mathcal{O}_J(t_{\gamma\alpha(p)}^*\Theta)^{-1}$,

$$d_1 = -(\alpha(C) \cdot \gamma^*(\mathcal{O}_J(\Theta)) \quad \text{and} \quad d_2 = -(\alpha(C) \cdot \mathcal{O}_J(\Theta)) \,.$$

Moreover, if $\Delta_C: C \to C \times C$ denotes the diagonal map,

$$\begin{aligned}(\Delta \cdot L_\gamma) = \deg \Delta_C^*(\gamma\alpha \times \alpha)^*\mu^*\mathcal{O}_J(\Theta)^{-1} &= -\deg \alpha^*(\gamma + 1_J)^*\mathcal{O}_J(\Theta) \\ &= -(\alpha(C) \cdot (\gamma + 1_J)^*\mathcal{O}_J(\Theta)) \,.\end{aligned}$$

Using this, Proposition 5.1.5 and Poincaré's Formula 2.1 imply

$$\begin{aligned}\operatorname{Tr}_r(\gamma) &= \tfrac{g}{(\Theta^g)}(\Theta^{g-1}\cdot((\gamma+1_J)^*\mathcal{O}_J(\Theta)\otimes\gamma^*\mathcal{O}_J(\Theta)^{-1}\otimes\mathcal{O}_J(\Theta)^{-1}))\\ &= \tfrac{1}{(g-1)!}(\Theta^{g-1}\cdot((\gamma+1_J)^*\mathcal{O}_J(\Theta)\otimes\gamma^*\mathcal{O}_J(\Theta)^{-1}\otimes\mathcal{O}_J(\Theta)^{-1}))\\ &= (\alpha(C)\cdot(\gamma+1_J)^*\mathcal{O}_J(\Theta)) - (\alpha(C)\cdot\gamma^*\mathcal{O}_J(\Theta)) - (\alpha(C)\cdot\mathcal{O}_J(\Theta))\\ &= -(\Delta\cdot L_\gamma)+d_1+d_2\,. \end{aligned}$$ □

Finally we express the Rosati-involution $\gamma \mapsto \gamma'$ on $\operatorname{End}(J)$ with respect to the canonical polarization Θ on J in terms of correspondences on $C\times C$. For this denote by $\tau: C\times C \to C\times C$ the canonical involution $(p,q)\mapsto(q,p)$.

(5.3) Proposition. $\gamma_{\tau^*L} = \gamma'_L$ *for every correspondence* L *on* $C\times C$.

Proof. By Theorem 5.1 we may assume $L = (\gamma_L\alpha\times\alpha)^*\mu^*\mathcal{O}_J(\Theta)^{-1}$. Thus $\tau^*L = (\alpha\times\gamma_L\alpha)^*\mu^*\mathcal{O}_J(\Theta)^{-1}$, and using Proposition 3.5

$$\begin{aligned}\gamma_{\tau^*L}\alpha(p) &= (\tau^*L)(p)\otimes(\tau^*L)(c)^{-1} = \alpha^*\gamma_L^*\mathcal{O}_J(t^*_{\alpha(p)}\Theta)^{-1}\otimes\alpha^*\gamma_L^*\mathcal{O}_J(\Theta)\\ &= -\alpha^*\widehat{\gamma}_L\phi_\Theta\alpha(p) = \phi_\Theta^{-1}\widehat{\gamma}_L\phi_\Theta\alpha(p) = \gamma'_L\alpha(p)\end{aligned}$$

for every $p\in C$. This implies the assertion, since $\alpha(C)$ generates the group J. □

§ 6 Endomorphisms Associated to Curves and Divisors

In this section we prove the criterion of Matsusaka [2] for numerical equivalence of 1-cycles respectively algebraic equivalence of divisors in terms of the endomorphisms associated to cycles introduced in Section 5.4. The criterion is valid for an arbitrary abelian variety, but the proof uses the theory of Jacobians.

Let X be an abelian variety of dimension g. Suppose D is a divisor and Γ an algebraic 1-cycle on X. In Section 5.4 we associated to the pair (Γ, D) an endomorphism $\delta(\Gamma, D)$ of X. Since by Corollary 5.4.5 the endomorphism $\delta(\Gamma,D)$ depends only on the divisor D up to linear equivalence, it makes sense to write $\delta(\Gamma, L) := \delta(\Gamma, D)$ for $L = \mathcal{O}_X(D)$.

Our first aim is to deduce a different expression for $\delta(\Gamma, L)$. Since $\delta(\Gamma, L)$ is additive in the first argument, we may assume that Γ is a reduced irreducible curve in X. Let $\varphi: C\to\Gamma = \varphi(C)$ be its normalization. Moreover we may assume that $\varphi(c) = 0$ for some $c\in C$. Let J denote the Jacobian of C and $\alpha = \alpha_c: C\to J$ the embedding with base point c. According to the Universal Property of the Jacobian 4.1 the morphism φ extends to a homomorphism $\widetilde{\varphi}$ fitting into the following commutative diagram

$$\begin{array}{ccc} & J & \\ {}^{\alpha}\nearrow & & \searrow^{\widetilde{\varphi}} \\ C & \xrightarrow[\varphi]{} & X \end{array}$$

To simplify notation, we identify J with its dual $\widehat{J}$ via the canonical isomorphism ϕ_Θ. Then we have

(6.1) Proposition. $\delta(\varphi(C), L) = -\widetilde{\varphi}\widehat{\widetilde{\varphi}}\phi_L$.

Proof. For all x out of an open dense set of X and a suitably chosen divisor D with $L = \mathcal{O}_X(D)$, we have, using Corollary 4.2 and the definition of the map S in Section 5.4,

$$\begin{aligned} -\widetilde{\varphi}\widehat{\widetilde{\varphi}}\phi_L(x) &= \widetilde{\varphi}\varphi^*(t_x^*L \otimes L^{-1}) = \widetilde{\varphi}\mathcal{O}_C(\varphi^*(t_x^*D - D)) \\ &= S(\varphi(C) \cdot (t_x^*D - D)) = \delta(\varphi(C), D)(x) . \end{aligned}$$ □

With the preceding notation:

(6.2) Proposition. *a)* $\operatorname{Tr}_r(\delta(\varphi(C), L)) = -2(\varphi(C) \cdot L)$
b) $\operatorname{Tr}_r(\delta(L, \varphi(C))) = -(2g-2)(\varphi(C) \cdot L)$

Proof. Note first that $\operatorname{Tr}_r(\delta(\varphi(C), L)) = -\operatorname{Tr}_r(\widetilde{\varphi}\widehat{\widetilde{\varphi}}\phi_L) = -\operatorname{Tr}_r(\widehat{\widetilde{\varphi}}\phi_L\widetilde{\varphi})$ and $-\widehat{\widetilde{\varphi}}\phi_L\widetilde{\varphi}$ is an endomorphism of J, since we identified $J = \widehat{J}$. The idea is to take a correspondence of $C \times C$ associated to $-\widehat{\widetilde{\varphi}}\phi_L\widetilde{\varphi}$ and use Proposition 5.2 in order to compute the trace. Define $M = (\varphi \times \varphi)^*\mu^*L$ (where as usual $\mu\colon X \times X \to X$ denotes the addition map). Then $\gamma_M = -\widehat{\widetilde{\varphi}}\phi_L\widetilde{\varphi}$, since $\gamma_M\alpha(p) = M(p) \otimes M(c)^{-1} = \varphi^*t_{\varphi(p)}^*L \otimes \varphi^*L^{-1} = \varphi^*\phi_L\varphi(p) = -\widehat{\widetilde{\varphi}}\phi_L\widetilde{\varphi}\alpha(p)$ for all $p \in C$ by Corollary 4.2.

The bidegree of M and its intersection number with the diagonal are given by

$$d_1(M) = d_2(M) = \deg M|\{p\} \times C = \deg \varphi^*L = (\varphi(C) \cdot L)$$

and
$$(\Delta \cdot M) = \deg \Delta_C^*M = \deg(2\varphi)^*L = 4(\varphi(C) \cdot L) .$$

So Proposition 5.2 implies

$$\operatorname{Tr}_r(\delta(\varphi(C), L)) = \operatorname{Tr}_r(\gamma_M) = d_1(M) + d_2(M) - (\Delta \cdot M) = -2(\varphi(C) \cdot L) .$$

Finally b) follows from a) and $\delta(\varphi(C), L) + \delta(L, \varphi(C)) = -(\varphi(C) \cdot L)1_X$ (see Lemma 5.4.4). □

(6.3) Corollary. $\delta(\varphi(C), L) = 0$ *if and only if* $\delta(L, \varphi(C)) = 0$.

Recall that an algebraic 1-cycle Γ on X is called *numerically equivalent to zero* if $(\Gamma \cdot L) = 0$ for every line bundle L on X. The following criterion is due to Matsusaka [2].

(6.4) Theorem. *a) Suppose L is a nondegenerate line bundle and Γ an algebraic 1-cycle on X. If $\delta(\Gamma, L) = 0$, then Γ is numerically equivalent to zero.*

b) Suppose $\Gamma \subset X$ is a curve generating X as a group and L a line bundle on X. Then $\delta(\Gamma, L) = 0$ if and only if L is algebraically equivalent to zero.

Note that in a) also the converse implication is valid (see Matsusaka [2])). We only proved that $\delta(\Gamma, L) = 0$, if Γ is algebraically equivalent to zero (see Proposition 5.4.3).

Proof. a) Suppose $\Gamma = \sum r_i \Gamma_i$ with irreducible reduced curves Γ_i. Let J_i denote the Jacobian of the normalization $\varphi_i: C_i \to \Gamma_i \subset X$ and define $\widetilde{\varphi}_i: J_i \to X$ as in (1). Then $0 = \delta(\Gamma, L) = \sum r_i \delta(\Gamma_i, L) = -\sum r_i \widetilde{\varphi}_i \widehat{\widetilde{\varphi}}_i \phi_L$ by Proposition 6.1. Since ϕ_L is an isogeny, this implies $\sum r_i \widetilde{\varphi}_i \widehat{\widetilde{\varphi}}_i = 0$.

Let L' be an arbitrary line bundle on X. We have to show $(\Gamma \cdot L') = 0$. But Proposition 6.1 and (1) imply $\delta(\Gamma, L') = -\sum r_i \widetilde{\varphi}_i \widehat{\widetilde{\varphi}}_i \phi_{L'} = 0$. So from Proposition 6.2 we get $(\Gamma \cdot L') = \sum r_i (\Gamma_i \cdot L') = -\sum r_i \frac{1}{2} \operatorname{Tr}_r(\delta(\Gamma_i, L')) = -\frac{1}{2} \operatorname{Tr}_r(\delta(\Gamma, L')) = 0$.

b) By assumption $\widetilde{\varphi}: J \to X$ is surjective and thus $\widehat{\widetilde{\varphi}}: \widehat{X} \to J$ has finite kernel. It follows that $M = \widehat{\widetilde{\varphi}}^* \mathcal{O}_J(\Theta)$ is an ample line bundle on $\widehat{X}$ and $\widetilde{\varphi}\widehat{\widetilde{\varphi}} = \phi_M: \widehat{X} \to X$ (see Corollary 2.4.6) is an isogeny. So $\delta(\Gamma, L) = -\widetilde{\varphi}\widehat{\widetilde{\varphi}}\phi_L = 0$ if and only if $\phi_L = 0$, i.e. $L \in \operatorname{Pic}^0(X)$. □

§ 7 Examples of Jacobians

Given a smooth projective curve C of genus g, it is difficult in general to compute a period matrix for its Jacobian $J(C) = H^0(\omega_C)^* / H_1(C, \mathbb{Z})$. However, if C admits a sufficiently large group of automorphisms, there is a method, due to Bolza [1], for doing this. We want to explain it, work out an example, and state Bolza's result.

Suppose $\varphi: C \to C$ is an automorphism. The Universal Property of the Jacobian 4.1 provides a unique automorphism $\widetilde{\varphi}$ of $J(C)$ such that for any $c \in C$ the following diagram commutes

$$\begin{array}{ccc} C & \xrightarrow{\varphi} & C \\ {\scriptstyle \alpha_c}\downarrow & & \downarrow{\scriptstyle \alpha_{\varphi(c)}} \\ J(C) & \xrightarrow{\widetilde{\varphi}} & J(C) \end{array} .$$

Choose a symplectic basis $\lambda_1, \dots, \lambda_g, \mu_1, \dots, \mu_g$ of the lattice $H_1(C, \mathbb{Z})$ (see the remark after Proposition 1.2) and a basis $\omega_1, \dots, \omega_g$ of $H^0(\omega_C)$ such that the corresponding period matrix is of the form $\Pi = (Z, \mathbb{1})$ for some $Z \in \mathfrak{H}_g$. Let ${}^tM = {}^t\begin{pmatrix} \alpha & \beta \\ \gamma & \delta \end{pmatrix} \in G_{\mathbb{1}} = \mathrm{Sp}_{2g}(\mathbb{Z})$ and $A \in M_g(\mathbb{C})$ denote the rational and analytic representation of $\widetilde{\varphi}$ with respect to these bases. Then, according to equation 8.1(6) and the fact that $\widetilde{\varphi}$ is an automorphism, we have $A(Z, \mathbb{1}) = (Z, \mathbb{1})\,{}^tM$ or equivalently

$$Z = (\alpha Z + \beta)(\gamma Z + \delta)^{-1} \quad \text{and} \quad A = {}^t(\gamma Z + \delta) \,. \tag{1}$$

Suppose now C admits a covering $C \to \mathbb{P}_1$ such that φ descends to an automorphism of $\mathbb{P}_1$. Realizing the covering as a concrete Riemann surface over $\mathbb{P}_1$ with the help of a system of canonical dissections, one can determine the action of φ on the fundamental group $\pi_1(J(C)) = H_1(C, \mathbb{Z})$, i.e. compute the matrix ${}^tM$ of the rational representation of $\widetilde{\varphi}$.

Suppose moreover that C is defined by an equation in $\mathbb{P}_2$ and we are given an explicit basis of $H^0(\omega_C)$ in terms of this equation, such that the analytic representation $\tilde{A}$ of $\widetilde{\varphi}$ with respect to this basis can be computed. Since A and $\tilde{A}$ are equivalent matrices, this gives us the eigenvalues of A and in particular its determinant. All this information gives restrictions on the matrix Z.

In the case that there is only one curve C with a given automorphism group, this procedure may be sufficient to determine a period matrix of $J(C)$, as the following example shows:

Let C be the smooth curve defined inhomogeneously by the equation

$$y^2 = x^6 - 1 \,.$$

C is a hyperelliptic curve of genus 2 admitting the automorphism

$$\varphi : \begin{cases} x & \mapsto x' = \zeta x \\ y & \mapsto y' = -y \end{cases}$$

with $\zeta = e(\frac{2\pi i}{6})$. The hyperelliptic involution $y \mapsto -y$ induces the double covering $C \to \mathbb{P}_1$, $(x, y) \mapsto x$, ramified in the 6th roots of unity. Consider the following picture

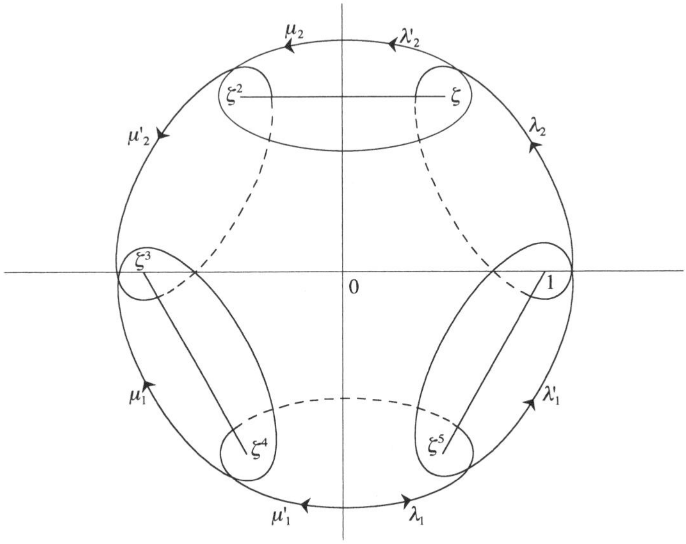

Here C is to be thought of consisting of two copies of the x-plane $\mathbb{P}_1$, glued together in the usual way via the three straight line dissections joining the ramification points $1, \zeta, \dots, \zeta^5$. Moreover, $\lambda_i, \mu_i, \lambda'_i, \mu'_i$, $i = 1, 2$, indicate 1-cycles on C with dotted segments lying on the lower sheet and full segments lying on the upper sheet. Obviously $\lambda_1, \lambda_2, \mu_1, \mu_2$ is a symplectic basis of $H_1(C, \mathbb{Z})$. The automorphism φ is induced by rotation in the x-plane by $\zeta = e(\frac{2\pi i}{6})$ with center 0, but exchanges the two sheets over the point 0. Hence $\lambda'_1, \lambda'_2, \mu'_1, \mu'_2$ are the images of the cycles $\lambda_1, \lambda_2, \mu_1, \mu_2$ under φ.

In order to compute ${}^tM$, we have to express the cycles $\lambda'_1, \lambda'_2, \mu'_1$ and μ'_2 in terms of the symplectic basis: obviously $\lambda'_2 = \mu_2$ and $\mu'_1 = -\lambda_1$. On the other hand, comparing their intersection numbers with the basis elements $\lambda_1, \lambda_2, \mu_1, \mu_2$ yields $\lambda'_1 = \mu_1 - \mu_2$ and $\mu'_2 = -\lambda_1 - \lambda_2$.

Hence
$$ {}^tM = \rho_r(\widetilde{\varphi}) = \begin{pmatrix} 0 & \begin{matrix} -1 & -1 \\ 0 & -1 \end{matrix} \\ \begin{matrix} 1 & 0 \\ -1 & 1 \end{matrix} & 0 \end{pmatrix} $$

and equations (1) read

$$ Z = \begin{pmatrix} 1 & -1 \\ 0 & 1 \end{pmatrix} \left(\begin{pmatrix} -1 & 0 \\ -1 & -1 \end{pmatrix} Z \right)^{-1} \quad \text{and} \quad A = {}^t\!\left(\begin{pmatrix} -1 & 0 \\ -1 & -1 \end{pmatrix} Z \right). \tag{2} $$

In order to determine the determinant of A, consider the basis $\frac{dx}{y}, \frac{x\,dx}{y}$ of $H^0(\omega_C)$ (for the fact that this is a basis of $H^0(\omega_C)$ see Shafarevich [1] III § 5.5). Recall that $J(C) = H^0(\omega_C)^*/H_1(C, \mathbb{Z})$. Thus the analytic representation of the induced endomorphism $\widetilde{\varphi}$ of $J(C)$ is given by the dual map of $\varphi^*: H^0(\omega_C) \to H^0(\omega_C)$. Since $\varphi^*(\frac{dx}{y}) = -\zeta \frac{dx}{y}$ and $\varphi^*(\frac{x\,dx}{y}) = -\zeta^2 \frac{dx}{y}$, it follows that $\tilde{A} = \operatorname{diag}(-\zeta, -\zeta^2)$ is the matrix of the analytic representation with respect to this basis. We get

$$\det A = \det \rho_a(\widetilde{\varphi}) = \det \tilde{A} = -1\ . \tag{3}$$

Now (2) implies $\det Z = -1$ and writing $Z = \begin{pmatrix} z_1 & z_2 \\ z_2 & z_3 \end{pmatrix}$ we obtain the following equations $z_1 = z_3 = -2z_2$ and $z_1 z_3 - z_2^2 = -1$. Since $\operatorname{Im} Z$ is positive definite, this gives $z_1 = z_3 = -2z_2 = \frac{2i}{\sqrt{3}}$ and we finally obtain the following period matrix for $J(C)$

$$\Pi = \begin{pmatrix} \frac{2i}{\sqrt{3}} & -\frac{i}{\sqrt{3}} & 1 & 0 \\ -\frac{i}{\sqrt{3}} & \frac{2i}{\sqrt{3}} & 0 & 1 \end{pmatrix}.$$

If one starts with the basis $\mu_1, \mu_2, \lambda_1, \lambda_2$ instead of the symplectic basis $\lambda_1, \lambda_2, \mu_1, \mu_2$, one ends up with the following period matrix for $J(C)$

$$\widetilde{\Pi} = \begin{pmatrix} \frac{2i}{\sqrt{3}} & \frac{i}{\sqrt{3}} & 1 & 0 \\ \frac{i}{\sqrt{3}} & \frac{2i}{\sqrt{3}} & 0 & 1 \end{pmatrix}.$$

This proves case IV of the following result of Bolza [1]. The proofs of the other cases are analogous.

type	equation	$\overline{\operatorname{Aut}}\,C$	$\widetilde{\Pi} = (Z, \mathbb{1})$
I	$y^2 = (x^2 - a^2)(x^2 - b^2)(x^2 - 1)$	$\mathbb{Z}/2\mathbb{Z}$	$Z = \begin{pmatrix} z & \frac{1}{2} \\ \frac{1}{2} & z' \end{pmatrix}$
II	$y^2 = x(x^2 - a^2)(x^2 - a^{-2})$	D_2	$Z = \begin{pmatrix} z & \frac{1}{2} \\ \frac{1}{2} & z \end{pmatrix}$
III	$y^2 = (x^3 - a^3)(x^3 - a^{-3})$	D_3	$Z = \begin{pmatrix} 2z & z \\ z & 2z \end{pmatrix}$
IV	$y^2 = x^6 - 1$	D_6	$Z = \begin{pmatrix} \frac{2i}{\sqrt{3}} & \frac{i}{\sqrt{3}} \\ \frac{i}{\sqrt{3}} & \frac{2i}{\sqrt{3}} \end{pmatrix}$
V	$y^2 = x(x^4 - 1)$	$\mathfrak{S}_4$	$Z = \begin{pmatrix} \frac{-1+i\sqrt{2}}{2} & \frac{1}{2} \\ \frac{1}{2} & \frac{-1+i\sqrt{2}}{2} \end{pmatrix}$
VI	$y^2 = x(x^5 - 1)$	$\mathbb{Z}/5\mathbb{Z}$	$Z = \begin{pmatrix} 1-\epsilon^4 & -\epsilon^2-\epsilon^4 \\ -\epsilon^2-\epsilon^4 & \epsilon \end{pmatrix}$, with $\epsilon = e(\frac{2\pi i}{5})$

Here $\overline{\operatorname{Aut}}\,C$ denotes the *reduced automorphism group* of C, that is $\operatorname{Aut} C$ modulo the hyperelliptic involution. Moreover D_n denotes the dihedral group of order $2n$. It is easy to give the equations as well as the period matrices in such a form that all types except VI are a specialization of type I. But here we follow Bolza. Note that the list is complete in the sense that every curve C of genus 2 with nontrivial reduced automorphism group appears.

Note moreover that the moduli space of curves of type I is of dimension two. Correspondingly the space of period matrices is of dimension two in this case. It is not known, which period matrix corresponds to a particular curve of this type. A similar remark is valid also for types II and III.

This method can be applied also to many hyperelliptic curves of higher genus as well as many nonhyperelliptic curves (see Exercises 15 to 18).

§ 8 The Criterion of Matsusaka-Ran

In this section we prove a criterion for a polarized abelian variety to be a Jacobian. We give here a modified version of Collino's proof (see Collino [1]).

Recall that a curve C on an abelian variety X is said to generate X, if X is the smallest abelian variety containing C. More generally, an effective algebraic 1-cycle $\sum r_\nu C_\nu$ on X, $r_\nu > 0$ for all ν, *generates* X, if the union of the curves C_ν generates X.

(8.1) Criterion of Matsusaka-Ran. *Suppose (X, L) is a polarized abelian variety of dimension g and $C = \sum_{\nu=1}^{n} r_\nu C_\nu$ is an effective 1-cycle generating X with $(C \cdot L) = g$. Then $r_1 = \cdots = r_n = 1$, the curves C_ν are smooth, and (X, L) is isomorphic to the product of the canonically polarized Jacobians of the C_ν's:*

$$(X, L) \simeq (J(C_1), \Theta_1) \times \cdots \times (J(C_n), \Theta_n) .$$

In particular, if C is an irreducible curve generating X with $(C \cdot L) = g$, then C is smooth and (X, L) is the Jacobian of C. Matsusaka's version of the criterion is a special case of this. For another proof see Remark 12.2.5.

(8.2) Corollary. *a) A principally polarized abelian surface is either the Jacobian of a smooth curve of genus 2 or the canonically polarized product of two elliptic curves.*

b) A principally polarized abelian threefold is either the Jacobian of a smooth curve of genus 3 or the principally polarized product of an abelian surface with an elliptic curve respectively three elliptic curves.

Proof. a) is an immediate consequence of the criterion. b) The moduli spaces of curves of genus 3 and of principally polarized abelian threefolds are both irreducible algebraic varieties of dimension 6. Hence by Torelli's Theorem a general principally polarized abelian threefold is a Jacobian. Since under specialization effective 1-cycles go to effective 1-cycles and the intersection numbers are preserved, Criterion 8.1 implies the assertion. □

The argument of the last sentence of the proof of Corollary 8.2 yields more generally

(8.3) Corollary. *Any specialization of a Jacobian is a product of Jacobians.*

Proof of 8.1. Step I: First we outline the general set up: for $\nu = 1, \dots, n$ denote by $\widetilde{C}_\nu$ the normalization of C_ν. According to the Universal Property of the Jacobian 4.1 the composed map $\iota_\nu : \widetilde{C}_\nu \to C_\nu \hookrightarrow X$ factorizes via a unique homomorphism $\psi_\nu : J_\nu = J(\widetilde{C}_\nu) \to X$ (of course we may assume that every C_ν passes through the origin in X). Writing $g_\nu = (C_\nu \cdot L)$, we have

$$g = \sum_{\nu=1}^{n} r_\nu g_\nu \ .$$

Fix a divisor $D \in |L|$. By Proposition 4.1.7 we may assume that D is reduced. By eventually passing to an algebraically equivalent line bundle, we may assume that its pullback $\iota_\nu^* D$ is a divisor and thus a divisor of degree g_ν on $\widetilde{C}_\nu$ for all ν. Recall the Abel-Jacobi map $\alpha_\nu := \alpha_{\iota_\nu^* D} : \widetilde{C}_\nu^{(g_\nu)} \to J_\nu$ defined by $\alpha_\nu(\sum x_i) = \mathcal{O}_{\widetilde{C}_\nu}(\sum x_i - \iota_\nu^* D)$.

Define rational maps $h_\nu : X \dashrightarrow \widetilde{C}_\nu^{(g_\nu)}$ by $h_\nu(x) = \iota_\nu^* t_x^* D$. Then we have the following diagram

$$\begin{array}{ccccc}
X & \xrightarrow{h=(h_1,\dots,h_n)} & \widetilde{C}_1^{(g_1)} \times \cdots \times \widetilde{C}_n^{(g_n)} & & \widetilde{C}_1 \cup \cdots \cup \widetilde{C}_n \\
{\scriptstyle \phi_L}\downarrow & & \downarrow{\scriptstyle \alpha=\alpha_1\times\cdots\times\alpha_n} & & \downarrow{\scriptstyle \iota=\iota_1+\cdots+\iota_n} \\
\operatorname{Pic}^0(X) & \xrightarrow{\iota^*=(\iota_1^*,\dots,\iota_n^*)} & J_1 \times \cdots \times J_n & \xrightarrow{\psi=\psi_1+\cdots+\psi_n} & X
\end{array}$$

The diagram is commutative, since for a general $x \in X$ we have

$$\iota_\nu^* \phi_L(x) = \iota_\nu^* \mathcal{O}_X(t_x^* D - D) = \mathcal{O}_{\widetilde{C}_\nu}(\iota_\nu^* t_x^* D - \iota_\nu^* D) = \alpha_\nu h_\nu(x) \ .$$

Step II: We claim that $r_\nu = 1$ and $\widetilde{C}_\nu$ is a curve of genus g_ν for $\nu = 1, \dots, n$, and that $\psi = \psi_1 + \cdots + \psi_n$ is an isogeny.

Identifying $J_\nu = \widehat{J}_\nu$ as usual, the map ι^* is dual to $-\psi$ by Exercises 2.11, 2.12, and Corollary 4.2. Since ψ is surjective by assumption, the homomorphism ι^* and thus also the composed map $\iota^* \phi_L$ has finite kernel. According to the commutativity of the diagram this gives

$$\dim X = \dim \overline{(\operatorname{im} \alpha h)} \leq \dim \overline{(\operatorname{im} h)} \leq \sum_{\nu=1}^{n} g_\nu \leq \sum_{\nu=1}^{n} r_\nu g_\nu = g = \dim X \ .$$

It follows that $r_1 = \cdots = r_n = 1$. Moreover, since X and $\widetilde{C}_1^{(g_1)} \times \cdots \times \widetilde{C}_n^{(g_n)}$ are both of the same dimension, the map h is dominant. Hence $\iota^* \phi_L(X) = \operatorname{im} \alpha$.

Since $\operatorname{im}\alpha$ generates the abelian variety $J_1 \times \cdots \times J_n$, this implies that α is surjective and $g_\nu = \dim J_\nu$ for $\nu = 1, \ldots, n$. So $\widetilde{C}_\nu$ is of genus g_ν and ψ is an isogeny.

Step III: We claim that ψ is an isomorphism and the line bundles $\psi^* L$ and $p_1^* \psi_1^* L \otimes \cdots \otimes p_n^* \psi_n^* L$ define the same principal polarization on $J_1 \times \cdots \times J_n$.

It suffices to show that $\psi^* L$ and $p_1^* \psi_1^* L \otimes \cdots \otimes p_n^* \psi_n^* L$ are algebraically equivalent line bundles inducing a principal polarization, since then $1 = h^0(\psi^* L) = \deg \psi \cdot h^0(L)$ implies that ψ is an isomorphism.

First we show that $\psi_i^* L$ defines a principal polarization on J_i for $i = 1, \ldots, n$. Assume $h^0(\psi_i^* L) \geq 2$. For any $y \in J_i$ consider the exact sequence

$$0 \longrightarrow H^0(t_y^* \psi_i^* L(-\widetilde{C}_i)) \longrightarrow H^0(t_y^* \psi_i^* L) \longrightarrow H^0(t_y^* \psi_i^* L | \widetilde{C}_i) \; .$$

For a general $y \in J_i$ we have $h^0(t_y^* \psi_i^* L(-\widetilde{C}_i)) = 0$, since not every linear system $|t_y^* L|$ contains $\widetilde{C}_i$, and thus $h^0(t_y^* \psi_i^* L | \widetilde{C}_i) \geq 2$. Together with $\deg t_y^* \psi_i^* L | \widetilde{C}_i = \deg L | C_i = (L \cdot C_i) = g_i$ and Riemann-Roch for curves, this implies that for a general $y \in J_i$ the divisor $t_y^* \psi_i^* D | \widetilde{C}_i$ is special (recall from Step I: $D \in |L|$). On the other hand, the divisors $t_y^* \psi_i^* D | \widetilde{C}_i$ form just the image of the rational map $h_i \colon X \dashrightarrow \widetilde{C}_i^{(g_i)}$. Since $\operatorname{im} h_i$ is dense in $\widetilde{C}_i^{(g_i)}$, this implies that a general divisor of degree g_i on $\widetilde{C}_i$ is special, a contradiction. Hence $h^0(\psi_i^* L) = 1$ and thus $\psi_i^* L$ defines a principal polarization on J_i.

Next we claim that $\widehat{\psi}_i \phi_L \psi_j = 0$ for all $i \neq j$. Note first that $\psi_i \colon J_i \to X$ is an embedding, since $\phi_{\psi_i^* L} = \widehat{\psi}_i \phi_L \psi_i$ is an isomorphism. Let N_{J_i} denote the corresponding norm-endomorphism (see Section 5.3), i.e. $N_{J_i} = \psi_i \phi_{\psi_i^* L}^{-1} \widehat{\psi}_i \phi_L$. Since $\psi = \psi_1 + \cdots + \psi_n$ is an isogeny, the subvarieties J_i and J_j of X intersect in finitely many points. This implies

$$0 = N_{J_i} N_{J_j} = (\psi_i \phi_{\psi_i^* L}^{-1}) \widehat{\psi}_i \phi_L \psi_j (\phi_{\psi_j^* L}^{-1} \widehat{\psi}_j \phi_L) \; .$$

But $\psi_i \phi_{\psi_i^* L}^{-1}$ is injective and $\operatorname{im}(\phi_{\psi_j^* L}^{-1} \widehat{\psi}_j \phi_L) = J_j$, which implies the claim.

Using this we obtain

$$\begin{aligned} \phi_{\psi^* L} &= \widehat{\psi} \phi_L \psi \\ &= (\widehat{\psi}_1, \ldots, \widehat{\psi}_n) \phi_L (\psi_1 + \cdots + \psi_n) \\ &= (\widehat{\psi}_1 \phi_L \psi_1) \times \cdots \times (\widehat{\psi}_n \phi_L \psi_n) \\ &= \phi_{p_1^* \psi_1^* L \otimes \cdots \otimes p_n^* \psi_n^* L} \; . \end{aligned}$$

Now Proposition 2.5.3 shows that $\psi^* L$ and $p_1^* \psi_1^* L \otimes \cdots \otimes p_n^* \psi_n^* L$ are algebraically equivalent line bundles on X, completing the proof of Step III.

Since ψ is an isomorphism, the curves $C_\nu = \widetilde{C}_\nu$ are smooth. Moreover it follows from Step III that $\psi_\nu^* L$ defines a principal polarization on J_ν. It remains to show that this is the canonical polarization of J_ν. So we are reduced to the irreducible case and set $C = C_1 = \widetilde{C}_1$, $J = J_1$ and $\psi = \psi_1$.

Step IV: By what we have seen above, $\alpha = \alpha_1$ is birational and $\iota^* = \iota_1^*$ and ϕ_L are isomorphisms. So $h^{-1} = \phi_L^{-1} \iota^{*-1} \alpha: C^{(g)} \to J$ is a birational morphism. It is defined as follows: for a general $(x_1 + \cdots + x_g) \in C^{(g)}$ there is a unique $x \in X$ such that $h(x) = t_x^* D|C = x_1 + \cdots + x_g$. Then $h^{-1}(x_1 + \cdots + x_g) = x$. Define a map $\beta: C^{(g-1)} \times C \to J$ by

$$\beta(x_1 + \cdots + x_{g-1}, x_g) = h^{-1}(x_1 + \cdots + x_g) + x_g$$

(the last + means addition in J, this makes sense, since $C \subset J$).

We claim $D = \operatorname{im} \beta$. In particular, D is an irreducible divisor. For the proof suppose $h^{-1}(x_1 + \cdots + x_g) = x$. By definition $x_g \in t_x^* D$ or equivalently $x + x_g \in D$, so $\operatorname{im} \beta \subseteq D$. Certainly $\operatorname{im} \beta$ is an irreducible divisor in J.

Assume there is a divisor $D' \neq 0$ such that $D = \operatorname{im} \beta + D'$. By eventually passing to a translate of C, we may assume that C intersects D' and $\operatorname{im} \beta$ properly, but not their intersection (note that D is a reduced divisor). So we may write $D|C = y_1 + \cdots + y_g$ with $y_g \in D'$. But then

$$\beta(y_1 + \cdots + y_{g-1}, y_g) = h^{-1}(y_1 + \cdots + y_g) + y_g = y_g \in \operatorname{im} \beta ,$$

a contradiction.

Step V: Now we are in a position to show: *$\psi^* L$ defines the canonical polarization on J.*

Recall that the canonical polarization on J is defined by the divisor $\Theta = \{x_1 + \cdots + x_{g-1} \mid x_\nu \in C \subset J\}$ in J. So it suffices to show that $D = \operatorname{im} \beta$ is a translate of the divisor $(-1)_J^* \Theta$ in J.

According to Corollary 4.9.3 the morphism β is of the form

$$\beta(x_1 + \cdots + x_{g-1}, x_g) = \gamma(x_1 + \cdots + x_{g-1}) + \delta(x_g)$$

with morphisms $\gamma: C^{(g-1)} \to J$ and $\delta: C \to J$. Since for every $x \in C$ the dimension of $\beta(C^{(g-1)} \times \{x\})$ is $g-1$, we have $\operatorname{im} \gamma + \delta(x) = \beta(C^{(g-1)} \times \{x\}) = \operatorname{im} \beta = D$, so δ is constant and we may assume $\delta \equiv 0$. This means

$$\beta(x_1 + \cdots + x_{g-1}, x_g) = \gamma(x_1 + \cdots + x_{g-1})$$

for all $x_1, \ldots, x_g \in C$. Fix $y_1, \ldots, y_{g-1} \in C$. For all $x_1, \ldots, x_g \in C$ we have

$$\begin{aligned} \beta(x_1 + \cdots + x_{g-1}, x_g) &= \beta(x_1 + \cdots + x_{g-1}, y_1) \\ &= h^{-1}(x_1 + \cdots + x_{g-1} + y_1) + y_1 \\ &= h^{-1}(x_2 + \cdots + x_{g-1} + y_1 + x_1) + y_1 \\ &= \gamma(x_2 + \cdots + x_{g-1} + y_1) + y_1 - x_1 . \end{aligned}$$

Repeating this process, we finally obtain

$$\beta(x_1+\cdots+x_{g-1},x_g)=\gamma(y_1+\cdots+y_{g-1})+y_1+\cdots+y_{g-1}-x_1-\cdots-x_{g-1}\,.$$

This implies the assertion and completes the proof of the theorem. □

§ 9 Trisecants of the Kummer Variety

Let C be a smooth algebraic curve of genus $g \geq 2$ and (J,Θ) its Jacobian. We may assume that the line bundle $\mathcal{O}_J(\Theta)$ is of characteristic zero with respect to some symplectic basis of $H_1(C,\mathbb{Z})$. According to Theorem 4.8.1 the Kummer variety K of J is the image of the map $\phi_{2\Theta}: J \to \mathbb{P}_{2^g-1}$. The aim of this section is to show that K admits a 4-dimensional family of trisecants in $\mathbb{P}_{2^g-1}$. First we show that the intersection of Θ with suitable translates of Θ is reducible. This will imply the existence of trisecants of K.

For simplicity we denote by $p-q$ the image of (p,q) under the difference map $\delta: C\times C \to J$. If C is hyperelliptic, we denote by ι the hyperelliptic involution.

(9.1) Proposition. *a)* $\Theta \cap t^*_{p-q}\Theta \subset t^*_{p-r}\Theta \cup t^*_{s-q}\Theta$ *for all* $p \neq q,r,s \in C$.
b) The intersection $\Theta\cap t^*_{p-q}\Theta$ *is reducible, except* C *is hyperelliptic and* $q=\iota(p)$. *In the remaining case* $\Theta \cap t^*_{p-\iota(p)}\Theta$ *is irreducible.*

Proof. a) According to Riemann's Theorem 2.4 there is a theta characteristic $\kappa \in \operatorname{Pic}^{g-1}(C)$ with $\alpha^*_\kappa\Theta = W_{g-1}$. In terms of W_{g-1} the assertion reads

$$W_{g-1}\cap(W_{g-1}+(q-p)) \subset (W_{g-1}+(r-p))\cup(W_{g-1}+(q-s))\,. \quad (1)$$

Consider the sets $V_q := \{L \in \operatorname{Pic}^{g-1}(C) \mid h^0(L(-q)) \geq 1\}$ and $V'_p := \{L \in \operatorname{Pic}^{g-1}(C) \mid h^0(L(p)) \geq 2\}$. It suffices to show that

$$W_{g-1}\cap(W_{g-1}+(q-p)) \subset V_q \cup V'_p\,,$$

since $V_q \subset (W_{g-1}+(q-s))$ for any $s\in C$ and $V'_p \subset (W_{g-1}+(r-p))$ for any $r \in C$. Suppose L is contained in $W_{g-1}\cap(W_{g-1}+(q-p))$, but not in V'_p. We have to show $L\in V_q$. The assumption implies $h^0(L(p)) = h^0(L) = h^0(L(p-q)) = 1$. This means that p and q are base points of the linear system $|L(p)|$. Hence q is a base point of $|L|$, i.e. $h^0(L(-q))>0$.

b) Suppose first that C is non-hyperelliptic or that $q\neq\iota(p)$, if C is hyperelliptic. It suffices to show that $W_{g-1}\cap(W_{g-1}+(q-p))$ is contained neither in V_q nor in V'_p. For this we apply the Geometric Riemann-Roch for the canonical map $\varphi=\varphi_{\omega_C}: C\to\mathbb{P}_{g-1}$ (see (1) in the proof of Proposition 2.9).

Choose points $p_1, \dots, p_{g-2} \in C$ such that the g points $\varphi(p_1), \dots, \varphi(p_{g-2})$, $\varphi(p), \varphi(q)$ span $\mathbb{P}_{g-1}$. Then $L = \mathcal{O}_C(p_1 + \dots + p_{g-2} + q)$ satisfies $1 = h^0(L(p)) = h^0(L) = h^0(L(p-q))$. So $L \in W_{g-1} \cap (W_{g-1} + (q-p))$ but $L \notin V'_p$.

On the other hand, let H be a general hyperplane in $\mathbb{P}_{g-1}$ containing $\varphi(p)$ but not $\varphi(q)$. Choose points $p_1, \dots, p_{g-1} \in C$, different from p such that $\varphi(p_1), \dots, \varphi(p_{g-1})$ span H. Then $L = \mathcal{O}_C(p_1 + \dots + p_{g-1})$ satisfies $h^0(L) = 1$, $h^0(L(p)) = 2$ and thus $h^0(L(-q)) = 0$ and $h^0(L(p-q)) = 1$. So $L \in W_{g-1} \cap (W_{g-1} + (q-p))$ but $L \notin V_q$.

Finally suppose C is hyperelliptic and $q = \iota(p)$. It suffices to show that $W_{g-1} \cap (W_{g-1} + (q - \iota(q))) = V_q$. This implies that the intersection is irreducible, since V_q is the image of the irreducible variety $C^{(g-2)}$ under the morphism $C^{(g-2)} \to W_{g-1}$, $p_1 + \dots + p_{g-2} \mapsto \mathcal{O}(p_1 + \dots + p_{g-2} + q)$.

We will only show $W_{g-1} \cap (W_{g-1} + (q - \iota(q))) \subset V_q$, the converse implication being obvious. So suppose $L \in \operatorname{Pic}^{g-1}(C)$ is a line bundle with $h^0(L) > 0$ and $h^0(L(\iota(q) - q)) > 0$. We have to show $h^0(L(-q)) > 0$. This is clear, if $h^0(L) \geq 2$. On the other hand, $h^0(L) = 1$ if and only if we have $L = \mathcal{O}_C(p_1 + \dots + p_{g-1})$ with uniquely determined points p_i with $p_i \neq \iota(p_j)$ for $i \neq j$. The assumption $h^0(L(\iota(q) - q)) > 0$ implies that $p_i = q$ for some i. This is obvious if $h^0(L(\iota(q))) = 1$, but also valid if $h^0(L(\iota(q))) = 2$. So $h^0(L(-q)) > 0$. □

Let $\pi: \mathbb{C}^g \to J$ be the canonical projection and ϑ the canonical theta function on $\mathbb{C}^g$ of Theorem 3.2.7 generating $H^0(\mathcal{O}_J(\Theta))$. As usual we write $\overline{v} = \pi(v)$ for any $v \in \mathbb{C}^g$.

(9.2) Proposition. *For any $\overline{x}, \overline{y}, \overline{z} \in J - \{0\}$ with $\overline{x} \neq \overline{y} + \overline{z}, \overline{y}, \overline{z}$ the following conditions are equivalent*

i) $$\Theta \cap t^*_{\overline{x}}\Theta \subset t^*_{\overline{y}}\Theta \cup t^*_{\overline{z}}\Theta \; ,$$

ii) $$c_1\vartheta(v)\vartheta(v+y+z) + c_2\vartheta(v+x)\vartheta(v-x+y+z) + c_3\vartheta(v+y)\vartheta(v+z) = 0$$

for some constants c_1, c_2, c_3 with $c_1c_2c_3 \neq 0$ and all $v \in \mathbb{C}^g$.

Note that Fay's trisecant identity, which will be proved in the next section, is a precise version of equation ii).

Proof. ii) $\Longrightarrow$ i): Suppose $\overline{v} \in \Theta \cap t^*_{\overline{x}}\Theta$, i.e. $\vartheta(v) = \vartheta(v + x) = 0$. Then ii) implies $\vartheta(v + y)\vartheta(v + z) = 0$, i.e. $\overline{v} \in t^*_{\overline{y}}\Theta \cup t^*_{\overline{z}}\Theta$.

i) $\Longrightarrow$ ii): By assumption the equation $\vartheta(v+y)\vartheta(v+z) = 0$ of $\pi^*(t^*_{\overline{y}}\Theta \cup t^*_{\overline{z}}\Theta)$ in $\mathbb{C}^g$ is contained in the ideal of $\pi^*(\Theta \cap t^*_{\overline{x}}\Theta)$ in $\mathbb{C}^g$. Since it is generated by the functions ϑ and $t^*_x\vartheta$, this means that there are holomorphic functions f and g on $\mathbb{C}^g$ such that

$$\vartheta(v+y)\vartheta(v+z) = f(v)\vartheta(v) + g(v)\vartheta(v+x)\ . \tag{2}$$

Restricting (2) to $\pi^*\Theta$ gives $\vartheta(v+y)\vartheta(v+z) = g(v)\vartheta(v+x)$ for all $v \in \pi^*\Theta$. This implies that g may be considered as an element of the vector space $H^0(\mathcal{O}_\Theta(t_{\overline{y}}^*\Theta + t_{\overline{z}}^*\Theta - t_{\overline{x}}^*\Theta)) = H^0(\mathcal{O}_\Theta(t_{\overline{y}+\overline{z}-\overline{x}}^*\Theta))$. For the last equation use the Theorem of the Square 2.3.3. The exact sequence

$$0 \longrightarrow \mathcal{O}_J(t_{\overline{y}+\overline{z}-\overline{x}}^*\Theta - \Theta) \longrightarrow \mathcal{O}_J(t_{\overline{y}+\overline{z}-\overline{x}}^*\Theta) \longrightarrow \mathcal{O}_\Theta(t_{\overline{y}+\overline{z}-\overline{x}}^*\Theta) \longrightarrow 0$$

induces an isomorphism $H^0(\mathcal{O}_J(t_{\overline{y}+\overline{z}-\overline{x}}^*\Theta)) \to H^0(\mathcal{O}_\Theta(t_{\overline{y}+\overline{z}-\overline{x}}^*\Theta))$, since the divisor $t_{\overline{y}+\overline{z}-\overline{x}}^*\Theta - \Theta$ is algebraically equivalent to zero, but not linearly equivalent to zero and thus its cohomology vanishes. It follows that there are constants c_2 and c such that $g(v) = c_2\vartheta(v-x+y+z) + c\vartheta(v)$ for all $v \in \mathbb{C}^g$. Inserting this into (2) gives

$$\vartheta(v+y)\vartheta(v+z) - c_2\vartheta(v+x)\vartheta(v-x+y+z) = (f(v) + c\vartheta(v+x))\vartheta(v)$$

for all $v \in \mathbb{C}^g$. Applying the Theorem of the Square 2.3.3, one checks that the left hand side is an element of $H^0(\mathcal{O}_J(t_{\overline{y}+\overline{z}}^*\Theta) \otimes \mathcal{O}_J(\Theta))$. This implies that $f + c\vartheta(\cdot + x) = c_1\vartheta(\cdot + y + z)$ for some constant c_1. Certainly $c_1c_2 \neq 0$, so this completes the proof. □

The following proposition gives a geometric interpretation of condition ii) in terms of the Kummer map $\varphi = \varphi_{2\Theta}: J \to \mathbb{P}_{2^g-1}$.

(9.3) Proposition. *Suppose the points $\overline{x}, \overline{y}, \overline{z} \in J - \{0\}$ with $\overline{x} \neq \overline{y} + \overline{z}, \overline{y}, \overline{z}$ satisfy the equivalent conditions of Proposition 9.2. Then the three points $\varphi(\overline{\tfrac{1}{2}y + \tfrac{1}{2}z})$, $\varphi(\overline{x - \tfrac{1}{2}y - \tfrac{1}{2}z})$ and $\varphi(\overline{\tfrac{1}{2}y - \tfrac{1}{2}z})$ are collinear.*

Note that the conditions $\overline{x}, \overline{y}, \overline{z} \neq 0$ and $\overline{x} \neq \overline{y} + \overline{z}, \overline{y}, \overline{z}$ are equivalent to the fact that the three points $\varphi(\overline{\tfrac{1}{2}y + \tfrac{1}{2}z})$, $\varphi(\overline{x - \tfrac{1}{2}y - \tfrac{1}{2}z})$ and $\varphi(\overline{\tfrac{1}{2}y - \tfrac{1}{2}z})$ are different from each other, i.e. the corresponding lines are trisecants of K.

Proof. By assumption the line bundle $\mathcal{O}_J(\Theta)$ is of characteristic zero with respect to the chosen symplectic basis of $H_1(C, \mathbb{Z})$. Let $K(\mathcal{O}(2\Theta)) = K_1 \oplus K_2$ be the decomposition induced by this basis and $\{\vartheta_w \mid w \in K_1\}$ the basis of canonical theta functions for $H^0(\mathcal{O}(2\Theta))$ of Theorem 3.2.7. Then the Kummer map is $\varphi(\overline{v}) = (\cdots : \vartheta_w(v) : \cdots)_{w \in K_1}$ for all $v \in \mathbb{C}^g$. So it suffices to show that there are constants $c_1, c_2, c_3 \in \mathbb{C}$ such that

$$c_1\vartheta_w(\tfrac{1}{2}y + \tfrac{1}{2}z) + c_2\vartheta_w(x - \tfrac{1}{2}y - \tfrac{1}{2}z) + c_3\vartheta_w(\tfrac{1}{2}y - \tfrac{1}{2}z) = 0 \tag{3}$$

for all $w \in K_1$.

To see this, apply the substitution $v \mapsto v - \tfrac{1}{2}y - \tfrac{1}{2}z$ to equation ii) in Proposition 9.2. We obtain

$$c_1\vartheta(v+\tfrac{1}{2}y+\tfrac{1}{2}z)\vartheta(v-\tfrac{1}{2}y-\tfrac{1}{2}z)+c_2\vartheta(v+x-\tfrac{1}{2}y-\tfrac{1}{2}z)\vartheta(v-x+\tfrac{1}{2}y+\tfrac{1}{2}z)+ \\ +c_3\vartheta(v+\tfrac{1}{2}y-\tfrac{1}{2}z)\vartheta(v-\tfrac{1}{2}y+\tfrac{1}{2}z)=0 \qquad (4)$$

Let $Q: H^0(\mathcal{O}(2\Theta)) \times H^0(\mathcal{O}(2\Theta)) \to \mathbb{C}$ be the symmetric bilinear form with matrix $\mathbb{1}$ with respect to the basis $\{\vartheta_w \mid w \in K_1\}$. The Addition Formula Exercise 7.1 implies $\vartheta(\cdot + a)\vartheta(\cdot - a) = \sum_{w\in K_1} \vartheta_w(a)\vartheta_w$ for all $a \in \mathbb{C}^g$. It follows that $Q(\vartheta(\cdot + a)\vartheta(\cdot - a), \vartheta_w) = \vartheta_w(a)$ for all $w \in K_1$. Hence applying $Q(\cdot, \vartheta_w)$ to equation (4) gives (3). □

According to Proposition 9.1 we have $\Theta \cap t^*_{p-q}\Theta \subset t^*_{p-r}\Theta \cup t^*_{s-q}\Theta$ for general points $p, q, r, s \in C$ and the assumptions of Proposition 9.2 are satisfied. So Proposition 9.3 implies that the Kummer variety of any curve C admits a 4-dimensional family of trisecants. Welters conjectured that this property characterizes Jacobians among all principally polarized abelian varieties. To be more precise he conjectured even that, if the Kummer variety of a principally polarized abelian variety X admits at least one trisecant, then X is a Jacobian.

§ 10 Fay's Trisecant Identity

Fay's trisecant identity is a precise version of equation ii) in Proposition 9.2, which provides trisecants of the associated Kummer variety. Its coefficients are expressed in terms of theta functions. The proof which we give here is due to Raina [1]. It uses only elementary methods of algebraic geometry. The idea is to interpret both sides of the identity as sections of a certain line bundle.

Let C be a smooth projective curve of genus $g \geq 2$ with Jacobian $J = \mathbb{C}^g/\Lambda$ and a theta divisor Θ of characteristic zero with respect to a symplectic basis of Λ. First we fix some notation. For any $c \in \mathbb{C}^g$ denote by

$$L_{\bar{c}} = t^*_{\bar{c}}\mathcal{O}_J(\Theta)$$

the line bundle of characteristic c in $\operatorname{Pic}^{\Theta}(J)$. Moreover recall the canonical theta function ϑ^c of Theorem 3.2.7 generating $H^0(L_{\bar{c}})$. Recall the difference map $\delta: C \times C \to J = \operatorname{Pic}^0(C)$, $\delta(x,y) = \mathcal{O}_C(x-y)$. Let $p: \widetilde{C} \to C$ denote the universal covering of C and choose a basis $\omega_1, \dots, \omega_g$ of $H^0(\omega_C)$. Obviously the map $\widetilde{\delta}: \widetilde{C} \times \widetilde{C} \to \mathbb{C}^g$, $\widetilde{\delta}(x,y) = \int_y^x \underline{\omega}$ with $\underline{\omega} = {}^t(p^*\omega_1, \dots, p^*\omega_g)$ is a lifting of δ.

Fix a theta characteristic $\lambda \in \operatorname{Pic}^{g-1}(C)$ with

$$h^0(C, \lambda) = 1 .$$

According to Exercise 12 such a theta characteristic exists. Choose a representative $c \in \mathbb{C}^g$ of the point

$$\bar{c} := \lambda \otimes \kappa^{-1} \in J = \operatorname{Pic}^0(C)$$

where κ is the Riemann constant for Θ. Finally let s denote a holomorphic section $\not\equiv 0$ of λ. Since $p^*\lambda$ is the trivial line bundle on $\widetilde{C}$ (any holomorphic line bundle on the unit disc is trivial), we may consider p^*s as a holomorphic function on $\widetilde{C}$. The *prime form* E on $\widetilde{C} \times \widetilde{C}$ is the meromorphic function on $\widetilde{C} \times \widetilde{C}$ defined by

$$E(x,y) = \frac{\vartheta^c(\int_y^x \omega)}{p^*s(x)p^*s(y)} .$$

A priori this definition depends on the choice of λ, its section s, the symplectic basis and the lifting of δ. We will see however that E depends only on the lifting.

(10.1) Fay's Trisecant Identity. *For all* $x_1, y_1, x_2, y_2 \in \widetilde{C}$ *and* $v \in \mathbb{C}^g$*:*

$$\vartheta^0\big(v + \int_{y_1}^{x_1} \omega + \int_{y_2}^{x_2} \omega\big)\vartheta^0(v)E(x_1,x_2)E(y_1,y_2) =$$
$$= \vartheta^0\big(v + \int_{y_2}^{x_1} \omega\big)\vartheta^0\big(v + \int_{y_1}^{x_2} \omega\big)E(x_1,y_1)E(x_2,y_2) -$$
$$- \vartheta^0\big(v + \int_{y_1}^{x_1} \omega\big)\vartheta^0\big(v + \int_{y_2}^{x_2} \omega\big)E(x_1,y_2)E(x_2,y_1) .$$

Setting $x = \int_{y_2}^{x_1} \omega$, $y = \int_{y_1}^{x_1} \omega$ and $z = \int_{y_2}^{x_2} \omega$ and $c_1 = E(x_1,x_2)E(y_1,y_2)$, $c_2 = -E(x_1,y_1)E(x_2,y_2)$ and $c_3 = E(x_1,y_2)E(x_2,y_1)$, Fay's Trisecant Identity coincides with equation ii) in Proposition 9.2. In particular, by Proposition 9.3, it implies the existence of trisecants of the corresponding Kummer variety. This explains the name Trisecant Identity.

For the proof of the Trisecant Identity we need some technical statements, which are collected in the following proposition. In order to state it, denote by $q_1, q_2: C^2 \to C$ and $p_i: C^4 \to C$, $i = 1, \dots, 4$ the natural projections. Moreover let Δ be the diagonal in C^2 and $\Delta_{ij} := (p_i, p_j)^*\Delta$ for $1 \le i < j \le 4$. In the notation we will not distinguish between Δ respectively Δ_{ij} and their pullbacks to the universal coverings. Finally define the map $\delta^{(2)}: C^4 \to J$ respectively its lifting $\tilde{\delta}^{(2)}: \widetilde{C}^4 \to \mathbb{C}^g$ by

$$\delta^{(2)}(x_1,y_1,x_2,y_2) = \delta(x_1,y_1) + \delta(x_2,y_2)$$

respectively

$$\tilde{\delta}^{(2)}(x_1,y_1,x_2,y_2) = \tilde{\delta}(x_1,y_1) + \tilde{\delta}(x_2,y_2) = \int_{y_1}^{x_1} \omega + \int_{y_2}^{x_2} \omega .$$

(10.2) Proposition. *For every* $\eta \in \mathrm{Pic}^{g-1}(C)$

a) $\delta^* L_{\eta\otimes\kappa^{-1}} = q_1^*(\omega_C \otimes \eta^{-1}) \otimes q_2^*(\eta) \otimes \mathcal{O}_{C^2}(\Delta)$,

b) $\delta^{(2)^*} L_{\eta\otimes\kappa^{-1}} = p_1^*(\omega_C \otimes \eta^{-1}) \otimes p_2^*(\eta) \otimes p_3^*(\omega_C \otimes \eta^{-1}) \otimes p_4^*(\eta) \otimes$
$\otimes \mathcal{O}_{C^4}(\Delta_{12} + \Delta_{14} + \Delta_{23} + \Delta_{34} - \Delta_{13} - \Delta_{24})$,

c) E *is a lifting of a holomorphic section of* $\mathcal{O}_{C^2}(\Delta)$ *to* $\widetilde{C} \times \widetilde{C}$, *alternating as a function of* $\widetilde{C} \times \widetilde{C}$,

d) $h^0(C^4, \delta^{(2)^*} L_{\eta\otimes\kappa^{-1}}) = 1$, *if* $h^0(C, \eta) = 0$.

Before we prove Proposition 10.2, let us deduce Fay's Trisecant Identity from it.

Proof of 10.1. The idea of the proof is to rewrite the Trisecant Identity in such a way that the variable v becomes a characteristic. Namely, by equation 3.2(3) the identity is equivalent to

$$\begin{aligned}\tilde{\delta}^{(2)*}\vartheta^v(x_1, y_1, x_2, y_2)\vartheta^v(0) &= \\ = \frac{1}{E(x_1,x_2)E(y_1,y_2)} \big(\tilde{\delta}^*\vartheta^v(x_1, y_2)\tilde{\delta}^*\vartheta^v(x_2, y_1)&E(x_1, y_1)E(x_2, y_2) - \\ \tilde{\delta}^*\vartheta^v(x_1, y_1)\tilde{\delta}^*\vartheta^v(x_2, y_2)&E(x_1, y_2)E(x_2, y_1)\big)\end{aligned} \tag{1}$$

for all $v \in \mathbb{C}^g$ and all $x_1, y_1, x_2, y_2 \in \widetilde{C}$. According to Proposition 10.2 c) both sides of the Trisecant Identity are holomorphic functions in v. Hence it suffices to show (1) for a general $v \in \mathbb{C}^g$. To be more precise, it suffices to show equation (1) for any $v \in \mathbb{C}^g - \pi^*\Theta$, i.e. those v with $h^0(C, \overline{v} \otimes \kappa) = 0$.

Note first that the left hand side of (1) is a (lifting of a) holomorphic section of $\delta^{(2)^*} L_{\overline{v}}$. We claim that also the right hand side is a holomorphic section of this line bundle. Writing $\overline{v} = \eta \otimes \kappa^{-1}$, Proposition 10.2 a), b) and c) imply that it is a meromorphic section of $\delta^{(2)^*} L_{\overline{v}}$. On the other hand, $E(x_1, x_2)E(y_1, y_2)$ is a lifting of a holomorphic section of $\mathcal{O}_{C^4}(\Delta_{13} + \Delta_{24})$ vanishing with multiplicity 1 on all points of $\Delta_{13} + \Delta_{24} - (\Delta_{13} \cap \Delta_{24})$. Moreover the second factor on the right hand side of (1) vanishes on all points of the form (x_1, y_1, x_1, y_2) and (x_1, y_1, x_2, y_1), i.e. on $\Delta_{13} + \Delta_{24}$. Since $\Delta_{13} \cap \Delta_{24}$ is of codimension 2, this implies that the right hand side is holomorphic, which completes the proof of the claim.

By assumption on $\overline{v}$ and Proposition 10.2 d) we have $h^0(\delta^{(2)^*} L_{\overline{v}}) = 1$. So both sides on (1) differ by a constant. But setting $x_2 = y_2$ and using the fact that E is alternating one easily sees that this constant is 1. □

It remains to prove Proposition 10.2, which will take up the rest of this section.

Proof of Proposition 10.2 a). $\delta^* L_{\eta\otimes\kappa^{-1}} = q_1^*(\omega_C \otimes \eta^{-1}) \otimes q_2^*(\eta) \otimes \mathcal{O}_{C^2}(\Delta)$ *for all* $\eta \in \mathrm{Pic}^{g-1}(C)$.

According to the Seesaw Principle A.9 it suffices to show that the restrictions to $C \times \{p\}$ and $\{p\} \times C$ of both sides of the equation coincide for all $p \in C$. But using Lemma 3.4 and the fact that Θ is symmetric

$$\begin{aligned}\delta^* L_{\eta\otimes\kappa^{-1}}|C \times \{p\} &= \alpha_p^* L_{\eta\otimes\kappa^{-1}} = \alpha_p^* \mathcal{O}_J(t^*_{\eta\otimes\kappa^{-1}}(-1)^*\Theta)\\ &= \omega_C \otimes \eta^{-1} \otimes \mathcal{O}_C(p) = q_1^*(\omega_C \otimes \eta^{-1}) \otimes q_2^*(\eta) \otimes \mathcal{O}_{C^2}(\Delta)|C \times \{p\}\end{aligned}$$

and similarly

$$\begin{aligned}\delta^* L_{\eta\otimes\kappa^{-1}}|\{p\} \times C &= \alpha_p^*(-1)^* L_{\eta\otimes\kappa^{-1}} = \alpha_p^* \mathcal{O}_J(t_{\eta^{-1}\otimes\kappa}(-1)^*\Theta)\\ &= \eta \otimes \mathcal{O}_C(p) = q_1^*(\omega_C \otimes \eta^{-1}) \otimes q_2^*(\eta) \otimes \mathcal{O}_{C^2}(\Delta)|\{p\} \times C\end{aligned}$$ □

Proof of Proposition 10.2 b). $\delta^{(2)^*} L_{\eta\otimes\kappa^{-1}} = p_1^*(\omega_C \otimes \eta^{-1}) \otimes p_2^*(\eta) \otimes p_3^*(\omega_C \otimes \eta^{-1}) \otimes p_4^*(\eta) \otimes \mathcal{O}_{C^4}(\Delta_{12} + \Delta_{14} + \Delta_{23} + \Delta_{34} - \Delta_{13} - \Delta_{24})$ *for all* $\eta \in \mathrm{Pic}^{g-1}(C)$.

We apply the Seesaw Principle A.9 to line bundles on $C^4 = C^2 \times C^2$. Using Proposition 10.2 a) we have for all $p, q \in C$.

$$\begin{aligned}\delta^{(2)^*} L_{\eta\otimes\kappa^{-1}}|C^2 \times \{(p,q)\} &= \delta^* t^*_{\mathcal{O}(p-q)} L_{\eta\otimes\kappa^{-1}} = \delta^* L_{\mathcal{O}(p-q)\otimes\eta\otimes\kappa^{-1}}\\ &= q_1^*\big(\omega_C \otimes \eta^{-1} \otimes \mathcal{O}(q-p)\big) \otimes q_2^*\big(\eta \otimes \mathcal{O}(p-q)\big) \otimes \mathcal{O}_{C^2}(\Delta)\,.\end{aligned}$$

But this is just the restriction of the right hand side to $C^2 \times \{(p,q)\}$. Similarly one checks that the restrictions of both sides to $\{(p,q)\} \times C^2$ coincide. □

Proof of Proposition 10.2 c). E *is a lifting of a holomorphic section of* $\mathcal{O}_{C^2}(\Delta)$ *to* $\widetilde{C} \times \widetilde{C}$, *alternating as a function on* $\widetilde{C} \times \widetilde{C}$.

By definition E is the pullback of a meromorphic section of the line bundle $\delta^* L_{\lambda\otimes\kappa^{-1}} \otimes q_1^*\lambda^{-1} \otimes q_2^*\lambda^{-1}$. Hence for the proof of the first assertion it suffices to show that the corresponding divisor on $C \times C$ is Δ.

By assumption $h^0(\lambda) = 1$, so $\lambda = \mathcal{O}_C(p_1 + \cdots + p_{g-1})$ with uniquely determined points $p_1, \ldots, p_{g-1} \in C$. Since $\delta^* L_{\lambda\otimes\kappa^{-1}} = \mathcal{O}_{C^2}(\delta^* t^*_{\lambda\otimes\kappa^{-1}}\Theta)$, it suffices to show

$$\delta^* t^*_{\lambda\otimes\kappa^{-1}}\Theta = (p_1 + \cdots + p_{g-1}) \times C + C \times (p_1 + \cdots + p_{g-1}) + \Delta\,.$$

By Proposition 10.2 a) the divisors on both sides of this equation are linearly equivalent, so we have only to show that a point $(p, q) \in C \times C$ is contained in the support of the left hand side if and only if it is contained in the support of the right hand side. But $(p, q) \in \delta^* t^*_{\lambda\otimes\kappa^{-1}}\Theta = \delta^* \alpha^*_{\lambda^{-1}} W_{g-1}$ if and only if $\lambda \otimes \mathcal{O}(p - q) \in W_{g-1}$ or equivalently $h^0(\lambda \otimes \mathcal{O}(p - q)) > 0$.

Note that $h^0(\lambda \otimes \mathcal{O}(p)) = 1$ or 2 by Riemann-Roch. If $h^0(\lambda \otimes \mathcal{O}(p)) = 1$, necessarily $|\lambda \otimes \mathcal{O}(p)| = p_1 + \cdots + p_{g-1} + p$. This implies $h^0(\lambda \otimes \mathcal{O}(p-q)) > 0$ if and only if $p = q$ or $p = p_\nu$ for some $1 \leq \nu \leq g-1$.

On the other hand, $h^0(\lambda \otimes \mathcal{O}(p)) = 2$ if and only if $h^0(\lambda \otimes \mathcal{O}(-p)) = h^1(\lambda \otimes \mathcal{O}(p)) = 1$ i.e. $p = p_\nu$ for some $1 \leq \nu \leq g-1$. This completes the proof of the first assertion.

Finally, E is alternating if and only if ϑ^c is an odd function. Clearly ϑ^c is either even or odd, since $\bar{c} = \lambda \otimes \kappa^{-1}$ is a 2-division point of J. But by Riemann's Singularity Theorem 2.5 we have $\operatorname{mult}_0(\vartheta^c) = \operatorname{mult}_{\lambda \otimes \kappa^{-1}}(\Theta) = \operatorname{mult}_\lambda(W_{g-1}) = h^0(\lambda) = 1$, so ϑ^c is odd. □

For the proof of Proposition 10.2 d) we need the following three Lemmas:

(10.3) Lemma. *a)* $\delta^* L_{\eta \otimes \kappa^{-1}} | \Delta \simeq \mathcal{O}_\Delta$ *for all* $\eta \in \operatorname{Pic}^{g-1}(C)$,

b) For $\eta \in \operatorname{Pic}^{g-1}(C)$ *with* $h^0(\eta) = 0$ *the natural restriction map* $\rho \colon H^0(C^2, \delta^* L_{\eta \otimes \kappa^{-1}}) \to H^0(\delta^* L_{\eta \otimes \kappa^{-1}} | \Delta)$ *is an isomorphism.*

Proof. a) follows from Proposition 10.2 a) and the fact that $\mathcal{O}_\Delta(\Delta) = \omega_\Delta^{-1}$ by the adjunction formula. b) Consider the exact sequence

$$0 \longrightarrow q_1^*(\omega_C \otimes \eta^{-1}) \otimes q_2^*(\eta) \longrightarrow \delta^* L_{\eta \otimes \kappa^{-1}} \longrightarrow \delta^* L_{\eta \otimes \kappa^{-1}} | \Delta \longrightarrow 0 .$$

The assertion follows from the fact that $h^i(q_1^*(\omega_C \otimes \eta^{-1}) \otimes q_2^*(\eta)) = 0$ for $i = 0, 1$ by Künneth's formula. □

(10.4) Lemma. *For any* $\eta \in \operatorname{Pic}^{g-1}(C)$ *with* $h^0(\eta) = 0$ *the restriction map yields an isomorphism*

$$H^0(C^4, \delta^{(2)^*} L_{\eta \otimes \kappa^{-1}}) \to H^0(\delta^{(2)^*} L_{\eta \otimes \kappa^{-1}} | \Delta_{12} + \Delta_{14} + \Delta_{23} + \Delta_{34}) .$$

Proof. Consider the exact sequence

$$\begin{aligned} 0 \to \delta^{(2)^*} L_{\eta \otimes \kappa^{-1}}(-\Delta_{12} - \Delta_{14} - \Delta_{23} - \Delta_{34}) \to \\ \to \delta^{(2)^*} L_{\eta \otimes \kappa^{-1}} \to \delta^{(2)^*} L_{\eta \otimes \kappa^{-1}} | \Delta_{12} + \Delta_{14} + \Delta_{23} + \Delta_{34} \to 0 . \end{aligned}$$

Denote for abbreviation $N_\eta = p_1^*(\omega_C \otimes \eta^{-1}) \otimes p_2^*(\eta) \otimes p_3^*(\omega_C \otimes \eta^{-1}) \otimes p_4^*(\eta)$. Then $\delta^{(2)^*} L_{\eta \otimes \kappa^{-1}}(-\Delta_{12} - \Delta_{14} - \Delta_{23} - \Delta_{34}) = N_\eta(-\Delta_{13} - \Delta_{24})$ by Proposition 10.2 b) and it suffices to show that $h^i(C^4, N_\eta(-\Delta_{13} - \Delta_{24})) = 0$ for $i = 0, 1$. For this consider the exact sequence

$$0 \longrightarrow N_\eta(-\Delta_{13} - \Delta_{24}) \longrightarrow N_\eta \longrightarrow N_\eta | \Delta_{13} + \Delta_{24} \longrightarrow 0 .$$

By Künneth's formula we have $h^i(C^4, N_\eta) = 0$ for $i = 0,1$ implying $h^0(N_\eta(-\Delta_{13} - \Delta_{24})) = 0$ and $H^1(N_\eta(-\Delta_{13} - \Delta_{24})) = H^0(N_\eta|\Delta_{13} + \Delta_{24})$. Again by Künneth's formula $h^0(N_\eta|\Delta_{13}) = 0 = h^0(N_\eta|\Delta_{24})$ which gives $h^0(N_\eta|\Delta_{13} + \Delta_{24}) = 0$. This implies the assertion. □

(10.5) Lemma. *Let Δ_{ij} and Δ_{kl} be any two distinct elements of the set $\{\Delta_{12}, \Delta_{14}, \Delta_{23}, \Delta_{34}\}$. For all $\eta \in \mathrm{Pic}^{g-1}(C)$ with $h^0(\eta) = 0$ the restriction maps in the commutative diagram*

$$\begin{array}{ccc} H^0(\delta^{(2)^*} L_{\eta\otimes\kappa^{-1}}|\Delta_{ij}) & \xrightarrow{r} & H^0(\delta^{(2)^*} L_{\eta\otimes\kappa^{-1}}|\Delta_{1234}) \\ & {\scriptstyle r_1}\searrow \quad \nearrow{\scriptstyle r_2} & \\ & H^0(\delta^{(2)^*} L_{\eta\otimes\kappa^{-1}}|\Delta_{ij} \cap \Delta_{kl}) & \end{array}$$

are isomorphisms. Here Δ_{1234} denotes the diagonal in C^4.

Proof. Without loss of generality we may assume $(i,j) = (1,2)$ and $(k,l) = (3,4)$ or $(2,3)$.

We first claim that r_1 is an isomorphism. The composition of the embedding $\Delta_{12} \hookrightarrow C^4$ with $\delta^{(2)}: C^4 \to \mathrm{Pic}^0(C)$ coincides with the composed map $\delta(p_3, p_4): \Delta_{12} \to \mathrm{Pic}^0(C)$. This gives $\delta^{(2)^*} L_{\eta\otimes\kappa^{-1}}|\Delta_{12} = (p_3, p_4)^* \delta^* L_{\eta\otimes\kappa^{-1}}$ and thus $H^0(\delta^{(2)^*} L_{\eta\otimes\kappa^{-1}}|\Delta_{12}) \simeq H^0(C^2, \delta^* L_{\eta\otimes\kappa^{-1}})$ by Künneth's formula. For $(k,l) = (3,4)$ we obtain the following commutative diagram:

$$\begin{array}{ccc} H^0(\delta^{(2)^*} L_{\eta\otimes\kappa^{-1}}|\Delta_{12}) & \xrightarrow{r_1} & H^0(\delta^{(2)^*} L_{\eta\otimes\kappa^{-1}}|\Delta_{12} \cap \Delta_{34}) \\ \wr| & & \wr| \\ H^0(C^2, \delta^* L_{\eta\otimes\kappa^{-1}}) & \xrightarrow{\rho} & H^0(\delta^* L_{\eta\otimes\kappa^{-1}}|\Delta) \end{array} \quad . \tag{2}$$

Since ρ is an isomorphism by Lemma 10.3 b), so is r_1. For $(k,l) = (2,3)$ we obtain the commutative diagram

$$\begin{array}{ccc} H^0(\delta^{(2)^*} L_{\eta\otimes\kappa^{-1}}|\Delta_{12}) & \xrightarrow{r_1} & H^0(\delta^{(2)^*} L_{\eta\otimes\kappa^{-1}}|\Delta_{12} \cap \Delta_{23}) \\ \wr| & & \wr| \\ H^0(C^2, \delta^* L_{\eta\otimes\kappa^{-1}}) & \xrightarrow{\mathrm{id}} & H^0(C^2, \delta^* L_{\eta\otimes\kappa^{-1}}) \end{array} \quad , \tag{3}$$

which completes the proof of the assertion for r_1.

It remains to show that r_2 is an isomorphism. This is clear for the pair $(k,l) = (3,4)$, since then r_2 corresponds to the identity map under the isomorphism $H^0(\delta^{(2)^*} L_{\eta\otimes\kappa^{-1}}|\Delta_{12} \cap \Delta_{34}) \simeq H^0(\delta^* L_{\eta\otimes\kappa^{-1}}|\Delta)$ of diagram (2). If $(k,l) = (2,3)$ the map r_2 corresponds under the isomorphism $H^0(\delta^{(2)^*} L_{\eta\otimes\kappa^{-1}}|\Delta_{12} \cap \Delta_{23}) \simeq H^0(C^2, \delta^* L_{\eta\otimes\kappa^{-1}})$ of (3) to the isomorphism ρ of Lemma 10.3 b). □

Now we are in the position to complete the proof of Proposition 10.2

Proof of Proposition 10.2 d). $H^0(C^4, \delta^{(2)^*} L_{\eta\otimes\kappa^{-1}}) = 1$ *for all* $\eta \in \mathrm{Pic}^{g-1}(C)$ *with* $h^0(C,\eta) = 0$.

According to Lemma 10.4 we have to show that $h^0(\delta^{(2)^*} L_{\eta\otimes\kappa^{-1}}|\Delta_{12} + \Delta_{14} + \Delta_{23} + \Delta_{34}) = 1$. For this it suffices to show that the natural map $H^0(\delta^{(2)^*} L_{\eta\otimes\kappa^{-1}}|\Delta_{12} + \Delta_{14} + \Delta_{23} + \Delta_{34}) \to H^0(\delta^{(2)^*} L_{\eta\otimes\kappa^{-1}}|\Delta_{1234})$ is an isomorphism, since $h^0(\delta^{(2)^*} L_{\eta\otimes\kappa^{-1}}|\Delta_{1234}) = 1$ by Lemma 10.3 a).

Suppose $\sigma \in H^0(\delta^{(2)^*} L_{\eta\otimes\kappa^{-1}}|\Delta_{12} + \Delta_{14} + \Delta_{23} + \Delta_{34})$ vanishes on Δ_{1234}. Since the map r of Lemma 10.5 is an isomorphism, $\sigma|\Delta_{ij}$ vanishes for each $\Delta_{ij} \in \{\Delta_{12}, \Delta_{14}, \Delta_{23}, \Delta_{34}\}$. Hence $\sigma \equiv 0$, and the map is injective. In particular $h^0(\delta^{(2)^*} L_{\eta\otimes\kappa^{-1}}|\Delta_{12} + \Delta_{14} + \Delta_{23} + \Delta_{34}) \leq 1$. For the surjectivity it suffices to show that $\delta^{(2)^*} L_{\eta\otimes\kappa^{-1}}|\Delta_{12} + \Delta_{14} + \Delta_{23} + \Delta_{34}$ admits a nonzero section. For this let s be a nonzero section of $\delta^{(2)^*} L_{\eta\otimes\kappa^{-1}}|\Delta_{1234}$ over Δ_{1234}. Let s_{ij} be the section of $\delta^{(2)^*} L_{\eta\otimes\kappa^{-1}}|\Delta_{ij}$ over Δ_{ij} which maps to s under the restriction map r of Lemma 10.5. Now $r = r_2 \circ r_1$ implies that s_{ij} and s_{kl} coincide on $\Delta_{ij} \cap \Delta_{kl}$. Hence the sections s_{ij} can be patched together to give a global nonzero section of $\delta^{(2)^*} L_{\eta\otimes\kappa^{-1}}|\Delta_{12} + \Delta_{14} + \Delta_{23} + \Delta_{34}$. □

Exercises

(1) *(The Dual Jacobian Variety)* Let C be a smooth projective curve of genus g.

a) Show that the composed map

$$H^1(C,\mathbb{Z}) \hookrightarrow H^1(C,\mathbb{C}) = H^0(\omega_C) \oplus \overline{H^0(\omega_C)} \to \overline{H^0(\omega_C)}$$

is injective. So $\overline{H^0(\omega_C)}/H^1(C,\mathbb{Z})$ is a complex torus.
Show that $\overline{H^0(\omega_C)}/H^1(C,\mathbb{Z}) = \widehat{J(C)}$, the dual Jacobian variety.

b) The canonical principal polarization on $\widehat{J(C)}$ is given by the hermitian form $\overline{H^0(\omega_C)} \times \overline{H^0(\omega_C)} \to \mathbb{C}$, $(\omega_1, \omega_2) \mapsto 2i \int_C \omega_1 \wedge \overline{\omega}_2$.

(2) *(Albanese Torus of a Compact Kähler Manifold)* Let X be a compact Kähler manifold.

a) Show that the map $H_1(X,\mathbb{Z})/\text{torsion} \to H^0(\Omega^1_X)^*$, $\gamma \mapsto \{\omega \mapsto \int_\gamma \omega\}$ is injective with image a lattice in $H^0(\Omega^1_X)^*$.

The complex torus $\mathrm{Alb}(X) := H^0(\Omega^1_X)^*/(H_1(X,\mathbb{Z})/\text{torsion})$ is called the *Albanese torus* of X. Fix a point $x_0 \in X$. The holomorphic map $\alpha_{x_0}\colon X \to \mathrm{Alb}(X)$, $x \mapsto \{\omega \mapsto \int_{x_0}^x \omega\}$ (mod $H_1(X,\mathbb{Z})/\text{torsion}$) is called the *Albanese map* of X.

b) The Albanese torus admits the following universal property:
Let $\varphi: X \to T$ be a holomorphic map into a complex torus T. There exists a unique homomorphism $\widetilde{\varphi}: \mathrm{Alb}(X) \to T$ of complex tori such that the following diagram is commutative

$$\begin{array}{ccc} X & \stackrel{\varphi}{\longrightarrow} & T \\ {\scriptstyle \alpha_{x_0}}\downarrow & & \downarrow{\scriptstyle t_{-\varphi(x_0)}} \\ \mathrm{Alb}(X) & \stackrel{\widetilde{\varphi}}{\longrightarrow} & T \end{array}$$

(3) *(The Picard Torus of a Compact Kähler Manifold)* Let X be a compact Kähler manifold.

a) Show that the composed map

$$H^1(X,\mathbb{Z})/\text{torsion} \to H^1(X,\mathbb{C}) = H^0(\Omega^1_X) \oplus \overline{H^0(\Omega^1_X)} \to \overline{H^0(\Omega^1_X)}$$

is injective with image a lattice in $\overline{H^0(\Omega^1_X)}$.

The complex torus $\mathrm{Pic}^0(X) := \overline{H^0(\Omega^1_X)}/(H^1(X,\mathbb{Z})/\text{torsion})$ is called the *Picard torus* of X.

b) There is a canonical isomorphism $\mathrm{Pic}^0(X) \simeq \ker(c_1: H^1(\mathcal{O}^*_X) \to H^2(X,\mathbb{Z}))$.
c) $\mathrm{Alb}(X)\widehat{\ } = \mathrm{Pic}^0(X)$.

(4) Let $X \subset \mathbb{P}_N$ be a smooth projective variety of dimension n. Denote by $\omega \in H^{1,1}(X) \cap H^2(X,\mathbb{Z})$ the first Chern class of the line bundle $\mathcal{O}_X(1)$. Show that the alternating form $E: H^1(X,\mathbb{Z})/\text{torsion} \times H^1(X,\mathbb{Z})/\text{torsion} \to \mathbb{Z}$, $E(\varphi_1,\varphi_2) = \int_X \left(\bigwedge^{n-1}\omega\right) \wedge \varphi_1 \wedge \varphi_2$ defines a polarization on $\mathrm{Pic}^0(X)$, called the *canonical polarization of* $\mathrm{Pic}^0(X)$. Hence by Exercise 3 c) the complex tori $\mathrm{Pic}^0(X)$ and $\mathrm{Alb}(X)$ are abelian varieties dual to each other.

(5) Let $X \subset \mathbb{P}_N$ be a smooth projective variety of dimension $n \geq 3$, and H a general hyperplane. Show that the embedding $X \cap H \hookrightarrow X$ induces an isomorphism of the canonically polarized Picard tori $\mathrm{Pic}^0(X) \to \mathrm{Pic}^0(X \cap H)$.

(6) Any polarized abelian variety is isomorphic to the canonically polarized Picard torus of a smooth projective surface. (Hint: use Exercise 5).

(7) Let C be a curve of genus 2 and $\alpha = \alpha_c: C \to J(C)$ the embedding with respect to the point $c \in C$.

a) Show that for any distinct points x and y in $J(C)$ there are exactly two translates of $\alpha(C)$ passing through x and y.
b) For any $x \in J$ and any tangent vector $t \neq 0$ of $J(C)$ at x, there are either one or two translates of $\alpha(C)$ passing through x and touching t. There are exactly 6 tangent directions such that there is only one such translate.

(8) *(Proof of Torelli's Theorem 1.7)* Let (J,Θ) be the Jacobian of a curve C. We have to reconstruct the curve C from the pair (J,Θ). Consider the Gauss map $G: \Theta_s \to \mathbb{P}_{g-1}$ (see Section 4.4). Let $\overline{\Theta}$ be the normalization of the closure of the graph of G and $\overline{G}: \overline{\Theta} \to \mathbb{P}_{g-1}$ the induced morphism.

a) Show that $\overline{G}$ is finite. Let $B \subset \mathbb{P}_{g-1}$ denote the branch locus of $\overline{G}$.
b) If C is not hyperelliptic, then $B = \varphi_{\omega}(C)^*$, the dual hypersurface of the canonical curve $\varphi_{\omega}(C)$ in $\mathbb{P}_{g-1}$.
c) If C is hyperelliptic, then $B = \varphi_{\omega}(C)^* \cup \bigcup_{\nu=1}^{2g+2} \varphi_{\omega}(x_\nu)^*$, where $x_\nu \in C$, $1 \leq \nu \leq 2g+2$, are the branch points of the hyperelliptic covering and $\varphi_{\omega}(x_\nu)^*$ denotes the dual hyperplane of the point $\varphi_{\omega}(x_\nu)$ in $\mathbb{P}_{g-1}$.
d) In the nonhyperelliptic case the Theorem follows by double duality $(\varphi_{\omega}(C)^*)^* = \varphi_{\omega}(C)$. In the hyperelliptic case the curve C is determined by the branch points.

(See Andreotti [1]. For other proofs see Torelli [1], Comessatti [1], Weil [2], Matsusaka [1], and Martens [1].)

(9) Let $\kappa \in \operatorname{Pic}^{g-1}(C)$ and Θ be the theta divisor on $J(C)$ with $\alpha_\kappa^*\Theta = W_{g-1}$. Since Θ and $(-1)^*\Theta$ are algebraically equivalent, there is an $x \in J(C)$ such that $(-1)^*\Theta = t_x^*\Theta$. Show that $x = \omega_C \otimes \kappa^{-2}$.

(10) Let C be a smooth algebraic curve, $\alpha_c : C \to J(C)$ the embedding with respect to the point $c \in C$ and Θ the theta divisor on $J(C)$ defined by $\alpha_L^*\Theta = W_{g-1}$ with $L = \omega_C \otimes \mathcal{O}_C((1-g)c)$. Show that $\alpha_c^*\mathcal{O}_{J(C)}(\Theta) = \mathcal{O}_C(g \cdot c)$. (Hint: use Lemma 3.4.)

(11) Let C be a smooth algebraic curve and $\alpha = \alpha_c : C \to J(C)$ the embedding with respect to the point $c \in C$. Suppose $\mathcal{P}_C$ is the Poincaré bundle of degree zero on $C \times J(C)$, normalized with respect to c, and let Δ denote the diagonal in C^2. Show that $(\mathrm{id}_C \times (-1)\alpha_c)^*\mathcal{P}_C \simeq \mathcal{O}_{C^2}(\{c\} \times C + C \times \{c\} - \Delta)$.

(12) Show that any curve C of genus $g \geq 1$ admits a theta characteristic κ with $h^0(\kappa) = 1$. (Hint: use Exercise 6.1 to show that the theta divisor Θ of $J(C)$ contains a 2-division point x with $\operatorname{mult}_x(\Theta) = 1$.)

(13) For a general smooth projective curve C of genus g we have $\operatorname{End} J(C) \simeq \mathbb{Z}$. (See Koizumi [2].)

(14) Let C be a smooth projective curve with Jacobian J. Recall the isomorphism of abelian groups $\operatorname{Corr}(C,C) \to \operatorname{End}(J)$ of Theorem 5.1. The ring structure of $\operatorname{End}(J)$ induces a ring structure on $\operatorname{Corr}(C,C)$ as follows. Let $l_1, l_2 \in \operatorname{Corr}(C,C)$.

a) Show that there are divisors D_1 and D_2 on $C \times C$ defining l_1 and l_2 such that $C \times D_1$ and $D_2 \times C$ intersect properly in $C \times C \times C$. Moreover the class of the correspondence in $\operatorname{Corr}(C,C)$ defined by the divisor $D = p_{13*}((C \times D_1) \cdot (D_2 \times C))$ does not depend on the choice of the divisors D_1 and D_2.
b) $\gamma_{\mathcal{O}(D_1)}\gamma_{\mathcal{O}(D_2)} = \gamma_{\mathcal{O}(D)}$.

(15) The plane quartic with equation $X_0^3X_1+X_1^3X_2+X_2^3X_0=0$ is called the *Klein quartic.* It is the unique plane quartic with automorphism group of order 168. Show that $\Pi=(Z,\mathbb{1}_3)$ with

$$Z=\frac{1}{14}\begin{pmatrix}-2 & 11 & -1\\ 11 & -22 & 2\\ -1 & 2 & 3\end{pmatrix}+\frac{\sqrt{7}}{14}i\begin{pmatrix}6 & -5 & 3\\ -5 & 10 & -6\\ 3 & -6 & 5\end{pmatrix}$$

is a period matrix of the Klein quartic. (Hint: use a modification of the method of Section 7).

(16) Let C be the hyperelliptic curve of genus $g\geq 2$ defined by the affine equation $y^2=x^{2g+2}-1$. It is the unique such curve with reduced automorphism group the dihedral group of order $4g+4$. Show that $\Pi=(Z,\mathbb{1}_g)$ with $Z=(z_{jk})$,

$$z_{jk}=\frac{1}{g+1}\sum_{\nu=1}^{j}\Big(\frac{1+\cos\frac{2\nu-1}{g+1}\pi}{\sin\frac{2\nu-1}{g+1}\pi}i+\frac{1+\cos\frac{2(k-\nu)+1}{g+1}\pi}{\sin\frac{2(k-\nu)+1}{g+1}\pi}i\Big)$$

for $j\leq k$, is a period matrix of the Jacobian $J(C)$.

(17) Let C be the hyperelliptic curve of genus $g\geq 2$ defined by the affine equation $y^2=x^{2g+2}-x$. It is the unique such curve with reduced automorphism group cyclic of order $2g+1$. Show that $\Pi=(Z,\mathbb{1}_g)$ with $Z=(z_{jk})$,

$$z_{jk}=1-\sigma_1^{-1}\sum_{\nu=1}^{j}\sigma_\nu\sigma_{k-j+\nu}\quad\text{for } 1\leq j\leq k\leq g \text{ and}$$

$$\sigma_1=e\Big(-\tfrac{g\pi i}{2g+1}\Big)\quad\text{and}$$

$$\sigma_{\nu+1}=\frac{\sigma_1}{1+e(\frac{2\nu\pi i}{2g+1})}\Big(1-\sum_{\mu=2}^{\nu}e((g-\nu+\mu-1)\tfrac{2\pi i}{2g+1})\sigma_\mu\sigma_{\nu-\mu+2}\Big)$$

for $\nu=1,\dots,g-1$, is a period matrix for the Jacobian $J(C)$.

(18) Let C be the hyperelliptic curve of genus $g\geq 2$ defined by the affine equation $y^2=x^{2g+1}-x$. It is the unique such curve with reduced automorphism group the dihedral group of order $4g$ for $g\geq 3$ and the symmetric group $\mathfrak{S}_4$ for $g=2$. Show that $\Pi=(Z,\mathbb{1}_g)$ with $Z=(z_{jk})$,

$$z_{1,k}=\alpha_k\quad\text{for } k=1,\dots,g,$$

$$z_{j,j}=-2\alpha_2\quad\text{for } j=2,\dots,g,$$

$$z_{j,k}=\alpha_{k-j+1}-\alpha_{k-j+2}\quad\text{for } 2\leq j<k\leq g \text{ and}$$

$$\alpha_j=\tfrac{2}{g}e\Big((2j-3)\tfrac{\pi i}{2g}\Big)\Big(e((2j-3)\tfrac{\pi i}{g})-1\Big)^{-1}\quad\text{for } j=2,\dots,g \text{ and}$$

$$\alpha_1=\tfrac{1}{2}\Big(-\alpha_2-\sum_{j=2}^{g}\alpha_j-1\Big)$$

is a period matrix for the Jacobian $J(C)$.

(For the last three exercises see Schindler [1]. For $g=2$ compare the results with cases IV, V and VI in Bolza's list in Section 7. Note that for $g=2$ the curve in the last exercise is isomorphic to the curve of case V in the list.)

Chapter 12

Prym Varieties

In the previous chapter we saw that to any smooth projective curve one can associate a principally polarized abelian variety, its Jacobian. This gives a map t from the moduli space $\mathcal{M}_g$ of smooth projective curves of genus g to the moduli space $\mathcal{A}_{\mathbb{1}_g}$ of principally polarized abelian varieties of dimension g, which by Torelli's Theorem is injective. We thus obtain a $3g-3$ dimensional subvariety $t(\mathcal{M}_g)$ of $\mathcal{A}_{\mathbb{1}_g}$. For every point of $t(\mathcal{M}_g)$ one can interpret the geometry of the theta divisor in terms of the corresponding curve (see for example Riemann's Singularity Theorem 11.2.5).

Wirtinger [1] and Mumford [5] showed that to any étale double covering of a smooth projective curve of genus $g+1$ one can associate an element of the moduli space $\mathcal{A}_{\mathbb{1}_g}$, called the Prym variety of the covering. This gives a $3g$-dimensional subvariety of $\mathcal{A}_{\mathbb{1}_g}$, with $t(\mathcal{M}_g)$ in its boundary. Again one can interpret the geometry of the theta divisor in terms of the corresponding double covering (see Section 6).

One would like to generalize this procedure. To be more precise, one would like to have an interpretation of the geometry of the theta divisor of *any* point of $\mathcal{A}_{\mathbb{1}_g}$ in terms of curve theory.

The first step in this direction was done by Tyurin. He showed in [1] that an abelian subvariety of the Jacobian $J(C)$ which is associated to a symmetric correspondence σ of a curve C, satisfying the equation

$$\sigma^2 + (m-2)\sigma - (m-1) = 0$$

for some positive integer m, is in many cases principally polarized. Later these varieties were called *Prym-Tyurin varieties* or *generalized Prym varieties*. Welters shows in [3] that any principally polarized abelian variety is a Prym-Tyurin variety (see Corollary 2.4). So this notion is a suitable candidate for the above mentioned interpretation of the theta divisor. However, it remains a problem to make this precise. The aim would be to find a stratification of $\mathcal{A}_{\mathbb{1}_g}$, such that for any stratum S there is an explicit family of curves with correspondences whose Prym-Tyurin varieties are exactly the elements of S.

The first invariant one could think of to construct the required stratification is the integer m appearing in the above equation. We will see that m is just the exponent of the abelian subvariety of the Jacobian $J(C)$ associated to σ (see Section 2). In particular we have $m = 1$ for Jacobians and $m = 2$ for Prym varieties. A first step in this program would be to compute for a given g the smallest integer M such that any principally polarized abelian variety of dimension g is a Prym-Tyurin variety of exponent $m \leq M$. In Corollary 2.4 we will see that $M \leq 3^{g-1}(g-1)!$.

This chapter should be considered as an introduction to the theory of (generalized) Prym varieties. It is organized as follows: the exponent $e(Y)$ of an abelian subvariety $Y \subset X$ is reviewed in Section 1. If X is principally polarized, $e(Y)$ has a pleasant interpretation in terms of endomorphisms of X (Proposition 1.1). A consequence is that complementary abelian subvarieties have the same exponent (Corollary 1.2). In Section 2 a *Prym-Tyurin variety for a curve* C is defined as a principally polarized abelian subvariety (Z, Ξ) of the Jacobian $(J(C), \Theta)$ such that $\iota_Z^*\Theta \equiv e(Z)\Xi$. We prove a universal property for Prym-Tyurin varieties as well as a criterion of Welters for a principally polarized abelian variety to be Prym-Tyurin. Suppose $f: C \to C'$ is a covering of a smooth projective curve C' of genus ≥ 1. If the complementary abelian variety Z of $f^*(J(C'))$ in $J(C)$ is a Prym-Tyurin variety for C, we call it the *Prym variety associated to the covering* f. In Section 3 we prove a theorem of Mumford's, saying that there are exactly 3 types of coverings $f: C \to C'$ leading to Prym varieties: étale double coverings, double coverings ramified in 2 points, and genus 2 coverings of an elliptic curve. In Section 4 we give a second, topological proof of the fact that a double covering ramified in at most 2 points defines a Prym variety. The main results of Sections 5 and 6 about the Abel-Prym map and the theta divisor of a Prym variety are due to Mumford [5]. The Universal Property (5.1) was given in Masiewicki [1]. In Section 7 we prove Recillas' Theorem relating Prym varieties associated to étale double coverings of trigonal curves to Jacobians of tetragonal curves (see Recillas [1]). Section 8 is dedicated to Donagi's tetragonal construction, which implies that the Prym-Torelli map is not injective (see Donagi [1]). Finally, in Section 9 we follow Kanev [2] to prove a criterion giving sufficient conditions for a correspondence to define a Prym-Tyurin variety.

As in the last chapter we use some results on algebraic curves, for which we refer to Arbarello et al. [1]. Let X be an abelian variety and Θ a divisor defining a principal polarization on X. If $f: Y \to X$ is a morphism of varieties, then according to the Moving Lemma 5.4.1 there is always a translate $t_x^*\Theta$ such that $f^*t_x^*\Theta$ is a divisor on Y. So, if we write $f^*\Theta$, we always assume the divisor Θ to be chosen in such a way that $f^*\Theta$ is defined.

§ 1 Abelian Subvarieties of a Principally Polarized Abelian Variety

In Section 5.3 we introduced the notion of complementary abelian subvarieties of a polarized abelian variety (X, L) and studied some first properties. In this section we derive further results on such subvarieties in the special case of a principal polarization $L = \mathcal{O}(\Theta)$.

Let (X, Θ) be a principally polarized abelian variety and $\iota = \iota_Y : Y \hookrightarrow X$ an abelian subvariety of X. In order to simplify notation, we identify X with its dual abelian variety $\widehat{X}$ via the isomorphism $\phi_\Theta : X \to \widehat{X}$, and write $\phi_Y := \phi_{\iota^*\Theta} : Y \to \widehat{Y}$ for the isogeny of the induced polarization. Recall the exponent $e(Y)$ of Y. It is defined as the exponent of the finite group $\ker \phi_Y$. According to Proposition 1.2.6 the map $\psi_Y = e(Y)\phi_Y^{-1}$ is an isogeny. With this notation the norm-endomorphism N_Y and the symmetric idempotent ε_Y of Y are

$$N_Y = \iota\psi_Y\widehat{\iota} \quad \text{and} \quad \varepsilon_Y = \iota\phi_Y^{-1}\widehat{\iota}\,. \tag{1}$$

As we saw in Theorem 5.3.2, the assignment $Y \mapsto \varepsilon_Y$ gives a bijection between the sets of abelian subvarieties Y of X and symmetric idempotents in $\mathrm{End}_{\mathbb{Q}}(X)$. In this way the involution $\varepsilon \mapsto 1 - \varepsilon$ on the set of symmetric idempotents of $\mathrm{End}_{\mathbb{Q}}(X)$ leads to the notion of complementary abelian subvarieties.

Let Z be the abelian subvariety of X complementary to Y, i.e.

$$Z = \mathrm{im}\big(e(Y)1_X - N_Y\big)\,.$$

Our first aim is to show that the exponents of Y and Z coincide. This is a consequence of the following

(1.1) Proposition. $e(Y) = \min\{n > 0 \mid n\varepsilon_Y \in \mathrm{End}(X)\}$ *for any abelian subvariety* Y *of a principally polarized abelian variety* (X, Θ).

Proof. Denote $e = \min\{n > 0 \mid n\varepsilon_Y \in \mathrm{End}(X)\}$. By definition of the exponent $e(Y)\varepsilon_Y = \iota\psi_Y\widehat{\iota}$ is an endomorphism, so $e \le e(Y)$.

On the other hand, since ι is a closed immersion, it follows that $e\phi_Y^{-1}\widehat{\iota}$ is a homomorphism. So its dual $\iota(e\phi_Y^{-1})\widehat{\ } = \iota(e\phi_Y^{-1})$ is also a homomorphism. Again since ι is a closed immersion, $e\phi_Y^{-1} \in \mathrm{Hom}(\widehat{Y}, Y)$. But it follows immediately from the definition of the exponent, that $e(Y)$ is the smallest positive integer such that $e(Y)\phi_Y^{-1} \in \mathrm{Hom}(\widehat{Y}, Y)$. Hence $e(Y) \le e$, which completes the proof. □

Since $\varepsilon_Z = 1 - \varepsilon_Y$ for any pair (Y, Z) of complementary abelian subvarieties, the proposition implies

(1.2) Corollary. *Complementary abelian subvarieties of a principally polarized abelian variety have the same exponent.*

Note that for an arbitrary polarization Proposition 1.1 and Corollary 1.2 are not valid. For example let (X_i, L_i), $i = 1, 2$ be principally polarized abelian varieties. Consider $X = X_1 \times X_2$ with polarization $L = p_1^* L_1 \otimes p_2^* L_2^2$. Then (X_1, X_2) is a pair of complementary abelian subvarieties with $e(X_1) = 1$ and $e(X_2) = 2$.

In the sequel we denote by e the common exponent of the complementary abelian subvarieties Y and Z. Then property 5.3(4) simplifies to

$$N_Z = e - N_Y \ . \tag{2}$$

The following proposition gives further possibilities for expressing Z in terms of Y.

(1.3) Proposition. $Z = (\ker N_Y)_0 = \ker \widehat{\iota} \simeq (X/Y)\widehat{}$.

Proof. We have $Z = \operatorname{im} N_Z \subset (\ker N_Y)_0$, since $N_Y N_Z = 0$ by 5.3(3). As Z and $(\ker N_Y)_0$ are abelian subvarieties of the same dimension, this gives the first equation. Moreover, $(\ker N_Y)_0 = (\ker \widehat{\iota})_0$ by (1), since ι a closed immersion and ψ_Y an isogeny.

In order to show that $(\ker \widehat{\iota})_0 = \ker \widehat{\iota}$, i.e. that $\ker \widehat{\iota}$ is connected, consider the exact sequence $0 \longrightarrow Y \hookrightarrow X \longrightarrow X/Y \longrightarrow 0$. By Proposition 2.4.2 the dual sequence is also exact: $0 \longrightarrow (X/Y)\widehat{} \longrightarrow \widehat{X} \xrightarrow{\widehat{\iota}} \widehat{Y} \longrightarrow 0$. So $\ker \widehat{\iota} \simeq (X/Y)\widehat{}$. In particular $\ker \widehat{\iota}$ is connected. □

(1.4) Corollary. $K(\iota^*\Theta) = \iota^{-1} Z \simeq Y \cap Z$

Proof. $K(\iota^*\Theta) = \ker \phi_{\iota^*\Theta} = \ker(\widehat{\iota}\iota) = \iota^{-1}\ker(\widehat{\iota}) = \iota^{-1} Z \simeq Y \cap Z$. □

By the symmetry of the situation Corollary 1.4 implies that $K(\iota_Y^*\Theta)$ and $K(\iota_Z^*\Theta)$ are isomorphic as abelian groups. Thus the types of the induced polarizations are related as follows:

(1.5) Corollary. *Let (Y, Z) be a pair of complementary abelian subvarieties of a principally polarized abelian variety with* $\dim Y \geq \dim Z = r$. *If the induced polarization $\iota_Z^*\Theta$ is of type $(d_1, \ldots, d_r)$, then $\iota_Y^*\Theta$ is of type $(1, \ldots, 1, d_1, \ldots, d_r)$.*

By definition the integer d_r is the exponent of the abelian subvariety Z. In particular we see again that the exponents of Y and Z coincide.

According to Corollary 5.3.6 the homomorphism $\mu = \iota_Y + \iota_Z : Y \times Z \to X$ is an isogeny. It is of exponent e, since by equation (2)

$$(N_Y, N_Z)(\iota_Y + \iota_Z) = e_X \ . \tag{3}$$

(1.6) Lemma. *The induced polarization splits:* $\phi_{(\iota_Y+\iota_Z)^*\Theta} = \phi_Y \times \phi_Z$.

Proof. The isogeny $\phi_{(\iota_Y+\iota_Z)^*\Theta}$ is of the form $\left(\begin{smallmatrix} \alpha & \beta \\ \gamma & \delta \end{smallmatrix}\right) : Y \times Z \to \widehat{Y} \times \widehat{Z}$. First we claim $\beta = 0$: by definition $\beta = \widehat{\iota}_Y \iota_Z$. Using 5.3(3) we have

$$0 = N_Y N_Z = \iota_Y \psi_Y \widehat{\iota}_Y \iota_Z \psi_Z \widehat{\iota}_Z \ .$$

But $Z = \operatorname{im} \psi_Z \widehat{\iota}_Z$, the homomorphism ι_Y is a closed immersion, and ψ_Y is an isogeny. So $\beta = \widehat{\iota}_Y \iota_Z = 0$. In the same way we obtain $\gamma = \widehat{\iota}_Z \iota_Y = 0$. Finally we have $\alpha = \widehat{\iota}_Y \iota_Y = \phi_Y$ and similarly $\delta = \phi_Z$. This completes the proof. □

Next we prove some auxiliary results on abelian subvarieties of X, which will be applied later. For a pair (Y, Z) of complementary abelian subvarieties of exponent e let $Y_e + Z_e$ denote the subgroup of e-division points in X generated by Y_e and Z_e.

(1.7) Lemma. $\ker N_Y \cap X_e = \ker N_Y \cap \ker N_Z = Y_e + Z_e \subset X_e$.

Proof. According to (3) the group $\ker N_Y \cap \ker N_Z = \ker(N_Y, N_Z)$ consists of e-division points. So the first equation follows from $N_Z = e - N_Y$. It remains to show the second equality. Using $\ker(\iota_Y + \iota_Z) = \{(x, -x) \mid x \in Y \cap Z\} \simeq Y \cap Z$, we obtain the following diagram:

$$\begin{array}{ccccccccc}
 & & & & 0 & & 0 & & \\
 & & & & \downarrow & & \downarrow & & \\
0 & \longrightarrow & \ker N_Y \cap \ker N_Z & \longrightarrow & X_e & \xrightarrow{(N_Y,N_Z)} & Y \cap Z & \longrightarrow & 0 \\
 & & \| & & \downarrow & & \downarrow & & \\
0 & \longrightarrow & \ker(N_Y, N_Z) & \longrightarrow & X & \xrightarrow{(N_Y,N_Z)} & Y \times Z & \longrightarrow & 0 \\
 & & & & {\scriptstyle e_X}\downarrow & & \downarrow{\scriptstyle \iota_Y+\iota_Z} & & \\
 & & & & X & = & X & & \\
 & & & & \downarrow & & \downarrow & & \\
 & & & & 0 & & 0 & &
\end{array}$$

By equations 5.3(2) and (4) we have $Y_e + Z_e \subset \ker(N_Y, N_Z)$. But even equality holds, since by the exactness of the upper row in the diagram:

$$\begin{aligned} \#(Y_e + Z_e) = \#(Y_e \times Z_e)\#(Y \cap Z)^{-1} &= \#(X_e)\#(Y \cap Z)^{-1} \\ &= \#(\ker N_Y \cap \ker N_Z) \ . \qquad \square \end{aligned}$$

Finally we will show that for any abelian subvariety $Y \subset X$ of exponent e there is an isogeny $f\colon \widetilde{Y} \to Y$ such that the induced polarization on $\widetilde{Y}$ is the e-fold of a principal polarization.

(1.8) Proposition. *There exist a principally polarized abelian variety $(\widetilde{Y}, \widetilde{\Sigma})$, homomorphisms $\tilde{\iota}$ and $\widetilde{N}$ and an isogeny f as in the following (noncommutative) diagram*

$$\begin{array}{ccc} \widetilde{Y} & & \\ {\scriptstyle f}\Big\downarrow & \overset{\widetilde{N}}{\nwarrow}\!\!\searrow^{\tilde{\iota}} & \\ Y & \underset{\iota}{\hookrightarrow} & X \end{array} \qquad \textit{such that} \qquad \begin{array}{rl} i) & \tilde{\iota} = \iota f \\ ii) & \tilde{\iota}^*\Theta \equiv e\widetilde{\Sigma} \\ iii) & \tilde{\iota}\widetilde{N} = N_Y \\ iv) & \widetilde{N}\tilde{\iota} = e_{\widetilde{Y}} \ . \end{array}$$

Proof. First we construct $\widetilde{Y}$. For this suppose $Y = V/\Lambda$ and denote by $E\colon \Lambda \times \Lambda \to \mathbb{Z}$ the alternating form corresponding to the polarization $\iota^*\Theta$. Define an alternating form on V by $\widetilde{E} = \frac{1}{e}E$. Since $\widetilde{E}$ is integer valued on the lattice $e^2\Lambda \subset \Lambda$ there is a lattice $\widetilde{\Lambda}$ with $e^2\Lambda \subset \widetilde{\Lambda} \subset \Lambda$ such that $\widetilde{E}$ is unimodular on $\widetilde{\Lambda}$.
Define $\widetilde{Y} = V/\widetilde{\Lambda}$. Then $\widetilde{E}$ defines a principal polarization on $\widetilde{Y}$. Choose $\widetilde{\Sigma}$ to be a theta divisor defining the polarization $\widetilde{E}$. Moreover define $f\colon \widetilde{Y} \to Y$ to be the isogeny induced by the inclusion $\widetilde{\Lambda} \subset \Lambda$, $\tilde{\iota} := \iota f$ and $\widetilde{N} := \phi_{\widetilde{\Sigma}}^{-1}\hat{\tilde{\iota}}$. Then by construction

$$\tilde{\iota}^*\Theta = f^*(\iota^*\Theta) \equiv e\widetilde{\Sigma} \ .$$

It remains to show iii) and iv): By i) and ii) we have $\hat{f}\phi_Y f = \phi_{f^*\iota^*\Theta} = e\phi_{\widetilde{\Sigma}}$ and equivalently $f\phi_{\widetilde{\Sigma}}^{-1}\hat{f} = \psi_Y$. Hence

$$\tilde{\iota}\widetilde{N} = \iota f \phi_{\widetilde{\Sigma}}^{-1}\hat{\tilde{\iota}} = \iota f \phi_{\widetilde{\Sigma}}^{-1}\hat{f}\hat{\iota} = \iota\psi_Y\hat{\iota} = N_Y \ .$$

Moreover, since $\hat{\tilde{\iota}}\tilde{\iota} = \phi_{\tilde{\iota}^*\Theta} = e\phi_{\widetilde{\Sigma}}$ by ii), we obtain: $\widetilde{N}\tilde{\iota} = \phi_{\widetilde{\Sigma}}^{-1}\hat{\tilde{\iota}}\tilde{\iota} = e\phi_{\widetilde{\Sigma}}^{-1}\phi_{\widetilde{\Sigma}} = e_{\widetilde{Y}}$. $\square$

§ 2 Prym-Tyurin Varieties

Let C be a smooth projective curve of genus g and (J, Θ) its Jacobian variety. The abelian subvarieties Z of J, for which the induced polarization is a multiple of a principal polarization Ξ, are of particular importance: one can study the geometry of the theta divisor Ξ in terms of the curve C. This leads to the notion of Prym-Tyurin varieties.

A principally polarized abelian variety (Z, Ξ) is called *Prym-Tyurin variety* or *generalized Prym variety* if there is a smooth projective curve C with Jacobian (J, Θ) such that Z is an abelian subvariety of J with

$$\iota_Z^* \Theta \equiv e \Xi \tag{1}$$

for some integer e. Necessarily e is the exponent of Z in J. We also say that Z is a Prym-Tyurin variety *for the curve* C and that e is the *exponent* of the Prym-Tyurin variety Z.

As we shall see later there may be several curves C for which Z is a Prym-Tyurin variety. The curve C turns out to be a tool for studying the geometry of the theta divisor Ξ of Z.

First we will show that our definition of a Prym-Tyurin variety coincides with the usual definition (as given for example in Bloch-Murre [1] and Kanev [1]): suppose σ is an endomorphism of J, symmetric with respect to the Rosati involution of (J, Θ) and satisfying

$$\sigma^2 + (m-2)\sigma - (m-1) = 0 \tag{2}$$

for some integer m. Then the abelian subvariety $Z = \operatorname{im}(\sigma - 1)$ of J is a generalized Prym variety in the sense of Bloch-Murre and Kanev, if the induced polarization is a multiple of some principal polarization of Z. Originally this definition was given in terms of correspondences, which can be done according to Theorem 11.5.1. Certainly a generalized Prym variety in the sense of Bloch-Murre and Kanev is also a Prym-Tyurin variety in the above sense. But also the converse holds: if Z is a Prym-Tyurin variety of exponent e for the curve C, then the endomorphism $\sigma = 1 - N_Z$ satisfies $Z = \operatorname{im}(1-\sigma)$ and equation (2) with $m = e$. In fact, equation (2) is just $N_Z(e - N_Z) = 0$ (see 5.3(4)), translated into terms of σ. Note that in terms of the symmetric idempotent ε_Z this is equivalent to $\varepsilon_Z(1-\varepsilon_Z) = 0$.

Suppose now (Z, Ξ) defines a Prym-Tyurin variety for C. Since Ξ defines a principal polarization, we can identify Z with its dual variety $\widehat{Z}$ via the isomorphism ϕ_Ξ. Then $\psi_Z = 1_Z$ by (1), and the equation defining the norm-endomorphism N_Z of Z reads

$$N_Z = \iota_Z \widehat{\iota}_Z \ .$$

Fix a point $c \in C$ and consider the Abel-Jacobi map $\alpha = \alpha_c: C \to J$. The composition

$$\pi = \pi_c: C \overset{\alpha_c}{\longrightarrow} J \overset{\widehat{\iota}_Z}{\longrightarrow} Z$$

is called the *Abel-Prym map* of Z. It satisfies the following

(2.1) Universal Property. *Let the notation be as above and let X be an abelian variety and $\varphi: C \to X$ a morphism. Assume that the induced homomorphism $\widetilde{\varphi}: J \to X$ satisfies*

$$\widetilde{\varphi} N_Y = 0 \ .$$

(Here N_Y is the norm-endomorphism of the complementary abelian subvariety Y of Z in J.) Then there is a unique homomorphism $\psi: Z \to X$ such that for any $c \in C$ the following diagram commutes

$$\begin{array}{ccc} C & \overset{\varphi}{\longrightarrow} & X \\ {\scriptstyle \pi_c}\downarrow & & \downarrow{\scriptstyle t_{-\varphi(c)}} \\ Z & \overset{\psi}{\longrightarrow} & X \end{array} \ .$$

Proof. As usual let e denote the exponent of Z in J. We first claim that $\widehat{\iota}_Z$ maps the e-division points in J surjectively onto the e-division points of Z: $\widehat{\iota}_Z(J_e) = Z_e$.
For the proof let $z \in Z_e$. Choose $x \in J$ with $\widehat{\iota}_Z(x) = z$. By Proposition 1.3

$$ex \in \ker \widehat{\iota}_Z = (\ker N_Z)_0 = \operatorname{im} N_Y \ .$$

So $ex = N_Y(x_1)$ for some $x_1 \in J$. Choose $x_2 \in J$ with $x_1 = ex_2$. Then $x - N_Y(x_2) \in J_e$ with $\widehat{\iota}_Z(x - N_Y(x_2)) = z$, since $\widehat{\iota}_Z N_Y = 0$ by equation 5.3(3). This completes the proof of the claim.

By assumption and since $N_Y = e - N_Z$, we have $e\widetilde{\varphi} = \widetilde{\varphi} N_Z$. This implies $\widetilde{\varphi}\iota_Z(Z_e) = \widetilde{\varphi}\iota_Z\widehat{\iota}_Z(J_e) = \widetilde{\varphi} N_Z(J_e) = e\widetilde{\varphi}(J_e) = 0$. Hence there is a homomorphism $\psi: Z \to X$ such that $\widetilde{\varphi}\iota_Z = e\psi$ and

$$e\widetilde{\varphi} = \widetilde{\varphi} N_Z = \widetilde{\varphi}\iota_Z\widehat{\iota}_Z = e\psi\widehat{\iota}_Z \ .$$

Consequently $\widetilde{\varphi} - \psi\widehat{\iota}_Z: J \to X$ is a homomorphism with $\operatorname{im}(\varphi - \psi\widehat{\iota}_Z) \subset (\ker e_Z)_0 = 0$. This implies $\widetilde{\varphi} = \psi\widehat{\iota}_Z$, hence $\varphi - \varphi(c) = \widetilde{\varphi}\alpha_c = \psi\widehat{\iota}_Z\alpha_c = \psi\pi_c$ for all $c \in C$. The uniqueness of the map ψ follows from the fact that the curve $\pi(C)$ generates Z as a group. □

Given a principally polarized abelian variety (Z, Ξ), it is an obvious problem to find a curve C such that (Z, Ξ) is a Prym-Tyurin variety for C. Welters' idea is to start with a curve D in Z, possibly singular, and consider a smooth covering $C \to D$. In order to formulate the result, recall that for any algebraic r-cycle Y of an n-dimensional variety X we denote by $[Y]$ its fundamental class in $H^{2n-2r}(X, \mathbb{Z})$.

(2.2) Welters' Criterion. *Let (Z, Ξ) be a principally polarized abelian variety of dimension g and C a smooth projective curve. Then (Z, Ξ) is a Prym-Tyurin variety of exponent e for the curve C if and only if there is a morphism $\varphi: C \to Z$ such that*

a) $\varphi^*: Z \to J = J(C)$ is an embedding,

b) $\varphi_*[C] = \frac{e}{(g-1)!} \bigwedge^{g-1}[\Xi]$ in $H^{2g-2}(Z, \mathbb{Z})$.

Note that we identified $Z = \widehat{Z} = \operatorname{Pic}^0(Z)$ via the isomorphism ϕ_Ξ, so φ^* is the pullback map of line bundles. According to Theorem 4.11.1 two algebraic r-cycles on an abelian variety are numerically equivalent if and only if their fundamental classes in $H^{2g-2r}(X, \mathbb{Z})$ coincide. So condition b) is equivalent to

b') the algebraic 1-cycles $\varphi_* C$ and $\frac{e}{(g-1)!}\Xi^{g-1}$ are numerically equivalent.

The main tool for the proof of Welters' Criterion is the following

(2.3) Lemma. *Let (Z, Ξ) be a principally polarized abelian variety of dimension g and (J, Θ) the Jacobian variety of a smooth projective curve C. Given a morphism $\varphi: C \to Z$ and an integer e the following statements are equivalent:*

i) $(\varphi^*)^*\Theta \equiv e\Xi$,

ii) $\varphi_*[C] = \frac{e}{(g-1)!} \bigwedge^{g-1}[\Xi]$ in $H^{2g-2}(Z, \mathbb{Z})$.

Proof. Since we identified $J = \widehat{J}$ via ϕ_Θ and $Z = \widehat{Z}$ via ϕ_Ξ, condition i) is equivalent to $\widehat{\varphi^*}\varphi^* = \phi_{(\varphi^*)^*\Theta} = \phi_{e\Xi} = e_Z$. According to Corollary 11.4.2 we have $-\widehat{\varphi^*} = \widetilde{\varphi}$. Let $f: \Gamma \to \varphi(C) \subset Z$ be the normalization of $\varphi(C)$ and $g: C \to \Gamma$ the induced morphism such that $\varphi = f \circ g$. If $\tilde{f}$ and $\tilde{g}$ denote the extensions of f and g to the corresponding Jacobians, then we have $\tilde{g}\widehat{\tilde{g}} = \deg g = \deg \varphi$. So, using Propositions 11.6.1 and 5.4.7, condition i) is equivalent to

$$\begin{aligned}\delta(\varphi_*[C], \Xi) &= \deg \varphi \; \delta(\varphi(C), \Xi) = -\deg \varphi \tilde{f}\widehat{\tilde{f}} = -\tilde{f}\tilde{g}\widehat{\tilde{g}}\widehat{\tilde{f}} \\ &= -\widetilde{\varphi}\widehat{\widetilde{\varphi}} = -e_Z = \delta\Big(\frac{e}{(g-1)!} \bigwedge^{g-1}[\Xi], \Xi\Big) .\end{aligned}$$

According to Theorems 11.6.4 and 4.11.1, this is equivalent to ii), since the line bundle $\mathcal{O}_Z(\Xi)$ is nondegenerate. □

Proof of 2.2. Assume (Z, Ξ) is a Prym-Tyurin variety for C. We claim that the Abel-Prym map $\pi = \widehat{\iota}_Z\alpha: C \to Z$ satisfies conditions a) and b).

By Proposition 11.3.5, $\pi^* = \alpha^* \iota_Z = -\iota_Z$, so π^* is an embedding. Moreover $(\pi^*)^*\Theta = (-\iota_Z)^*\Theta \equiv \iota_Z^*\Theta \equiv e\Xi$ by assumption, so b) follows from Lemma 2.3. The converse implication is also an immediate consequence of the lemma. □

(2.4) Corollary. *Every principally polarized abelian variety of dimension g is a Prym-Tyurin variety of exponent $3^{g-1}(g-1)!$.*

Proof. Let (X,Θ) be a principally polarized abelian variety of dimension g. According to the Theorem of Lefschetz 4.5.1 the map $\varphi_{3\Theta}: X \hookrightarrow \mathbb{P}_N$ is an embedding. By Bertini's Theorem there is a $\mathbb{P}_{g-1}$ in $\mathbb{P}_N$ such that $C = X \cap \mathbb{P}_{g-1}$ is a smooth irreducible curve. Since C generates X as an abelian group, the embedding $\varphi: C \hookrightarrow X$ extends to a surjective homomorphism $\widetilde{\varphi}: J(C) \to X$. We claim that the kernel of $\widetilde{\varphi}$ is connected, i.e. $\ker \varphi$ is an abelian variety. For this it suffices to show that the rational representation of $\widetilde{\varphi}$ is surjective. But $\varrho_r(\widetilde{\varphi})$ is just the canonical map $H_1(C,\mathbb{Z}) \to H_1(X,\mathbb{Z})$, which is surjective according to the Lefschetz Hyperplane Theorem (see Griffiths-Harris [1] p. 156 and the Remark on p. 159).

Now Proposition 2.4.2 implies that the dual map $\varphi^* = \widehat{\widetilde{\varphi}}: X = \widehat{X} \to J(C) = \widehat{J}(C)$ is an embedding. On the other hand by construction

$$\varphi_*[C] = \overset{g-1}{\bigwedge}[3\Theta] = \frac{3^{g-1}(g-1)!}{(g-1)!} \overset{g-1}{\bigwedge}[\Theta] \,.$$

So Welters' Criterion 2.2 implies the assertion. □

The curve C for which (X,Θ) is a Prym-Tyurin variety is not uniquely determined. One can show that there are even infinitely many different curves C and different integers $e > 0$ such that (X,Θ) is a Prym-Tyurin variety for C of exponent e.

(2.5) Remark. An easy consequence is *Matsusaka's Criterion* for a principally polarized abelian variety to be a Jacobian:

Let (X,Θ_X) be a principally polarized abelian variety of dimension g and $C \subset X$ an irreducible curve with $[C] = \frac{1}{(g-1)!}\bigwedge^{g-1}[\Theta_X]$ in $H^{2g-2}(X,\mathbb{Z})$. Then C is smooth and $(X,\Theta_X) \simeq (J,\Theta_J)$, the Jacobian variety of C.

Note that this is a weaker version of the Criterion of Matsusaka-Ran 11.8.1.

Proof. Let $\varphi: \widetilde{C} \to C \subset X$ be the normalization of C. Identify $X = \widehat{X}$ as usual and consider the pullback map $\varphi^*: X \to J = \mathrm{Pic}^0(\widetilde{C})$. According to Lemma 2.3 the assumption on C gives $(\varphi^*)^*\Theta_J \equiv \Theta_X$. In particular φ^* is injective and X is of exponent $e = 1$ as abelian subvariety of J. Let Y be the complementary abelian subvariety of X in J. According to Corollary 1.5 the induced polarization on Y is a principal polarization Θ_Y. Moreover by Lemma 1.6

$$\iota_X + \iota_Y : (X \times Y, p_X^* \Theta_X + p_Y^* \Theta_Y) \to (J, \Theta_J)$$

is an isomorphism of polarized abelian varieties. But Θ_J is irreducible, so $Y = 0$ and we get the assertion. □

A consequence of Remark 2.5 and Welters' Criterion is

(2.6) Corollary. *For a principally polarized abelian variety (X, Θ_X) and a smooth irreducible curve C the following conditions are equivalent:*

i) (X, Θ_X) is a Prym-Tyurin variety of exponent 1 for the curve C.
ii) $(X, \Theta_X) \simeq (J, \Theta_J)$, the Jacobian of C.

§ 3 Prym Varieties

In the last section we saw that every principally polarized abelian variety is a Prym-Tyurin variety for some curve C. Now we consider the problem of finding Prym-Tyurin varieties for a given curve C. One method is the following: to a morphism $f: C \to C'$ of curves one can associate in a natural way a subvariety Z of the Jacobian J of C. It turns out that there are exactly three types of morphisms f such that Z is a Prym-Tyurin variety for the curve C. These varieties are called Prym varieties.

Let $f: C \to C'$ be a morphism of degree n of smooth projective curves. In order to avoid trivialities, we assume that C' is of genus ≥ 1. Denote by (J, Θ) and (J', Θ') respectively the corresponding Jacobians. Recall the norm map $N_f : J \to J'$, $\mathcal{O}_C(\sum r_\nu p_\nu) \mapsto \mathcal{O}_{C'}(\sum r_\nu f(p_\nu))$ of Section 11.4. Identifying $J = \widehat{J}$ and $J' = \widehat{J'}$ as usual, the pullback map f^* is a homomorphism of J' into J.

(3.1) Lemma. $(f^*)^* \Theta \equiv n\Theta'$.

Proof. By definition $N_f f^*$ is multiplication with n on J'. So using equation 11.4(1) we obtain $\phi_{n\Theta'} = n_{J'} = N_f f^* = \widehat{f^*} f^* = \phi_{(f^*)^*\Theta}$, and Proposition 2.5.3 gives the assertion. □

Denoting $Y = \operatorname{im} f^*$, the map f^* factorizes into an isogeny j and the canonical embedding ι_Y. With $\phi_Y = \phi_{\iota_Y^* \Theta}$ as above the following diagram commutes

$$\begin{array}{ccccc} & & \overset{f^*}{\frown\!\!\longrightarrow} & & \\ J' & \xrightarrow{j} & Y & \xrightarrow{\iota_Y} & J \\ {\scriptstyle n_{J'}}\downarrow & & \downarrow{\scriptstyle \phi_Y} & & \Vert \\ J' & \xleftarrow{\widehat{j}} & \widehat{Y} & \xleftarrow{\widehat{\iota_Y}} & \widehat{J} \\ & & \underset{N_f}{\longleftarrow\!\!\smile} & & \end{array} .$$

The norm map N_f and the norm-endomorphism N_Y are related as follows

(3.2) Proposition. $\qquad f^*N_f = \frac{n}{e(Y)}N_Y$.

Proof. From the diagram we deduce $\phi_Y = n\widehat{j}^{-1}j^{-1}$, since j is an isogeny. So

$$\psi_Y = e(Y)\phi_Y^{-1} = \tfrac{e(Y)}{n} j\widehat{j}\,,$$

implying $\qquad f^*N_f = \iota_Y j\widehat{j}\widehat{\iota_Y} = \frac{n}{e(Y)}\iota_Y\psi_Y\widehat{\iota_Y} = \frac{n}{e(Y)}N_Y$. □

The aim is to associate to the covering f a Prym-Tyurin variety. A natural candidate would be Y. In fact, according to Proposition 11.4.3 and Lemma 3.1 it is a Prym-Tyurin variety, provided f does not factorize via an étale covering. However, more important as a candidate is the complementary abelian subvariety Z of Y in J. If Z is a Prym-Tyurin variety for C, we call Z the *Prym variety* associated to the covering $f: C \to C'$. The following theorem, due to Wirtinger [1] and Mumford [5], gives a list of all coverings determining Prym varieties in this way.

(3.3) Theorem. *Let $f: C \to C'$ be a finite morphism of degree $n \geq 2$ of smooth projective curves of genus g and $g' \geq 1$. Then the abelian subvariety Z of $J(C)$, as defined above, is a Prym variety if and only if f is of one of the following types:*
a) f is étale of degree 2,
b) f is of degree 2 and ramified in 2 points,
c) C is of genus 2 and C' is of genus 1.

From Proposition 3.2 one easily deduces that the Prym variety Z is of exponent 2 in the cases a) and b). As for c), consider the factorization $f = f_e g$ of Corollary 11.4.4. Here C_e is an elliptic curve and $e(Z) = \deg g$.

Proof. Step I: Suppose Z is a Prym variety. Necessarily Z is of exponent $e \geq 2$ in J, since otherwise the canonical polarization on J would split by Lemma 1.6 and Corollary 1.5. Since $\iota_Z^*\Theta$ is of type $(e, \dots, e)$, the polarization on Y defined by $\iota_Y^*\Theta$ is of type $(1,\dots,1,e,\dots,e)$ again by Corollary 1.5. This implies $g' = \dim Y \geq \dim Z = g - g'$, i.e.

$$g \leq 2g'. \tag{1}$$

Using Hurwitz's formula we get

$$2g' - 1 \geq g - 1 = n(g'-1) + \tfrac{\delta}{2} \geq n(g'-1) \tag{2}$$

with δ the degree of the ramification divisor of f. Hence

$$(n-2)g' \leq n-1. \tag{3}$$

We consider the following four cases separately:
Case 1: $n \geq 3$, $g' \geq 3$: On the one hand we have $6 \leq 2n$, on the other hand (3) implies $2n \leq 5$, a contradiction.
Case 2: $n \geq 3$, $g' = 2$: Here (3) gives $n = 3$ implying $\delta = 0$ and $g = 4$ by (2). So f is étale and $\dim Y = \dim Z = 2$. Since the exponent e divides $n = 3$ by Proposition 3.2, we have also $e = 3$ and the polarization $\iota_Y^*\Theta$ is of type $(3,3)$. But f being étale of degree 3 implies that $(f^*)^*\Theta = \varphi^*\iota_Y^*\Theta$ is of type $(3,9)$, contradicting Lemma 3.1.
Case 3: $n \geq 2$, $g' = 1$: By (1) the curve C is of genus $g = 2$ and we are in case c) of the theorem.
Case 4: $n = 2$, $g' \geq 2$: Inequality (2) gives: $2g' - 1 \geq 2g' - 2 + \frac{\delta}{2}$. So $\delta \leq 2$ and we are either in case a) or b) of the theorem.
Step II: We have to show that in the cases a), b) and c) the abelian subvariety Z is a Prym variety. It suffices to show that the induced polarization is of type $(e,\dots,e)$. This is clear in case c) the subvariety Z being an elliptic curve.

In case b) the morphism f is ramified and of degree 2, so $Y = J'$ by Proposition 11.4.3. On the other hand $\iota_Y^*\Theta$ defines the square of a principal polarization by Lemma 3.1. Hence, according to Corollary 1.5, the line bundle $\iota_Z^*\Theta$ is of type $(2,\dots,2)$, since $\dim Y = \dim Z$.

As for case a): since $(f^*)^*\Theta = j^*\iota_Y^*\Theta$ is of type $(2,\dots,2)$ and $j\colon J' \to Y$ is an isogeny of degree 2, the line bundle $\iota_Y^*\Theta$ is of type $(1,2,\dots,2)$. But $\dim Z = \dim Y - 1$, so $\iota_Z^*\Theta$ is of type $(2,\dots,2)$ by Corollary 1.5. □

Finally we prove a formula relating the theta divisors of J, J' and the Prym variety Z in cases a) and b) of the theorem.

(3.4) Proposition. *Suppose $f\colon C \to C'$ is a double covering, ramified in at most two points, and (Z,Ξ) the associated Prym variety. Then*

$$2\Theta \equiv N_f^*\Theta' + \widehat{\iota}_Z^*\Xi\ .$$

Proof. In terms of divisors Lemma 1.6 reads $(\iota_Y + \iota_Z)^*\Theta \equiv q_Y^*\iota_Y^*\Theta + q_Z^*\iota_Z^*\Theta$ with q_Y and q_Z the natural projections of $Y \times Z$. Recall that $N_Y + N_Z = (\iota_Y + \iota_Z)(N_Y, N_Z) = 2_J$. So by Proposition 3.2

$$\begin{aligned} 4\Theta \equiv 2_J^*\Theta &\equiv (N_Y, N_Z)^*(q_Y^*\iota_Y^*\Theta + q_Z^*\iota_Z^*\Theta) \\ &= N_Y^*\Theta + N_Z^*\Theta = N_f^*(f^*)^*\Theta + \widehat{\iota}_Z^*\iota_Z^*\Theta \,. \end{aligned}$$

But $(f^*)^*\Theta \equiv 2\Theta'$ by Lemma 3.1 and $\iota_Z^*\Theta \equiv 2\Xi$. This gives the assertion, since the Néron-Severi group of J is torsion free. □

§ 4 Topological Construction of Prym Varieties

In Theorem 3.3 we saw that there are two types of double coverings determining Prym varieties: namely those ramified in none or two points. In this section we study these coverings from the topological point of view. This gives a second proof of the fact that they determine Prym varieties.

Let $f: C \to C'$ be a double covering of smooth projective curves, étale or ramified in two points. As in the last section denote by Z the abelian subvariety of the Jacobian $J = J(C)$ complementary to the abelian subvariety $Y = \operatorname{im} f^*$. Let $\iota: C \to C$ be the involution corresponding to the double covering f. It extends to an involution $\tilde{\iota}$ on J. In terms of $\tilde{\iota}$ the norm-endomorphisms of the abelian subvarieties Y and Z can be described as follows.

(4.1) Lemma. $\quad N_Y = 1 + \tilde{\iota}$ *and* $N_Z = 1 - \tilde{\iota}$.

In particular, using Proposition 1.3 we get

$$Y = \operatorname{im}(1 + \tilde{\iota}) = \ker(1 - \tilde{\iota})_0 \quad \text{and} \quad Z = \operatorname{im}(1 - \tilde{\iota}) = \ker(1 + \tilde{\iota})_0 \,. \tag{1}$$

Proof. According to Proposition 3.2 we have $N_Y = f^*N_f$, so for any $x = \mathcal{O}_C(\sum r_\nu p_\nu) \in J$:

$$N_Y(x) = f^*N_f\big(\mathcal{O}_C(\textstyle\sum r_\nu p_\nu)\big) = \mathcal{O}_C\big(\textstyle\sum r_\nu(p_\nu + \iota p_\nu)\big) = x + \tilde{\iota}x = (1 + \tilde{\iota})(x) \,.$$

Consequently $N_Z = 2 - N_Y = 1 - \tilde{\iota}$. □

Recall that $J = H^0(\omega_C)^*/H_1(C, \mathbb{Z})$. In these terms the induced action of ι on $H^0(\omega_C)^*$ respectively $H_1(C, \mathbb{Z})$ is just the analytic respectively rational representation of $\tilde{\iota}$. Denote by $H^0(\omega_C)^-$ and $H_1(C, \mathbb{Z})^-$ the (-1)-eigenspaces in $H^0(\omega_C)$ and $H_1(C, \mathbb{Z})$ with respect to the action of the involution ι. An immediate consequence of (1) is

(4.2) Proposition. $Z = \left(H^0(\omega_C)^-\right)^* / H_1(C, \mathbb{Z})^-$.

Suppose now that f is an étale covering. Setting $g = \dim Z$, the curves C' and C are of genus $g+1$ and $2g+1$. Choose a symplectic basis $\lambda_0, \lambda_1, \ldots, \lambda_g, \mu_0, \mu_1, \ldots, \mu_g$ of $H_1(C', \mathbb{Z})$ (see the remark after Proposition 11.1.2). From the topological point of view $f: C \to C'$ is a connected degree 2 covering of the topological space C' and as such determined by a nonzero element of $H_1(C', \mathbb{Z}/2\mathbb{Z})$. We may assume that this element is the image of the cycle λ_0 in $H_1(C', \mathbb{Z}/2\mathbb{Z})$. Topologically f can be realized as follows: cut the surface C' along μ_0 and glue two copies of it with upper and lower boundary of μ_0 reversed, so that the orientations fit together.

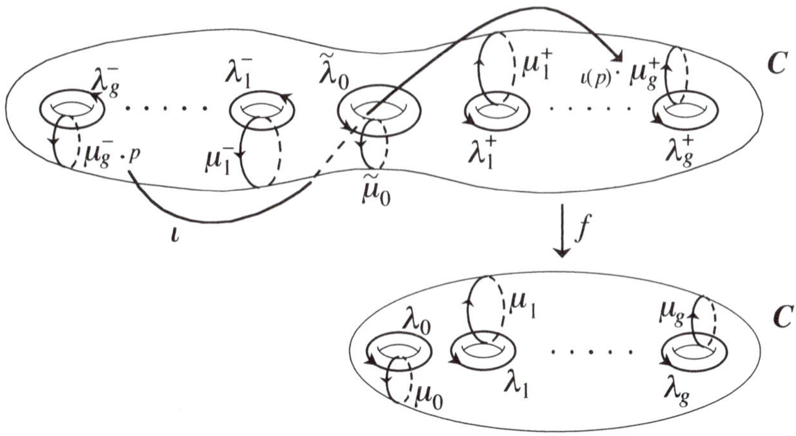

We obtain cycles $\widetilde{\lambda}_0, \lambda_i^+, \lambda_i^-, \tilde{\mu}_0, \mu_i^+, \mu_i^-$ which obviously form a symplectic basis of $H_1(C, \mathbb{Z})$. The involution ι corresponding to the covering is just the map interchanging the copies. It is clear from the picture, how ι acts on the lattice $H_1(C, \mathbb{Z})$. Obviously a basis of the (-1)-eigenspace $H_1(C, \mathbb{Z})^-$ is given by the skew symmetric cycles

$$\alpha_i := \lambda_i^+ - \lambda_i^- \ , \quad \beta_i := \mu_i^+ - \mu_i^- \ , \qquad i = 1, \ldots, g \ .$$

Let $E: H_1(C, \mathbb{Z}) \times H_1(C, \mathbb{Z}) \to \mathbb{Z}$ denote the alternating form associated to the canonical polarization on $J(C)$. In order to show that Z is a Prym variety, we compute its restriction to the basis α_i, β_i:

$$E(\alpha_i, \beta_j) = 2\delta_{ij} \ , \quad E(\alpha_i, \alpha_j) = E(\beta_i, \beta_j) = 0 \ , \quad 1 \le i, j \le g \ .$$

So the induced polarization is twice a principal polarization and Z is a Prym variety.

(4.3) Remark. *Every Jacobian (J, Θ) of a smooth projective curve C of genus g is a limit of a family of Prym varieties (Z_t, Ξ_t) of dimension g associated to étale double coverings.*

This was shown in Wirtinger [1] (see also Beauville [1]). We want to sketch his argument: choose distinct points p and q of C and identify them to obtain a singular curve C_0 with one double point whose normalization is C. Now choose a one-parameter family C_t of smooth genus $g+1$ curves degenerating to C_0. There is a family of 1-cycles μ_t on C_t shrinking to the singular point in C_0. The 1-cycle μ_t determines an étale double covering $f_t: \widetilde{C}_t \to C_t$ as described above. The curves $\widetilde{C}_t$ degenerate to a singular curve $\widetilde{C}_0$. The two singular points of $\widetilde{C}_0$ arise from shrinking the vanishing cycles μ_t' and μ_t'' of $\widetilde{C}_t$. So $\widetilde{C}_0$ is a reducible curve obtained by identifying the distinguished points $p_1 = q_2$ and $q_1 = p_2$ of two copies C' and C'' of C. The curve $\widetilde{C} = C' \cup C''$, the normalization of $\widetilde{C}_0$, is the induced double covering of C.

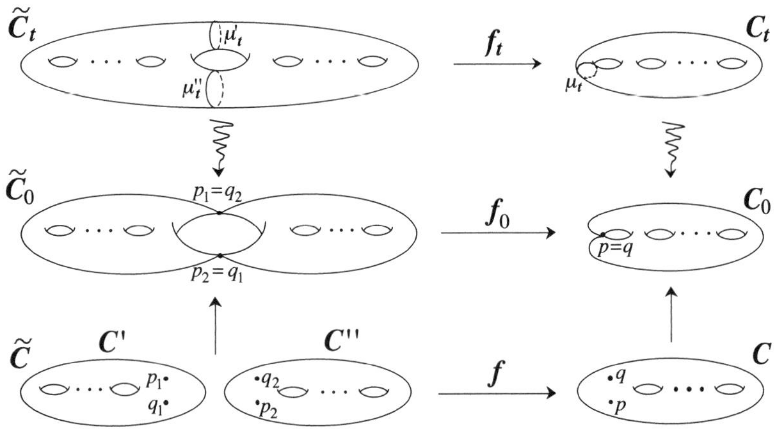

Recall that the Jacobian of $\widetilde{C}$ is the product of the Jacobians of its components, i.e. $J(\widetilde{C}) = J \times J$. For the reducible double covering $f: \widetilde{C} \to C$ one can define a Prym variety Z in the same way as in the irreducible case: the involution $\tilde{\imath}$ on $J(\widetilde{C}) = J \times J$ induced by the involution corresponding to the double covering f is obviously given by $\tilde{\imath}(x_1, x_2) = (x_2, x_1)$. Analogously as above, the Prym variety Z for the covering f is

$$Z = \operatorname{im}(1 - \tilde{\imath}) = \{(x, -x) \mid x \in J\}$$

(see also equation (1)). So Z is isomorphic to the Jacobian J of C. On the other hand, one can show, considering explicitly the degenerations of differential 1-forms and 1-cycles on C_t, that the family of Prym varieties (Z_t) associated to the coverings $f_t: \widetilde{C}_t \to C_t$ degenerates to Z. □

Finally, suppose that $f: C \to C'$ is a double covering ramified in two points p_0 and q_0 of C'. Setting $g = \dim Z$, the curves C' and C are of genus g and $2g$. As before, choose a symplectic basis $\lambda_1, \dots, \lambda_g, \mu_1, \dots, \mu_g$ of $H_1(C', \mathbb{Z})$. Topologically the covering can be realized as follows: let γ be a path joining p_0 and q_0 which does not intersect any of the cycles λ_i, μ_i. Cut the surface C' along γ and glue two copies of it with upper and lower boundaries reversed, such that the orientations fit together.

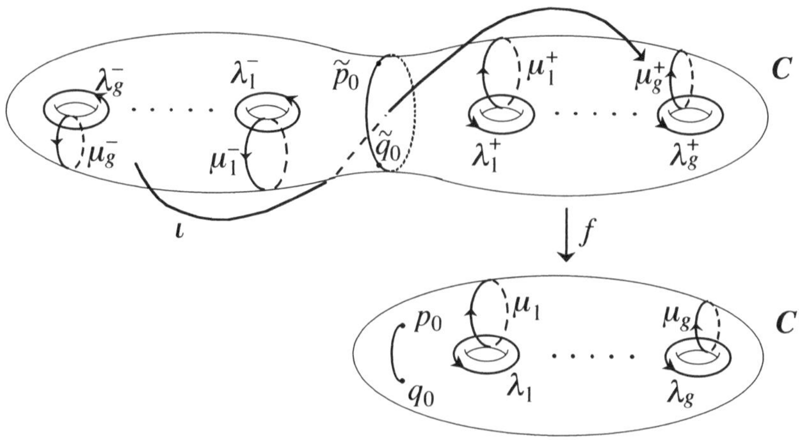

We obtain cycles $\lambda_i^+, \lambda_i^-, \mu_i^+, \mu_i^-$, which obviously form a symplectic basis of $H_1(C, \mathbb{Z})$. As in the étale case one sees that the cycles

$$\alpha_i := \lambda_i^+ - \lambda_i^- , \quad \beta_i := \mu_i^+ - \mu_i^- , \qquad i = 1, \dots, g$$

form a basis of the (-1)-eigenspace $H_1(C, \mathbb{Z})^-$ of the involution ι. The restriction to $H_1(C, \mathbb{Z})^-$ of the alternating form $E: H_1(C, \mathbb{Z}) \times H_1(C, \mathbb{Z}) \to \mathbb{Z}$ of the canonical polarization on $J(C)$ is given by

$$E(\alpha_i, \beta_j) = 2\delta_{ij} , \quad E(\alpha_i, \alpha_j) = E(\beta_i, \beta_j) = 0 , \quad 1 \leq i, j \leq g .$$

So the induced polarization on Z is twice a principal polarization and Z is a Prym variety.

§ 5 The Abel-Prym Map

In Section 2 we defined the Abel-Prym map $\pi : C \to Z$ associated to any Prym-Tyurin variety Z for a curve C. In this section we want to study in more detail the Abel-Prym map of Prym varieties.

Let us fix the notation for this section: suppose $f: C \to C'$ is a double covering of smooth projective curves of genus ≥ 1, étale or ramified in two points $\tilde{p}_0$ and $\tilde{q}_0$ of C (with images p_0 and q_0 in C'), determining a Prym variety (Z, Ξ). Identify $J = \widehat{J}$ and $Z = \widehat{Z}$ as usual and let $\pi = \pi_c: C \to Z$ be the Abel-Prym map with respect to a point $c \in C$. The following theorem, due to Masiewicki [1], is a translation of the Universal Property 2.1 into terms of the involution ι, corresponding to the double covering f.

(5.1) Universal Property. *Suppose $\varphi: C \to X$ is a morphism of C into an abelian variety X. If $\varphi\iota = -\varphi$, then there exists a unique homomorphism $\psi: Z \to X$, such that for any $c \in C$ the following diagram commutes*

$$\begin{array}{ccc} C & \xrightarrow{\varphi} & X \\ {\scriptstyle \pi_c}\downarrow & & \downarrow {\scriptstyle t_{-\varphi(c)}} \\ Z & \xrightarrow{\psi} & X \end{array} .$$

For the proof we only note that $N_Y = 1 + \tilde{\iota}$ by Lemma 4.1 and thus the condition $\widetilde{\varphi}N_Y = 0$ is the extension of $\varphi + \varphi\iota = 0$ to the Jacobian. □

The following proposition exhibits the degree of the Abel-Prym map $\pi: C \to Z$.

(5.2) Proposition. *a) Suppose C is not hyperelliptic. Then $\pi(p) = \pi(q)$ for distinct points $p, q \in C$ if and only if f is ramified in p and q. In particular $\pi: C \to Z$ is injective in the étale case.*

b) Suppose C is hyperelliptic. Then $\pi: C \to Z$ is of degree 2 onto its image and $\pi(p) = \pi(q)$ for distinct points $p, q \in C$ if and only if $p + \iota(q)$ is in the unique linear system of degree 2 and dimension 1 on C.

Proof. By definition of the Abel-Prym map and Lemma 4.1 $\pi(p) = \pi(q)$ for $p \neq q$ in C if and only if $(1 - \tilde{\iota})\mathcal{O}_C(p - c) = (1 - \tilde{\iota})\mathcal{O}_C(q - c)$. This is the case if and only if $p - \iota(p) \sim q - \iota(q)$, i.e.

$$\pi(p) = \pi(q) \Longleftrightarrow p + \iota(q) \sim q + \iota(p) . \tag{1}$$

If C is not hyperelliptic, the right hand side is equivalent to $p + \iota(q) = q + \iota(p)$. Since $p \neq q$ this proves a).

Let C be hyperelliptic with unique linear system g^1_2 of degree 2 and dimension 1. The hyperelliptic involution differs from the involution ι, since it has a rational quotient. So for all but a finite number of points $p \in C$ there is a unique $q \in C$ such that $p + \iota(q) \in g^1_2$. Since $\iota^* g^1_2 = g^1_2$ by the uniqueness of g^1_2, it follows that $p + \iota(q) \sim q + \iota(p)$. So (1) implies that for any $p \in C$ there is a unique $q \in C$ such that $\pi(p) = \pi(q)$. □

In order to determine the cases, in which the Abel-Prym map π is an embedding, we analyze its differential $d\pi$. For this we recall the scheme theoretic construction of the double covering $f: C \to C'$: there is a nontrivial line bundle η on C' with $\eta^2 \simeq \mathcal{O}_{C'}$, if f is étale, and $\eta^2 \simeq \mathcal{O}_{C'}(p_0 + q_0)$ otherwise. In fact, this isomorphism defines an algebra structure on the sheaf $\mathcal{O}_{C'} \oplus \eta$ and the covering C can be defined as $C = \mathbf{Spec}(\mathcal{O}_{C'} \oplus \eta)$ (see Hartshorne [1] Ex. IV.2.7). Moreover we need the following two properties, for the proof of which we refer to Barth et. al. [1] p.42:

$$1) \quad f_*\mathcal{O}_C = \mathcal{O}_{C'} \oplus \eta^{-1}\,, \qquad \text{and} \qquad 2) \quad \omega_C = f^*(\omega_{C'} \otimes \eta)\,.$$

Recall that $d\pi$ is a holomorphic map from the tangent bundle $\mathcal{T}_C$ of C to the tangent bundle $\mathcal{T}_Z$ of Z. According to Lemma 1.4.2 and Proposition 4.2 there is a canonical isomorphism $\mathcal{T}_Z = Z \times (H^0(\omega_C)^-)^*$. The projectivization of the composed map $\mathcal{T}_C \to \mathcal{T}_Z = Z \times (H^0(\omega_C)^-)^* \to (H^0(\omega_C)^-)^*$ is a priori a rational map $C \to P\big((H^0(\omega_C)^-)^*\big)$, called the *projectivized differential of the Abel-Prym map* π.

(5.3) Proposition. *The projectivized differential of the Abel-Prym map* $\pi: C \to Z$ *is the composed map* $\varphi_{\omega_{C'}\otimes\eta} \circ f: C \to C' \to P(H^0(\omega_{C'} \otimes \eta)^*)$.

Proof. Since $Z = \operatorname{im}(1 - \tilde{\iota})$ by Equation 4(1), we may consider $1 - \tilde{\iota}$ as a homomorphism of J into Z. According to Lemma 4.1 the Abel-Prym map may be written as $\pi = (1 - \tilde{\iota})\alpha$. By definition the differential $d(1 - \tilde{\iota})_0$ of $(1 - \tilde{\iota})$ at the point $0 \in J$ is the analytic representation $\rho_a(1 - \tilde{\iota})$. So the following diagram commutes

$$\begin{array}{ccccc}
\mathcal{T}_C & \xrightarrow{d\alpha} & \mathcal{T}_J = J \times H^0(\omega_C)^* & \xrightarrow{d(1-\tilde{\iota})} & \mathcal{T}_Z = Z \times (H^0(\omega_C)^-)^* \\
& \searrow & \downarrow & & \downarrow \\
& & H^0(\omega_C)^* & \xrightarrow{\rho_a(1-\tilde{\iota})} & (H^0(\omega_C)^-)^* \quad .
\end{array}$$

Using Proposition 11.1.4 this implies that the composed map $P(\rho_a(1-\tilde{\iota}))\varphi_{\omega_C}$ is the projectivized differential of π. But $\rho_a(1-\tilde{\iota})$ is just the projection onto the (-1)-eigenspace of $H^0(\omega_C)^*$ with respect to $\tilde{\iota}$. On the other hand we have $H^0(\omega_C)^- = H^0(\omega_{C'} \otimes \eta)$, since obviously $H^0(\omega_C) = H^0(f_*f^*(\omega_{C'} \otimes \eta)) = H^0((\omega_{C'} \otimes \eta) \otimes f_*\mathcal{O}_C) = H^0(\omega_{C'} \otimes \eta) \oplus H^0(\omega_{C'})$ is the decomposition of $H^0(\omega_C)$ into the $\tilde{\iota}$-eigenspaces. So the projectivized differential is the composed map

$$C = P(\mathcal{T}_C) \xrightarrow{\varphi_{\omega_C}} P(H^0(\omega_C)^*) \longrightarrow P(H^0(\omega_{C'} \otimes \eta)^*)\,.$$

But this map is given by the sublinear system $f^*|\omega_{C'} \otimes \eta|$ of $|\omega_C|$, which completes the proof. □

(5.4) Remark. In the case of an étale covering $f: C \to C'$ the global sections of $\omega_{C'} \otimes \eta$ are called *Prym differentials*, in honour of Prym who studied them in Prym [1]. Correspondingly in the theory of curves the map $\varphi_{\omega_{C'}\otimes\eta}$ is called *Prym canonical map.* This was apparently the reason for Mumford, to invent the name Prym variety (see Mumford [5]).

As a consequence of the proposition we obtain that the differential $d\pi_p$ of π at a point p is not injective if and only if p is a base point of the linear system $f^*|\omega_{C'} \otimes \eta|$ or equivalently $f(p)$ is a base point of $|\omega_{C'} \otimes \eta|$. This leads to the following

(5.5) Corollary. *The differential $d\pi_p$ at a point $p \in C$ is injective unless one of the following two cases holds:*

a) f is étale, the curve C' is hyperelliptic, $\eta = \mathcal{O}_{C'}(f(p) - q)$ for some $q \in C'$, and $f(p)$ and q are distinct ramification points of the hyperelliptic covering.

b) f is ramified in $p_0, q_0 \in C'$, the curve C' is hyperelliptic, and $\eta = \mathcal{O}_{C'}(f(p))$ with $2f(p) \sim p_0 + q_0$.

Proof. By what we have seen above, the map $d\pi_p$ is not injective if and only if $f(p)$ is a base point of the linear system $|\omega_{C'} \otimes \eta|$ or equivalently if and only if

$$h^0(\omega_{C'} \otimes \eta) = h^0\left(\omega_{C'} \otimes \eta(-f(p))\right) .$$

Suppose first that f is étale. Then we have $\deg(\omega_{C'} \otimes \eta) = 2g(C') - 2$ and $h^0(\omega_{C'} \otimes \eta) = g(C') - 1$, since $\eta \neq \mathcal{O}_C$. Hence $d\pi_p$ is not injective if and only if $h^0\left(\eta^{-1}(f(p))\right) = 1$, i.e. $\eta \simeq \mathcal{O}_{C'}(f(p) - q)$ for some $q \neq f(p)$. Squaring gives $\mathcal{O}_{C'} \simeq \eta^2 = \mathcal{O}_{C'}(2f(p) - 2q)$. So $\mathcal{O}_{C'}\left(2f(p)\right) \simeq \mathcal{O}_{C'}(2q)$ and C' is hyperelliptic with $f(p)$ and q distinct ramification points of the hyperelliptic covering.

In case f is ramified we have $\deg(\omega_{C'} \otimes \eta) = 2g(C') - 1$ and $h^0(\omega_{C'} \otimes \eta) = g(C')$. Hence $d\pi_p$ is not injective if and only if $\omega_{C'} \otimes \eta(-f(p)) \simeq \omega_{C'}$, i.e. $\eta = \mathcal{O}_{C'}(f(p))$. Squaring gives $\mathcal{O}_{C'}(p_0 + q_0) \simeq \eta^2 = \mathcal{O}_{C'}(2f(p))$ and C' is hyperelliptic. □

Note that in case a) of Corollary 5.5 the curve C necessarily is also hyperelliptic, since $\mathcal{O}_C = f^*\mathcal{O}_{C'}(f(p) - q)$ and thus $f^{-1}(f(p)) \sim f^{-1}(q)$. So combining Corollary 5.5 and Proposition 5.2, we obtain in the nonhyperelliptic case

(5.6) Corollary. *Suppose C is not hyperelliptic.*

a) If $f: C \to C'$ is étale, the Abel-Prym map $\pi: C \to Z$ is an embedding.

b) If $f: C \to C'$ is ramified in $\tilde{p}_0, \tilde{q}_0 \in C$, then $\pi: C \to Z$ embeds $C - \{\tilde{p}_0, \tilde{q}_0\}$ and $\pi(\tilde{p}_0) = \pi(\tilde{q}_0)$ is an ordinary double point unless C' is hyperelliptic and $\eta = \mathcal{O}_{C'}(p)$ for some $p \in C'$ with $2p \sim p_0 + q_0$.

In the hyperelliptic case we have for both types of coverings $f: C \to C'$:

(5.7) Corollary. *Suppose C is hyperelliptic. Then $D = \pi(C)$ is a smooth hyperelliptic curve and the Prym variety (Z, Ξ) is the Jacobian of D.*

Proof. According to Welter's Criterion (2.2) we have $\pi_*[C] = \frac{2}{(g-1)!} \bigwedge^{g-1}[\Xi]$ in $H^{2g-2}(Z, \mathbb{Z})$. Since π is of degree 2, this implies $[D] = \frac{1}{(g-1)!} \bigwedge^{g-1}[\Xi]$, and Matsusaka's Criterion (see Remark 2.5) says that D is smooth with Jacobian (Z, Ξ). Moreover D is hyperelliptic, since the norm map of the double covering $\pi: C \to D$ sends the hyperelliptic linear system g_2^1 on C to a linear system of degree 2 and dimension 1 on D. □

§ 6 The Theta Divisor of a Prym Variety

Let $f: C \to C'$ be an étale double covering of smooth projective curves of genus $g(C) = 2g+1$ and $g(C') = g+1 \geq 2$, and let (Z, Ξ) denote the associated Prym variety. By definition, a suitable translate of the theta divisor Θ of the Jacobian J of C restricts to a divisor on Z algebraically equivalent to 2Ξ. In this section we will show how one can choose Θ and Ξ within their algebraic equivalence classes to get the equality $\iota_Z^*\Theta = 2\Xi$. Moreover we will study the singularities of Ξ.

Recall the notation of Section 3: as usual we identify the Jacobians J of C and J' of C' with their dual varieties via the canonical polarizations. The morphism f induces $N_f: J \to J'$, the norm map and $f^*: J' \to J$, the pullback map of line bundles, which are related by $\widehat{N_f} = f^*$. The homomorphism f^* factorizes into an isogeny $j: J' \to Y = \operatorname{im} f^*$ of degree 2 and the canonical embedding $\iota_Y: Y \hookrightarrow J$. This leads to the following diagram

$$\begin{array}{ccccccccc}
 & & & & 0 & & 0 & & \\
 & & & & \downarrow & & \downarrow & & \\
0 & \longrightarrow & (J/Y)\widehat{} & \longrightarrow & \ker N_f & \longrightarrow & \mathbb{Z}/2\mathbb{Z} & \longrightarrow & 0\\
 & & \| & & \downarrow & & \downarrow & & \\
0 & \longrightarrow & (J/Y)\widehat{} & \longrightarrow & J & \xrightarrow{\widehat{\iota_Y}} & \widehat{Y} & \longrightarrow & 0\\
 & & & & {\scriptstyle N_f}\downarrow & & \downarrow{\scriptstyle \widehat{j}} & & \\
 & & & & J' & = & J' & & \\
 & & & & \downarrow & & \downarrow & & \\
 & & & & 0 & & 0 & &
\end{array}$$

Since $Z = (J/Y)\widehat{}$ by Proposition 1.3, this proves

(6.1) Proposition. $\ker N_f$ *consists of two connected components* Z *and* Z_1.

Our next aim is to distinguish the components Z and Z_1 in terms of line bundles on C: mapping Z and Z_1 bijectively onto subvarieties of $\mathrm{Pic}^{g(C)-1}(C)$, these components can be distinguished by the cohomology of the corresponding line bundles.

Fix an even theta characteristic κ on C which is the pullback of a theta characteristic κ' on C'. Note that such a theta characteristic exists. To see this, consider the sets S^+ of even theta characteristics on C', and S_0^+ of theta characteristics κ' on C' such that $\kappa' \otimes \eta$ is even (here η denotes as usual the 2-division point of $\mathrm{Pic}^0(C')$ corresponding to the covering $f\colon C \to C'$). Since $\#S^+ = \#S_0^+ = 2^g(2^{g+1}+1)$ by Proposition 11.2.7, the sets S^+ and S_0^+ have a nonempty intersection. Any $\kappa' \in S^+ \cap S_0^+$ satisfies $h^0(f^*\kappa') = h^0(f_*f^*\kappa') = h^0(\kappa') + h^0(\kappa' \otimes \eta) \equiv 0 \,(\mathrm{mod}\, 2)$. With this notation we have

(6.2) Theorem. $Z = \{L \in \ker N_f \subset \mathrm{Pic}^0(C) \mid h^0(L \otimes \kappa) \equiv 0 \,(\mathrm{mod}\, 2)\}$ *and* $Z_1 = \{L \in \ker N_f \mid h^0(L \otimes \kappa) \equiv 1 \,(\mathrm{mod}\, 2)\}$.

Another description of the components Z and Z_1 is (see Exercise 6):

$$Z = (1 - \tilde{\iota})\,\mathrm{Pic}^0(C) \quad \text{and} \quad Z_1 = (1 - \tilde{\iota})\,\mathrm{Pic}^1(C)\ .$$

Proof. Step I: The parity of $h^0(L \otimes \kappa)$ *is constant on* Z *and* Z_1*:*
For the proof we apply a theorem of Mumford [4] on families of vector bundles on curves with quadratic form (see also Arbarello et al. [1]). Recall that a nondegenerate quadratic form on a vector bundle E on C' with values in a line bundle M on C' is the composition of the diagonal map $E \to E \otimes E$ with a symmetric bilinear form $B\colon E \otimes E \to M$, which is nondegenerate on every fibre. For every $L \in \ker N_f$ we will construct such a quadratic form on the vector bundle $E = f_*L$. Note that $\mathcal{O}_C = f^*N_f(L) = L \otimes \iota^*L$. So for every open $U \subset C'$ and $\sigma, \sigma' \in \Gamma(U, f_*L) = \Gamma(f^{-1}U, L)$ the section $\sigma \otimes \iota^*\sigma'$ is a section of $\mathcal{O}_C$ over $f^{-1}U$. Let Nm denote the usual norm map of the function fields of C and C', then $B(\sigma, \sigma') = Nm(\sigma \otimes \iota^*\sigma')$ is a section of $\mathcal{O}_{C'}$ over U. Certainly this defines a nondegenerate quadratic form on the vector bundle f_*L with values in $\mathcal{O}_{C'}$. Tensoring f_*L with the theta characteristic κ' of above, we obtain a quadratic form on $f_*L \otimes \kappa'$ with values in $\omega_{C'}$.

This construction can be easily generalized to families of line bundles on C, namely the restriction of the Poincaré bundle on $C \times \mathrm{Pic}^0(C)$ to $C \times Z$ and $C \times Z_1$. In this way we obtain a family of vector bundles on C', endowed with a quadratic form with values in $\omega_{C'}$ parametrized by the components Z and Z_1. Now Mumford's theorem says that the map $L \mapsto h^0(f_*L \otimes \kappa') (\mathrm{mod}\, 2)$ is constant on Z and Z_1. This implies the assertion, since $h^0(f_*L \otimes \kappa') = h^0(f_*(L \otimes f^*\kappa')) = h^0(L \otimes \kappa)$.

Step II: There is an $L \in \ker N_f$ *with* $h^0(L \otimes \kappa) \neq 0$*:*
For the proof let $\sum p_\nu$ be a divisor in the linear system $|\omega_{C'}|$. For every ν choose a point $\tilde{p}_\nu \in C$ over p_ν, then

$$N_f(\mathcal{O}_C(\sum \tilde{p}_\nu) \otimes \kappa^{-1}) = \omega_{C'} \otimes N_f f^* \kappa'^{-1} = \omega_{C'} \otimes \kappa'^{-2} = \mathcal{O}_{C'} .$$

So $L = \mathcal{O}_C(\sum \tilde{p}_\nu) \otimes \kappa^{-1}$ is an element of $\ker N_f$. In particular $L \otimes \kappa = \mathcal{O}_C(\sum \tilde{p}_\nu)$ has a non-trivial section.
Step III: In order to complete the proof, note first that Z is contained in the set $\{L \in \ker N_f \mid h^0(L \otimes \kappa) \equiv 0 \ (\mathrm{mod}\, 2)\}$, since the trivial line bundle is contained in Z and $h^0(\kappa) \equiv 0 \ (\mathrm{mod}\, 2)$ by assumption. We have to show that $\{L \in \ker N_f \mid h^0(L \otimes \kappa) \equiv 1 \ (\mathrm{mod}\, 2)\}$ is nonempty. For this it suffices to show that for $L \in \ker N_f$ with $h^0(L \otimes \kappa) \neq 0$ there is a $p \in C$ with

$$h^0\big(L(p - \iota(p)) \otimes \kappa\big) = h^0(L \otimes \kappa) - 1 . \tag{1}$$

Let $r = h^0(L \otimes \kappa)$. Choose a point $p \in C$ such that $\iota(p)$ is not a base point of the complete linear system $|L \otimes \kappa|$. Then $h^0\big(L(-\iota(p)) \otimes \kappa\big) = r - 1$ and thus $h^0\big(L(p - \iota(p)) \otimes \kappa\big) = r$ or $r - 1$.

Assume $h^0\big(L(p - \iota(p)) \otimes \kappa\big) = r$. The Riemann-Roch Theorem for curves implies $h^1\big(L(p - \iota(p)) \otimes \kappa\big) = r = h^1\big(L(-\iota(p)) \otimes \kappa\big)$ or equivalently $h^0\big(L^{-1}(\iota(p) - p) \otimes \kappa\big) = h^0\big(L^{-1}(\iota(p)) \otimes \kappa\big)$. So p is a base point of the linear system $|L^{-1} \otimes \kappa|$. But since $h^0(L^{-1} \otimes \kappa) = h^1(L^{-1} \otimes \kappa) = h^0(L \otimes \kappa) > 0$ not every $p \in C$ is a base point of $|L^{-1} \otimes \kappa|$. Hence a general $p \in C$ satisfies (1). □

Recall the canonical theta divisor $W_{g(C)-1}$ in $\mathrm{Pic}^{g(C)-1}(C)$. Choose the theta divisor Θ of $J = \mathrm{Pic}^0(C)$ as follows

$$\Theta = W_{g(C)-1} - \kappa$$

with κ the particular theta characteristic chosen above. From Riemann's Singularity Theorem 11.2.5 we deduce:

(6.3) Proposition. *a) There is a theta divisor* Ξ *defining the principal polarization of* Z *such that* $\iota_Z^* \Theta = 2\Xi$.
b) $Z_1 \subset \Theta$.

Proof. b) follows immediately from the definition of Θ and the previous proposition. As for a): by Theorem 6.2 we have $h^0(L \otimes \kappa) \geq 2$ for any $L \in \Theta \cap Z$. So by Riemann's Singularity Theorem L is a singular point of Θ. Hence $\iota_Z^* \Theta$ consists entirely of multiple components. On the other hand, for an arbitrary theta divisor $\widetilde{\Xi}$ defining the principal polarization of Z we have $\iota_Z^*\Theta \equiv 2\widetilde{\Xi}$. Since the homomorphism $\phi_{2\widetilde{\Xi}}: Z \to \widehat{Z}$ is surjective, there is a $z \in Z$ such that

$$\iota_Z^*\Theta - 2\widetilde{\Xi} \sim \phi_{2\widetilde{\Xi}}(z) \sim 2t_z^*\widetilde{\Xi} - 2\widetilde{\Xi}$$

So $\Xi := t_z^*\widetilde{\Xi}$ satisfies the assertion. □

Case a) of the proposition says that every point in Ξ is of even multiplicity considered as a point in Θ. In particular, every singular point x of Ξ is also a singular point of Θ. Translating this into terms of the tangent cone $TC_x\Theta$ of Θ in x, one of the following two cases occurs: the tangent cone $TC_x\Theta$ either intersects the tangent space T_xZ transversely or contains it. This gives

(6.4) Corollary. $\operatorname{sing}\Xi = \{x \in Z \mid \operatorname{mult}_x\Theta \geq 4\} \cup \{x \in Z \mid {\operatorname{mult}_x\Theta=2 \atop T_xZ\subset TC_x\Theta}\}$

A singular point x of Ξ is called *stable* if $\operatorname{mult}_x\Theta \geq 4$, it is called *exceptional* otherwise. We will see in Remark 6.7 that exceptional singularities exist only for special coverings $f: C \to C'$. The importance of exceptional singularities lies in the fact that they provide special line bundles of low degree on C'.

(6.5) Proposition. *A singularity L of Ξ (considered as a line bundle on C) is exceptional if and only if $L \otimes \kappa = f^*M \otimes \mathcal{O}_C(B)$ with $M \in \operatorname{Pic}(C')$ such that $h^0(M) = 2$ and B an effective divisor on C.*

The proof uses a result of Kempf, which is part of his generalization of Riemann's Singularity Theorem, namely

(6.6) Proposition (Kempf). *With the notation as above, suppose $L \in J = \operatorname{Pic}^0(C)$ with $h^0(C, L \otimes \kappa) = r \geq 1$. Let $\{s_1, \dots, s_r\}$ and $\{t_1, \dots, t_r\}$ be bases of $H^0(L \otimes \kappa)$ and $H^0(L^{-1} \otimes \kappa)$ respectively. Consider $H^0(\omega_C)$ as the cotangent space T_L^*J to J at the point L. Then*

$$\det(s_i \otimes t_j) = 0$$

is an equation of the tangent cone $TC_L\Theta$ of Θ at L.

Of course the proposition is valid not only in our situation, but for every smooth curve C and every theta characteristic on C. For the proof we refer to Arbarello et al. [1] p. 240.

Proof of Proposition 6.5. For an exceptional singularity $L \in \operatorname{sing}\Xi$ we have $h^0(L \otimes \kappa) = 2$ by Riemann's Singularity Theorem 11.2.5. Choose a basis s_1, s_2 of $H^0(L \otimes \kappa)$. Since L is an element of $\ker N_f$, we have $\iota^*L = L^{-1}$. Now $\kappa = f^*\kappa'$ implies $L^{-1} \otimes \kappa = \iota^*(L \otimes \kappa)$. Thus ι^*s_1, ι^*s_2 is a basis for $H^0(L^{-1} \otimes \kappa)$. By Proposition 6.6 an equation for the tangent cone $TC_L\Theta$ is given by $\det(s_i \otimes \iota^* s_j) = 0$. Since $Z = \operatorname{im}\{(1-\tilde{\iota}): J \to J\}$ by Lemma 4.1, the cotangent space $T_L^* Z$ of Z in L is the (-1)-eigenspace in $H^0(\omega_C) = T_L^* J$ under the action of ι. Moreover, $\omega \mapsto \frac{1}{2}(\omega - \iota^*\omega)$ defines the projection $T_L^* J \to T_L^* Z$. This shows that the restriction of $TC_L\Theta$ to $T_L Z$ is given by $(\det(s_i \otimes \iota^* s_j - s_j \otimes \iota^* s_i)^{\frac{1}{2}}) = 0$, i.e. by $s_1 \otimes \iota^* s_2 - s_2 \otimes \iota^* s_1 = 0$.

Hence $T_L Z \subset TC_L\Theta$ if and only if $s_1 \otimes \iota^* s_2 = s_2 \otimes \iota^* s_1$ as sections in $H^0(\omega_C)$, i.e. if and only if $\iota^*(\frac{s_1}{s_2}) = \frac{s_1}{s_2}$ as meromorphic functions on C. That is, $\frac{s_1}{s_2}$ is the pullback of a meromorphic function on C'. In terms of linear systems this means $T_L Z \subset TC_L\Theta$ if and only if the map $\varphi_{L\otimes\kappa}: C \to \mathbb{P}_1$ factorizes via f:

$$\begin{array}{ccc} C & \xrightarrow{\varphi_{L\otimes\kappa}} & \mathbb{P}_1 \\ & {}_f\searrow \quad \nearrow_{\psi} & \\ & C' & . \end{array}$$

Define $M = \psi^* \mathcal{O}_{\mathbb{P}_1}(1)$. Then $h^0(M) = 2$ and $L \otimes \kappa = f^* M \otimes \mathcal{O}_C(B)$ where B denotes the base locus of $L \otimes \kappa$. This completes the proof. □

(6.7) Remark. *Let the notation be as above. If C' is a general curve of genus $g+1$ (in the sense of Brill-Noether), then the theta divisor Ξ of Z does not admit exceptional singularities.*

Proof. Suppose $L \in \Xi$ is exceptional. According to Proposition 6.5 we have $L \otimes \kappa = f^* M \otimes \mathcal{O}_C(B)$ with $h^0(M) = 2$ and B effective on C. Consider the Petri map $\mu: H^0(M) \otimes H^0(\omega_{C'} \otimes M^{-1}) \to H^0(\omega_{C'})$ (see Arbarello et al. [1] p. 215).

By the base point free pencil trick (see Arbarello et al. [1] p.126) the kernel of μ is isomorphic to $H^0(\omega_{C'} \otimes M^{-1} \otimes M^{-1}(B'))$ where B' is the base locus of the linear system $|M|$. On the other hand, since $L \in \ker N_f$,

$$\begin{aligned} \omega_{C'} = \kappa'^2 &= N_f(L) \otimes N_f f^* \kappa' = N_f(L \otimes \kappa) \\ &= N_f(f^* M \otimes \mathcal{O}_C(B)) = M^2 \otimes N_f \mathcal{O}_C(B)) \ . \end{aligned}$$

So $h^0(\omega_{C'} \otimes M^{-2}(B')) = h^0(N_f(\mathcal{O}_C(B)) \otimes \mathcal{O}_{C'}(B')) > 0$, contradicting Petri's theorem (see Arbarello et al. [1] p.215), which says that μ is injective for a general curve C'. □

§ 7 Recillas' Theorem

By a *d-gonal curve* we mean a smooth projective curve C together with a base point free linear system g_d^1 or, equivalently, a morphism $C \to \mathbb{P}_1$ of degree d. In this section we present Recillas' Theorem (see Recillas [1]) saying that the Prym variety associated to an étale double covering of a general trigonal curve is the Jacobian of a general tetragonal curve and conversely.

Let X be a general tetragonal curve of genus g. Here *general* means that X is not hyperelliptic and any fibre of the 4-fold covering $k: X \to \mathbb{P}_1$, given by the g_4^1, admits at most one ramification point and this is of index ≤ 3. First we describe, how one can associate to X an étale double covering $f: C \to C'$ of a trigonal curve C'.

We start with a simple geometric description: the Geometric Riemann-Roch Theorem (see proof of Proposition 11.2.9) says that there is a pencil of planes $\{P_t \mid t \in \mathbb{P}_1\}$ in $\mathbb{P}_{g-1}$ intersecting the canonical model of X in $\mathbb{P}_{g-1}$ in the given g_4^1. But four different coplanar points determine three pairs of diagonals, whose three intersection points form the "diagonal triangle" of the four points. The vertices of the diagonal triangles, associated to the pencil $\{X \cap P_t \mid t \in \mathbb{P}_1\}$, sweep out a trigonal curve C'. The double covering $f: C \to C'$ is given by the space of diagonals.

Now we give an abstract and more detailed version of this construction: as usual, the g_4^1 will be considered as a curve isomorphic to $\mathbb{P}_1$ in the fourth symmetric product $X^{(4)}$ of the curve X. Let $s: X^{(2)} \times X^{(2)} \to X^{(4)}$ denote the sum map and $q_1: X^{(2)} \times X^{(2)} \to X^{(2)}$ the projection onto the first factor. Define a curve C in $X^{(2)}$ by

$$C = q_1 s^{-1}(g_4^1) = \{p_1 + p_2 \in X^{(2)} \mid p_1 + p_2 + p_3 + p_4 \in g_4^1 \text{ for some } p_3, p_4 \in X\}$$

with reduced subscheme structure. There is an obvious map $h: C \to \mathbb{P}_1$. If $k^{-1}(x) = \{p_1, \dots, p_4\}$ for some $x \in \mathbb{P}_1$, then

$$h^{-1}(x) = \{p_1 + p_2, p_1 + p_3, p_1 + p_4, p_2 + p_3, p_2 + p_4, p_3 + p_4\} .$$

So h is a morphism of degree 6.

(7.1) Lemma. *C is a smooth irreducible curve of genus $2g + 1$.*

Proof. Recall the well known 1-1 correspondence between the set of (smooth) degree n coverings of $\mathbb{P}_1$ modulo isomorphisms which are ramified at most in a finite subset δ of $\mathbb{P}_1$, and the set of representations of the fundamental group $\rho : \pi_1(\mathbb{P}_1 - \delta, x_0) \to \mathfrak{S}_n$ modulo conjugation. Intuitively, if $1, \dots, n$ denote the sheets of the covering over $\mathbb{P}_1$ and σ is an element of $\pi_1(\mathbb{P}_1 - \delta, x_0)$

which "goes once around a point $x \in \delta$", then above x the sheets $1, \ldots, n$ are glued according to the permutation $\rho(\sigma)$. From this one immediately sees that a covering is irreducible if and only if the corresponding representation has a transitive image in $\mathfrak{S}_n$.

Let δ denote the ramification divisor of the 4-fold covering $k: X \to \mathbb{P}_1$ and let $\rho: \pi_1(\mathbb{P}_1 - \delta, x_0) \to \mathfrak{S}_4$ be a corresponding representation. By the assumption made on $k: X \to \mathbb{P}_1$, the image of ρ is a transitive subgroup of $\mathfrak{S}_4$, generated by cycles of length ≤ 3. But any such subgroup contains the alternating group $\mathfrak{A}_4$.

The representation ρ induces a representation $\rho^{(2)}: \pi_1(\mathbb{P}_1 - \delta, x_0) \to \mathfrak{S}_6$: the action of $\pi_1(\mathbb{P}_1 - \delta, x_0)$ on the set $\{1, \ldots, 4\}$ induces an action on the second symmetric product $\{1, \ldots, 4\}^{(2)} = \{\{1,2\}, \{1,3\}, \{1,4\}, \{2,3\}, \{2,4\}, \{3,4\}\}$. By construction the representation $\rho^{(2)}$ corresponds to the covering $h: C \to \mathbb{P}_1$. Since by assumption $\operatorname{im} \rho$ contains $\mathfrak{A}_4$, the group $\pi_1(\mathbb{P}_1 - \delta, x_0)$ acts transitively on the set of pairs $\{i, j\}$. So C is smooth and irreducible.

It remains to compute its genus: obviously $h: C \to \mathbb{P}_1$ is ramified exactly over the divisor δ and the degree of the ramification divisor of h is twice the degree of δ. Hence by Hurwitz's formula $g(C) = 6(g(\mathbb{P}_1) - 1) + 1 + \deg \delta = 2g + 1$. □

The curve C admits a natural involution ι, sending an element $p_1 + p_2$ of $h^{-1}(x)$ to the complementary element $p_3 + p_4$. By the assumption on $k: X \to \mathbb{P}_1$ the involution ι acts fix point free. Let

$$f: C \to C' := C/\iota$$

denote the natural projection. The covering h induces a map $h': C' \to \mathbb{P}_1$ of degree 3. In the above notation

$$h'^{-1}(x) = \{(p_1+p_2) \sim (p_3+p_4), (p_1+p_3) \sim (p_2+p_4), (p_1+p_4) \sim (p_2+p_3)\}.$$

Clearly h' is ramified exactly over δ. Summing up, $f: C \to C'$ is an étale double covering of a trigonal curve C' of genus $g+1$.

(7.2) Theorem. *Let X be a general tetragonal curve as above and $f: C \to C'$ the associated étale double covering of a trigonal curve C'. There is an isomorphism between the Prym variety (Z, Ξ) associated to $f: C \to C'$ and the Jacobian (J, Θ) of X.*

Proof. Choose a line bundle $L \in \operatorname{Pic}^2(X)$ with $L^2 = g^1_4$ and denote as usual by $\alpha_L: X^{(2)} \to J(X)$ the map $\alpha_L(p_1 + p_2) = \mathcal{O}_X(p_1 + p_2) \otimes L^{-1}$, as well as its restriction $\alpha = \alpha_L: C \to J(X)$ to C. By the choice of L we have $\alpha\iota = -\alpha$ on C. According to the Universal Property 5.1 the morphism $\alpha_L: C \to J(X)$ factorizes via the Abel-Prym map $\pi = \pi_c$ for any point $c \in C$

$$\begin{array}{ccc} C & \xrightarrow{\alpha} & J(X) \\ {\scriptstyle \pi_c}\downarrow & & \downarrow{\scriptstyle t_{-\alpha(c)}} \\ Z & \xrightarrow{\tilde{\alpha}} & J(X) \end{array} .$$

We have to show that $\tilde{\alpha}: Z \to J(X)$ is an isomorphism, compatible with the natural principal polarizations. We claim that it suffices to show that

$$\alpha_*[C] = \tfrac{2}{(g-1)!} \bigwedge^{g-1} [\Theta] \quad \text{in } H^{2g-2}(J(X), \mathbb{Z}). \tag{1}$$

To see this, note first that $\alpha(C)$ generates $J(X)$ as an abelian variety. Consequently $\tilde{\alpha}$ is an isogeny, Z and $J(X)$ being of the same dimensions. According to Welters' Criterion 2.2 we have $\pi_*[C] = \frac{2}{(g-1)!} \bigwedge^{g-1}[\Xi]$ in $H^{2g-2}(Z, \mathbb{Z})$. So $\tilde{\alpha}_* \bigwedge^{g-1}[\Xi] = \bigwedge^{g-1}[\Theta]$ and $\tilde{\alpha}$ is an isomorphism of polarized abelian varieties by Lemma 2.3.

For the proof of (1) let $q: X^2 \to X^{(2)}$ be the canonical map and $\Delta_{\mathbb{P}_1}$ respectively Δ_X the diagonals in $\mathbb{P}_1 \times \mathbb{P}_1$ respectively $X \times X$. Using that $[\Delta_{\mathbb{P}_1}] = [\mathbb{P}_1 \times \{x\}] + [\{x\} \times \mathbb{P}_1]$ in $H^2(\mathbb{P}_1 \times \mathbb{P}_1, \mathbb{Z})$ we have

$$[\Delta_X] + q^*[C] = (k \times k)^*[\Delta_{\mathbb{P}_1}] = 4[X \times \{p\}] + 4[\{p\} \times X] \quad \text{in } H^2(X^2, \mathbb{Z})$$

for some $p \in X$. Denote $\delta := q(\Delta_X)$ and $X + p := q(X \times \{p\}) = q(\{p\} \times X)$. Applying q_* we obtain

$$[\delta] + 2[C] = 8[X + p] \quad \text{in } H^2(X^{(2)}, \mathbb{Z}) \,, \tag{2}$$

since $q: \Delta_X \to \delta$ and $q: X \times \{p\} \to X + p$ are of degree 1. For $M \in \operatorname{Pic}^1(X)$ with $M^2 = L$ we have $\alpha(2p) = 2\alpha_M(p)$ for all $p \in X$ with $\alpha_M: X \to J(X)$ the embedding associated to M. This implies

$$\alpha_*[\delta] = 2_{J(X)_*}[X] = 4[X] \quad \text{in} \quad H^{2g-2}(J(X), \mathbb{Z}) \,.$$

Here the push forward homomorphisms $\alpha_*: H^2(X^{(2)}, \mathbb{Z}) \to H^{2g-2}(J(X), \mathbb{Z})$ and $2_{J(X)*}$ on $H^{2g-2}(J(X), \mathbb{Z})$ are defined by the corresponding push forward homomorphism in homology via Poincaré duality. In particular $2_{J(X)*}$ is multiplication by $2^2 = 4$ on $H^{2g-2}(J(X), \mathbb{Z})$, since $H_2(J(X), \mathbb{Z})$ is generated by 2-cycles $\lambda_\mu \star \lambda_\nu$ with $1 \le \mu < \nu \le 2g$ (see Lemma 4.10.1) and by definition $2_{J(X)*}(\lambda_\mu \star \lambda_\nu) = (2\lambda_\mu) \star (2\lambda_\nu) = 4\lambda_\mu \star \lambda_\nu$.

On the other hand, we have

$$\alpha_*[X + p] = [X] \quad \text{in} \quad H^{2g-2}(J(X), \mathbb{Z}) \,.$$

Since $[X] = \frac{1}{(g-1)!} \bigwedge^{g-1}[\Theta]$, the assertion follows by applying α_* to equation (2). □

Theorem 7.2 has a counterpart, also due to Recillas [1]

(7.3) Theorem. *The Prym variety* (Z, Ξ) *associated to an étale double covering* $f: C \to C'$ *of a trigonal, non hyperelliptic curve* C' *is isomorphic to the Jacobian* $(J(X), \Theta)$ *of a tetragonal curve* X.

Since the proof is very similar to Donagi's tetragonal construction, which we give in the next section, we omit it here. Beauville [2] proves a more general theorem of which Theorems 7.3 and 8.2 are special cases.

§ 8 Donagi's Tetragonal Construction

The tetragonal construction (see Donagi [1]) associates to any étale double covering of a tetragonal curve two more such double coverings, all with the same Prym variety: let C' be a general tetragonal curve of genus $g+1$. Here *general* means that in any fibre of the morphism $k: C' \to \mathbb{P}_1$, given by the g_4^1, there is at most one ramification point and this is of index ≤ 3. Let $f: C \to C'$ be an étale double covering. We will see that the set of divisors of degree 4 on C which push down via f to divisors of the g_4^1 define a curve $\widetilde{C}$ with a map of degree 16 onto $g_4^1 \simeq \mathbb{P}_1$. This curve has two components C_1 and C_2 which again are étale double coverings of tetragonal curves C_1' and C_2' respectively. The aim is to show that the three associated Prym varieties coincide. Parts of the proof below are due to Beauville [2] and Welters [1].

For the precise definition of the curve $\widetilde{C}$ consider the induced morphism $f^{(4)}: C^{(4)} \to C'^{(4)}$. Define $\widetilde{C}$ to be the pullback of the g_4^1

$$\begin{array}{ccccc} \widetilde{C} & = & {f^{(4)}}^{-1}(g_4^1) & \hookrightarrow & C^{(4)} \\ {\scriptstyle \tilde{f}}\downarrow & & \downarrow & & \downarrow {\scriptstyle f^{(4)}} \\ \mathbb{P}_1 & = & g_4^1 & \overset{j}{\hookrightarrow} & C'^{(4)} \end{array} \qquad (1)$$

Obviously the induced map $\tilde{f}: \widetilde{C} \to g_4^1 = \mathbb{P}_1$ is of degree $2^4 = 16$, étale over the open set of smooth divisors of the linear system g_4^1. We call two divisors in $\widetilde{C}$ equivalent, if they push down to the same divisor in g_4^1 and share an even number of points of C. Denote by C_1 and C_2 the corresponding equivalence classes. We claim that the monodromy of the map $\tilde{f}: \widetilde{C} \to \mathbb{P}_1$ preserves these equivalence classes. To see this, note first that $\tilde{f}$ is ramified exactly over the ramification locus of $k: C' \to \mathbb{P}_1$. Going around a branch point in $\mathbb{P}_1$ causes a divisor $p_1 + \cdots + p_4 \in \widetilde{C}$ to exchange either zero or two of its points p_i for their conjugates with respect to the involution ι on

C corresponding to the covering f. The following picture illustrates this in the case of a simple ramification point: the divisor D remains fixed, whereas in D' two points are exchanged by their conjugates

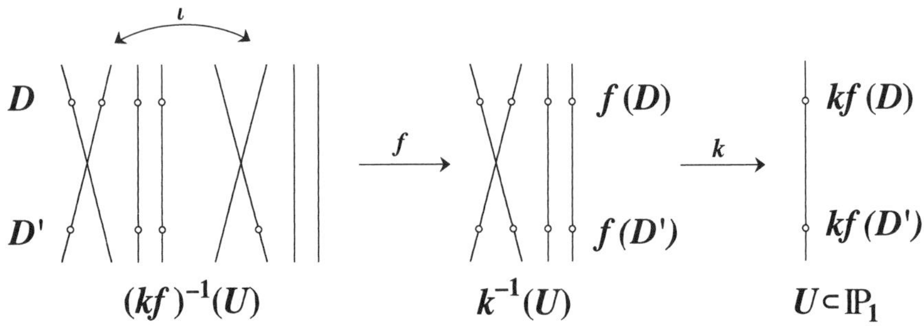

So the map $\tilde{f}: \widetilde{C} \to \mathbb{P}_1$ splits into two maps $C_\nu \to \mathbb{P}_1$ of degree 8 with disjoint curves C_1 and C_2.

(8.1) Lemma. *The curves C_1 and C_2 are smooth and irreducible of genus $2g+1$ if and only if the g_4^1 is general in the above sense.*

We postpone the proof of the lemma until the end of this section, in order not to interrupt the construction. For a generalization see Exercise 14.

The involution ι on C induces an involution on $\widetilde{C}$ defined by $p_1 + \cdots + p_4 \mapsto \iota(p_1) + \cdots + \iota(p_4)$. Clearly this gives fixed point free involutions ι_ν on the components C_ν. Denoting $C'_\nu = C_\nu / \iota_\nu$ we get the following factorizations

$$\begin{array}{ccc} C_\nu & \longrightarrow & \mathbb{P}_1 \\ & {\scriptstyle f_\nu} \searrow \quad \nearrow {\scriptstyle k_\nu} & \\ & C'_\nu & \end{array}$$

with étale double coverings f_ν and tetragonal coverings k_ν for $\nu = 1, 2$.

Summing up, we constructed out of the étale double covering $f: C \to C'$ and the tetragonal covering $k: C' \to \mathbb{P}_1$ two more such data $f_\nu: C_\nu \to C'_\nu$ and $k_\nu: C'_\nu \to \mathbb{P}_1$. The main result is the following theorem, due to Donagi.

(8.2) Theorem. *The Prym variety (Z_ν, Ξ_ν) associated to the double covering $f_\nu: C_\nu \to C'_\nu$ is isomorphic to the Prym variety (Z, Ξ) associated to the double covering $f: C \to C'$ for $\nu = 1, 2$.*

Proof. It suffices to prove the assertion for C_1.
Step I: Choose a point $c \in C_1 \subset C^{(4)}$ and denote by $\alpha = \alpha_c: C^{(4)} \to J = J(C)$ and $\alpha' = \alpha_{f^{(4)}(c)}: C'^{(4)} \to J' = J(C')$ the corresponding morphisms. Then the following diagram is commutative

$$\begin{array}{ccccc} \widetilde{C} = C_1 \cup C_2 & \hookrightarrow & C^{(4)} & \xrightarrow{\alpha} & J \\ & & {\scriptstyle f^{(4)}}\big\downarrow & & \big\downarrow{\scriptstyle N_f} \\ & & C'^{(4)} & \xrightarrow{\alpha'} & J' \end{array} \qquad (2)$$

It shows that $\alpha(\widetilde{C}) \subset \ker N_f = Z \cup Z'$, since α' maps $f^{(4)}(C_\nu) = |g^1_4| = \mathbb{P}_1$ onto 0 in J' for $\nu = 1, 2$. By choice of c we have $\alpha(C_1) \subset (\ker N_f)_0 = Z$. Define $\varphi = \alpha|C_1 : C_1 \to Z$. For $p_1 + \cdots + p_4 \in C_1$ we have

$$\begin{aligned}(\varphi + \varphi\iota_1)(p_1 + \cdots + p_4) &= \mathcal{O}_C(p_1 + \cdots + p_4 + \iota(p_1) + \cdots + \iota(p_4) - 2c) \\ &= \mathcal{O}_C(f^* g^1_4 - 2c) \ .\end{aligned}$$

But this is a constant in Z. So replacing φ by a suitable translate (which we also denote by φ), we have $\varphi\iota_1 = -\varphi$. According to the Universal Property 5.1 the map φ factorizes via the Abel-Prym map $\pi_1 : C_1 \to Z_1$

$$\begin{array}{ccc} C_1 & \xrightarrow{\varphi} & Z \\ {\scriptstyle \pi_1}\big\downarrow & & \big\downarrow{\scriptstyle t_{-\varphi(c)}} \\ Z_1 & \xrightarrow[\widetilde{\varphi}]{} & Z \ . \end{array}$$

We have to show that $\widetilde{\varphi}$ is an isomorphism, compatible with the natural principal polarizations on Z and Z_1. Similarly as in the proof of Theorem 7.2 it suffices to show

$$\alpha_*[C_1] = \tfrac{2}{(g-1)!} \bigwedge^{g-1} [\Xi] \quad \text{in } H^{2g-2}(Z, \mathbb{Z}). \qquad (3)$$

The proof of (3) proceeds in three steps.

Step II: Denote $V_i = \alpha(C_i)$ for $i = 1, 2$. As we saw above, $V_1 \subset Z$ and $V_2 \subset Z \cup Z'$. After a suitable translation we may assume that also $V_2 \subset Z$. *We claim* $[V_1] = [V_2]$ *in* $H^{2g-2}(Z, \mathbb{Z})$.

For this we show first that $[C_1] = [C_2]$ in $H^6(C^{(4)}, \mathbb{Z})$. Let $p : C^4 \to C^{(4)}$ denote the natural map and $T_i := p^{-1}(C_i)$ for $i = 1, 2$. It suffices to show $[T_1] = [T_2]$ in $H^6(C^4, \mathbb{Z})$, since the map $p^* : H^6(C^{(4)}, \mathbb{Z}) \to H^6(C^4, \mathbb{Z})$ is injective. Consider the Künneth decomposition

$$H^6(C^4, \mathbb{Z}) = \bigoplus_{\substack{0 \le \nu_i \le 2 \\ \nu_1 + \cdots + \nu_4 = 6}} H^{\nu_1}(C, \mathbb{Z}) \otimes \cdots \otimes H^{\nu_4}(C, \mathbb{Z}) \ .$$

For every multi-index $(\nu_1, \ldots, \nu_4)$ there is an i with $\nu_i = 2$. By the symmetry of the situation, it suffices to show that in the decomposition of $[T_1] - [T_2]$ all components with $\nu_4 = 2$ are 0. In other words, we have to show that the projection of $[T_1] - [T_2]$ onto the first factor of the decomposition

$$\begin{aligned}H^6(C^4, \mathbb{Z}) = \big(H^4(C^3, \mathbb{Z}) \otimes H^2(C, \mathbb{Z})\big) \oplus \big(H^5(C^3, \mathbb{Z}) \otimes H^1(C, \mathbb{Z})\big) \oplus \\ \big(H^6(C^3, \mathbb{Z}) \otimes H^0(C, \mathbb{Z})\big)\end{aligned} \qquad (4)$$

is 0. For this consider the projection $q: C^4 \to C^3, (p_1,p_2,p_3,p_4) \mapsto (p_1,p_2,p_3)$. For every $(p_1,p_2,p_3) \in q(T_1 \cup T_2)$ there is exactly one point $p_4' \in C'$ such that $f(p_1) + f(p_2) + f(p_3) + p_4' \in g_4^1$. Let $f^{-1}(p_4') = \{p_4, \iota(p_4)\}$, then $q^{-1}(p_1,p_2,p_3) \cap (T_1 \cup T_2) = \{(p_1,p_2,p_3,p_4),(p_1,p_2,p_3,\iota(p_4))\}$. Since one of these points is contained in T_1 and the other in T_2, this implies that q restricts to isomorphisms $q: T_i \to q(T_i)$ for $i = 1,2$. Hence $q_*([T_1]-[T_2]) = 0$ in $H^4(C^3,\mathbb{Z})$. But $q_*: H^6(C^4,\mathbb{Z}) \to H^4(C^3,\mathbb{Z})$ coincides with the projection onto the first factor in (4), if we identify as usual $H^2(C,\mathbb{Z}) = \mathbb{Z}$. This shows that $[C_1] = [C_2]$ in $H^6(C^{(4)},\mathbb{Z})$.

Denoting as usual by $\iota_Z: Z \hookrightarrow J$ the canonical embedding, we obtain $\iota_{Z*}[V_1] = \alpha_*[C_1] = \alpha_*[C_2] = \iota_{Z*}[V_2]$. Since $2_Z = \phi_{\iota_Z^*\Theta} = \widehat{\iota}_Z\iota_Z$, this gives $2_{Z*}[V_1] = \widehat{\iota}_{Z*}\iota_{Z*}[V_1] = \widehat{\iota}_{Z*}\iota_{Z*}[V_2] = 2_{Z*}[V_2]$. So $2^2[V_1] = 2^2[V_2]$, which implies the assertion, since $H^*(Z,\mathbb{Z})$ is torsion free.

Step III: $\widehat{\iota}_{Z*}\alpha_*[\widetilde{C}] = \frac{16}{(g-1)!}\bigwedge^{g-1}[\Xi]$.

The proof applies a formula of Macdonald's for the class of the curve g_4^1 in $C'^{(4)}$ in $H^6(C'^{(4)},\mathbb{Z})$ (see Arbarello et al. [1] Lemma 8.3.2):

$$[g_4^1] = \sum_{k=0}^{3} \binom{2-g}{k}\eta'^k \cdot \frac{\bigwedge^{3-k}[\alpha'^*\Theta']}{(3-k)!} \qquad \text{in } H^6(C'^{(4)},\mathbb{Z}) .$$

Here $\eta'^k \in H^{2k}(C^{(4)},\mathbb{Z})$ denotes the fundamental class of the image of $C'^{(4-k)}$ in $C'^{(4)}$ under the embedding $p_1+\dots+p_{4-k} \mapsto p_1+\dots+p_{4-k}+kp$ with some point $p \in C'$. Similarly define classes $\eta^k \in H^{2k}(C^{(4)},\mathbb{Z})$. It is well known and easy to see that $\eta'^k = \bigwedge^k \eta'^1$ and $\eta^k = \bigwedge^k \eta^1$ and both are related by

$$f^{(4)^*}\eta'^k = 2^k\eta^k . \tag{5}$$

Moreover by Poincaré's Formula 11.2.1

$$\alpha_*\eta^k = \frac{\bigwedge^{2g+k-3}[\Theta]}{(2g+k-3)!} . \tag{6}$$

Applying $f^{(4)^*}$ to Macdonald's formula, we get with (5) and the commutativity of diagram (2)

$$[\widetilde{C}] = f^{(4)^*}[g_4^1] = \sum_{k=0}^{3} \binom{2-g}{k}2^k\eta^k \cdot \frac{\alpha^* N_f^* \bigwedge^{3-k}[\Theta']}{(3-k)!} .$$

Applying α_*, equation (6) and the projection formula yield

$$\alpha_*[\widetilde{C}] = \sum_{k=0}^{3} 2^k\binom{2-g}{k}\frac{\bigwedge^{2g+k-3}[\Theta]}{(2g+k-3)!} \cdot \frac{N_f^*\bigwedge^{3-k}[\Theta']}{(3-k)!} .$$

According to Proposition 3.4, $2[\Theta] = N_f^*[\Theta'] + \widehat{\iota}_Z^*[\Xi]$, so applying the binomial formula

$$\alpha_*[\widetilde{C}] = 2^{3-2g} \sum_{k=0}^{3} \binom{2-g}{k} \sum_{n=0}^{2g+k-3} \binom{2g+k-3}{n} \frac{N_f^* \wedge^{n-k+3}[\Theta']}{(3-k)!} \cdot \frac{\widehat{\iota}_Z^* \wedge^{2g-n+k-3}[\Xi]}{(2g+k-3)!} \qquad (7)$$

Before we proceed, we claim that $\widehat{\iota}_{Z*} N_f^* \wedge^m \frac{[\Theta']}{m!} = \begin{cases} 2^{2g+1} & \text{if} \quad m=g+1 \\ 0 & \quad\;\; m \neq g+1. \end{cases}$

To see this, consider the composed map $\widehat{\iota}_{Z*} N_f^* : H^{2m}(J', \mathbb{Z}) \to H^{2m}(J, \mathbb{Z}) \to H^{2m-2g-2}(Z, \mathbb{Z})$. By trivial reasons $\widehat{\iota}_{Z*} N_f^* \equiv 0$ for $m \neq g+1$. Note moreover that $\widehat{\iota}_Z$ is multiplication by 2 on Z. So the push forward homomorphism $\widehat{\iota}_{Z*}$ on $H^0(Z, \mathbb{Z}) = H^{2g-2g}(Z, \mathbb{Z})$ is multiplication by 2^{2g} as we saw in the proof of Theorem 7.2. Applying this and Poincaré's Formula 11.2.1, we obtain

$$\widehat{\iota}_{Z*} N_f^* \frac{\wedge^{g+1}[\Theta']}{(g+1)!} = \widehat{\iota}_{Z*} N_f^*[0] = 2\widehat{\iota}_{Z*}[Z] = 2^{2g+1} \; .$$

So from equation (7) we get

$$\begin{aligned} \widehat{\iota}_{Z*} \alpha_*[\widetilde{C}] &= 2^{3-2g} \sum_{k=0}^{3} \binom{2-g}{k} \binom{2g+k-3}{g+k-2} \frac{2^{2g+1}(g+1)!}{(3-k)!} \cdot \frac{\wedge^{g-1}[\Xi]}{(2g+k-3)!} \\ &= 2^4 \frac{\wedge^{g-1}[\Xi]}{(g-1)!} \sum_{k=0}^{3} \binom{2-g}{k} \binom{g+1}{3-k} = 2^4 \frac{\wedge^{g-1}[\Xi]}{(g-1)!} \end{aligned}$$

since the sum is 1. In fact, it can be interpreted as the coefficient of x^3 in the series of $(1+x)^{2-g}(1+x)^{g+1} = (1+x)^3$.

Step IV: From Step II we deduce

$$\widehat{\iota}_{Z*} \alpha_*[\widetilde{C}] = \widehat{\iota}_{Z*}([V_1] + [V_2]) = 2\widehat{\iota}_{Z*}[V_1] = 2 \cdot 2_{Z*}[V_1] = 8[V_1] = 8\alpha_*[C_1] \; ,$$

since $V_1 \subset Z$ and $\widehat{\iota}_Z$ restricted to Z is multiplication by 2. Combining this with the result of Step III and using the fact that $H^{2g-2}(Z, \mathbb{Z})$ is torsion free we obtain (3), which completes the proof of the theorem. □

Proof of Lemma 8.1. Step I: $\widetilde{C} = C_1 \cup C_2$ is smooth.

The curve $\widetilde{C}$ is smooth in a point $\widetilde{D}$ if and only if the Zariski tangent space $T_{\widetilde{D}}\widetilde{C}$ is of dimension 1. If $D := \tilde{f}(\widetilde{D})$, then diagram (1) yields a pullback diagram

$$\begin{array}{ccc} T_{\widetilde{D}}\widetilde{C} & \hookrightarrow & T_{\widetilde{D}}C^{(4)} \\ {\scriptstyle d\tilde{f}}\downarrow & & \downarrow{\scriptstyle df^{(4)}} \\ T_D\mathbb{P}_1 & \overset{dj}{\hookrightarrow} & T_D C'^{(4)} \end{array} \; .$$

But $C^{(4)}$ and $C'^{(4)}$ are smooth of dimension 4, hence

$$\begin{aligned} \dim T_{\widetilde{D}}\widetilde{C} &= \dim \ker df^{(4)} + \dim(\operatorname{im} dj \cap \operatorname{im} df^{(4)}) \\ &= 4 - \dim \operatorname{im} df^{(4)} + \dim(\operatorname{im} dj \cap \operatorname{im} df^{(4)}) \\ &= 4 + \dim \operatorname{im} dj - \dim(\operatorname{im} dj + \operatorname{im} df^{(4)}) \\ &= 1 + (4 - \dim(\operatorname{im} dj + \operatorname{im} df^{(4)})) \; . \end{aligned}$$

So $\widetilde{C}$ is smooth in $\widetilde{D}$ if and only if $\operatorname{im} dj + \operatorname{im} df^{(4)} = T_D C'^{(4)}$ or equivalently if and only if the composition

$$T_D \mathbb{P}_1 \overset{dj}{\hookrightarrow} T_D C'^{(4)} \longrightarrow \operatorname{coker} df^{(4)}$$

is surjective.

In order to analyze this composition, recall the following canonical identifications: $T_D C'^{(4)} = H^0(\mathcal{O}_D(D))$ and $T_{\widetilde{D}} C^{(4)} = H^0(\mathcal{O}_{\widetilde{D}}(\widetilde{D}))$ (see Arbarello et al. [1] Ex. IV B-2). Moreover $T_D \mathbb{P}_1 = T_D g_4^1 = H^0(\mathcal{O}_{C'}(D))/H^0(\mathcal{O}_{C'})$. Recall also that the sheaf $\mathcal{O}_{C'}(D)$ may be considered as a subsheaf of the sheaf of rational functions on C' by setting $\mathcal{O}_{C'}(D)_q = m_{C',q}^{-\nu_q(D)}$ for all $q \in C'$. Then $H^0(\mathcal{O}_{C'}(D))$ is a subspace of the function field $\mathbb{C}(C')$ of the curve C' and moreover $\mathcal{O}_D(D)_q = m_{C',q}^{-\nu_q(D)}/\mathcal{O}_{C',q}$.

Suppose $D = \sum_i n_i q_i$ with $n_i > 0$ and points $q_i \neq q_j$ of C'. Then the map $dj: H^0(\mathcal{O}_{C'}(D))/H^0(\mathcal{O}_{C'}) \to H^0(\mathcal{O}_D(D)) = \bigoplus_i m_{C',q_i}^{-n_i}$ is the natural one: if $h \in H^0(\mathcal{O}_{C'}(D))$, then the i-th component of $dj(h + H^0(\mathcal{O}_{C'}))$ is the image of h in $m_{C',q_i}^{-n_i}/\mathcal{O}_{C',q_i}$.

In order to describe the map $df^{(4)}$, denote by q' and q'' the points of C over q. Then $\widetilde{D} = \sum_i n_i' q_i' + \sum_i n_i'' q_i''$ with $n_i' + n_i'' = n_i$, where we assume $n_i' \geq n_i''$. So $H^0(\mathcal{O}_{\widetilde{D}}(\widetilde{D})) = \bigoplus_i (m_{C,q_i'}^{-n_i'}/\mathcal{O}_{C,q_i'} \oplus m_{C,q_i''}^{-n_i''}/\mathcal{O}_{C,q_i''})$. The map $df^{(4)}: \bigoplus_i (m_{C,q_i'}^{-n_i'}/\mathcal{O}_{C,q_i'} \oplus m_{C,q_i''}^{-n_i''}/\mathcal{O}_{C,q_i''}) \to \bigoplus_i (m_{C',q_i}^{-n_i}/\mathcal{O}_{C',q_i})$ again is the natural one: if t_i is a local parameter of C' in q_i and t_i' and t_i'' the induced local parameters of C in q_i' and q_i'' respectively, then $df^{(4)}(t_i'^{\nu}) = df^{(4)}(t_i''^{\nu}) = t_i^{\nu}$ for all $\nu \in \mathbb{Z}$.

Combining everything, we see that $\widetilde{C}$ is smooth if and only if the natural map

$$H^0(\mathcal{O}_{C'}(D)) \to \bigoplus m_{C',q_i}^{-n_i}/m_{C',q_i}^{-n_i'} \tag{8}$$

is surjective, where the sum is to be taken over all i for which $n_i > n_i'$. Now, if $D = \tilde{f}(\widetilde{D})$ is a smooth divisor, the right hand side of (8) is zero and $\widetilde{C}$ is smooth at $\widetilde{D}$. Suppose now $D = 2q_1 + q_2 + q_3$. If $n_1' = 2$, the same argument as above works. If $n_1' < 2$, then the right hand side of (8) is $m_{C',q_1}^{-2}/m_{C',q_1}^{-1}$. Let $h \in H^0(\mathcal{O}_{C'}(D)) \subset \mathbb{C}(C')$ correspond to a divisor $D_h \neq D$ in g_4^1. Then $D_h = \operatorname{Div}(h) + D$, so $0 = \nu_{q_1}(D_h) = \nu_{q_1}(h) + 2$, i.e. $\nu_{q_1}(h) = -2$. So h maps to a generator of $m_{C',q_1}^{-2}/m_{C',q_1}^{-1}$ and $\widetilde{C}$ is smooth at $\widetilde{D}$. The same argument works for $D = 3q_1 + q_2$. Moreover, if $D = 4q_1$ respectively $D = 2q_1 + 2q_2$, then at $\widetilde{D} = 2q_1' + 2q_1''$ respectively $\widetilde{D} = q_1' + q_1'' + q_2' + q_2''$ the cokernel of $df^{(4)}$ is of dimension 2. So the map (8) cannot be surjective and $\widetilde{C}$ is not smooth at $\widetilde{D}$.

Step II: Fix a point $x \in \mathbb{P}_1$, such that $k^{-1}(x)$ is a smooth divisor on C'. Let $p, q \in C$ be points in $(kf)^{-1}(x)$ with $q \neq \iota(p)$. We claim:

There exists a path $\gamma: [0,1] \to \mathbb{P}_1$ connecting x with a branch point of k, and paths $\gamma_p, \gamma_q: [0,1] \to C$ over γ with $\gamma_p(0) = p$, $\gamma_q(0) = q$ and $\gamma_p(1) = \gamma_q(1)$, such that $f\gamma_p(t) \neq f\gamma_q(t)$ for all $t \in]0,1[$, for which $\gamma(t)$ is not a branch point of k.

We leave the elementary topological proof of this statement to the reader. The idea is as follows: since C is connected, there exists a path $\widetilde{\gamma}$ in C connecting p and q. Choose an open disc U in $\mathbb{P}_1$ containing the path $kf(\widetilde{\gamma})$ and consider $(kf)^{-1}(U)$ as 8 sheets, i.e. copies of U, glued together along dissections connecting ramification points. Then construct γ_p out of $\widetilde{\gamma}$ starting at p and γ_q as a path over $(kf)(\gamma_p)$ with $\gamma_q(0) = q$. Show that γ_p and γ_q meet in a ramification point.

Step III: C_i is connected for $i = 1, 2$.

Let $x \in \mathbb{P}_1$, not a branch point of $k: C' \to \mathbb{P}_1$. It suffices to connect two arbitrary points D_1 and D_2 of C_i in the fibre $f^{-1}(x)$ by a path in C_i. We may assume that $D_1 = p_1 + p_2 + p_3 + p_4$ and $D_2 = \iota(p_1) + \iota(p_2) + p_3 + p_4$.

By definition of C_i a path in C_i is equivalent to paths $\gamma_1, \ldots, \gamma_4: [0,1] \to C$ and a path $\gamma: [0,1] \to \mathbb{P}_1$ such that

a) $kf\gamma_i(t) = \gamma(t)$ for all $t \in [0,1]$ and $i = 1, \ldots, 4$,

b) if $\gamma(t)$ is not a branch point of k, then $f\gamma_1(t), \ldots, f\gamma_4(t)$ are distinct points in C'.

Apply Step II to $p = p_1$ and $q = \iota p_2$. Let r be the ramification point $r = \gamma_p(1) = \gamma_q(1)$. Define $\gamma_1 = \gamma_p$ and $\gamma_2 = \iota\gamma_q$. Then γ_1 connects p_1 with r and γ_2 connects p_2 with ιr. Certainly we can add paths γ_3 and γ_4 over γ such that $\gamma_i(0) = p_i$ and $\gamma_1, \ldots, \gamma_4, \gamma$ satisfy condition b) above. So we get a path in C_i between D_1 and the divisor $D = r + \iota r + \gamma_3(1) + \gamma_4(1)$. The paths $(\iota\gamma_2)^{-1}, (\iota\gamma_1)^{-1}, \gamma_3^{-1}, \gamma_4^{-1}, \gamma^{-1}$ yield a path in C_i connecting D with D_2. So adding both paths, we obtain a path in C_1 connecting D_1 with D_2. This completes the proof of Step III.

Step IV: To compute the genus, note that the degree of the ramification divisor of $C_i \to \mathbb{P}_1$ is twice the degree of the ramification divisor of the covering $k: C' \to \mathbb{P}_1$. Then the assertion follows from Hurwitz's formula.

□

§ 9 Kanev's Criterion

Let C be a smooth projective curve of genus g and (J, Θ) its Jacobian. To any abelian subvariety Z of J one can associate in a natural way a class of correspondences of C. We want to prove a criterion due to Kanev [2], saying that Z is a Prym-Tyurin variety under the assumption that the associated class of correspondences contains a fixed point free representative. This generalizes the case of Prym varieties associated to an étale double covering. At the end of this section we will give some examples.

Let L be a correspondence on $C \times C$ of bidegree (d_1, d_2). Recall from Proposition 11.5.1 that $L \mapsto \gamma_L$ induces an isomorphism $\operatorname{Corr}(C \times C) \to \operatorname{End}(J)$. The correspondence is called *fixed point free*, if $(L \cdot \Delta) = 0$ for the intersection of L with the diagonal Δ of $C \times C$.

(9.1) Theorem. *Let Z be an abelian subvariety of exponent e of the Jacobian J of C. Suppose there is an effective fixed point free correspondence L on $C \times C$ of bidegree (d, d) with $\gamma_L = 1_J - N_Z$. Then Z is a Prym-Tyurin variety for the curve C. Moreover there are theta divisors Θ on J and Ξ on Z such that $\iota_Z^* \Theta = e\Xi$.*

According to Proposition 11.5.2 the fixed points of L are related to the rational trace of γ_L by the formula $\operatorname{Tr}_r(\gamma_L) = d_1 + d_2 - (L \cdot \Delta)$. So L is fixed point free and of bidegree (d, d) if and only if $\operatorname{Tr}_r(\gamma_L) = 2d$. On the other hand by Corollary 5.3.10, $\operatorname{Tr}_r(\gamma_L) = \operatorname{Tr}_r(1_J - N_Z) = 2g - 2e \dim Z$. Hence

$$d = g - e \dim Z \tag{1}$$

Consider the special case of an étale double covering $f: C \to C'$ with involution ι on C. Let Z be the abelian subvariety $\operatorname{im}(1_J - \tilde{\iota})$ of J. The correspondence $L = \mathcal{O}_{C \times C}(D)$ with $D = \{(p, \iota(p)) \mid p \in C\}$ is fixed point free, of bidegree $(1, 1)$ and satisfies $\gamma_L = \tilde{\iota} = 1_J - N_Z$. So Z is a Prym variety of exponent 2, and we obtain a third proof of Theorem 3.3 a). Furthermore the last assertion in Theorem 9.1 is a generalization of Proposition 6.3 a).

The idea of the proof is to show the equality of divisors $\iota_Z^* \Theta = e\Xi$ directly. For this we introduce auxiliary principally polarized abelian varieties $\widetilde{Y}$ and $\widetilde{Z}$. Denote by Y the abelian subvariety of J complementary to Z. Applying Proposition 1.8 to both abelian subvarieties Y and Z of J, one sees that there are principally polarized abelian varieties $(\widetilde{Y}, \widetilde{\Sigma})$ and $(\widetilde{Z}, \widetilde{\Xi})$, homomorphisms $\tilde{\iota}_Y$, $\tilde{\iota}_Z$, $\widetilde{N}_Y$, $\widetilde{N}_Z$ and isogenies μ_Y and μ_Z as indicated in the following noncommutative diagram

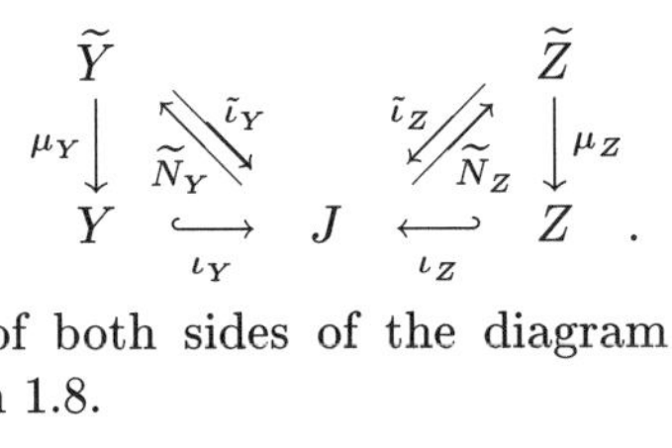

The homomorphisms of both sides of the diagram satisfy the equations i),..., iv) of Proposition 1.8.

(9.2) Proposition. *There are theta divisors* Θ *on* J *and* $\widetilde{\Xi}$ *on* $\widetilde{Z}$ *such that*

$$\tilde{\iota}_Z^*\Theta = e\widetilde{\Xi} \ .$$

From the proposition the theorem is an easy consequence.

Proof of Theorem 9.1. If Θ is the theta divisor of Proposition 9.2, the divisor $\iota_Z^*\Theta$ is well defined, since $\tilde{\iota}_Z = \iota_Z \mu_Z$ and μ_Z is an isogeny by Proposition 1.8. Write $\iota_Z^*\Theta = \sum_j r_j \Xi_j$ with $r_j > 0$ and $\mu_Z^*\Xi_j = \sum_k \widetilde{\Xi}_{jk}$ with pairwise different irreducible divisors Ξ_j on Z and $\widetilde{\Xi}_{jk}$ on $\widetilde{Z}$. Then $\tilde{\iota}_Z^*\Theta = \mu_Z^*\iota_Z^*\Theta = \sum_{j,k} r_j \widetilde{\Xi}_{jk}$. Proposition 9.2 implies that $r_j = e$ for all j and $\widetilde{\Xi} = \sum_{j,k} \widetilde{\Xi}_{jk}$. Define $\Xi = \sum_j \Xi_j$. Since $\mu_Z^*\Xi = \widetilde{\Xi}$ defines a principal polarization, the isogeny μ_Z is an isomorphism and Ξ defines also a principal polarization. Hence $\iota_Z^*\Theta = e\Xi$, completing the proof of the theorem. □

For the proof of the proposition we need several lemmas.

(9.3) Lemma. $e\Theta \sim \widetilde{N}_Y^*\widetilde{\Sigma} + \widetilde{N}_Z^*\widetilde{\Xi}$ *for a suitable theta divisor* Θ *on* J.

Proof. Let $\widetilde{\Theta}$ be any theta divisor on J. Using properties ii) and iii) of Proposition 1.8, Lemma 1.6 and $(\iota_Y + \iota_Z)(N_Y, N_Z) = e_J$ we get

$$\begin{aligned} e(\widetilde{N}_Y^*\widetilde{\Sigma} + \widetilde{N}_Z^*\widetilde{\Xi}) &\equiv \widetilde{N}_Y^*\tilde{\iota}_Y^*\widetilde{\Theta} + \widetilde{N}_Z^*\tilde{\iota}_Z^*\widetilde{\Theta} \\ &= N_Y^*\widetilde{\Theta} + N_Z^*\widetilde{\Theta} \\ &= (N_Y, N_Z)^*(\iota_Y + \iota_Z)^*\widetilde{\Theta} = e_J^*\widetilde{\Theta} \equiv e^2\widetilde{\Theta} \ . \end{aligned}$$

Since $\mathrm{NS}(J)$ is torsion free, this implies $\widetilde{N}_Y^*\widetilde{\Sigma} + \widetilde{N}_Z^*\widetilde{\Xi} - e\widetilde{\Theta} \equiv 0$. But the homomorphism $\phi_{e\widetilde{\Theta}}: J \to \widehat{J}$ is surjective, so $\widetilde{N}_Y^*\widetilde{\Sigma} + \widetilde{N}_Z^*\widetilde{\Xi} - e\widetilde{\Theta} \sim \phi_{e\widetilde{\Theta}}(x) = et_x^*\widetilde{\Theta} - e\widetilde{\Theta}$ for some $x \in J$. Setting $\Theta = t_x^*\widetilde{\Theta}$ gives the assertion. □

In the sequel let Θ denote the theta divisor of Lemma 9.3. As usual denote by $\alpha_c: C \to J$ the embedding $p \mapsto \mathcal{O}_C(p - c)$. Without loss of generality we may assume that the theta divisors $\widetilde{\Sigma}$ and $\widetilde{\Xi}$ are symmetric. According to Lemma 9.3 we may also assume that Θ is symmetric.

(9.4) Lemma. *There is an* $\varepsilon \in \mathrm{Pic}(C)$ *such that for all* $z \in \widetilde{Z}$ *and* $c \in C$

$$\alpha_c^*\widetilde{N}_Z^*\mathcal{O}_{\widetilde{Z}}(t_z^*\widetilde{\Xi}) = \tilde{\iota}_Z(z)^{-1} \otimes \mathcal{O}_C(c) \otimes L(c)^{-1} \otimes \varepsilon \ .$$

Proof. Choose $x \in \widetilde{Z}$ with $ex = z$. From Proposition 1.8 and the facts that $N_Y|Z = 0$ and $N_Z|Z = e$ we deduce $\widetilde{N}_Y = \widetilde{N}_Y t_{\tilde{\iota}_Z(x)}$ and $t_z\widetilde{N}_Z = \widetilde{N}_Z t_{\tilde{\iota}_Z(x)}$. Hence by Lemma 9.3 and Lemma 11.3.4 and using that Θ is symmetric

$$\begin{aligned}\alpha_c^* \widetilde{N}_Y^* \mathcal{O}_{\widetilde{Y}}(\widetilde{\Sigma}) \otimes \alpha_c^* \widetilde{N}_Z^* \mathcal{O}_{\widetilde{Z}}(t_z^* \widetilde{\Xi}) &= \\ &= \alpha_c^* t_{\tilde{\imath}_Z(x)}^* (\widetilde{N}_Y^* \mathcal{O}_{\widetilde{Y}}(\widetilde{\Sigma}) \otimes \widetilde{N}_Z^* \mathcal{O}_{\widetilde{Z}}(\widetilde{\Xi})) \\ &= e\alpha_c^* \mathcal{O}_J(t_{\tilde{\imath}_Z(x)}^* \Theta) \\ &= \tilde{\imath}_Z(z)^{-1} \otimes \kappa^e \otimes \mathcal{O}_C(ec) \, .\end{aligned}$$

Here κ is the uniquely determined line bundle on C with $W_{g-1} = \alpha_\kappa^* \Theta$. Now consider this equation with c replaced by a point $a \in C$. Since $\widetilde{N}_Z \alpha_c = t_{\widetilde{N}_Z \alpha_c(a)} \widetilde{N}_Z \alpha_a$, this gives

$$\begin{aligned}\alpha_c^* \widetilde{N}_Z^* \mathcal{O}_{\widetilde{Z}}(t_z^* \widetilde{\Xi}) &= \alpha_a^* \widetilde{N}_Z^* \mathcal{O}_{\widetilde{Z}}(t_{z+\widetilde{N}_Z \alpha_c(a)}^* \widetilde{\Xi}) \\ &= \tilde{\imath}_Z(z + \widetilde{N}_Z \alpha_c(a))^{-1} \otimes \kappa^e \otimes \mathcal{O}_C(ea) \otimes \alpha_a^* \widetilde{N}_Y^* \mathcal{O}_{\widetilde{Y}}(-\widetilde{\Sigma}) \, .\end{aligned}$$

But $\tilde{\imath}_Z \widetilde{N}_Z \alpha_c(a) = N_Z \alpha_c(a) = (1_J - \gamma_L)\alpha_c(a) = \mathcal{O}_C(a-c) \otimes L(a)^{-1} \otimes L(c)$ by definition of γ_L. We obtain the assertion with $\varepsilon = \kappa^e \otimes \mathcal{O}_C((e-1)a) \otimes L(a) \otimes \alpha_a^* \widetilde{N}_Y^* \mathcal{O}_{\widetilde{Y}}(-\widetilde{\Sigma})$. □

Let τ denote the involution $(p,q) \mapsto (q,p)$ of $C \times C$. We may apply Lemma 9.4 to the correspondence $\tau^* L$, since $\gamma_{\tau^* L} = \gamma_L' = (1_J - N_Z)' = 1_J - N_Z$ by Proposition 11.5.3. Hence there is an $\eta \in \operatorname{Pic}(C)$ such that for all $z \in \widetilde{Z}$ and $c \in C$

$$\alpha_c^* \widetilde{N}_Z^* \mathcal{O}_{\widetilde{Z}}(t_z^* \widetilde{\Xi}) = \tilde{\imath}_Z(z)^{-1} \otimes \mathcal{O}_C(c) \otimes \tau^* L(c)^{-1} \otimes \eta \, . \tag{2}$$

(9.5) Lemma. $\varepsilon \otimes \eta = \omega_C$.

Proof. Let $\mathcal{P}$ denote the Poincaré bundle of degree zero on $C \times J$ normalized with respect to the point $c \in C$, i.e. $\mathcal{P}|\{c\} \times J = \mathcal{O}_J$.

Step I: $\left(\mathrm{id}_C \times (N_Z \alpha_c)\right)^* \mathcal{P} =$
$\mathcal{O}_{C^2}(\Delta - \{c\} \times C - C \times \{c\}) \otimes \tau^* L^{-1} \otimes p_1^* L(c) \otimes p_2^*(\tau^* L)(c)$.

For the proof we apply the Seesaw Principle A.9: on the one hand, we have for every $p \in C$

$$\begin{aligned}\left(\mathrm{id}_C \times (N_Z \alpha_c)\right)^* \mathcal{P}|C \times \{p\} &= \mathcal{P}|C \times \{N_Z \alpha_c(p)\} = N_Z \alpha_c(p) \\ &= (1_J - \gamma_L)\alpha_c(p) = \mathcal{O}_J(p-c) \otimes L(p)^{-1} \otimes L(c) \, .\end{aligned}$$

But this equals obviously the left hand side restricted to $C \times \{p\}$. On the other hand, both sides restricted to $\{c\} \times C$ are trivial.

Step II: $(\widetilde{N}_Z \alpha_c \times 1_{\widetilde{Z}})^* \mathcal{O}_{\widetilde{Z} \times \widetilde{Z}}(\mu^* \widetilde{\Xi} - q_1^* \widetilde{\Xi} - q_2^* \widetilde{\Xi}) = (\mathrm{id}_C \times \tilde{\imath}_Z)^* \mathcal{P}^{-1}$ *with the addition map* μ *and the natural projections* q_i *of* $\widetilde{Z} \times \widetilde{Z}$.

For all $z \in \widetilde{Z}$ we have by Lemma 9.4

$$(\widetilde{N}_Z\alpha_c \times 1_{\widetilde{Z}})^*\mathcal{O}_{\widetilde{Z}\times\widetilde{Z}}(\mu^*\widetilde{\Xi} - q_1^*\widetilde{\Xi} - q_2^*\widetilde{\Xi})|C \times \{z\} =$$
$$= \alpha_c^*\widetilde{N}_Z^*\mathcal{O}_{\widetilde{Z}}(t_z^*\widetilde{\Xi}) \otimes \left(\alpha_c^*\widetilde{N}_Z^*\mathcal{O}_{\widetilde{Z}}(\widetilde{\Xi})\right)^{-1}$$
$$= \tilde{\iota}_Z(z)^{-1} = \mathcal{P}^{-1}|C \times \{\tilde{\iota}_Z(z)\} = (\mathrm{id}_C \times \tilde{\iota}_Z)^*\mathcal{P}^{-1}|C \times \{z\} .$$

Moreover, the restrictions of both sides to $\{c\} \times \widetilde{Z}$ are trivial on $\widetilde{Z}$, so the assertion follows again from the Seesaw Principle.
Step III: Combining Steps I and II and using $N_Z = \tilde{\iota}_Z\widetilde{N}_Z$ we get

$$(\widetilde{N}_Z\alpha_c \times \widetilde{N}_Z\alpha_c)^*\mathcal{O}_{\widetilde{Z}\times\widetilde{Z}}(\mu^*\widetilde{\Xi} - q_1^*\widetilde{\Xi} - q_2^*\widetilde{\Xi}) =$$
$$= \mathcal{O}_{C^2}(-\Delta + \{c\}\times C + C\times\{c\}) \otimes \tau^*L \otimes p_1^*L(c)^{-1} \otimes p_2^*(\tau^*L)(c)^{-1} .$$

By assumption L is fixed point free, so restricting to the diagonal $\Delta \simeq C$ gives
$$2\alpha_c^*\widetilde{N}_Z^*\mathcal{O}_{\widetilde{Z}}(\widetilde{\Xi}) = \omega_C \otimes \mathcal{O}_C(2c) \otimes L(c)^{-1} \otimes (\tau^*L)(c)^{-1} ,$$
since $\mathcal{O}_\Delta(-\Delta) = \omega_C$ by the adjunction formula. On the other hand, by Lemma 9.4 and (2) with $z = 0$ we have

$$2\alpha_c^*\widetilde{N}_Z^*\mathcal{O}_{\widetilde{Z}}(\widetilde{\Xi}) = \mathcal{O}_C(2c) \otimes L(c)^{-1} \otimes (\tau^*L)(c)^{-1} \otimes \varepsilon \otimes \eta .$$

This implies the assertion. □

By assumption $d = \deg L(c) = \deg(\tau^*L)(c)$, so $\deg\varepsilon = \deg\eta$, and Lemma 9.5 implies $\deg\varepsilon = \deg\eta = g - 1$.

Proof of Proposition 9.2. We claim that $\Theta = W_{g-1} - \varepsilon$ satisfies $\tilde{\iota}_Z^*\Theta = e\widetilde{\Xi}$.
Step I: $\tilde{\iota}_Z^{-1}(\Theta) \subset \widetilde{\Xi}$.
Suppose $z \in \widetilde{Z} - \widetilde{\Xi}$. We have to show that $\tilde{\iota}_Z(z) \notin \Theta$. For any $c \in C$ we have, by the choice of z and since $\widetilde{\Xi}$ is symmetric, $\widetilde{N}_Z\alpha_c(c) = 0 \notin t_{-z}^*\widetilde{\Xi}$. Hence $c \notin \alpha_c^*\widetilde{N}_Z^*(t_{-z}^*\widetilde{\Xi})$. Consider the line bundle

$$M := \mathcal{O}_C(\alpha_c^*\widetilde{N}_Z^*(t_{-z}^*\widetilde{\Xi}) - c) \otimes L(c) .$$

By Lemma 9.4 we have $M = \tilde{\iota}_Z(z) \otimes \varepsilon$. In particular M does not depend on c. Since L is effective and fixed point free, c is not a base point of the line bundle $L(c)$. By what we have shown above, this implies $h^0(M \otimes \mathcal{O}_C(c)) = h^0(M) + 1$.

Assume $h^0(M) = r > 0$. Since $\deg M = \deg\varepsilon = g - 1$, Riemann-Roch gives $h^0(\omega_C \otimes M^{-1} \otimes \mathcal{O}_C(-c)) = h^1(M \otimes \mathcal{O}_C(c)) = h^0(M \otimes \mathcal{O}_C(c)) - 1 = r$. On the other hand, $h^0(\omega_C \otimes M^{-1}) = h^1(M) = r$, so c is a base point of $\omega_C \otimes M^{-1}$. This is a contradiction, since c is arbitrary.

It follows that $h^0(M) = 0$, that is $M \notin W_{g-1}$ or equivalently $\tilde{\iota}_Z(z) = M \otimes \varepsilon^{-1} \notin W_{g-1} - \varepsilon = \Theta$.

Step II: Since $\widetilde{\Xi}$ is a divisor defining a principal polarization on $\widetilde{Z}$, it is of the form $\widetilde{\Xi} = \sum_i \widetilde{\Xi}_i$ with irreducible divisors $\widetilde{\Xi}_i$ which are linearly independent in $\mathrm{NS}(\widetilde{Z})$. By what we have just proved, $\tilde{\iota}_Z^*\Theta = \sum_i r_i \widetilde{\Xi}_i$ with $r_i \geq 0$. By Proposition 1.8.ii) $\tilde{\iota}_Z^*\Theta \equiv \sum e\widetilde{\Xi}_i$, hence $e = r_i$ for all i. □

Finally we give two examples of correspondences of curves, where the theorem applies, both due to Kanev. The first example is a generalization of Recillas' Theorem to d-gonal curves with $d \geq 4$: let X be a smooth curve with a base point free g_d^1 and let $k: X \to \mathbb{P}_1$ be the corresponding morphism. Assume that any fibre of k contains at most one ramification point, and this is of index ≤ 3. Define a curve

$$C = \{p_1 + p_2 \in X^{(2)} \mid |g_d^1 - p_1 - p_2| \neq \emptyset\}$$

with reduced subscheme structure. With similar arguments as used in the proof of Lemma 7.1 one can show that C is a smooth and irreducible curve. Define a correspondence L on $C \times C$ by $L = \mathcal{O}_{C\times C}(D)$ with D the reduced divisor

$$D = \{(p_1 + p_2, q_1 + q_2) \in C \times C \mid |g_d^1 - p_1 - p_2 - q_1 - q_2| \neq \emptyset\} \ .$$

It is easy to see that L is fixed point free with degrees $d_1 = d_2 = \binom{d-2}{2}$. Moreover, one can show that $1-\gamma_L$ is the norm endomorphism of the abelian subvariety $Z = \mathrm{im}(1-\gamma_L)$ of $J(C)$ (see Kanev [2]). So the correspondence D satisfies the assumptions of Theorem 9.1 and (Z, Ξ) is a Prym-Tyurin variety of exponent $d-2$ for C.

(9.6) Proposition. $\quad (Z, \Xi) \simeq \big(J(X), \Theta\big)$.

We omit the proof, since it is very similar to the proof of Recillas' theorem, which is a special case of it (see Kanev [2]).

The last example is an analogue of Donagi's tetragonal construction in the case of pentagonal curves. Let C' be a smooth curve with a base point free g_5^1. Let $k: C' \to \mathbb{P}_1$ be the corresponding morphism. We assume that any fibre of k contains at most one ramification point and this is of index ≤ 3. Let $f: C \to C'$ be an étale double covering with involution ι. Consider the induced map $f^{(5)}: C^{(5)} \to C'^{(5)}$. Then ${f^{(5)}}^{-1}(g_5^1)$ consists of two components C_1 and C_2 which are smooth irreducible curves, isomorphic under the action induced by ι (see Exercise 14). Define a correspondence L on $C_1 \times C_1$ by $L = \mathcal{O}_{C_1\times C_1}(D)$ with D the reduced divisor

$$D = \big\{\big(p_1 + \cdots + p_5, \iota(p_1) + \cdots + \iota(p_4) + p_5\big) \in C_1^2 \mid f^{(5)}(p_1 + \cdots + p_5) \in g_5^1\big\}$$

It is easy to see that D is fixed point free with degrees $d_1 = d_2 = 5$. Moreover, one can show that $1-\gamma_L$ is the norm-endomorphism of the abelian subvariety $Z = \operatorname{im}(1-\gamma_L)$ of $J(C_1)$ (see Kanev [2]). So the correspondence L satisfies the assumptions of Theorem 9.1 and (Z, Ξ) is a Prym-Tyurin variety of exponent 4 for C.

(9.7) Proposition. *(Z, Ξ) is isomorphic to the Prym variety (Z_f, Ξ_f) associated to the double covering $f: C \to C'$.*

We omit the proof since it is very similar to the proof of Theorem 8.2.

Exercises

(1) Suppose (X, L) is a polarized abelian variety of dimension g.

a) Show that for any positive integer n there are only finitely many abelian subvarieties of X of exponent less or equal to n.

b) Conclude that for any smooth projective curve C and any positive integer n there are up to isomorphisms only finitely many morphisms of degree less or equal to n of C onto curves of genus ≥ 1.

A special case of b) is the *Theorem of de Franchis* [1]: any smooth projective curve C admits up to isomorphisms only finitely many morphisms onto curves of genus ≥ 2.

(2) Suppose $f: (X, L) \to (X', L')$ is an isogeny of polarized abelian varieties. Let Y be an abelian subvariety of X and d the exponent of the finite group $\ker f|Y$. Show that $\frac{1}{d^2} e(Y) \leq e(f(Y)) \leq e(Y)$.

(3) Let C be a smooth projective curve of genus 3.

a) The following conditions are equivalent:

i) C is a double covering of an elliptic curve,

ii) C admits an embedding into an abelian surface X.

b) Let $f: C \to C'$ be a double covering of an elliptic curve and $C \hookrightarrow X$ a corresponding embedding into an abelian surface. Let Z be the abelian subvariety of $J(C)$, complementary to $\operatorname{im} f^* \subset J(C)$. Both abelian surfaces X and Z admit natural $(1,2)$-polarizations, identifying one with the dual of the other (see Barth [1]).

(4) Let $f: C \to C'$ be a double covering of smooth projective curves, ramified in $2n$ points for some $n \geq 2$. Compute the type of the abelian subvariety Z of $J(C)$ complementary to $\operatorname{im} f^*$.

(5) Let $f\colon C \to C'$ be a finite morphism of smooth projective curves. Consider the abelian subvariety $Y = \operatorname{im} f^*$ of $J(C)$, and let k denote the largest integer such that $\ker(f^*N_f)$ contains the group $J(C)_k$ of k-division points of $J(C)$.
a) Show that $e(Y) = \frac{\deg f}{k}$.
b) Give an example of a morphism f with $e(Y) < \deg f$.

(6) Let $f\colon C \to C'$ be an étale double covering of smooth projective curves. Recall that $\ker N_f$ consists of two connected components, the Prym variety Z associated to f and Z_1. Show that $Z = \{\sum_{\nu=1}^{2N} \mathcal{O}_C(p_\nu - \iota p_\nu) \mid N \geq 0\,,\, p_\nu \in C\}$ and $Z_1 = \{\sum_{\nu=1}^{2N+1} \mathcal{O}_C(p_\nu - \iota p_\nu) \mid N \geq 0\,,\, p_\nu \in C\}$. A set of generators for Z is given by $\{2\mathcal{O}_C(p - \iota p) \mid p \in C\}$. Conclude $Z = (1 - \tilde{\iota})\operatorname{Pic}^0(C)$ and $Z_1 = (1 - \tilde{\iota})\operatorname{Pic}^1(C)$.

(7) *(Schottky-Jung-Relation)* Let $f\colon C \to C'$ be an étale double covering of smooth projective curves corresponding to a 2-division point $\eta \in \operatorname{Pic}^0(C')$. Choose an even theta characteristic κ' on C' and denote $\kappa = f^*\kappa'$. Define the theta divisors $\Theta = W_{g(C)-1} - \kappa$ on $J(C)$ and $\Theta' = W_{g(C')-1} - \kappa'$ as in Section 6. Show that $(f^*)^*\Theta = \Theta' + t_\eta^*\Theta'$ (see Mumford [5]).

(8) Let C' be a smooth hyperelliptic curve of genus ≥ 2 and $B \subset \mathbb{P}_1$ the set of branch points of the hyperelliptic double covering.
a) A connected étale double covering $f\colon C \to C'$ corresponds uniquely to a decomposition of B into two non-empty disjoint subsets $B = B_1 \cup B_2$, both of even cardinality.
b) Let $B = B_1 \cup B_2$ be a decomposition inducing $f\colon C \to C'$ and $C_i \to \mathbb{P}_1$ the double covering branched over B_i. The Prym variety (Z, Ξ) associated to f is isomorphic to the product $(J(C_1), \Theta_1) \times (J(C_2), \Theta_2)$.
c) Conclude that the Jacobian of any hyperelliptic curve is the Prym variety associated to some étale double covering (see Mumford [5]).

(9) Let X be a smooth projective curve of genus 5 neither hyperelliptic nor trigonal.
a) The canonical model of X is the complete intersection of three quadrics $\{Q_0 = 0\} \cap \{Q_1 = 0\} \cap \{Q_2 = 0\}$ in $\mathbb{P}_4$. The discriminant curve
$$C' = \{(x_0 : x_1 : x_2) \in \mathbb{P}_2 \mid \det(x_0Q_0 + x_1Q_1 + x_2Q_2) = 0\}$$
is a plane quintic depending only on X but not on the choice of the quadrics Q_i.
b) C' is smooth if and only if $\operatorname{rank}(x_0Q_0 + x_1Q_1 + x_2Q_2) = 4$ for all $(x_0 : x_1 : x_2) \in C'$.
c) Suppose C' is smooth. Then every quadric $x_0Q_0 + x_1Q_1 + x_2Q_2$ corresponding to a point of C' has two different rulings. The two rulings define an étale double covering $f\colon C \to C'$.

d) With the notation of b) the Prym variety associated to f is isomorphic to the Jacobian of X.

e) Let $\eta \in \operatorname{Pic}^0(C)$ be the 2-division point associated to f. Show that $h^0(\mathcal{O}_{C'}(1) \otimes \eta) = 0$ (see Masiewicki [1]).

(10) Let C' be a smooth projective curve of genus $\neq 4$ and $f: C \to C'$ an étale double covering corresponding to a 2-division point $\eta \in \operatorname{Pic}^0(C')$. Show that, if the Prym variety associated to f is a Jacobian, then C' is either hyperelliptic or trigonal or a plane quintic with $h^0(\mathcal{O}_{C'}(1) \otimes \eta) = 0$ (see Shokurov [1], the idea is to apply Exercise 13).

(11) Let (Z, Ξ) be the Prym variety of dimension g associated to an étale double covering $f: C \to C'$.

a) $\dim \operatorname{sing} \Xi \geq g - 6$.

b) If C' is a general curve of genus $g + 1$, then $\operatorname{sing} \Xi$ is irreducible of dimension $g - 6$ for $g \geq 7$, finite for $g = 6$ and empty for $g \leq 5$ (see Welters [2] and Debarre [1]).

(12) Let (Z, Ξ) be as in Exercise 11. Any component $V \subset \operatorname{sing} \Xi$ of dimension $\geq g - 4$ contains an open dense subset of exceptional singularities (see Mumford [5]).

(13) Let (Z, Ξ) be as in Exercises 10 and 11 with $g \geq 4$. If $\dim \operatorname{sing} \Xi \geq g - 4$, then we are in one of the following cases

a) C' hyperelliptic,

b) C' trigonal,

c) C' a plane quintic with $h^0(\mathcal{O}_{C'}(1) \otimes \eta) = 0$,

d) C' a double covering of an elliptic curve,

e) C' of genus 5 admitting an even theta characteristic κ with $h^0(\kappa) \neq 0$ and $h^0(\kappa \otimes \eta)$ even.

(See Mumford [5]. Hint: use Exercise 12)

(14) Let $f: C \to C'$ be an étale double covering of smooth projective curves and g^r_d a complete base-point free linear system on C'. Define $V = {f^{(d)}}^{-1}(g^r_d)$, the scheme theoretic preimage of g^r_d under $f^{(d)}: C^{(d)} \to C'^{(d)}$. As a set V consists of the divisors of degree d on C, which push down to divisors of the g^r_d.

a) V consists of two connected components V_1 and V_2.

b) If d is odd, the involution ι, corresponding to the double covering f, induces an isomorphism $V_1 \simeq V_2$.

c) Suppose the following conditions are fulfilled

i) every fibre of the morphism $k: C' \to \mathbb{P}_r$ associated to the g^r_d contains at most one ramification point and this is of index ≤ 3,

ii) if $r > 1$, the map k is birational onto its image.

Then V_1 and V_2 are normal and irreducible varieties.

(See Welters [1] and Beauville [2]).

(15) Let the notation be as in Exercise 14. Consider the map $\alpha = \alpha_D : C^{(d)} \to J(C)$ corresponding to the divisor $D \in C^{(d)}$. After suitable translations we may assume that $\alpha(V_1)$ and $\alpha(V_2)$ are contained in the Prym variety Z associated to f. If $d > 2r$, then $\alpha(V_1)$ and $\alpha(V_2)$ have the same class in $H^{2(g-r)}(Z, \mathbb{Z})$. To be more precise

$$[\alpha(V_i)] = \frac{2^{d-2r-1}}{(g-r)!} \bigwedge^{g-r} [\Xi] \ ,$$

where Ξ is a theta divisor on Z and $g = \dim Z$. (See Beauville [2]. Hint: generalize Steps II and III of the proof of Theorem 8.2.)

(16) Let $\mathcal{R}_g$ denote the moduli space of nontrivial étale double coverings of curves of genus g. It is a finite covering of the moduli space $\mathcal{M}_g$ of curves of genus g of degree $2^{2g} - 1$. By associating to the étale double covering $f: C \to C'$ of $\mathcal{R}_{g+1}$ its Prym variety $Z \in \mathcal{A}_{\mathbb{1}_g}$ we get a morphism

$$p_g : \mathcal{R}_{g+1} \to \mathcal{A}_{\mathbb{1}_g} \ ,$$

called the *Prym map*.

a) The Prym map p_g is not everywhere injective. (Hint: show that the three double coverings of Donagi's tetragonal construction in Section 8 are not isomorphic in general. For another case in which p_g is not injective see Verra [1].)

b) (*Torelli Theorem for the Prym map*) The Prym map p_g is generically injective for $g \geq 6$. This implies that the image of p_g is of dimension $3g$ for $g \geq 6$. (See Friedmann-Smith [1], Kanev [1], Welters [4] and Debarre [1].)

c) The general fibre of p_5 consists of 27 elements. The Galois group of the field extension $\mathbb{C}(\mathcal{R}_6)|\mathbb{C}(\mathcal{A}_{\mathbb{1}_5})$ is isomorphic to the Galois group of the 27 lines on a cubic surface. (See Donagi-Smith [1].)

d) The general fibre of p_4 is a double covering of a Fano surface. (See Donagi [1].)

e) For a general $(X, \Theta) \in \mathcal{A}_3$ the fibre $p_3^{-1}(X)$ is isomorphic to the Kummer variety of X.

Appendix

A. Algebraic Varieties and Complex Analytic Spaces

In this Appendix we compile some results about algebraic varieties over $\mathbb{C}$, complex analytic spaces and their interrelations, which are frequently used in this book.

To any algebraic variety X over the complex numbers one can associate a complex analytic space X_{hol} in a natural way (see Serre [1]). This construction is functorial: for any morphism $f : Z \to X$ of algebraic varieties over $\mathbb{C}$ there is a natural morphism of complex analytic spaces $f_{\text{hol}} : Z_{\text{hol}} \to X_{\text{hol}}$. Moreover, many of the important notions in both categories are preserved under $X \to X_{\text{hol}}$. For example: an algebraic variety X is complete if and only if X_{hol} is compact, and X is smooth if and only if X_{hol} is a complex manifold. Conversely, a complex analytic space $\mathfrak{X}$ is called *algebraic*, if there is an algebraic variety X over $\mathbb{C}$ such that $\mathfrak{X} \simeq X_{\text{hol}}$.

In both, the algebraic and the complex analytic category one has the notion of a vector bundle and its associated locally free sheaf. For our purposes it is convenient not to distinguish between these objects: we always speak of vector bundles, even if we mean the associated locally free sheaf. In particular, for a line bundle L on a complex analytic space (respectively algebraic variety) X we denote by $H^i(X, L)$, or $H^i(L)$ if there is no ambiguity about the base space X, the i-th cohomology group of X with values in the locally free sheaf associated to L.

With a similar construction as above one can associate to an algebraic vector bundle (or more generally an algebraic coherent sheaf) F on the algebraic variety X over $\mathbb{C}$ a holomorphic vector bundle (respectively a holomorphic coherent sheaf) F_{hol} over the complex analytic space X_{hol}. Again this process is functorial: given algebraic vector bundles (respectively coherent sheaves) F and G on X, one can associate to any algebraic homomorphism $f : F \to G$ a holomorphic homomorphism $f_{\text{hol}} : F_{\text{hol}} \to G_{\text{hol}}$.

Not every holomorphic vector bundle on an arbitrary algebraic complex analytic space comes from an algebraic one. However in the case of complete algebraic varieties we have the following comparison theorems due to Serre [1].

(A.1) Theorem. *Let X be a complete algebraic variety over $\mathbb{C}$.*

a) For any holomorphic vector bundle (respectively holomorphic coherent sheaf) $\mathcal{F}$ on X_{hol} there exists a unique algebraic vector bundle (respectively algebraic coherent sheaf) F such that $F_{\text{hol}} \simeq \mathcal{F}$.

b) For every homomorphism $\mathfrak{f}: \mathcal{F} \to \mathcal{G}$ of holomorphic coherent sheaves $\mathcal{F}$ and $\mathcal{G}$ on X_{hol} there exists a unique homomorphism $f : F \to G$ of algebraic coherent sheaves on X such that $f_{\text{hol}} = \mathfrak{f}$.

(A.2) Theorem. *Let X be a complete algebraic variety over $\mathbb{C}$. For any coherent sheaf F on X the natural maps $H^i(X, F) \to H^i(X_{\text{hol}}, F_{\text{hol}}), i \in \mathbb{Z}$, are isomorphisms of $\mathbb{C}$-vector spaces.*

Thus in the case of a complete algebraic variety X we need not distinguish between algebraic vector bundles on X and holomorphic vector bundles on X_{hol}.

Recall the following theorem due to Chow (see Griffiths-Harris [1]).

(A.3) Theorem. *Suppose X is a complete algebraic variety over $\mathbb{C}$ and $\mathfrak{Z}$ a closed analytic subset of X_{hol}, then there is an algebraic subvariety Z of X, such that $Z_{\text{hol}} \simeq \mathfrak{Z}$.*

A consequence of Chow's theorem is

(A.4) Corollary. *Suppose X and Z are complete algebraic varieties over $\mathbb{C}$ and $\mathfrak{f} : Z_{\text{hol}} \to X_{\text{hol}}$ is a holomorphic map. Then there is an algebraic map $f : Z \to X$ with $f_{\text{hol}} = \mathfrak{f}$.*

Moreover we need the following theorem due to Grauert-Remmert [1].

(A.5) Theorem. *Suppose X is a normal algebraic variety over $\mathbb{C}$ and $\mathfrak{Z}$ is a normal complex analytic space. If $\mathfrak{f}: \mathfrak{Z} \to X_{\text{hol}}$ is a finite morphism, then there exists a unique normal algebraic variety Z over $\mathbb{C}$ and a finite morphism $f : Z \to X$ such that $Z_{\text{hol}} = \mathfrak{Z}$ and $f_{\text{hol}} = \mathfrak{f}$.*

Next let us recall a result about quotients of complex analytic spaces (see Cartan [1]). For this let G be a group acting as a group of isomorphisms on the complex analytic space $\mathfrak{X}$. The quotient $\mathfrak{X}/G$, endowed with the quotient topology, admits the structure of a ringed space in a natural way: denote by $\pi : \mathfrak{X} \to \mathfrak{X}/G$ the canonical projection. Then by definition $\mathcal{O}_{\mathfrak{X}/G}(U)$, for $U \subseteq \mathfrak{X}/G$ open, is the set of functions $f : U \to \mathbb{C}$, for which $f\pi$ is an element of $\mathcal{O}_{\mathfrak{X}}(\pi^{-1}U)$. Moreover recall that the group G *acts properly and discontinuously* on the complex analytic space $\mathfrak{X}$, if for any pair of compact subsets K_1, K_2 of $\mathfrak{X}$ the set $\{g \in G | gK_1 \cap K_2 \neq \emptyset\}$ is finite. Then

(A.6) Theorem. *Suppose $\mathfrak{X}$ is a complex analytic space and G is a group, acting properly and discontinuously on $\mathfrak{X}$. The quotient $\mathfrak{X}/G$ is also a complex analytic space. Moreover, if $\mathfrak{X}$ is normal, so is $\mathfrak{X}/G$.*

In the special case that $\mathfrak{X}$ is a complex manifold there is a criterion for the quotient $\mathfrak{X}/G$ to be also a complex manifold. Note that the action of G on $\mathfrak{X}$ is said to be *free*, if $gx = x$ for some $x \in X$ and $g \in G$ implies $g = \mathrm{id}$.

(A.7) Corollary. *Let $\mathfrak{X}$ be a complex manifold and suppose G is a group acting freely and properly discontinuously on $\mathfrak{X}$. Then $\mathfrak{X}/G$ is also a complex manifold.*

Finally we give a proof of the so-called *Seesaw Theorem*, which we need in the following version. The same proof works also in the algebraic category.

(A.8) Seesaw Theorem. *Let $\mathfrak{X}$ be a connected compact complex manifold, $\mathfrak{Z}$ a reduced analytic space, and $\mathcal{L}$ a holomorphic line bundle on $\mathfrak{X} \times \mathfrak{Z}$.*

a) The set $\mathfrak{Z}_0 = \{z \in \mathfrak{Z} | \quad \mathcal{L}|\mathfrak{X} \times \{z\} \quad \text{is trivial}\}$ is Zariski-closed in $\mathfrak{Z}$.

b) If $q : \mathfrak{X} \times \mathfrak{Z}_0 \to \mathfrak{Z}_0$ denotes the projection map, then there is a holomorphic line bundle $\mathcal{M}$ on $\mathfrak{Z}_0$ such that

$$\mathcal{L}|\mathfrak{X} \times \mathfrak{Z}_0 \simeq q^*\mathcal{M}$$

Proof. a) Note first that a holomorphic line bundle $\mathcal{N}$ on a connected compact complex manifold is trivial if and only if $h^0(\mathcal{N}) > 0$ and $h^0(\mathcal{N}^{-1}) > 0$. Consequently

$$\mathfrak{Z}_0 = \{z \in \mathfrak{Z} \mid h^0(\mathcal{L}|\mathfrak{X} \times \{z\}) > 0 \quad \text{and} \quad h^0(\mathcal{L}^{-1}|\mathfrak{X} \times \{z\}) > 0\}$$

and the assertion follows from the Semicontinuity Theorem (see Grauert-Remmert [2] Theorem 10.5.4).

b) According to the Base Change Theorem (see Grauert [1] p.2 (2)) the sheaf $\mathcal{M} := q_*(\mathcal{L}|\mathfrak{X} \times \mathfrak{Z}_0)$ is invertible on $\mathfrak{Z}_0$ and the canonical base change homomorphism

$$\varphi(z) : \mathcal{M}(z) \to H^0\left(\mathcal{L}|\mathfrak{X} \times \{z\}\right) \cong \mathbb{C}$$

is an isomorphism for every $z \in \mathfrak{Z}_0$. We have to show that the canonical map

$$q^*\mathcal{M} = q^*q_*\left(\mathcal{L}|\mathfrak{X} \times \mathfrak{Z}_0\right) \to \mathcal{L}|\mathfrak{X} \times \mathfrak{Z}_0$$

is an isomorphism. For this it suffices to show that the induced map

$$H^0\left(q^*\mathcal{M}|\mathfrak{X} \times \{z\}\right) \to H^0\left(\mathcal{L}|\mathfrak{X} \times \{z\}\right)$$

is surjective for every $z \in \mathfrak{Z}_0$, since $\mathcal{L}|X \times \{z\}$ is trivial. But this is a consequence of the surjectivity of $\varphi(z)$. □

The following statement is an immediate consequence of the Seesaw Theorem. It is often called the *Seesaw Principle.*

(A.9) Corollary. *Let $\mathfrak{X}$ and $\mathfrak{Z}$ be compact complex manifolds and $\mathcal{L}$ a holomorphic line bundle on $\mathfrak{X} \times \mathfrak{Z}$. If $\mathcal{L}|\mathfrak{X} \times \{z\}$ is trivial for all z out of an open dense subset of $\mathfrak{Z}$ and $\mathcal{L}|\{x_0\} \times \mathfrak{Z}$ is trivial for some $x_0 \in \mathfrak{X}$, then $\mathcal{L}$ is trivial.*

B. Line Bundles and Factors of Automorphy

In this section we outline how to describe certain line bundles by factors of automorphy. Though the results are valid in greater generality, we restrict ourselves to the cases needed in this book.

Suppose X is a complex manifold (for us X will be a complex torus or the universal abelian variety over the Siegel upper half space), and let $\pi: \widetilde{X} \to X$ be the universal covering. The topological space $\widetilde{X}$ inherits the structure of a complex manifold in a natural way. We denote by $\pi_1(X)$ the fundamental group of $\widetilde{X} \to X$. In the notation we omit the base point of the fundamental group. Our aim is to describe those holomorphic line bundles L on X, whose pullback $\pi^* L$ is trivial, in terms of the cohomology of the fundamental group $\pi_1(X)$ acting from the left on $\widetilde{X}$.

Let us recall the definition of the cohomology group $H^1(\pi_1(X), H^0(\mathcal{O}^*_{\widetilde{X}}))$ of $\pi_1(X)$ with values in the multiplicative group $H^0(\mathcal{O}^*_{\widetilde{X}})$ of nonvanishing holomorphic functions on $\widetilde{X}$: the action of the fundamental group $\pi_1(X)$ on $\widetilde{X}$ induces a $\pi_1(X)$-module structure on $H^0(\mathcal{O}^*_{\widetilde{X}})$. A holomorphic map $f: \pi_1(X) \times \widetilde{X} \to \mathbb{C}^*$ satisfying the *cocycle relation*

$$f(\lambda\mu, \tilde{x}) = f(\lambda, \mu\tilde{x})\, f(\mu, \tilde{x}) \tag{1}$$

for all $\lambda, \mu \in \pi_1(X)$ and $\tilde{x} \in \widetilde{X}$ is called *1-cocycle* of $\pi_1(X)$ with values in $H^0(\mathcal{O}^*_{\widetilde{X}})$. Under multiplication these 1-cocycles form an abelian group $Z^1(\pi_1(X), H^0(\mathcal{O}^*_{\widetilde{X}}))$. For historical reasons we call the elements of the group $Z^1(\pi_1(X), H^0(\mathcal{O}^*_{\widetilde{X}}))$ *factors of automorphy* or simply *factors.* The factors of the form $(\lambda, \tilde{x}) \mapsto h(\lambda\tilde{x})h(\tilde{x})^{-1}$ for some $h \in H^0(\mathcal{O}^*_{\widetilde{X}})$ are called *boundaries.* They form the subgroup $B^1(\pi_1(X), H^0(\mathcal{O}^*_{\widetilde{X}}))$ of $Z^1(\pi_1(X), H^0(\mathcal{O}^*_{\widetilde{X}}))$. The group $H^1(\pi_1(X), H^0(\mathcal{O}^*_{\widetilde{X}}))$ is defined to be the quotient

$$H^1(\pi_1(X), H^0(\mathcal{O}^*_{\widetilde{X}})) = Z^1(\pi_1(X), H^0(\mathcal{O}^*_{\widetilde{X}}))/B^1(\pi_1(X), H^0(\mathcal{O}^*_{\widetilde{X}}))\ .$$

Any f in $Z^1(\pi_1(X), H^0(\mathcal{O}^*_{\widetilde{X}}))$ defines a line bundle on X as follows: consider the holomorphic action of $\pi_1(X)$ on the trivial line bundle $\widetilde{X} \times \mathbb{C} \to \widetilde{X}$

$$\lambda \circ (\tilde{x}, t) = (\lambda \tilde{x}, f(\lambda, \tilde{x}) t)$$

for all $\lambda \in \pi_1(X)$. This action is free and properly discontinuous, so the quotient

$$L = \widetilde{X} \times \mathbb{C} / \pi_1(X)$$

is a complex manifold (see Corollary A.7). Considering the projection $p: L \to X$ induced by $\widetilde{X} \times \mathbb{C} \to \widetilde{X}$ one easily checks that L is a holomorphic line bundle on X.

By definition the group of holomorphic line bundles on X is canonically isomorphic to the group $H^1(X, \mathcal{O}_X^*)$. Hence the construction above gives a map $Z^1(\pi_1(X), H^0(\mathcal{O}_{\widetilde{X}}^*)) \to H^1(X, \mathcal{O}_X^*)$. The following proposition shows, that this map induces an injective homomorphism of cohomology groups.

(B.1) Proposition. *There is a canonical isomorphism*

$$\phi_1: H^1(\pi_1(X), H^0(\mathcal{O}_{\widetilde{X}}^*)) \to \ker\bigl(H^1(X, \mathcal{O}_X^*) \xrightarrow{\pi^*} H^1(\widetilde{X}, \mathcal{O}_{\widetilde{X}}^*)\bigr) .$$

Proof. Step I: Existence of ϕ_1.
It is not difficult to see that the map $Z^1(\pi_1(X), H^0(\mathcal{O}_{\widetilde{X}}^*)) \to \ker(H^1(X, \mathcal{O}_X^*) \to H^1(\widetilde{X}, \mathcal{O}_{\widetilde{X}}^*))$ defined above is a homomorphism and factorizes via the group $H^1(\pi_1(X), H^0(\mathcal{O}_{\widetilde{X}}^*))$. However in Chapter 2 we need its explicit description in terms of cocycles in $Z^1(X, \mathcal{O}_X^*)$. For this let $\{U_i\}_I$ be an open covering of X, such that for every $i \in I$ there is a connected set $W_i \subseteq \pi^{-1}(U_i)$ with $\pi_i := \pi | W_i : W_i \to U_i$ biholomorphic. For every pair $(i,j) \in I^2$ there is a unique $\lambda_{ij} \in \pi_1(X)$ such that

$$\pi_j^{-1}(x) = \lambda_{ij} \pi_i^{-1}(x) . \tag{2}$$

for all $x \in U_i \cap U_j$.

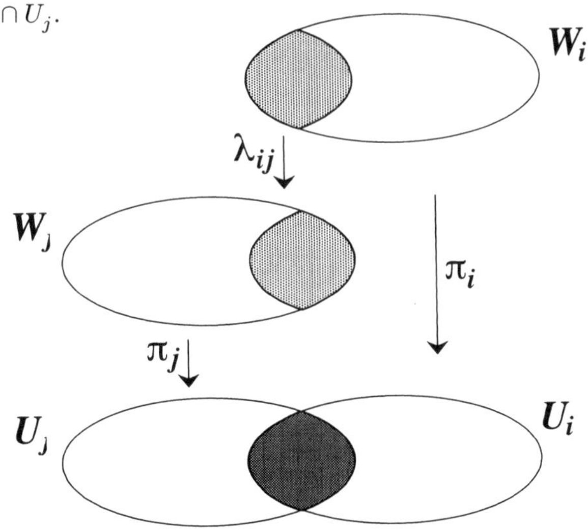

This implies that

$$\lambda_{ij}\lambda_{jk} = \lambda_{ik} \tag{3}$$

for all $i, j, k \in I$.

Suppose $f \in Z^1(\pi_1(X), H^0(\mathcal{O}^*_{\widetilde{X}}))$. For $i, j \in I$ and $x \in U_i \cap U_j$ define $g_{ij}(x) = f(\lambda_{ij}, \pi_i^{-1}(x))$. Then $\{U_i, g_{ij}\}_I$ is a 1-cocycle in $Z^1(X, \mathcal{O}^*_X)$, since $g_{ij}\, g_{jk} = g_{ik}$ for all $i, j, k \in I$ which is an immediate consequence of (3) and the cocycle relation (1) for f. So we get a map $Z^1(\pi_1(X), H^0(\mathcal{O}^*_{\widetilde{X}})) \to H^1(X, \mathcal{O}^*_X)$. Obviously this is a homomorphism of groups.

In order to define ϕ_1 it remains to show, that $B^1(\pi_1(X), H^0(\mathcal{O}^*_{\widetilde{X}}))$ is in the kernel of this homomorphism. Let $h \in H^0(\mathcal{O}^*_{\widetilde{X}})$. Our homomorphism maps the factor $h(\lambda\tilde{x})h(\tilde{x})^{-1} \in B^1(\pi_1(X), H^0(\mathcal{O}^*_{\widetilde{X}}))$ to the cocycle $\{g_{ij}\}_I$ with $g_{ij}(x) = h(\lambda_{ij}\pi_i^{-1}(x))h(\pi_i^{-1}(x))^{-1}$ for every $i, j \in I$ and $x \in U_i \cap U_j$. By (2) we have $g_{ij} = h(\pi_j^{-1})h(\pi_i^{-1})^{-1}$, so $\{g_{ij}\}_I$ is an element of $B^1(X, \mathcal{O}^*_X)$. Since the definition of ϕ_1 does not depend on the choice of U_i and π_i, this completes the proof of Step I.

Step II: The inverse map.

Suppose $L \in \ker(H^1(X, \mathcal{O}^*_X) \to H^1(\widetilde{X}, \mathcal{O}^*_{\widetilde{X}}))$, in other words π^*L is the trivial line bundle on $\widetilde{X}$. Let $\alpha: \pi^*L \to \widetilde{X} \times \mathbb{C}$ be a trivialization. The action of $\pi_1(X)$ on $\widetilde{X}$ induces holomorphic automorphisms of π^*L over this action. Via α we get for every $\lambda \in \pi_1(X)$ an automorphism ϕ_λ of the trivial line bundle $\widetilde{X} \times \mathbb{C}$. Necessarily ϕ_λ is of the form $\phi_\lambda(\tilde{x}, t) = (\lambda\tilde{x}, f(\lambda, \tilde{x})t)$ with a map $f: \pi_1(X) \times \widetilde{X} \to \mathbb{C}$, holomorphic in $\widetilde{X}$. The equation $\phi_{\lambda\mu} = \phi_\lambda\phi_\mu$ implies that f is a 1-cocycle in $Z^1(\pi_1(X), H^0(\mathcal{O}^*_{\widetilde{X}}))$.

Suppose $\alpha': \pi^*L \to \widetilde{X} \times \mathbb{C}$ is a second trivialization. Then there is an $h \in H^0(\mathcal{O}^*_{\widetilde{X}})$, such that $\alpha'\alpha^{-1}(\tilde{x}, t) = (\tilde{x}, h(\tilde{x})t)$ for all $(\tilde{x}, t) \in \widetilde{X} \times \mathbb{C}$. Denoting by ϕ'_λ the automorphism of $\widetilde{X} \times \mathbb{C}$ associated to $\lambda \in \pi_1(X)$ with respect to the trivialization α', we obtain:

$$\phi'_\lambda(\tilde{x}, t) = (\alpha'\alpha^{-1})\phi_\lambda(\alpha'\alpha^{-1})^{-1}(\tilde{x}, t) = (\lambda\tilde{x}, h(\lambda\tilde{x})f(\lambda, \tilde{x})h^{-1}(\tilde{x})t) \ .$$

Hence the class of the cocycle f in $H^1(\pi_1(X), H^0(\mathcal{O}^*_{\widetilde{X}}))$ does not depend on the trivialization $\pi^*L \to \widetilde{X} \times \mathbb{C}$ and we get a canonical map $\ker(H^1(X, \mathcal{O}^*_X) \to H^1(\widetilde{X}, \mathcal{O}^*_{\widetilde{X}})) \to H^1(\pi_1(X), H^0(\mathcal{O}^*_{\widetilde{X}}))$. It is easy to see that this map is the inverse of ϕ_1, and this completes the proof. □

(B.2) Remark. With an analogous proof one can show that there are canonical homomorphisms

$$\phi_n: H^n(\pi_1(X), H^0(\widetilde{X}, \pi^*\mathcal{F})) \longrightarrow H^n(X, \mathcal{F})$$

for all sheaves $\mathcal{F}$ of abelian groups on X and all $n \geq 0$ (see Mumford [2] Appendix to § 2). In particular, using the notation of Step I in the proof of Proposition B.1 the homomorphisms ϕ_1 and ϕ_2 are given as follows:

$$(\phi_1 f)_{ij} = f(\lambda_{ij}, \pi_i^{-1}) \quad \text{and} \quad (\phi_2 F)_{ijk} = F(\lambda_{ij}, \lambda_{jk}, \pi_i^{-1})$$

for $f \in Z^1(\pi_1(X), H^0(\widetilde{X}, \pi^*\mathcal{F})), F \in Z^2(\pi_1(X), H^0(\widetilde{X}, \pi^*\mathcal{F}))$ and $i, j, k \in I$. □

Next we want to describe the group of global sections of a line bundle L in $\ker(\pi^*: H^1(X, \mathcal{O}_X^*) \to H^1(\widetilde{X}, \mathcal{O}_{\widetilde{X}}^*))$ using a factor of automorphy. For any line bundle L on X there is a natural isomorphism $H^0(X, L) \xrightarrow{\sim} H^0(\widetilde{X}, \pi^*L)^{\pi_1(X)}$. Suppose that π^*L is trivial. A trivialization $\alpha: \pi^*L \to \widetilde{X} \times \mathbb{C}$ induces an isomorphism $H^0(\widetilde{X}, \pi^*L)^{\pi_1(X)} \xrightarrow{\sim} H^0(\widetilde{X}, \widetilde{X} \times \mathbb{C})^{\pi_1(X)}$. If $f \in Z^1(\pi_1(X), H^0(\mathcal{O}_{\widetilde{X}}^*))$ is the factor of automorphy associated to L with respect to the same trivialization $\alpha: \pi^*L \to \widetilde{X} \times \mathbb{C}$ (see proof of Proposition B.1), the elements in $H^0(\widetilde{X}, \widetilde{X} \times \mathbb{C})^{\pi_1(X)}$ are just the holomorphic functions $\vartheta: \widetilde{X} \to \mathbb{C}$ satisfying

$$\vartheta(\lambda\tilde{x}) = f(\lambda, \tilde{x})\vartheta(\tilde{x}) \tag{3}$$

for all $\tilde{x} \in \widetilde{X}$ and $\lambda \in \pi_1(X)$. Via the composed isomorphism $H^0(X, L) \xrightarrow{\sim} H^0(\widetilde{X}, \pi^*L)^{\pi_1(X)} \xrightarrow{\sim} H^0(\widetilde{X}, \widetilde{X} \times \mathbb{C})^{\pi_1(X)}$ the sections of L over X may be considered as holomorphic functions on $\widetilde{X}$, satisfying this functional equation. Note that the isomorphism depends on the trivialization: choosing another trivialization exactly means identifying the elements of $H^0(X, L)$ with holomorphic functions satisfying (3) with respect to an equivalent factor of automorphy (to see this, look at the definition of the maps in Step II of the proof of Proposition B.1).

Finally we show that the homomorphism of Proposition B.1 is functorial. For this suppose $\varphi: Y \to X$ is a holomorphic map of complex manifolds. Let $\widetilde{Y}$ and $\widetilde{X}$ denote the universal covering spaces of Y and X and $\pi_1(Y)$ and $\pi_1(X)$ the corresponding fundamental groups. The map φ induces a natural homomorphism $\varphi_*: \pi_1(Y) \to \pi_1(X)$ (of course the base points for $\pi_1(X)$ and $\pi_1(Y)$ have to be chosen compatibly) and a holomorphic map $\widetilde{\varphi}: \widetilde{Y} \to \widetilde{X}$ such that the following diagram commutes

$$\begin{array}{ccc} \widetilde{Y} & \xrightarrow{\widetilde{\varphi}} & \widetilde{X} \\ {\scriptstyle \pi_Y}\downarrow & & \downarrow{\scriptstyle \pi_X} \\ Y & \xrightarrow{\varphi} & X \end{array} .$$

Pulling back cocycles we obtain homomorphisms between the cohomology groups of X and Y: the map $f \mapsto (\varphi_* \times \widetilde{\varphi})^* f$ induces the homomorphism $H^1(\pi_1(X), H^0(\mathcal{O}_{\widetilde{X}}^*)) \to H^1(\pi_1(Y), H^0(\mathcal{O}_{\widetilde{Y}}^*))$, whereas $H^1(X, \mathcal{O}_X^*) \to H^1(Y, \mathcal{O}_Y^*)$ is given by $\{g_{ij}\}_I \mapsto \{\varphi^* g_{ij}\}_I$. One easily verifies that the diagram

$$\begin{array}{ccc} H^1(\pi_1(X), H^0(\mathcal{O}^*_{\widetilde{X}})) & \longrightarrow & H^1(\pi_1(Y), H^0(\mathcal{O}^*_{\widetilde{Y}})) \\ {\scriptstyle \phi_1}\downarrow & & \downarrow{\scriptstyle \phi_1} \\ H^1(X, \mathcal{O}^*_X) & \longrightarrow & H^1(Y, \mathcal{O}^*_Y) \end{array}$$

commutes. In particular, if L is any line bundle on X defined by a 1-cocycle f in $Z^1(\pi_1(X), H^0(\mathcal{O}^*_{\widetilde{X}}))$, the line bundle φ^*L on Y is determined by the cocycle $(\varphi_* \times \widetilde{\varphi})^* f : (\lambda, \tilde{y}) \mapsto f(\varphi_*(\lambda), \widetilde{\varphi}(\tilde{y}))$ of $Z^1(\pi_1(Y), H^0(\mathcal{O}^*_{\widetilde{Y}}))$.

Bibliography

Albert, A.A.

[1] *A solution of the principal problem in the theory of Riemann matrices.* Ann. Math. 35, 500–515 (1934)

[2] *On the construction of Riemann matrices II.* Ann. Math. 36, 376–394 (1935)

Andreotti, A.

[1] *On a theorem of Torelli.* Am. J. Math. 80, 801–828 (1958)

Appell, P.

[1] *Sur les fontions périodiques de deux variables.* Journ. de Math. Sér. IV, 7, 157–219 (1891)

Arbarello, E. - Cornalba, M. - Griffiths, P.A. - Harris, J.

[1] Geometry of Algebraic Curves. Vol. 1. Grundlehren 267, Springer-Verlag (1985)

Bagnera, G. - de Franchis, M.

[1] *Le nombre ρ de M. Picard pour les surfaces hyperelliptiques et pour les surfaces irrégulières de genre zéro.* Rend. Circ. Matem. Palermo 30, 185–238 (1910)

Baker, H.

[1] Abel's theorem and the theory of theta functions. Cambridge Univ. Press (1897)

Barth, W.

[1] *Abelian Surfaces with* $(1,2)$*-Polarization.* Adv. Studies in Pure Math. 10, Algebraic Geometry, 41–84, Sendai (1985)

Barth, W. - Peters, C. - van de Ven, A.

[1] *Compact Complex Surfaces.* Ergebnisse der Mathematik. Springer-Verlag, Berlin (1984)

Beauville, A.

[1] *Prym varieties and the Schottky problem.* Invent. Math. 41, 149–196 (1977)

[2] *Sous-variétés spéciales des variétés de Prym.* Comp. Math. 45, 3, 357–383 (1982)

[3] *Prym Varieties: A Survey.* Proc. Symp. in Pure Math. 49 Part 1, 610–620, (1989)

Bertrand, D.

[1] *Minimal heights and polarizations on abelian varieties.* Math. Sciences Research Inst., Berkeley, 06220-87 (1987)

Birkenhake, Ch. - Lange, H.
[1] *Cubic theta relations.* Journal reine angew. Math. 407, 167–177 (1990)
[2] *Symmetric Theta Structures.* manuscripta math. 70, 67–91 (1990)
[3] *The exponent of an abelian subvariety.* Math. Ann. 290, 801–814 (1991)
[4] *Norm-endomorphisms of abelian subvarieties.* To appear in the Proc. of the Trento conference on Alg. Geom., December 1990.

Birkenhake, Ch. - Lange, H. - v. Straten, D.
[1] *Abelian surfaces of type* $(1,4)$. Math. Ann. 285, 625–646 (1989)

Bloch, S. - Murre, J.P.
[1] *On the Chow group of certain types of Fano threefolds.* Comp. Math. 39, 47–105 (1979)

Bolza, O.
[1] *On Binary Sextics with Linear Transformations onto themselves.* Am. J. Math. 10, 47–70 (1888)

Borel, A.
[1] Introduction aux groupes arithmétiques. Hermann, Paris (1969)

Bourbaki, N.
[1] Algébre. Chap IX, Hermann, Paris (1958)

Brasch, H.-J.
[1] *Lifting Level D-Structures of Abelian Varieties.* Schriftenreihe Schwerpunkt Kompl. Mannigf. Nr. 130, Erlangen (1991)

Cartan, H.
[1] *Quotients of complex analytic spaces.* Contributions to function theory. 1–15, Bombay (1960)

Castelnuovo, G.
[1] *Sulle funzioni abeliane I, II, III, IV.* Rom. Acc. L. Rend. (5) 30, 50–55, 99–103, 195–200, 355–359 (1921)

Clebsch, A. - Gordan, P.
[1] Theorie der Abelschen Funktionen. Teubner, Leipzig (1866)

Coble, A.B.
[1] Algebraic Geometry and Theta Functions. Am. Math. Soc. Coll. Publ. 10, Providence (1961)

Collino, A.
[1] *A New Proof of the Ran-Matsusaka Criterion for Jacobians*, Proc AMS 92, 329–331 (1984)

Comessatti, A.
[1] *Sulle trasformazione hermitiane delle varietá di Jacobi.* Atti. R. Accad. Sci. Torino, 50, 439–455 (1914–15)
[2] *Sulle superficie de Jacobi semplicemente singolari.* Mem. Soc. ital. delle Szienze (dei XL) (3) 21, 45–71 (1919)

Conforto, F.
[1] Abelsche Funktionen und algebraische Geometrie. Grundlehren 84, Springer-Verlag (1956)

Debarre, O.
[1] *Sur le problème de Torelli pour les variétés de Prym.* Amer. J. Math. 111, 111–134 (1989)

Deligne, P.
[1] La théorie de Kodaira, Exposé 5 in Séminaire de Géométrie Algébrique, Orsay. (1967–1968)

Donagi, R.
[1] *The tetragonal construction.* Bull. AMS 4, 181–185 (1981)

Donagi, R. - Smith, R.
[1] *The structure of the Prym map.* Acta Math. 146, 25–102 (1981)

Enriques, F. - Severi, F.
[1] *Mémoire sur les Surfaces Hyperelliptiques.* Acta Math. 32, 283–392 (1909)

Fay, J.
[1] Theta Functions on Riemann Surfaces. Lecture Notes in Math. 352, Springer-Verlag, Heidelberg (1973)

de Franchis, M.
[1] *Un Teorema sulle Involuzioni irrazionali.* Rend. Cir. Mat. Palermo 36, 368 (1913)
[2] *Matrici di Riemann e varietà abeliane.* Reale Acc. It., Ford. A. Volta, Atti Conv. 9, 43–84 (1943)

Friedman, R. - Smith, R.
[1] *The generic Torelli theorem for the Prym map.* Invent. Math. 67, 473–490 (1982)

Frobenius, F. G.
[1] *Theorie der linearen Formen mit ganzen Coeffizienten.* Journal reine angew. Math. 86, 146–208 (1879); 88, 96–116 (1880)
[2] *Über die Grundlagen der Theorie der Jacobischen Functionen.* Journal reine angew. Math. 97, 16–48 and 188–224 (1984)

Godement, R.
[1] Topologie Algébrique et Théorie des Faisceaux. Hermann, Paris (1958)

Grauert, H.
[1] *Ein Theorem der Analytischen Garbentheorie und die Modulräume komplexer Strukturen.* Institut Des Hautes Études Scientifiques, Publications Mathematiques, N° 5 (1960)

Grauert, H. - Remmert, R.
[1] *Komplexe Räume.* Math. Ann. 136, 245–318 (1958)
[2] Coherent Sheaves. Grundlehren der math. Wiss. Bd. 265, Springer-Verlag, Berlin-Heidelberg-New York-Tokyo (1984)

Greenberg, M.J. - Harper, J.R.
[1] Algebraic Topology. A First Course. Benjamin (1981)

Griffiths, Ph. - Harris, J.
[1] Principles of Algebraic Geometry. John Wiley & Sons, New York (1978)

Grothendieck, A.
[1] *Fondements de la Géométrie Algébrique*, Sém. Bourbaki 1957–1962, Secrétariat Math., Paris (1962)
[2] *Eléments de Géométrie Algébriques III*. Publ. Math. IHES 11 (1961), and 17 (1963)

Grothendieck, A. – Dieudonné, J.
[1] Eléments de Géometrie Algébrique, I. Grundlehren 166, Springer-Verlag, Heidelberg (new ed. 1971)

Hartshorne, R.
[1] Algebraic Geometry. Grad. texts math 52, Springer-Verlag, Berlin-Heidelberg-New York (1977)

Hirzebruch, F.
[1] Topological Methods in Algebraic Geometry. Grundlehren 131, Springer-Verlag, New York (1974)

Hochschild, G.
[1] The structure of Lie groups. Holden Day, San Francisco 1965.

Hudson, R.W.H.T.
[1] Kummer's quartic surface. Cambridge Univ. Press (1905)

Hulek, K.
[1] Projective Geometry of Algebraic Curves. Astérisque 137 (1986)

Hulek, K. - Lange, H.
[1] *Examples of abelian surfaces in* $\mathbb{P}^4$. Journal reine angew. Math. 363, 200–216 (1985)

Humbert, G.
[1] *Théorie générale des surfaces hyperelliptiques*. Journ. de Math. Sér. IV, 9, 29–170 and 361–475 (1893)

Hurwitz, A.
[1] *Über algebraische Korrespondenzen und das verallgemeinerte Korrespondenzprinzip*. Math. Ann. 28, 561–585 (1887)

Igusa, J.
[1] Theta Functions. Die Grundlehren der mathem. Wiss. Bd. 194, Springer-Verlag, Berlin-Heidelberg-New York (1972)

Jacobson, N.
[1] Basic Algebra I, II. Freeman, New York (1980, 1985)

Jessop, C.M.
[1] A Treatise on the Line Complex. Cambridge Univ. Press (1903)

Kanev, V.
[1] *The global Torelli theorem for Prym varieties at a generic point.* Math. USSR-Izv. 20, 235–258 (1983)
[2] *Principal polarizations of Prym-Tjurin varieties.* Comp. Math. 64, 243–270, (1987)

Kempf, G.
[1] Appendix to *Varieties defined by quadratic equations*, by D. Mumford, in "Questions on Algebraic Varieties" CIME, Roma, 95–100 (1970)
[2] *Projective coordinate rings of abelian varieties*, in: Algebraic Analysis, Geometry and Number Theory. Edited by J.I. Igusa. 225-236 The John Hopkins Press (1989)
[3] Complex Abelian Varieties and Theta Functions. Springer Verlag (1991)

Kleiman, S.L.
[1] *Relative Serre duality for quasi-coherent sheaves.* Comp. Math. 41, 39–60 (1980)

Koizumi, S.
[1] *Theta relations and projective normality of abelian varieties.* Am. J. Math. 98, 865–889 (1976)
[2] *The Ring of Algebraic Correspondences on a Generic Curve of Genus g.* Nagoya Math. J. 60, 173–180 (1976)

Krazer, A.
[1] Lehrbuch der Thetafunktionen. Teubner, Leipzig (1903)

Krazer, A. - Wirtinger, W.
[1] *Abelsche Funktionen und allgemeine Thetafunktionen.* Enzyklopädia Math. Wiss. II.B.7, 604–873

Lang, S.
[1] Abelian Varieties. Interscience Pub., New York (1959)
[2] Introduction to Algebraic and Abelian Functions. Springer-Verlag, New York (1982)

Lange, H.
[1] *Jacobian surfaces in* $\mathbb{P}_4$. Journ. reine ang. Math. 372, 71–86 (1986)
[2] *Abelian varieties with several principal polarizations.* Duke Math. Journ. 55, 617–628 (1988)
[3] *Equations for Endomorphisms of Abelian Varieties.* Math. Ann. 280, 613–623 (1988)

Lange, G. - Narasimhan, M. S.
[1] *Squares of ample line bundles on abelian varieties.* Expo. Math. 7, 275–287 (1989)

Lazarsfeld, R.

[1] *Projectivité normale des surfaces abéliennes.* Rédigé par O. Debarre. Prépublication No. 14, Europroj – C.I.M.P.A., Nice (1990)

Lefschetz, S.

[1] *On certain numerical invariants of algebraic varieties with application to Abelian varieties.* Trans. Amer. Math. Soc. 22, 327–482 (1921)

[2] *Hyperelliptic surfaces and Abelian varieties.* Chap. 17, 349–395, Vol. 1 of *Selected Topics in Algebraic Geometry*, Washington (1928)

Lieberman, D.

[1] *Numerical and homological equivalence of algebraic cycles on Hodge manifolds.* Amer. J. Math. 90, 366–374 (1968)

Manolache, N.

[1] *Syzygies of abelian varieties embedded in* $\mathbb{P}_4(\mathbb{C})$. Journal reine angew. Math. 384, 180–191 (1988)

[2] *The equations of the abelian surfaces embedded in* $\mathbb{P}_4(\mathbb{C})$. Journal reine angew. Math. 394, 196–202 (1989)

Martens, H.H.

[1] *A new proof of Torelli's theorem.* Ann. of Math. 78, 107–111 (1963)

Masiewicki, L.

[1] *Universal properties of Prym varieties with an application to algebraic curves of genus five.* Trans AMS 222, 221–240 (1976)

Matsusaka, T.

[1] *On a theorem of Torelli.* Amer. J. Math. 80, 784–800 (1958)

[2] *On a characterization of a Jacobian variety.* Mem. Coll. Science, Kyoto, Series A. 32, 1–19 (1959)

Morikawa, H.

[1] *Cycles and Endomorphisms of Abelian Varieties.* Nag. Math. J. 7, 95–102 (1954)

Mukai, S.

[1] *Duality between* $D(X)$ *and* $D(\widehat{X})$ *with its application to Picard sheaves.* Nag. Math. J. 81, 153–175 (1981)

[2] *Symplectic structure of the moduli space of sheaves on an abelian or* $K3$ *surface.* Invent. Math. 77, 101–116 (1984)

[3] *Fourier Functor and its Application to the Moduli of Bundles on an Abelian Variety.* Adv. Studies Pure Math. 10, 515–550 (1987)

Mumford, D.

[1] *On the equations defining Abelian varieties I.* Invent. math. 1, 287–354 (1966)

[2] Abelian varieties. Oxford Univ. Press (1970)

[3] *Varieties defined by quadratic equations.* in: "Questions on Algebraic Varieties" CIME, Roma, 29–94 (1970)

[4] *Theta characteristics of an algebraic curve.* Ann. scient. Éc. Norm. Sup., 4.4, 181–192 (1971)

[5] *Prym varieties I*, in: Contributions to Analysis. Academic Press, New York, 325–350 (1974)
[6] Algebraic Geometry I. Complex Projective Varieties. Grundlehren der math. Wiss. Bd.221, Springer-Verlag, Berlin-Heidelberg-New York (1976)
[7] Tata Lectures on Theta I, II. Progr. in Math. 28, 43, Birkhäuser (1983,1984)

Narasimhan, M.S. - Nori, M.V.
[1] *Polarizations on an abelian variety.* Proc Indian Acad. Sci. Math. Sci. 90, 125–128 (1981)

Ohbuchi, A.
[1] *Some remarks on simple line bundles on abelian varieties.* Manuscr. Math. 57, 225–238 (1987)
[2] *A note on the normal generation of ample line bundles on abelian varieties.* Proc Japan Acad. 64, 119–120 (1988)

Poincaré, H.
[1] *Sur les fonctions abéliennes.* Acta Math. 26, 43–98 (1902)

Poincaré, H. - Picard, E.
[1] *Sur un théorème de Riemann relatif aux fonctions de n variables indépendentes admettant 2n systèmes de périodes.* Comptes Rendus 97, 1284–1287 (1883)

Pontryagin, L.
[1] *Homologies in Compact Lie Groups.* Mat. Sbornik, Novaya Seriya 6(3), 389–422 (1939)

Prym, F. - Rost, G.
[1] Theorie der Prym'schen Funktionen erster Ordnung im Anschluß an die Schöpfung Riemann's. Teubner, Leipzig (1911)

Raina, A.K.
[1] *Fay's trisecant identity and conformal field theory.* Comm. Math. Phys. 122, 625–641 (1989)

Ramanan, S.
[1] *Ample divisors on abelian surfaces.* Proc. London Math. Soc. 51, 231–245 (1985)

Ran, Z.
[1] *On Subvarieties of Abelian Varieties.* Inv. Math. 62, 459–479 (1981)
[2] *The Structure of Gauss-like Maps.* Comp. Math. 52, 171–177 (1984)

Recillas, S.
[1] *Jacobians of curves with a* g^1_4 *are Prym varieties of trigonal curves.* Bol. Soc. Mat. Méxicana 19, 9–13 (1974)

Reider, I.
[1] *Vector bundles of rank* 2 *and linear systems on algebraic surfaces.* Ann. Math. 127, 309–316 (1988)

Riemann, B.

[1] *Theorie der Abel'schen Funktionen.* Journal reine angew. Math. 54, 115–155 (1857)

Rosati, C.

[1] *Sulle corrispondenze fra i punti di due curve algebrique.* Ann. di Matem. (3), 28, 35–61 (1918)

Scharlau, W.

[1] Quadratic and Hermitian Forms. Grundlehren 270, Springer-Verlag, (1985)

Schindler, B.

[1] *Jacobische Varietäten hyperelliptischer Kurven und einiger spezieller Kurven vom Geschlecht 3.* Dissertation, Erlangen (1991)

Schoen, C.

[1] *Produkte Abelscher Varietäten und Moduln über Ordnungen.* Preprint (1991)

Scorza, G.

[1] *Intorno alla teoria generale delle matrici di Riemann e ad alcune sue applicatione.* Circ. Matem. di Palermo 41, 263–380 (1916)

[2] *Le algebre di ordine qualunque e le matrici di Riemann.* Circ. Matem. di Palermo 45, 1–204 (1921)

Serre, J.P.

[1] *Géométrie algébrique et géométrie analytique.* Ann. Inst. Fourier 6, 1–42 (1955)

[2] Groupes algébriques et corps de classes. Hermann (1959)

Severi, F.

[1] Trattato di Geometrica Algebraica. Bologna (1926)

Shafarevich, I.R.

[1] Basic Algebraic Geometry. Grundlehren 213, Springer-Verlag, Heidelberg (1974)

Shimura, G.

[1] *On analytic families of polarized abelian varieties and automorphic functions.* Ann. of Math. 78, 149–193 (1963)

Shimura, G. - Taniyama, Y.

[1] Complex Multiplication of Abelian Varieties. Math. Soc. Japan, Tokyo (1961)

Shioda, T.

[1] *The period map of abelian surfaces.* J. Fac. Sci. Univ. Tokyo, Sect. I A 25, 47–59 (1978)

Shioda, T. - Mitani, N.

[1] *Singular abelian surfaces and binary quadratic forms.* in: Classification of algebraic varieties and compact complex manifolds. Lecture Notes in Math. 412, 255–287, Springer-Verlag, Berlin-Heidelberg-New York (1974)

Shokurov, V.

[1] *Distinguishing Prymians from Jacobians.* Invent. Math. 65, 209–219 (1981)

Siegel, C.L.

[1] *Symplectic geometry.* Am. J. Math. 65, 1–86 (1943)

[2] Topics in Complex Function Theory. Vol. III. John Wiley & Sons, New York (1972)

Stahl, H.

[1] Abelsche Funktionen. Teubner, Berlin (1896)

Swinnerton-Dyer, H.P.F.

[1] Analytic Theory of Abelian Varieties. London Math. Soc. L. N. S. 14, Cambridge Univ. Press (1974)

Torelli, R.

[1] *Sulle varietà di Jacobi.* Rend. R. Accad. Lincei 22, 98–103 (1913)

Tyurin, A.N.

[1] *Five lectures on three-dimensional varieties.* Uspeki Mat. Nauk 29, 5, 3–50 (1972), Russian Math. Surv. 27, 5, 1–53 (1972)

Ueno, K.

[1] Classification Theory of Algebraic Varieties and Compact Complex Spaces. Lecture Notes in Math. 439, Springer-Verlag, Heidelberg (1975)

Umemura, H.

[1] *Some results in the theory of vector bundles.* Nagoya Math. J. 52, 97–128 (1973)

Verra, A.

[1] *The Prym map has degree two on plane sextics.* Preprint, Genova (1992)

v.d. Waerden, B.L.

[1] Algebra I, II. Springer-Verlag, Berlin-Heidelberg-New York (1971)

Weil, A.

[1] Courbes algébriques et variétés abéliennes. Hermann, Paris (1948)

[2] *Zum Beweis des Torellischen Satzes.* Nachr. Akad. Wiss. Göttingen, Math. Phys. Kl. 11a, 33–53 (1957)

[3] Variétés kähleriennes. Hermann, Paris (1958)

Wells, R.O.

[1] Differential Analysis on Complex Manifolds. Grad. texts math. 65, Springer-Verlag, New York (1980)

Welters, G.
[1] *Divisor Varieties, Prym Varieties and a Conjecture of Tyurin.* Preprint Nr. 139, University Utrecht (1980)
[2] *A theorem of Gieseker-Petri type for Prym varieties.* Ann. Sci. Ecole Norm. Sup. 18, 671–683 (1985)
[3] *Curves of twice the minimal class on principally polarized abelian varieties.* Indig. Math. 94, 87–109 (1987)
[4] *Recovering the curve data from a general Prym variety.* Amer. J. Math. 109, 165–182 (1987)

Weyl, H.
[1] *On generalized Riemann matrices.* Ann. Math. 35, 714–729 (1934)

Wirtinger, W.
[1] Untersuchungen über Thetafunctionen. Teubner, Berlin (1895)
[2] *Über einige Probleme in der Theorie der Abelschen Funktionen.* Acta Math. 26, 133–156 (1902)

Glossary of Notation

$\pi: V \to X$	Universal covering map of a complex torus	7
T_0X	Tangent space at X at 0	7
t_{x_0}	Translation on X by the point x_0	9
ρ_a	Analytic representation of $\mathrm{Hom}_{\mathbb{Q}}(X, X')$	10
ρ_r	Rational representation of $\mathrm{Hom}_{\mathbb{Q}}(X, X')$	10
$(\ker f)_0$	Component of $\ker f$ containing 0	11
$\deg f$	Degree of a homomorphism	12
n_X	Multiplication on X by the integer n	12
X_n	n-Division points in X	12
$e(f)$	Exponent of the isogeny f	12
$IF^n(X)$	Vector space of translation invariant n-forms on X	15
Ω^p_X	Sheaf of holomorphic p-forms on X	15
Ω	$= \mathrm{Hom}_{\mathbb{C}}(V, \mathbb{C})$	15
$\bar{\Omega}$	$= \mathrm{Hom}_{\bar{\mathbb{C}}}(V, \mathbb{C})$	15
$IF^{p,q}(X)$	Vector space of translation invariant (p,q)-forms on X	16
$\mathcal{A}^{p,q}_X$	Sheaf of $\mathcal{C}^\infty$-(p,q)-forms on X	17
$\mathrm{NS}(X)$	Néron-Severi group of X	29
$L(H, \chi)$	Line bundle with first Chern class H and semicharacter χ	30
$\mathrm{Pic}^0(X)$	Group of line bundles on X algebraically equivalent to zero	31
a_L	Canonical factor for the line bundle L	32
$\widehat{X}$	Dual complex torus	35
ϕ_L	The homomorphism $X \to \widehat{X}$, $x \mapsto t^*_x L \otimes L^{-1}$	37
$K(L)$	Kernel of ϕ_L	38
$\Lambda(L)$	$= \{v \in V \mid \mathrm{Im}\, H(v, \Lambda) \subseteq \mathbb{Z}\}$	38
$K(H)$	Kernel of $\phi_{L(H,\chi)}$	38
$\Lambda(H)$	$= \Lambda(L(H,\chi))$	38
$\mathcal{P}$	Poincaré bundle	39
$\mathrm{Pic}^H(X)$	Line bundles L in $\mathrm{Pic}(X)$ with $c_1(L) = H$	41
χ_0	Semicharacter of a line bundle of characteristic zero	48
L_0	$= L(H, \chi_0)$, the line bundle of characteristic zero in $\mathrm{Pic}^H(X)$	48
$Pf(E)$	Pfaffian of E	52
ϑ^c	Theta function with characteristic c	53
ϑ^c_w	$= a_L(w, \cdot)^{-1} \vartheta^c(\cdot + w)$	55
$Pfr(E)$	Reduced Pfaffian	58
$\mathcal{A}^{0,q}_X(L)$	Sheaf of $\mathcal{C}^\infty$-$(0,q)$-forms with values in the line bundle L	58

$i(L)$	Index of the line bundle L	63
$\chi(L)$	Euler-Poincaré characteristic of the line bundle L	66
(L^g)	Self-intersection number of the line bundle L	67
φ_L	Rational map $X \to \mathbb{P}_n$ associated to L	73
$(L_1 \cdot \ldots \cdot L_g)$	Intersection number of the line bundles $L_1, \ldots, L_g$	78
D_s	Smooth part of the divisor D	82
$a(X)$	Algebraic dimension of the complex torus X	86
$\mathrm{Pic}_s^H(X)$	Symmetric line bundles in $\mathrm{Pic}^H(X)$	90
$(-1)_L$	Normalized isomorphism of L over $(-1)_X$	91
$H^0(L)_+$	$(+1)$-Eigenspace in $H^0(L)$	91
$H^0(L)_-$	(-1)-Eigenspace in $H^0(L)$	91
$\mathrm{mult}_x(D)$	Multiplicity of the divisor D in x	94
e^H	Alternating form on X_2 associated to H	95
q_L	Quadratic form on X_2 associated to the line bundle L	95
$X_2^+(D)$	$= \{x \in X_2 \mid \mathrm{mult}_x(D) \equiv 0 \pmod 2\}$	97
$X_2^-(D)$	$= \{x \in X_2 \mid \mathrm{mult}_x(D) \equiv 1 \pmod 2\}$	97
$\star$	Pontryagin product $H_p(X,\mathbb{Z}) \times H_q(X,\mathbb{Z}) \to H_{p+q}(X,\mathbb{Z})$	104
I^o	Ordered multi-index complementary to I in $\{1,\ldots,2g\}$	105
$\{V\}$	Homology class of the cycle V	106
$*$	Star-operator	108
$e(L)$	Exponent of the polarization induced by L	116
ψ_L	The homomorphism $e(L)\phi_L^{-1}: \widehat{X} \to X$	116
$'$	Rosati involution	116
P_f^r	Characteristic polynomial of the rational representation	117
P_f^a	Characteristic polynomial of the analytic representation	117
$\mathrm{Tr}_r(f)$	Rational trace	118
$\mathrm{Tr}_a(f)$	Analytic trace	118
$N_r(f)$	Rational norm	118
$N_a(f)$	Analytic norm	118
$e(Y)$	Exponent of the abelian subvariety Y of X	125
N_Y	Norm-endomorphism of the abelian subvariety Y of X	125
ε_Y	Symmetric idempotent of the abelian subvariety Y of X	125
X^ε	Abelian subvariety associated to the symmetric idempotent ε	126
$\delta(\mathsf{V},\mathsf{W})$	Endomorphism associated to the algebraic cycles V and W	131
$(\mathsf{V} \cdot \mathsf{W})$	Intersection number of the algebraic cycles V and W	131
$\mathcal{G}(L)$	Theta group of the line bundle L	149
$G(L)$	Theta group of the line bundle L	150
$[\alpha, w]$	Element of the theta group $G(L)$	150
e^L	Commutator map $K(L) \times K(L) \longrightarrow \mathbb{C}^*$	154

$\mathcal{H}(D)$	Heisenberg group of type D 163
$K(D)$	$= \mathbb{Z}^g/D\mathbb{Z}^g \oplus \mathbb{Z}^g/D\mathbb{Z}^g$ 163
e^D	Commutator map $K(D) \times K(D) \to \mathbb{C}^*$ 163
$\mathrm{Sp}(D)$	Symplectic group of type D 166
$G^e(L)$	Extended theta group 174
$\mathcal{G}^e(L)$	Extended theta group 174
$\mathcal{H}^e(D)$	Extended Heisenberg group 175
$\mathfrak{H}_g$	Siegel upper half space 214
Λ_Z	The lattice $(Z, D)\mathbb{Z}^{2g}$ 214
X_Z	The abelian variety $\mathbb{C}^g/\Lambda_Z$ 214
H_Z	Hermitian form on X_Z 214
$\mathrm{Sp}_{2g}(\mathcal{R})$	Symplectic group 215
G_D	$= \{M \in \mathrm{Sp}_{2g}(\mathbb{Q}) \mid {}^tM\Lambda_D \subseteq \Lambda_D\}$ 215
$M(Z)$	$= (\alpha Z + \beta)(\gamma Z + \delta)^{-1}$ 215
$\mathcal{A}_D$	Moduli space for polarized abelian varieties of type D 219
$\mathrm{Sp}_{2g}^D(\mathcal{R})$	$= \left\{R \in M_{2g}(\mathcal{R}) \mid R\begin{pmatrix} 0 & D \\ -D & 0 \end{pmatrix} {}^tR = \begin{pmatrix} 0 & D \\ -D & 0 \end{pmatrix}\right\}$ 219
Γ_D	The symplectic group $\mathrm{Sp}_{2g}^D(\mathbb{Z})$ 219
$\widetilde{\mathcal{A}}_D$	$\simeq \mathcal{A}_D$ 219
$\widetilde{\mathcal{A}}_D(D)$	Moduli space for polarized abelian varieties of type D with level D-structure 222
$\Gamma_D(n)$	$= \{R \in \Gamma_D \mid R \equiv \mathbb{1}_{2g} \pmod{n}\}$ 223
$\widetilde{\mathcal{A}}_D(n)$	Moduli space for polarized abelian varieties of type D with level n-structure 223
$\widetilde{\mathcal{A}}_D^{\Delta}$	Moduli space for polarized abelian varieties of type D with a decomposition 224
$(S)_0$	$= {}^t(s_{11}, \ldots, s_{gg}) \in \mathbb{R}^g$ 224
$\vartheta\begin{bmatrix} c^1 \\ c^2 \end{bmatrix}(v, Z)$	Classical theta function with characteristic $\begin{bmatrix} c^1 \\ c^2 \end{bmatrix}$ 227
$X_{(Z,D)}$	The abelian variety $\mathbb{C}^g/(Z\mathbb{Z}^g \oplus D\mathbb{Z}^g)$ 229
j_Z	The isomorphism $\mathbb{R}^{2g} \to \mathbb{C}^g, \quad x \mapsto (Z, \mathbb{1})x$ 233
$\mathfrak{X}_D$	$= (\mathbb{C}^g \times \mathfrak{H}_g)/\Lambda_D$ 234
$G_D(D)$	$= \left\{\begin{pmatrix} \mathbb{1}+D\tilde{a} & D\tilde{b}D \\ \tilde{c} & \mathbb{1}+\tilde{d}D \end{pmatrix} \in G_D \mid \tilde{a}, \tilde{b}, \tilde{c}, \tilde{d} \in M_g(\mathbb{Z})\right\}$ 237
$\mathcal{A}_D(D)$	$\simeq \widetilde{\mathcal{A}}_D(D)$ 237
$\mathcal{A}_D(D)_0$	Moduli space for polarized abelian varieties of type D with orthogonal level D-structure 238
$A \otimes B$	Kronecker product of matrices 249
$\mathcal{H}_m$	$= \{Z \in M_m(\mathbb{C}) \mid {}^tZ = -Z,\ \mathbb{1}_m - {}^t\overline{Z}Z > 0\}$ 263
$\mathcal{H}_{r,s}$	$= \{Z \in M(r \times s, \mathbb{C}) \mid \mathbb{1}_s - {}^t\overline{Z}Z > 0\}$ 269
$U_{r,s}$	$= \left\{M \in M_{r+s}(\mathbb{C}) \mid {}^tM\begin{pmatrix} \mathbb{1}_r & 0 \\ 0 & -\mathbb{1}_s \end{pmatrix}\overline{M} = \begin{pmatrix} \mathbb{1}_r & 0 \\ 0 & -\mathbb{1}_s \end{pmatrix}\right\}$ 274

$\widetilde{U}_{m,m}$	$=\left\{M=\begin{pmatrix}\alpha & \beta\\ -\bar\beta & \bar\alpha\end{pmatrix}\in M_{2m}(\mathbb{C}) \mid {}^tM\begin{pmatrix}\mathbb{1}_m & 0\\ 0 & -\mathbb{1}_m\end{pmatrix}\overline{M}=\begin{pmatrix}\mathbb{1}_m & 0\\ 0 & -\mathbb{1}_m\end{pmatrix}\right\}$	274
$p_a(C)$	Arithmetic genus of the curve C	289
$J(C)$	Jacobian variety of the curve C	322
$(J(C),\Theta)$	Canonically polarized Jacobian	323
$\mathrm{Div}^n(C)$	Divisors of degree n on C	325
$C^{(n)}$	n-th symmetric product of C	325
α_{D_n}	Abel-Jacobi map $C^{(n)}\to J(C)$	325
$\mathrm{Pic}^n(C)$	Line bundles of degree n on C	325
$\mathcal{T}_C$	Tangent bundle of C	326
$[Y]$	Fundamental cohomology class of the subvariety Y	328
$\mathrm{sing}\,\Theta$	Singularity locus of Θ	331
$\mathcal{P}_C^n$	Poincaré bundle of degree n of C	334
N_f	Divisor norm map	337
γ_L	Homomorphism defined by the correspondence L	340
$\mathrm{Corr}(C_1,C_2)$	Group of equivalence classes of correspondences	340
$\overline{\mathrm{Aut}\,C}$	reduced automorphism group of a hyperelliptic curve	347
$\mathrm{Alb}(X)$	Albanese variety of X	361
ϕ_Y	The homomorphism ϕ_{ι^*Y}	367
ψ_Y	The homomorphism $e(Y)\phi_Y^{-1}$	367

Index

2